S£85.00
9G

Economic Modelling
in the OECD
Countries

INTERNATIONAL STUDIES IN ECONOMIC MODELLING

Series Editor
Homa Motamen
Director, Institutional Investment
(Grenfell & Colegrave)
Canadian Imperial Bank of Commerce
London

Input-Output Analysis
M. Ciaschini
Modelling the Labour Market
M. Beenstock

In preparation

Models of Disequilibrium and Shortage in the Centrally Planned Economies
C. Davis and W. Charemza
Economic Models of Latin America
D.E. Hojman
Economic Models of Trade Unions
P. Garonna

Economic Modelling in the OECD Countries

Edited by
Homa Motamen
Executive Director, Canadian
Imperial Bank of Commerce,
Securities Europe Ltd, London

London
CHAPMAN AND HALL
New York

First published in 1988 by
Chapman and Hall Ltd
11 New Fetter Lane, London EC4P 4EE
Published in the USA by
Chapman and Hall
29 West 35th Street, New York, NY 10001

© *1988 Chapman and Hall Ltd*

Printed in Great Britain by
J. W. Arrowsmith Ltd, Bristol

ISBN 0 412 29770 1

All rights reserved. No part of this book may be reprinted, or reproduced or utilized in any form or by any electronic, mechanical or other means, now known or hereafter invented, including photocopying and recording, or in any information storage and retrieval system, without permission in writing from the publisher.

British Library Cataloguing in Publication Data

Economic modelling in the OECD countries.—
 (International studies in economic modelling)
 1. Macroeconomics—Mathematical models
 2. Organisation for Economic Co-operation
 and Development countries—Economic
 policy—Mathematical models
 I. Motamen, Homa II. Series
 339'.0724   HD87

ISBN 0-412-29770-1

Library of Congress Cataloging-in-Publication Data

Economic modelling in the OECD countries/edited by Homa Motamen.
 p. cm.—(International studies in economic modelling)
 Bibliography: p.
 Includes index.

ISBN 0-412-29770-1

 1. Economic policy—Econometric models. 2. Economic forecasting—Econometric models. I. Motamen, Homa. II. Series.
 HD75.5.E27 1987
 338.9'00724—dc19   87–27006

To my children
NICHOLAS and ANTONIA

Contents

Contributors		xi
Introduction to the Series		xv
Preface and Acknowledgements		xvii

1 The LINK model and its use in international scenario analysis 1
  Lawrence R. Klein

2 Projections of the OECD economies in the global perspective, 1986–2000: policy simulations by the FUGI global macroeconomic model 11
  Akira Onishi

3 Supply-side policies in four OECD countries 31
  Anthonie Knoester

4 An empirical analysis of policy co-ordination in the United States, Japan and Europe 53
  Hali J. Edison and Ralph Tryon

5 How much could the international co-ordination of economic policies achieve? An example from US–EEC policy-making 71
  Andrew J. Hughes Hallett

6 Capital risk and models of investment behaviour 103
  Robert S. Pindyck

7 Adjustment costs and mean-variance efficiency in UK financial markets 119
  Christopher J. Green

8 The macroeconomic and sectoral effects of the Economic Recovery Tax Act: some simulation results 141
  Flint Brayton and Peter B. Clark

9	Use of anticipations data in the anticipations model Walter Naggl	165
10	An endogenously time-varying parameter (TVP) model of investment behaviour: theory and application to Belgian data Marcel Gérard and Carine Vanden Berghe	183
11	Budget consolidation, effective demand and employment Wulfheinrich von Natzmer	203
12	Interaction between economic growth and financial flows: presentation of a model analysing the impact of short-term financial disturbances on economic growth Hasse Ekstedt and Lars Westberg	219
13	Asymmetry in conservation: a capital stock analysis Jonathan V. Greenman	245
14	Adjustment options for the US economy Jan C. Siebrand and Job Swank	265
15	Model building for decision aid in the agri-economic field Patrick Anglard, Francoise Gendreau and A. Rault	283
16	Estimated optimal lags for the optimization models: a method for estimating the optimal lag between economic variables Kaoru Ichikawa	313
17	Macroeconomic policy and aggregate supply in the UK Michael Beenstock and Paul Lewington	327
18	Two recent trends combined in an econometric model for the Netherlands: the supply-side and sectoral approach Johan P. Verbruggen	353
19	The supply-side of RIKMOD: short-run producer behaviour in a model of monopolistic competition Michael Hoel and Ragnar Nymoen	381
20	Direct interventions, interest rate shocks and monetary disturbances in the Canadian Foreign Exchange market: a simulation study Kanta Marwah and Halldor P. Palsson	407

Contents ix

21	Effects of a fall in the price of oil: the case of a small oil-exporting country Kjell Berger, Ådne Cappelen, Vidar Knudsen and Kjell Roland	457
22	Modelling the effects of investment subsidies W. Driehuis and P. J. van den Noord	473
23	Collective bargaining and macroeconomic performance: the case of West Germany Ullrich Heilemann	491
24	A cost–push model of galloping inflation: the case of Yugoslavia Davorin Kračun	507
25	Short-term forecasting of wages, employment and output in Barbados Daniel O. Boamah	539
26	Reducing working time for reducing unemployment? A macroeconomic simulation study for the Belgian economy Joseph Plasmans and Annemie Vanroelen	561
27	An econometric model for the determination of banking system excess reserves José Luis Escrivá and Antoni Espasa	609
28	Forecasting versus policy analysis with the ORANI model Peter B. Dixon, Brian R. Parmenter and Mark Horridge	653
29	An applied general equilibrium model of the United States economy John V. Colias	667
30	A quarterly econometric model for the Spanish economy Ignacio Mauleón	683
31	Macroeconomic policies and adjustment in Yugoslavia: some counter-factual simulations Fahrettin Yagci and Steven Kamin	713
	Index	731

Contributors

Patrick Anglard — ADERSA, Verrières-le-Buisson, France

Michael Beenstock — City University Business School, London, UK

Kjell Berger — Central Bureau of Statistics, Oslo, Norway

Daniel O. Boamah — Central Bank of Barbados, Barbados, West Indies

Flint Brayton — Division of Research and Statistics, Federal Reserve Board, Washington DC, USA

Ådne Cappelen — Central Bureau of Statistics, Oslo, Norway

Peter B. Clark — Intenational Monetary Fund, Washington DC, USA

John V. Colias — BellSouth Services, Birmingham, Alabama, USA

Peter B. Dixon — Institute of Applied Economic and Social Research, University of Melbourne, Australia

W. Driehuis — Faculty of Economics, University of Amsterdam, The Netherlands

Hali J. Edison — Division of International Finance, Federal Reserve Board, Washington DC, USA

Hasse Ekstedt — Department of Economics, Göteborg University, Sweden

José Luis Escrivá — Banco de España, Madrid, Spain

Antoni Espasa — Banco de España, Madrid, Spain

Francoise Gendreau	ADERSA, Verrières-le-Buisson, France
Marcel Gérard	Institut Catholique des Hautes Études Commerciales, Brussels, Belgium
Christopher J. Green	Department of Business and Economics, University of Wales Institute of Science and Technology, Cardiff, UK
Jonathan V. Greenman	Corporate Planning Department, British Petroleum plc, London, and Department of Mathematics, University of Essex, Colchester, UK
Ullrich Heilemann	Rhine–Westphalian Institute for Economic Research, Essen, FRG
Michael Hoel	Department of Economics, University of Oslo, Norway
Mark Horridge	IMPACT Project, IMPACT Research Centre, University of Melbourne, Australia
Andrew J. Hughes Hallett	Department of Economics, University of Newcastle, Newcastle upon Tyne, UK
Kaoru Ichikawa	Department of Economics, Chiba University, Japan
Steven Kamin	Division of International Finance, Federal Reserve Board, Washington DC, USA
Lawrence R. Klein	University of Pennsylvania, Philadelphia, USA
Anthonie Knoester	Economic Institute, Nijmegen University, The Netherlands
Vidar Knudsen	Central Bureau of Statistics, Oslo, Norway
Davorin Kračun	School of Economics and Commerce, University of Maribor, Yugoslavia
Paul Lewington	City University Business School, London, UK
Kanta Marwah	Carleton University, Ottawa, Canada

Ignacio Mauleón	Bank of Spain, Madrid, Spain
Walter Naggl	University of Munich, FRG
Wulfheinrich von Natzmer	University of Freiburg, FRG
P. J. van den Noord	SEO, Foundation for Economic Research, University of Amsterdam, The Netherlands
Ragnar Nymoen	Bank of Norway, Oslo, Norway
Akira Onishi	Institute of Applied Economic Research, Soka University, Tokyo, Japan
Halldor P. Palsson	Carleton University, Ottawa, Canada
Brian R. Parmenter	Institute of Applied Economic and Social Research, University of Melbourne, Australia
Robert S. Pindyck	Sloan School of Management, Massachusetts Institute of Technology, Cambridge MA, USA
Joseph Plasmans	Studiecentrum voor Economisch en Sociaal Onderzoek, UFSIA, University of Antwerp, Belgium
A. Rault	ADERSA, Verrières-le-Buisson, France
Kjell Roland	Central Bureau of Statistics, Oslo, Norway
Jan C. Siebrand	Erasmus University, Rotterdam, The Netherlands
Job Swank	Erasmus University, Rotterdam, The Netherlands
Ralph Tryon	Division of International Finance, Federal Reserve Board, Washington DC, USA
Carine Vanden Berghe	Institut Catholique des Hautes Études Commerciales, Brussels, Belgium
Annemie Vanroelen	Studiecentrum voor Economisch en Sociaal Onderzoek, UFSIA, University of Antwerp, Belgium

Johan P. Verbruggen	Ministry of Economic Affairs, The Hague, The Netherlands
Lars Westberg	Department of Economics, Göteborg University, Sweden
Fahrettin Yagci	The World Bank, Washington, DC, USA

Introduction to the Series

There has been a growing dependence in the past two decades on modelling as a tool for better understanding of the behaviour of economic systems, and as an aid in policy and decision making. Given the current state of the art globally, the introduction of a series such as this can be seen as a timely development. This series will provide a forum for volumes on both the theoretical and applied aspects of the subject.

International Studies in Economic Modelling is designed to present comprehensive volumes on modelling work in various areas of the economic discipline. In this respect one of the fundamental objectives is to provide a medium for ongoing review of the progression of the field.

There is no doubt that economic modelling will figure prominently in the affairs of government and in the running of the private sector, in efforts to achieve a more rational and efficient handling of economic affairs. By formally structuring an economic system, it is possible to simulate and investigate the effect of changes on the system. This in turn leads to a growing appreciation of the relevance of modelling techniques. Our aim is to provide sufficient space for authors to write authoritative handbooks, giving basic facts with an overview of the current economic models in specific areas and publish a useful series which will be consulted and used as an accessible source of reference.

The question may arise in some readers' minds as to the role of this series *vis-à-vis* other existing publications. At present, no other book series possesses the characteristics of *International Studies in Economic Modelling* and as such cannot fill the gap that will be bridged by it. Those journals which focus in this area do not present an exhaustive and comprehensive overview of a particular subject and all the developments in the field. Other journals which may contain economic modelling papers are not sufficiently broad to publish volumes on all aspects of modelling in a specific area which this series is designed to cover.

A variety of topics will be included encompassing areas of both micro and macroeconomics, as well as the methodological aspects of model construction. Naturally, we are open to suggestions from all readers of, and contributors to, the series regarding its approach and content.

Finally, I would like to thank all those who have helped the launch of this series. The encouraging response received from authors who have contributed the forthcoming volumes and from the subscribers to the series has indicated the need for such a publication.

Homa Motamen
London, Dec 1987

Preface and Acknowledgements

This volume aims to bring together the work of those involved in the development and use of macroeconomic models in the OECD economies. The main areas covered in the book are: international policy co-ordination, risk, supply-side, exchange rates, equilibrium models, use of resources, the United States economy, wages and employment, international modelling, country models, modelling techniques and domestic economic management.

I am grateful to John Edmonston, Lyndon Driscoll, Sean Holly and Brian Scobie for their help and support in preparing this volume.

Homa Motamen
London, Dec 1987

1
The LINK model and its use in international scenario analysis

LAWRENCE R. KLEIN

The main ideas and structure of the world econometric model developed by project LINK are known, but the system is always undergoing change, and this fact, together with the need for substantive analysis in the light of most recent developments prompts me to give a few general indications about the total system. Originally, the project was conceived as an attempt to model the international transmission mechanism. The initial focus was on models of major industrial market economies where short run, cyclical-like fluctuations were transmitting disturbance from one country to another. Also, a guiding principle of the project has been that every resident model builder knows his or her own country best; so the principal emphasis has been on linking together prevailing national models, to operate in an international mode, rather than on constructing look-alike systems at some research centre.

What was first an attempt to study the international transmission mechanism evolved into a world model. At first, regional area models for developing and centrally planned economies were added in order to close the system, but gradually individual models for major developing and centrally planned economies have been introduced. Wherever possible, these have been prepared by resident model builders and entered into the system, but completeness has been an overriding aim, and the LINK centre has prepared or commissioned several models for cases in which a resident model builder could not be readily found. At present there are several resident model builders who are readying their models for formal entry into the system, but they help to monitor and guide inputs for the interim models that have been constructed at the LINK centre. A Chinese and an Indian team are presently designing 'own' systems for the world's two largest countries, and these will replace interim models in due course. Development of world telecommunications networks is making it ever more possible to maintain a single world model at the centre, consisting of components that are supplied from the corners of the globe.

The LINK system now consists of 71 individual country models and 8 regional models of residual groupings of countries. This is a large heterogeneous system of nearly 20 000 equations. It poses a considerable management problem, but it can be done, especially with many new facilities and procedures that have only recently become available.

There is much to be said for uniform systems. Notation, documentation and computer handling are vastly simplified. It is easier to understand the workings of a homogeneous system and to trace the causal structure of economic events. Also, many homogeneous systems tend to be more compact and smaller than LINK, and the single-handed researcher definitely prefers a small over a large system. More sophisticated statistical treatment of the former is possible. While I fully appreciate these arguments in favour of 'small is beautiful', I am fully committed to the continued development of a large heterogeneous system. Actually, the Wharton world model is a homogeneous system, while LINK is heterogeneous, and I have worked with both systems. The former is more viable from a commercial point of view, but my true love for both research and substance remains the LINK system. It is not a matter of 'bigger is better', but it must, of necessity, be a large system if it is to be used for the kinds of applications that

Table 1.1 LINK models

North America:
    Canada, USA
Developed East:
    Australia, Japan, New Zealand
EEC:
    Belgium/Luxemburg, Denmark, France, Germany (FR), Greece, Ireland, Italy, Netherlands, Portugal, Spain, UK
Rest of Industrialized Countries:
    Austria, Finland, Iceland, Norway, Sweden, South Africa, Switzerland
South and East Asia:
    Hong Kong, India, Indonesia, Korea, Malaysia, Pakistan, Philippines, Singapore, Taiwan, Thailand, Other South-East Asia, South-East Asia Least Developed
West Asia:
    Iran, Iraq, Kuwait, Saudi Arabia, Israel, West Asia Oil Importers, Other West Asia Oil Exporters
Mediterranean:
    Cyprus and Malta, Turkey, Yugoslavia
Centrally Planned Economies (CPE):
    Bulgaria, China, Czechoslovakia, Germany (GDR), Hungary, Poland, Romania, USSR
Latin America:
    Argentina, Bolivia, Brazil, Chile, Colombia, Eucador, Mexico, Paraguay, Peru, Uruguay, Venezuela, Caribbean
Africa:
    Algeria, Egypt, Ethiopia, Gabon, Ghana, Kenya, Libya, Morocco, Nigeria, Sudan, Tunisia, Africa Least Developed, Other African Countries
Rest of the World

will be described in later sections of this chapter. Also, there is much merit to the argument that the far-flung and diversified information system that goes with the dispersed structure of LINK has a peculiar richness of detail and can be insightful even if cumbersome and very expensive to maintain (Table 1.1).

Each resident model builder manages an 'own' system for domestic use, in an unlinked mode. Results from such systems are sent to the LINK centre, incorporating informed knowledge about exogenous inputs. All these dispersed models are solved simultaneously and consistently, according to the well-known LINK algorithm, at the University of Pennsylvania, and made available to all project participants in hard copy, by machine retrieval on a network, and by face-to-face oral communication at semi-annual project meetings. By feeding in unlinked model results, the participants get back linked model results with many relevant world aggregates. In this co-operative research mode, the project has functioned since 1969.

Over the years the system has evolved by incorporating more financial market information, shifting from fixed (Bretton Woods) to floating exchange rates, paying more attention to energy economics,[1] adding models (smaller OECD, individual LDC and individual CPE models), paying more attention to international debt, lengthening the projection horizon, improving computation (cost, speed, interaction, communication).

1.1 THE 1986 BASELINE CASE

The baseline solution is an attempt at preparing a world economic forecast and is, therefore, of limited interest for scenario or response analysis, but since the world economy has undergone such large policy and other external changes in early 1986, a careful analysis of the present baseline case, tells us a great deal about the working of the system.

The drop in oil prices, which is viewed as the disintegration of a cartel, is treated as a major exogenous event. This has great impact on the system's outcome and raises the interesting question whether price decline (1986) has effects that are symmetrical with a price rise (1974).

The export price deflator for Saudi Arabia provides a good numerical estimate of the course assumed in this forecast for oil prices (Table 1.2).

Table 1.2 Percentage change in export price deflator for Saudi Arabia

1985	1986	1987	1988	1989	1990
−6.0	−35.0	0.1	5.2	5.6	4.5

[1] From the very beginning, international bilateral trade in mineral fuels (SITC 3), has been separately treated in the external relationship for all countries.

Table 1.3 Some main economic policy magnitudes, 1985–1990: LINK assumptions and estimates

	1985	1986	1987	1988	1989	1990
USA Treasury bill rate (%)	7.5	6.8	7.2	7.2	7.2	7.0
Average exchange rate (%)	2.9	−18.9	−5.3	−2.5	−2.2	−2.1
Deficit (National Income Account, NIA, $ billion)	−189.7	−145.9	−88.4	−68.1	−56.1	−37.3
UK interest rate (%)	12.2	10.8	9.0	8.7	8.9	9.0
Exchange rate, £/$	0.78	0.65	0.62	0.60	0.59	0.59
Germany money market rate (%)	6.5	6.4	6.1	5.9	5.7	5.6
Exchange rate, DM/$	2.95	2.24	2.04	1.93	1.84	1.76
Japan interest rate (%)	6.43	6.20	6.21	6.08	6.11	6.10
Exchange rate, ¥/$	238.5	178.9	168.0	162.2	157.4	152.6

Coincident with the oil price decline, there is a result associated with monetary policy which takes place on a co-ordinated basis across many countries. Scenarios of co-ordinated monetary easing, with lower interest rates, have long been studied with the LINK system, and we have a fresh instance of that policy at this time (Klein et al., 1981).

Since 22 September 1985, there has been a co-ordinated move to bring down the exchange value of the US dollar. The dollar started falling endogenously, before 22 September, but its decline was probably accelerated and stretched out after the major finance ministers set out to attack it.

Finally, there is a change in US fiscal policy towards deficit reduction, either according to the Gramm–Rudman–Hollings legislation or as a result of the public debate about the bill. The baseline case assumes that a conscious effort is being made to reduce the deficit, and this internal matter has great repercussions for the whole international economy. These four major developments are all factored into the latest LINK simulation (Table 1.3).

The lower oil price, averaging at $18 per barrel, imparts an upward thrust to the world economy. It lowers the growth of producer and consumer prices in importing countries and encourages monetary authorities to be more stimulative. For major world economic areas, the outcome is a downward drift in unemployment and a low inflation rate, accompanying a robust real growth rate. The unemployment rate shows slight evidence of decline in Europe for the first time in many forecast exercises, and inflation projections do not turn up after a year or two. What was formerly an expected slowdown or growth recession for 1986 has been smoothed over and even improved for 1987. A slight slowdown appears to have taken place in the industrial world in 1985 (Table 1.4).

If China were not in the LDC group, their growth rate would not look so impressive. Inflation results are not comprehensive for LDC and CPE groups; therefore, they are reported here for the OECD group alone.

The recent world developments in oil markets and in policy co-ordination among the large industrial countries has, according to this model, imparted a

Table 1.4 Growth and inflation 1985–1990: LINK estimates

	1985	1986	1987	1988	1989	1990
GNP (1970 US $) (%)						
OECD	2.8	3.1	3.6	2.9	2.9	2.7
LDC[a]	5.3	4.1	4.9	5.0	4.7	4.9
CPE	3.4	3.7	3.7	3.5	3.3	3.1
World	3.4	3.4	3.8	3.4	3.3	3.2
Unemployment (%) OECD	8.9	8.7	8.5	8.4	8.3	8.1
Inflation (%) OECD	4.2	3.2	3.9	4.3	4.5	4.9

[a] China is included in the LDC–Asia group and not in CPE.

Table 1.5 Mexican debt, interest service, and current account 1985–1990

	1985	1986	1987	1988	1989	1990
Debt (US $ billion)[a]	96.7	100.8	102.6	106.7	111.5	116.6
Interest service (US $ billion)	9.3	8.5	9.8	11.9	13.3	14.9
Current account (US $ billion)	0.5	−1.7	−2.3	−2.4	−1.9	−3.1

[a] One billion equal to one thousand million.

favourable pattern of economic performance and ironed out cyclical and other disturbing results.

Consider the results for Mexico under the terms of these developments assumed for oil markets (Table 1.5).

This is obviously a projection of a worsening situation. An already large debt continuous to rise; interest service, except for some relief, through falling world rates, in 1986 also rises; and the current account slips from surplus into increasing deficit. The model estimate for export growth, dollar denominated, is negative in 1986 and less than 7% for the rest of this decade. The US interest rate, on the other hand, is estimated to be more than 7.0%. The prime rate is somewhat higher. Given a risk premium, in the case of Mexico, this model provides a very unfavourable combination of export growth and interest cost.

The asymmetry of the situation is not only that falling oil prices give this poor picture for Mexico, and other oil exporters with large debt burdens, but also that there are potential troubles for bank creditors, and this has implications for world financial stability. The model can provide estimates of endogenous variables as indicated in Table 1.5; it cannot determine a 'flash point', at which poor indicators would trigger panic conditions in financial markets. It could, however, estimate some quantitative magnitudes associated with North–South transfers of financial capital to alleviate this situation for Mexico.

1.2 SENSITIVITY TO OIL PRICE CHANGES, DOLLAR DEPRECIATION AND PROTECTIONISM: SCENARIO ANALYSIS

The baseline solution does have an assumption of lower oil prices; the Saudi Arabian export price index is 35% lower in 1986 than in 1985. But the drop could be even greater. We have, accordingly, estimated a case where the oil price averages $10 per barrel against the $18 figure used in the baseline (Tables 1.6 and 1.7). By looking at the results of this projection we can isolate the model's sensitivity to oil price change alone.

The industrial countries would gain in terms of trade balance, largely at the expense of OPEC nations, and to some extent the Soviet Union and Eastern Europe. Real world trade would gain another 1.5% over the baseline value. All major areas would gain in terms of GDP, but the OECD countries would gain the most – more than another full percentage point above baseline values.

Early on, the inflation rate would be a full percentage point lower and the European unemployment rate would be marginally lower, say by 0.2%. Mexico,

Table 1.6 Effect of projected oil price drop of 45% 1986–1990 ($10/barrel in 1986)

	OECD	LDC	OPEC	CPE	OECD	North America	EEC
	GDP percentage changes				Unemployment rate (difference)		
1986	0.89	0.83	1.95	0.07	−0.2	−0.5	0
1987	1.23	0.84	1.71	0.12	−0.5	−1.3	−0.1
1988	1.22	0.88	1.33	0.12	−0.5	−1.2	−0.2
1989	1.11	0.91	1.63	0.11	−0.4	−0.9	−0.2
1990	1.10	0.89	1.80	0.14	−0.3	−0.5	−0.2
	Trade balance (fob) in bill$				Inflation rate (consumer deflator)		
1986	42.04	−26.99	−34.77	−14.00	−1.2	−1.1	−1.1
1987	39.87	−24.38	−33.66	−13.03	−0.4	+0.1	−0.9
1988	44.04	−26.58	−36.45	−13.85	−0.2	+0.1	−0.6
1989	48.35	−28.98	−40.58	−15.35	0	+0.3	−0.3
1990	54.53	−34.73	−47.01	−16.62	0	+0.2	−0.1

Table 1.7 Effect of projected oil price drop of 45%, 1986–1990

	Change in trade balance in bill US $				
	1986	1987	1988	1989	1990
Mexico	−3.49	−3.27	−3.44	−3.79	−4.12
Venezuela	−3.30	−3.30	−3.51	−3.83	−4.47
Nigeria	−2.90	−2.30	−2.26	−2.31	−4.14
Indonesia	−2.80	−3.22	−3.35	−3.64	−3.90
USSR	−16.57	−13.26	−13.88	−15.20	−16.73
UK	−4.13	−4.28	−2.81	−1.74	−0.42
Norway	−1.98	−2.16	−2.40	−2.64	−2.87

Scenario analysis

where the problems are severe at $18 oil would be much worse off at $10. After an initial gain – presumably by increasing oil production in order to recoup lost revenues – the growth rate of Mexico's GDP would fall by more than another one-quarter point. All the major oil exporters, including Saudi Arabia, would suffer losses in the trade accounts. Of course, the balance of the USSR deteriorates.

All the summit countries, with the exception of the UK, gain through lower oil costs. Canada's results are hardly significant; they start out slightly negative and then turn positive, i.e. above the baseline values. Any asymmetry in this calculation, in comparison with an oil price rise, would have to come through nonmodel reaction in the Mexican and similar cases to the buildup of a large external deficit.

To isolate the effects of currency realignment, we made a simulation with the dollar depreciating by 20%, over the baseline values, where there is already a programmed decline of 20% (Tables 1.8 and 1.9). In this present scenario we have the dollar falling only against other OECD currencies. The baseline solution estimates a current account deficit of $69 billion in 1990. That figure is reduced by another $48 billion in this scenario; so the US current account is estimated to be still in deficit after a 40% fall in the dollar.

Scenario GNP is up by almost 0.5% in 1990, but inflation is also up. The trade gains by the United States in this case come at the expense of GDP losses in major OECD countries. The dollar is not depreciated against developing countries; so they also gain in this scenario.

Table 1.8 Change in trade balance for US $ depreciation of 20%

	1986	1987	1988	1989	1990
US	−2.95	0.86	4.45	10.11	12.95
Japan	3.41	3.65	4.41	2.89	3.00
Germany	0.17	−1.25	−3.54	−6.52	−7.63
UK	−1.22	−3.17	−3.92	−2.20	1.44
France	−0.09	−2.78	−4.48	−6.74	−8.14
Italy	−2.28	−2.43	0.03	0.05	−0.58
Canada	1.92	1.44	1.96	1.10	0.68

Table 1.9 Change in current account balance for US $ depreciation of 20%

	1986	1987	1988	1989	1990
US	10.76	23.94	38.87	44.79	47.97
Japan	1.98	1.52	1.82	0.78	0.28
Germany	0.04	−3.35	−8.40	−13.23	−16.37
UK	0.98	0.83	2.35	4.61	7.48
France	0.93	−1.00	−1.66	−3.30	−4.09
Italy	−3.33	−4.58	−3.30	−4.88	−6.33
Canada	1.67	2.14	3.13	3.36	3.05

Table 1.10 Effect of projected imposition of US tariffs 1987–88, 20%

	1987	1988	1989	1990
	Index of real world exports, deviation from baseline levels			
World exports real				
Percentage difference of baseline	−1.1	−1.6	−0.7	−0.3
	Percentage change in GDP, deviation from baseline rates			
OECD	−0.47	−0.96	−1.28	−1.14
LDC	0.02	−0.11	−0.64	−0.69
CPE	0	−0.01	−0.03	−0.05
	Change in inflation rate (consumption deflators)			
OECD	+0.2%	+0.2%	−0.2%	−0.2%
North America	+0.5%	+0.4%	−0.5%	−0.5%
EEC	+0.1%	0.1%	0.2%	0.2%
Developed/East	−0.1%	0 %	+0.1%	0%

Finally, let us consider a unilateral protectionist move by the United States, in frustration with having a large external deficit especially in the early years of a pattern of dollar depreciation. Quite arbitrarily we impose a 20% surcharge on all US imports, and our gains come at the expense of everyone else (Table 1.10).

World trade falls off by 1.6%, and world GDP is also down, but by only 0.25%. Although the United States gain on the side of current account and also the federal budget deficit, we lose as much as 0.4% of real GNP.

This alternative simulation was introduced along the lines of the popular case, last year, known as the 'Motorola surcharge', which looked for a temporary surcharge for 3 years on a sliding scale of severity.

The principal reason for introducing a comprehensive surcharge once again was to use it for a provocative first-round effort in an international telecommunications experiment. Using a conversational mode on the BITNET–EARNET system, the members of project LINK, assembled in the computer centre of the City University of New York (CUNY), sent the main tables for the unilateral import surcharge to colleagues in Madrid, Rome, Bologna, Paris and London. After exchanging messages, with virtually no time delay, about the provocative solution, we learned that our European colleagues did not want to recommend retaliation by imposition of import duties against products of the United States. Instead they recommended some domestic counter action, namely, to lower interest rates by one full percentage point in the EEC and by 0.5 percentage point in Japan.

By adopting an expansionary monetary policy in response to an American duty, the EEC switch from a GDP loss of almost 0.2% to a gain of 0.4%. Unemployment and the inflation rate are not significantly changed.

The United States lose some of their surcharge gains for internal and external deficits by about $1 billion or less. The other summit countries have noticeably better production levels after having compensated the unilateral surcharge by a cut in interest rates, but the gains are modest, all less than 1% over baseline. If added negotiating rounds were to be introduced in the telecommunications experiment, either the domestic policy response would have to be larger, say by lowering interest rates by 1.5 or 2.0 percentage points, or by using more policy instruments in retaliation.

On the heels of the 'Motorola surcharge' proposals last year to try to shore up the American external and internal deficits through protectionism, there arose a more specific proposal to impose a 3-year duty of 25% on US imports from Japan, South Korea, Taiwan and Brazil. This was uniquely suited for analysis with the LINK system because each of these four target countries is separately modelled

Table 1.11 Effects of surcharge (25%) against Japan, South Korea, Taiwan and Brazil

	Deviation from baseline (% in own currency units)				
	1986	1987	1988	1989	1990
Canada exports	2.6	1.7	1.2	−1.6	−0.9
Imports	1.2	1.6	1.5	0.5	−0.1
GNP	0.8	0.5	0.3	−0.7	−0.6
France	0.6	0.5	0.3	−0.4	−0.5
	0.5	0.4	0.0	−0.8	−1.2
	0.1	0.1	0.1	0.0	0.1
Germany	0.7	0.5	0.5	−0.1	0.0
	0.4	0.5	0.3	−0.2	−0.3
	0.3	0.2	0.2	0.0	0.1
Italy	1.0	1.0	1.1	0.4	0.6
	0.4	0.2	−0.4	−1.6	−2.3
	0.3	0.4	0.5	0.6	0.8
UK	0.7	0.6	0.6	0.0	−0.1
	0.2	0.1	−0.2	−0.7	−1.0
	0.2	0.3	0.2	−0.1	−0.3
US	0.0	−0.3	−0.7	−0.8	−0.5
	−1.2	−2.0	−2.3	−1.0	0.0
	0.1	−0.1	−0.3	−0.4	−0.1
Japan	−3.7	−2.9	−2.0	2.3	2.8
	−0.6	−1.4	−2.1	−2.2	−2.0
	−0.9	−0.8	−0.6	0.6	0.9
Brazil	−6.7	−6.9	−7.5	−1.6	−1.7
	−1.5	−1.8	−1.7	−0.6	−0.4
	−0.8	−1.0	−1.2	−0.6	−0.5
South Korea	−3.9	−3.9	−3.9	−0.2	−0.1
	−3.0	3.1	−3.3	−0.4	−0.3
	−2.0	2.1	−2.3	−0.2	0.0
Taiwan	−3.6	−3.7	−3.7	−0.3	−0.1
	−2.9	−3.4	−3.8	−1.2	−0.9
	−1.8	−1.8	−1.7	0.0	0.0

in LINK. According to our calculations, US imports from these four target countries would be reduced by about $21 billion and increased by about $10 billion from the rest of the world. World trade would have been cut by about 0.4% off baseline values.

The results in Table 1.11 show that Western European countries gain at the expense of the target countries, all of which lose exports and GDP growth. Their imports fall because their activity levels fall, but the export decrease dominates the import decrease. Other countries, besides those in Western Europe or Canada, gain as well. The United States does not gain, in the sense that protectionism does not pay, although the twin deficits would improve.

The calculations were used, together with national interindustry and regional US models, to estimate the impact of such proposals on the Port of New York. It is an example of how a large detailed model like LINK, kept on line, can be used for study of a variety of problems that come up unannounced.

REFERENCE

Klein, L. R., Voisin, Pascal and Simes, Richard (1981) Co-ordinated monetary policy and the world economy. *Prévison et Analyse Economique*, **2** (3), 75–105.

2

Projections of the OECD economies in the global perspective, 1986–2000: policy simulations by the FUGI global macroeconomic model

AKIRA ONISHI

This chapter makes projections of the OECD economies for the period 1986–2000, using the latest version of the FUGI global macroeconomic model, i.e. type IV 011–62, and giving due consideration to the complex international linkages which make for global interdependence. Directing attention to both (1) a baseline future scenario and (2) alternative policy mixes, the model forecasts that the real average economic growth rate of the OECD countries as a whole during 1986–2000 will likely be from around 2.8% (in the baseline projection) to 3.5% (with alternative policies), while the world economy will likely sustain annual growth rates of between 3.3% and 3.8% for the period.

2.1 INTRODUCTION: CURRENT DEVELOPMENT AND FUTURE CHALLENGE

The trend toward greater interdependence of the OECD economies on the global level is progressing, and is not limited to the exports and imports of goods and capital, but has meant a strengthening of international relationships of interdependence also through exchanges of information, culture and individual talents. This in turn has pushed the scope of economics to transcend traditional frameworks which encompassed primarily the economy of one or another single country. In a similar way, the appearance of complex and interrelated problems having to do with such things as population, environment, food, resources, energy, development, arms competitions, displaced persons, etc., has posed problems whose solutions are quite impossible within the old frameworks of an economics dealing for the most part with only one country. These facts help explain the popularity of 'global modelling', especially in the years since 1970.

The reader is probably familiar with the *Limits to Growth* report presented to the Club of Rome in 1972. This study, put together mainly by Dennis and Donella Meadows, students of Professor J. Forrester of the Sloan School of Management at Massachusetts Institute of Technology, used a System Dynamics model developed by Forrester. Using the system dynamics (SD) method, this report made a pioneering analysis of global systems, attempting to see how such factors as population, environment, energy, resources and food production are related to one another.

This 1972 study, whose computer computations required about 380 software steps, required only a very short computation time to produce forecasts for the year 2000, or even for as far into the future as the year 3000. Computations up to the year 2000 could be made for the most part in just a few seconds. Even though it was for the most part a simple model that we might compare, so to speak, to a baby's teething ring, what it had to say was nevertheless impressive. Also, it was one of the first global models made available to the public (Forrester, 1971; Meadows *et al.*, 1972).

Afterwards, the World Multilevel System model of Drs Mesarović and Pestel made its appearance, as well as the United Nations World model developed mainly by Professor Wassily Leontief and the model used by the OECD for its *Interfutures* report, based on a model produced by the UK's Department of the Environment. More recent is the GLOBUS model developed in West Germany by a research group centring on Drs Karl Deutsch and Stuart A. Bremer, in collaboration with Professor Harold Guetzkow. The latter model is rather closer to an international political model than to an economic model (Mesarović and Pestel, 1974; Leontief *et al.*, 1977; OECD, 1979; Bremer, 1985).

Among systems that centre on economic models, there is the Project LINK system, developed primarily by Professor Lawrence Klein, a winner of the Nobel Prize for economics. At present it is being used in the Department of International Economic and Social Affairs of the United Nations for short-term global economic forecasting (Klein *et al.*, 1982).

The model which we have developed is called the FUGI (Future of Global Interdependence) model. It made its first public appearance at the same time and place as the first public report on the Leontief-oriented UN World model, namely at the 1977 Symposium of the International Institute for Applied Systems Analysis (IIASA), which during the 1970s hosted in Austria yearly symposia on global modelling (Bruckmann, 1980).

The model presented in 1977 was a first-generation model, whereas we are now in the process of building a fifth-generation model. The fourth-generation model which is operative at present is primarily an economic model. It establishes dynamic international links among the 62 countries and regions into which it classifies the global economy. It is for the most part the relatively larger countries among the 159 member countries of the United Nations Organization which are treated as country units, while smaller countries are grouped together in regions.

The number of equations used is approximately 13 700, while the number of software steps for computation is approximately 100 000. Computation, including tabulation, can nevertheless be performed very rapidly, and only about 20 minutes is required to make forecasts from the present up to the year 2000.

The FUGI model is at present being used by the Projections and Perspectives Studies Branch, Department of International Economic and Social Affairs of the United Nations, for simulations of United Nations medium- and long-term international development strategies, while the Project LINK model is being used for short-term forecasts (Onishi, 1985).

Stimulated by our latest joint research with the United Nations University on a 'global early warning system for displaced persions', we have felt the need for our FUGI model to go beyond its present capacities centred on an 'economic' model (in the rather traditional, restricted sense of the term) and to develop into a model that can in the future analyse 'global problematiques' or 'global complexes of symptoms' and complicated questions including various types of environmental problems and the sorts of displaced persons issues to which we are now directing our attention. We are thus expanding the scope of our fifth-generation FUGI model, presently under development, to deal with such issues.

The fifth-generation model treats all countries, regardless of how large or small, as having the possibility of being dealt with as country units. It is designed to be a comprehensive system model which can deal not only with economic problems but also with environmental issues, population, energy, food, indicators of social welfare and issues of human rights.

As our model was originally researched and developed principally around economic issues, the task which we have put before us is to devise and incorporate subsystems that deal with environmental issues and such areas of concern as human rights and peace and security (Onishi, 1986).

Although our methodology is first and foremost based on various country or regional studies, we have felt it desirable, using these country or regional studies as a base, to adopt an orientation that further gives consideration to a highly sophisticated global modelling system.

We call this a dynamic soft systems approach, and it reflects the fact that in the fields of so-called present-day high technology, particularly in the astounding development of computers during the 1970s and 1980s, extraordinarily sophisticated handling of information has become possible. In this regard, too, the software which computers use, that is to say, utilization techniques, have made very notable forward strides, so that approaches to many problems have been made easier for us not only with respect to hardware systems but also with respect to software.

The dynamic soft systems approach is, we feel, supported not only by the so-called soft sciences but also by developments in a number of interrelated fields of the natural sciences.

For example, our understanding of the human brain has greatly advanced through recent developments in brain physiology. As a result, it is seen that the 'right brain' perceives images of reality, while the 'left brain' analyses these in a logical and conceptual way and constructs logical models. As a part of its own division of labour, the brain's central ridge facilitates high-level flows or exchanges of information between the left and right brains. Through a skilful treatment of the organically linked functions of the left and right brains, a soft system model can, we feel, be produced. In a similar way, that we have tried to develop for our present purpose is not a model that merely collects information but a model that collects information skilfully, analyses it, and provides a sophisticated global information system.

It is also worth noting that recent developments in life sciences are making ever clearer the connectedness between individual cells of the human body and the human body as a whole organic entity. Individual cells contain, it has been found, information pertaining to the entire body. Thus, at times of special stress, the individual cells dispose of a regulatory mechanism by which they pool their forces, working together in the face of difficulties. This is an extraordinarily important capacity which living things possess, and we in fact need to incorporate this very sort of capacity into any global modelling system in order to prevent or mitigate, through international co-operation, appearances of undesirable phenomena in our global human society.

In my view, the first-generation modern economics was based upon Newtonian dynamics and Darwinism. The second-generation economics is econometrics, which has been greatly developed through progress with statistics and economic modelling. Third-generation economics seems likely to come to be called 'humanomics', with methodology based upon a dynamic soft systems approach reflecting progress in life sciences, biotechnology, ecology and soft system science. Toward the 21st century, it is expected that economic models will come to have much 'softer' dynamic systems.

2.2 MODEL STRUCTURE

The FUGI global macroeconomic model type IV 011–62 which we currently employ classifies the world economy, which it treats as an amalgam of developed market economies (or OECD economies), developing market economies, and centrally planned economies, into 62 countries and regional groups.

With the FUGI global macroeconomic model, each national and regional economy has eleven sub-blocs, namely: (1) sectoral production with energy constraints, (2) expenditures on GDP at constant prices, (3) profits–wages, (4) prices, (5) expenditures on GDP at current prices, (6) money, (7) interest rates, (8) government finance, (9) international balance of payments, (10) international finance, and (11) foreign exchange rates. Each country and regional model was designed so as to reflect the special characteristics of each country and region.

Each national and regional economy is interlinked with other economies through both direct and indirect complex informational networks having to do with trade, capital flows, government finance, money, interest rates, foreign exchange rates, export–import prices, etc.

The model allows complex computations that show, for example, the various modalities in which major sectoral production is carried out under conditions of energy constraints, and systems by which supply and demand operate in their relation to labour force, etc.

The model also indicates the composition of expenditures on GDP at constant prices. It can elucidate not only such complex trade structures as, for example, Japanese exports to and imports from the United States, but also such matters as private consumption, governmental consumption expenditures, and investments.

There are also elucidated such aspects of the distribution side of income as profits and wages. Structures are presented which show the peculiarities by which, through the intermediary of profits and wages, the net product of a given country is distributed.

There are then price systems to be considered. In spite of the common word 'price', there are in fact a variety of different indicators, e.g. wholesale price indices, consumer price indices, various types of price deflators for GDP components, etc. A feature of the systems used in the model is their ability to indicate the ways in which various types of prices – export and import prices, oil prices, prices of primary products, etc. – are decided upon. There is furthermore a system that computes the composition of expenditures on GDP at current prices.

In addition, there is a system for elucidating the determination of money supply and interest rates. Given the problem of the present-day high interest rates in the United States, it is possible, for example, to throw light on the phenomenon by which long-term interest rates rise as a result of increases in the United States budgetary deficit, or to understand better the types of impacts produced when the United States Federal Reserve Board (FRB) intervenes by setting money supply target zones or by manipulating official discount rate.

This system elucidates, for example, what is not a negligible background causative factor in the generation of current debt problems of developing countries, namely, the system by which increased American military expenditures give rise to government deficits, which in turn given rise to high interest rates which push up interest rates on the global level, causing a serious worsening of the debt problems of developing countries. This is most certainly a background factor linked to cases of failures in development in the developing countries.

Details of government finance, revenues and expenditures can also be clearly shown. In this way it can be shown, for example, how the US military expenditures are calculated, and how American and Soviet military expenditures constitute a scenario of retaliatory reactions in which, for example, any increase in Soviet military expenditures brings an increase in American military expenditures. Nevertheless, it is seen that if America's budgetary deficit should worsen beyond a certain point, military expenditures are cut back somewhat from earlier projections and are thus not completely a holy ground devoid of all controls. The same is true of the Soviet Union.

The model also includes an extremely meticulous system for examining international balances of payments. Not only can we elucidate circumstances and issues related to the import and export of goods, but also those related to the import and export of services. These issues include the realities of decision-making with respect to private foreign direct investments and portfolio investments in stocks and bonds.

In the field of international finance, one can see the ways in which official development assistance (ODA), both bilateral and multilateral, is being carried out, the ways in which private foreign direct investments are distributed, and the sorts of impacts produced on developing countries as a result.

In the field of international finance, the model exhibits a very detailed system with respect to foreign indebtedness, using World Bank data, and thus throws much new light on the ways in which each country's external indebtedness may be expected to develop in the future.

And lastly, there is a system for determining foreign exchange rates.

The above are the model's general specifications. It is important to keep in mind that the model is not an abstraction but an already existing tool which can make immediate and relevant computations. It is already fully in operation, and after we compile some additional policy information and give this information a somewhat more sophisticated treatment, the model as a whole should be in good shape for practical application of alternative policy simulations.

In brief, by virtue of its global system with complex linkages, the FUGI macroeconomic model can analyse complex phenomena including the impacts of alternative policy mixes on the economic growth of both the OECD economies and the rest of the world. For instance, the use of this model permits analyses not merely of the influence which the OECD countries' economic policies exert on these countries' economic relations with one another, but also of globally interdependent relationships having a complex structure within the framework of the world economy as a whole.

If one should try to consider, for example, only the topic of a possible American surcharge imposed on imports from Japan, the dimensions of the impact which this would produce on the Japanese economy or on Japan–US economic relations cannot in the absence of such a large-scale global model be directly perceived with any reliability. In the domain of international economics, which has a dynamic system structure that is complex to a degree that is in fact a little hard to believe at first sight, one simply cannot cut the ice with any small-scale, single-country model. If, forcing matters, one should nevertheless try to make a small-scale country model apply to the complex international domain, this is only likely to lead to mistaken results in the work of analysis. It is precisely in these types of complex system simulations that a very large-scale global computer model wields its strength.

For reference purposes, the theoretical outline of the FUGI macroeconomic model and an example of the modelling of the Japanese economy are given in an Appendix (not included in the present book, but available on application to the author – for address see Acknowledgements).

2.3 FUTURE SCENARIOS FOR THE OECD COUNTRIES

In our attempts at making projections of the OECD economies in a global perspective for the period 1986–2000, we hypothesized two cases as follows: (1) the baseline case, and (2) alternative policy simulation.

2.3.1 The baseline scenario

The baseline scenario is constructed upon the hypothesis that although economic policies in both the OECD and the rest of the world's countries are subject to certain changes which have been taking place in the economic environment, the scope for further policy choices does not indicate the likelihood of any dramatic changes. In other words, the parameters for the model's policy response functions and structural behavioural equations as estimated from past data (up to and including the present and current information) are taken as being nearly unchanging.

2.3.2 Alternative policies

The alternative policy scenario is constructed upon the hypothesis that the future course of the world economy can be changed as the result of human wisdom, appropriate policies and international co-operation. This policy scenario combines the following suppositions, starting from 1986:

(a) Co-ordinated lowering of interest rates among the OECD countries

The major OECD countries make new efforts at international co-ordination of monetary policies. More specifically, Japan, West Germany and the United States will take the initiative, in co-operation with the other major OECD countries, in lowering their official discount rates by 1 percentage point (0.5 percentage point in the case of West Germany) below the rate that would prevail in the baseline scenario. Following the example of Japan, West Germany and the United States, the other OECD countries carry out monetary policies that likewise make possible the achievement of lower interest rates.

(b) Global disarmament

In the global disarmament scenario, we hypothesize that military expenditures in all countries of the world are expected to be progressively cut by 5% annually (in nominal terms), beginning in 1986. We also suppose that in the case of the OECD economies 50% of the financial resources made available by these reductions in defence expenditures will be transferred to official development assistance (ODA) to the developing countries, while the remaining 50% will be applied to improving the welfare of the people of their own countries (e.g. reducing fiscal deficits or increasing expenditures on R and D). In the case of the developing countries and centrally planned economies, we considered that the financial resources which are released from the burden of national defence expenditures will be used principally for increases in domestic capital formation.

(c) Increased R and D in the OECD countries

The major OECD countries would increase their R and D by an amount equivalent to 0.5% of GDP in addition to the baseline, with the aim of evaluating the impact of R and D on investment, trade and economic growth of the OECD countries in an interdependent world nearing the 21st century.

(d) Achievement of ODA equal to 0.7% of GNP

OECD/DAC member countries gradually increase official development assistance (ODA) to developing countries, and by 2000 (in Japan by 1995) achieve the goal of ODA equivalent to 0.7% of GDP.

(e) Japanese policies for expanding domestic demand and opening markets

In this scenario, it is hypothesized that Japan will put into practice the following policies, beginning in 1986:

(1) Japan will introduce fiscal policies to stimulate domestic demand by reducing both personal income texes (by 2 trillion yen) and corporate taxes (by 1 trillion yen), as well as to increase government expenditures by 1 percentage point compared with the annual increase rate of the baseline
(2) Japan will progressively eliminate tariffs and nontariff barriers, achieving markets that are in fact completely open by 1990
(3) Japan's technology transfers to developing countries and industrial co-operation with the OECD countries will be promoted by raising annual private foreign direct investments between 1986 and 2000 by an additional amount equivalent to 0.5% of GDP in comparison with the baseline. In particular, it is expected that Japanese private enterprises will encourage multinationaliza-tion and adopt strategies of importing some 30% of overseas production in which they are directly involved in order to contribute to the expansion of Japan's imports not only from the United States and Europe but also from the developing countries.

2.4 THE PROJECTION RESULTS

2.4.1 The baseline projection

(1) According to the baseline projection, oil prices will move away from a nominal $27 per barrel in 1985 to approximately $22 in 1990 and approximately $25 in 2000. The real price of oil may be expected to slightly decrease when deflated by the OECD countries' weighted average export price indices.

(2) The annual average growth rate of the world economy (including the centrally planned economies) is estimated at 3.2% during 1986–1990, 3.4% during 1990–1995, and 3.2% during 1995–2000. This represents an increase in the yearly average for the entire decade of the 1980s from 3.0% to a yearly average of 3.3% for the decade of the 1990s. The annual average real growth rate of GDP of the OECD economies as a whole during the 1980s would be expected to grow from 2.5% to 2.9% during the 1990s. In the case of the developing market economies, average 2.8% real growth during the 1980s would be expected to accelerate to 4.2% during the 1990s. Through the stabilization of the price of oil, one may expect an improvement in the real economic growth of the non-oil-producing developing countries, namely an increase of the yearly average growth of 3.6% for the 1980s to 4.3% in the following decade. In the centrally planned economies, one may expect a slightly decreasing trend, namely an average yearly real growth of 4.8% in the 1980s but only 3.7% in the 1990s. This is partly because in the course of the two decades the real growth rate of the economy of the People's Republic of China may be expected to decrease from 8.1% to 4.9%.

Among the major OECD economies, the highest rate of economic growth between the years 1986 and 2000 is forecast for Japan (the yearly average being 4.1%), followed by France (3.5%), Italy (2.9%), Canada (2.5%), the United States (2.5%), the UK (2.4%) and West Germany (2.3%), in descending order (Table 2.1).

(3) The ratio of industrial activities to GDP in the OECD economies, i.e. the so-called industrialization rate, will in keeping with the trend towards an increasingly high-information society fall slightly, from 32.1% in 1985 to 30.8% in 2000. The proportion of the GDP in the OECD economies represented by services will probably rise slightly from 65.0% in 1985 to 66.0% in 2000. Conversely, the industrialization rate in the developing market economies is forecast to rise from 29.8% (of which 18% accounts for manufacturing industries) in 1985 to 35.2% (22.1% for manufacturing) in 2000. Although the ratio of manufacturing industry to GDP in the developing market economies will not come to equal by 2000 the OECD economies' corresponding figure of 26.3% in that year, their overall industrialization rate is seen as catching up with and surpassing that of the OECD economies.

In the year 2000, the world's highest ratio of manufacturing industry to GDP will be found in the East Asian NICs (47.7%). By 2000, this ratio in the East Asian newly industrializing countries (NICs) will have far surpassed the

Table 2.1 Projections of the OECD economies in the global perspective: baseline on annual average growth rates of real gross domestic product (GDP); unit: %

	70-75	75-80	70-80	80-85	85-90	80-90	90-95	95-2000	90-2000	85-2000
World	3.9	3.9	3.9	2.7	3.2	3.0	3.4	3.2	3.3	3.3
Pacific Basin	3.5	4.3	3.9	3.7	3.5	3.7	3.7	3.2	3.5	3.5
World (excluding CPEs)	3.6	3.8	3.7	2.3	2.8	2.5	3.2	3.1	3.2	3.1
Pacific Basin market economies	3.3	3.9	3.6	3.1	2.8	3.0	3.3	3.0	3.2	3.1
Developed market economies (AMEs)	3.2	3.4	3.3	2.3	2.6	2.5	3.0	2.9	2.9	2.8
OECD	3.1	3.4	3.3	2.3	2.6	2.5	3.0	2.8	2.9	2.8
The major seven	3.0	3.5	3.3	2.4	2.7	2.6	3.1	2.9	3.0	2.9
Japan	4.6	5.1	4.9	4.2	3.1	3.6	4.4	4.8	4.6	4.1
Canada	5.0	2.9	4.0	2.1	2.7	2.4	2.4	2.4	2.4	2.5
United States	2.5	3.4	2.9	2.6	2.6	2.7	2.8	2.1	2.5	2.5
France	4.0	3.2	3.6	1.1	2.4	1.7	3.7	4.2	4.0	3.5
Germany (Federal Republic of)	2.1	3.6	2.8	0.9	2.7	1.8	2.3	2.0	2.1	2.3
Italy	2.4	3.8	3.1	0.6	2.9	1.9	2.9	2.8	2.9	2.9
United Kingdom	2.2	1.7	1.9	1.9	2.4	2.2	2.5	2.2	2.4	2.4
EC	2.8	3.1	2.9	1.1	2.5	1.8	2.7	2.7	2.7	2.6
Other developed market economies	6.1	2.6	4.4	1.8	2.1	1.7	2.9	4.2	3.5	3.0
Developing market economies (DMEs)	5.9	5.6	5.8	2.2	3.4	2.8	4.2	4.2	4.2	3.9
Oil-exporting countries	6.2	5.6	5.9	1.2	1.8	1.3	4.2	4.1	4.1	3.3
OPEC in Middle East	8.2	5.1	6.7	0.3	1.8	0.8	3.3	3.3	3.2	2.7
Nonoil-exporting countries	5.8	5.6	5.7	2.8	4.3	3.6	4.3	4.3	4.3	4.3
NICs	8.7	6.9	7.8	2.2	4.7	3.6	5.4	5.2	5.3	5.1
Asian NICs	8.7	9.3	9.0	6.4	6.5	6.4	5.8	5.3	5.5	5.8
Latin America NICs	8.7	6.3	7.5	0.9	4.0	2.7	5.2	5.2	5.2	4.8
Asia and Pacific	5.0	6.0	5.5	5.0	4.2	4.6	4.5	4.4	4.4	4.3
East Asia	8.6	9.4	9.0	6.4	7.0	6.7	5.8	5.2	5.5	6.0
ASEAN	7.4	7.4	7.4	4.2	2.5	3.2	4.4	4.4	4.4	3.8
Other Asia and Pacific	2.6	3.6	3.1	5.0	3.9	4.4	3.8	3.9	3.9	3.9
Middle East	8.1	5.2	6.7	0.8	2.0	1.2	3.3	3.5	3.4	2.9
Africa	3.3	5.8	4.6	2.3	2.9	2.4	3.5	3.1	3.3	3.2
Latin America and Caribbean	6.8	5.4	6.1	0.5	3.5	2.1	4.7	4.8	4.7	4.3
Centrally planned economies (CPEs)	5.7	4.5	5.1	4.6	4.7	4.8	4.1	3.2	3.7	4.0
USSR and East Europe	5.8	3.5	4.6	3.0	3.4	3.2	3.3	2.5	2.9	3.1
China and other Asian CPEs	5.4	7.3	6.3	7.8	6.8	7.8	5.4	4.3	4.8	5.5

70-84 actual; 85 estimate; 86-2000 projection.

corresponding ratio for Japan, which has already peaked at 40.5% in 1985 and is expected to follow a declining course, to stand at about 34.1% in 2000. Though not as spectacular as in the case of the East Asian NICs group, there is also a noteworthy rise in the ratio of manufacturing industry to GDP in the ASEAN group, (Association of South East Asian Nations) most likely growing from 20.3% in 1985 to 25.6% in 2000.

(4) If we adopt an index of 1 for average real per capita GDP in the developing market economies as a whole (calculated in 1980 US $), the OECD economies had a level that was 10.1 times greater in 1980, and will in 2000 have a level that is 13.0 times greater. It should thus be noted that the north–south income gap will tend to increase rather than decrease. The above-defined per capita GDP index in the oil-producing developing market economies will fall from 2.4 (6.5 for those in the Middle East only) in 1980 to 2.0 (5.1 for Middle East countries) in 2000. While the index for non-oil-producing developing countries as a group records the same figure, 0.7, in 2000 as in 1980, in the case of the Asian NICs the index may be expected to rise from 2.2 (2.1 in the case of the East Asian NICs only) in 1980 to 4.6 (4.4 for East Asian NICs) in 2000.

Among the OECD countries, a rise in the above-defined index is especially noteworthy in Japan, going from 9.8 in 1980 to 16.6 in 2000, within a relatively small margin of equalling the corresponding figure of 15.0 for the United States in the latter year. Looking at real per capita income calculated in 1980 US $, the $912 figure for the developing market economies as a whole will increase to $1121 by 2000, while the $9599 figure for the OECD economies as a whole in 1980 would be expected to increase to $14 607 in 2000.

2.4.2 Alternative projection

What are the results of the alternative policy mix projection which presupposes greater OECD co-operation aimed at additionally encouraging economic development in the developing countries?

(1) According to this optimistic projection, the oil price will in 1990 be $24 per barrel, or a nominal $2 higher than in the baseline projection. In the year 2000, it will be $33 per barrel, or $8 higher than in the baseline projection. It is seen that as the real oil price will remain nearly constant, this is in keeping with improved conditions for real growth in the world economy.

(2) The average yearly growth rate of the OECD economies as a whole is 0.66 percentage point above that of the baseline projection for the period 1986–2000, averaging nearly 3.5% in comparison with the 2.8% seen in the baseline projection. The baseline of 2.9% for the decade of the 1990s is similarly upgraded (by 0.72 percentage point), with the expectation that it would maintain a level of at least 3.6% (Fig. 2.1). The average yearly growth rate for the world economy as a whole is forecast as approximately 3.7% during the latter half of the 1980s and 3.9% during the decade of the 1990s. Although this rate is during both periods 0.53 percentage point higher than

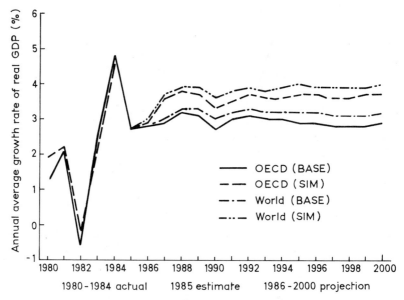

Fig. 2.1 Projections of the OECD economies in the global perspective: annual average growth rates of real GDP. BASE and SIM denote baseline and alternative policy simulation

in the baseline forecast, in the case of the developing market economies (taken as a whole) average annual growth may be expected to be only 0.46 percentage point higher than in the baseline projection for the latter half of the 1980s (reaching 3.8%), but thereafter an average 0.68 percentage point higher for the decade of the 1990s, registering an average 4.9%. Since average yearly economic growth in the developing market economies is estimated as having been only 2.2% during the first half of the 1980s, even with this optimistic scenario's forecast improvement during the latter half of the decade, average annual growth during the 1980s as a whole would be only 3.0%, falling short of the 7% goal set for the United Nations Third Development Decade, 1981–1990 (Tables 2.2 and 2.3).

The yearly average growth of 1.4% in the economies of the oil-producing developing countries during the 1980s increases to 4.5% in the 1990s, and the average 3.9% growth in the nonoil-producing developing countries during the 1980s rises to 5.1% during the subsequent decade.

(3) As a result of increased growth rates for the world economy as a whole, world trade will definitely follow a path of expansion and is seen to be greater than in the baseline projection by approximately $119 billion nominal in 1990 and greater by approximately $693 billion in the year 2000. In real terms (calculated in constant 1975 dollars), this represents an improvement, i.e. expansion, in the world trade figures of approximately $39 billion in 1990 and of approximately $210 billion in 2000.

Table 2.2 Projection of the OECD economies in the global perspective: effect of alternative policy-mix on annual average growth rates of real GDP; unit: %

	70–75	75–80	70–80	80–85	85–90	80–90	90–95	95–2000	90–2000	85–2000
World	3.9	3.9	3.9	2.7	3.7	3.2	3.9	3.8	3.9	3.8
Pacific Basin	3.5	4.3	3.9	3.7	4.1	4.0	4.4	4.2	4.3	4.2
World (excluding CPEs)	3.6	3.8	3.7	2.3	3.3	2.7	3.9	3.9	3.9	3.7
Pacific Basin market economies	3.3	3.9	3.6	3.1	3.5	3.4	4.2	4.1	4.2	4.0
Developed market economies (AMEs)	3.2	3.4	3.3	2.3	3.2	2.8	3.6	3.7	3.6	3.5
OECD	3.1	3.4	3.3	2.3	3.2	2.8	3.6	3.7	3.6	3.5
The major seven	3.0	3.5	3.3	2.4	3.4	2.9	3.8	3.8	3.8	3.7
Japan	4.6	5.1	4.9	4.2	4.3	4.2	5.1	6.0	5.5	5.1
Canada	5.0	2.9	4.0	2.1	3.1	2.6	3.1	3.4	3.2	3.2
United States	2.5	3.4	2.9	2.6	3.3	3.0	3.8	3.3	3.6	3.5
France	4.0	3.2	3.6	1.1	2.6	1.8	4.2	4.5	4.4	3.8
Germany (Federal Republic of)	2.1	3.6	2.8	0.9	2.9	1.9	2.6	2.3	2.5	2.6
Italy	2.4	3.8	3.1	0.6	3.0	1.9	3.0	3.0	3.0	3.0
United Kingdom	2.2	1.7	1.9	1.9	2.6	2.3	2.7	2.4	2.5	2.5
EC	2.8	3.1	2.9	1.1	2.7	1.9	3.0	3.0	3.0	2.9
Other developed market economies	6.1	2.6	4.4	1.8	2.3	1.8	3.3	4.5	3.9	3.4
Developing market economies (DMEs)	5.9	5.6	5.8	2.2	3.8	3.0	5.0	4.9	4.9	4.5
Oil-exporting countries	6.2	5.6	5.9	1.2	2.1	1.4	4.7	4.4	4.5	3.7
OPEC in Middle East	8.2	5.1	6.7	0.3	2.1	1.0	3.5	3.7	3.6	3.1
Nonoil-exporting countries	5.8	5.6	5.7	2.8	4.8	3.9	5.1	5.1	5.1	5.0
NICs	8.7	6.9	7.8	2.2	5.6	4.0	6.8	6.3	6.5	6.2
Asian NICs	8.7	9.3	9.0	6.4	6.9	6.7	6.6	6.1	6.4	6.6
Latin America NICs	8.7	6.3	7.5	0.9	5.0	3.2	6.8	6.3	6.6	6.1
Asia and Pacific	5.0	6.0	5.5	5.0	4.5	4.8	4.9	4.9	4.9	4.8
East Asia	8.6	9.4	9.0	6.4	7.4	6.9	6.6	6.0	6.3	6.7
ASEAN	7.4	7.4	7.4	4.2	2.7	3.3	4.8	4.9	4.9	4.2
Other Asia and Pacific	2.6	3.6	3.1	5.0	4.1	4.5	4.0	4.2	4.1	4.1
Middle East	8.1	5.2	6.7	0.8	2.3	1.4	3.8	3.9	3.8	3.3
Africa	3.3	5.8	4.6	2.3	3.1	2.4	3.8	3.5	3.6	3.4
Latin America and Caribbean	6.8	5.4	6.1	0.5	4.4	2.5	5.9	5.6	5.8	5.3
Centrally planned economies (CPEs)	5.7	4.5	5.1	4.6	4.8	4.9	4.2	3.3	3.7	4.1
USSR and East Europe	5.8	3.5	4.6	3.0	3.6	3.3	3.4	2.6	3.0	3.2
China and other Asian CPEs	5.4	7.3	6.3	7.8	6.8	7.8	5.4	4.3	4.8	5.5

70–84 actual; 85 estimate; 86–2000 projection.

Table 2.3 Impacts of alternative policy-mix on the OECD economies in the global perspective: changes of alternative simulation over the baseline for the year 1990 (D = difference, R = ratio)

	GDP (D) (%)	CPI (D) (%)	WPI (D) (%)	UNEMPR (D) (%)	ETFOB (R) (%)	MTFOB (R) (%)	TB (D) MDS	CBP (D) MDS	FERSI (R) (%)
World	0.512	0.100	0.137	−0.064	5.116	5.116	2	−9365	−0.000
World (excluding CPEs)	0.609	0.099	0.147	−0.105	5.356	5.369	−583	−9951	−0.000
Asia and Pacific Basin	0.854	−0.112	0.254	−0.282	4.764	6.382	−18650	−21980	0.091
Developed market economies (AMEs)	0.583	−0.035	0.089	−0.262	4.880	5.782	−15847	−23253	−0.169
OECD	0.588	−0.035	0.092	−0.275	4.870	5.828	−17302	−24787	−0.203
The major seven	0.694	−0.088	0.087	−0.355	5.389	6.747	−20154	−26391	−2.771
Japan	1.198	−0.473	2.016	−0.149	−2.633	16.803	−31956	−34839	16.266
Canada	0.371	−0.075	−0.282	−0.009	5.366	5.295	1071	922	−2.693
United States	0.843	0.012	−0.109	−0.651	9.887	5.437	4454	4500	0.0
France	0.362	−0.713	−1.052	−0.330	9.060	6.456	3881	3094	−15.217
Germany (Federal Republic of)	0.287	−0.074	−0.480	−0.278	4.915	5.975	−139	−785	−5.744
Italy	0.065	0.899	0.260	−0.098	6.893	4.021	1651	1376	0.0
United Kingdom	0.123	0.157	−0.054	−0.002	5.877	4.544	882	−659	−1.456
EC	0.186	0.005	−0.371	−0.131	5.596	4.538	8824	6215	−1.641
Other developed market economies	0.294	0.067	0.018	−0.003	5.338	1.806	1454	1534	0.0
Developing market economies (DMEs)	0.730	0.659	0.404	−0.035	6.917	3.972	15264	13302	−0.000
Oil-exporting countries	0.364	0.247	0.194	−0.001	8.366	4.803	7048	5517	0.808
OPEC in Middle East	0.527	0.043	−0.242	−0.000	9.162	3.744	3382	3014	−0.300
Nonoil-exporting countries	0.926	0.800	0.450	−0.043	6.140	3.575	8216	7784	−0.000
NICs	1.314	0.978	0.615	−0.102	6.194	3.853	5792	5848	−0.938
Asian NICs	0.933	−0.675	−0.779	−0.114	4.597	1.951	3887	3992	0.156
Latin America NICs	1.458	1.008	0.602	−0.098	9.684	9.525	1904	1856	−0.939
Asia and Pacific	0.465	0.001	−0.077	−0.026	5.309	2.561	5875	5439	0.407
East Asia	0.921	−0.749	−0.884	−0.106	3.801	1.459	2829	2706	−0.310
ASEAN	0.382	−0.086	−0.167	−0.072	7.171	3.226	3109	3017	−0.047
Other Asia and Pacific	0.270	0.341	0.288	−0.007	5.958	4.042	−63	−285	1.545
Middle East	0.497	0.040	−0.273	−0.000	8.494	3.751	3566	3134	0.030
Africa	0.281	0.263	0.181	−0.020	8.295	5.345	1349	1115	0.866
Latin America and Caribbean	1.240	0.922	0.490	−0.089	7.925	6.256	4473	3613	−0.000
Centrally planned economies (CPEs)	0.140	−0.012	−0.007	0.0	3.360	3.223	585	585	−0.719
USSR and East Europe	0.226	0.006	0.004	0.000	3.022	3.125	284	284	−0.600
China and other Asian CPEs	0.007	−0.045	−0.025	0.0	7.732	3.965	301	301	−0.787

(4) Although rates of price increases in the world economy as a whole will be somewhat higher than in the case of the baseline projection due to a higher rate of growth in the world economy together with somewhat higher prices for oil and other primary products, this rise in costs will probably not be so great as some people might imagine since it will be to a considerable extent absorbed by improvements in productivity that will accompany appropriate currency management. With the exception of the centrally planned economies, the average yearly increase in the wholesale price index for the rest of the world economy as a whole, for example, will rise slightly from 10.1% in the baseline projection 1986–1990 to 10.3%, and from 7.6% to 8.0% for 1990–2000. The average yearly price increase in the OECD economies will remain unchanged during 1980–1990 at 4.8% in both the baseline forecast and the alternative policies forecast, but is expected to increase under the alternative policy conditions by 0.2 percentage point from an average 5.6% to an average 5.8% during the period 1990–2000. In the developing countries, likewise, there would not seem to be any great danger that serious global inflation might be induced due to the acceleration of global economic growth.
(5) Among the OECD countries as a whole, the real per capita GDP (in 1980 dollars) is $11 898 in 1990 and $16 002 in 2000, which represents an improvement over the baseline projection of $301 and $1415, respectively. The corresponding figures for the developing market economies as a whole are $953 in 1990 and $1211 in 2000. Japan's per capita income is seen as rising from $13 202 in 1990 to $21 547 in 2000, approximately on a par with the $19 387 figure (year 2000) forecast for the United States.

2.5 IMPACTS OF ALTERNATIVE POLICIES ON THE JAPANESE ECONOMY

Let us ask, what might be the future course of the Japanese economy, which recently has received the double impact of a sudden rise in the value of the yen and a lowering of the price of oil. The area of extremely high oil prices induced by OPEC and of extraordinarily high exchange values for the US dollar has now ended as the world economy, holding possibilities for new development led by sophisticated technological innovations, gropes towards the 21st century.

The new growth in the world economy is becoming centred on the Pacific Basin, and Japan is very much being expected to show leadership in economic management in a global context. Precisely now is the time that Japan ought to show the world blueprints for economic development grounded in a global perspective.

In the face of the rapid rise in the value of the yen and drop in oil prices, Japan should not respond passively, but rather should actively carry out appropriate and comprehensive policies. In that way, then, to what extent can the Japanese economy contribute to the welfare of the world as a whole?

Market economic systems are constantly being subjected to alteration as a result of changes in information and in people's psychology, and may thus be said to be extremely delicate. However, such systems are by no means led only by what Adam Smith called the unseen hand. They are system whose courses of development can, if found to be off course, be put back on the right track through the interplay of human intelligence mediated by large-scale informational systems. The opportunity, that is to say the real possibility, has come for Japan and the world economy to change the economic environment and to cast away the traps of low growth and high unemployment into which they have fallen as a result of the oil shock.

How are things likely to develop if present conditions continue? Indeed, the rapid rise in the value of the yen and the drop in oil prices have begun to have a large impact on Japan's economy. For example, an upward revaluation of 40 yen with respect to the US dollar may be expected to reduce real exports by 0.5 percentage point in 1986 and by 5.5 percentage points in 1987, in comparison with what they would have been otherwise. Real imports, on the other hand, may be expected to increase by 0.13 percentage point in 1986 and by 1.6 percentage point in 1987.

If, however, we look at the values of imports and exports on a nominal dollar basis, the trend toward improving the trade imbalance in real terms is cancelled out in nominal terms, with the result that Japan's overall trade surplus will actually grow by $19.8 billion in 1986 and the Japan–US trade imbalance will rather increase in dollar terms. The Japanese economy can be expected to be faced with a strong-yen recession with a drop in real economic growth rate of 1.6 percentage points in 1986 in comparison with what it would otherwise have been if all other factors were assumed to be unchanged. This will probably have the effect of applying a brake to the expansion of imports, and may be called the paradox of the stronger yen. This stronger value of the yen may be expected to lower wholesale prices by 5.8 percentage points and to lower consumer prices by 1.2 percentage points during 1986.

As an additional factor, the lowering of the price of oil by $5 per barrel will in 1986 tend to depress wholesale prices by an additional 0.4 percentage point. This drop in the price of oil can at the same time be expected, however, to bring about increased real economic growth by 0.4 percentage point in 1986 and by 0.2 percentage point in 1987 in comparison with what it would be if oil prices had remained $5 per barrel higher.

Consequently, with the continuation of present trends whereby the effects of the higher yen and cheaper oil in large part reinforce one another, Japan's wholesale and consumer prices are forecast to undergo the extraordinary situation in 1986 of falling by 10.1% and of rising by 0.8%, respectively, in comparison with the previous year. According to our estimates, this year's growth rate is said to be most likely in the neighbourhood of 2.5%. In the baseline projection, it is predicted to rise in 1987, but only to 3.4%, this stronger yen recession being mitigated by the recent co-ordinated lowering of interest rates in the international arena. The growth rate is in our forecast predicted to recover in 1988, and to register a yearly average of 3.1% for the period 1986–1990, and 4.6% for 1990–2000.

Japan's current balance of payments surplus will in the baseline projection decline somewhat in 1986 and 1987, to 88.9 and $84.3 billion respectively; but beginning in 1989 it will again begin to increase, and will reach $109.2 billion in 1990 and $119.0 billion in 2000. Japan's FOB-base trade surplus with the United States will decrease from $53.5 billion in 1986 to $49.7 billion in 1988. However, it will increase again in 1989 and reach $61.0 billion in 1990 and $82.8 billion in 2000, with the result that Japan–US trade friction will probably not be improved (Fig. 2.2).

If the above-described alternative policy mix were adopted, the real growth of the Japanese economy could be expected to average 4.3% during 1986–1990 and 5.5% during 1990–2000. By opening the market and expanding direct overseas investment by private enterprises, the rate of nominal import growth would increase (over and above the baseline projection) by an average of 3.3 percentage points during the coming 5-year period to an average level of 7.4%, while nominal

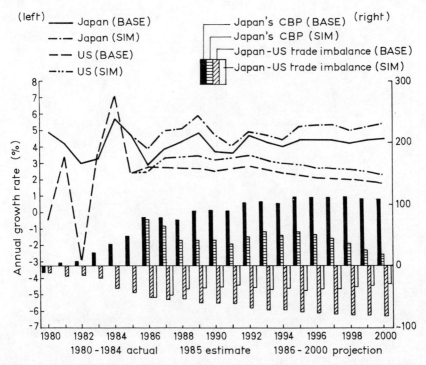

Fig. 2.2 Projections of Japan's and the US economic growth, Japan's current balance of payments and the US–Japan trade imbalance. BASE and SIM denote baseline and alternative policy simulation.

exports would on the other hand decrease by an average −0.56 percentage point to an average annual growth rate of 4.9%. In this way, an economic growth would be achieved which encompasses an active readjustment of patterns of production, imports and exports.

As a result, Japan's surplus in the current balance of payments would be less than in the baseline projection by $13.9 billion in 1987, less by $34.8 billion in 1990, and less by $74.2 billion in 2000. Thus the absolute size of the surplus would peak at $70.4 billion in 1987, declining to $45.5 billion in 1990 and $44.7 billion in 2000.

There is a possibility that Japan's surplus *via-à-vis* the United States could decrease from $53.5 billion in 1986 to a level of $42.2 billion in 1990, though in that case it would likely again rise to reach a level of some $73.4 billion in 2000. A greater decrease in the short term would be more problematic because further improving the Japan–US trade imbalance would be not so much the responsibility

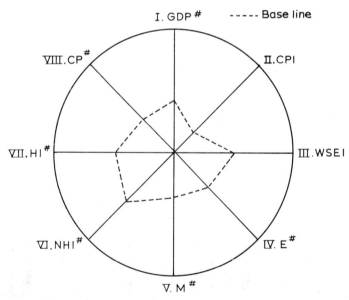

Fig. 2.3 Impact of alternative policy on the Japanese economy.
GDP#   annual average economic growth rate 86–2000
CPI   consumer price index (average percentage changes 86–2000)
WSEI   nominal wage rate per employee (average percentage changes 86–2000)
E#   exports of goods and services (const.) (average percentage changes 86–2000)
M#   imports of goods and services (const.) (average percentage changes, 86–2000)
NHI#   non-housing investment (const.) (average percentage changes 86–2000)
HI#   housing investment (const.) (average percentage changes 86–2000)
CP#   private final consumption expenditure (const.) (average percentage changes 86–2000)

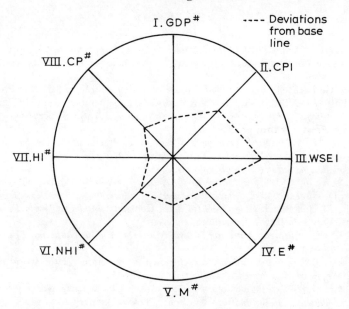

Fig. 2.4 Impact of alternative policy on the OECD economies.

of Japan, but would rather involve the structure of American exports and imports as they relate to the activities of US-based multinational corporations.

With the above-described alternative policy mix, Japan's wholesale prices would rise by an average 2.2% yearly during the period 1986–2000, while the rise in consumer prices could, given the opening of the market to foreign competitors, be expected not to exceed a yearly average of approximately 2.4%. With the application of the above-outlined policy package, we could expect a future in which the Japanese economy operates in greater harmony with the economy of the world as a whole (Figs 2.3, 2.4).

ACKNOWLEDGEMENTS

The computations were done by HITAC M-260H with GEMS software at the Center for Global Modelling, Soka University. The author is grateful to his global Modelling staff members including Kazuo Aoki, Osamu Nakamura, Kimihiro Onishi and Norihiro Tsuruta for their helpful co-operation in the computer modelling work. Assistance in computation work made by Hitachi Co. is also gratefully acknowledged. Please address correspondence to Professor Akira Onishi, Director, Center for Global Modelling, Soka University, Hachioji-shi, Tokyo, Japan 192.

REFERENCES

Bremer, Stuart A. (1985) *The GLOBUS Model: History, Structure and Illustrative Results*, Institute for Comparative Social Research (Wissenschaftszentrum), Steinplatz 2, D-1000 Berlin 12, May.

Bruckman, Gerhart (ed.) (1980) FUGI: Future of global interdependence, in Proceedings of the Fifth IIASA Symposium on Global Modelling, September 26–28, 1977, reprinted in *Input–Output, Approaches in Global Modelling*, IIASA Proceedings Series 9, Pergamon Press, Oxford, pp. 91–360.

Forrester, Jay W. (1971) *World Dynamics*, Wright-Allen Press, Cambridge, Massachusetts.

Klein, Lawrence R., Pauly, Peter and Voisin, Pascal (1982) The world economy, a global model, in *Perspectives in Computing*, No. 2, IBM, Armonk, New York, pp. 4–17.

Leontief, W. et al. (1977) *The Future of the World Economy*, Oxford University Press.

Meadows, D. H. et.al. (1972) *Limits to Growth*, Potomac Associates, Washington DC.

Mesarović, M. and Pestel, E. (1974) *Multilevel Computer Model of World Development Systems*, Vols 1–6, IIASA, Laxenburg, Austria.

OECD (1979) *Interfutures Facing the Future: Mastering the Probable and Managing the Unpredictable*, Paris.

Onishi, Akira (1985) North–south interdependence: projections of the world economy, 1985–2000. *Journal of Policy Modelling*, 7 (4), pp. 1–18.

Onishi, Akira (1986) A Supplementary Report on the Global Early Warning System for Displaced Persons, United Nations University, Tokyo, January.

3

Supply-side policies in four OECD countries

ANTHONIE KNOESTER

3.1 INTRODUCTION

A key message of supply-side economics is that tax cuts could be a major help in increasing economic growth and lowering unemployment. However, up till now, there are hardly any quantitative foundations for such policy recommendations. This forms a serious limitation for the credibility of supply-side economics.

This chapter suggests a general framework for analysing supply-side economics with the help of macroeconomic models. The starting point is the contention of Klein (1978), Tinbergen (1982) and Tobin (1981) that in compiling a model – whether large or small – it is wrong to stress any theoretical issue unilaterally. We follow the synthetic approach which they recommend, meaning that all theoretical questions should be given in principle their full weight, being arranged according to schools, if desired. At the same time, it is best to keep a model as simple as possible. This is the more important because this chapter will deal with a macroeconomic model not for one but for four OECD countries, namely, Germany, the Netherlands, the United Kingdom and the United States.

As a result we have focused on the empirical elaboration of those behavioural equations which are essential for any analysis which pretends to deal with supply-side economics. For this purpose equations describing the supply-side of the economy have been added to a standard Keynesian income–expenditure block. These supply-side equations are first of all equations for production capacity and the utilization rate as can be derived from neoclassical growth theory. It should be emphasized, however, that we have also focused on the analysis of the phenomenon that higher taxes can be shifted forward into higher wages. As we will make clear the implications of such forward shifting are of fundamental importance for a proper view on supply-side policies. In particular, two elements seem to be relevant.

First, shifting forward of taxation provides an important explanation for the fact that in the past in almost all industrialized countries, real wages have expanded substantially. In Knoester and Van der Windt (1985) this has been investigated for ten OECD countries. It has been found that shifting forward of taxation explains 40 to 50 percentage points of real wage growth in the 1960s and 1970s in Australia, Canada, the Netherlands and Sweden. For Germany, Italy and the United States it explains about 25 percentage points and for France, Japan and the United Kingdom about 15 percentage points. These findings give an additional explanation for the occurrence of classical unemployment as pointed out by for example Malinvaud (1977).

A second important implication of forward shifting of taxation is that it suggests a completely different view on fiscal policy compared with the Keynesian and classical views. As shown in Knoester (1983, 1984) forward shifting results in lower economic growth and more unemployment as a consequence of a simultaneous increase in public spending and taxes or social security contributions. Thus, in the present situation, a *negative balanced-budget multiplier* seems to be more relevant than the classical balanced-budget multiplier of zero or the positive Keynesian one. In this chapter we will show that in reverse a simultaneous decrease in public spending and direct taxes provides a quantitative basis for tax cuts as advocated by supply-sides.

We will end with a discussion of the different ways of financing tax cuts. Most supply-siders either remain silent on the financial consequences of tax cuts or argue that such cuts can be made at the expense of the government budget deficit. Therefore the different ways of financing lower taxes will be analysed systematically for all four countries. We will discuss lower taxation financed by cutting public spending, higher indirect taxes and borrowing in the capital market. Also some aspects of international co-ordination of tax-cutting policies are discussed.

The plan of this chapter is as follows. Section 3.2 deals with the structure of the general model that has been used as a starting point for the empirical part of our analysis. In Section 3.3 empirical evidence is presented for four OECD countries, namely, for Germany, the Netherlands, the United Kingdom and the United States. Discussed are estimated equations for private consumption and investments, imports, exports, wages, employment and the inflation rate. Section 3.4 deals with the quantitative effects of supply-side policies. In particular the different ways of financing lower taxes are analysed. The final section ends with the main conclusions of our analysis.

3.2 A GENERAL MODEL FOR FOUR COUNTRIES

This section contains the general model developed for four countries, the values for fixed weights in the definition equations for these countries and a list of symbols used. This model must be completed with the estimated behavioural equations as discussed in Section 3.3. A dot on top of a symbol represents percentage changes, otherwise levels are used. A dash above symbols suggests that these symbols have been given a fixed value (in general the 1977 value). If needed, the model has been linearized; differences on behalf of this procedure are

A general model for four countries

not mentioned explicitly. The general model can be divided into six interdependent blocks.

The first block describes the supply side of the economy by means of a clay–clay type vintage model. The capital–output ratio is considered to be constant, while productivity increases per vintage because of the embodied labour-augmenting technological progress. The life span of the vintages is shortened when the rise in real labour costs exceeds the rise in productivity. Production capacity can be computed on the basis of these starting points. The details can be found in Knoester and Van Sinderen (1984). Besides production capacity, in addition actual output, productivity and the utilization rate are computed in this block.

The second block comprises private-sector employment. The structure of this equation is likewise grafted on the clay–clay type of vintage model which is used to compute production capacity. This block comprises moreover the definition of unemployment, assuming public sector employment and labour supply to be exogenous.

The third block covers effective demand. Foreign trade is determined by world trade and relative prices. Domestic consumption depends on disposable income. Net investment is determined by profits, the utilization rate, the rate of inflation and government debt.

In the fourth block wages and prices are determined. Prices depend on wage and import costs and on a demand–pull effect symbolized by the utilization rate. The wage equation comprises GDP prices and productivity. A high level of unemployment causes a persistent negative pressure on wages. Finally, wages are determined by social security charges and taxes on wages. As will be shown later, this forward shifting of direct taxes and social security contributions into wages is of prime importance for the occurrence of supply-side effects as a result of more taxation.

The fifth block is concerned with local and central government and the social security system. Government consumption, government investment and investment in housing are treated exogenously, as is the number of public sector employees. Wage rates in the public sector are linked to private sector wages. Tax revenues are defined in the standard way. The applied progression factor is based on OECD figures. Both social security benefits and contributions are exogenous. The sixth block, finally, comprises the definitions of income distribution.

In short, this model comprises a mixture of Keynesian income–expenditure analysis and neoclassical growth theory. In addition, it takes into account the effect of issuing government debt on the investment level. Equations which are essential for an adequate analysis of supply-side policies will be estimated.

3.2.1 The general model

(a) Supply

$$\dot{\text{cap}} = \alpha_1 (i/y)_{t-x} - \alpha_2 (\dot{p}_1 - \dot{p}_y)_{t-x} \tag{3.1}$$

$$\dot{y} = \overline{(V/Y)}_{t-1} \dot{v} - \overline{(M/Y)}_{t-1} \dot{m} \tag{3.2}$$

$$\dot{Y} = \dot{y} + \dot{p}_y \tag{3.3}$$

$$\Delta q = \dot{y} - \dot{\text{cap}} \tag{3.4}$$

$$q = \Delta q + q_{t-1} \tag{3.5}$$

$$\dot{h} = \dot{y} - \dot{\text{ab}} \tag{3.6}$$

(b) Labour market

$$\dot{a}_b = \beta_1 \dot{a}_{b_{t-1}} + \beta_2(\dot{p}_l - \dot{p}_y)_{t-x} + \beta_3(\Delta q)_{t-x} + \beta_4(i/y)_{t-x} + \beta_5 \tag{3.7}$$

$$\Delta u = \dot{p}_a - \overline{(LB/L)}_{-1} \dot{\text{ab}} \tag{3.8}$$

$$u = \Delta u + u_{t-1} \tag{3.9}$$

$$U^{-1} = (u^{-1})_{t-1} - \left(\frac{1}{u_{t-1}}\right)^2 \Delta u \tag{3.10}$$

(c) Demand

$$i/y = \gamma_1(i/y)_{t-1} + \gamma_2(Z/Y)_{t-x} + \gamma_3(\Delta q)_{t-x} + \gamma_4(S/Y)_{t-x} + \gamma_5 \dot{p}_{y_{t-x}} + \gamma_6 \tag{3.11}$$

$$\dot{i} = \overline{(y/i)}_{t-1} \Delta(i/y) + \dot{y} \tag{3.12}$$

$$\dot{c} = \varepsilon_1 \overline{(LB_{t-x}/C_{-1})} LD_{t-x} + \varepsilon_2 \overline{(ZB_{t-x}/C_{-1})} ZD_{t-x} + \varepsilon_3 \dot{p}_c + \varepsilon_4 \tag{3.13}$$

$$\dot{e} = \eta_1 \dot{m}w_{t-x} + \eta_2(\dot{p}_e - \dot{p}'_e)_{t-x} + \eta_3 q_{t-x} + \eta_4 \tag{3.14}$$

$$\dot{m} = \lambda_1 \dot{v}_{t-x} + \lambda_2(\dot{p}_m - \dot{p}_{vd})_{t-x} + \lambda_3(\Delta n/v_{-1})_{t-x} + \lambda_4 \tag{3.15}$$

$$\dot{v} = \overline{(c/v)}_{t-1} \dot{c} + \overline{(i/v)}_{t-1} \dot{i} + \overline{(e/v)}_{t-1} \dot{e} + \overline{(x/v)}_{t-1} \dot{x} + 1.0(\Delta n/v_{-1}) + ex \tag{3.16}$$

(d) Prices and wages

$$\dot{p}_l = \mu_1 \dot{p}_{y_{t-x}} + \mu_2 \dot{h}_{t-x} + \mu_3(\Delta DR^{wn})_{t-x} + 1.0 \Delta DR^{wg} + \mu_4 u_{t-x}^{-1} \tag{3.17}$$

$$\dot{p}_v = o_1(\dot{p}_l - \dot{h})_{t-x} + o_2 \dot{p}_{m_{t-x}} + o_3 \Delta q_{t-x} + o_4 \overline{(V_d/V)}_{-1} \Delta(TKS/V_d)_{t-x} \tag{3.18}$$

$$\dot{p}_y = \overline{(v/y)}_{t-1} \dot{p}_v - \overline{(m/y)}_{t-1} \dot{p}_m \tag{3.19}$$

$$\dot{p}_e = \rho_1 \dot{p}'_{e_{t-x}} + \rho_2[\dot{p}_{vd} - \Delta(TK/V_d) - \dot{p}'_e]_{t-x} + \rho_3 \tag{3.20}$$

$$\dot{p}_c = 1.0 \dot{p}_{vd} \tag{3.21}$$

$$\dot{p}_{vd} = \overline{(V/VD)}_{-1} \dot{p}_v - \overline{(E/V_d)}_{-1} \dot{p}_e \tag{3.22}$$

(e) Public sector

$$\dot{TL} = p_f L \dot{TL} \tag{3.23}$$

A general model for four countries

$$FT/Y = 1.0(FT/Y)_{t-1} + \overline{(LOV/Y)}_{t-1}\dot{LOV} + \overline{(X/Y)}_{t-1}(\dot{x} + \dot{p}_v) - \overline{(TL/Y)}_{t-1}\dot{TL}$$
$$+ \overline{(SU/Y)}_{t-1}\dot{SU} + \overline{(SVO/Y)}_{t-1}\dot{SVO} - \overline{(PR/Y)}_{t-1}\dot{PR} \qquad (3.24)$$

(f) Income distribution

$$\dot{LB} = \dot{p}_l + \dot{a}_b \qquad (3.25)$$

$$\dot{LOV} = \dot{p}_l + a\dot{o}v \qquad (3.26)$$

$$\dot{L} = \overline{(LB/L)}_{t-1}\dot{LB} + \overline{(LOV/L)}_{t-1}\dot{LOV} \qquad (3.27)$$

$$\dot{LTL} = \overline{(L/LTL)}_{t-1}\dot{L} + \overline{(SU/LTL)}_{t-1}\dot{SU} + \overline{(SVO/LTL)}_{t-1}\dot{SVO}$$
$$- \overline{(PR/LTL)}_{t-1}\dot{PR} \qquad (3.28)$$

$$\dot{LD} = \overline{(L/LD)}_{t-1}\dot{L} - \overline{(TL/LD)}_{t-1}\dot{TL} + \overline{(SU/LD)}_{t-1}\dot{SU}$$
$$+ \overline{(SVO/LD)}_{t-1}\dot{SVO} - \overline{(PR/LD)}_{t-1}\dot{PR} \qquad (3.29)$$

$$\dot{Z} = \overline{(Y/Z)}_{t-1}\dot{Y} - \overline{(L/Z)}_{t-1}\dot{L} \qquad (3.30)$$

$$\dot{ZD} = \overline{(Z/ZD)}_{t-1}\dot{Z} - \overline{(TZ/ZD)}_{t-1}\dot{TZ} \qquad (3.31)$$

$$Z/Y = (Z/Y)_{t-1} + 0.3(\dot{Z} - \dot{Y}) \qquad (3.32)$$

$$\Delta DR = \overline{(TL/L)}_{t-1}(\dot{TL} - \dot{L}) + \overline{(PR/L)}_{t-1}(\dot{PR} - \dot{L}) \qquad (3.33)$$

$$\Delta DR^{wn} = [\overline{(TL/(SU+L_b))}]_{t-1}\dot{TL} + [\overline{(PR/SU+L_b)}]_{t-1}\dot{Pr}^{wn} \qquad (3.34)$$
$$- [\overline{(SU/(SU+L_b))}]_{t-1}[\overline{(TL+PR)/(SU+L_b)}]_{t-1}\dot{SU}$$
$$- [\overline{(L_b/(SU+L_b))}]_{t-1}[\overline{(TL+PR)/(SU+L_b)}]_{t-1}\dot{L}_b$$

$$\Delta DR^{wg} = \overline{(PR^{wg}/L_b)}_{t-1}(\dot{PR}^{wg} - \dot{L}_b) \qquad (3.35)$$

$$\dot{L}_b = \dot{LB} - \dot{PR}^{wg} \qquad (3.36)$$

Endogenous: $c\dot{a}p$, $\dot{y}$, $\dot{Y}$, Δq, q, $\dot{h}$, $\dot{a}_b$, Δu, u, u^{-1}, i/y, $\dot{i}$, $\dot{c}$, $\dot{e}$, $\dot{m}$, $\dot{v}$, $\dot{p}_l$, $\dot{p}_v$, $\dot{p}_y$, $\dot{p}_e$, $\dot{p}_c$, $\dot{p}_{vd}$, $\dot{TL}$, FT/Y, $\dot{LB}$, $\dot{LOV}$, $\dot{L}$, $\dot{LTL}$, $\dot{LD}$, $\dot{Z}$, $\dot{ZD}$, Z/Y, ΔDR, ΔDR^{wn}, ΔDR^{wg}, L_b

Exogenous: S/Y, $\dot{p}'_e$, $(\Delta n/v_{-1})$, TKS/V_d, $\dot{p}_b$, $\dot{SU}$, $\dot{SVO}$, $\dot{PR}$, $\dot{PR}^{wn}$, $\dot{PR}^{wg}$, $a\dot{o}v$

In general capitals refer to nominal values and small letters to volumes or prices.

(g) List of symbols

Endogenous:

a_b = private sector employment
c = volume of private consumption

Table 3.1 Values for fixed weights in the models

Symbol	Occurrence in equation	Germany	Netherlands	United Kingdom	United States
LB/L	8, 27	0.80	0.77	0.77	0.83
$(1/u)^2$	10	0.06	0.06	0.03	0.02
LOV/L	27	0.20	0.23	0.23	0.17
L/LTL	28	1.01	0.91	0.95	0.96
SU/LTL	28	0.20	0.31	0.10	0.12
SVO/LTL	28	0.05	0.06	0.05	0.04
PR/LTL	28	0.26	0.28	0.10	0.12
L/LD	29	1.26	1.12	1.18	1.14
LT/LD	29	0.26	0.23	0.24	0.19
SU/LD	29	0.25	0.37	0.12	0.14
SVO/LD	29	0.07	0.07	0.07	0.05
PR/LD	29	0.32	0.33	0.13	0.14
Y/Z	30	2.88	3.12	3.77	3.43
L/Z	30	1.74	1.80	2.25	2.10
y/i	12	8.53	8.33	7.82	8.52
c/v	16	0.50	0.44	0.52	0.66
i/v	16	0.14	0.12	0.12	0.18
e/v	16	0.22	0.35	0.23	0.08
x/v	16	0.10	0.05	0.06	0.08
v/y	2, 19	1.23	1.51	1.29	1.09
m/y	2, 19	0.23	0.51	0.29	0.09
TL/L	32	0.21	0.21	0.20	0.17
PR/L	32	0.26	0.31	0.12	0.12
p_f	23	1.64	1.73	1.48	1.55
LOV/y	24	0.11	0.14	0.14	0.11
X/Y	24	0.12	0.07	0.07	0.08
TL/Y	24	0.10	0.11	0.11	0.09
SU/Y	24	0.11	0.20	0.06	0.08
SVO/Y	24	0.03	0.04	0.03	0.02
PR/Y	24	0.14	0.18	0.07	0.07
V_d/V	18	0.789	0.675	0.764	0.928
V/V_d	22	1.267	1.481	1.309	1.078
e/V_d	22	0.267	0.481	0.309	0.078
$TL/(SU+L_b)$	34	0.185	0.172	0.198	0.168
$PR/(SU+L_b)$	34	0.131	0.120	0.043	0.045
$SU/(SU+L_b)$	34	0.250	0.334	0.158	0.148
$(TL+PR)/(SU+L_b)$	34	0.315	0.292	0.241	0.213
$L_b/(SU+L_b)$	34	0.750	0.666	0.842	0.852
$PRWG/L_b$	34	0.216	0.282	0.153	0.184
LW/L_b		1.216	1.282	1.153	1.184
LB/C	13	0.74	0.74	0.76	0.75
ZB/C	13	0.41	0.42	0.23	0.26

A general model for four countries

C = nominal private consumption
cap = production capacity
FT = public sector deficit
e = exports
h = labour productivity
i = gross private investments
L = compensation of employees
L_b = gross wages
LB = compensation of employees in enterprises
LD = disposable income
LOV = compensation of civil servants
LTL = total income before taxes
m = imports
n = stock formation
p_i = investment price deflator
p_l = wage rate
p_e = prices of exports
p_v = total expenditure price deflator
p_{vd} = domestic expenditure price deflator
p_y = GDP price deflator
q = utilization rate
TL = direct taxes on wages and transfer income
u = unemployment rate
v = total expenditure
V_d = domestic expenditure
y = gross domestic product (GDP)
Y = nominal GDP
Z = profits
ZD = disposable profits
DR^{wn} = forward shifting by employees
DR^{wg} = forward shifting by employers
DR = forward shifting of direct taxes and social security contributions

Exogenous:

aov = civil servants
d_{74} = dummy 1974
m_w = world trade
p_a = labour supply
p'_e = prices of competitors on foreign markets
p_f = progression factor of taxes
p_m = imports price deflator
p_{vv} = price deflator of competitors in foreign markets
PR = social security contributions
S = debt of general government

SU = social security benefits
SVO = social assistance grants
X = government expenditure excluding wages and salaries
TZ = taxes on profits
TKS = indirect taxes

3.3 ESTIMATED EQUATIONS

For Germany, Holland, the United Kingdom and the United States eight equations have been estimated each time for the wage rate, prices of domestic expenditure, prices of exports, private sector employment, consumption and investments and exports and imports. Those used are yearly figures, with 1960–1982 as the sample period. The employed estimation technique is ordinary least squares (OLS). We have started from the following hypotheses.

For nominal wages we assume that they are fixed in a bargaining process of which roots can be found in Phillips (1958) and in Dicks-Mireaux and Dow (1959). The bargaining model has been treated more explicitly especially in recent times. For example, see Johnston (1974), Johnston and Timbrell (1973), OECD (1978), Corden (1981), Brandsma and Van der Windt (1983) and Knoester and Van der Windt (1985). According to these papers wage-earners are primarily concerned with increasing or at least maintaining real net earnings. As a result, wage-earners will try to shift higher direct taxes and social security contributions forward into higher nominal wages. In addition to this shifting 'traditional' determining factors are relevant. Operation of the Phillips curve mechanism can be seen in this respect as the factor representing the relative forces of employers' and workers' claims in the bargaining process. Inclusion of productivity and the inflation rate are, of course, standard elements in wage bargaining.

In addition to nominal wages, the inflation rate must be explained, which is done in the standard way. Apart from cost factors such as wage and import costs, the inflation rate depends on a demand–pull factor, the utilization rate. Prices of exports depend partly on export prices of foreign competitors.

Private sector employment growth depends largely on structural factors as suggested by the clay–clay vintage supply block used to quantify production capacity. As shown in Knoester and Van Sinderen (1984), this model can make it clear that the investment ratio, real wage costs and technical progress are determinants of private sector employment growth. In addition the utilization rate is included as a demand factor. As the desired employment growth in one period need not necessarily correspond with actual developments, room has been made also for the one-year lagged private sector employment growth.

Also, we have estimated a behavioural equation for investment as a percentage of GDP. The following determining factors have been investigated: gross profit ratio, government debt as a percentage of GDP, changes in the utilization rate, and the inflation rate. The one year lagged gross investment ratio can be included

Estimated equations

in principle. Most of these terms are fully in line with the standard ingredients of investment behaviour. The profit ratio, for instance, represents the internal finance possibilities of new net investment. In addition, this term can be interpreted as a proxy for anticipated profits. Besides profits, external finance possibilities can be important for net investment. Economic literature mentions many possibilities, such as real interest rates, some money or liquidity ratio, a real cash-balance term, a credit-rationing term and so on. All these terms have in common that they represent in one way or another a relevant transmission channel for monetary impulses to the real sector of the economy. We have approximated such channels by the inclusion of government debt as a percentage of GDP.

The inclusion of changes in the utilization rate represents the impact of demand on gross investment. The combination of the *level* of the gross investment with *changes* in the utilization rate has been chosen because this connection can be derived from the hypothesis that the desired size of the capital stock depends on the level of the utilization rate. The inflation rate has been included to quantify the negative effects of inflation-induced uncertainties on gross investment.

The equations for private consumption, imports and exports have been estimated in the standard way. Determinants are real disposable income, world trade, relative prices and home pressure of demand. Tables 3.2–3.9 show the estimated results for the equations under discussion. Most of the results speak for themselves, so that it suffices to end this section with a few notes only. First, it is striking that for investment, private-sector employment, wages and prices in Germany, the Netherlands, the United Kingdom and the United States, an acceptable fit is obtainable with the aid of the determining factors that are identical across the board. As for the estimated price equations, note that in all four countries the coefficients for wage costs and import costs add up to approximately unity. This is in line with the price theory – as pointed out in Eckstein and Fromm (1968) among others – in which fixed costs come on top of the variable costs via a fixed mark-up factor. The size of the coefficient for import costs varies moreover with the openness of the economy in question.

The estimated wage equations show interesting outcomes. In all countries employees' forward shifting of direct taxes and social security contributions into higher nominal wages is significant. The coefficient found for the Netherlands points to an almost complete forward shifting, whereas in Germany, the United Kingdom and the United States nearly three-quarters and one-half is shifted forward respectively. It is noted that an OECD study of 1978 on public expenditure trends showed that there was a certain forward shifting of direct taxes. As we shall see later on, the established forward shifting is of decisive importance for a correct appreciation of supply-side policies.

The estimated equations for private sector employment show interesting results too. It is possible to trace a similar determining structure for the four countries, having marked neoclassical characteristics such as the included investment ratio and real wage growth. Taking the established partial adjustment coefficient into account, we find that elasticity between private sector employment and real wages

Table 3.2 Estimated equations for private employment growth, sample 1960–1982

	Private employment	One-year lagged private employment	Real wages	Utilization rate	Gross investment ratio	Constant	R^2	DW	SEE
Germany	$\dot{a}_b =$	$0.50(t-1)$ (4.80)	$-0.43t_{333}$ (−3.49)	$+0.54t$ (8.34)	$+0.25(t-1)$ (5.67)	-1.50 (−)	0.82	2.73	0.46
Netherlands	$\dot{a}_b =$	$0.39(t-1)$ (2.33)	$-0.29t_{4444}$ (−2.14)	$+0.46(t-1/3)$ (4.34)	$+0.25(t-1)$ (5.21)	-1.80 (−)	0.73	1.68	0.53
United Kingdom	$\dot{a}_b =$	$0.35(t-1)$ (2.50)	$-0.49t_{333}$ (−2.67)	$+0.89(t-\frac{1}{2})$ (5.43)	$+0.25(t-1)$ (6.05)	-2.00 (−)	0.67	2.16	1.1
United States	$\dot{a}_b =$	$0.33(t-1)$ (2.47)	$-0.77t_{333}$ (−2.33)	$+1.02(t-1/4)$ (7.49)	$+0.50(t-1)$ (11.95)	-3.10 (−)	0.77	1.88	1.1

Annual figures. Numbers in parentheses are t-values, lags are symbolized by $t-x$, where $x=1$ is one year lag; t_{333} represents weighted lags, namely, $0.3t + 0.3(t-1) + 0.3(t-2)$. R^2 = Squared correlation coefficient, SEE = Standard error of estimation, DW = Durbin–Watson statistic.

Table 3.3 Estimated equations for the gross investment ratio, sample 1960–1982

	Gross investment ratio	One-year lagged gross investment ratio	Gross profit ratio	Utilization rate	Government debt ratio	Inflation rate	Constant	R^2	DW	SEE
Germany	$i/y =$	$0.52(t-1)$ (4.87)	$+0.16(t-1\frac{1}{2})$ (2.80)	$+0.10(t-1)$ (2.94)	$-1.01t$ (−5.32)	$-0.96t_{55555}$ (−4.38)		0.91	1.68	0.41
Netherlands	$i/y =$	$0.58(t-1)$ (6.00)	$+0.32(t-1)$ (4.05)	$+0.22(t-\frac{1}{4})^a$ (3.50)	$-0.16(t-1\frac{1}{2})$ (−2.68)	$-0.18(t-1)$ (−2.24)		0.93	2.19	0.62
United Kingdom	$i/y =$	$0.42(t-1)$ (3.03)	$+0.18(t-1)$ (4.65)	$+0.073t$ (6.00)	$-0.02(t-1)$ (−4.73)		-1.82 (−0.81)	0.81	1.50	0.25
United States	$i/y =$	$0.68(t-1)$ (5.34)	$+0.22t$ (2.68)	$+0.10(t-1)^a$ (2.99)	$-0.08(t-1)$ (−2.63)			0.86	1.81	0.36

Estimated equations

Table 3.4 Estimated equations for the inflation rate of domestic expenditure, sample 1960–1982

	Inflation rate	Unit labour costs	Import costs	Utilization rate	Indirect taxes	R^2	DW	SEE
Germany	$p_v =$	$0.71(t-\frac{1}{4})$ (14.84)	$+0.25(t-\frac{1}{4})$ (5.50)	$+0.28t$ (3.44)	$+1.00\delta_1 t$ (—)	0.79	1.17	0.90
Netherlands	$p_v =$	$0.51t$ (10.29)	$+0.46(t-\frac{1}{4})$ (9.40)	$+0.34t$ (2.91)	$+1.00\delta_2 t$ (—)	0.85	1.54	1.30
United Kingdom	$p_v =$	$0.73(t-\frac{1}{2})$ (11.34)	$+0.27(t-\frac{1}{4})$ (5.12)	$+0.35(t-1)$ (2.04)	$+1.00\delta_3 t$ (—)	0.96	2.26	1.35
United States	$p_v =$	$0.87(t-\frac{1}{4})$ (28.25)	$+0.14(t-\frac{1}{2})$ (8.65)	$+0.27t$ (5.66)	$+1.00\delta_4 t$ (—)	0.98	2.07	0.51

Annual figures. Numbers in parentheses are t-values, lags are symbolized by $t-x$, where $x=1$ is one year lag. Indirect taxes have been included by $\delta\Delta(TKS/V_d)_{t-1/4}$ where $\delta_1 = 0.79$, $\delta_2 = 0.68$, $\delta_3 = 0.76$ and $\delta_4 = 0.93$. R^2 = Squared correlation coefficient, etc.

Table 3.5 Estimated equations for real export growth, sample 1960–1982

	Real export	World trade	Relative price of exports	Home pressure of demand	Constant	R^2	DW	SEE
Germany	$\dot{e} =$	$1.0(t-\frac{1}{4})$ (—)	$-0.66t_{532}$ (−3.12)		$+0.85$ (1.45)	0.60	2.04	2.67
Netherlands	$\dot{e} =$	$1.0(t-\frac{1}{4})$ (—)	$-1.88t_{22222}$ (−4.16)		$+1.06$ (2.01)	0.85	2.13	2.41
United Kingdom	$\dot{e} =$	$1.0(t-\frac{1}{2})$ (—)	$-1.44t_{22222}$ (−3.90)	$-0.60(t-\frac{1}{2})$ (−3.63)	-1.60 (−2.74)	0.61	2.64	2.52
United States	$\dot{e} =$	$1.0(t-\frac{1}{4})$ (—)	$-1.43(t-\frac{1}{2})$ (−5.98)	$-0.22(t-\frac{1}{4})$ (−2.06)	-0.84 (−1.38)	0.77	2.45	2.93

Annual figures. Numbers in parentheses are t-values, lags are symbolized by $t-x$, where $x=1$ is one year lag; t_{532} represents weighted lags, namely, $0.5t + 0.3(t-1) + 0.2(t-2)$. $R^2 = \ldots\ldots\ldots$

Supply-side policies

Table 3.6 Estimated equations for real import growth, sample 1960–1982

	Real imports	Total expenditure	Relative prices	Stock formation	Constant	R^2	DW	SEE
Germany	$\dot{m} =$	1.0 (—)	$-0.32(t-\tfrac{1}{4})$ (−2.37)	$+2.63(t-\tfrac{1}{4})$ (2.81)	$+2.99$ (5.22)	0.80	1.80	2.68
Netherlands	$\dot{m} =$	1.0 (—)	$-0.44t_{4321}$ (−5.23)	$+2.28(t-\tfrac{1}{4})$ (4.89)	$+1.17$ (3.21)	0.94	2.17	1.53
United Kingdom	$\dot{m} =$	$1.0(t-\tfrac{1}{2})$ (—)	$-0.27(t-\tfrac{1}{2})$ (−2.42)	$+3.23(t-\tfrac{1}{4})$ (4.17)	$+1.01$ (1.79)	0.72	1.86	2.70
United States	$\dot{m} =$	1.0 (—)	$-0.37(t-\tfrac{3}{4})$ (−3.15)	$+3.01t$ (2.54)	$+3.07$ (3.67)	0.74	2.05	3.91

Annual figures. Numbers in parentheses are t-values, lags are symbolized by $t-x$, where $x=1$ is one year lag. $\bar{R}^2 =$ ....

Table 3.7 Estimated equations for the nominal wage rate, sample 1960–1982 (Netherlands 1958–1982)

	Nominal wages	GDP deflator	Productivity	Forward shifting of direct taxes and social security contributions (employees)	Forward shifting of social security contributions (employers)	Unemployment	Constant	R^2	DW	SEE
Germany	$\dot{p}_l =$	$1.0t$ (—)	$+0.87(t-\tfrac{1}{2})$ (12.85)	$+0.69(t-\tfrac{1}{2})$ (2.48)	$+1.00t$ (—)	$+0.80(t-1)$ (2.36)		0.91	2.27	0.92
Netherlands	$\dot{p}_l =$	$1.0t$ (—)	$+0.80t$ (9.23)	$+0.93(t-\tfrac{1}{2})$ (2.03)	$+1.00t$ (—)	$+1.43t$ (2.93)	$3.41t^a$ (2.18)	0.87	1.75	1.54
United Kingdom	$\dot{p}_l =$	$1.0t$ (—)	$+0.80(t-\tfrac{1}{4})$ (8.19)	$+0.62t$ (2.58)	$+1.00t$ (—)	$-0.68(t-\tfrac{1}{2})^b$ (−2.28)	$-3.74t^c$ (−5.03)	0.97	2.05	1.24
United States	$\dot{p}_l =$	$1.0(t-\tfrac{1}{2})$ (—)	$+0.53t$ (5.67)	$+0.53(t-\tfrac{3}{4})$ (2.28)	$+1.00t$ (—)	$+1.45t$ (3.98)		0.83	2.09	0.91

Annual figures. Numbers in parentheses are t-values, lags are symbolized by $t-x$, where $x=1$ is one year lag. $\bar{R}^2 =$ ....
[a] Dummy 1975.

Estimated equations

	Export prices	Export prices of competitors	Relative costs	Constant	R^2	DW	SEE
Germany	$\hat{p}_e =$	$1.01(t - \frac{1}{4})$ (6.89)	$+0.72(t - \frac{1}{2})$ (4.28)	-0.89 (-1.13)	0.75	1.72	1.67
Netherlands	$\hat{p}_e =$	$1.00(t - \frac{3}{4})$ (—)	$+0.47(t - 1)$ (2.92)	$-1.50 + 19.20t^a$ (-1.78) (5.43)	0.78	1.45	3.45
United Kingdom	$\hat{p}_e =$	$1.05(t - \frac{1}{4})$ (24.05)	$+0.63(t - \frac{1}{4})$ (7.30)		0.92	1.98	2.05
United States	$\hat{p}_e =$	$1.06(t - \frac{3}{4})$ (12.71)	$+0.64t_{0532}$ (3.41)		0.84	2.30	2.26

Annual figures. Numbers in parentheses are t-values, lags are symbolized by $t - x$, where $x = 1$ is one year lag, t_{0532} represents a weighted lag, namely, $0.5(t - 1) + 0.3(t - 2) + 0.2(t - 3)$. $R^2 = $ ....
[a] Dummy 1974.

Table 3.9 Estimated equations for real consumption growth

	Consumption	Total disposable income	Total disposable profits	Consumption price	Constant	R^2	DW	SEE
Germany	$\dot{c} =$	$1.01\alpha_1(t - \frac{1}{4})$ (7.88)	$+0.29\beta_1 t$ (2.03)	$-1.03t$ (-6.56)	$+1.08$ (1.15)	0.87	1.34	1.06
Netherlands	$\dot{c} =$	$0.97\alpha_2 t$ (9.11)	$+0.24\beta_2(t - \frac{1}{4})$ (2.51)	$-1.02t$ (-9.22)	$+1.06$ (1.32)	0.87	2.01	1.10
United Kingdom	$\dot{c} =$	$0.78\alpha_3(t - \frac{1}{4})$ (8.47)	$+0.15\beta_3 t$ (2.89)	$-0.71t$ (-10.92)	$+1.49$ (3.82)	0.89	2.28	0.79
United States	$\dot{c} =$	$0.71\alpha_4(t - \frac{1}{4})$ (3.36)	$+0.53\beta_4 t$ (4.24)	-0.59 (-5.54)	$+1.06$ (1.08)	0.75	2.39	0.94

Annual figures. Numbers in parentheses are t-values, lags are symbolized by $t - x$, where $x = 1$ is one year lag. Also $\alpha_1 = LB_{t-1\frac{1}{4}}/C_{t-1} = 0.74$, $\alpha_2 = LB_{t-1}/C_{t-1} = 0.74$, $\alpha_3 = LB_{t-1\frac{1}{4}}/C_{t-1} = 0.76$, $\alpha_4 = LB_{t-1\frac{1}{4}}/C_{t-1} = 0.75$, $\beta_1 = ZB_{t-1\frac{1}{4}}/C_{t-1} = 0.41$, $\beta_2 = ZB_{t-1\frac{1}{4}}/C_{t-1} = 0.42$, $\beta_3 = ZB_{t-1}/C_{t-1} = 0.23$, $\beta_4 = ZB_{t-1}/C_{t-1} = 0.26$. $R^2 = $ ....

ranges between 0.5 and 2.0 for the four countries. This suggests strong support for the occurrence of classical unemployment is suggested by Malinvaud (1977), den Hartog and Tjan (1976), and Dreze and Modigliani (1981) among others.

Our analysis confirmed that investment is traditionally the hardest equation to estimate in empirical analysis. We found, for instance, that for Germany and the United Kingdom the level of the utilization rate was significant, and for the other countries, it was changes in this rate. A relatively strong result was, however, that in all four countries the terms representing the internal and external finance possibilities of the investment ratio are significant and have the correct signs. Obviously, even in investments that are hard to quantify, there are still some important determining factors that can be traced internationally.

3.4 SUPPLY-SIDE POLICIES

Following for example Bartlett and Roth (1983) the key message of supply-side economics is that lower taxation would be a major help in restoring economic growth and in lowering unemployment. Thus, in a sense this new school breaks with the monetarist-inspired fixation on government spending and government budget deficits. However, supply-siders either remain silent on the financial consequences of tax cuts or argue that such cuts can be made at the expense of the budget deficit, since its shortfall will be made good in due course by the resultant economic upturn. So a systematic analysis of supply-side economics within a quantitative framework seems to be urgently needed. This holds the more as far as an international comparison is under discussion.

The models suggested in Sections 3.2 and 3.3 may serve as such a framework. It goes without saying that these models do not pretend to be the ultimate wisdom but are merely a starting-point for further elaboration. However, this chapter should be seen as an attempt to demonstrate which elements should, in any event, be included in models dealing with supply-side economics.

We have focused on analysing the effects of lower direct taxes financed in different ways, namely, by cutting public spending, by increasing direct taxes and by extra borrowing in the capital market. In addition we have looked for what happens when direct taxes are lowered accompanied by international policy co-ordination. Each time we have simulated the short and long run effects of a 1% GDP decrease in direct taxes under the said different regimes for Germany, the Netherlands, the United Kingdom and the United States. The results can be summarized as follows.

3.4.1 Lower direct taxes financed by cutting public spending

As a result of a 1% GDP decrease in direct taxes financed by a simultaneous decrease in public spending the following pictures arises. In the short run, i.e. after one year, production rises in Germany, the Netherlands and the United

Supply-side policies

States. Only in the United Kingdom does production fall off slightly. In all four countries the unemployment rate falls in the short run (Table 3.10).

In the long run, i.e. after 5 years, we find in all four countries a substantial rise in production and employment and a subsequent decrease in the unemployment rate. This is accompanied by an increase in the investment ratio, the profit ratio and the volume of exports and an improvement in the utilization rate. These results can be explained as follows.

It is important to distinguish between the negative and positive effects of this policy mix. The negative effects are the result of the decrease in public spending. As public sector wages, government consumption and social security benefits fall, private consumption and investments will fall. Yet we do not find these negative effects of less public spending reflected in our multiplier tables.

The reason is, of course, that these negative effects are outweighed by the positive effects of lower direct taxes. Less taxation increases disposable income and hence private consumption. More important is, however, that lower direct taxes will moderate the wage rate because of negative forward shifting. Negative forward shifting is exactly the opposite to what happens when direct taxes and social security contributions are increasing.

As pointed out in Knoester (1983, 1984) higher taxes and social security contributions will be shifted forward into higher (real) wages because employees are bargaining for *real net* income and not for nominal income. In Knoester and Van der Windt (1985) a general wage-bargaining model is shown, in which claims and offers of employees and employers determine the wage rate. This model shows that, besides traditional determinants of the wage rate such as the inflation rate, productivity growth and the Phillips curve, there is an additional and in most cases underestimated determinant, namely, the forward shifting of direct taxes and/or social security contributions. It should be emphasized that the roots of such a wage bargaining model can already be found in Zeuthen (1930), Phillips (1958) and Dicks-Mireaux and Dow (1959).

Anyhow, this shifting forward of taxation has important policy implications. As shown in Knoester (1983, 1984) it suggests that the economic consequences of a simultaneous increase in public spending and taxation are neither neutral nor positive, but negative. In other words, the balanced-budget multiplier shows neither the classical value of zero nor the Keynesian positive value as suggested by Haavelmo (1945). For this reason we call this negative balanced-budget multiplier *the inverted Haavelmo effect*.

The positive effects found on economic growth and employment as a result of a decrease in direct taxes financed by a simultaneous decrease in public spending are completely consistent against this background. The underlying mechanism – negative shifting forward – is that lower taxation will lower the wage rate resulting in an increase of the profit ratio, so that investment and economic growth increases. In addition the lower wage rate will improve the competitive position by which exports will grow. An essential element for these results is the said negative forward shifting of lower taxation into a lower wage rate which follows from

Table 3.10 Effects of a once-and-for-all 1% GDP decrease in direct taxes financed by a simultaneous decrease in public spending

Change in levels after year	Germany		The Netherlands		United Kingdom		United States	
	1	5	1	5	1	5	1	5
Profit ratio (% GDP)	0.6	1.1	1.0	1.2	1.0	1.0	0.6	0.8
Investment ratio (% GDP)	0.0	0.7	0.1	0.8	-0.0	0.4	0.1	0.5
Volume of production (%)	0.4	2.6	0.7	2.6	-0.1	1.5	0.5	2.1
Inflation rate (%)	-0.6	-1.9	-1.6	-2.2	-0.6	-4.0	-0.2	-1.9
Export volume (%)	0.1	0.7	0.0	0.9	0.2	1.3	-0.1	0.3
Utilization rate (%)	0.3	1.3	0.6	0.7	-0.3	0.4	0.5	0.4
Private employment (%)	0.2	2.7	0.2	1.7	0.1	1.7	0.4	2.4
Unemployment rate (%)	-0.2	-2.1	-0.2	-1.3	-0.1	-1.3	-0.3	-2.0
Public sector deficit (% GDP)	0.0	-0.4	0.1	-0.2	0.0	-0.2	0.0	-0.2

Table 3.11 Effects of a once-and-for-all 1% GDP decrease in direct taxes financed by a simultaneous decrease in public spending accompanied by international policy co-ordination

Change in levels after year	Germany		The Netherlands		United Kingdom		United States	
	1	5	1	5	1	5	1	5
Profit ratio (% GDP)	0.6	0.9(1.1)	0.8	0.9(1.3)	0.9	0.9(1.0)	0.6	0.7(0.8)
Investment ratio (% GDP)	0.0	0.7(0.7)	0.1	0.6(0.9)	-0.0	0.3(0.4)	0.1	0.5(0.5)
Volume of production (%)	0.3	2.1(2.9)	0.5	1.5(3.1)	-0.2	0.5(1.5)	0.5	1.8(2.2)
Inflation rate (%)	-0.6	-2.3(-1.7)	-1.6	-2.7(-2.1)	-0.6	-5.6(-3.6)	-0.2	-2.0(-1.7)
Export volume (%)	0.0	0.0(1.6)	0.0	0.0(1.8)	0.0	0.0(1.5)	0.0	0.0(1.1)
Utilization rate (%)	0.2	0.9(1.5)	0.4	-0.2(0.9)	-0.4	-0.4(0.4)	0.5	0.1(0.4)
Private employment (%)	0.2	2.2(3.0)	0.2	1.0(2.1)	0.0	0.8(1.8)	0.4	2.1(2.6)
Unemployment rate (%)	-0.1	-1.7(-2.4)	-0.1	-0.8(-1.6)	-0.0	-0.6(-1.4)	-0.3	-1.8(-2.2)
Public sector deficit (% GDP)	0.1	-0.3(-0.5)	0.1	0.0(-0.3)	0.0	0.1(-0.2)	0.0	-0.2(-0.3)

Figures without parentheses are for the effects produced by the exclusion of relative prices in the equations for the volume of imports and exports. Figures in parentheses are for the effects obtained by the exclusion of relative prices and the inclusion of a once-and-for-all 1% increase in foreign trade.

Supply-side policies

the chosen wage bargaining model. In that case such a policy mix would improve the economic performance substantially in all four countries.

3.4.2 Lower direct taxes and public spending accompanied by international policy co-ordination

Suppose a 1% GDP decrease in direct taxes and a simultaneous decrease in public spending are pursued not by one but by all industrialized countries. In that case the effect of lower taxes on the wage rate will not improve the economic performance in so far as it would have been improved by the competitiveness in the case of a single country action. Table 3.11 shows the results obtained under such international policy co-ordination. It means in technical terms that the elasticities for relative prices in the equations for imports and exports have been given a zero value.

Indeed, our results show a less attractive outcome compared with that of the single country action. Even so, the ultimate outcome remains positive for economic growth as well as for employment. The reason is, of course, that the positive domestic effects of the chosen policy mix are substantial. Our results make perfectly clear that for such a policy the openness of an economy would not be of decisive importance. The effects of lower taxation on the wage rate, the profit and investment ratio and on economic growth provide sufficient grounds for pursuing this policy.

One could even argue that international policy co-ordination with respect to lower direct taxes and public spending would increase world trade because of the pressure on imports as a result of the increased domestic economic growth. In this chapter we have quantified such a possibility tentatively by combining the minimum outcome of a concerted action with a once-and-for-all increase in foreign trade by 1%, being about half of the domestic effect on economic growth. As a result, the positive effects of the policy mix appear to be in the same range as would be the case in a single country action. This outcome suggests that such combined action at international level could be a major help in solving the economic problems of the 1980s in the OECD area.

3.4.3 Lower direct taxes financed by higher indirect taxes

Another possibility for financing lower direct taxes is to increase indirect taxes. The results can be found in Table 3.12. It appears that, at first glance, such a policy mix seems to be a serious alternative for lower direct taxes financed by cutting public spending. Here again economic growth and employment increases whereas the unemployment rate falls, albeit to a lesser extent.

It should be emphasized, though, that these results are based on the absence of any forward shifting of indirect taxes into nominal wages. As shown in Section 3.3, we have supposed – for reasons of simplicity – that employees will compensate a higher inflation rate into higher nominal wages only for the increase

48   Supply-side policies

Table 3.12 Effects of a once-and-for-all 1% GDP decreases in direct taxes financed by a simultaneous increase in indirect taxes

Change in levels after year	Germany		The Netherlands		United Kingdom		United States	
	1	5	1	5	1	5	1	5
Profit ratio (% GDP)	0.0	0.3(−0.3)	0.4	0.4(−0.3)	0.2	0.2(−0.2)	−0.2	−0.0(−0.2)
Investment ratio (% GDP)	0.0	0.4(−0.2)	0.1	0.3(−0.2)	0.0	0.1(−0.0)	−0.0	0.0(−0.1)
Volume of production (%)	0.5	2.1(0.3)	0.9	1.7(−0.1)	0.2	1.3(0.4)	0.2	1.0(0.2)
Inflation rate (%)	0.1	−1.1(1.1)	−0.9	−1.1(0.6)	0.1	−2.8(−0.2)	0.7	−1.6(0.8)
Export volume (%)	0.1	0.8(−0.0)	0.0	1.0(0.2)	0.1	0.8(−0.2)	−0.0	0.9(−0.6)
Utilization rate (%)	0.5	1.3(0.6)	0.8	0.7(0.1)	0.1	0.6(0.3)	0.2	0.4(0.3)
Private employment (%)	0.3	2.3(0.3)	0.8	1.1(0.0)	0.2	1.6(0.6)	0.2	1.3(0.5)
Unemployment rate (%)	−0.2	−1.9(−0.2)	−0.2	−0.9(−0.0)	−0.1	−1.2(−0.5)	−0.1	−1.1(−0.4)
Public sector deficit (% GDP)	0.1	−0.3(0.0)	0.0	−0.1(0.1)	0.1	−0.0(0.0)	0.1	−0.0(0.1)

Figures in parentheses are for effects obtained by the inclusion of consumer prices instead of GDP prices in the wage equation.

Table 3.13 Effects of a once-and-for-all 1% GDP decrease in direct taxes financed by simultaneous increase in government debt

Change in levels after year	Germany		The Netherlands		United Kingdom		United States	
	1	5	1	5	1	5	1	5
Profit ratio (%) GDP	0.4	0.2	0.9	0.4	0.6	0.4	0.4	−0.2
Investment ratio (% GDP)	−0.1	−0.8	0.2	−0.5	0.0	0.1	0.1	−0.6
Volume of production (%)	0.9	0.1	1.5	0.9	0.7	1.1	1.2	−0.4
Inflation rate (%)	−0.5	−1.3	−1.6	−1.7	−0.8	−2.6	−0.0	−1.1
Export volume (%)	0.1	0.5	0.0	0.7	−0.1	−0.0	−0.2	0.1
Utilization rate (%)	0.8	0.3	1.4	−0.1	0.5	0.3	1.1	−1.2
Private employment (%)	0.5	0.7	0.5	0.8	0.4	1.5	0.9	0.2
Unemployment rate (%)	−0.4	−0.5	−0.4	−0.6	−0.3	−1.2	−0.7	−0.2
Public sector deficit (% GDP)	0.8	0.9	0.8	0.8	0.7	0.6	0.8	1.0

Conclusions

in GDP prices. Following the wage bargaining model of Knoester and Van der Windt (1985) consumer prices seem to be the more relevant price deflator for employees. Instead of GDP prices consumer prices include the pressure on inflation of higher indirect taxes. If employees were to succeed in compensating their wages for this price deflator, they would also succeed in shifting forward higher indirect taxes into higher nominal wages. In Knoester and Van der Windt (1985) it has been shown that such behaviour seems to be relevant for a number of OECD countries.

For this reason, we have repeated the policy mix under discussion, but now with consumer prices instead of GDP prices in the wage equation. As a result the picture changes substantially. In such an event the suggested policy mix no longer has positive effects but has negative effects on economic growth, whereas the effects on the unemployment rate are negligible. Thus, if employees succeed in shifting forward higher indirect taxes into higher nominal wages — for which empirical evidence is available — lower direct taxes financed in this way will not bring the required improvement in economic performance, which means that this policy mix is inferior to the financing of lower direct taxes by cutting public spending.

3.4.4 Lower direct taxes financed by issuing government debt

Lower direct taxes financed by increasing the government budget deficit have been experienced to a large extent by the Reagan administration. So far, it does not seem to be overwhelmingly successful, which is reason enough for looking at this policy mix with the models developed in the study reported in this chapter. A 1% GDP decrease in direct taxes financed by borrowing in the capital market gives the following result.

In the short run in all four countries production increases whereas unemployment falls. In the long run in Germany production hardly improves, whereas it falls in the United States. For the Netherlands and the United Kingdom it was found that a positive effect on economic growth also remains in the long run (Table 3.13). It should be emphasized, however, that these outcomes are highly tentative, because the models used were not developed to analyse budget deficit financing but balanced-budget financing. But even so, lower direct taxes financed by cutting public spending seems to be for all four countries a much better alternative. This can also be deduced from the outcomes for the public sector deficit. Our findings for all four countries do not support the claim that the budget deficit will decrease in due course as a result of the economic upturn produced by this policy mix.

3.5 CONCLUSIONS

This chapter has concentrated on the analysis of supply-side policies in four OECD countries, namely, Germany, the Netherlands, the United Kingdom and the United States. For this purpose we have developed a general model in which the supply-side of the economy has been added to the standard Keynesian income–expenditure block. The inclusion of supply-side equations has not been limited to equations for production capacity and the utilization rate. We have also focused on the analysis of the forward shifting of higher taxes into higher real wages. Such forward shifting provides a quantitative basis for tax cuts as advocated by supply-siders. Our findings can be summarized as follows.

(1) In modelling the demand and supply side of the economy we found a general structure for all four countries.
(2) An interesting result is that the shifting forward of taxation into higher wages seems to be an international phenomenon.
(3) Such forward shifting implies a completely different view of the effects of balanced-budget financing, i.e. of a simultaneous increase in public spending and direct taxes. It appears that as a result of this policy mix neither the classical balanced-budget multiplier of zero, nor the positive Keynesian value is valid, but rather a negative balanced-budget multiplier.
(4) In reverse, the results found provide a quantitative basis for direct tax cuts as advocated by supply-siders. A condition is, though, that these tax cuts should be accompanied by a simultaneous cut in public spending.
(5) Lower direct taxes accompanied by international policy co-ordination may diminish the positive effects on economic growth and employment in so far as the underlying transmission channel runs through an improvement in competitiveness in the case of single country action. However, such international policy co-ordination will also have positive effects on the growth rate of world trade. These positive effects can overcompensate the 'loss' resulting from unchanged competitiveness.
(6) At first glance lower direct taxes financed by higher indirect taxes seems to be a serious alternative for lower taxes financed by cutting public spending. This is a result of a complete absence of any forward shifting of higher indirect taxes into higher wages. If employees succeed in compensating consumer prices instead of GDP prices in wages – for which empirical evidence is available – the forward shifting of indirect taxes will diminish the positive effects of lower direct taxes on economic growth and employment. In that case, this would hardly be a fruitful policy mix.
(7) Lower direct taxes financed by issuing extra government debt is no serious alternative to lower direct taxes financed by cutting public spending. In two of the four countries under discussion, this will even lead to zero or a lower economic growth. Also, in all four countries the increase in the government budget deficit will not vanish in due course because of an economic upturn.

REFERENCES

Bartlett, B. and Roth, T. P. (1983) *The Supply-Side Solution*, Macmillan, London.
Brandsma, A. S. and Windt, N. van der (1983) Wage bargaining and the Phillips-curve: a macroeconomic view. *Applied Economics*, 15, No. 1, February, 61–71.

References

Corden, W. M. (1981) Taxation, real wage rigidity and employment. *The Economic Journal*, 91, No. 362, June, 309–30.

Dicks-Mireaux, L. A. and Dow, J. C. R. (1959) The determinants of wage inflation: United Kingdom, 1946–56. *Journal of the Royal Statistical Society*, Series A, 122, Part 2, 145–74.

Drèze, J. H. and Modigliani, F. (1981) The trade-off between real wages and employment in an open economy (Belgium). *European Economic Review*, 15, No. 1, January, 1–40.

Eckstein, O. and Fromm, G. (1968) The price equation, *American Economic Review*, 58, 1159–83.

Haavelmo, T. (1945) Multiplier effects of a balanced budget. *Econometrica*, 13, 311–8.

Hartog, H. den and Tjan, H. S. (1976) Investments, wages, prices and demand for labour; a clay–clay vintage model for the Netherlands. *De Economist*, 124, No. 1/2, 32–55.

Johnston, J. (1974) A model of wage determination under bilateral monopoly, in *Inflation and Labour Markets* (eds D. Laidler and D. L. Purdy), Manchester University Press, 61–78.

Johnston, J. and Timbrell, M. (1973) Empirical tests of a bargaining theory of wage rate determination. *The Manchester School of Economic and Social Studies*, 41, No. 2, June, 141–67.

Klein, L. R. (1978) The supply-side. *American Economic Review*, 68, 1–7.

Knoester, A. (1983) Stagnation and the inverted haavelmo effect: some international evidence. *De Economist*, 131, No. 4, 548–84.

Knoester, A. (1984) *Negative consequences of public sector expansion in the US and Europe*, Occasional paper, November, American Enterprise Institute for Public Policy Research, Washington, DC.

Knoester, A. and Van Sinderen, J. (1984) A simple way of determining the supply side in macroeconomic models. *Economic Letters*, 16, Nos 1–2, 83–91.

Knoester, A. and Van der Windt, N. (1985) *Real wages and taxation in ten OECD countries*, Discussion paper 8501 G/M, Institute for Economic Research, Erasmus University, Rotterdam. A revised version of this paper has been published in *Oxford Bulletin of Economics and Statistics*, 49 (1), 151–69, 1987.

Malinvaud, E. (1977) *The Theory of Unemployment Reconsidered*, Basil Blackwell, Oxford.

OECD (1978) *Public Expenditure Trends*, Paris.

Phillips, A. W. (1958) The relation between unemployment and the rate of change of money wage rates in the United Kingdom, 1861–1957. *Economica*, 25, No. 100, November, 283–99.

Tinbergen, J. (1982) De noodzaak van een synthese. *Economisch Statistische Berichten*, 67, 1284–5.

Tobin, J. (1981) Comment on Albert Ando: On a theoretical and empirical basis of macroeconomic models, in *Large-scale macroeconometric models* (eds J. Kmenta and J. B. Ramsey), North-Holland, Amsterdam, pp. 391–2.

Zeuthen, F. (1930) *Problems of Monopoly and Economic Warfare*, George Routledge & Sons, London.

4

An empirical analysis of policy co-ordination in the United States, Japan and Europe

HALI J. EDISON and RALPH TRYON

Co-ordination of macroeconomic policy has been a major topic at recent summit meetings, and has been the subject of a number of theoretical studies. However, relatively little *empirical* research exists on policy co-ordination. This chapter is an attempt to help fill this gap. The chapter considers the quantitative importance of the co-ordination of fiscal and monetary policy under flexible exchange rates. We also evaluate the mechanisms by which the effects of macroeconomic policy are transmitted abroad. The nature of the equilibrium reached in the absence of co-ordination is also analysed, and the empirical results are related to the theoretical literature. The analysis is based on simulations with the Multicountry Model (MCM) developed at the Federal Reserve Board.

In the postwar period, trading and financial ties have increased dramatically among the industrialized countries. These ties imply that one country's economic policies have spillover effects on other countries' welfare and consequently have implications for their economic policies. This interdependence suggests that co-ordination of economic policy between countries is important. However, as Oudiz and Sachs (1984) put it, 'the advocacy of international co-ordination has been far more plentiful than its actual implementation'. This chapter aims to shed some light on the potential gains from co-ordination.

Several important theoretical papers have been written on policy co-ordination, but there is relatively little empirical work in the area. Jeffrey Sachs and Gilles Oudiz, in the paper just referred to, made a pathbreaking effort to measure the potential welfare gains that could be realized from policy co-ordination among Japan, Germany and the United States. (Their paper also contains a good list of references to the theoretical literature. See, for example, Hamada (1974, 1976), Canzoneri and Gray (1983), Miller and Salmon (1985) and Buiter and Marston (1985).) In our chapter we work within the general framework of Oudiz and Sachs, but we use a somewhat simpler empirical methodology, and focus on a different aspect of the gains from policy co-ordination.

Over the past five years, US policies have been subject to much criticism by its major trading partners. One of the major sources of friction has been the 'mix' of US fiscal and monetary policies. The United States has been running a large government deficit the inflationary effects of which have been offset domestically by a restrictive monetary policy. It is generally argued that the result has been higher real interest rates in the United States, a stronger dollar, and a large US current account deficit. The authorities have been under substantial pressure both at home and from abroad to change the policy mix, and recent months have indeed seen several steps in this direction.

However, if the US government lowers spending there would be a significant tendency to lower real income in the United States, and, through lower US imports, in the rest of the world. This could potentially lead to a worldwide recession. On the other hand, part of this recession could be offset by the expansion of other OECD economies, Japan and Germany in particular. An important question is how other OECD economies can act effectively in response to changes in US economic policy. In this chapter we consider this as a problem in the empirical analysis of policy co-ordination.

We examine the impact of the US Balanced Budget and Deficit Act of 1985 (also referred to as the Gramm-Rudman Act) on the US economy and on Germany and on Japan. The chapter seeks to evaluate different policy responses to this change in US policy. In particular, we consider first an independent response by Germany and Japan, made in the absence of complete information about the extent of US monetary accommodation, and a similar response for the United States. We then analyse some of the implications of a fully co-ordinated response. The aim of the chapter is to provide an empirical evaluation of the gains from co-ordination using simulations with the Federal Reserve Board Multicountry Model (MCM).

The plan of the chapter is as follows: Section 4.1 gives a brief overview of the MCM model properties. Section 4.2 describes the basic features of the Gramm-Rudman Act and Section 4.3 discusses our framework for the analysis of co-ordination. Section 4.4 presents and analyses the empirical results from our model simulations; the conclusions follow in Section 4.5.

4.1 THE FRB MULTICOUNTRY MODEL

The Multicountry Model (MCM) is a system five of quarterly national macro-economic models at the centre of which is a medium-sized model of the US economy.[1] Linked to this US model and to each other are models of Canada, West Germany, Japan and the United Kingdom.[2] The single models vary in size

---

[1] For complete details on the theoretical basis of the MCM see Stevens et al. (1984). For an update on the MCM theoretical basis see Edison, Marquez and Tyron (1987).
[2] Henceforth we refer to West Germany as Germany.

The FRB multicountry model

from 150 to 250 behavioural equations and identities; also included in the system is an abbreviated section representing the rest of the world.

The system has three salient features relating to the international scope of the MCM. The first is the endogeneity of the bilateral dollar exchange rates[3]. In the current version of the model exchange rates are determined by the open parity condition with the expected exchange rate a function of relative price differentials. The second noteworthy feature of the model is the use of bilateral, rather than aggregate, goods trade equations. These bilateral import demand equations are used to explain each country's imports from each of the other countries. The third feature of interest is the oil sector: the MCM models explicitly the consumption of and trade in oil; oil also enters the supply side, as a factor of production.

In the typical country model, prices and quantities are determined by the behaviour of four classes of economic agents: the monetary authorities, the government, commercial banks, and firms and households. The actions of these agents are modelled in the goods market, the labour market and the asset market. Each country is assumed to produce a composite consumption–investment commodity. By assumption, goods produced in the different countries are imperfect substitutes.

Aggregate demand is divided into six major components: personal consumption, fixed investment, inventory investment, government spending, exports, and imports. Consumption depends upon private disposable income and net worth following the life-cycle hypothesis. The fixed investment equations are based on neoclassical investment theory, being positively related to changes in income and negatively to changes in the user cost of capital. Inventories act as a buffer stock and absorb any discrepancy between production and sales. Real government spending is assumed to be exogenous. Imports and exports of goods and services are broken down into merchandise trade, investment income, and other services.

The supply side of the prototype country model models potential GNP as a function of the capital stock and labour force using a Cobb–Douglas production function. Capacity utilization is defined as the ratio of actual to potential GNP. The labour market is assumed not to clear completely in any one period, allowing for the existence of labour unions and minimum wage laws. The wage equation follows the familiar Phillips–Lipsey–Friedman approach: the change in wages is a function of unemployment and the expected inflation rate. The expected inflation rate is represented by a distributed lag on past price changes. Prices themselves are determined by a markup over average costs, which include wage costs adjusted for changes in labour productivity, and the cost of imports (including oil).

In the prototype model, the money market focuses on the role of reserves in the system. For a given level of base money the short-term interest rate adjusts to clear money market. Money demands by the public (currency, demand deposits, and time deposits) and by the banking sector (free reserves) are modelled explicitly. Long-term interest rates are modelled as distributed lags on the short-terms rate.

---

[3] For more details on the exchange rate in the MCM see Hooper (1986).

The MCM is in many ways a conventional demand-oriented macroeconometric model – the innovations are chiefly in the modelling of trade and exchange rates, and in the multicountry structure. In particular, the treatment of expectations and aggregate supply is quite conventional, which does raise some questions about the appropriate way to model policy regime changes. Expectations of future prices and exchange rates are both modelled explicitly in the MCM country model; however, expectations are determined by an adaptive structure, and are not 'forward-looking'. Exchange rate expectations are used in the determination of the spot exchange rate, while price expectations offset nominal wages and the real interest rate, which appear throughout the model.

To provide a rough indication of the importance of expectations and supply effects in the MCM for the sorts of policy exercises reported in this chapter, we present some summary multipliers in Table 4.1. The table shows the effects on real GNP and prices of standardized fiscal and monetary shocks, for the US and foreign countries. The results can be interpreted as a measure of the degree of

Table 4.1 Fiscal and monetary multipliers in the MCM (amounts are cumulative percentage deviations from baseline)

Year:	1	2	3	4	5	6
US fiscal shock[a]						
US GNP	1.6	1.8	1.4	0.9	0.5	0.1
US prices	0.1	0.4	0.9	1.4	1.9	2.3
Foreign GNP	0.3	0.7	0.9	0.9	1.0	1.0
Foreign prices	0.2	0.4	0.6	0.7	1.0	1.2
US monetary expansion[b]						
US GNP	0.4	1.5	2.2	2.0	1.4	0.9
US prices	0.1	0.4	0.8	1.4	2.1	2.6
Foreign GNP	−0.2	−0.7	−0.8	−0.8	−0.9	−1.1
Foreign prices	−0.4	−0.6	−0.7	−0.7	−0.9	−1.1
Foreign fiscal shock[c]						
US GNP	0.3	0.5	0.4	0.2	0.1	0.0
US prices	0.0	0.2	0.3	0.4	0.6	0.7
Foreign GNP	1.1	1.4	1.3	1.2	1.1	1.1
Foreign prices	0.0	0.3	0.6	0.9	1.2	1.6
Foreign monetary expansion[d]						
US GNP	0.0	0.0	0.0	0.1	0.1	0.0
US prices	−0.1	−0.2	−0.2	−0.1	−0.1	0.0
Foreign GNP	0.3	1.5	2.3	2.2	1.8	1.5
Foreign prices	0.5	0.6	0.7	1.1	1.5	2.0

This table is adapted from the results presented at the recent Brookings Institutions Conference on Empirical Macroeconomics for Open Economies (March 1986). (Money growth rates are held exogenous in all countries unless noted.)
[a] Permanent increase of US real government expenditure of 1% of baseline GNP.
[b] One-time increase in US money supply of four percentage points.
[c] Permanent increase in foreign real government expenditures of 1% of baseline GNP.
[d] One-time increase in foreign money supply of four percentage points.

crowding out and neutrality of money in the model. The results also show the extent of foreign linkages in the MCM.

The table shows that crowding-out of the US fiscal shock is virtually complete after six years. Money is not neutral in the United States over this simulation period, but the positive effects of the temporary increase in money growth are substantially reduced. The foreign models, taken together, exhibit noticeably less crowding out and neutrality of money than does the US model. Finally, the spillover effects of US policy on foreign economies are very strong, but foreign policy has much smaller effects on the US. (The transmission effects are primarily through exchange rate changes, which affect foreign economies more than the United States.)

4.2 THE GRAMM-RUDMAN ACT

This section briefly outlines some of the institutional aspects of the Balanced Budget and Deficit Reduction Act of 1985, also known as the Gramm-Rudman Act. In addition it gives some indication how this will impact government spending. The Gramm-Rudman Act targets are now binding on the US administration and congressional budget proposals. Table 4.2 shows what these limits are. Under the new law the administration may not propose a budget with deficits that exceeds these specific limits nor may Congress adopt budget resolution which proposes higher deficits.

The Gramm-Rudman Act utilizes two mechanisms to enforce its deficit targets. First, it amends the regular congressional budget process. This has included making the reconciliation process earlier in the budget cycle and the provision for only one annual congressional budget resolution. Other changes mandate new procedures that are designed to make enforcement of budget targets stricter and more certain.

The second mechanism that the Gramm-Rudman Act uses to enforce deficit targets is a procedure for automatic outlay cuts called sequestration. This process is triggered if the estimate of the deficit for the coming fiscal year prepared by the Congressional Budget Office and the Office of Management and Budget just before the fiscal year begins exceeds the limit for that year specified in the Act. Even after this process is triggered Congress is given the opportunity to assert

Table 4.2 Gramm-Rudman deficit limits (fiscal years, billions of current dollars)

	1986	1987	1988	1989	1990	1991
Maximum allowable deficit (in budget deliberations)	172	144	108	72	36	0
Trigger for automatic outlay cut mechanism	172	154	118	82	46	0

priorities other than those embodied in the automatic formula provided that the Congressional alternative brings the deficit within the requirement limits.

In modelling the Gramm-Rudman Act in this chapter, it is assumed that the US macroeconomic authorities lose control over one of their policy instruments, namely government spending. (It is assumed that tax rates do not change, according to current administration policy.) Thus government spending is effectively predetermined because the government deficit is limited by the Gramm-Rudman Act. We assume that the constraint imposed by the act is always binding. US government spending is forced to be cut from the baseline in accordance with the numbers on Table 4.2.

4.3 STRATEGIC CONSIDERATIONS IN ECONOMIC POLICY-MAKING

The term 'co-ordination' can refer to several aspects of the way in which countries jointly formulate their economic policies. Here we consider three possibilities.

The first concerns the *timing* of joint policy actions. Suppose, for example, that two countries agree to intervene in the foreign exchange markets to support a particular currency. The outcome might be different depending on whether or not the intervention occurred simultaneously (that is, was co-ordinated). It is entirely conceivable that nonlinearities in the system could cause the exchange rate to respond if both countries intervened together, than if each intervened, by the same amount, a month apart. (The G-5 intervention in September 1985 might be an example of such a case.)

A second type of co-ordination involves exchanging information among countries about the policy stance of each. Following, say, an unexpected change in oil prices, individual countries would face the problem of determining the appropriate policy response. The best policy for one country might depend on what action was taken by another. For example, if one country responded by easing its monetary policy, that country's trading partners might want to ease their own monetary policies to keep real exchange rates unchanged.

Finally, countries might engage in co-operative policy 'trading' in which one country undertakes a policy action which lowers its own welfare but helps others, in return for similar concessions by foreign countries. An obvious example is in negotiating tariff reductions, where one country might agree to stop protecting an important domestic industry in return for access to a foreign market.

Or, as in the cases analysed by Oudiz and Sachs, the trade might occur in macroeconomic policy. One country might inflate its own economy slightly past the optimal point, raising income but at the cost of a deterioration of the exchange rate. The expansions would, however, benefit foreign economies by raising demand for their exports. In exchange for this, the foreign economies could inflate their own economies, offsetting the exchange rate change and raising demand in the first country.

For clarity we term these three types of policy interaction 'synchronization', 'co-ordination', and 'co-operation' respectively. In this chapter we focus on the gains from co-ordination, as defined here. This is in contrast to much of the recent work in this area, which considers the question of policy co-operation, as we use the term. Our interest in this aspect stems not from any presumption that the exchange of information is either more or less important than policy trading, but rather from our desire to explore another side of the question.

The basic theoretical framework we use adopts the classical convention of Tinbergen of targets and instruments in a multicountry environment (see Oudiz and Sachs (1984), also Hamada (1974, 1976), Canzoneri and Gray (1983), Miller and Salmon (1985) and Buiter and Marston (1985)). In the general framework an n country world economy is considered in which each country has k targets. For each country i there is a vector of targets $\mathbf{T}^i = (T_1, \ldots, T_k)$. The ith country has m controls or policy instruments $P^i = (P_1^i, \ldots, P_m^i)$. The macroeconomic authorities of each country choose P^i in a way which will maximize their own utility, $U^i(\mathbf{T}^i)$. In an interdependent world the $\mathbf{T}^i$ will be a function of all n countries' controls. Therefore to maximize its utility, country 1's authority will have to condition on what the other $n - 1$ countries are doing.

In this chapter three equilibria are considered. In the first policy-makers must form expectations under uncertainty about what foreign policy is. The second is a Nash equilibrium, or the unco-ordinated policymaking equilibrium. In this instance no trades are being made; rather country 1 formulates its policy taking the actions of the $n - 1$ countries as known and given. In the Nash equilibrium each country knows what every other country is doing and that they all know and use the 'true' model. The last equilibrium considered is a co-operative one in which some form of explicit co-operation takes place, moving some or all countries to a higher level of utility.

These concepts can be illustrated for the case of two countries, each with one policy instrument, using the standard indifference curve framework, as shown in Fig. 4.1. On the horizontal axis the control variable is measured for country 1 and the vertical axis depicts that for country 2. The curves in the diagram represent the utility for each country – actually a family of these curves could be drawn in.

Suppose country 1 expects country 2 to set its control variable at $E_1(C_2)$, where E_1 denotes expectations by country 1. Then country 1 chooses policy C_1^a, expecting to reach point A_1. Country 2 does likewise, expecting to reach A_2. The actual point reached by both countries is A. In this example each country sets a higher value for the control variable than expected by the other, so utility in each country is higher than expected. But point A is clearly suboptimal – each country, had it known the other would select a higher value for its control, would have selected a still higher value for its own.

At the Nash equilibrium, point N, country 1's expectation for country 2's policy, $E_1(C_2^n)$, is equal to that policy, C_2^n, and similarly, $E_2(C_1^n) = C_1^n$. At this point each side is optimizing given the actual policy of the other player – this is not true at point A. The difference between the two points is simply that at A, expected foreign policy is not equal to actual foreign policy. Because countries

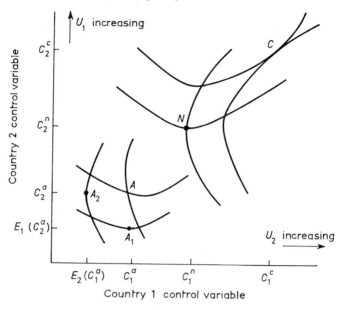

Fig. 4.1

must act without knowledge of actual policy abroad, welfare is reduced.

In other words, while the Nash equilibrium N is not a co-operative equilibrium, it is *co-ordinated* in the sense that each side has full knowledge of what the other is doing. (This knowledge might be communicated directly, or it might be obtained by predicting foreign behaviour given knowledge of the underlying economic structure.)

Of course, the Nash equilibrium is not optimal either, because welfare could still be improved by a trade. Raising C_1 above its Nash value would change U_1 only slightly, because U_1 is maximized with respect to C_1 at N. But U_2 would clearly rise. A similar argument applies for C_2, and by raising both C_1 and C_2 an efficient point such as C can be reached. Notice that reaching C requires more than an exchange of information – it requires an agreement in which each party agrees to make a sacrifice for the good of the other. This is what we term policy 'co-operation'.

In their pioneering article, Oudiz and Sachs make use of the properties of the Nash equilibrium (together with other assumptions) to infer the policy-makers' utility functions. This, together with the parameters of the structural model, enables them not only to find an efficient point such as C but to measure the welfare gain from moving from N to C. The major difficulty with this analysis is that it requires some very stringent assumptions. One needs to be able to identify a particular period in time as a Nash equilibrium and to specify the exact form of the policy-makers' utility functions.

Empirical simulations of economic policy co-ordination 61

In this chapter we use a less rigorous but, we think, equally informative approach to study the difference between points such as *A* and *N*, and *N* and *C*. We assume first that following a particular event (or 'shock') each country's expectation about the stance of foreign policy remains unchanged, and we consider various policy responses in this case. We then consider the same type of policy response when policy-makers assume that a foreign policy reaction occurs. Finally, we give a simple example of explicit policy co-operation, as defined above. In order to evaluate gains and losses we simply offer a description of the changes in income, prices, and the real exchange rate, rather than trying to compute a single measure of utility.

4.4 EMPIRICAL SIMULATIONS OF ECONOMIC POLICY CO-ORDINATION

This section presents the results of a series of simulations of the Gramm-Rudman law in the United States. The simulations were performed with the Federal Reserve Board (FRB) Multicountry Model (MCM) over the period 1986–1990 Q4. In the empirical implementation of our analysis of policy co-ordination we make several simplifying assumptions. First, it is assumed that economic structure of the MCM is the true model of the world, and that policy-makers believe this. Second, we consider policy co-ordination among only three countries in the MCM: the United States, Japan and Germany. Finally, each country is assumed to have two policy instruments – monetary and fiscal policy – and three targets – GNP, inflation, and the real exchange rate.

We start by describing the underlying baseline projection used in the analysis. The baseline used for these simulations is essentially similar to that used for the Brookings Conference on Empirical Macroeconomics (March, 1986). The common elements in the baseline are the components of GNP, prices, exchange rates, and interest rates in each country. For the most part, these paths are simply extrapolations of forecasts obtained from the OECD Economic Outlook for December 1984. For the period beyond which forecast have been prepared by the OECD, we extrapolated at the growth rates projected by the OECD for the first half of 1986. Exchange rates and interest rates are held constant over the projection period using 1985 Q4 data. This baseline path is treated as the initial Nash equilibrium. That is, each country is assumed to have achieved its optimal policy mix given the other countries' policy actions, prior to the passage of the Gramm-Rudman Act.

4.4.1 The initial effects

Table 4.3 summarizes the effects of the Gramm-Rudman law in the absence of any other policy reaction either in the United States or abroad. (All countries are assumed to keep their money supplies unchanged.) In this model, the reduction

Table 4.3 US fiscal contraction with no monetary accommodation

	1986 Q4	1987 Q4	1988 Q4	1989 Q4	1990 Q4
United States					
Real GNP (%)	−1.5	−2.3	−2.5	−1.6	−1.9
Growth rate	−1.5	−0.8	−0.3	0.9	−0.3
Price level (%)	−0.1	−0.6	−1.3	−2.1	−3.0
Inflation rate (+/−)	−0.1	−0.4	−0.7	−0.8	−0.9
Interest rate (+/−)	−1.2	−2.3	−3.2	−3.4	−4.5
Money supply − M1 (%)	0.0	0.0	0.0	0.0	0.0
Growth rate	0.0	0.0	0.0	0.0	0.0
Trade balance ($B)	9.3	16.9	24.1	24.1	32.7
Current account ($B)	11.2	26.1	45.4	59.8	92.7
Exchange rate − FX/$ (%)	−1.8	−3.6	−5.2	−5.5	−6.8
Government deficit ($B)	−8.9	−22.6	−47.6	−72.6	−114.9
Germany					
Real GNP (%)	−0.4	−0.9	−1.3	−1.4	−1.8
Growth rate	−0.4	−0.4	−0.5	−0.1	−0.3
Price level (%)	−0.3	−0.7	−1.2	−1.7	−2.4
Inflation rate (+/−)	−0.3	−0.4	−0.5	−0.5	−0.7
Interest rate (+/−)	−0.5	−0.8	−1.1	−1.2	−1.8
Money supply − CBM (%)	0.0	0.0	0.0	0.0	0.0
Growth rate	0.0	0.0	0.0	0.0	0.0
Trade balance ($B)	−1.8	−3.9	−7.0	−9.1	−12.9
Current account ($B)	−1.3	−2.9	−5.3	−7.2	−11.3
Exchange rate − $/DM (%)	1.7	3.6	5.2	5.4	6.4
Japan					
Real GNP (%)	−0.7	−1.5	−2.1	−2.2	−2.8
Growth rate	−0.7	−0.8	−0.6	−0.1	−0.5
Price level (%)	−0.1	−0.2	−0.4	−0.6	−0.9
Inflation rate (+/−)	−0.1	−0.1	−0.2	−0.2	−0.3
Interest rate (+/−)	−0.2	−0.4	−0.6	−0.6	−0.7
Money supply − M2 (%)	0.0	0.0	0.0	0.0	0.0
Growth rate	0.0	0.0	0.0	0.0	0.0
Trade balance ($B)	−4.2	−8.8	−13.2	−14.8	−20.4
Current account ($B)	−4.1	−9.7	−15.4	−19.0	−26.9
Exchange rate − $/yen (%)	2.1	3.6	5.1	5.4	7.3

Note: Amounts shown are deviations from baseline.
(+/−) = absolute deviation.
Dollar values converted at baseline rates.
B = billion (one billion equal to one thousand million).

Empirical simulations of economic policy co-ordination

of government spending in the United States leads directly to a reduction in real income. With a fixed path for the money supply, interest rates fall leading to a depreciation of the dollar. (The reduction in real income tends to reduce imports and therefore to cause the dollar to depreciate; this effect is offset by the lower interest rates.) Prices fall (relative to the baseline) as output falls because excess capacity in the economy reduces profit margins. Again, this effect offsets the inflationary effect of the dollar depreciation.

All of these factors tend to cause the US current account position to improve. Falling income lowers goods imports, while the reduction in interest rates reduces investment income payments. The dollar depreciation and the fall in prices tend to increase exports although on balance exports fall because foreign income is reduced. The net result is a substantial improvement in the US current account. We might note that because the United States has a large net external debt in the baseline simulation changes in US interest rates have an important direct effect on the current account. This effect cumulates as current account improvements reduce the stock of liabilities to foreigners.

In general, the effects on the US economy appear to be of moderate magnitudes. Real income falls by 1–1/2% in the first year and by 2–1/2% after three years before recovering somewhat. (In the long run these decreases would be almost completely reversed.) The inflation rate falls slightly, by 0.5–0.9%, and the exchange rate depreciates by only 5–7%. Nominal interest rates fall substantially, by 1.2 percentage points after four quarters and by 4.5 percentage points at the end of the simulation. As noted, the current account improves quite substantially by the end of this period.

In Germany and Japan the effects of the US fiscal contraction are felt as a direct reduction in demand for exports and as an appreciation of their exchange rates which further lowers exports. Prices and interest rates fall, as in the US, but the current account in both countries goes into deficit.

Initially, the impact on real income is smaller abroad than in the United States, but over time the effects of the dollar depreciation reduces exports sharply. In Japan the percentage fall in output actually exceeds that in the United States by the end of the period. Nevertheless the magnitude of the effects is generally moderate – the growth rate of real income is reduced by about 1/2% per year, while the inflation rate falls by a little less. Interest rates decline by less than one percentage point in Japan, and somewhat more in Germany; the two countries' exchange rates appreciate by only 5–7% against the dollar.

4.4.2 Foreign response

Table 4.4 shows the results from a second simulation in which foreign countries respond to the US fiscal contraction by using monetary policy to hold their real exchange rates constant. This expansionary policy offsets one of the major sources of deflationary pressure abroad, with the result that the fall in foreign income is greatly reduced. In Germany, real GNP falls by 0.9% at the end of the period,

Table 4.4 US fiscal contraction with foreign monetary accommodation

	1986 Q4	1987 Q4	1988 Q4	1989 Q4	1990 Q4
United States					
Real GNP (%)	−1.5	−2.3	−2.5	−1.6	−1.9
Growth rate	−1.5	−0.8	−0.3	1.0	−0.3
Price level (%)	−0.2	−0.6	−1.4	−2.2	−3.1
Inflation rate (+/−)	−0.2	−0.5	−0.7	−0.8	−0.9
Interest rate (+/−)	−1.3	−2.3	−3.3	−3.5	−4.6
Money supply – M1 (%)	0.0	0.0	0.0	0.0	0.0
Growth rate	0.0	0.0	0.0	0.0	0.0
Trade balance ($B)	9.3	16.5	23.4	23.2	32.0
Current account ($B)	11.1	26.0	45.5	60.6	94.4
Exchange rate – FX/$ (%)	−1.1	−2.3	−3.2	−3.3	−3.8
Government deficit ($B)	−9.0	−23.0	−48.7	−74.8	−118.6
Germany					
Real GNP (%)	−0.4	−0.7	−0.9	−0.7	−0.9
Growth rate	−0.4	−0.3	−0.2	0.2	−0.2
Price level (%)	−0.1	−0.4	−0.7	−1.1	−1.6
Inflation rate (+/−)	−0.1	−0.3	−0.3	−0.4	−0.5
Interest rate (+/−)	−0.9	−1.5	−2.2	−2.3	−3.1
Money supply – CBM (%)	0.3	0.8	1.4	1.8	2.3
Growth rate	0.3	0.5	0.6	0.4	0.5
Trade balance ($B)	−1.9	−3.6	−5.5	−5.8	−8.2
Current account ($B)	−2.0	−3.7	−5.9	−6.8	−10.5
Exchange rate – $/DM (%)	0.7	1.4	1.8	1.7	1.9
Japan					
Real GNP (%)	−0.5	−0.7	−0.6	0.2	0.7
Growth rate	−0.5	−0.2	0.1	0.7	0.5
Price level (%)	−0.0	−0.1	−0.2	−0.2	−0.1
Inflation rate (+/−)	−0.0	−0.1	−0.1	0.0	0.1
Interest rate (+/−)	−0.8	−1.4	−1.9	−2.0	−2.6
Money supply – M2 (%)	0.9	2.0	3.5	5.0	7.7
Growth rate	0.9	1.1	1.6	1.4	2.8
Trade balance ($B)	−3.5	−6.5	−9.4	−9.3	−11.9
Current account ($B)	−3.7	−8.3	−13.7	−17.2	−24.5
Exchange rate – $/yen (%)	0.6	1.1	1.2	0.7	0.3

Note: Amounts shown are deviations from baseline.
(+/−) = absolute deviation.
Dollar values converted at baseline rates.
B = billion (one billion equal to one thousand million).

instead of 1.8%, while in Japan output actually rises slightly. In contrast, during the first two years of the two simulations the output paths are very similar for both countries, since at that point the chief influence is the loss of exports to the United States.

The inflationary price paid for these gains is very modest, which reflects in part the 'stickiness' of the price determination mechanism in the MCM. The increase in the money growth rate required to maintain constant real exchange rates is about 1/2% per year in Germany (for central bank money), and $1-2\frac{1}{2}\%$ per year in Japan (for M2). Interest rates fall by 2–3 percentage points.

What is striking about this policy response is that it has essentially no effect on the United States economy: the paths for income, prices, interest rates, and the current account are virtually identical to the two simulations. The implication of this result is that, at least for policy changes of this order of magnitude, the United States can effectively ignore the foreign response in calculating its own optimal policy.

4.4.3 US response

In Table 4.5 we show the results of the opposite case, in which the United States does ease its monetary policy to accommodate the fiscal contraction, but foreign countries do not. We assume that the US monetary reaction is aimed at stabilizing real GNP, rather than the real exchange rate. This choice of target is motivated by the commonplace observation that the United States is a less open and outward-looking economy than is Germany or Japan.

We should note that in our model it is not generally possible to use monetary policy to hold real GNP in the United States to an arbitrary path in the face of a shock. (This is because of the nature of the lagged response of demand to interest rate changes.) Therefore we have simply selected a path for monetary growth which substantially reduces the decline in US GNP, without trying to eliminate the decline entirely.

In this simulation, US GNP returns (approximately) to the baseline path after the first two years. The monetary expansion reduces nominal interest rates, which fall by about $4\frac{1}{2}$ percentage points. This leads to a depreciation of the dollar of 12% by the end of the period. Despite the dollar's depreciation, the inflation rate is very slightly lower throughout the simulation.

These benefits come, however, at the expense of the other countries. The additional depreciation of the dollar (as compared with the first simulation with no policy response) leads to a further reduction in exports. In Germany this is offset by the positive effects of stronger income in the US, but in Japan exports and output both fall sharply. Japanese real GNP falls by over 3% by the end of the simulation period. Clearly, US policies do offset foreign countries.

Table 4.5 US fiscal contraction with US monetary accommodation

	1986 Q4	1987 Q4	1988 Q4	1989 Q4	1990 Q4
United States					
Real GNP (%)	−1.0	−0.9	−0.3	0.7	−0.1
Growth rate	−1.0	0.1	0.7	1.0	−0.8
Price level (%)	−0.0	−0.2	−0.4	−0.6	−0.8
Inflation rate (+/−)	−0.0	−0.1	−0.2	−0.2	−0.2
Interest rate (+/−)	−2.0	−4.4	−4.4	−4.6	−4.7
Money supply − M1 (%)	1.5	3.5	4.5	4.5	4.0
Growth rate	1.5	2.0	1.0	−0.0	−0.5
Trade balance ($B)	6.9	11.6	16.0	18.0	27.3
Current account ($B)	8.1	26.3	45.0	64.1	93.1
Exchange rate − FX/$ (%)	−2.7	−8.0	−9.5	−11.3	−12.0
Government deficit ($B)	−20.7	−56.8	−104.9	−137.6	−168.8
Germany					
Real GNP (%)	−0.3	−0.5	−0.9	−1.2	−1.7
Growth rate	−0.3	−0.2	−0.5	−0.2	−0.5
Price level (%)	−0.5	−1.2	−1.7	−2.1	−2.6
Inflation rate (+/−)	−0.5	−0.7	−0.6	−0.3	−0.5
Interest rate (+/−)	−0.6	−1.2	−1.2	−1.4	−1.8
Money supply − CBM (%)	0.0	0.0	0.0	0.0	0.0
Growth rate	0.0	0.0	0.0	0.0	0.0
Trade balance ($B)	−1.0	−2.4	−5.8	−7.4	−12.0
Current account ($B)	−0.4	−0.2	−2.4	−2.6	−7.3
Exchange rate − $/DM (%)	3.2	8.6	10.9	12.5	12.6
Japan					
Real GNP (%)	−1.0	−2.2	−2.9	−3.1	−3.4
Growth rate	−1.0	−1.3	−0.7	−0.2	−0.3
Price level (%)	−0.1	−0.4	−0.6	−0.9	−1.2
Inflation rate (+/−)	−0.1	−0.3	−0.2	−0.3	−0.3
Interest rate (+/−)	−0.3	−0.6	−0.7	−0.7	−0.7
Money supply − M2 (%)	0.0	0.0	0.0	0.0	0.0
Growth rate	0.0	0.0	0.0	0.0	0.0
Trade balance ($B)	−6.3	−11.1	−17.2	−17.6	−24.1
Current account ($B)	−6.3	−12.4	−20.6	−22.6	−31.0
Exchange rate − $/yen (%)	3.4	9.2	10.3	12.6	13.8

Note: Amounts shown are deviations from baseline.
(+/−) = absolute deviation.
Dollar values converted at baseline rates.
B = billion (one billion equal to one thousand million).

Table 4.6 US fiscal contraction with US and foreign monetary accommodation

	1986 Q4	1987 Q4	1988 Q4	1989 Q4	1990 Q4
United States					
Real GNP (%)	−1.0	−1.0	−0.3	0.7	0.0
Growth rate	−1.0	0.0	0.7	1.0	−0.7
Price level (%)	−0.1	−0.3	−0.6	−0.8	−1.0
Inflation rate (+/−)	−0.1	−0.2	−0.3	−0.2	−0.2
Interest rate (+/−)	−2.0	−4.5	−4.6	−4.7	−4.8
Money supply − M1 (%)	1.5	3.5	4.5	4.5	4.0
Growth rate	1.5	2.0	1.0	−0.0	−0.5
Trade balance ($B)	6.9	10.6	14.4	15.5	25.5
Current account ($B)	7.8	25.4	45.2	64.9	96.0
Exchange rate − FX/$ (%)	−1.0	−4.8	−5.9	−7.3	−7.9
Government deficit ($B)	−20.9	−57.3	−106.9	−141.4	−175.5
Germany					
Real GNP (%)	−0.2	−0.1	−0.0	0.3	−0.0
Growth rate	−0.2	0.1	0.1	0.3	−0.3
Price level (%)	−0.2	−0.5	−0.8	−1.0	−1.2
Inflation rate (+/−)	−0.2	−0.3	−0.3	−0.1	−0.3
Interest rate (+/−)	−1.6	−3.0	−3.3	−3.5	−3.6
Money supply − CBM (%)	0.7	1.7	2.8	3.5	4.0
Growth rate	0.7	1.1	1.1	0.7	0.5
Trade balance ($B)	−1.9	−2.5	−3.0	−1.7	−2.7
Current account ($B)	−2.5	−2.9	−3.9	−2.6	−4.5
Exchange rate − $/DM (%)	0.6	3.2	3.8	4.6	5.1
Japan					
Real GNP (%)	−0.3	−0.3	0.5	1.8	2.5
Growth rate	−0.3	0.0	0.8	1.3	0.8
Price level (%)	−0.0	−0.1	−0.0	0.1	0.5
Inflation rate (+/−)	−0.0	−0.1	0.1	0.2	0.3
Interest rate (+/−)	−1.4	−2.8	−3.0	−3.0	−3.1
Money supply − M2 (%)	2.1	5.4	8.0	10.7	13.4
Growth rate	2.1	3.3	2.5	2.8	2.7
Trade balance ($B)	−3.7	−6.7	−9.5	−9.8	−12.7
Current account ($B)	−4.3	−10.2	−17.2	−22.2	−29.8
Exchange rate − $/yen (%)	0.4	2.7	3.0	3.5	3.3

Note: Amounts shown are deviations from baseline.
(+/−) = absolute deviation.
Dollar values converted at baseline rates.
B = billion (one billion equal to one thousand million).

4.4.4 Joint policy response

Finally, we consider a simulation in which both the United States and the foreign countries use monetary policy to offset the effects of the initial fiscal contraction. The United States is assumed to follow the same monetary growth path as in the previous simulation, while Germany and Japan again use monetary policy to fix their real exchange rates. Table 4.6 presents the results.

As in the earlier example, the foreign expansion maintains GNP roughly unchanged in the face of the US contraction, except in the last two years for Japan, when income actually rises. Again, the foreign policy response has only minor effects on US income, prices, interest rates, and current account. Thus we still conclude that US policies can be set independently of the foreign response.

However, the *foreign* monetary policy needed to stabilize real exchange rates is very different in the last simulation (with US monetary accommodation) from the earlier simulation (without US accommodation). The foreign money growth rates shown in Table 4.6 are more than twice as large as those in Table 4.4. Therefore we conclude that in responding to a given event, foreign economic policy-making should, in general, take into account the response of US economic policy to the same event.

This result is an example of an important, and obvious, asymmetry in the present-day world economy, that the United States is a much larger and substantially less open economy than its major trading partners. Therefore other countries must take into account US policy actions, but the US need not take theirs into account. (Of course, we have given an analysis of only one possible policy-making problem – it does not follow that in response to *any* shock the United States could ignore the actions for foreign countries.)

This example also illustrates the scope for gains from sharing information about foreign economic policies. In the terms of our earlier discussion, the Nash equilibrium (loosely) corresponds to the last simulation, in which all parties optimize given full knowledge of other countries' policies.

4.5 CONCLUSION

We can represent these simulations in terms of our theoretical framework using Fig. 4.2. After the implementation of the Gramm-Rudman Act the world economy finds itself at a point such as A_0, which corresponds to our first simulation (Table 4.3). At this point policy in all countries is suboptimal given the changed conditions, which we model as the imposition of a constraint on US fiscal policy.

We assume that foreign fiscal policy remains unchanged throughout, so that each country has only one policy instrument, namely the money supply. We have further simplified the problem by assuming that each country has only one target – real GNP in the United States, and the real exchange rate abroad. This assumption allows each country to hit its target exactly, which greatly simplified the task of computing the optimal policies.

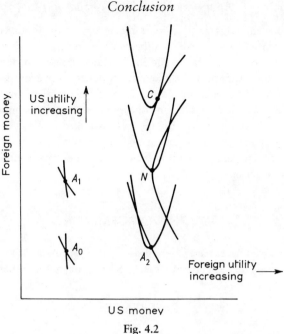

Fig. 4.2

We first suppose that policy-makers do not know where the new Nash equilibrium (point N) is. Indeed, they naively assume that policy abroad will remain unchanged while their own countries take action. This gives us points A_1 (corresponding to Table 4.4) and A_2 (Table 4.5). The Nash equilibrium is the point at which countries optimize given full knowledge of others' responses. This is shown at point N, which corresponds to the simulation in Table 4.6. Our results suggest that for United States policy, there is only a slight difference between A_2, where the foreign response is ignored, and the Nash equilibrium, where the foreign reaction is taken into account. This is shown by drawing point A_2 on a vertical line beneath N. On the other hand, the optimal foreign policy response changes considerably depending on whether or not the US action is taken into account. Thus, point A_1 is well below N on the vertical axis.

With this framework we are not able to obtain any explicit measures of the welfare gain associated with obtaining information about foreign policy. However, the difference in real income corresponding to the difference between points N and A_1 is of the order of 1%, and the implied monetary policy differs by a factor of two. Therefore it seems fair to describe the differences as important.

With our framework we are also unable to say much about the location of the co-operative point C. (Indeed, when we reduce the policy problem to one target and one control the Nash equilibrium is pareto-efficient and there are no further gains from co-operation to be realized.) Sachs and Oudiz analyse the gains to be realized in money from N to C, and find them to be surprisingly small.

We conclude from our analysis that the information about other countries' policies needed to support a Nash equilibrium is quantitatively important. Another way of putting this is that the Nash equilibrium appears to differ significantly from other possible equilibria in the world economy. (For example, one in which countries hold static expectations about foreign policy.) The implication for the modeller is that the choice of solution concept is quantitatively important in doing policy analysis. The implication for the policy-maker is that institutions which ensure the interchange of information about economic policy are in important form of policy co-ordination, perhaps at least as important as explicit bargaining over joint policy actions.

ACKNOWLEDGEMENTS

The views expressed herein are solely those of the authors and do not necessarily represent the views of the Federal Reserve System or any members of its staff.

REFERENCES

Buiter, W. H. and Marston, R. C. (1985) *International Economic Policy Coordination*, Cambridge University Press, Cambridge.

Canzoneri, M. E. and Gray, J. (1983) Two essays on monetary policy in an interdependent world. *International Economic Review* (to be published).

Edison, H. J., Marquez, J. and Tryon, R. (1987) The Structure and Properties of the Federal Board, *Economic Modelling*, 4(2), 115–315.

Hamada, K. (1974). 'Alternative exchange rate systems and the interdependence of monetary policies, *National Monetary Policies and the International Financial System*, (ed. R. Z. Aliber) University of Chicago Press, Chicago, pp. 13–33.

Hamada, K. (1976) A strategic analysis of monetary interdependence. *Journal of Political Economy*, 84, August, 77–99.

Hooper, P. (1986) Exchange Rate Simulation Properties of the MCM, *European Economic Review*, 30, pp. 121–98.

Miller, M. and Salmon, M. (1985) Dynamic games and time inconsistency of optimal policies in Open Economies, in *International Economic Policy Co-ordination*, (eds W. Buiter and R. Marston) Cambridge University Press, Cambridge.

Oudiz, G. and Sachs, J. (1984) Macroeconomic policy coordination among industrial economies, Brookings Papers on Economic Activity, No. 1, pp. 1–76.

Stevens, G. V. G., Berner, R. B., Clark, P. B., Hernandez-Cuta, E., Howe, H. T. and Kwack, S. Y. (1984) *The US Economy in an Interdependent World: A Multicountry Model*, Board of Governors of the Federal Reserve System, Washington, DC.

5

How much could the international co-ordination of economic policies achieve? An example from US–EEC policy-making

A. J. HUGHES HALLETT

The economic recessions since 1973 have emphasized the interdependence of the major economies and their policy choices. This chapter examines whether decentralized control of the world economy effectively limits our ability to steer individual economies. It is well known that non-co-operative policies are Pareto inefficient; but there is, as yet, no empirical evidence on the costs of unco-ordinated policies, or the potential gains and risks in co-operation. In contrast to recent theoretical work in the area, this chapter uses an estimated multicountry model in a dynamic game framework to estimate those costs and benefits. Policy design depends crucially on the asymmetries between economies. Successful co-ordination depends on anticipations, and on timing the fiscal and monetary policy impacts correctly. The gains from sustainable co-operation are relatively small and benefit Europe.

5.1 INTRODUCTION

Economic interdependence poses problems of both measurement and management. Official concern at the increasing complexity of the problems faced by policy-makers in open economies is well known. Persistent recessions since 1974, two oil price shocks, and an international debt crisis, have made policy-makers all too aware of the links between their economies, and that mutual dependence through trade and capital movements also makes their policy choices interdependent. The problem is serious; the average OECD country now exports about

30% of its GNP, so that 'spillover' effects of policy changes from one country to another (and back again) can be very powerful. Even the United States is more than 20% dependent on foreign activity. How should policy be designed in these circumstances, and could internationally co-ordinated policies lead to better results?

The realization that foreign reactions often interfere with domestic policies has pursuaded many politicians to call for co-ordinated policies. Competitive devaluations used to be the standard example, but nowadays few countries can divorce their monetary policies from foreign monetary conditions. Similarly budget reductions abroad may frustrate domestic reflation plans, while foreign budget deficits can crowd out domestic investment. Thus not all countries can reduce inflation simultaneously by tight money, high interest rates, or currency appreciation. But if all countries do tighten their policies simultaneously, the losses of output and employment are likely to be larger than any of them planned because of the spillovers between economies. For example, given the degree of their interdependence, the European economies have an obvious incentive to export their inflation and unemployment – yet if they all do that, no country will benefit. Similarly, if they all expand together the inflation gains may well be larger, and employment gains smaller, than expected because of the spillovers. Such surprises can be avoided only if policy-makers anticipate the consequences of current and future foreign policies correctly.

The problem here is clear enough; unco-ordinated policies may actually limit our ability to control individual economies. Therefore it seems odd that, after a decade of annual economic summit meetings, there are so few results on how interdependence should affect policy design, and so little evidence on the policy spillover effects between economies. Similarly, although economic theory has shown that co-ordinated policies can bring gains, there is no evidence on either whether these gains will be significant over a period of time or on how they would be distributed between countries. The agenda for this chapter is therefore:

(1) to evaluate the policy spillovers between the major economies;
(2) to incorporate rational expectations of foreign actions into the policy selections; and
(3) to determine the likely gains (their size and distribution) from sustainable co-operative or bargaining solutions.

Previous work in this area has depended on simple analytic models of identically symmetric economies and/or static decision rules. For example, Cooper (1969), Hamada and Sakuri (1978) and Johansen (1982) have identically symmetric economies and static decisions. Currie and Levine (1984), Miller and Salmon (1985), and Taylor (1985) consider identically symmetric economies but dynamic decisions. Finally Canzoneri and Gray (1985) allow asymmetric spillovers (in sign but not size) under static decisions, while the only empirical study so far (Oudiz and Sachs, 1984) considers asymmetric economies and a static decision process.

This chapter considers policies for the 1974–78 recession, which led to the proposal of co-ordinated policies at the 1978 Bonn economic summit. Estimates of the expected benefits under different policy strategies for the US and EEC economies are made based on one of the multicountry models which the policy makers themselves have used. Our results confirm Cooper's (1969) theoretical arguments for international co-operation: co-ordination restores policy effectiveness, and also cuts the cost of intervention by speeding up policy responses. These gains actually arise from co-ordinated monetary actions – so that the 'G5' group of governments (US, Japan, France, Germany and UK) are right to try to harmonize their interest rates as a first step.

However, the central result is that asymmetries between countries play a vital role. Co-ordination produces better results both because it allows governments greater freedom to specialize in those policy instruments which have comparative advantage, and because it allows governments to co-ordinate the *timing* of their policy impacts. Hence co-ordination should aim to exploit the differences between economies, and to organize the sequencing of policy actions rather than promote parallel policies. This implies that the European Economic Commission (EEC) is wrong to make the convergence of economic structures the goal of co-ordination (EEC, 1984). Better results will follow by exploiting the asymmetry of policy responses between countries. This finding is important because it shows that previous studies, which have used models of identically symmetric economies and/or static decisions, could not pick up a major part of the gains from policy co-ordination. This, of course, includes the earlier empirical work of Oudiz and Sachs who did not consider the gains from co-ordinating the timing of interventions or the question of bargaining power. On the other hand, this chapter is deliberately set in that same empirical mould. It is not always practical to use theoretical models in this context because there are three spillover channels to consider (via income, monetary, and price changes) and we need to evaluate their *net* spillover effects. No theoretical model has yet managed to treat all three channels (and very few consider more than one) because it is too difficult to sign the combined effect of all the partial derivatives of different instruments, activated in different periods, on each target in each period. Consequently theoretical models usually show an extreme sensitivity to changes in their underlying assumptions and restrictions. For example, Mundell's (1963) model yields totally opposite results if flexible exchange rates become fixed; in effect by restricting just one parameter to zero. Similarly Corden and Turnovsky (1983) can reverse Hamada and Sakuri's (1978) conclusions simply by considering two production sectors per economy instead of one. These examples are a striking illustration of Malinvaud's (1981) warning that the untested restrictions assumed in theoretical models are often more extreme in their consequences than those in empirical models.

5.2 ECONOMIC CO-OPERATION AND WORLD RECESSION: A REVIEW

The severity of the recessions which followed the 1973/74 oil price rise has provided good reason to examine the case for internationally co-ordinated policies. With unemployment rising and output falling worldwide, the US in fact chose to pursue a policy of deficit financing (tax cuts and tight money since 1975; increased public expenditures since 1980), while European governments followed contractionary policies (mainly public expenditure cuts and tight monetary control). Even within these two different strategies, there were many policy changes. Budgetary policy started out by being cautious in 1974; but by 1975 several countries had expanded their public expenditures to compensate for the external deficits caused by the rise in oil prices. In some countries monetary restraint was also eased. Then in 1976–77 most countries cut back on their fiscal activity again, so that output continued to stagnate while unemployment increased and inflationary expectations persisted. In particular the record shows that US and EEC policies, from 1975 up to the second oil price shock, were increasingly divergent. The EEC economies, in particular, resorted to unstable fiscal and monetary actions. A large cut of government expenditures in 1976 was followed by a rise in 1977 and a shift from social outlays to government consumption in 1978. During these years changes in the discount rate were also alternating. Such an erratic and unsynchronized pattern of policy adjustments suggests that co-ordination of policy effort between the US and the various EEC governments might bring important benefits; an argument explicitly endorsed by the declarations made during the Bonn Summit meeting of 1978.

Nevertheless many policy-makers and economists appear to believe in the virtues of co-ordinated policies. In 1974 finance ministers of the industrialized economies agreed to avoid competitive deflations which would merely pass current account deficits between partners. In 1978 the US administration called for joint action to expand the major economies as a 'locomotive' for world recovery. However, Germany's solo attempt to carry out that programme rapidly ran into trouble – as did later attempts by France and Sweden. Policy-makers called again for joint reflation in the 1980–2 recession, although co-ordinated policies were by then opposed by the US government. More recently it has been argued that European governments should accept greater fiscal expansion in return for reduced US deficits, and that co-ordinated policies would have helped by reducing exchange rate fluctuations. See Helmut Schmidt, V. Giscard d'Estaing, Martin Feldstein, and Lawrence Klein in *The Economist* (between February and June 1983); also Bergsten *et al.* (1982), Begg (1983), and Layard *et al.* (1986) for the economic arguments.

However, advocating co-operation has led to very little positive action, precisely because there are no estimates of either how large the gains are likely to be, or how they might be distributed between countries. Similarly there is little agreement on what exactly determines the size of those gains, so that detailed arguments for co-operation are difficult to establish in practice. In these circumstances, it would have been very hard for any sovereign government to justify surrendering some of its own objectives for the incompletely specified and unquantified benefits of international action; benefits which might, in any case, largely go to the other participants.

Table 5.1 The objectives of economic policy 1974–78

	US Ideal (%)	Priority	EEC Ideal (%)	Priority	Symbol	Units
Targets:						
Inflation	−3	4	−3	1	P	% annual growth
Output	3	4	3	4	X	% annual growth
Balance of trade	1	1	1	4	B	% of GNP
Employment	1½	1	1½	1	E	% of labour force
Investment ratio	3	1	3	1	IR	% of GNP
Profit ratio	3	1	3	1	PR	% of GNP
Instruments:						
Government expenditure	2	1	4	1	G	% annual growth
Social security outlays	15	1	16	1	S	% annual growth
Direct tax, households	10	1	17	1	TH	% annual growth
Direct tax, companies	9	1	14	1	TC	% annual growth
Discount rate	0.4	1	−¼	1	R	Level
Central bank loans	5	1	19	1	LC	% annual growth
Commercial bank loans	5	1	19	1	LB	% annual growth

The exercises which follow therefore examine the advantages in promoting economic recovery through concerted action by the US and EEC policy-makers. The US and EEC are assumed to have the same target variables, but the priorities which governments attach to these targets differ to match their observed policy stances. The choice of the 1974–78 planning period implies that we are looking for improved policy reactions to a common unanticipated shock – the oil price rise of 1973. However, the planners are also assumed to be interested in governing supply conditions in 1978 and beyond. Hence the targets combine conventional stabilization objectives (unemployment, inflation, production growth, the balance of trade) with indicators of potential supply (the real investment/production ratio, and the ratio of profits to national income). The difference in policy stance is represented by giving inflation and production growth a relatively high weight in the US objective function, but weighting production and the balance of trade more heavily for the EEC objectives. Exchange rates float freely throughout.

The following instruments were used by the US and EEC to improve economic performance during this period: government expenditures; social security expenditures; direct taxes on households and corporations; discount rates; net government borrowing from central and commercial banks. The policy variables, priorities, and ideal values, are summarized in Table 5.1. All instruments are defined here as discretionary *adjustments* to the historically chosen policy values. We found by experimentation that the penalties on deviations of these instruments from their ideal path should be roughly equally distributed.

5.3 THE MODEL AND ITS ASYMMETRIC POLICY RESPONSES

The policy dependence between the US and EEC economies has been explored here using a version of the COMET multicountry econometric model (Barten *et al.*, 1976). COMET is typical of the empirical models which are currently used for analysing economic interdependence. It describes the economies of the EEC countries, the US, the remaining OECD countries, OPEC, the Socialist and the developing countries. The version used here has been re-estimated and contains the following changes:

(1) aggregation into three economic blocks, to give a systematic specification across the US and EEC economies;
(2) consistent accounts for the rest of the world;
(3) the introduction of financial sectors, international capital movements, and endogenous exchange rates. (The model, as used, is described in Van der Windt *et al.* (1984).)

The importance of having a rest of the world block is that one country's trade deficit is not automatically the other country's surplus. One can then see if policy improvements in the US and EEC are obtained at the expense of other countries.

The US and EEC blocks, some 200 equations in total, have similar structures based on a conventional Keynesian demand system covering consumption, investment, and foreign trade. Cobb–Douglas production functions determine labour demand and investment. The supply side of both blocks are modelled by similarly specified potential output functions; and the potential and actual outputs are reconciled by a capacity utilization index. In each block prices are related to import prices and the GDP deflator. The GDP deflator itself is determined by unit labour, capital costs and the utilization index, while wages depend on consumption prices, productivity, unemployment, taxes and social security contributions.

The monetary sectors describe the financial relations between the central and commercial banks, the government, the private sector, and the foreign sector. Private and government financial surpluses are the main input from the real sector. Interest rates, which affect both real spending and international borrowing, are determined by those surpluses and by the behaviour of the banks. There is an important difference between the US and the EEC here. In either case government expenditures are financed, to a large extent, by loans from the private sector. In the EEC, this involves selling bonds to the banking system which then adjusts its portfolio and offsets any upward pressure on interest rates. Meanwhile the extra government expenditures also pass through the banking system, so that credit creation effects tend to reduce interest rates and increase domestic activity until inflation and wage increases set in to reverse that some periods later. Government expenditures in the US, however, are financed by borrowing savings directly from the private sector and from abroad, or by tax and other fiscal measures. With a low propensity to save and greater interest elasticity of investment, this leads to the crowding out of private investment followed by higher interest rates and falling activity levels. At the same time prices rise due to higher capital costs. This crowding out process sets in after one year. It is sustained because there is no offsetting expansionary pressure or credit creation.

The model and its asymmetric policy responses

An analysis of the model's dynamic properties revealed that policy changes in the US generated quicker and stronger responses than those in the EEC (compare Tables A2 and A3 of Appendix A). This stems principally from reactions of real quantities to price, wage or interest rate changes, which are both larger and faster in the US structural equations. These differences are accentuated on the US side by international capital movements (which depend on relative interest rates, domestic savings, the trade balance and relative growth rates). For example, both economies react in a similar way to a rise in the domestic interest rate, but the US balance of payments alone is affected significantly (and positively). Again the impact of a rise in US interest rates on European targets is roughly equal to that for an equivalent rise in the EEC interest rate; but the EEC rate does not have the equivalent effects on US targets. Hence internal US policies for monetary control, or for financing the budget, tend to attract capital from abroad in a way that EEC policies do not.

Thus, in this model, US policy instruments have more powerful short-run multipliers, but less powerful long-run multipliers, than their EEC counterparts. These differences of impacts and timing show up clearly in the target responses to changes in interest rates, money supply and (less clearly) in government and social security expenditures. Only taxation on households generates a larger response in Europe. In the longer run the US multipliers for government and social security expenditures, taxation and the interest rate weaken to the extent that sign changes appear in all of them, while that for money supply strengthens. The main European multipliers (government expenditures, social security payments and taxation) also weaken slightly, while the monetary instruments (the interest rate and money supply) strengthen. That leaves the US multipliers weaker for all instruments except the interest rate (whose relative strength over its EEC counterpart has fallen) and the money supply (whose relative strength over its EEC counterpart remains about the same).

The spillover effects from one economy to the other show a similar pattern in that the effects of EEC policy changes on the US build up over time, whereas the effects of US policies on the EEC are immediate and do not build up (Tables A4 and A5 of Appendix A). Moreover, the US spillover multipliers are nearly all stronger than the corresponding EEC multipliers. Thus rising public expenditures and high interest rates in the US would only temporarily increase European output, profits and trade balances. This result is broadly consistent with most theoretical models, and with recent history. In contrast, public expenditure cuts and low interest rates in Europe would first add to a US trade deficit (while expanding US output only slightly) but would lead to a reduction of that deficit and extra inflation later on. Recent events, at least, have been consistent with the first part of that finding.

However, the monetary instruments are the only variables with strong spillover effects, and here the impact of US monetary policy on the EEC far exceeds that of EEC monetary policy on the US. Moreover, that asymmetry becomes more marked the longer the time lag since a policy change. It is important to identify these spillover effects since, as Canzoneri and Gray (1985) point out, the appropriate policy responses for interdependent economies depend on knowing the type of the policy regime faced. Canzoneri and Gray consider three regimes defined by the signs of the spillovers:

(1) the 'beggar-thy-neighbour' case, where money supply expansion in each economy has negative effects on output in the other economy;
(2) the 'locomotive' case, where money supply expansion has positive effects on output in the other country;
(3) the 'asymmetric' case, where the money supply spillovers may be of different sizes but must have different signs.

The present model points to the 'asymmetric' case for short-run effects of US monetary policy on the EEC and vice versa. Expansion of the US money supply has a positive impact on EEC output and employment, whereas expanding the EEC money supply has a negative effect on US output and employment. Thus the tight American monetary policy of recent years, backed up by a relatively tight European money supply, have both worked to the American short-term advantage. However, the spillovers from such policies will eventually tend to increase output abroad while depressing it domestically; the spillover multipliers are negative, the domestic multipliers positive. Therefore monetary policy turns into a 'beggar-thy-neighbour' regime in the longer term; but, being dominated by the US money supply, it is certainly not symmetric. That poses an awkward problem of selecting the timing of the interventions so as to get the desired *net* effects, in each period, while the policy regime shifts with the passage of time. This implies:

(1) a dynamic analysis of policy spillover effects is essential since those effects can, like any policy multipliers, change sign over time and the dynamic consequences of different strategies may be quite different from their impact effects; and
(2) the *net* impacts of current and past policy changes (at home and abroad) will probably have to be evaluated numerically.

Extending this classification to interest rates, we find that the spillovers from lower interest rates on output and employment are all positive while domestic effects are also positive.[1] This variable therefore introduces a 'locomotive' policy regime, which will allow an expansion of the international money supply. Once again the US policy spillovers outweigh those from the EEC, particularly those which arise from interest rate changes. Finally, public expenditures also offer a 'locomotive' regime in the short run which becomes ineffective with respect to output and employment in the longer term. But, as explained above, the US

---

[1] The Federal Reserve's MCM model and the Japanese EPA model both corroborate these findings (Oudiz and Sachs, 1984) if Germany is taken to represent the EEC. The same holds for the government expenditure multipliers. The conclusions of this section may therefore be reasonably robust.

public expenditure domestic multipliers change sign after one year (while their EEC counterparts do not) so that the underlying policy regime now shifts for domestic, rather than international, reasons.

Hence, taking these points together, numerical policy evaluations seem unavoidable because we have to estimate the *net* spillover effects from:

(1) a combination of multiple transmission channels between an instrument in one economy and a target in another;
(2) the combined effects of different instruments on a foreign target variable (spillovers may be small individually but substantial in sum);
(3) spillover multipliers which are asymmetric between economies;
(4) dynamic spillover multipliers which may change sign at successive intervals.

5.4 NON-CO-OPERATIVE POLICIES WITH ANTICIPATIONS

In order to measure the potential gains from co-operative policy-making, we must first determine how far each country can expect to satisfy its own objectives when all countries follow non-co-operative strategies. To obtain that benchmark each government must condition its decisions on the actions to be expected from rational policy-makers abroad. They must also expect that foreign decision-makers will, at the same time, be choosing their policies in the same way, so that the consequences of foreign reactions must be allowed for in domestic decisions. This means each decision rule must evaluate the expected foreign responses to domestic policy adjustments, simultaneously with the usual joint determination of the decision values themselves. Otherwise surprises cannot be avoided since it is inconsistent to determine country 2's policies, $x^{(2)}$ say, using a rule which specifies $\partial x^{(2)}/\partial x^{(1)} \neq 0$ but $\partial x^{(1)}/\partial x^{(2)} = 0$, while $x^{(1)}$ depends on a rule which maintains $\partial x^{(1)}/\partial x^{(2)} \neq 0$ but $\partial x^{(2)}/\partial x^{(1)} = 0$. A more general solution may not involve zero restrictions, but inconsistencies will appear whenever the computed $\partial x^{(2)}/\partial x^{(1)}$ value in the rule for $x^{(2)}$ differs from its conjectured value in the rule for $x^{(1)}$ (and vice versa for $\partial x^{(1)}/\partial x^{(2)}$). If the assumed policy responses differ from actual policy reactions, then the anticipations generated within the policy game must be incomplete. Consequently the Lucas critique will apply and policy makers will make systematic 'errors' because they fail to base their decisions on rational expectations of future developments.

One solution method which does evaluate foreign reactions jointly with domestic decisions is discussed in Hughes Hallett (1984) and Brandsma and Hughes Hallett (1984). It works as follows. Let there be two countries which have m_i domestic targets, $y_t^{(i)}$, n_i policy instruments, $x_t^{(i)}$, and which are subject to noncontrollable (random) events $s_t^{(i)}$ over periods $t = 1 \ldots T$. The decision variables are therefore $y^{(i)'} = (y_1^{(i)'} \ldots y_T^{(i)'})$ and $x^{(i)'} = (x_1^{(i)'} \ldots x_T^{(i)'})$ for $i = 1, 2$, and $s^{(i)'} = (s_1^{(i)'} \ldots s_T^{(i)'})$ represent external shocks. Suppose each country aims for ideal values $y^{(i)d}$ and $x^{(i)d}$, so that $\tilde{y}^{(i)} = y^{(i)} - y^{(i)d}$ and $\tilde{x}^{(i)} = x^{(i)} - x^{(i)d}$ define its decision 'failures'. The national objectives of each country are represented by the loss functions:

$$w^{(i)} = \tfrac{1}{2}[\tilde{y}^{(i)'} B^{(i)} \tilde{y}^{(i)} + \tilde{x}^{(i)'} A^{(i)} \tilde{x}^{(i)}] \qquad i = 1, 2 \qquad (5.1)$$

where $B^{(i)}$ and $A^{(i)}$ are positive definite and symmetric matrices. Let the two economies be linked via the world economic system

$$y_t = f(y_t, y_{t-1}, x_t^{(1)}, x_t^{(2)}, e_t) \qquad (5.2)$$

where e_t represents any noncontrollable variables. The constraints on each country's targets can then be condensed to

$$y^{(i)} = R^{(i,1)} x^{(1)} + R^{(i,2)} x^{(2)} + s^{(i)} \qquad i = 1, 2 \qquad (5.3)$$

where $R^{(i,j)}$, $j = 1, 2$, are $(m_i T \times n_j T)$ matrices containing submatrices of dynamic multipliers $R_{tk}^{(i,j)} = \partial y_t^{(i)}/\partial x_k^{(j)}$ if $t \geq k$, and zeros elsewhere. Hence $R^{(i,i)}$ measures the responses of country i's targets to its own instruments, and $R^{(i,j)}$ the spillover effects from country j's decisions.

Each policy-maker will find his own decision variables constrained by $\tilde{y}^{(i)} = R^{(i,i)} \tilde{x}^{(i)} + [R^{(i,j)} \tilde{x}^{(j)} + c^{(i)}]$ where $c^{(i)} = s^{(i)} - y^{(i)d} + \Sigma_j R^{(i,j)} x^{(j)d}$. Each country's optimal strategy therefore depends on, and must be determined simultaneously with, the decisions to be expected abroad. In the absence of any co-operation, the optimal decisions $(x^{(1)*}, x^{(2)*})$ will satisfy $w^{(i)}(x^{(i)*}, x^{(j)*}) \leq w^{(i)}(x^{(i)}, x^{(j)*})$ for $i = 1$ and 2, and all feasible $x^{(i)} \neq x^{(i)*}$. This equilibrium holds only when both countries perceive that no further gains can be made by varying their reactions to the decisions currently expected from their opponent, because to do so would trigger optimal counter-reactions (in the opponent's interest) which more than offset any gains made by the first country. (See Bresnahan (1981), Ulph (1983), Holt (1985) for the conjectural variations method. Brandsma and Hughes Hallett (1984) extend it for any number of targets and periods, and for asymmetric decision-makers.) But whenever net gains could be made, despite foreign responses, a further round of policy adjustments must be expected. The necessary conditions for jointly optimal decisions by both countries are therefore satisfied by solving

$$\partial w^{(i)}/\partial x^{(i)} + [(\partial y^{(i)}/\partial x^{(j)}) \partial x^{(j)}/\partial x^{(i)} + \partial y^{(i)}/\partial x^{(i)}]' \partial w^{(i)}/\partial y^{(i)} = 0 \qquad (5.4)$$

for $i = 1$ and 2 simultaneously. These equations are nonlinear in the unknowns $(\partial x^{(1)}/\partial x^{(2)}, \partial x^{(2)}/\partial x^{(1)})$ and $(x^{(1)}, x^{(2)})$, but a solution may be identified by numerical search. Let $D^{(i)} = \partial x^{(i)}/\partial x^{(j)}$. Inserting trial values, $D_s^{(1)}$ and $D_s^{(2)}$, into Equation 5.4 automatically generates the iteration:[2]

$$\begin{bmatrix} I & -D_{s+1}^{(1)} \\ -D_{s+1}^{(2)} & I \end{bmatrix} \begin{bmatrix} \tilde{x}_{s+1}^{(1)} \\ \tilde{x}_{s+1}^{(2)} \end{bmatrix} = \begin{bmatrix} F_{s+1}^{(1)} c^{(1)} \\ F_{s+1}^{(2)} c^{(2)} \end{bmatrix} \qquad (5.5)$$

---

[2] To save introducing new notation, s is an iteration index for the entire $\mathbf{x}^{(i)}$ vector, while $\mathbf{x}_t^{(i)}$ defines the subvector of $\mathbf{x}^{(i)}$ which is active in period t.

where $F^{(i)}_{s+1} = -(G^{(i)'}_s B^{(i)} R^{(i,i)} + A^{(i)})^{-1} G^{(i)'}_s B^{(i)}$ with $G^{(i)}_s = R^{(i,i)} + R^{(i,j)} D^{(j)}_s$, and where

$$D^{(i)}_{s+1} = F^{(i)}_{s+1} R^{(i,j)} \neq D^{(i)}_s. \qquad i = 1, 2 \quad \text{and} \quad j \neq i \tag{5.6}$$

Unfortunately this iteration may fail to converge and, in any case, it typically yields multiple solutions. However, $D^{(i)}_{s+1}$ can be replaced by $\gamma_i D^{(i)}_{s+1} + (1-\gamma_i) D^{(i)}_s$ where $0 \leq \gamma_i \leq 1$ is chosen (at each step) so as to force $x^{(1)}_{s+1}$ and $x^{(2)}_{s+1}$ 'downhill' – i.e. so that $w^{(i)}(x^{(1)}_{s+1}, x^{(2)}_{s+1}) \leq w^{(i)}(x^{(1)}_s, x^{(2)}_s)$ for $i = 1, 2$. This modification introduces a directed search such that at least one country is better off (and neither worse off) at each step. An exhaustive search will then identify the optimal decisions $(x^{(1)*}, x^{(2)*})$ as defined by the joint minima specified above.[3]

This solution procedure explicitly recognizes that a player should only alter his conjectures about an opponent's reactions in a way which is Pareto improving for both players, since otherwise the opponent simply will not react in the way conjectured. In this context, one cannot expect any sovereign government to adjust its responses *unilaterally* against its own interests because that would amount to imposing intercountry comparisons without any compensating policy bargain. In any case, to establish an equilibrium position requires a solution yielding the best available objective function values for both countries simultaneously. Finally, the purpose of Equation 5.6 is to eliminate inconsistencies between conjectured responses, $D^{(i)}_s$, and optimized reactions, $D^{(i)}_{s+1}$. But, as Basar (1985) shows, consistency with real valued policy reactions is not always possible. Nevertheless, the Pareto improving property will continue to eliminate surprises, and any incentive to deviate unilaterally from the final decisions, by picking the *best* real valued solution subject to no country being worse off than it would be with alternative conjectures.

On completing this 'downhill' search, the optimal decisions can be written as (Hughes Hallett, 1984)

$$\tilde{x}^{(i)*} = -[G^{(i)'}_* B^{(i)} G^{(i)}_* + A^{(i)}]^{-1} G^{(i)'}_* B^{(i)} E_{1i}(d^{(i)}_*) \qquad i = 1, 2 \tag{5.7}$$

where $G^{(i)}_* = R^{(i,i)} + R^{(i,j)} D^{(j)}_*, d^{(i)}_* = c^{(i)} + R^{(i,j)} F^{(j)}_* c^{(j)}$, and where $F^{(j)}_*$ is defined by Equation 5.5 but evaluated at the terminal values $(D^{(1)}_*, D^{(2)}_*)$. Certainty equivalence has been applied to each objective, $E_{ti}(w^{(i)})$, where $E_{ti}(\cdot) = E(\cdot|\Omega_{ti})$ denotes an expectation conditional on country i's information at t. The computed policies are therefore conditioned on a single information set representing either one or the other government's view of the past and future at a given moment. They would be revised by explicit reoptimizations conditional on subsequent information sets (Ω_{ti}, for $t = 2 \ldots T$ and $i = 1, 2$; see Hughes Hallett (1984).) Note that the decisions in Equation 5.7 are identical to the decisions for a closed economy whose targets behave as

$$y^{(i)} = (R^{(i,i)} + R^{(i,j)} D^{(j)}_*) \tilde{x}^{(i)} + d^{(i)}_* \tag{5.8}$$

---

[3] The starting point is arbitrary, but could be the zero conjectures (open loop Nash) solution. Since the latter exists (Aubin, 1979), this search will terminate.

Thus the *net* effect of domestic action and foreign reactions is to change the policy responses from those in Equation 5.3 to those in Equation 5.8.

One objection to standard game theory solutions is that the conjectures which agents are presumed to hold in the standard case are usually wrong (Ulph, 1983). The open-loop Nash solution, for instance, requires each policy maker to base his actions on the assumption that the other will not react to any policy adjustments, although both economies will be affected. That assumption nevertheless generates nonzero reactions (i.e. $D_1^{(1)}, D_1^{(2)}$ given $D_0^{(1)} = 0$, $D_0^{(2)} = 0$) which falsify the original conjectures. This is true for all conjectures except those at the point where the conjectures turn out to equal the optimal reactions. The inconsistencies at earlier steps of Equation 5.6 therefore just a demonstration of the *Lucas critique*. The policies of one country, based on certain conjectured responses abroad, automatically generate different reactions from those conjectured; and that discrepancy invalidates the original policy selection because the spillovers from the new reactions alter the responses of the first country's targets to its own instruments. For instance, the open loop Nash solution generates policies for country 1 assuming that $R^{(1,1)}$ describes the target responses to domestic policy changes, whereas $R^{(1,1)} + R^{(1,2)} D_1^{(2)}$ will actually determine those responses. Country 1 then has to modify its proposed action to account for the changed dynamics induced by foreign reactions. More generally the responses change from $R^{(i,i)} + R^{(i,j)} D_s^{(j)}$ to $R^{(i,i)} + R^{(i,j)} D_{s+1}^{(j)}$. Only at termination is the Lucas critique satisfied because, if one country then tries to gain by altering its reactions, the other will be made worse off and would retreat to a position which makes them both worse off.[4] Notice that neither $D_*^{(j)}$ nor $D_s^{(j)}$ are block lower triangular, even when $D_0^{(j)} = 0$, which implies that the decisions of different dates are mutually rather than recursively dependent. This means that the conjectures are intertemporal and hence that, in satisfying the Lucas critique, the current decisions will allow for any foreign policy threats which may lie in the future.

5.5 CO-OPERATIVE DECISION-MAKING

Non-co-operative decisions are known to be socially inefficient in the absence of side payments. Da Cuhna and Polak (1967), for example, showed that the entire set of nondominated (Pareto efficient) decisions can be generated by minimizing

$$w = \alpha w^{(1)} + (1 - \alpha) w^{(2)} \qquad 0 < \alpha < 1 \qquad (5.9)$$

subject to Equation 5.3. Hence, once a collective objective (or α value) is agreed, both countries could gain by solving one global policy problem. But whether those gains would be worthwhile compared with the loss of individual sovereignty is an empirical issue for which evidence has yet to be produced. Moreover,

---

[4] Note that the same argument applies should the search terminate without consistent conjectures because the latter imply imaginary policy responses. Basar (1985) suggests nonlinear reaction functions for this case; but that situation did not arise in any of the exercises below.

Empirical results   83

co-operation may entail redistributing the policy gains in a way which agrees poorly with national priorities or bargaining power.[5] It is therefore important to estimate both the size and distribution of the gains from co-operation, compared with those of a non-co-operative strategy, in order to say anything about the value or sustainability of co-operation in practice.

A second set of questions concerns the qualitative differences between optimal co-operative and non-co-operative policies. Do the gains from co-operation derive from smaller interventions, or from better target realizations, or from the fact that policy goals can be reached earlier under co-ordinated policies? We must also examine the proposition that co-operation allows countries to exploit any asymmetries between their policy responses more efficiently. Three types of asymmetry are relevant: (1) asymmetric domestic impacts of different instruments (leading to policy specialization); (2) asymmetric spillovers between countries (leading to unequal bargaining strengths); and (3) asymmetric distribution of responses over time (leading to different speeds of adjustment). Such asymmetries could increase a country's freedom of choice through its relative size or openness; or, more likely, because greater flexibility in its product, labour, and financial markets permits the consequences of international events to be absorbed more easily. This was conjectured, but not tested, by Koromzay *et al.* (1984).

5.6 EMPIRICAL RESULTS: CO-OPERATIVE AND NON-CO-OPERATIVE STRATEGIES

Table 5.2 presents optimized decisions (averaged over the period 1974–78) for the US and EEC under three different strategies:

(1) Isolationist policies which ignore interdependence; i.e. decisions by Equation 5.4, with the restriction $\partial y^{(i)}/\partial x^{(j)} = R^{(i,j)} = 0$, but target outcomes by Equation 5.3.
(2) Non-co-operative policies, including anticipated reaction terms, by Equation 5.7.
(3) Co-operative (Pareto efficient) policies, on an equal shares basis, by minimizing Equation 5.9 with $\alpha = \frac{1}{2}$ subject to Equation 5.3.[6]

These three strategies represent the three fundamental choices facing policy-makers at the start of the planning exercise. They can decide to ignore interdependence, or to set their policies competitively, or to accept an equitable compromise by co-operating. Detailed questions about the exact intervention

---

[5] A 'real world' illustration is provided by the rise in inflation and current account deficit suffered by Germany after the co-operative policies, agreed at the Bonn Summit meeting in 1978, were enacted. Those policies were rapidly abandoned.
[6] This co-operative solution is a convenient point of reference since it corresponds to weighting national interests by relative GNP size (1978 figures). Bargaining solutions, and alternative α values, are considered in Section 5.8.

paths can then be settled once the fundamental strategy has been selected. The non-co-operative strategy provides the benchmark against which the gains from co-operation, and the characteristics of a co-operative policy, may be measured. The isolationist policies are used here to establish the *scale* of the changes resulting from co-operative instead of competitive policy-making; although they also show some of the costs of ignoring interdependence. All the values reported in Table 5.2 are computed using the initial (1974) information set and are defined, in each case, as deviations from their simulated values based on that information. They therefore represent the options as they would have appeared when the policy-makers had to choose their fundamental strategy.

A popular objection to policy analysis based on non-co-operative decisions is that those decisions are often time-inconsistent; with the passage of time policy-makers may find that they can improve their economy's performance, for the planning periods which remain, by reneging on their previously announced policies. If that can be predicted in advance, the non-co-operative policies will cease to be credible. However, as far as the non-co-operative decisions of this chapter are concerned, explicit reoptimization of $E_{ti}(w^{(i)})$ shows that the revised decisions for $x_t^{(i)} \ldots x_T^{(i)}$, for each t in turn, remain identical to those expected earlier if there are no changes in the information set (Hughes Hallett, 1984). There is nothing here to stop agents attempting to make extra gains by reoptimizing the purely future elements in $E_{ti}(w^{(i)})$, but if they manage to do so the performance index for the planning interval *as a whole* will be made worse.[7] Therefore the

Table 5.2 Mean interventions under isolationist, optimal non-co-operative, and co-operative strategies (1974 information set) (relative to the historical values)

Instruments	Isolationist		Non-co-operative		Co-operative	
	US	EEC	US	EEC	US	EEC
G	− 0.96	+ 0.52	+ 0.18	+ 0.68	+ 0.23	+ 1.01
S	− 2.38	+ 0.26	− 1.11	+ 0.13	− 0.61	+ 0.32
TH	− 0.73	+ 0.41	− 0.67	− 0.62	− 0.40	− 0.59
TC	− 0.44	− 0.03	− 0.23	− 0.10	− 0.14	− 0.06
R	− 0.21	− 0.81	− 0.49	− 0.51	− 0.43	− 0.15
LC	+ 0.10	+ 0.01	− 0.05	+ 0.03	+ 0.03	0.00
LB	− 0.24	− 0.18	− 0.72	− 0.04	− 0.27	− 0.19
Targets						
P	− 1.6	− 2.2	− 2.2	− 2.3	− 2.3	− 2.1
X	+ 1.9	+ 2.4	+ 2.3	+ 2.3	+ 2.7	+ 2.4
B	− 2.1	− 0.1	− 0.5	+ 0.3	− 0.2	+ 0.4
E	+ 0.9	+ 0.8	+ 1.2	+ 0.8	+ 1.2	+ 0.9
IR	+ 1.5	+ 3.8	+ 1.8	+ 3.3	+ 2.0	+ 3.7
PR	+ 1.2	+ 3.4	+ 1.5	+ 3.6	+ 1.3	+ 3.6

[7] This happens because the constrained objective has an intertemporal rather than the usual time-recursive structure, so the decisions cannot be separated into a time-nested sequence of actions where both countries can try to gain by reoptimizing (Hughes Hallett, 1984).

Empirical results

non-co-operative decisions defined by Equation 5.8 will be credible since any decisions on fundamental strategy will be taken with respect to their currently expected value for the entire planning period, and not on the basis of potential improvements to the later periods which would yield a worse performance taking all periods together. This property may seem rather specialized, but it is consistent with the aim of establishing a reputation over a period of time and with the fact that policy-makers will be judged on the sum of their achievements over a given period.

5.6.1 Optimal policies ignoring interdependence

The isolationist policies for the US involve a cut of 1% per annum in government expenditures, and of $2\frac{1}{2}$% per annum in social security expenditures, while both quantities rise in the EEC. This US fiscal disengagement is due to the fact that domestic deficit financing in the US induces crowding out effects which lead to contractions in output, followed by upward pressure on both prices and interest rates. This was pointed out in Section 5.3. Consequently, if one ignores spillovers from abroad, the US budget deficit should be cut as the economy moves into recession. This is accomplished here by expenditure cuts combined with tax cuts of $1\frac{1}{4}$% per annum. The US government did in fact introduce such a strategy (defence expenditures excepted) for the 1979–81 recession.

In contrast, deficit financing in the EEC has expansionary effects on output and inflation in the first few years, and adverse effects only later. So expenditures are increased to start with, while taxes are sharply reduced. This pattern is, however, reversed in the final two years.[8] Thus the EEC budget deficit would increase, while that in the US would decrease, over this five year period.

These fiscal changes mainly affect the household sector in both the US and the EEC; and the fact that social security cuts are more intensively used in the US is the product of real wage rigidity found in the US economy compared with the real disposable wage rigidity of the European economies (Branson and Rotemberg, 1980). Hence real wage flexibility is crucial for analysing not only an economy's responses to shocks from abroad, but also for determining what kind of policies are appropriate.

The monetary variables show that the change in European budgets is not financed by loans from the central banks (LC) and that the loans from commercial banks (LB) cause no extra money creation. The main European instrument, the discount rate, is lowered throughout the period. In the US the dependence of budget financing on savings requires a more sophisticated strategy. First a lowering of the discount rate has to induce a fall in interest changes to stimulate economic activity. A rise of the discount rate (R) is then used to acquire savings until the loans from the banks to the government start to decline, after which the discount rate is lowered again.

---

[8]The year-by-year policy changes are reported in Appendix A, Table A1.

The basic position is therefore one of standard Keynesian deficit financing in Europe, to stimulate aggregate demand and output, with monetary control and falling interest rates to reduce inflation. Meanwhile policies which have come to be associated with supply-side economics operate in the US; a reduction in government intervention, especially in social security outlays and taxes, is coupled with a tight monetary policy.

5.6.2 Competitive strategies

Turning to the non-co-operative policies, recognition of their interdependence should lead both economies to introduce rising government expenditures, although the change is only marked in the US. The US social security cuts are halved, and in Europe they turn positive in 1975–1977. Once again tax cuts appear in both economies, although they are more variable than in the previous case. The US fiscal position is now broadly neutral with reduced, but more flexible, interest rates to offset any crowding out of the expansionary impulse. The EEC, on the other hand, switches to fiscal expansion with an accommodating monetary policy.

The changes in monetary policy are more complicated because policy-makers must now allow for the possibility that their budget deficits will be financed from abroad. An initial rise in European discount rates attracts American capital, as witnessed by lower loans to the US government. The US has to resort to raising interest rates; but rational expectations of the US monetary policy then lead to larger adjustments in both monetary and fiscal policy in Europe. Indeed, Table 5.2 shows that the main changes here are in the monetary variables; the US follows a more restrictive policy but reduces interest rates, while the EEC now intervenes more actively both with loans and the discount rate.

Finally, recognizing interdependence has had a greater effect on US policies than on EEC policies. The US fiscal disengagement is halved, while its monetary policy more than doubles in overall impact. Meanwhile monetary expansion is reduced in the EEC and fiscal neutrality becomes fiscal expansion. Thus policy interactions have led to some convergence in the national policies. The US has dropped its supply-side stance in favour of some demand creating measures; and the EEC intervenes with a more flexible monetary policy, using alternately the discount rate and money supply.

5.6.3 Co-operative policies

Co-operation between the US and the EEC takes the optimal policies a step further towards convergence; but the changes are now larger for the EEC than the US, and appear in government and social expenditures and in the monetary instruments. Government expenditures rise faster in both economies, and the US social security cuts are halved once again while those in the EEC lose their vigour. Tax cuts are still in evidence, although less prominent in the US.

Overall the co-operative policies call for reduced interventions in the US, and

Empirical results

for more stable policies in both economies compared to the non-co-operative strategies. Indeed, Fig. 5.1 shows that the US interventions are smoother than their EEC counterparts, and that (for both countries) the degree of smoothness increases with co-ordination. But for the EEC it is the optimal non-co-operative interventions which are least smooth. Thus the EEC has to work harder, and has

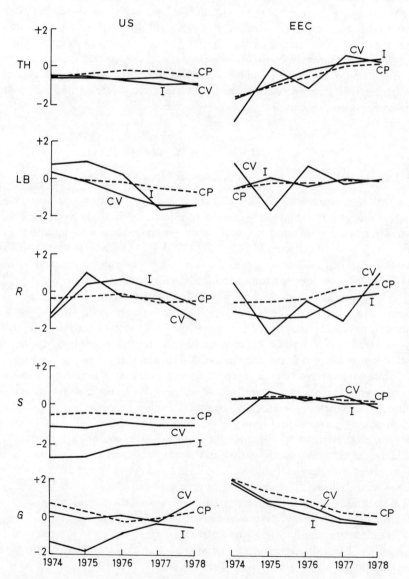

Fig. 5.1 The main instrument adjustments (percentage growth per annum) (I isolationist, CV non-co-operative, CP co-operative)

greater difficulty in controlling the spillovers, in a non-co-operative environment. Moreover, the activism of monetary policy has now vanished – most noticeably the sharp restrictions of money supply and interest rate policy which appeared in the US non-co-operative solution. Both loans to the government and the discount rates follow either constant or steadily changing paths. This suggests that it is important to co-ordinate the *timing* of monetary changes, and that the gains to co-operation may be significant only when that is done. Indeed, the dominance of US monetary instruments among the spillover multipliers, and the asymmetry of the strategies which anticipate policy actions abroad (the US specializing in monetary control, the EEC in fiscal expansion), indicate that the important spillovers were caused by monetary rather than fiscal action. Therefore international policy co-ordination should perhaps concentrate on monetary policy, leaving fiscal instruments free for domestic purposes.

5.7. THE INCENTIVES TO CO-OPERATE

5.7.1 Comparative advantage and policy specialization

One striking feature of the results in Table 5.2 is the increasing specialization in the non-co-operative and co-operative solutions. Both the US and the EEC concentrate on the instruments with more powerful domestic multipliers; the monetary instruments (R and LB) for the US, and the fiscal instruments (G and TH) for the EEC. This specialization can be explained by the comparative advantage of the policy instruments under each strategy.

In the co-operative solution each country uses its instruments to 'produce' several target values in the global objective of Equation 5.9, where $w^{(i)} \geqslant 0$. It therefore pays each country to put most effort into those instruments with the lowest opportunity costs, i.e. those with the largest (absolute) multipliers. Specialization according to comparative advantage always makes it possible to reallocate the intervention effort between countries so as to get improved outcomes for both components in a global objective, since both countries are directly or indirectly 'producing' some of the changes in both $w^{(1)}$ and $w^{(2)}$.

In non-co-operative problems there is no global objective, so comparative advantage is restricted to the different instruments in the home country. That explains the different specialization patterns in the non-co-operative and the co-operative solutions of Table 5.2. Tables A2 and A3 of Appendix A show that government and social expenditures, interest rates and loans to the government (G, S, R and LB) have comparative advantage among the US instruments, while government expenditures, taxes and interest rates (G, TH and R) have it among the EEC instruments. But the comparative advantage of G and S in the US, and of R in the EEC, disappears when compared with the same instrument in the other economy. Hence the non-co-operative solutions show specialization in all those instruments; but those without *international* comparative advantage (i.e. G and S for the US, and R for the EEC) are phased out in the co-operative solution.

The incentives to co-operate

Hence the opportunity to reallocate instruments according to international (in addition to national) comparative advantage is one important reason why a co-operative solution can always produce a *better* outcome than the optimal non-co-operative solution. Policy-makers can then exploit the comparative advantage of the same instrument in different countries, as well as that of different instruments within one country. This comparative advantage argument is closely related to Mundell's 'principle of effective market classification'. But we have

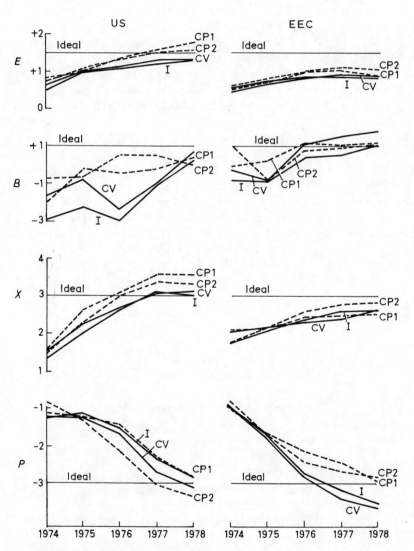

Fig. 5.2 The main target adjustments (percentage growth per annum) (I isolationist, CV non-co-operative, CP1, CP2 co-operative) ($\alpha = \frac{1}{2}, \frac{2}{3}$)

had to distinguish the co-operative from the non-co-operative case here, and also to provide for policy specialization rather than for a strict allocation of instruments.

5.7.2 The expected target values

Table 5.2 shows the average expected changes in the target variables compared with the values which could have been expected by simulating the historical instrument choices. These expected outcomes are better for all variables in both economies, excepting the US balance of trade. However, the EEC does relatively badly with the employment target, indicating that European unemployment will be hard to eliminate. Overall the US targets improve as we move from the isolationist to the co-operative solution (especially for production, balance of trade, and investment), but the EEC targets are not much affected by the type of strategy chosen. Once again we have evidence of asymmetry in US–EEC economic relations.

Figure 5.2 sketches those target values in more detail, and makes clear where co-operation is advantageous. Under co-operation the US in fact does uniformly better only in employment, output and the trade balance. The EEC does uniformly better only in employment and the trade balance; and both countries do worse with inflation. Nevertheless gains arise in all cases because co-operation *both* advances the timing *and* increases the speed with which the target variables approach their ideal values. Tables 5.3 and 5.4 illustrate these two characteristics

Table 5.3 The number of years by which the co-operative solution ($\alpha = \frac{1}{2}$) reaches its best result in advance of the non-co-operative solution

	US	EEC
P	3/4	$-1\frac{1}{2}$
X	3/4	3
B	3	0
E	$1\frac{1}{4}$	0

Table 5.4 The average rates of approach (% per year) to the ideal

	US		EEC	
	Non-co-operative	Co-operative ($\alpha = \frac{1}{2}$)	Non-co-operative	Co-operative ($\alpha = \frac{1}{2}$)
P	0.38	0.55	0.61	0.40
X	0.37	0.75	0.32	0.42
B	0.72	0.63	0.68	0.40
E	0.16	0.29	0.18	0.25

Table 5.5 The average target failures under optimal non-co-operative and co-operative strategies

	US			EEC		
	Non-co-operative	Co-operative ($\alpha = \frac{1}{2}$)	Co-operative ($\alpha = 2/3$)	Non-co-operative	Co-operative ($\alpha = \frac{1}{2}$)	Co-operative ($\alpha = 2/3$)
P	0.8	0.7	0.9	0.7	0.9	0.9
X	−0.7	−0.3	−0.3	−0.7	−0.6	−0.5
B	−1.5	−1.2	−1.1	−0.7	−0.6	−0.5
E	−0.3	−0.3	−0.3	−0.7	−0.6	−0.1
IR	−1.2	−1.0	−1.0	0.3	0.7	0.8
PR	−1.5	−1.7	−1.7	0.6	0.6	0.6

for the main target variables of each country. The upshot is better expected outcomes for most targets in terms of smaller average deviations from their ideal paths (see Table 5.5). These results therefore confirm one of Cooper's (1969) arguments: policy co-ordination damps out the transitory effects of shocks more rapidly and hence lowers the costs of reaching a given performance level in terms of the targets.

5.7.3 Externalities

The results in Table 5.2 showed that the gains from co-operation would take the form of both smaller instrument interventions and better expected target values in the US; but in the EEC those gains arise from a reduction in intervention effort rather than from improved target values. Thus the EEC would exploit the US economy as a locomotive; improvements in the US would enable the EEC to reach the same target values with less effort, both because the favourable spillovers are enhanced and because the unfavourable spillovers are blocked. Similarly an improved EEC economic performance means that the US instruments need to intervene less strongly to start a recovery following the oil price shock, with the result that the US instruments switch to achieving better outcomes for their own targets.

Co-operation therefore reduces certain policy externalities; in this case the cost (to the EEC) of blocking spillovers from US policy, and the cost (to the US) of generating a recovery, are both reduced. Consequently the US and EEC target responses to domestic instrument changes appear to be stronger under co-operative policies than under competitive policies. That result illustrates Cooper's other argument for co-operation: co-ordination restores the policy impacts weakened by interdependence.

5.7.4 The gains from co-ordination

The objective function evaluations in Table 5.6 show that accommodating economic interdependence brings larger gains to the US than the EEC, but the

Table 5.6 The performance indicators under different strategies

	Strategy			
	Isolationist	Non-co-operative	Co-op($\alpha = 1/2$)	Co-op($\alpha = 2/3$)
US objectives: w^*_{US}	188.9	52.2	47.5	44.7
EEC objectives: w^*_{EEC}	66.9	40.7	27.2	29.1

gains to explicit co-operation are significantly smaller and mainly benefit the EEC. Thus the US can expect to gain nearly twice as much as the EEC from competitive policy-making (72% against 39%), whereas the subsequent gains from co-operation are much smaller (9% against 33% respectively) and favour the EEC by 3 to 1. To place these results in context, the gains to co-operation can be expressed in units of equivalent GNP growth; that is the average rate of GNP growth, all other decision variables fixed, which would yield the same objective function gains for each economy. Here the gains to co-operation for the US is equivalent to an extra 0.47% annual GNP growth for 5 years, while for the EEC it is worth an extra 1.35% annual GNP growth for 5 years. This is a comparison advocated by Oudiz and Sachs (1984) who obtained the equivalent of 0.2% and 0.3% GNP growth for the US and West Germany in a static decision framework. This suggests again that co-ordinating the *timing* of policy effects can be at least as important as co-ordinating the impacts of contemporaneous instruments. It also suggests that the lion's share of the EEC's gains will go to the weaker European economies. It would therefore be important to extent this study to analysing the distribution of gains within the EEC.

5.8 BARGAINING STRATEGIES

The incentives to co-operate must be analysed in a way which does not depend on comparisons with one particular co-operative solution. The difficulty here is to identify co-operative strategies which yield worthwhile gains to each country, and which also imply an acceptable distribution of gains between participating countries. That distribution depends on the value of α; and to determine the value of α which is most likely to be sustained in a co-operative agreement requires a bargaining theory which recognizes the relative power of the participants. Economic theory contains several different models of bargaining behaviour, each with rather different theoretical properties. In this case, however, they all imply virtually identical values for α.

The first requirement of a negotiated co-operative solution is that it should dominate the optimal non-co-operative solution, since the latter represents the best outcome either country could expect from unilateral action. Section 5.5 pointed out that the set of nondominated (Pareto efficient) solutions is generated by minimizing $w = \alpha w_{US} + (1 - \alpha) w_{EEC}$ for $\alpha \varepsilon (0, 1)$. The national objective function values, w^*_{US} and w^*_{EEC}, implied by that set of solutions have been plotted in

Fig. 5.3, together with the threat point N (the best non-co-operative outcome). The shaded segment of LL' represents the set of co-operative solutions which dominate that threat point. Although other points on LL' offer one country an even better result, they require the other country to accept an outcome which is worse than that available under unilateral action. The losing country would therefore revert to its non-co-operative solution, forcing a return to N.

We can now consider picking co-operative solutions from this nondominated set. The standard bargaining theory is represented by Nash's co-operative equilibrium model (Nash, 1953). This model takes the measure of 'relative power', which determines the outcome of the bargaining process, to be given by the relative utilities at the threat point. This is plausible since each country will be willing to bargain only in so far as it expects to get a pay-off better than that attained at the threat point, and both countries should be willing to accept a division of the net incremental gains in a proportion directly related to the losses incurred by not making an agreement. Nash demonstrated his solution to be the only one that satisfies axioms of rationality, feasibility, pareto-optimality, independence of irrelevant alternatives, symmetry, and independence with respect to linear

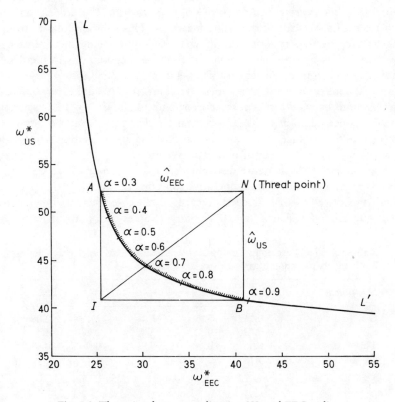

Fig. 5.3 The gains from co-ordinating US and EEC policy

transformations of the set of pay-offs. That solution is obtained at the point where the product of the co-operation gains from the threat point, say $(w^*_{US} - w^N_{US})(w^*_{EEC} - w^N_{EEC})$, is maximized. In this case it yields $\alpha = 0.68$.

The Nash solution has been criticised on the grounds that the independence of irrelevant alternatives axiom is inappropriate in a bargaining problem (Kalai and Smorodinski, 1975; Kalai, 1977). No individual would agree to ignore the opportunities to improve his position if new options should become available to him. If the independence of irrelevant alternatives axiom is replaced by the monotonicity in individual utility functions over possible outcomes, then a unique bargaining solution exists in which the distribution of utility gains is proportional to the maximum individual gains when the other country is no worse off than at the threat point. This solution corresponds to picking the minimum product of the losses from the ideal point, I, which represents the maximum gain each country could make while the other maintains his threat point outcome (Kalai and Smorodinski, 1975). That yields $\alpha = 0.67$ here.

Another possibility is pick α in proportion to the maximum possible gains either country could make without triggering a non-co-operative response by the other, i.e. $\alpha = \hat{w}_{US}/(\hat{w}_{US} + \hat{w}_{EEC})$ This is the proportional solution put forward by Kalai (1977), but expressed in terms of utility gains from I in order to maintain independence of the scale of the utility measures. This solution follows from the axiom of 'step-by-step' negotiations, in which the point of agreement at each step is used as a threat point in a subsequent round of negotiations to increase the utility of one or both countries. In this case it implies $\alpha = 0.65$.

Lastly, the maximum sum of co-operative gains from N implies $\alpha = 0.67$. Since we are minimising loss functions, this corresponds to Harsayni's (1956) suggestion considered once again as utility gains from I rather than from the origin. It is equivalent to the producer plus consumer surplus welfare measures used elsewhere for evaluating the net benefit of policy interventions.

Bargaining power is therefore distributed 2 to 1 in favour of the US, and the decisions associated with that co-operative solution are summarized in Table 5.7. The instrument values are scarcely different from the case where national interests are weighted by NGP shares ($\alpha = \frac{1}{2}$). The target values turn out to be a bit more

Table 5.7 The bargaining co-operative solution, average percentage adjustments per annum to the historical choices

Instruments	US	EEC	Targets	US	EEC
G	0.26	0.99	P	−2.12	−2.06
S	−0.73	0.33	X	2.68	2.49
TH	−0.48	−0.56	B	−0.07	0.44
TC	−0.15	−0.07	E	1.23	0.90
R	−0.47	−0.17	IR	2.01	3.80
LC	0.03	0.00	PR	1.33	3.61
LB	−0.34	−0.16			

volatile (mainly in the balance of trade and inflation figures) but they are otherwise broadly similar to the equal shares solution. Hence variations in bargaining power are likely to affect only the expected target values; this is in sharp contrast to the gains from co-operation (over no co-operation) which came from modifying the instrument values. Bargaining power does however affect the distribution of the gains from co-operation. Table 5.6 shows that the US makes a 6% gain in w^*_{US} over the equal shares solution, while the EEC suffers a 7% loss in w^*_{EEC}. Those changes are 'worth' an extra 0.3% annual GNP growth to the US (bringing the total gain from co-operation to an equivalent of a 0.75% increase in the GNP growth rate), but a loss of 0.28% in the EEC's annual GNP growth rate (leaving the total gain from co-operation at 1.16% extra GNP growth). Thus, despite having less bargaining power, the EEC still gains by more.

5.9 CONCLUSIONS

This chapter has identified three important empirical arguments for co-ordinating policies internationally:

(1) Co-ordination restores policy effectiveness, which is weakened by interdependence and by expectations of foreign reactions. This happens because co-operation involves trading reductions in foreign externalities caused by domestic action against reduced externalities from foreign reactions. As a result, smaller interventions can achieve better expected outcomes on average.
(2) Co-ordination speeds up an economy's responses to policy action, and damps out the transitory effects of external shocks more rapidly.
(3) Co-ordination enables governments to extend the range of comparative policy advantage, and hence the policy specializations, which they can exploit. Thus, paradoxically, co-ordination helps restore a degree of autonomy in policy choice to the policy-makers and the loss of sovereignty may be more apparent than real.

The first two arguments are empirical confirmation of the theoretical results which Cooper (1969) derived from the comparative statics of a small analytic model: co-ordination cuts the cost of intervention since one needs to intervene less and for shorter periods. But the third argument shows that co-ordination brings *additional* benefits to countries which have asymmetric policy responses. These additional gains cannot be detected in Cooper's analysis since that was based on a model of two identically symmetric economies.

Asymmetries between countries play a crucial role in determining the gains from co-operation. Three kinds of asymmetry are of interest: asymmetric domestic policy responses (which determine policy specializations); asymmetric spillover effects (which determine bargaining strengths); and asymmetric adjustment speeds (which determine the cost of policy responses). In our US–EEC example, the important asymmetries were larger and faster domestic policy responses in the

US, and the dominance of US monetary policy among the spillover effects. Thus the US should specialize in monetary policy and the EEC in fiscal policy. For the same reasons the US can make greater gains from competitive policies, while the EEC, despite its low bargaining power, gains more from any sustainable co-operative arrangement.

However, the incentives to co-operate were not in themselves large, which probably explains why the Americans officially favour competitive decision-making with a full exchange of information (Feldstein, 1983). The gains from co-operation are nevertheless significantly larger here, for any sustainable bargain, than previous estimates which have been based on static decision procedures (Oudiz and Sachs, 1984). That suggests a major part of the gain comes from co-ordinating the *timing* of policy impacts – particularly with respect to monetary policy.

There are a number of important issues which cannot be resolved without further research:

(a) This analysis is based on unified EEC policies. Are the gains from co-ordination *within* the EEC larger than those from co-ordination with the US? How are those intra-EEC gains distributed? Would an unequal distribution of those gains make it difficult for the EEC to co-ordinate its own policies?

(b) Can international co-operation in fact be sustained? So far we have relied on models of rational bargaining behaviour; but are the incentives to cheat sufficiently small, or must punishment schemes be introduced? To answer such questions one first needs estimates of the size and distribution of the potential gains from co-operation, and that is the point of this chapter.

(c) What role do (asymmetric) priorities play? Johansen (1982) argues that the cost of unco-ordinated policies varies with national preferences, and we know that co-operative and non-co-operative equilibria can coincide when the target priorities are identical (Hughes Hallett and Rees, 1983). But economic performance is not usually very sensitive to changes in relative priorities. Taylor (1985) and Hughes Hallett confirm this, showing that the objective function values are less sensitive than the policy values to variations in priorities.

(d) To what extent are these results dependent on one particular model? This question must always be asked since no model is perfect. However, the model used here is a conventional one, with rational expectations being generated by game theoretic behaviour rather than imposed on certain variables from outside. We also pointed out that the sensitivity of results obtained from small theoretical models to small changes in the underlying restrictions is often much greater than that of empirical models – even when the latter are based on the same theory (Malinvaud, 1981). One reason for this is the difficulty of accommodating the net cumulative effects of several policy instruments activated in different periods and operating through several channels simultaneously. This suggests a twin-track research strategy: simple theoretical models to establish the essential elements of policy design, but

APPENDIX A

Table A1 Optimal isolationist, competitive and co-operative policy adjustments for the US and EEC economies (1974 information set)

Variables		Isolationist		Optimal non-co-operative		Co-operative $\alpha = 1/2$		Co-operative $\alpha = 2/3$	
		US	EEC	US	EEC	US	EEC	US	EEC
1974:	G	−1.33	1.88	0.31	1.99	0.78	2.04	0.86	2.11
	S	−2.83	0.29	−1.20	−0.86	−0.62	0.28	−0.75	0.38
	TH	−0.44	−1.74	−0.62	−2.91	−0.53	−1.60	−0.64	−1.54
	TC	−0.87	−0.21	−0.70	−0.50	−0.37	−0.19	−0.45	−0.17
	R	−1.44	−1.01	−1.33	0.60	−0.39	−0.57	−0.36	−0.76
	LC	0.39	−0.02	0.22	0.18	0.10	−0.01	0.11	−0.01
	LB	0.76	−0.53	0.35	0.88	0.28	−0.53	0.22	−0.45
1975:	G	−1.78	0.84	−0.07	0.87	0.37	1.44	0.35	1.50
	S	−2.80	0.39	−1.22	0.75	−0.55	0.42	−0.69	0.48
	TH	−0.56	−0.94	−0.59	0.00	−0.42	−1.05	−0.48	−1.05
	TC	−0.62	−0.03	−0.35	0.01	−0.26	−0.06	−0.29	−0.05
	R	0.37	−1.41	0.99	−2.22	−0.29	−0.55	−0.17	−0.64
	LC	0.19	0.03	0.11	−0.12	0.04	0.01	0.05	0.01
	LB	0.89	0.09	−0.20	−1.61	−0.18	−0.20	−0.18	−0.14
1976:	G	−0.88	0.31	0.06	0.81	−0.24	1.00	−0.25	0.90
	S	−2.20	0.39	−0.95	0.28	−0.56	0.39	−0.66	0.36
	TH	−0.78	−0.08	−0.63	−1.14	−0.24	−0.47	−0.30	−0.40
	TC	−0.36	0.03	−0.15	−0.02	−0.08	−0.03	−0.07	−0.03
	R	0.67	−1.32	−0.27	−0.44	−0.13	−0.38	−0.17	−0.31
	LC	0.09	−0.01	0.07	0.09	0.07	−0.01	0.08	−0.01
	LB	0.19	−0.33	−0.92	0.71	−0.20	−0.20	−0.28	−0.18
1977:	G	−0.32	−0.19	−0.25	−0.03	−0.14	0.38	−0.16	0.27
	S	−2.13	0.11	−0.09	0.54	−0.61	0.31	−0.70	0.26
	TH	−0.97	0.24	−0.58	0.63	−0.29	0.07	−0.33	0.13
	TC	−0.26	0.03	0.02	0.07	−0.01	0.00	0.01	0.00
	R	0.04	−0.28	−0.35	−1.47	−0.78	0.29	−0.97	0.42
	LC	−0.11	0.02	−0.07	−0.02	−0.01	0.01	−0.01	0.01
	LB	−1.64	0.02	−1.43	−0.23	−0.53	−0.00	−0.63	0.01
1978:	G	−0.49	−0.26	0.86	−0.22	0.38	0.20	0.50	0.17
	S	−1.96	0.10	−1.08	−0.06	−0.72	0.22	−0.85	0.19
	TH	−0.91	0.46	−0.93	0.32	−0.53	0.10	−0.63	0.07
	TC	−0.11	0.02	0.03	−0.04	0.01	−0.01	0.03	−0.01
	R	−0.69	−0.01	−1.49	0.97	−0.58	0.44	−0.69	0.45
	LC	−0.09	0.00	−0.10	0.00	−0.05	−0.00	−0.06	0.00
	LB	−1.39	−0.02	−1.40	0.03	−0.71	−0.02	−0.82	−0.02

Legend and units: see Table 5.1.

Table A2 Selected US instruments, domestic multipliers

Target	Instrument G	S	TH	TC	R	LB
Impact multipliers:						
P	−0.07	0.04	0.04	0.04	0.48	−0.21
X	0.42	0.14	−0.07	−0.03	1.14	0.25
B	−0.89	−0.38	0.06	0.02	2.10	−1.00
E	0.21	0.05	−0.05	−0.02	−0.50	0.09
IR	0.58	0.10	−0.16	−0.05	−0.78	−0.06
PR	0.52	−0.24	−0.29	−0.01	−1.63	0.21
Dynamic (5 year lag) multipliers:						
P	0.23	0.11	−0.07	−0.01	0.17	0.59
X	−0.13	−0.02	0.05	0.01	−0.94	2.22
B	0.66	0.13	−0.17	−0.10	−0.39	−3.72
E	−0.06	−0.02	0.02	0.00	−0.46	0.98
IR	−0.37	−0.09	0.12	0.02	1.29	1.52
PR	−0.14	−0.04	0.05	0.02	1.49	−1.27

Table A3 Selected EEC instruments, domestic multipliers

Target	Instrument G	S	TH	TC	R	LB
Impact multipliers:						
P	−0.07	0.11	0.35	0.03	0.04	0.00
X	0.49	0.21	−0.19	−0.01	−0.38	−0.01
B	−0.20	−0.03	0.13	0.02	−0.09	−0.01
E	0.14	0.06	−0.06	−0.00	−0.07	−0.00
IR	0.61	0.16	−0.50	−0.02	−0.46	−0.01
PR	1.00	0.27	−0.66	0.03	−1.36	−0.02
Dynamic (5 year lag) multipliers:						
P	−0.16	0.14	0.19	0.02	0.59	0.18
X	0.30	0.06	0.03	−0.01	−0.31	0.15
B	−0.11	0.13	0.41	0.04	−0.21	−0.33
E	0.12	−0.01	−0.03	−0.01	−0.14	−0.01
IR	0.39	0.02	0.09	−0.03	−0.82	−0.06
PR	0.60	0.23	0.18	0.01	−0.28	−0.27

Table A4 Selected US instruments, spillover multipliers

Target	Instrument G	S	TH	TC	R	LB
Impact multipliers:						
P	−0.04	−0.01	0.01	0.01	0.13	−0.02
X	0.23	0.06	−0.05	−0.02	−0.68	0.04
B	0.30	0.06	−0.08	−0.03	−0.87	0.08
E	0.07	0.02	−0.02	−0.01	−0.18	0.01
IR	0.28	0.08	−0.06	−0.03	−0.83	0.05
PR	0.46	0.12	−0.11	−0.04	−1.52	0.10
Dynamic (5 year lag) multipliers:						
P	0.10	−0.01	−0.00	0.00	0.60	−0.94
X	0.06	0.02	0.00	−0.01	−0.47	−0.04
B	0.28	0.09	−0.05	−0.04	−1.77	1.10
E	−0.02	−0.01	0.01	0.00	−0.00	−0.08
IR	−0.01	−0.01	0.03	−0.00	−0.33	−0.12
PR	0.39	0.10	−0.07	−0.06	−1.89	−0.04

Table A5 Selected EEC instruments, spillover multipliers

Target	Instrument G	S	TH	TC	R	LB
Impact multipliers:						
P	−0.06	−0.01	0.04	0.00	0.31	0.00
X	0.11	0.06	−0.03	−0.00	−0.50	−0.01
B	0.55	0.26	−0.30	−0.00	0.95	−0.02
E	0.05	0.03	−0.01	0.00	−0.20	−0.00
IR	0.09	0.06	−0.00	0.00	−0.17	−0.00
PR	0.14	0.06	−0.07	−0.00	−0.65	−0.01
Dynamic (5 year lag) multipliers:						
P	−0.49	−0.04	0.23	0.03	0.63	−0.03
X	0.25	0.08	0.04	−0.01	−0.43	−0.00
B	−0.45	−0.02	0.55	0.05	−0.12	−0.45
E	0.09	0.04	0.04	−0.00	−0.18	−0.01
IR	0.06	0.04	0.12	0.00	0.37	−0.07
PR	−0.00	−0.03	0.05	0.00	0.78	−0.11

empirical models to evaluate different policy strategies in practice. Unless some empirical analysis of interdependence is done, it seems unlikely that such fundamental questions as whether to co-operate or not can be answered reliably. Indeed, although the detailed numerical results reported here may be sensitive to the model, the conclusions which have been drawn are all qualitative and are therefore necessarily less sensitive.

Note: The present chapter presents an alternative version of the arguments put forward in Hughes Hallett (1986).

REFERENCES

Aubin, J. P. (1979) *Mathematical Methods of Game and Economic Theory*, North Holland, Amsterdam.
Barten, A. P., d'Alcantara, G. and Carrin, C.J. (1976) COMET: A medium-term macroeconomic model for the European Economic Community. *European Economic Review*, 7, 63–115.
Basar, T. (1985) *A Tutorial on Dynamic and Differential Games*, Co-ordinated Science Laboratory, University of Illinois, Champaign, Ill.
Basevi, G., Blanchard, O., Buiter, W., Dornbusch, R. and Layard, R. (1984) *Europe: the Case for Unsustainable Growth*, Centre for European Policy Studies, Brussels.
Begg, D. (1983) The economics of floating exchange rates: the lessons of the 70s and the research programme for the 80s, Memorandum in *International Monetary Arrangements*, Vol. 3, HMSO, London, pp. 4–56.
Bergstein, C. Fred *et al.* (1982) *Promoting World Recovery* (a statement on global economic strategy by 26 economists from 14 countries), Institute for International Economics, Washington, DC.
Brandsma, A. S. and Hughes Hallett, A. J. (1984) Economic conflict and the solution of dynamic games. *European Economic Review*, 26, 13–32.
Branson, W. H. and Rotemberg, J. (1980) International adjustments with wage rigidity. *European Economic Review*, 13, 309–32.
Bresnahan, T. F. (1981) Duopoly models with consistent conjectures. *American Economic Review*. 71, 934–45.
Canzoneri, M. and Gray, J. A. (1985) Monetary policy games and the consequences of non-co-operative behaviour. *International Economic Review* 26, 547–64.
Cooper, R. N. (1969) Macroeconomic policy adjustment in interdependent economies. *Quarterly Journal of Economics*, 83, 1–24.
Corden, W. M. and Turnovsky, S. J. (1983) Negative transmission of economic expansion. *European Economic Review*, 20, 289–310.
Currie, D. and Levine, P. L. (1984) Macroeconomic policy design in an interdependent world, PRISM research paper No. 23, Queen Mary College, University of London.
Da Cuhna, N. and Polak, E. (1967) Constrained minimisation of vector-valued criteria in finite dimensional spaces. *Journal of Mathematical Analysis and Applications*, 19, 103–24.
EEC (1984) The Annual Economic Report 1984–5, *The European Economy*, No. 23, The European Economic Commission, Brussels.
Feldstein, M. (1983) The world economy today. *Economist*, June 11.
Hamada, K. and Sakuri, M. (1978) International transmission of stagflation under fixed and flexible exchange rates. *Journal of Political Economy*, 86, 877–95.

References

Harsayni, J. C. (1956) Approaches to the bargaining problem before and after the theory of games: a critical discussion of Zeuthen's, Hicks' and Nash's Theories. *Econometrica*, **24**, 144–57.

Holt, C. A. (1985) An experimental test of the consistent conjectures hypothesis. *American Economic Review*, **75**, 314–25.

Hughes Hallett, A. J. (1986) Autonomy and the choice of policy in asymmetrically dependent economies, *Oxford Economic Papers*, **38**, 516–44.

Hughes Hallett, A. J. (1988) How robust are the gains to policy co-ordination to variations in the model and objectives? *Ricerche Economiche* (special issue on Game theory and Policy Co-ordination).

Hughes Hallett, A. J. (1984) Non-co-operative strategies for dynamic policy games and the problem of time inconsistency. *Oxford Economic Papers*, **36**, 381–99.

Hughes Hallett, A. J. and Rees, H. J. B. (1983) *Quantitative Economic Policies and Interactive Planning*, Cambridge University Press, Cambridge.

Johansen, L. (1982) The possibility of international equilibrium at low levels of activity. *Journal of International Economics*, **13**, 257–65.

Kalai, E. (1977) Proportional solutions to bargaining situations: interpersonal utility comparisons. *Econometrica*, **45**, 1623–30.

Kalai, E. and Smorodinski, M. (1975) Other solutions to Nash's bargaining problem. *Econometrica*, **43**, 513–8.

Koromzay, V., Llewellyn, J. and Potter, S. (1984) Exchange rates and policy choices: some lessons from interdependence in a multilateral perspective. *American Economic Review*, 311–15.

Malinvaud, E. (1981) Econometrics faced with the needs of macroeconomic policy. *Econometrica*, **49**, 1363–75.

Miller, M. and Salmon, M. (1985) Policy co-ordination and the time inconsistency of optimal policy in open economies. *Economic Journal* (supplement, 124–35).

Mundell, R. A. (1963) Capital mobility and stabilisation policy under fixed and flexible exchange rates. *Canadian Journal of Economics*, **29**, 475–85.

Nash, J. F. (1953) Two person co-operative games. *Econometrica*, **21**, 128–40.

Oudiz, G. and Sachs, J. (1984) Policy co-ordination in industrialised countries. *Brookings Economic Papers* (1), 1–64.

Taylor, J. B. (1985) International co-ordination in the design of macroeconomic policy rules. *European Economic Review*, **28**, 53–82.

Ulph, D. (1983) Rational conjectures in the theory of oligopoly. *International Journal of Industrial Organisation*, **1**, 131–54.

Van der Windt, N., Siebrand, J. C., Swank, J. and Pijpers, J. R. (1984a, b) Rasmus: an annual model of the US and EEC economies, discussion papers 8405/G and 8411/G, Erasmus University, Rotterdam.

6

Capital risk and models of investment behaviour

ROBERT S. PINDYCK

Most investment expenditures are at least partly irreversible – although capital in place can be sold from one firm to another, its scrap value is often small because it has no alternative use other than that originally intended for it. An emerging literature has shown that this makes investment decisions highly sensitive to uncertainty over future market conditions, and in theory changes in risk levels should strongly affect investment spending. However, explicit measures of risk are usually missing from empirical investment models.

This chapter discusses the effects of risk on investment and capacity choice, and explains why q theory and related investment models, based on the rule 'invest when the marginal value of a unit of capital is at least as large as the cost of the unit', are theoretically flawed. It argues that the inclusion of explicit measures of risk can help explain and predict investment spending. The results of causality tests and a set of simple regressions are presented that strongly support this argument.

6.1 INTRODUCTION

It is obvious to any business person that economic decisions often depend critically on the nature and extent of market risk. Clearly a decision to invest in new capital will depend not only on projections of market demand, but also on the degree to which future demand is uncertain. Indeed, much of corporate finance theory, as taught in business schools, deals with methods for properly taking risk into account when making capital budgeting decisions. Yet most econometric models of aggregate economic activity ignore the role of risk, or deal with it only implicitly. The point of this chapter is that a more explicit treatment of risk may help to better explain and forecast economic fluctuations, and especially movements in investment spending.

Consider, for example, the recessions of 1975 and 1980. The sharp jumps in world energy prices that occurred in 1974 and 1979–80 clearly contributed to those recessions, and they did so in a number of ways. First, they caused a reduction in the real national incomes of oil importing countries. Second, they led to 'adjustment effects' – inflation and a further drop in real income and output resulting from the rigidities that prevented wages and nonenergy prices from coming into equilibrium quickly. (For a discussion of these effects, see Pindyck (1980) and Pindyck and Rotemberg (1984).)

But those energy shocks also caused greater uncertainty over future economic conditions. For example, it was unclear whether energy prices would continue to rise or later fall, what the impact of higher energy prices would be on the marginal product of various types of capital, how long-lived the inflationary impact of those shocks would be, etc. Other events also contributed to what became a more uncertain economic environment, especially in 1979–82: much more volatile exchange rates, and (at least in the United States) more volatile interest rates. This increased uncertainty must have contributed to the decline in investment spending that occurred during these periods.[1]

This more volatile economic environment was reflected in an increase in the volatility of US stock prices. From 1970 to 1981, the average monthly variance of the New York Stock Exchange Index was about 2.5 times as large as for the period 1950–1969. This increase in stock market volatility can be explained in part by the more volatile economic conditions described above. Elsewhere I have argued that it corresponds to an increase in the variance of the real gross marginal return on capital: see Pindyck (1984).[2] Again, this should have affected investment spending, and economic performance in general.

In what follows I will focus on investment spending, and argue that a more explicit treatment of risk is needed to better model and forecast investment at the aggregate or sectoral level. This is particularly true when investment is irreversible, as most investment is, at least in part. In such a case the decision to invest involves an additional opportunity cost – installing capital today forecloses the possibility of installing it instead at some point in the future (or never installing it at all). Put differently, a firm has options to install capital at various points in the future (options that can be exercised at the cost of purchasing the capital), and if the firm installs capital now, it closes those options. If uncertainty over future market conditions increases, the opportunity cost associated with closing these options increases, and current investment spending becomes less attractive.

In the next two sections I explain this aspect of firms' investment decisions in more detail, and discuss the implications for q-theory models of investment.

---

[1] This point was made by Bernanke (1983), particularly with respect to changes in oil prices. Also, see Evans (1984) and Tatom (1984) for a discussion of the depressive effects of increased interest rate volatility.
[2] That paper also argues that the increase in the variance of stock returns may have been partly responsible for the decline in real stock prices over the period 1965–81.

Section 4 presents the results of some simple (and preliminary) empirical tests. These results indicate that risk does seem to help explain and predict aggregate investment spending.

6.2 THE DETERMINANTS OF INVESTMENT SPENDING

The explanation of aggregate and sectoral investment behaviour remains especially problematic in the development of macroeconometric models. Existing models have had, at best, limited success in explaining or predicting investment. The problem is not simply that these models have been able to explain and predict only a small portion of the movements in investment. In addition, constructed quantities that in theory should have strong explanatory power – e.g. Tobin's q, or various measures of the cost of capital – in practice do not, and leave much of investment spending unexplained. (See Kopcke (1985) for an overview, as well as examples and comparisons of traditional approaches to modelling investment spending.)

It is easy to think of reasons for the failings of these models. For example, even leaving aside problems with their theoretical underpinnings, there are likely to be formidable estimation problems resulting from aggregation (across firms, and also across investment projects of different gestations). I will not attempt to survey these problems here, nor in any way provide a general overview of the state of investment modelling. Instead I want to focus on one special aspect of investment – the role of risk, and in particular the effects on investment spending of uncertainty over future values of the marginal revenue product of capital.

Of course nondiversifiable risk plays a role in even the simplest models of investment, by affecting the cost of capital. But there is an emerging literature that suggests that risk may be a more crucial explanator of investment. The thrust of this literature begins with the fact that much of investment spending is *irreversible* – i.e. a widget factory, once constructed, can be used to make widgets, but not much else. Given this irreversibility, one must view an investment expenditure as essentially the exercising of an option (an option to productively invest). But once such an option is exercised, it is 'dead', i.e. one cannot decide to exercise it instead at some point in the future, or never at all. In other words, one gives up the option of waiting for new information (about evolving demand and cost conditions), and using that information to re-evaluate the desirability and/or timing of the expenditure.[3]

---

[3] This is developed in the recent papers by Bernanke (1983) and McDonald and Siegel (1986). Other examples of this literature include Cukierman (1980), Brennan and Schwartz (1985), and Majd and Pindyck (1985). In the papers by Bernanke and Cukierman, uncertainty over future market conditions is reduced as time passes, so that firms have an incentive to delay investing when markets are volatile (e.g. during recessions). In the other papers, future market conditions are *always* uncertain. But as with a call option on a dividend-paying stock, an investment expenditure should be made only when the value of the resulting project exceeds its cost by a positive amount, and again, increased uncertainty will increase the incentive to delay the investment.

This lost option value must be included as part of the cost of the investment. Doing so leads to an investment rule that is different from the standard rule: 'Invest when the marginal value of a unit of capital is at least as large as the purchase and installation cost of the unit'. Instead, the marginal value of the unit must *exceed* the purchase and installation cost *by an amount equal to the value of keeping the firm's option to invest alive.*

To see this more clearly, consider a firm that has some degree of monopoly power (i.e. faces a downward demand curve), and must decide how much initial output capacity to install. (To keep things simple we will ignore any lumpiness in investment, and assume that the firm can chose any output capacity it wants.) Again, a key assumption is that the firm's investment is irreversible – although capital in place can be sold by one firm to another, its scrap value is small because it has no alternative use other than that originally intended for it (e.g. to produce widgets).

Let $\Delta V(K)$ denote the value to the firm of an *incremental unit of capacity*, given that the firm already has capacity K. Determining $\Delta V(K)$ in practice is a nontrivial matter, because it will be a highly nonlinear function of (unknown) future demands. For example, if future demand falls, this incremental unit of capacity might not be used by the firm. As a result, methods other than discounted cash flow techniques are needed to calculate $\Delta V(K)$.[4] We will not be concerned with this problem here, and instead simply assume that $\Delta V(K)$ can indeed be calculated. It will of course be a declining function of K.

Given that the firm knows $\Delta V(K)$, it must now decide whether to install the incremental unit. Although the firm has the option to install the unit, it might not want to exercise this option.[5] Deciding whether or not to exercise this option is the key to the firm's investment decision.

In order to determine *when* to exercise its option to invest, the firm must also *value* the option. Again, we are concerned here with an incremental unit of capacity, given a current capacity K, i.e. a unit that has value $\Delta V(K)$. Denote by $\Delta F(K)$ the *value of the firm's option* to install this incremental unit. Note that this option has value even if it is currently not optimal for the firm to exercise it (just as a call option on shares of a common stock has value even if it will not be exercised today). Also, the greater the uncertainty over future demand, the greater will be the value of this option (just as a call option has greater value the greater the price volatility of the stock on which it is written).

Assume the cost of a unit of capital is constant, and denote this cost by k. Then k is the 'exercise price' of the firm's option on the incremental unit of

---

[4] As noted by McDonald and Siegel (1985), once a unit of capital is in place, the firm has the *option* of whether or not to utilize it as market conditions evolve. Thus option pricing methods can be used to value the unit. Implications for marginal investment decisions and capacity choice is examined in Pindyck (1986).

[5] What gives a firm this option? It may be that the firm owns a patent on a particular production technology. More generally, the firm's managerial resources and expertise, reputation and market position enable it to productively undertake investments that individuals or other firms cannot undertake.

capital, i.e. the price the firm must pay to exercise the option. Suppose demand conditions are such that it is rational for the firm to exercise the option. In that case, $\Delta F(K) = \Delta V(K) - k$. Now suppose instead that $\Delta F(K) > \Delta V(K) - k$. This might be the case, for example, if current K is large and future demand is extremely uncertain, in which case the firm will want to hold its option to invest, rather than exercise it. Then what is the firm's optimal choice of capacity K? It is the largest level K^* such that:

$$\Delta V(K^*) = k + \Delta F(K^*) \tag{6.1}$$

In other words, the firm should keep adding capacity until it reaches the point where the value of the marginal unit is just equal to its purchase cost *plus* the value of the option to install the unit (an option that is closed, yielding $\Delta V(K) - k$, once the unit is indeed installed).

Now, what is the effect of an *increase in uncertainty* over future demand conditions? The immediate effect, whatever the current capacity K, is to increase the value of the firm's option to invest in the marginal unit, i.e. to increase $\Delta F(K)$. But this means that all else equal, the firm will want to hold *less capacity* than it would otherwise. The reason is that it is now worth more to the firm to keep its option to invest alive. Thus an increase in uncertainty will reduce firms' desired capital stocks, and have a *depressive effect on investment spending*.

If there is considerable uncertainty over future demand conditions the value of the firm's options to invest will be large, and an investment rule that ignores this will be grossly in error. In fact, as the important paper by McDonald and Siegel (1986) has shown, for even moderate levels of uncertainty the effect can be quite large; an investment in an individual project might require that the present value of the project be at least *double* the cost of the project. Also, changes in the level of uncertainty can have a major effect on the critical present value needed for a positive investment decision, and thus such changes should have a major effect on investment spending.

In most modelling work, effects of risk are handled by assuming that a risk premium can be added to the discount rate used to calculate the present value of a project. That discount rate is typically obtained from the Capital Asset Pricing Model (CAPM). But as we have learned from the theory of financial option pricing, the correct discount rate cannot be obtained without actually solving the option valuation problem, generally will not be constant over time, and will not equal the average cost of capital for the firm. As a result, simple cost of capital measures, based on rates of return (simple or adjusted) to equity and debt, may be poor explanators of investment spending.

6.3 MARGINAL q AND INVESTMENT

In essence, the q theory of investment says that firms have an incentive to invest whenever marginal q – the present value of a marginal unit of capital divided by the cost of that unit – exceeds one. But models based on this theory have not been very successful in explaining investment.[6] There may be several reasons for this failing, but one possibility is that if risk is significant, the model is theoretically flawed. In fact marginal q should *not* equal one in equilibrium. Instead it should exceed one, because as we have seen, investment should occur only when the present value of a marginal unit of capital exceeds the cost of the unit by an amount equal to the value of keeping the firm's option to invest alive.

To see this more clearly, let us return to the problem of a firm deciding whether to invest in an incremental unit of capacity. Recall that the optimal investment decision is to invest up to the point K^* where $\Delta V(K^*) = k + \Delta F(K^*)$. Marginal q is the present value of the marginal unit of capital – in my notation, $\Delta V(K^*)$ – divided by the cost of the unit, k. Then clearly $q = \Delta V(K^*) > 1$ in equilibrium.

A correct measure for marginal q could be defined as:

$$q^* = [\Delta V(K^*) - \Delta F(K^*)]/k \qquad (6.2)$$

which will equal 1 in equilibrium. The problem is that this measure cannot be observed directly, nor can it easily be computed from other firm or industry-wide data. Furthermore, the use of an *average* measure of q (the market value of the firm divided by the replacement cost of its capital) will be even more misleading. The reason is that the market value of the firm will include the value of capital in place *plus* the value of the firm's options to install more capital in the future.[7] What is needed instead is the value of capital in place *less* the value of those options.

Since the q theory has become prominent as a basis for structural models of investment spending, it is useful to discuss it in somewhat more detail. I do this with reference to the recent paper by Abel and Blanchard (1983). The model that they developed is one of the most sophisticated attempts to explain investment in a q theory framework; it uses a carefully constructed measure for marginal rather than average q, incorporates delivery lags and costs of adjustment, and explicitly models expectations of future values of explanatory variables.

The model is based on the standard discounted cash flow rule, 'invest in the marginal unit of capital if the present discounted value of the expected flow of profits resulting from the unit is at least equal to the cost of the unit'. Let $\pi_t(K_t, I_t)$ be the maximum value of profits at time t, given the capital stock K_t and investment level I_t, i.e. it is the value of profits assuming that variable factors are used optimally. It depends on I_t because of costs of adjustment; $\partial \pi/\partial I < 0$, and

---

[6] For example, Abel and Blanchard (1983) find that even when it is properly measured, marginal q 'is a significant explanator of investment, but, leaves unexplained a large, serially correlated fraction of investment'. Also, see Kopcke (1985).

[7] Assuming the firm has chosen its capacity optimally, its value is given by:

$$\text{Value} = \int_0^{K^*} \Delta V(K) \, dK + \int_{K^*}^{\infty} \Delta F(K) \, dK,$$

i.e. the value of capital in place plus the value of options to productively install more capital in the future.

$\partial^2 \pi / \partial I^2 < 0$, i.e. the more rapidly new capital is purchased and installed, the more costly it is. Then the present value of current and future profits is given by:

$$V_t = E_t \left[\sum_{j=0}^{\infty} \left[\prod_{i=0}^{j} (1 + R_{t+i})^{-1} \right] \pi_{t+j}(K_{t+j}, I_{t+j}) \right] \quad (6.3)$$

where E_t denotes an expectation, and R is the discount rate. Maximizing this with respect to I_t, subject to the condition $K_t = (1 - \delta)K_{t-1} + I_t$ (where δ is the rate of depreciation), gives the following marginal condition:

$$-E_t(\partial \pi_t / \partial I_t) = q_t, \quad (6.4a)$$

where

$$q_t = E_t \left[\sum_{j=0}^{\infty} \left[\prod_{i=0}^{j} (1 + R_{t+i})^{-1} \right] (\partial \pi_{t+j} / \partial K_{t+j})(1 - \delta)^j \right] \quad (6.4b)$$

In other words investment should occur up to the point where the cost of an additional unit of capital is just equal to the present value of the expected flow of incremental profits resulting from the unit. Abel and Blanchard estimate both linear and quadratic approximations to q_t, and use vector autoregressive representations of R_t and $\partial \pi_t / \partial K_t$ to model expectations of future values. Their representation of R_t is based on a weighted average of the rates of return on equity and debt.

If the correct discount rates R_{t+i} were known, Equations 6.4a and 6.4b would indeed accurately represent the optimal investment decision of the firm. *The problem is that these discount rates are usually not known, and generally will not be equal to the average cost of capital of the firm, or some related variable.* Instead, these discount rates can only be determined as part of the solution to the firm's optimal investment problem. This involves valuing the firm's options to make (irreversible) marginal investments (now or in the future), and determining the conditions for the optimal exercise of those options. Thus the solution to the investment problem is more complicated than the first-order condition given by Equations 6.4a and 6.4b would suggest.

As an example, consider a project that has zero systematic (nondiversifiable) risk. The use of a risk-free interest rate for R would lead to much too large a value for q_t, and might suggest that an investment expenditure should be made, whereas in fact it should be delayed. Furthermore, there is no simple way to adjust R properly. The problem is that the calculation ignores the opportunity cost of exercising the option to invest. This may be why Abel and Blanchard conclude that 'our data are not sympathetic to the basic restrictions imposed by the q theory, even extended to allow for simple delivery lags'.

6.4 DOES RISK HELP EXPLAIN INVESTMENT SPENDING?

We have seen that when investment is irreversible, there is an especially strong link between risk and investment spending. An increase in uncertainty over future demand raises the value of the firm's option to invest in a marginal unit of capital, and therefore raises the incentive to keep that option alive (by not investing) rather than exercising it (by investing), so that other things equal, investment spending will fall. Furthermore, recent studies have shown that in theory, this effect can be quantitatively important. See McDonald and Siegel (1986), Brennan and Schwartz (1985), Majd and Pindyck (1987), and Pindyck (1986). But do the data support the theory?

A major problem with testing the theory is finding a good measure of market uncertainty. One possibility is to use survey data. An alternative, which I pursue here, is to use data on the stock market. When product markets become more volatile, we would expect stock prices to also become more volatile, so that the variance of stock returns will be larger. This was indeed the case, for example, during the recessions of 1975 and 1980, and most dramatically during the Great Depression. Thus the variance of aggregate stock returns should be a reasonable measure of aggregate product market uncertainty.

Stock returns themselves are also a predictor of aggregate investment spending, as was illustrated by the recent work of Fischer and Merton (1984). We would expect stock returns to have predictive power because they reflect new information about economic variables. That new information can imply revised expectations about future corporate earnings, and changes in the discount rates used to capitalize those earnings. Fischer and Merton show that the predictive power of stock returns is strong, not only with respect to aggregate investment spending, but with respect to other components of GNP as well.

Our concern here is whether the *variance* of stock returns – my measure of aggregate product market uncertainty – also has predictive power with respect to investment, and whether that predictive power goes beyond that of stock returns themselves, as well as other variables that would usually appear in an empirical investment equation. Ideally, this should be examined in the context of a structural model of investment derived from firms' optimizing decisions. However, such a model would be quite complicated, especially if it were to take account of construction lags, capital of different vintages, etc. Instead, I conduct two related tests of an exploratory nature.

First, I test whether the variance of stock returns can be said to 'cause' the real growth rate of investment, in the sense of Granger (1969) and Sims (1972). To say that 'X causes Y', two conditions should be met. First, X should help to predict Y, i.e. in a regression of Y against past values of Y, the addition of past values of X as independent variables should contribute significantly to the explanatory power of the regression. Second, Y should *not* help to predict X. (If X helps to predict Y *and* Y helps to predict X, it is likely that one or more other variables are in fact 'causing' both X and Y.)

In each case I test the null hypothesis that one variable does not help predict the other. For example, to test the null hypothesis that 'X does not cause Y', I regress Y against lagged values of Y and lagged values of X (the 'unrestricted' regression), and then regress Y only against lagged values of Y (the 'restricted' regression). A simple F test is used to determine whether the lagged values of X contribute significantly to the explanatory power of the first regression.[8] If they do, I can reject the null hypothesis and conclude that the data are consistent with X causing Y. The null hypothesis that 'Y does not cause X' is then tested in the same manner.

A weakness of this test of causality is that a third variable, Z, might in fact be causing Y, but might be contemporaneously correlated with X. For example, lagged values of the variance of stock returns might be a significant explanator of the growth of investment in a bivariate regression, but might become insignificant when other variables are added, such as stock returns themselves, interest rates, etc. Therefore as a second exploratory test I also run a set of simple regressions of the growth rate of investment against stock returns, the variance of stock returns, and a set of additional explanatory variables that usually appear in empirical investment equations.

6.4.1 Data

These causality tests and multivariate regressions require a series for the variance of stock returns. I constructed such a series using daily data for the total logarithmic return on the combined value-weighted New York and American Stock Exchange Index, obtained from the CRISP tape. From these daily data I constructed nonoverlapping estimates of the monthly variance by calculating the sample variance, corrected for nontrading days. I then sum the monthly estimates for each quarter, to obtain a series of nonoverlapping estimates of the quarterly variance, denoted by VAR. The returns themselves were also summed over each quarter and then adjusted for inflation, to yield a quarterly series for real stock returns (RTRN).

Because the daily stock return data were available only from the fourth quarter of 1962 through the fourth quarter of 1983, and because the causality tests and regressions require lags of several quarters, the sample is limited to the 21 year period 1963-4 to 1983-4.

I test whether the variance of stock returns can help predict the growth rate of real nonresidential fixed investment (INGR). I also examine the two major components of INGR, the growth rate of real investment in structures (ISGR), and the growth rate of real investment in durable equipment (IEGR). These multivariate regressions also include the following additional explanatory variables: the quarterly change in the BAA corporate bond rate (DRBAA), the

---

[8] The F-statistic is as follows:

$$F = (N - k)(SSR_r - SSR_u)/r(SSR_u)$$

where SSR_r and SSR_u are the sums of squared residuals in the restricted and unrestricted regressions respectively, N is the number of observations, k is the number of estimated parameters in the unrestricted regression, and r is the number of parameter restrictions. This statistic is distributed as $F(r/N - k)$.

change in the 3-month Treasury bill rate (DRTB3), the change in the rate of inflation as measured by the rate of growth of the Producer Price Index (DINF), and the growth rate of real GNP (GNPGR). Series for real investment and its components, GNP, the Producer Price Index, and the interest rates were all obtained from the Citibank database.

6.4.2 Causality tests

Granger causality tests are conducted as follows. For each investment series (INGR, ISGR, IEGR), we test the pair of null hypotheses: (i) VAR does not help predict the growth rate of investment, and (ii) the growth rate of investment does not help predict VAR. If we reject the first hypothesis and accept the second, then the data are consistent with variance of stock returns causing investment. Each hypothesis is tested by running the unrestricted regression,

$$Y_t = a + \sum_{i=1}^{n} b_i Y_{t-i} + \sum_{i=i}^{n} c_i X_{t-i}, \tag{6.5}$$

and the restricted regression,

$$Y_t = a + \sum_{i=1}^{n} b_i Y_{t-i}, \tag{6.6}$$

and testing the parameter restrictions $c_i = 0$ for all i (see footnote 8). The tests are done for the number of lags, n, equal to 4 and 6.

Table 6.1 Causality tests
(quarterly data: 1963-4 to 1983-4)

Regression of: $Y_t = a + \sum_{i=1}^{n} b_i Y_{t-1} + \sum_{i=1}^{n} c_i X_{t-i}$

No.	Lag (n)	Regression pair Y	X	R_r^2/R_u^2	$F^a(H_0: b_i = 0)$
1	4	INGR	VAR	0.355/0.495	4.98[c]
2	4	VAR	INGR	0.298/0.356	1.63
3	4	ISGR	VAR	0.201/0.326	3.33[b]
4	4	VAR	ISGR	0.298/0.347	1.38
5	4	IEGR	VAR	0.277/0.442	5.33[c]
6	4	VAR	IEGR	0.298/0.337	1.09
7	6	INGR	VAR	0.388/0.548	3.87[c]
8	6	VAR	INGR	0.299/0.410	2.05
9	6	ISGR	VAR	0.225/0.361	2.35[b]
10	6	VAR	ISGR	0.299/0.404	1.94
11	6	IEGR	VAR	0.304/0.473	3.52[c]
12	6	VAR	IEGR	0.299/0.388	1.58

[a] $F(4/75)$ for $n = 4$, $F(6/66)$ for $n = 6$.
[b] Significant at 5% level.
[c] Significant at 1% level.

Does risk help explain investment spending?

The results of these causality tests are summarized in Table 6.1, which shows the R^2s for the restricted and unrestricted regressions, and the F-statistic for the restrictions $c_i = 0$. Observe that for every investment variable and for both 4 and 6 lags, we reject the null hypothesis 'variance does not help predict the growth of investment', and we fail to reject the opposite null hypothesis 'the growth of investment helps to predict variance'. This is at least preliminary evidence that market risk, as measured by the variance of stock returns, plays a significant role in the determination of investment spending.

6.4.3 Multivariate regressions

As a further test, I run a set of regressions similar to those used by Fischer and Merton (1984) in their study of the predictive power of stock returns. Each investment variable is regressed first against lagged values of variance, then against lagged values of variance and lagged values of real stock returns, and finally against lagged variance, lagged stock returns, and the lagged values of four additional variables: the change in the BAA corporate bond rate, the change in the 3-month Treasury bill rate, the change in the inflation rate, and the rate of growth of real GNP.

The results are summarized in Table 6.2. Note that three lagged values of each independent variable appear in the regressions, but only the sum of the estimated coefficients is shown for each variable, together with an associated t-statistic.

As one would expect from the results of the causality tests, variance is highly significant when it appears as the only independent variable. Variance continues to be highly significant after adding real stock returns to the regression, and this second independent variable is also a significant explanator of the growth of total nonresidential investment, as Fischer and Merton found, as well as investment in equipment, but it is not significant for investment in structures. When the remaining explanatory variable are added, variance continues to be significant in the regressions for nonresidential investment and structures, but not for equipment. (The only significant explanator of equipment investment is the BAA bond rate.) This may reflect the fact that the irreversibility of investment is greater for structures than for equipment.

But the importance of our risk measure is also evident from the magnitudes of the variance coefficients. The quarterly variance of stock returns went from about 0.01 in the 1960s to about 0.02 in the mid-1970s. From regressions 3, 6 and 9 we see that this implies an approximately 4.5 percentage point decline in the growth rate of investment in structures (a drop from around 5% real growth during the 1960s to only slightly positive real growth), a 2.5 percentage point drop in the growth rate of investment in equipment, and a 3 percentage point drop in the growth rate of total investment.

Table 6.2 Variance of stock returns as a predictor of investment
(Quarterly data, 1963–4 to 1983–4)

Independent variable	Dependent variable								
	INGR	INGR	INGR	INGR	ISGR	ISGR	IEGR	IEGR	IEGR
Reg. No.	1	2	3	4	5	6	7	8	9
Constant	0.0314 (7.72)	0.0264 (6.30)	0.0199 (2.38)	0.0261 (5.36)	0.0228 (4.28)	0.0190 (1.75)	0.0349 (6.96)	0.0291 (5.67)	0.0201 (2.01)
$\sum_{i=1}^{3} \text{VAR}_{-i}$	−6.307 (−5.98)	−5.306 (−4.92)	−3.282 (−2.32)	−6.262 (−4.98)	−5.470 (−4.20)	−4.786 (−2.59)	−6.478 (−4.99)	−5.378 (−4.08)	−2.563 (−1.51)
$\sum_{i=1}^{3} \text{RTRN}_{-i}$		0.1738 (3.46)	0.1229 (1.91)		0.0769 (1.22)	0.0887 (1.05)		0.2150 (3.49)	0.1388 (1.80)
$\sum_{i=1}^{3} \text{DRBAA}_{-i}$			−0.0308 (−2.93)			−0.0056 (−0.41)			−0.0409 (−3.25)
$\sum_{i=1}^{3} \text{DRTB3}_{-i}$			0.0167 (2.44)			0.0199 (2.23)			0.0144 (1.76)
$\sum_{i=1}^{3} \text{DINF}_{-i}$			−0.2827 (−0.41)			−0.9405 (−1.05)			0.1495 (0.18)
$\sum_{i=1}^{3} \text{GNPGR}_{-i}$			0.3505 (0.66)			0.0816 (0.18)			0.5428 (0.86)

	(1)	(2)	(3)	(4)	(5)	(6)	(7)	(8)	(9)
RHO	0.0222	0.0187	0.0108	0.0226	0.0202	0.0126	0.0215	0.0174	0.0096
	(2.95)	(2.59)	(1.46)	(2.51)	(2.20)	(1.32)	(2.33)	(1.97)	(1.11)
R^2	0.425	0.501	0.627	0.320	0.335	0.477	0.350	0.442	0.596
SER	0.0204	0.0194	0.0184	0.0244	0.0247	0.0239	0.0252	0.0238	0.0216
DW	1.40	1.53	1.84	1.65	1.67	1.88	1.60	1.78	2.19

Variables:
- INGR = Quarterly growth rate of real business fixed investment.
- ISGR = Growth rate of real investment in structures.
- IEGR = Growth rate of real investment in durable equipment.
- VAR = Quarterly variance of real return on NYSE index.
- RTRN = Real return on NYSE index.
- DRBAA = Change in BAA corporate bond rate.
- DRTB3 = Change in 3-month Treasury bill rate.
- DINF = Change in inflation rate, as measured by PPI.
- GNPGR = Quarterly growth rate of real GNP.
- RHO = Primary order serial correlation coefficient.
- SER = Standard error of regression.
- DW = Durbin-Watson statistic.

Note: t-statistics in parentheses.

6.5 CONCLUSIONS

I argue in the beginning of the chapter that there are good theoretical reasons to expect market risk to have a major role in the determination of investment spending. This idea is not new; it has been elaborated upon in a number of articles during the past few years. However, it seems to be missing from most empirical work on investment. This may be a reflection of the fact that most theoretical models of irreversible investment under uncertainty are quite complicated, so that their translation into well-specified empirical models represents a formidable task.

This chapter has not attempted to make that translation. Instead I have sought and found only a rough empirical verification for the importance of risk. The tests and regressions reported in the previous section should be viewed as exploratory, and limited in their implications. For example, they are based on highly aggregated data, and what is probably a very imperfect measure of market risk.

Nonetheless, these findings support recent theoretical results regarding the effects of risk on irreversible investment, and they suggest that the explicit inclusion of market risk measures can improve our ability to explain and predict investment spending. The development of structural models that include such measures should be an important research priority.

ACKNOWLEDGEMENT

This chapter was written while the author was Visiting Professor at Tel-Aviv University. Research support from the National Science Foundation, under grant No. SES-8318990, is gratefully acknowledged.

REFERENCES

Abel, Andrew B., and Blanchard, Olivier J. (1983) The present value of profits and cyclical movements in investment, discussion paper No. 983, May, Harvard Institute of Economic Research.

Bernanke, Ben S. (1983) Irreversibility, uncertainty, and cyclical investment. *Quarterly Journal of Economics*, 98, February, 85–106.

Brennan, Michael J. and Schwartz, Eduardo S. (1985) Evaluating natural resource investments. *Journal of Business*, January.

Cukierman, Alex (1980) The Effects of uncertainty on investment under risk neutrality with endogenous information. *Journal of Political Economy*, 88, June 462–75.

Evans, Paul (1984) The effects on output of money growth and interest rate volatility in the United States. *Journal of Political Economy*, 92, April, 204–22.

Fischer, Stanley and Merton, Robert C. (1984) Macroeconomics and finance: The role of the stock market. *Carnegie-Rochester Conference Series on Public Policy*, 57–108.

Granger, Clive W. J. (1969) Investigating causal relations by econometric models and cross-spectral methods. *Econometrica*, 37, July, 429–38.

References

Kopcke, Richard W. (1985) The determinants of investment spending. *New England Economic Review*, July, 19–35.

Majd, Saman and Pindyck, Robert S. (1987) Time to build, option value, and investment decisions, 4. *Financial Economics*, **18**, 7–27.

McDonald, Robert and Siegel, Daniel (1985) Investment and the valuation of firms when there is an option to shut down. *International Economic Review*, October.

McDonald, Robert and Siegel, Daniel (1986) The value of waiting to invest. *Quarterly Journal of Economics* **101**, 707–27.

Pindyck, Robert S. (1980) Energy price increases and macroeconomic policy. *The Energy Journal*, October.

Pindyck, Robert S. (1984) Risk, Inflation, and the Stock Market. *American Economic Review*, **74**, June, 335–51.

Pindyck, Robert S. (1986) Irreversible investment and capacity choice, unpublished, March.

Pindyck, Robert S. and Rotemberg, Julio J. (1984) Energy shocks and the macroeconomy, in *Managing Oil Shocks*, (eds A. Alm and R. Weiner), Ballinger, Cambridge.

Sims, Christopher (1972) Money, income, and causality. *American Economic Review*, **62**, 540–52.

Tatom, John A. (1984) Interest rate variability: its link to the variability of monetary growth and economic performance, Federal Reserve Bank of St Louis *Review*, November, pp. 31–47.

7

Adjustment costs and mean-variance efficiency in UK financial markets

CHRISTOPHER J. GREEN

7.1 INTRODUCTION

The mean-variance model has, for a long time, appeared to offer the most attractive approach to understanding the workings of financial markets.[1] Beginning with the seminal works of Tobin (1958), Markovitz (1959) and Sharpe (1964), a vast body of theory has been constructed to explain a wide variety of phenomena related to portfolio choice and asset pricing.

Tests of the mean-variance model have typically been conducted in one of two settings. In the finance literature, the emphasis has been on studying the determination of the prices of individual securities and their stochastic properties relative to that of the market. A seminal example is Black, Jensen and Scholes (1972). In the economics literature, more attention has been given to the estimation at an aggregative level of asset demand functions belong the lines advocated by Tobin and Brainard (1968). These functions can then, in principle, be inverted to determine asset prices and yields, see Backus, Brainard, Smith and Tobin (1980).

Neither approach has proved reassuring for the mean-variance model. In the finance literature, estimated betas frequently have counterintuitive properties. Moreover, Roll (1977) has argued that a proper test of the mean-variance model requires data on market portfolios in a form more associated with aggregate economic modelling than with the disaggregated approach. However, the aggregative approach has not fared noticeably better. It has proven difficult to find simple relationships between the parameters of the asset demand functions on the one hand, and the coefficient of risk aversion and the subjective covariance matrix on the other (Blanchard and Plantes, 1977). The latter represent the most important parameters in the underlying mean-variance utility function. The asset

---

[1] See Ross (1978) for a dissenting view.

demand functions have therefore to be estimated directly and, in these functions, estimated interest rate substitution effects frequently have implausible signs and magnitudes or are statistically insignificant. This is not necessarily unexpected in a setting in which interest rates are highly correlated with one another and in which interest rates and asset holdings are determined simultaneously in continuouly clearing markets.

A more serious methodological problem with asset demand systems is posed by the way in which such systems are invariably used to determine interest rates: the asset demands are inverted and the result is described as showing the effects on interest rates of exogenous shocks to asset supplies in the context of a freely clearing market. It is my contention that this way of proceeding is fundamentally incorrect: if asset supplies (and private preferences) are the exogenous processes which determine interest rates, then the regression of an asset quantity on interest rates is not meaningful, since it amounts to regressing an exogenous variable on a collection of endogenous variables. The contrast with models of consumer behaviour is instructive (see, for example, Anderson and Blundell, 1982). In that area it is plausible to think of prices as predetermined at a point in time, and regressions of expenditure shares on consumer prices can therefore often be given a meaningful interpretation. In asset markets which continuously clear, it is far from obvious which variables can be regarded as predetermined within a regression framework.

In a recent paper (Green, 1987a), I developed the idea of *asset price expectation formation (APEX) equations*. The idea owes a great deal to pioneering work by Frankel (1985) and Frankel and Dickens (1984). It aims at utilizing insights from both the finance literature and the economics literature. Put simply, APEX equations are derived by inverting a set of mean-variance-based asset demand functions under the maintained assumptions of continuous market clearing and unbiased expectations. Under these assumptions, APEX equations turn out to be regression equations which can be estimated by standard procedures. Restrictions implied by the mean-variance model can likewise be tested in particularly straightforward ways. In principle, the APEX technique offers a constructive resolution of *all* the major methodological problems listed above.

In Green (1987a), the focus of interest was in using the technique to test mean-variance efficiency and estimate the parameters of the underlying utility function, notably the coefficient of relative risk aversion. It turned out that UK data covering the period 1972–77 did not unambiguously support the hypothesis of mean-variance efficiency. A number of associated hypotheses were accepted by the data, notably: symmetry of agents' subjective covariance matrix and equality of agents' subjective covariance matrix and the true covariance matrix of the data. The latter hypothesis requires a particularly strong set of restrictions. However, the estimated coefficient of relative risk aversion was found to be large in magnitude, significant and negative. This is inconsistent with mean-variance efficiency. It implies either that agents are minimizing not maximizing utility, or that agents are risk seekers. However, if agents literally were risk seekers, this would be inconsistent with observed portfolio diversification, apart from the constraints provided by legal restrictions on all-or-nothing portfolios.

In the present paper, I use the same APEX methodology, the same dataset and a mean-variance framework, but the focus of interest is on the importance of adjustment costs in determining UK asset demands and prices.

Adjustment costs have been offered as one of the principal rationales for explaining why markets which are informationally efficient should nevertheless give rise to time series data which are often highly autocorrelated (see Sargent, 1979). Aggregate private sector asset holdings typically are highly autocorrelated and traditional portfolio demand functions generally require lagged endogenous variables to explain fully the time series behaviour of asset stocks. In contrast, asset prices are much less highly autocorrelated. Some asset prices approximate a white noise process; most retain some persistence which is, however, much weaker than that to be observed in asset stocks. For some examples see Green (1984) on portfolio demand functions, and Dickens (1987) on asset prices.

Autocorrelation in asset stocks could be caused by a number of factors apart from adjustment costs in asset demands. In particular, it is possible that the rules determining asset supplies could be autocorrelated. If it is true that asset supplies are the exogeneous processes driving asset prices, then estimated portfolio demand functions are potentially non-informative about the presence or absence of adjustment costs, as they could equally well be interpreted as (possibly misspecified) asset supply functions. The APEX technique offers an alternative method of examining the impact of adjustment costs in asset markets, under the assumption that such markets are continuously cleared. Though it is not foolproof, the method appears less vulnerable to confusing low-order autocorrelations in asset demands and asset supplies. Moreover, as the APEX technique provides a new framework for studying asset demands, it seems worthwhile to examine the impact of adjustment costs within this framework.

In this chapter I examine and compare the results of estimating APEX equations with three different specifications of adjustment costs. Conveniently, two of these specifications are each nested in the third and hypothesis testing can therefore be carried out in a relatively unambiguous manner.

7.2 ASSET PRICE EXPECTATION FORMATION (APEX) EQUATIONS

The notation is as follows:

$x_t = n \times 1$ vector of portfolio shares

$W_t =$ wealth at market value (a scalar)

hence $x_{jt} W_t =$ market value of holdings of asset j at time t

$\mu = W_{t-1}/W_t$

$y_{t+1} = n \times 1$ vector of actual gross asset returns

each $y_{jt+1} = (d_{jt} + p_{jt+1})/p_{jt}$

with d_{jt} = the coupon on asset j at time t

p_{jt} = the price of asset j at time t

${}_t y_{jt+1} = (d_{jt} + {}_t p_{jt+1})/p_{jt}$ is the expectation at time t of y_{jt+1}

$q_{jt} = p_{jt}/p_{jt-1}$ is the gross rate of capital gain on asset j

q_t is a diagonal *matrix* of q_{jt}

i is a conformable vector of ones

I is the identity matrix.

I assume away aggregation problems by adopting the convenient fiction of a representative agent who chooses a vector of portfolio shares (x_t) to:

$$\text{maximize } U = U(M, V, C) \tag{7.1}$$

$$\text{subject to } i'x_t = 0 \tag{7.2}$$

$U(\;)$ is the usual second-order approximation to the expected utility of wealth. The arguments of $U(\;)$ consist of:

Subjective expectation of wealth $\quad M = {}_t y'_{t+1} x_t W_t$

Subjective variance of wealth $\quad V = x'_t \Sigma x_t W_t^2$

Adjustment Costs $\quad C = (x_t - \mu x_{t-1})' \Psi (x_t - \mu x_{t-1}) W_t^2$

Σ is an $n \times n$ subjective covariance matrix of asset returns

Ψ is an $n \times n$ matrix of adjustment costs.

I treat this as a one-period optimization problem though strictly speaking, the presence of adjustment costs (with $\partial U/\partial C < 0$) makes this approach questionable. It is however widely used, and a main purpose of the present chapter is to examine adjustment costs within a framework which can easily be compared to more traditional portfolio models. For expositional purposes, I assume in this section that adjustment costs are proportional to (the square of) the change in asset holdings. The nature and implications of different specifications of adjustment costs are spelled out in the next section.

Agents' attitudes towards risk are assumed to be characterized by constant relative risk aversion, implying that:

$$\frac{-2W_t \partial U/\partial V}{\partial U/\partial M} = \gamma \quad \text{(a constant)} \tag{7.3}$$

Largely as a convenient normalization, I also assume 'constant relative aversion to adjustment costs' implying that:

$$\frac{-2W_t \partial U/\partial C}{\partial U/\partial M} = \delta \quad \text{(a constant)} \tag{7.4}$$

Maximizing (7.1) subject to (7.2) by choice of x and imposing constant relative risk aversion and aversion to adjustment costs yields as asset demands:

$$x_t = Z_t y_{t+1} + \delta Z \Psi \mu x_{t-1} + \Omega^{-1} i (i' \Omega^{-1} i)^{-1} \quad (7.5)$$

with $Z \equiv \Omega^{-1} - \Omega^{-1} i (i' \Omega^{-1} i)^{-1} i' \Omega^{-1}$

$$\Omega = \gamma \Sigma + \delta \Psi$$

It is in this general form that portfolio equations are most commonly estimated; see for example Green (1984).

The next step is to assume that asset markets are continuously cleared by freely moving prices. Under the assumptions which have been made, setting asset demands equal to asset supplies and solving for prices is now equivalent to inverting equations (7.5). To do this, note first that the matrix Z is singular; this is just another way of stating that an n market model can determine only $n-1$ asset prices. Therefore, we first partition (7.5) into the first $(n-1)$ rows and the remaining nth row:

$$\begin{pmatrix} x_{1t} \\ x_{nt} \end{pmatrix} = \begin{pmatrix} Z_{11} & Z_{1n} \\ Z_{n1} & z_{nn} \end{pmatrix} \begin{pmatrix} {}_t y_{1t+1} \\ {}_t y_{nt+1} \end{pmatrix} + \delta \mu \begin{pmatrix} Z_{11} & Z_{1n} \\ Z_{n1} & z_{nn} \end{pmatrix} \begin{pmatrix} \Psi_{11} & \Psi_{1n} \\ \Psi_{n1} & \psi_{nn} \end{pmatrix} \begin{pmatrix} x_{1t-1} \\ x_{nt-1} \end{pmatrix} + \begin{pmatrix} \theta_1 \\ \theta_n \end{pmatrix}$$

(7.6)

Here $x_t = \begin{pmatrix} x_{1t} \\ x_{nt} \end{pmatrix}$ is $\begin{pmatrix} n-1 \times 1 \\ 1 \times 1 \end{pmatrix}$

$Z = \begin{pmatrix} Z_{11} & Z_{1n} \\ Z_{n1} & z_{nn} \end{pmatrix}$ is $\begin{pmatrix} n-1 \times n-1 & n-1 \times 1 \\ 1 \times n-1 & 1 \times 1 \end{pmatrix}$

$\theta \equiv \Omega^{-1} i (i' \Omega^{-1} i)^{-1} = \begin{pmatrix} \theta_1 \\ \theta_n \end{pmatrix}$ is $\begin{pmatrix} n-1 \times 1 \\ 1 \times 1 \end{pmatrix}$

and other vectors and matrices are partitioned conformably.

Dropping the last row of (7.6), premultiplying by Z_{11}^{-1} (which is non-singular) and rearranging gives:

$${}_t y_{1t+1} - i_t y_{nt+1} = (\Omega_{1n} - i\omega_{nn}) - \delta(\Psi_{1n} - i\psi_{nn})\mu + \hat{\Omega} x_{1t} - \delta \hat{\Psi} \mu x_{1t-1} \quad (7.7)$$

with $\hat{\Sigma} = \Sigma_{11} - i\Sigma_{n1} - \Sigma_{1n} i' + ii' \sigma_{nn}$

$\hat{\Psi} = \Psi_{11} - i\Psi_{n1} - \Psi_{1n} i' + ii' \psi_{nn}$

$\hat{\Omega} = \gamma \hat{\Sigma} + \delta \hat{\Psi}$ (7.8)

Equation (7.7) is the fundamental theoretical result of the analysis. It shows that the vector of expected asset returns relative to the return on the numeraire can be expressed as linear combinations of current and lagged portfolio shares, the inverse of the rate of change of wealth, and a vector of constants reflecting the risk and costs of adjustment associated with the numeraire. Note that the share of the numeraire asset does not appear on the right-hand side as it has been eliminated using the wealth identity (7.2).

To give an economic interpretation to (7.7) recall that the main assumptions of the model are:

1. One period optimization.
2. Constant relative aversion to risk and to adjustment costs.
3. Continuous market clearing.

Assumption (3) is particularly important as it legitimizes the inversion of the asset demands. In contrast to traditional portfolio models therefore, continuous market clearing is a maintained (untestable) hypothesis. Given these three assumptions and some hypotheses about expectations and about asset supplies, equation (7.7) provides an implicit set of solutions for $n-1$ current asset prices: $p_{1t}, \ldots, p_{n-1t}$. However, in a regression context, these solutions are no more useful than the asset demands (7.5). I therefore make the additional assumption:

4(i) Asset price expectations are unbiased and (by virtue of the representative agent assumption) homogeneous.

This implies that the subjective expectation of wealth (M) is identical to the mathematical expectation of wealth. It implies too that actual asset returns (which contain asset prices) provide information about expected returns, subject to a measurement error. Thus:

$$_t y_{1t+1} - i_t y_{nt+1} = y_{1t+1} - i y_{nt+1} - (\varepsilon_{1l+1} - i\varepsilon_{nt+1}) \tag{7.9}$$

where $\varepsilon'_{t+1} = (\varepsilon'_{1t+1} : \varepsilon_{nt+1})$ is a vector of white noise errors with covariance matrix defined as:

$$\Lambda = E_t(\varepsilon_{t+1}\varepsilon'_{t+1})$$
$$\hat{\Lambda} = E_t(\varepsilon_{1t+1} - i\varepsilon_{nt+1})(\varepsilon_{1t+1} - i\varepsilon_{nt+1})'$$
$$\text{Hence} \quad \hat{\Lambda} = \Lambda_{11} - i\Lambda_{nl} - \Lambda_{1n}i' + ii'\lambda_{nn}$$

Here, E_t is the mathematical expectation conditional on information at time t; and Λ is partitioned to conform with the definitions (7.8).

Combining (7.7) and (7.9) gives:

$$y_{1t+1} - iy_{nt+1} = (\Omega_{1n} - i\omega_{nn}) - \delta(\Psi_{1n} - i\psi_{nn})\mu + \hat{\Omega}x_{lt} - \delta\hat{\Psi}\mu x_{1t-1} + (\varepsilon_{it+1} - i\varepsilon_{nt+1}) \tag{7.10}$$

If the expectations $(_t y_{t+1})$ were rational, the errors in (7.10) would be uncorrelated with the right-hand variables, whatever the nature of the process determining asset supplies. If expectations are merely unbiased, the error orthogonality property may not hold. I therefore assume:

4(ii) Expected asset returns and their errors are orthogonal.

Together with assumption 4(i), this implies that (7.10) constitutes a set of regression equations which can be estimated by ordinary least squares.

Note that rational expectations would imply that assumptions 4(i) and 4(ii) were true, but the reverse implication does not hold. If expectations were rational, it would, in addition, be true that agents' subjective covariance matrix would coincide with the true covariance matrix of the data. The subjective covariance matrix of relative returns is $\hat{\Sigma}$; the objective covariance matrix, conditional on information at time t, is found by operating on (7.9) to get:

$$E_t(y_{1t+1} - iy_{nt+1} - {}_t y_{1t+1} + i_t y_{nt+1})(y_{1t+1} - iy_{nt+1} - {}_t y_{1t+1} + i_t y_{nt+1})' = \hat{\Lambda}$$

Equality of $\hat{\Sigma}$ and $\hat{\Lambda}$ amounts to constraints between the coefficient matrix difference $(\hat{\Omega} - \delta\hat{\Psi}) = \gamma\hat{\Sigma}$ and the covariance matrix of regression residuals $\hat{\Lambda}$. These two matrices should be proportional, with the factor of proportionality equal to the coefficient of relative risk aversion. Clearly, these are testable, though non-standard restrictions.

7.3 THE SPECIFICATION OF ADJUSTMENT COSTS IN PORTFOLIO DEMAND FUNCTIONS AND APEX EQUATIONS

The specification of adjustment costs in portfolio demand functions has typically been based on Tobin and Brainard's generalized partial adjustment model. In this model, the change in an investor's holdings of an asset is modelled as a linear combination of deviations of (lagged) actual holdings from (current) desired holdings. Tobin and Brainard have emphasized that each asset demand function must typically include as arguments the deviations of actual from desired holdings of *all* assets in the portfolio, otherwise the balance sheet constraint may impose a nonsensical structure on asset demands which are not explicitly modelled. The generalized partial adjustment model can be justified in terms of optimizing behaviour by assuming that agents face quadratic costs of adjustment and of being away from their desired (long-run) position. Recent developments have shown that this basic framework can be interpreted as providing the rationale for a wide class of error correction models (Nickell, 1985).

Relating this discussion back to the previous section, we see there that adjustment costs (C) are associated with the change in asset holdings: $(x_t - \mu x_{t-1})W_t$. The portfolio demand functions (7.5) therefore consist of two components: the terms in ${}_t y_{t+1}$ reflecting the 'desired' portfolio and determined by expected returns, and the terms in μx_{t-1} reflecting a partial adjustment towards that desired portfolio.[2]

One criticism of this formulation of adjustment costs is apparent from an

---

[2] A relatively minor point is that partial adjustment models are sometimes derived as the outcome of two optimization problems: the one determining the 'desired' portfolio and the other determining the 'actual' portfolio. This approach may be pedagogically convenient but it is logically rather unsatisfactory. Why should agents optimize to choose their actual portfolio, treating as fixed their desired portfolio which is itself the outcome of an optimization problem? The status of such a desired portfolio is unclear. In this paper therefore, I assume that agents carry out a single optimization problem subject to various forms of adjustment cost. This leads to a single dynamic decision rule of the partial adjustment form, albeit with some differences in the exact interpretation of coefficients.

inspection of the cost function (C). The total change in asset holdings consists of two components: net acquisitions, and capital gains or losses on existing holdings.

$$(x_t - \mu x_{t-1}) W_t = (x_t - \mu q_t x_{t-1}) W_t + (q_t - I) \mu x_{t-1} W_t \qquad (7.11)$$

total change = net acquisitions + capital gains

Costs are therefore incurred either when agents make transactions, or when there is a change in asset prices. It is not apparent that a change in asset prices *per se* is likely to give rise to costs of adjustment. In financial markets, such costs are thought of as being associated largely with bid-asked spreads and other brokerage costs of buying and selling securities. Asset price changes will, in general, provoke some transactions, but it is the transactions rather than the price changes which generate the costs of adjustment. Changes in asset holdings as a result of capital gains and losses are essentially incurred in a passive way. This suggests modifying the cost function to eliminate capital gains and losses:

$$C_1 = (x_t - \mu q_t x_{t-1})' \Psi (x_t - \mu q_t x_{t-1}) W_t^2 \qquad (7.12)$$

Friedman (1977) has argued that this formulation is still deficient in that it fails to differentiate between the costs of reallocating the existing portfolio and the costs of allocating net new inflows of cash. The latter, he argued, would be cheaper to allocate than the former, essentially because brokerage charges would be paid only once for a sale *or* purchase, rather than twice for a sale *and* purchase. Friedman modelled these differential adjustment costs by introducing the 'optimal marginal adjustment model' in which net new cash inflows are allocated differently from the reallocation of the existing portfolio.

Although it looks intuitively appealing, the optimal marginal adjustment model is not altogether easy to rationalize in terms of maximization of a 'sensible' objective function. One possibility is to add to the cost function an extra term representing the total net acquisition of assets. This can be developed as follows. The total change in wealth is given by:

$$W_t - W_{t-1} = F_t + G_t \qquad (7.13)$$

$$\text{with } F_t = i'(x_t - \mu q_t x_{t-1}) W_t = \text{total net acquisitions} \qquad (7.14)$$

$$G_t = i'(q_t - I) \mu x_{t-1} W_t = \text{total capital gains} \qquad (7.15)$$

The extra term in the cost function is:

$$C^* = (x_t - \mu q_t x_{t-1})' \Psi^* (F_t) W_t \qquad (7.16)$$

Ψ^* is a vector of negative elements so that, insofar as acquisitions of any particular asset are positively correlated with net cash inflows, adjustment costs are reduced. However, this proposal is not satisfactory as it does not add any new information to the cost function. Because of the identity (7.14), total flows (F_t) can be replaced by their individual components in (7.16). Summing (7.12) and (7.16) to arrive at total costs produces an expression which is essentially identical to (7.12), and in which the two components of costs, Ψ and $\Psi^* i'$, cannot be separately identified.

A better proposal for modelling differential adjustment costs is to utilize the identity (7.13) and work in terms of capital gains and losses. To simplify the argument, suppose that agents' desired and actual portfolios are equal. Then, unless desired portfolios change, there are just two possible sources of further transactions. The first arises from net inflows or outflows of new funds as we have seen. The second arises because capital gains or losses, even if they have no effect on desired portfolios (which is unlikely), will still, in general, unbalance actual portfolios and cause transactions aimed at rebalancing. Moreover, if transactions costs differ among assets, capital gains or losses will produce costs which will differ depending on where the gains or losses were incurred. As new inflows can be allocated at lower cost, they can be thought of as reducing investors' costs of portfolio adjustment. Capital gains or losses, on the other hand, impose a particular pattern of transactions on the portfolio and therefore can be thought of as increasing investors costs of adjustment. These arguments suggest defining an extra term in the cost function as:

$$C_2 = (x_t - \mu q_t x_{t-1})' \Phi (q_t - I) \mu x_{t-1} W_t^2 \qquad (7.17)$$

with Φ being a positive semi-definite matrix. Thus, if transactions are positively correlated with capital gains, costs are increased, and vice versa.

Under this hypothesis, total adjustment costs are the sum of the two components $C_1 + C_2$. C_1 are the costs associated with transactions in general; C_2 are the additional costs incurred when transactions are associated with capital gains or losses rather than net new inflows. It should be emphasized that these cost functions do not replicate Friedman's optimal marginal adjustment model. They do, however, provide an alternative method, grounded in optimizing behaviour, of differentiating between adjustment costs arising from two different sources.

The next step is to derive portfolio demand functions and APEX equations using the differential adjustment cost function $(C_1 + C_2)$ in place of the adjustment cost function (C) which was used in section 7.2.

Omitting details, the asset demand functions (7.5) are now specified as:

$$x_t = Z_t y_{t+1} + \delta Z \Psi \mu q_t x_{t-1} - \delta Z \Phi (q_t - I) \mu x_{t-1} + \Omega^{-1} i (i' \Omega^{-1} i)^{-1} \qquad (7.5')$$

with Z defined as before
and $\Omega \equiv \gamma \Sigma + \delta \Psi$
Likewise, the APEX equations (7.10) are:

$$\begin{aligned}
y_{1t+1} - i y_{nt+1} &= (\Omega_{1n} - i\omega_{nn}) - \delta(\Pi_{1n} - i\pi_{nn})\mu q_n - \delta(\Phi_{1n} - i\phi_{nn})\mu \\
&\quad + \hat{\Omega} x_{1t} - \delta \hat{\Pi} \mu q_{1t} x_{1t-1} - \delta \hat{\Phi} \mu x_{1t-1} \\
&\quad - (\Pi_{1n} i' - i i' \pi_{nn})(q_{1t} - I q_{nt}) \mu x_{1t-1} \qquad (7.10')
\end{aligned}$$

with $\hat{\Pi}' \equiv \hat{\Psi} - \hat{\Phi}$
$\hat{\Omega} \equiv \gamma \hat{\Sigma} + \delta \hat{\Psi}$
$\hat{\Sigma} \equiv \Sigma_{11} - i\Sigma_{n1} - \Sigma_{1n} i' + i i' \sigma_{nn}$
$\hat{\Psi} \equiv \Psi_{11} - i\Psi_{n1} - \Psi_{1n} i' + i i' \psi_{nn}$
$\hat{\Phi} \equiv \Phi_{11} - i\Phi_{n1} - \Phi_{1n} i' + i i' \phi_{nn}$ as before

As compared with the original specification of adjustment costs, the new specification introduces additional terms in lagged portfolio shares and in capital gains and losses. However, a more important point stands out from a comparison of the modified portfolio demand functions (7.5′) with the modified APEX equations (7.10′). A major drawback of generalized partial adjustment models is that they lead to portfolio demand functions in which the underlying behavioural content is obscured rather than clarified. In equation (7.5′), the economic content is contained in the matrices Σ, Ψ and Φ (which should be symmetric and positive semi-definite) and in the coefficients γ and δ (which should be positive). However, the observable coefficients are unfriendly non-linear functions of these underlying parameters. Symmetry of Z is easily tested though its rejection could be due either to non-symmetry of Σ or Ψ. The products $Z\Psi$ and $Z\Phi$ are not, in general, symmetric and it is exceedingly difficult to test separately the symmetry of Σ, Ψ or Φ. The same general problem appears in Friedman's optimal marginal adjustment model where a high proportion of the coefficients cannot be signed a priori. In short, portfolio demand functions incorporating adjustment costs produce few economic restrictions which can be tested easily.

A central advantage of the APEX approach is that it delivers behavioural restrictions in a much more easily testable form than in asset demand functions. Thus in equation (7.10′) the modified matrix $\hat{\Phi}$ appears directly (up to the factor of proportionality δ), and $\hat{\Sigma}$ and $\hat{\Psi}$ can easily be computed from identities using straightforward linear operations. Thus, separate tests for the symmetry of $\hat{\Sigma}$, $\hat{\Psi}$ and $\hat{\Phi}$ can be carried out easily by the imposition of simple linear (cross-equation) restrictions.

The form of equation (7.10′) also makes it easy to test hypotheses about the exact nature of adjustment costs. If adjustment costs are differential as between transactions induced by new flows and those induced by capital gains and losses, then the principal restrictions acceptable by the adjustment cost coefficients in equation (7.10′) will be those of symmetry of the matrices $\hat{\Psi}$ and $\hat{\Phi}$. In addition, the elements of the vector $(q_1 - Iq_n)\mu x_{1t-1}$ should have parameters which are all identical to that on μq_n in any given equation. (This follows from the structure of $II_{1n}i' - ii'\pi_{nn}$.) These latter restrictions, I will label 'numeraire restrictions' as they are associated with costs of adjusting the numeraire asset.

Other hypotheses about adjustment costs can now be interpreted as special cases of the differential adjustment cost model. There are three such hypotheses which are likely to be of most interest. Consider again the rationale for differential adjustment costs. They are thought to arise because transactions can be segregated into two kinds: one involving one-way costs (associated with a sale or a purchase) and the other involving two-way costs (associated with a sale and a purchase). In the general model, the costs associated with one-way transactions may differ from the additional costs associated with two-way transactions (i.e. $\Psi \neq \Phi$). The first special case therefore is the hypothesis that these two sets of costs are, in fact, equal. I call this the 'double-cost' hypothesis. It implies that $\Psi = \Phi$; $\hat{\Psi} = \hat{\Phi}$; and hence, $II = \hat{II} = 0$. Note that, in this case, the model reverts to the basic generalized partial adjustment model which can thus be reinterpreted as defensible against our earlier criticism. The second special case is that which asserts that there are, in fact, no differential adjustment costs. I call this the 'unified-cost' hypothesis; it implies that $\Phi = \hat{\Phi} = 0$, and the model reverts to the partial adjustment kind with a cost function given by C_1 (7.12). The third special case is that of no adjustment costs at all, i.e. $\Psi = \Phi = \hat{\Psi} = \hat{\Phi} = 0$. In this case, the terms involving lagged portfolio shares and wealth (x_{1t-1}, μ) do not contribute significantly to explaining asset returns.

In summary, different hypotheses about adjustment costs introduce into APEX equations different variables involving lagged portfolio shares rescaled in various ways by changes in asset prices and in wealth.

7.4 EMPIRICAL METHODOLOGY

My procedure for examining the empirical significance of adjustment costs is, as far as possible, that of nested hypothesis testing. This involves first estimating a general 'maintained' model and then examining successively more restricted versions of this model. In the present context, the maintained hypothesis is provided by equations (7.10') but ignoring restrictions implied by any particular interpretation of the coefficients. These equations state that asset returns relative to the numeraire can be expressed as a vector of constants plus linear combinations of current and lagged portfolio shares and the inverse of the change in wealth. The first step is to estimate these equations without restrictions on the coefficients. Subsequent steps involve testing the restrictions on coefficients implied by the mean-variance model with adjustment costs.

Adjustment cost specifications were examined in detail in Section 7.3. In brief, I test for the numeraire restrictions associated with the costs of adjusting the numeraire asset. I also test for the symmetry of Ψ and Φ and (separately) for the restrictions that $\Phi = \Psi$; $\Phi = 0$; and $\Psi = \Phi = 0$.

Turning now to the mean-variance aspects of the model, $\hat{\Sigma}$ is a subjective covariance matrix (of the relative returns $y_{1t+1} - iy_{nt+1}$) and should therefore be symmetric and positive semi-definite. In addition, I can, as indicated earlier, test whether agents' subjective covariance matrix $(\hat{\Sigma})$ coincides with the true covariance matrix of the data $(\hat{\Lambda})$. These matrices are equal if the covariance matrix of regression residuals $(\hat{\Lambda})$ is proportional to the coefficient matrix $(\gamma \hat{\Sigma} = \hat{\Omega} - \delta \hat{\Psi})$, with the factor of proportionality equal to the coefficient of relative risk aversion. This may be interpreted as a test for rationality of expectations. Since $\hat{\Lambda}$ is a covariance matrix, these restrictions are nested in the symmetry of $\hat{\Sigma}$. Thus, logically, a test for mean-variance efficiency *and* rationality of expectations is more restrictive than one for mean-variance efficiency alone. Note that, in this framework, rejection of rationality does not necessarily imply rejection of unbiasedness (which is untestable). A finding that agents were not using the true covariance matrix in their decisions, would, in my view, impel some doubts about the assumption of unbiasedness, but it would by no means be fatal to this assumption. Similar considerations apply to the error orthogonality assumption underlying (7.10′): non-proportionality of $\gamma\hat{\Sigma}$ and $\hat{\Lambda}$ is disturbing but not fatal to this assumption. These considerations explain why we chose to use assumptions 4(i) and 4(ii) in Section 7.2 in preference to that of rational expectations.

The final phase of the tests consists of checking that agents are *not* risk-neutral ($\gamma \neq 0$); that is, that portfolio shares do contribute significantly (albeit perhaps, in a restricted way) to the explanation of asset returns. This is a particularly important test. It corresponds to the test on portfolio demand functions that interest rates contribute significantly and in a finite way to explaining asset demands. To assert that agents are risk-averse is to assert that assets with imperfectly correlated (rationally expected) returns are less than perfect substitutes for one another.

There are a large number of hypotheses to be tested and they cannot be nested in any unique way. This gives rise to the possibility that test results may differ depending on the route chosen from least restrictive to most restrictive hypothesis. This possibility is inherent in testing multiple hypotheses whenever the tests cannot

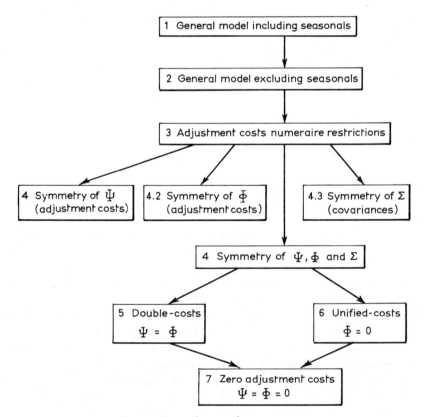

Fig. 7.1 Test schema: adjustment costs

Empirical methodology

be nested in a unique way. I therefore carried out the tests along a variety of different routes. It soon became clear however that there was no serious ambiguity in the test results. For any given hypothesis test, the accept/reject decision at the 95% significance level produced the same result no matter which particular test route was chosen. Accordingly, only a very limited set of test results are reported in this paper.

The first block of tests to be reported are set out in Fig. 7.1 and are concerned primarily with adjustment costs. I started by estimating the maintained (unrestricted) model and, as the data are seasonally adjusted, these regressions include seasonal dummies. To reduce the size of the problem, I tested (zero) restrictions on the seasonals before testing any other hypothesis. I next tested the numeraire restrictions and all the symmetry restrictions. The latter are fundamental to the optimizing basis of the model and are generally regarded as relatively weak sets of restrictions. Finally, I tested the three restricted adjustment cost hypotheses: double-costs; unified costs; and zero costs.

In an earlier paper (Green, 1987a), I reported in detail on the results of estimating APEX equations implied by the mean-variance model with the unified costs of adjustment hypothesis. In this present chapter therefore, I summarize only briefly the results concerned with mean-variance aspects of the model. As it happens, the data are not able to discriminate very clearly among the different adjustment cost hypotheses. For illustrative purposes therefore, in reporting the mean-variance results, I concentrate on the version of the model which incorporates the double costs of adjustment hypothesis.

The nesting of this second block of tests is set out in Fig. 7.2. I begin by imposing

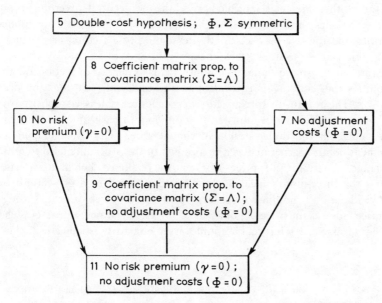

Fig. 7.2 Test schema: mean-variance efficiency and rational expectations

symmetry on the matrix of adjustment costs ($\hat{\Phi}$) and the subjective covariance matrix ($\hat{\Sigma}$). This corresponds to model 5 in Fig. 7.2. I then test successively for the rational expectations restrictions (proportionality of the coefficient matrix, $\gamma\hat{\Sigma}$, and the residual covariance matrix, $\hat{\Lambda}$) and the existence of a risk premium ($\gamma \neq 0$). For completeness, I include the tests for zero adjustment costs.

7.5 THE DATA

The data is described in detail in Green (1982, 1984). It consists of comprehensive monthly integrated balance sheet and flow data for the UK together with monthly rates of return. The estimation period covers 1972:7–1977:11. The first date follows immediately after the floating of sterling; the final date marks the end of this particular dataset.

There are four assets: equity claims on fixed capital, foreign currency denominated, government securities, and liquid assets. The latter is a consolidation of all short-term paper: currency, bank deposits and loans, bills and national savings. As far as possible these are all valued at market prices. Thus, equities are valued using Tobin's q; foreign currency using the dollar-sterling spot exchange rate; and bonds using a specially constructed price index which is consistent with the underlying maturity structure of the bond stock. Banks' holdings of securities are adjusted from book value to market value using formulae derived from their accounting procedures. For further details see Green (1982).

The portfolio shares are those of the consolidated private sector consisting of banks and other financial intermediaries, the non-central-government public sector, companies, and households. The supplies of assets by the monetary authorities and the overseas sector (the remaining two sectors) are assumed to be exogenous, as is the stock of fixed capital.

The rates of return are actual one-month yields as defined in Section 7.2 above. For equities the 'coupon' is equivalent to the dividend yield and the price is Tobin's q. The return on foreign currency denominated assets consists of the three-month Eurodollar rate adjusted to a one-month basis and modified by the actual change in the dollar-sterling exchange rate.[3] The average coupon on government bonds is generated as a by-product of the procedure used to estimate the average valuation ratio (price) of the stock of bonds. For liquid assets, the return is the three-month local authority rate adjusted to a one-month basis. These yields are all modified to be expressed as real rates of return using the retail price index as the price of goods and services. The numeraire asset is chosen to be liquid assets. I interpret the liquid asset rate to be stochastic within the

---

[3] Green (1982) explains why it is preferable to use the dollar-sterling rate rather than a weighted average exchange rate.

one-month decision period of the model. Accordingly, the numeraire is a risky asset.[4]

In summary then, the model consists of three regression equations with left-hand side variables consisting of the difference between the one-month return on bonds and that on liquid assets; the one-month return on foreign currency denominated assets and that on liquid assets; the one-month return on equities and that on liquid assets. Implicitly, the regressions seek to explain the one-period ahead expectations of the price of bonds, the spot exchange rate, and the price of equity. The explanatory variables in each equation in the maintained model amount to fifteen in number consisting of: the shares of bonds, foreign-currency denominated assets, and equities in private sector wealth; these shares lagged one period and rescaled by various combinations of the gross rate of capital gains (q_t), the inverse of the change in wealth (μ), and terms in the excess rate of capital gains over those on liquid assets $(q_i - q_n)$; the product of μ and the gross rate of capital gains on liquid assets; μ; and a constant.

7.6 ESTIMATION PROCEDURES

The restrictions associated with the various forms of adjustment costs are all linear, though some are cross-equation restrictions calling for a systems estimator under the null hypothesis. The main estimation and testing problem concerns the rational expectations restrictions which require that the difference between two matrices of coefficients be proportional to the error covariance matrix of the equations. These are non-standard non-linear restrictions which are awkward to impose directly. It turns out however, that maximum likelihood estimation under these restrictions can be carried out by a simple three-step iterative procedure which can be implemented in most 'canned' regression packages. For details see Green (1987b).

In carrying out estimation and hypothesis tests therefore, I adopted the following strategy. First, the model was estimated under all possible null hypotheses by Full Information Maximum Likelihood (FIML) which, in several cases, was simply equivalent to OLS on each equation separately. Next, likelihood ratio tests were carried out in accordance with the test structures given in Figs. 7.1 and 7.2. Likelihood ratio tests are justified by asymptotic considerations and are known to be too large on occasion in small samples. Monte Carlo evidence suggests that a degrees of freedom adjustment is sufficient to make the likelihood ratio test 'about' the correct size. (For further details see Bewley, 1983.) At the final stage therefore, the $(T - K)/T$ small sample correction of Bohm, Rieder and Tintner

---

[4] This is not unreasonable given that a high proportion of liquid assets as defined here have a maturity in excess of one month. It is possible to test whether the numeraire is safe. However such tests involve a considerable increase in the complexity of the nesting of hypotheses and they are therefore deferred to a subsequent paper.

(1980) was made to all the likelihood ratio statistics. (Here, T = number of observations, K = number of regressors in each equation.) Only in one case did this small-sample correction affect the accept/reject decision of the hypothesis to be tested and this was in the relatively unimportant test of the significance of the seasonal dummies. In the next section, I report only the results of the small-sample adjusted likelihood ratio tests.

7.7 EMPIRICAL RESULTS

The maintained model is (deliberately) overfitted and the parameter estimates for this model are not of much interest. A few summary statistics are given in Table 7.1. The most important of these are the LM tests for residual autocorrelation. These suggest that there remains some autocorrelation in the residuals of the equity yield and, to a lesser extent, the bond yield equations. Subsequent tests on more restricted models suggested that the estimated residual autocorrelations may in part have been caused by overfitting. LM tests on the double-costs model showed an absence of simple third-order autocorrelation. Conversely, the unified-costs model broadly shared the properties of the maintained model: some evidence of simple third-order autocorrelation but none of vector first-order autocorrelation. I did not attempt to estimate models with longer lags because of the difficulty in finding a rigorous economic justification for such lags. The addition of extra lags to the maintained model would make it considerably more difficult to find a sensible route by which to simplify the model, and the use of purely statistical criteria can easily give rise to nonsensical results. The advantage of the set-up in the present paper is that economic theory provides clear (if still not unique) routes by which to simplify

Table 7.1 Estimated APEX equations: summary statistics of unrestricted model

	\multicolumn{6}{c}{Equations}
	IG_{t+1}		IE_{t+1}		ID_{t+1}	
	Statistic	Significance level	Statistic	Significance level	Statistic	Significance level
R^2	0.56		0.59		0.38	
$LM_1; \chi^2(3)$	8.57	96.4	18.47	99.9	2.46	51.7
$LM_2; \chi^2(3)$	3.56	68.7	3.85	72.2	1.83	39.2

$IG_{t+1}, IE_{t+1}, ID_{t+1}$ = One month returns on bonds, equities and foreign currency assets (respectively) less the one period return on liquid assets.

LM_1 = Lagrange multiplier test for simple third-order autocorrelation and significance level of χ^2 statistic.

LM_2 = Lagrange multiplier test for vector first-order autocorrelation and significance level of χ^2 statistic.

Empirical results

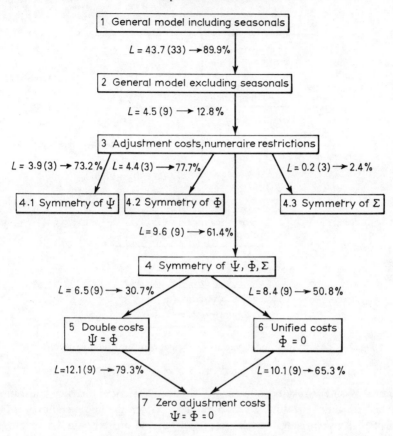

Fig. 7.3 Test results: adjustment costs

the maintained model. Moreover, the maintained model is already quite sizeable, with 78 parameters, including seasonals, to be estimated.

The results of the hypothesis tests are shown in Figs. 7.3 and 7.4 which are drawn on the same basis as Figs. 7.1 and 7.2 (respectively). Now however, continuous lines linking hypotheses refer to hypotheses which are accepted at the 95% level. Broken lines refer to hypotheses which are rejected at this level. Figures 7.3 and 7.4 show the results of each test in the form: $L = V(N) \to S$. V is the value of the chi-squared statistic; N is the degrees of freedom of the test; and S is the significance level of the test in percent.

Considering the adjustment costs first (Fig. 7.3), it is clear that, though the data easily accept symmetry restrictions, they also accept the restrictions that adjustment costs as specified here do not contribute significantly to explaining asset returns. No matter which test route is chosen the data easily accept these restrictions. In terms of the regression framework, this implies that lagged portfolio shares do not help explain relative asset returns.

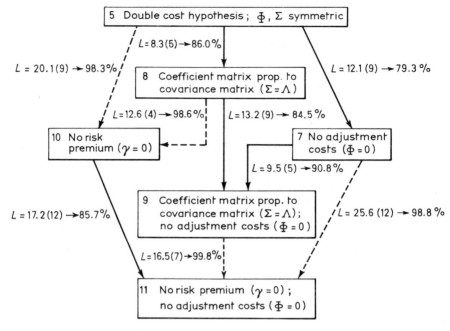

Fig. 7.4 Test results: mean-variance efficiency and rational expectations

Figure 7.4 shows results concerned with mean-variance efficiency and rational expectations under the double-cost hypothesis. Given the findings on adjustment costs, it is not surprising that the results in Fig. 7.4 are very similar to those in Green (1987a) in which the unified-cost hypothesis was the main object of analysis. In summary, the data are able to accept the restrictions of no adjustment costs and rational expectations, but reject the hypothesis that there is no systematic risk premium. The latter finding implies that current (time t) portfolio shares do contribute significantly to explaining asset returns (between time t and $t+1$). Moreover, the data are consistent with the rational expectations hypothesis of equality between agents' subjective covariance matrix and the objective covariance matrix of the data.

Some parameter estimates are provided in Table 7.2. The parameters of principal interest are the matrix of adjustment costs ($\hat{\Phi}$), the subjective covariance matrix ($\hat{\Sigma}$), and the coefficient of relative risk aversion (γ).[5] Two sets of estimates of these parameters are shown in Table 7.2. They correspond to the models numbered 5 and 8 in Fig. 7.4, i.e. the double-cost hypothesis with symmetry restrictions, and then with rational expectations restrictions.

[5] The parameter δ, 'the coefficient of aversion to adjustment costs' is essentially a normalizing factor which cannot be separately identified. To present the estimates of the matrix of adjustment costs, δ is set to unity.

Table 7.2 Estimated APEX equations: double-cost hypothesis

	Matrix of adjustment (costs = Φ)			Subjective covariance matrix (risk-aversion coefficient = $\gamma\hat{\Sigma}$)			Coefficient of relative risk aversion = γ	True covariance matrix = $\hat{\Lambda}$		
	Coefficient on:			Coefficient on:				Residual covariance of:		
	$-\dfrac{G_{t-1}}{W_t}$	$-\dfrac{E_{t-1}}{W_t}$	$-\dfrac{D_{t-1}}{W_t}$	$\dfrac{G_t+G_{t-1}}{W_t}$	$\dfrac{E_t+E_{t-1}}{W_t}$	$\dfrac{D_t+D_{t-1}}{W_t}$		IG	IE	ID
2.1 Symmetry-constrained estimates										
IG$_{t+1}$	0.339			−0.105				0.00235		
IE$_{t+1}$	1.330	0.807		−0.206	−1.179			0.00190	0.00803	
ID$_{t+1}$	−1.230	−1.183	0.523	0.221	0.210	−0.558		−0.00035	−0.00052	0.00049
Log likelihood = 335.6										
2.2 Rational expectations (residual-) constrained estimates: $\hat{\Sigma} = \hat{\Lambda}$										
IG$_{t+1}$	0.076			−0.343				0.00243		
IE$_{t+1}$	0.004	−0.040		−0.268	−1.219		−141.1	0.00190	0.00800	
ID$_{t+1}$	−0.134	0.066	−0.219	0.052	0.072	−0.076		−0.00037	−0.00051	0.00054
Log likelihood = 330.9										

IG$_{t+1}$, IE$_{t+1}$, ID$_{t+1}$ = One period return on bonds, equities, and foreign currency assets (respectively) less the one period return on liquid assets.
G$_t$, E$_t$, D$_t$ = Market values of private sector holdings of government bonds, equities, and net foreign currency assets (respectively).
W$_t$ = Private net worth = G$_t$ + E$_t$ + D$_t$ + liquid assets.

It is immediately apparent that the central problem in these results is that to which I drew attention in Green (1987a). This is the problem that the second-order conditions for a maximum are simply not satisfied. Enforcing the rational expectations restrictions that the subjective and objective covariance matrix are equal produces an estimate of the coefficient of relative risk aversion of -141.1. This is larger in size but still the same sign as my estimate of -101.9 in a model without adjustment costs. There are fewer restrictions to tie down the matrix of adjustment costs and, as estimated, it is neither positive nor negative semi-definite. Under the rational expectations restrictions, two out of three of the on-diagonal coefficients in $\hat{\Phi}$ are wrong-signed (negative). However, three out of six of the adjustment cost parameters change sign in moving from model 5 to model 8. Indeed, given the fact that these parameters are, collectively, not significant, not much importance can be attached to the exact magnitudes of the estimates of the elements in the adjustment cost matrix.

7.8 CONCLUDING REMARKS

Arising out of this investigation of the role of adjustment costs in determining portfolio demands and asset prices are two questions. The first reiterates that raised at the conclusion of Green (1987a): what explains the finding of a significant negative coefficient of relative risk aversion? This is clearly not entirely an accident; it is corroborated by the work of Wills (1984) on the UK in a different setting.

Of more interest in the present chapter however is the second question. Why are lags important in estimated portfolio demand functions and yet, when parameterized as adjustment costs in APEX equations, appear to have negligible importance? A part of the answer to this question has to do with the manner in which the adjustment costs are parameterized in the regression equations. In standard portfolio demand functions, adjustment costs appear as autoregressions in asset *holdings*. If asset markets continuously clear therefore, estimates of these equations cannot readily discriminate between autoregression in asset demands and autoregression in asset supplies. In contrast, APEX equations parameterize adjustment costs as introducing a moving average into the process determining asset *returns*. If markets clear and expectations are rational (which, as we have seen, is partly testable), then, at time t, all the systematic information about the process determining asset supplies is incorporated into actual portfolios and asset prices at time t. This in turn constitutes the information set on which expectations of time $t+1$ asset prices are based, and it is this information set which appears as the explanatory variables in the APEX equations. Since the systematic part of asset supplies is incorporated into current portfolios and prices, that part of the APEX equations which contains terms in lagged portfolios must reflect the costs of adjustment. It follows that APEX equations cannot get confused between asset demands and supplies. I would argue therefore that APEX equations constitute a very much clearer test for the presence of adjustment costs than do standard portfolio demand functions.

As against these considerations, it is also clear that APEX equations depend on strong assumptions. If these fail then the rest of the argument is put in question. In this respect, the present study is certainly far from providing the last word on the subject. Among other issues: I have considered only first-order adjustment costs; I have not explored fully the apparent autocorrelation in asset returns; I have not considered how the maintained assumptions might themselves be tested. Nevertheless, the evidence presented here suggests quite strongly that it cannot be inferred that autocorrelation in asset holdings implies that asset demands are autocorrelated; it may be that it is autocorrelation in asset supplies which is responsible for the persistence.

ACKNOWLEDGEMENT

I thank Mark Goode for energetic research assistance and helpful comments. Helpful comments on earlier versions of this chapter were also provided by Jeffrey Frankel, and participants in the 1986 Economic Modelling Conference. The manuscript was excellently typed by Mrs J. Thomas. The research reported in this paper was funded in part by ESRC grant no. B0023-2151 and in part by a grant from the Jane Hodge foundation.

REFERENCES

Anderson, G. J. and Blundell, R. W. (1982) Testing restrictions in a flexible dynamic demand system: an application to consumers' expenditure in Canada, *University of Manchester Discussion Papers in Econometrics*, ES124.

Backus, D., Brainard, W. C., Smith, G. and Tobin, J. (1980) A model of US financial and non-financial economic behaviour, *Journal of Money, Credit and Banking*, **XII**:2, May, 259–93.

Bewley, R. A. (1983) Tests of restrictions in large demand systems, *European Economic Review*, 20, 257–69.

Black, F., Jensen, M. C. and Scholes, M. (1972) The capital asset pricing model: some empirical tests, in *Studies in The Theory of Capital Markets* (ed. M. C. Jensen) Praeger, New York, pp. 79–121.

Blanchard, O. J. and Plantes, M. K. (1977) A note on gross substitutability of financial assets, *Econometrica* **45**:3, April, 769–71.

Bohm, B., Rieder, B. and Tintner, G. (1980) A system of demand equations for Austria, *Empirical Economics*, 5, 129–42.

Dickens, R. (1987) Variability in some major UK asset markets since the mid-1960s: an application of the ARCH model, in *The Operation and Regulation of Financial Markets* (eds C. A. E. Goodhart, D. Currie and D. T. Llewellyn) MacMillan, London, pp. 231–70.

Frankel, J. A. (1985) Portfolio shares as 'beta-breakers': a test of CAPM, *Journal of Portfolio Management*, (to be published).

Frankel, J. A. and Dickens, W. T. (1984) Are asset demand functions mean-variance efficient?, *NBER Working Paper No. 1113* (revised).

Friedman, B. M. (1977) Financial flow variables and the short-run determination of long-term interest rates, *Journal of Political Economy*, **85**:4, August, 661–89.

Green, C. J. (1982) Monetary policy and the structure of interest rates in the United Kingdom: a flow of funds model 1971–77, Ph.D. Dissertation, Yale University (unpublished).

Green, C. J. (1984) Preliminary results from five sector flow of funds model of the United Kingdom 1972–77, *Economic Modelling*, July, 304–26.

Green, C. J. (1987a) Asset demands and asset prices in the UK: is there a risk premium?, *Mimeo*.

Green, C. J. (1987b) Maximum likelihood estimation of the coefficient of relative risk aversion, *Mimeo*.

Markowitz, H. M. (1959) *Portfolio Selection*, Yale University Press, New Haven.

Nickell, S. J. (1985) Error correction, partial adjustment, and all that: an expository note, *Bulletin of The Oxford University Institute of Economics and Statistics*, May, pp. 119–29.

Roll, R. (1977) A critique of the asset pricing theory's tests: Part I, *Journal of Financial Economics*, **4**, 129–76.

Ross, S. (1978) The current status of the capital asset pricing model, *Journal of Finance*, **33**:3, June, pp. 885–901.

Sargent, T. J. (1979) *Macroeconomic Theory*, Academic Press, New York.

Sharpe, W. F. (1964) Capital asset prices: a theory of market equilibrium under conditions of risk, *Journal of Finance*, **19**:3, September, pp. 425–42.

Tobin, J. (1958) Liquidity preference as behaviour towards risk, *Review of Economic Studies*, **25**:67, February, pp. 65–86.

Tobin, J. and Brainard, W. C. (1968) Pitfalls in financial model building, *Amencal Economic Review Papers and Proceedings*, **58**, May, 99–122.

Wills, H. (1984) Inferring expectations, *Mimeo*.

8

The macroeconomic and sectoral effects of the Economic Recovery Tax Act: some simulation results

FLINT BRAYTON and PETER B. CLARK

The unprecedented size of federal deficits in recent years has stimulated interest in the effects of fiscal policy. The effects of fiscal policy can be classified broadly into two categories:

(1) those relating to the short-run level of economic activity through the effect on aggregate demand and
(2) those relating to the long-run level of economic activity through the effect on aggregate supply.

Almost by definition the long-run consequences are the more important ones, and, at least implicitly, they appear to be the ones at the centre of the current debate on the effects of the deficit. However, the short- and long-run effects of the fiscal changes made in the last several years are difficult to disentangle. The initial manifestation of the long-run consequences of the new fiscal stance – changes in the composition of output – may have been obscured thus far by the short-run changes in composition induced by the fiscal effects on aggregate demand. In fact, it is entirely possible – and the simulation results presented below bear this out – that the fiscal policy adopted in the 1980s has stimulated output in the short run but is likely to reduce potential output in the long run.

In this chapter the focus is on the longer-term effects; the fiscal policy actions we examine are the Economic Recovery Tax Act of 1981 (ERTA) and the changes in ERTA stipulated in the Tax Equity and Fiscal Responsibility Act of 1982 (TEFRA). The main features of ERTA/TEFRA were a cumulative 23% cut in personal income tax rates and a substantial acceleration in depreciation allowances.[1] To estimate the impact of these fiscal measures we used the Federal

---

[1] We do not investigate the impact of the other provisions of ERTA/TEFRA, such as the indexing of personal income taxes, all savers certificates, the liberalization of individual retirement accounts and Keogh plans, and changes in the rules governing the eligibility of leased equipment for the investment tax credit.

Reserve Board's MPS (MIT–Penn–Social Science Research Council) model for simulations covering the period from mid-1981 to mid-1988 with and without these changes in fiscal policy. The MPS model is a quarterly econometric model of the US economy that is especially appropriate for the task because careful attention has been paid to the proper specification of its long-run properties. These properties are described briefly below.

In the MPS model, long-term changes in potential output arising from ERTA/TEFRA depend most importantly on what happens to business fixed investment. A priori, the net long-term change in business investment under ERTA/TEFRA is ambiguous: on the one hand, ERTA/TEFRA contained provisions, mainly accelerated depreciation allowances, that provide direct incentives to business fixed investment. On the other hand, the additional consumption generated by the personal tax cuts raises interest rates (according to the simulation results presented below), and higher interest rates have a negative effect on business investment.

In standard macro models such as the MPS, the short-run effects of fiscal policy actions (given conventional monetary policy assumptions) typically involve substantial changes in the level of aggregate demand. These variations in economic activity make it difficult to interpret the path to the long-run equilibrium position and to distinguish between the transitory and permanent effects of the fiscal action. For example, a positive shock to aggregate demand resulting from a cut in personal taxes will generate new investment via the accelerator mechanism, but the positive effect on investment of such a tax cut is necessarily short lived.

In order to minimize these problems associated with fluctuations in output arising from changes in fiscal policy, estimates of the effects of ERTA/TEFRA were obtained with simulations in which the employment rate was held constant; this constraint was imposed by altering monetary policy to keep the unemployment rate in the simulation with ERTA/TEFRA removed the same as in the baseline simulation, which includes ERTA/TEFRA. We adopted the constraint on unemployment because, in the long-run equilibrium position of the economy, the unemployment rate is not affected by fiscal policy; therefore, keeping the unemployment rate unchanged eliminates the short-run output effects of fiscal policy and thereby facilitates analysis of the impact of the policy change in the long run.[2] The simulations extend through the first half of 1988, and although the US economy does not reach long-run equilibrium at the point, the shifts in the composition of output that occur by then can be used as good indicators of the long-run consequences of ERTA/TEFRA.

The organization of the chapter is as follows. First, we describe the key parts of the MPS model, including the manner in which the changes in fiscal policy associated with ERTA/TEFRA directly affect consumption and investment in the

---

[2] The long-run unemployment rate is sensitive to one policy variable – the ratio of unemployment benefits to after-tax wage income – but this is of no consequence to the analysis reported in this chapter.

model. We then discuss the key simulation results: the impact of the fiscal measures on the composition of output and on such variables as interest rates and exchange rates. Keeping the unemployment rate unchanged, we find that the effect of ERTA/TEFRA is to skew the composition of output sharply toward consumption and way from housing. Under the influence of direct incentives, mainly in the form of accelerated depreciation allowances, business fixed investment at first increases as a share of output despite a rise in nominal and real interest rates. However, the impact of the higher interest rates eventually more than offsets the effects of the investment incentives so that, in the long run, business capital formation also gets crowded out by the additional consumption spending caused by the personal tax cuts. Last, we simulate the effects of ERTA/TEFRA under the more commonly used monetary policy assumption in which the path of M1 is held invariant to the fiscal change. In this case ERTA/TEFRA induces a short-run cyclical upturn; through the accelerator, the upturn generates an increase in business fixed investment that is significantly larger than when the unemployment rate is held fixed. Nevertheless, higher interest rates again eventually reduce capital formation.

8.1 THE MPS MODEL

The theoretical core of the MPS model consists of a production function, the first-order profit-maximizing conditions for labour and capital, the life-cycle model of consumption, and a money demand equation.[3] These equations determine the most important steady-state properties of the model, all of which are quite standard in the context of growth theory. The rate of inflation equals the excess of the growth rate of the money supply over that needed to support the growth rate of real activity; a permanent change in the level of the money supply results in a proportional change in the price level and has no impact on real activity. The growth rate of real output equals the growth rate of the labour force plus the exogenous rate of change of labour input measured in efficiency units (time trend). The level of output per unit of labour input depends on the time trend and on the level of the capital stock. The interaction of the supply of saving of individuals, governments, and the foreign sector with the demand for capital on the part of firms determines the real interest rate and the capital stock. Absent from the model are two potential determinants of output in the long run – the impact of the real interest rate on the supply of private saving and the impact of the real after-tax wage on labour supply. The consumption function is such that the private saving rate in the long run is essentially independent of the rate of interest. The labour force in the long run depends only on demographic variables.

Later in this section we show how the changes brought about by ERTA/TEFRA affect several key sectors in the model, with particular attention to the impacts

[3] A detailed description of the model is presented in Brayton and Mauskopf (1985).

on the supply of saving and the demand for investment. A final subsection brings these pieces of the model together and draws general conclusions about the long-run effects of ERTA/TEFRA.

A brief discussion of the short-run properties of the model is required because the simulated effects of ERTA/TEFRA are also presented for the case of an unchanged path for M1, and the short-run properties are especially important for understanding these results. Lags associated with adjustment costs and expectation formation are imposed on the steady-state structure of the model, and they generate short-run behaviour that is Keynesian.

Wages and prices are sticky, and thus the goods and labour markets do not always clear. Aggregate demand largely determines output. Additional divergences between short- and long-run behaviour are introduced through a 'putty–clay' characterization of producers' durable equipment and through the modelling of firms as oligopolists. In the short run, changes in the supply of money affect real activity through interest rates. Fiscal policy affects real output in the short run through the direct impact of government spending on aggregate demand and less directly through the impact of tax policy on disposable income and investment incentives.

8.1.1 Consumption

The specification of consumption is based on the life-cycle model. Aggregate real consumption, defined as spending on nondurables and services plus the gross rental value of the stock of consumer durables, is explained as a function of income and wealth. Additional equations determine consumer expenditure on new autos and other durables.

The life-cycle model of consumption, augmented to include transfer income, yields a consumption equation of the form

$$\text{CON} = a\text{YLABOUR} + b\text{YTRANS} + cW, \qquad (8.1)$$

where CON is real consumption, YLABOUR and YTRANS are real, permanent, after-tax labour income and transfer income respectively, and W is real wealth. The coefficients a, b, and c are in theory not constants but functions of such variables as the real after-tax rate of interest and the elasticity of substitution between consumption in different periods of time.[4] In estimating the equation, simplifying assumptions have been made about the functional forms of the coefficients. In particular, a and b are assumed to be constants, while c is assumed to be a linear function of the real after-tax rate of interest, R:

$$c = c_0 + c_1 R.$$

The consumption function is further simplified by the restriction $a = c_1$, which

---

[4] For a more detailed discussion of the life-cycle model, see Ando and Modigliani (1963); and Steindel (1981).

is equivalent to assuming that there is zero elasticity of substitution between consumption in different time periods.[5] The following example will make this point clearer. Consider a tax change that reduces the tax rate on property income (thus increasing R) and raises the tax rate on labour income (thus reducing YLABOUR) so that total taxes and after-tax income are both unchanged. The experiment is thus structured so that there is no income effect on consumption; only the substitution effect (from the change in R) is at issue. If $a = c_1$, there is no substitution effect, and the net impact on consumption and saving is zero.[6] Making use of this restriction, the consumption function becomes

$$\text{CON} = a(\text{YLABOUR} + \text{YPROP}) + b\text{YTRANS} + c_0 W,$$

where YPROP (which equals RW) is property income. This is the consumption function in the MPS model, except that most of the explanatory variables are entered in the form of distributed lags to capture expectational and adjustment lags, and wealth has two components – corporate equity and other wealth. The definition of property income includes earnings retained by corporations. The consumption function determines saving as the difference between income (as defined in the equation) and consumption; therefore the important concept of saving in the model is net private saving, which includes retained earnings.

The effects of fiscal policy – especially the effects of a cut in personal taxes financed by bond sales – depend importantly on the extent to which changes in government saving may be offset by changes in other sources of saving. In the case of the response of private saving, two points are important. First, the definition of wealth in the MPS consumption equation includes government debt held by the private sector. It is assumed that future tax liabilities to service government debt are not anticipated by consumers. Because there is no offset in the form of future tax payments, government bonds are treated as net wealth. Therefore the so-called 'Richardian equivalence theorem' does not hold: a cut in personal tax rates that generates a deficit financed by the sale of government bonds does not

---

[5] For a justification of this restriction, see Ando (1974). When the version of the MPS model used for the simulations reported here was estimated, the restriction was supported by the data. The consumption equation has since been estimated using revised data on wealth. As described in Brayton and Mauskopf (1985), the estimates now show that $c_1 < a$, indicating a nonzero but small elasticity of substitution (saving responds positively to R). This change is probably not substantial enough to have much effect on the results presented here.

[6] This point can be shown more formally by combining the consumption function (Equation 8.1) with the identity that saving, or the change in wealth, equals income less consumption, and by assuming that all real magnitudes grow at the rate g in equilibrium. Let s be the ratio of saving to income, where income includes that from labour, transfers, and property, and define k as the ratio of transfer income to total income. Then

$$s = \frac{g[1 - a + k(a - b)]}{g - (a - c_1)R + c_0}.$$

If $a = c_1$, s is independent of R. After the simulations presented here had been performed, the MPS consumption function was re-estimated with new wealth data (see note 5).

lead to an equivalent increase in private saving.[7] Second, as discussed above, the specification of the main consumption function provides no substitution between consumption today and consumption tomorrow. Thus, changes in the real after-tax rate of interest (the relative price of consumption today and consumption tomorrow) have no effect on the private saving rate in the long run.[8]

However, there are several channels in the MPS model through which an increase in real interest rates increases private saving in the short run. First, purchases of consumer durables (which are not included in CON) vary inversely with the real short-term interest rate.[9] Second, the market value of corporate equity, which is a component of household wealth, also varies inversely with the real rate of interest. An increase in the rate of interest, for example, causes the market value of equity to fall, which in turn lowers consumption and increases saving. This effect dies away in the long run as individuals rebuild their wealth to the point where the ratio of wealth to income, which is proportional to the saving rate, has returned to its desired value. A third mechanism generating a positive relationship between saving and the interest rate in the short run operates through the price deflator for CON (PCON), which is a function of the real rate of interest. PCON varies directly with the real rate of interest because the price associated with the component of CON corresponding to the use of the stock of consumer durables is proportional to the real rental rate. This dependence of PCON on the rate of interest affects consumption and saving in the short run by changing the value of real wealth. The latter is calculated as nominal wealth divided by PCON. In the long run this effect will disappear as the short-run adjustments in saving return real wealth to its desired level.

ERTA/TEFRA affects consumption mainly through the cumulative reduction of 23% in personal income tax rates. For a given level of before-tax income, after-tax income is increased, and both consumption and saving will rise to a level that, in the long run, will leave the saving rate unchanged regardless of what happens to the rate of interest. Thus the increase in private saving will offset only a fraction of the reduction in government saving that was due to the personal income tax reduction. In the short run, however, the saving rate should be expected to rise as the higher rate of interest temporarily reduces the value of wealth.

---

[7] The Ricardian equivalence theorem has been popularized by Barro (1974; 1984). For critical comments on this theory, see Tobin (1980); and Haberler (1985). As Haberler points out, Ricardo's view is more accurately described as a nonequivalence theorem because he did not believe that people would fully anticipate future taxes and save an amount equal to the discounted value of future tax liabilities.

[8] The empirical literature varies in the importance it gives to the relationship between saving and changes in the real after-tax rate of interest. Boskin (1978), for example, finds a significantly positive and fairly substantial estimate of the interest elasticity of household saving in the United States. For a countary view, see Friend and Hasbrouck (1983).

[9] Note that changes in purchases of consumer durables affect only the definition of saving that is in the national income and product accounts. The concept of saving consistent with the definition of CON, in which an expenditure on durables is treated as an investment rather than as consumption, does not depend on the level of these purchases. The latter affect only the composition of CON, not its total. In addition, the construction of CON is such that changes in spending on consumer durables will affect saving in the NIPA much less in the long run than in the short run.

8.1.2 Investment and potential output

In the MPS model there are three components of fixed investment – producers' durable equipment, nonresidential structures, and residential structures. The investment equation for each component follows the neoclassical approach in which the desired capital stock is equal to the optimal capital–output ratio times the expected output. (In the case of housing, consumption rather than output is the scale variable.) The optimal capital–output ratio is in turn a function of the price of the investment good relative to the price of output and the cost of capital. It is the lattter variable that has been affected by ERTA/TEFRA. In general terms, the cost of capital, C, is given by

$$C = \left[(1-t)R_n + \delta - \frac{\dot{p}}{p} \right] \frac{[1 - \phi - t(1 - \gamma\phi)\text{PVD}]}{(1-t)}, \qquad (8.2)$$

except in the case of owner-occupied housing, C_H, for which it is

$$C_H = (1-t)R_n + \delta - \frac{\dot{p}}{p}, \qquad (8.3)$$

where

$t =$ tax rate (corporate or personal)
$R_n =$ nominal interest rate
$\delta =$ rate of economic depreciation
$\phi =$ investment tax credit rate
$p =$ price of the investment good
PVD $=$ present value of the depreciation allowance
$\gamma =$ amount by which the tax depreciation base must be reduced if the investment tax credit is taken.[10]

A dot over a variable denotes a time derivative. The direct effects exerted by ERTA/TEFRA on investment are reflected in the values of the exogenous tax parameters in Equations 8.2 and 8.3: t, ϕ, and PVD. Abstracting from changes in output, indirect effects of ERTA/TEFRA on investment occur through changes in the rate of interest.

The major direct change in the cost of capital stemming from ERTA/TEFRA was an increase in the present value of the depreciation allowances provided by the accelerated cost recovery system (ACRS).[11] ERTA/TEFRA also affected the

---

[10] For this exposition we have simplified the model's formulae for the cost of capital by abstracting from the choice between debt and equity finance.

[11] The provisions of ACRS have been incorporated in the model as follows: tax service lives are shortened from 10.5 years to 4.6 years for equipment, from 40 years to 15 years for nonresidential structures, and from 22.5 years to 15 years for rental housing. Declining-balance depreciation rates are increased (increasing PVD) from 150% to 175% for nonresidential structures and from 175% to 200% for rental housing, but they are reduced (reducing PVD) from 200% to 150% for equipment.

cost of capital by increasing the effective rate of the investment tax credit, increasing the amount by which the depreciable base must be reduced if an investment tax credit is taken, and, in housing, reducing the (personal) tax rate at which interest payments can be deducted.[12]

These changes have different effects on the cost of capital for the various types of capital investment. Using values from the third quarter of 1983 for nominal interest rates and other variables in the expressions for the cost of capital, we find that the direct effects of ERTA/TEFRA are a 3.7% reduction for business equipment, a 17.4% reduction for nonresidential structures, a 3.3% increase for rental housing, and an 11.2% increase for owner-occupied housing. Thus the cost of business fixed investment is reduced while that for housing is increased.

These differential effects of ERTA/TEFRA reflect several factors. First, the tax changes raise the cost of capital for housing investment because of the decrease in the personal income tax rate. Most of the housing stock is owned by households and noncorporate businesses, and thus the personal income tax rate is used for t. In the case of business fixed investment, the corporate profits tax rate, which is not affected by ERTA/TEFRA, is the relevant tax rate. Second, owner-occupied housing is more adversely affected than rental housing because the former did not receive the benefits of accelerated depreciation enjoyed by the latter. Third, the changes in depreciation allowances have a larger impact on nonresidential structures than on business equipment and hence lower the cost of structures more than that of equipment.

The results discussed above show that the direct effects of ERTA/TEFRA lower the cost of business capital and raise the cost of housing capital. However, the simulation results presented below indicate that the indirect effects of ERTA/TEFRA tend to increase the cost of capital by raising interest rates. An increase in interest rates is needed to reduce interest-sensitive expenditures and thereby bring saving and investment back into equality. In the case of business capital, the net effect of the direct and indirect influences depends on their relative strengths because they are of opposite sign; in the case of housing, both the direct and indirect effects act to increase the cost of capital.

To understand the long-run effects of ERTA/TEFRA on potential output, a brief description of the relationship between investment and the production function is necessary. The function for labour demand in the model is an inverted form of a Cobb–Douglas production function, in which output depends on labour, capital, and energy inputs. There are several other terms in the equation to capture short-run dynamics. In the long run this equation determines the level of output consistent with the capital stock and full employment of labour. Inputs of both capital and energy are measured not as the observed stocks or flows but rather as average capital–output and energy–output ratios for the existing stock of

---

[12] The effective rate of the investment tax credit was increased from 8.8% to 9.2% because of the shortening of the holding period required to keep the full amount of the investment tax credit. The depreciable tax base must now be reduced by 50% of the investment tax credit. Previously there had been no reduction, except in the early 1960s.

The MPS model

productive capacity, which is assumed to be putty–clay. The average ratios of capital to output and energy to output change over time as new investment is made using the then-current optimal values of the ratios. Because of the putty–clay assumption, the characteristics (ratios) of the old capital stock cannot be altered. The production function relates output and labour input in the nonfarm business sector excluding housing. The measure of capital stock – more precisely, the capital–output ratio – corresponds to producers' durable equipment. The structure of the model is such that the stock of producers' structures does not affect the production function.[13] The stock of residential structures determines housing output, which by assumption requires no labour input. In the discussion below of the long-run effects of ERTA/TEFRA on potential output, the concept of output is nonfarm business output less housing. To the extent that ERTA/TEFRA reduces housing investment, housing output (and gross national product) will also be reduced.

8.1.3 Foreign sector

The external repercussions of fiscal policy actions influence the magnitude of the impact of such actions on the domestic economy; the unprecedented appreciation of the dollar in the early 1980s has focused attention on the importance of this interaction.[14] The personal and business tax cuts under investigation generate an increased demand for resources. In an open economy, part of these additional resources can be supplied from the rest of the world. The greater the elasticity of world demand for dollar-denominated assets with respect to interest rate differentials, the smaller the extent to which interest-sensitive sectors of the economy are squeezed by the additional spending generated by the tax cuts.[15] Also, the larger is this elasticity, the more likely it will be that initially the exchange value of the dollar appreciates in response to an expansionary fiscal policy. While the direction of the effect of fiscal stimulus on a country's exchange rate (holding the quantity of money fixed) is theoretically ambiguous, the fact that the massive tax cuts under ERTA/TEFRA coincided with such substantial real appreciation of the dollar suggests that the elasticity of demand for dollar assets may be quite high.

---

[13] This feature of the model stems from the difficulty in describing how producers' structures enter into the production process. As an approximation, having fixed proportions between structures and the other inputs does not seem too unappealing. Such a view would leave output unaffected by the size of the stock of producers' structures as long as it is above a minimum level.

[14] For a more extensive discussion of the external effects of fiscal policy, see Cohen and Clark (1984), and the references cited therein. Explicit quantitative estimates of the impact of recent US fiscal actions on individual foreign countries are contained in Hooper (1985) and Symansky (1985).

[15] The Mundell–Fleming model assumes that bonds denominated in different currencies are perfect substitutes and that exchange rate expectations are static; in that limiting case, the domestic interest rate remains unaffected by a change in fiscal policy and the interest-sensitive sectors of the economy are not crowded out. Total domestic expenditures change by the amount of the fiscal shift, but real output remains unchanged as the traded goods sector gets squeezed. For a discussion of the Mundell–Fleming model, see Stevens et al. (1984).

In the MPS model, the US exchange rate, which is a weighted average of 10 bilateral exchange rates, is determined endogenously as the value that equilibrates the capital account with the current account. The capital account is generated by equations for domestic investors' holdings of foreign assets and foreign investors' holdings of domestic assets. The demands for these assets depend mainly on the differential between the short-term interest rate on domestic assets and on foreign assets, as adjusted for the expected rate of change of the exchange rate. Several proxies represent the latter: the exchange rate forward premium, past changes in the exchange rate, and the deviation of the real exchange rate from a level assumed to be consistent with purchasing power parity. The parameter values are such that expectations for the exchange rate are regressive: a movement of the exchange rate in one direction yields the expectation that it will then tend to move back in the opposite direction. The current account is primarily determined by the levels of real activity here and abroad and by the relative price of domestic and foreign goods.

The response of the exchange rate to a particular change in fiscal policy depends on the relative magnitudes of the resulting movements in the current account and in the capital account. In the case of a stimulative fiscal policy, for example, the capital inflow will tend to be larger, and the exchange value of the dollar will be more likely to appreciate, the larger is the resulting interest rate differential favouring dollar-denominated instruments and the larger is the responsiveness of asset preferences to that differential. The exchange value of the dollar will be more likely to depreciate the more sensitive are imports to the level of domestic activity. In our main simulation of ERTA/TEFRA, however, monetary policy is adjusted to keep the level of domestic activity approximately unchanged. There is, therefore, no income effect on imports and, consequently, the exchange rate can be expected unambiguously to increase. The appreciation of the dollar will produce a deficit in the current account that is consistent with the capital inflow, and it will set up the expectation of a depreciation that dampens to some extent the desired shift of capital.

Foreign interest rates may, of course, vary with US interest rates. In fact, the response of the dollar's exchange value to US policy actions depends crucially on how foreign policymakers react to economic conditions in the United States.[16] The average historical behaviour of several key foreign variables is embodied in the MPS model in rather elementary equations that include US macroeconomic variables as determinants. In the case of the foreign short-term interest rate, the estimated response of the foreign rate to the US short-term interest rate is approximately one-to-one in the long run. However, because the one-to-one ratio seemed to be at the upper bound of a plausible elasticity, we adjusted the equation so that the foreign rate moves by only half the amount that the US rate moves. Thus it was assumed that US policy actions can open up a permanent differential between US and foreign interest rates.

---

[16] This point is brought out strongly by Hooper (1985) and by Symansky (1985).

The MPS model

In the portfolio equations that determine the capital flows, the stock demand for assets depends upon the interest rate differential as adjusted for the expected change in the exchange rate. Thus a permanent change in the differential yields a change in the net capital flow primarily along the adjustment path to the new equilibrium level of the stocks. Once the stocks are in balance, the change in the net flow will continue only to the extent that the new level of the stocks and the new pattern of interest rates require a slightly different net flow to maintain the new stock equilibrium. As a result, the foreign sector will not be a permanent source of saving at a constant interest rate differential.[17] However, the specification of these portfolio equations linking *stock* asset demands to interest rates is parallel to the structure of the rest of the model, in which the impact of fiscal policy on the level of domestic interest rates ultimately depends on the magnitude of the *stock* of outstanding debt that must be held in domestic portfolios. The latter equals the amount of private capital that is displaced by government debt for a given level of wealth. Thus, the greater the interest rate sensitivity of world demand for dollar-denominated assets, the less will be the change in US interest rates and the smaller the impact on the private capital stock in the United States of a permanent shift of fiscal policy, defined as a permanent change in the stock of government debt.

8.1.4 The long-run consequences of ERTA/TEFRA

Before describing the simulation results, it is useful to bring together the previous sections and present some general conclusions about the long-run effects of ERTA/TEFRA. The fiscal shift generated by these tax laws results in a stock of government debt that is permanently higher than otherwise would have been the case.[18] The real interest rate must rise to reduce the size of the private capital stock so that a larger share of domestic wealth will be available for investment in government securities. The rise in the real interest rate does not permanently increase private saving, and thus it does not increase the amount of wealth that individuals desire to hold. The increase in the interest rate does generate an inflow of foreign wealth that can absorb part of the increase in the government debt. Although the size of the private capital stock must decline, the impact on the stock of business fixed capital is ambiguous. The cost of business fixed capital will tend to rise because of the higher interest rate; it will tend to fall because of the direct effects of the new tax laws on the present value of depreciation allowances and on the other tax parameters. The housing stock is unambiguously

---

[17] However, experiments with the MPS model indicate that the period over which the asset stocks respond to a permanent shift in the interest rate differential is quite long – at least 10 years.

[18] In this discussion the references to stocks of debt and wealth should be interpreted as stocks relative to some measure of the scale of the economy such as GNP or income. We postpone a discussion of the long-run stability of the fiscal shift embodied in ERTA/TEFRA except to note that if the policy change pushes the real after-tax interest rate above the real rate of growth of output, then to achieve a stable ratio of government debt to GNP, the government budget deficit exclusive of after-tax interest payments must be in surplus.

reduced; both the higher interest rate and the direct effects of the tax changes raise the cost of housing capital.

8.2 EMPIRICAL ESTIMATES

In this section we focus on changes in the composition of real GNP resulting from ERTA/TEFRA. Table 8.1 shows recent data for the shares of the major components of expenditure in real GNP. In certain cases these data appear to be consistent with the hypothesis that ERTA/TEFRA had a substantial impact on the composition of expenditures. For example, in the years 1982–84 there was a significant increase, relative to recent historical experience, in the proportion of real GNP devoted to consumption. This higher proportion may have been caused in part by the cuts in personal taxes that were part of ERTA/TEFRA. Also, between 1981 and 1984 there was a substantial drop in the share of real GNP represented by net exports of goods and services. This decline is consistent with the hypothesis that the fiscal stimulus associated with ERTA/TEFRA opened up an interest rate differential that induced a capital inflow and an appreciation of the dollar to yield a corresponding current account deficit.

Other data shown in Table 8.1, however, are not obviously consistent with the hypothesis that ERTA/TEFRA had a major impact on the composition of output. Relative to 1980 and 1981, gross and net investment first fell as a share of output but then increased. In addition, the share of output going to residential investment rose in 1983 and 1984 above the proportion in 1980 and 1981 despite the continued high level of real interest rates, which are associated in part with the fiscal stimulus. Obviously, short-run cyclical developments, and possibly other factors too, are at work that mask the longer-run effects of ERTA/TEFRA on these two sectors and probably on the consumption and the foreign sectors as well.

Table 8.1 Percentage composition of real gross national product[a]

Component	Average, 1960–79	1980	1981	1982	1983	1984
Personal consumption	61.3	63.2	62.9	65.1	65.8	64.8
State and local government purchases	12.3	12.1	11.7	11.9	11.4	11.0
Federal government purchases	10.0	7.2	7.3	7.9	7.6	7.5
Business fixed investment						
Gross	10.1	11.2	11.6	11.3	11.1	12.5
Net	3.0	2.8	2.9	2.0	1.9	3.6[b]
Residential investment	4.4	3.2	2.9	2.6	3.5	3.7
Net exports of goods and services	1.1	3.4	2.9	2.0	0.8	−0.9

[a] Excludes inventory changes.
[b] Estimate.

8.2.1 Effects with the unemployment rate held constant

Quantitative estimates of the effects of the change in fiscal policy can be obtained only through a simulation exercise. Using the MPS model, we computed the effects of ERTA/TEFRA as the difference between two simulations: a baseline case and a simulation assuming ERTA/TEFRA had not been enacted. The baseline simulation used historical values for variables through the third quarter of 1983 and then projected values through the first half of 1988. The projection portion of the baseline simulation has the following broad characteristics: the unemployment rate gradually falls from 9% in the second half of 1983 to 7% by the end of the period; the inflation rate is fairly stable in the range of 4 to 5%; and the federal funds rate generally lies in the range of 9 to 11%.[19] In the second simulation, we removed the cuts in personal taxes as well as accelerated depreciation and the other parts of ERTA/TEFRA related to business taxation. The federal funds rate was adjusted to keep the unemployment rate the same in both simulations.[20] This holds labour input constant; however, the level of real GNP can change through the production function to the extent that the capital stock is altered by the change in fiscal policy.

Conceptually, then, the experiment that was undertaken to measure the empact of ERTA/TEFRA at an unchanged unemployment rate necessarily involved a change in the mix of policies; in the second, counterfactual simulation the removal of the fiscal stimulus was counteracted by an expansionary monetary policy in order to keep the unemployment rate the same as in the baseline simulation. This procedure allows us to look at the compositional effects of the fiscal shift in the absence of short-run cyclical movements in output and to draw conclusions about the long-run consequences of ERTA/TEFRA.

It is important to note that the adjustment to the new equilibrium is still far from complete at the end of the first half of 1988 even though the simulation period extends for nearly seven years and even though the constant unemployment rate puts the model on a path that converges more rapidly to the long-run equilibrium than otherwise would be the case. The time required for all capital stocks to converge to the new equilibrium is quite lengthly, in excess of twenty years.

Table 8.2 reports the impact on the composition of output generated by ERTA/TEFRA with the compensating monetary policy. The results indicate that the fiscal shift has had, and will continue to have, a significant effect on the composition of real GNP. Not surprisingly, the proportion of output devoted to consumption (Table 8.2, line 1) rises as a result of the cut in personal taxes

---

[19] The characteristics of the baseline simulation are not too important to the multiplier results presented except to the extent that nonlinearities in the model cause the size of the multipliers to be dependent on the initial conditions. The MPS model is only moderately nonlinear, with the unemployment and interest rates being the principal variables behaving in a nonlinear fashion.

[20] The structure of the model is such that the quantity of money has little impact on the rest of the model except through interest rates. Thus the monetary policy assumed in this second simulation is most clearly characterized by the pattern of the federal funds rate.

Table 8.2 Simulated changes in the composition of real GNP generated by ERTA/TEFRA and a compensating monetary policy[a] (percentage points except as noted)

Component	1982	1983	1984	1985	1986	1987	1988[b]
1 Consumption	0.2	0.4	0.7	1.1	1.4	1.6	1.7
2 Business fixed investment	0.3	0.4	0.2	−0.1	−0.3	−0.4	−0.6
3 Residential investment	−0.3	−0.9	−1.0	−1.1	−1.3	−1.2	−1.1
4 Net exports of goods and services	−0.1	−0.2	−0.2	−0.2	−0.2	−0.2	−0.2
Memo: Impact on other variables							
5 Corporate bond rate	1.3	1.3	1.8	1.9	1.9	1.9	2.0
6 Inflation rate	−0.2	−0.2	−0.6	−0.4	−0.5	−0.5	−0.5
7 Federal budget (national income and product accounts, billions of dollars) (deficit, −)	−51	−106	−147	−177	−209	−237	−259
8 Exchange value of the dollar (%) (appreciation, +)	3.8	2.3	6.1	3.5	5.8	6.9	8.4
Capital stock							
9 Business (billions of 1972 dollars)	2.8	8.2	12.3	11.2	5.9	−1.3	−7.9
10 Percentage	0.2	0.6	0.9	0.8	0.4	−0.1	−0.5
11 Residential (billion of 1972 dollars)	−1.9	−11.5	−24.6	−39.5	−56.1	−74.0	−86.8
12 Percentage	−0.2	−1.0	−2.2	−3.4	−4.3	−5.9	−6.7

[a] Monetary policy was adjusted to hold the level of unemployment unchanged between simulations. The net change in shares for a year does not sum to zero because inventories and the government sector are not included.
[b] Average of values for first and second quarters.

under ERTA/TEFRA. The share of consumption gradually rises, reaching an increase of 1.7 percentage points at the end of the simulation period.[21] The continuous increase in the consumption share reflects two things: the multistep nature of the personal tax cut and the fact that, in the consumption equation, the response of consumption to a permanent increase in after-tax labour income rises gradually from the initial response as additional wealth is accumulated.

The share of output going to business fixed investment first increases but then declines starting in 1985. Thus when accelerator effects are removed (as was done by design in this simulation experiment), the MPS model indicates that the liberalized depreciation allowances and other features of ERTA/TEFRA affecting the business sector more than offset the negative impact of higher nominal and real interest rates generated by the personal tax cuts in the first few years of the ERTA/TEFRA regime. However, this positive effect is short lived; the longer-run impact on business fixed investment is negative as the increase in interest rates progresses. The growing impact on the corporate bond rate (Table 8.2, line 5). reflects the fact that with an increasing share of GNP going to consumption,

---

[21] To obtain some perspective as to what the figures in Table 8.2 imply for real dollar magnitudes, it is useful to know that 1984 GNP in 1972 dollars was about $1640 billion at a seasonally adjusted annual rate. Consequently a change in the composition of output of 1 percentage point would amount to roughly $16 billion in real terms in that year.

spending in the other sectors must be reduced by ever larger amounts.[22]

Housing is the sector affected most adversely by ERTA/TEFRA (Table 8.2, line 3). As described above, the cut in personal tax rates raised the after-tax cost of capital for residential investment. Much more important, the higher level of interest rates substantially reduces expenditures for housing. The estimate that investment in housing gets crowded out – reduced by roughly 1% of GNP in this case – confirms the widespread view that housing is very sensitive to interest rates. It also reflects the fact that in the MPS model, residential investment has the highest estimated interest elasticity of all categories of expenditures.

The foreign sector – net exports of goods and services – is only slightly affected by ERTA/TEFRA, according to the estimates presented in Table 8.2. Recall again that the level of real GNP is essentially unchanged in this experiment, and therefore the simulation lacks the usual deterioration in the balance of goods and services associated with higher imports generated by a fiscal stimulus. The decline in net exports in line 4 is caused by the appreciation of the exchange rate, shown in line 8, which in turn results from the rise in US interest rates relative to those abroad.

Lines 9–12 show how much the change in policies affects capital stocks. Although business fixed investment is curtailed beginning in 1985, the stock of business capital falls below that in the baseline solution only in 1987. Up until that point the stock of plant and equipment would be raised by the ERTA/TEFRA fiscal package and compensating monetary policy. The stock of residential housing, however, is always adversely affected, and the negative impact becomes increasingly severe. Thus the main result of the simulation experiment reported in Table 8.2 is that, abstracting from effects on the level of output, ERTA/TEFRA significantly skews the composition of output toward consumption and away from housing. Business capital formation is helped in the short run but eventually it too gets crowded out.

Further insight into the consequences of ERTA/TEFRA can be gained by looking at the effects on the magnitude and composition of net saving and investment shown in Table 8.3. Federal saving (line 2) is reduced directly because of the revenue losses from the cuts in personal and business taxes and also indirectly because of the higher interest expense resulting from a higher interest rate applied to a larger stock of outstanding debt. The increase in the federal deficit in the first half of 1988 – $259 billion – consists of $162 billion in direct tax losses and $97 billion in higher interest expenses. Because of higher private saving (line 5), the reduction in federal saving does not yield an equivalent reduction in total net saving and net investment (lines 1 and 6). A fraction of the increase in private saving is due to the higher level of disposable income generated by the personal

---

[22] The real interest rate (the nominal interest rate less the rate of inflation) actually rises somewhat more than does the nominal interest rate because the rate of inflation is reduced (Table 8.2, line 6). The appreciation of the dollar lowers the markup of prices over unit labour costs, and this causes a reduction in the inflation rate that will persist as long as the appreciation is maintained, given the fact that the unemployment rate is not allowed to change.

Table 8.3 Simulated changes in net saving and investment generated by ERTA/TEFRA and a compensating monetary policy[a] (billions of dollars)

Category	1982	1983	1984	1985	1986	1987	1988[b]
1 Net saving	−7.5	−21.6	−37.3	−51.4	−70.9	−82.9	−90.2
2 Federal	−50.8	−105.8	−146.7	−177.4	−209.4	−236.5	−258.9
3 State and local	−0.2	−1.4	−1.3	0.6	2.2	2.6	1.6
4 Foreign	1.7	2.6	1.1	3.0	2.6	1.8	1.8
5 Private	41.8	83.0	109.5	122.3	133.7	149.3	165.3
Undistributed profits	−2.5	0.1	−7.8	1.3	4.6	−1.5	−4.9
Personal	44.3	83.0	117.3	121.1	129.2	150.7	170.2
6 Net investment	−7.5	−21.6	−37.3	−51.4	−70.9	−82.9	−90.2
7 Gross business fixed	9.0	11.3	4.0	−8.0	18.6	−27.9	−39.1
8 Inventory	−6.1	0.4	1.3	2.1	2.2	−3.9	−3.8
9 Gross residential	−11.2	−34.0	−44.5	−50.6	−65.3	−70.0	−72.8
10 Less: depreciation	−0.6	−0.8	−2.0	−5.1	−10.7	−18.5	−25.6

[a] Monetary policy was adjusted to hold the unemployment rate unchanged between simulations. The components may not add to the totals because of rounding.
[b] Average of values for the first and second quarters.

tax cut; part of the increase is also due to the lagged response of consumption to the tax change. In addition, as discussed above, there are several reasons why the private saving rate rises in response to a higher real interest rate along the transition path to a new long-run equilibrium. The simulation results show that this transition path is quite lengthy; nevertheless, in the long run the private saving rate is invariant to the real rate of interest, and ultimately private saving will increase only to the extent that disposable income is higher. Over the period from 1985 to 1988 the increase in private saving offsets nearly two-thirds of the reduction in federal saving. The remaining difference is reflected in lower net investment – the saving of the state and local government sector and the foreign sector is not altered very much.[23]

The foreign sector supplies more saving (Table 8.3, line 4) but the size of the increase, $1 billion to $3 billion, is minuscule compared with the shift in federal saving. The simulated increase is small because capital flows in the model are not very sensitive to the differential between the return on dollar-denominated assets and the return on foreign-currency assets. Furthermore, the induced appreciation of the dollar causes an expectation of a depreciation, which tends to offset the impact of the differential in interest rates on capital flows. There is some evidence suggesting that the foreign sector supplied more saving to the US economy as a result of ERTA/TEFRA than is shown by these simulations; the

---

[23] In a paper stressing the effects of government deficits on the composition of output, Lawrence H. Summers (1984) presents some informal estimates of the impact of deficits on saving and investment. In some respects, his findings are similar to those reported in Table 8.3, namely, an enlarged deficit is greatly offset by private saving and has a fairly low negative impact on private fixed investment. However, he finds a much larger impact on net foreign saving than suggested by the results in Table 8.3. See Summers (1984).

Empirical estimates

evidence lies in the fact that for the last several years, the model's equations have underestimated foreign holdings of US assets and overestimated US holdings of foreign assets. Thus the net capital inflow that actually occurred was stronger than predicted, given the actual values of the explanatory variables, and this may have been associated with ERTA/TEFRA through channels not fully captured by the model.

Before drawing conclusions about the long-run consequences of ERTA/TEFRA from these simulation results, we briefly contrast our short-run findings with those of Throop. In an experiment directly comparable to ours, Throop investigates the consequences of ERTA/TEFRA through a simulation with a macroeconomic model that keeps the level of real GNP unaltered by the policy change (Throop, 1985). In comparison with our results, Throop finds that over the 1981/84 period, ERTA/TEFRA had a much larger impact on interest rates, consumption, and net exports and a much smaller impact on business fixed investment and housing. His most striking result is that, even for the relatively short period he examines, ERTA/TEFRA had no significant positive effect on business fixed investment. Thus the higher interest rates generated by the cut in personal tax rates have a dampening effect that offsets the stimulative effect of the business tax provisions.

Throop's result for business fixed investment contrasts with our finding and appears to be due to a smaller response of investment, especially in the residential category, to the cost of capital in Throop's model than in the MPS model. For residential investment, the lower sensitivity in Throop's model is evidenced by the smaller simulated impact on residential investment even though the change in the rate of interest is larger. In the case of business fixed investment, a smaller interest rate response is apparent from Throop's result that, in this category, the increase arising from only the business portion of ERTA/TEFRA is smaller than the gain we compute for the entire tax package. Because investment in Throop's model is less sensitive to the interest rate, his simulation requires a larger increase in the rate of interest to hold real GNP unchanged with ERTA/TEFRA. Consequently he finds that business fixed investment is not stimulated in the short run (the larger increase in the interest rate outweighs the lower interest sensitivity), that there is a substantial appreciation of the dollar, and that the stimulus to consumption is largely matched by a decline in net exports.

We now return to the MPS simulation results and address the question of the long-run consequences of ERTA/TEFRA. The increase in the real interest rate that emerges by 1988 – about $2\frac{1}{2}$ percentage points – would, if sustained, definitely have a negative impact on potential GNP. However, it is difficult to estimate precisely the change in the real interest rate beyond 1988. The increase in the real interest rate in 1988, combined with the provisions of ERTA/TEFRA, yields a 9% reduction in the optimal capital–output ratio for producers' durable equipment. As discussed above, new equipment is assumed to be installed at the optimal capital–output ratio existing at the time of the investment decision. Changes in the optimal ratio do not affect the use of old equipment because of its assumed putty–clay nature. Thus the full impact on potential output of a change in the optimal capital–output ratio does not occur until the entire capital stock has been replaced. In this case, the 9% reduction in the ratio, if sustained, would cause a $1\frac{1}{2}$% reduction in potential output, given the capital–input elasticity of 0.16 in the model's Cobb–Douglas production function. At the end of the simulation period, the reduction in (potential) output is only 1/2%; thus the adjustment process is only one-third complete by 1988.

It seems likely that the impact of ERTA/TEFRA on the real rate of interest would continue to grow if the simulations were extended beyond 1988. In the simulation containing ERTA/TEFRA, the ratio of federal debt to GNP is still increasing in 1988. If we knew how much further the debt/GNP ratio might increase, we could estimate the ultimate increase in the real interest rate. To project the ratio of federal debt to GNP, we analyse the budget identity of the federal government:

$$\dot{D} = G - T + RD, \qquad (8.4)$$

where

D = real federal debt
G = real federal expenditures excluding interest payments
T = real federal tax receipts net of tax on the interest on federal debt
R = real after-tax rate of interest.

For long-run analysis it is more convenient to rewrite Equation 8.4 with all quantities expressed as ratios to real GNP (denoted by the subscript x); this form of the budget identity is

$$\dot{D}_x = G_x - T_x + (R - g)D_x,$$

where g is the growth rate of real GNP. In a growth equilibrium, $\dot{D}_x = 0$; thus the equilibrium ratio of federal debt to GNP is

$$\bar{D}_x = \frac{\bar{G}_x - \bar{T}_x}{(g - \bar{R})}, \qquad (8.5)$$

where a bar over a variable denotes an equilibrium quantity.

Assuming that the values for $(G_x - T_x)$ and $(g - R)$ simulated for 1988 under ERTA/TEFRA (0.012 and 0.007, respectively) were sustained indefinitely, Equation 8.5 shows that D_x would approach 1.7. This would be a very large increase in the value of the debt ratio – as of the fourth quarter of 1984, it was 0.33. Results of long-run simulations of the MPS model reported by Anderson *et al.* indicate that an increase of this magnitude in D_x would cause the real interest rate eventually to rise more than the $2\frac{1}{2}$ percentage point increase simulated here for 1988 (Anderson et al., 1984).[24] They find that a change in fiscal policy that

---

[24] The version of the MPS model used by Anderson and others differs a little from the one on which the simulations in this chapter are based; in the former case some adjustments have been made to make the structure of the model more consistent with long-run steady-state growth.

increases the ratio of federal debt to GNP from zero to one increases the real interest rate by 4.5 percentage points. Furthermore, the level of the real after-tax interest rate, R, exceeds the real growth rate, g, at the higher debt ratio, and thus the federal government must run a surplus (excluding real after-tax interest payments) to maintain the ratio of the debt to GNP at unity. These findings indicate that if the fiscal policy associated with ERTA/TEFRA were continued after 1988, the real interest rate would continue to rise, and at some point the ratio of debt to GNP would become unstable, growing explosively. This outcome could be avoided only by shifting to a more restrictive fiscal policy.

8.2.2 Effects with the growth rate of M1 held constant

The short-run effects of the fiscal changes embodied in ERTA/TEFRA depend crucially on the accompanying monetary policy.[25] In the experiment described above it was assumed that, without ERTA/TEFRA, monetary policy was conducted to achieve the same level of unemployment that occured with the new fiscal programme in place. No claim was made that this was the most plausible or likely policy that would have been implemented in the absence of the fiscal initiative. Rather, it was adopted in order to facilitate examination of the long-run impact of the fiscal policy action.

To look at economic activity in the short run under ERTA/TEFRA, one is forced to ask what monetary policy would have been pursued if the fiscal change had not been implemented. One possibility is to construct a feedback rule that would describe the behaviour of the monetary authorities under alternative economic conditions. Such an approach would have the advantage of providing a flexible response to departures from desired targets for output, employment, and inflation; moreover, the approach avoids the instabilities that sometimes arise if the monetary authorities are assumed to follow a rigid monetary rule.[26]

However, in the absence of any consensus on the appropriate characterization of monetary policy by a feedback rule, we have chosen to investigate the short-run consequences of ERTA/TEFRA using a standard, rigid monetary policy rule: in the absence of the fiscal change, the monetary authorities would have achieved the same path of M1 as in the baseline simulation. This rule has the virtue of simplicity, and it has often been used in experiments of this type. It also has the advantage that it holds monetary policy constant in a well-defined way. However, it is probably unreasonable to suppose that the Federal Reserve would have operated in such a way as to achieve exactly the same growth rate in the money supply with and without ERTA/TEFRA.

The macroeconomic effects of the fiscal initiatives, computed under this assumption of unchanged monetary policy just described, are shown in Tables 8.4

---

[25] For a discussion of some of the issues involved in the specification of monetary policy in conducting counterfactual fiscal policy experiments, see Blinder (1984).

[26] For examples of feedback rules in the simulation of monetary policy, see Anderson and Enzler (1987); Anderson et al. (1984); and Clark (1984).

Table 8.4 Simulated effects of ERTA/TEFRA and a given M1

Item	1981		1982		1983		1984		1985		1986		1987	
	Q3	Q4	Q2	Q4	Q2	Q4	Q2	Q4	Q2	Q4	Q2	Q4	Q2	Q4
GNP (billions of 1972 dollars)	0.1	6.6	14.1	37.6	33.4	53.0	41.3	27.0	5.1	−25.0	−48.7	−59.1	−50.1	−27.6
Business fixed investment (billions of 1972 dollars)	0.0	2.2	6.7	13.0	17.2	21.5	19.7	14.5	6.3	−4.8	−16.0	−22.8	−22.4	−15.3
Unemployment rate (percentage points)	0.0	−0.1	−0.3	−0.8	−0.9	−1.3	−1.2	−1.0	−0.6	0.0	0.6	1.1	1.2	1.0
Federal budget deficit (billions of current dollars)	−2.7	−15.1	−14.4	−38.3	−51.2	−78.4	−85.1	−103.6	−130.3	−155.7	−186.8	−201.4	−204.7	−199.7
Treasury bill rate (percentage points)	0.0	0.2	0.6	1.6	1.8	3.0	3.4	3.4	3.1	2.4	1.5	0.9	0.7	0.9
Corporate bond rate (percentage points)	0.0	0.1	0.3	0.6	0.8	1.2	1.5	1.8	2.2	2.3	2.2	2.0	1.7	1.4
GNP implicit deflator (percentage points)	0.0	−0.2	0.1	0.1	0.5	0.9	1.6	1.7	1.6	1.1	0.4	−0.5	−1.2	−1.6
Net exports of goods and services (1972 dollars)	0.0	0.0	−1.3	−2.2	−1.5	−1.4	0.4	2.3	2.5	1.0	−0.5	−2.9	−5.6	−7.3
Net exports of goods and services (billions of current dollars)	0.0	−1.6	−4.9	−10.7	−8.4	−11.8	−6.7	2.0	8.7	12.8	15.9	13.8	4.8	−6.9
Foreign exchange value of the dollar (%) (appreciation, +)	0.0	0.0	−0.7	−2.2	−4.0	−4.1	−3.9	−1.5	1.1	4.3	5.8	7.1	5.7	2.2
Residential construction (billions of 1972 dollars)	0.1	0.2	−0.6	−2.3	−7.5	−9.3	−10.6	−10.7	−12.1	−14.0	−15.3	−15.3	−14.6	−14.5

Empirical estimates 161

Table 8.5 Simulated changes in the composition of real GNP generated by ERTA/TEFRA and a given M1[a] (percentage points)

Component	1982	1983	1984	1985	1986	1987	1988[b]
Consumption	−0.1	0.1	0.3	0.8	1.4	1.5	1.3
Business fixed investment	0.4	0.9	0.9	0.2	−0.7	−0.8	−0.3
Residential investment	−0.1	−0.6	−0.8	−0.8	−0.8	−0.7	−0.8
Net exports of goods and services	−0.1	−0.1	*	0.1	−0.1	−0.3	−0.4

[a] The net change in shares for a year does not sum to zero because inventories and the government sector are not included.
[b] Average of values for first and second quarters.

and 8.5. Note in the first line of Table 8.4 that ERTA/TEFRA is estimated to provide a very powerful stimulus to real GNP through 1984. The stimulus is generated by a large increase in consumption spending (not shown) resulting from the cuts in personal taxes. In addition, the fiscal package gives a short-run boost to business fixed investment (second line of Table 8.4) that is larger than in the fixed-unemployment-rate simulation because the accelerator effect is operating in this case and not in the previous one. The large initial impact can be seen by comparing business fixed investment in Tables 8.2 and 8.5. In Table 8.2, for example, the share of output going to business fixed investment in 1984 is raised only 0.2 percentage point by ERTA/TEFRA, but when GNP is allowed to vary, as in Table 8.5, the share devoted to business investment rises by 0.9 percentage point. Taken at face value, the results in Table 8.5 suggests that a major fraction of the boom in business fixed investment in 1983 and 1984 can be attributed to the direct and indirect effects of ERTA/TEFRA.

The results reported in Table 8.4 also indicate that with fixed M1 the rise in GNP and investment under ERTA/TEFRA is also transitory. Starting in late 1985 the economy is considerably weaker than it would have been in the absence of ERTA/TEFRA, a situation that reflects the strong cyclical properties of the MPS model. Typically the response of the model to shocks, given the assumption of fixed M1, is a path that oscillates toward the new equilibrium. In large part these cycles arise because the lags between changes in interest rates and in the interest-sensitive components of final demand are much longer than those between changes in interest rates and in money demand. In addition, the dynamics of the wage and price sectors result in a tendency for the real wage to oscillate, which introduces lower-frequency cycles in the model's response to shocks.[27] Even if the model did not have this cycling behaviour, the assumption of fixed M1 requires the eventual elimination of the initial boost to the price level that was generated

---

[27] For analysis of the oscillatory properties of a small, simplified version of the MPS model see Enzler and Johnson (1981).

by the additional aggregate demand.[28] The cancellation of the rise in price level occurs through higher unemployment: interest rates rise to choke off aggregate demand until the unemployment rate increases to a level sufficient to bring wages and prices back down.

The tendency of the MPS model to cycle so strongly under a fixed path for M1 makes it very difficult to discern the long-run consequences of a change in fiscal policy; thus one should not use such a monetary policy when simulating the long-run effects of ERTA/TEFRA, as discussed above. Nevertheless, the long-run effects, except for those on the price level, will be the same under either choice of monetary policy.

Other macroeconomic models have been used with a fixed path for a monetary aggregate to quantify the impacts of ERTA/TEFRA. However, because these other models do not cycle as strongly as the MPS model, the simulation results derived from these models have been used to draw conclusions about the effects of the new fiscal policy not only in the short run but also in the long run. In a recent paper, Blinder estimated the effects of the Reagan fiscal package using the Data Resources, Inc. (DRI) model and the Wharton Econometric Forecasting Associates (WEFA) model (as well as the MPS model) (Blinder, 1984).[29] He found that the fiscal package generated a continuously increasing level of real GNP (relative to the baseline) in both the DRI and WEFA models, reaching about 2% in 1989, the last year for which he reports results. Also, through 1989 the level of business fixed investment is always above the baseline path in these two models and hence gets 'crowded in' rather than crowded out.

These results differ dramatically from those obtained using the MPS model. In the case of the WEFA model, the difference appears to arise from a Phillips curve in which changes in the unemployment rate within some range have no effect on wage behaviour; thus fiscal policy can cause the economy to permanently expand by increasing labour input as long as the unemployment rate does not fall below this range. The results for the DRI model are harder to understand. The higher level of business investment requires that some other type of investment be reduced or that saving be higher. Interest rates are not appreciably higher; thus residential investment is not crowded out, and saving from the foreign sector (which is not reported) should not increase. Federal saving is reduced – the budget deficit is higher as a consequence of ERTA/TEFRA. The only remaining saving-investment component is private saving, and although it is not reported, it must be higher (as a ratio to output). If private saving is higher, then the difference between the DRI and MPS results arises from different specifications of consumption behaviour.

[28] Prices are also higher because of an initial depreciation of the dollar; the depreciation occurs because the widening deficit on the balance on goods and services has a negative influence on the dollar that more than offsets the positive influence of higher US interest rates. The dollar eventually appreciates relative to the baseline, however, because the decline in real GNP and the initial dollar depreciation both cause a subsequent improvement in the balance of goods and services compared to its baseline path.

[29] Blinder's simulations are done with M2 held fixed, and they also include small changes in federal expenditures. Nevertheless, his quantitative results from the MPS model are similar to ours with fixed M1.

Estimates of the effects of ERTA/TEFRA have also been made by Coen and Hickman (1984). Using the Hickman–Coen annual growth (HCAG) model, they simulate separately (among other experiments) the effects of the third round (July 1983) cut of 10% in personal income taxes and the impact of the accelerated depreciation allowances. In each case they held the growth rate of M1 unchanged at its baseline path. The simulation experiments were run out to 1989, when, according to the results, convergence to the new equilibrium is apparently achieved.

In the HCAG model, the effects of the cut in personal income taxes are different from those reported here for the MPS model. Business fixed investment in the HCAG model is insensitive to nominal and real interest rates, and as a result the increase in consumption is matched almost entirely by a crowding out of residential investment. There is essentially no change in business investment and potential and actual output. Another disturbing feature of these results is that the inflation rate is continuously above that in the baseline, and nominal interest rates are steadily rising. One may therefore question whether the simulation results in 1989 do in fact represent a long-run equilibrium position.

The simulated effects of the accelerated depreciation allowances are also surprising. Coen and Hickman find that this part of ERTA/TEFRA produces such a large change in real GNP – about 1% in 1983 – that the cut in tax liabilities actually *reduces* the federal budget deficit in 1983. Furthermore, by 1989 the tax change results in an apparently *permanent* increase in real consumption, investment, and real GNP, and a *permanent* reduction in nominal interest rates, the inflation rate, and the budget deficit. One must question whether a cut in business taxes would generate such a large effect.

ACKNOWLEDGEMENTS

This chapter was published as a summary in the *Federal Reserve Bulletin* for December 1985. The analyses and conclusions set forth are those of the authors and do not necessarily indicate concurrence by the Board of Governors of the Federal Reserve System, Washington, DC, by the Federal Reserve Banks, or by the members of their staffs. The authors thank Eileen Mauskopf and John Sturrock for comments.

REFERENCES

Anderson, Robert, Ando, Albert and Enzler, Jared (1984) Interaction between fiscal and monetary policy and the real rate of interest. *American Economic Review*, 74, May, 55–60.

Anderson, Robert and Enzler, Jared J. (1987) Policy design: policy rules that use forecasts, in *Macroeconomics and Finance: Essays in Honour of Franco Modigliani* (eds R. Dornbush and S. Fisher), MIT Press, Cambridge, MA.

Ando, Albert (1974) Some aspects of stabilization policies, the monetarist controversy, and the MPS model. *International Economic Review*, 15, October, 547.

Ando, Albert and Modigliani, Franco (1963) The 'life cycle' hypothesis of saving: aggregate implications and tests. *American Economic Review*, 53, March 55–84.

Barro, Robert J. (1974) Are government bonds net wealth? *Journal of Political Economy*, 82, November/December, 1095–117.

Barro, Robert J. (1984) *Macroeconomics*, Wiley, New York, pp. 380–93.

Blinder, Alan S. (1984) Reaganomics and growth: the message in the models, in *The Legacy of Reaganomics* (eds Charles R. Hulten and Isabel V. Sawhill), Urban Institute, Washington DC, pp. 210–12.

Boskin, Michael J. (1978) Taxation, saving and the rate of interest. *Journal of Political Economy*, 86, April, part 2, S3–S27.

Brayton, Flint and Mauskopf, Eileen (1985) The Federal Reserve Board MPS quarterly econometric model of the US economy. *Economic Modelling*, 2, July, 170–292.

Clark, Peter B. (1984) Inflation and unemployment in the United States: recent experience and policies, Working paper series 33, January, Board of Governors of the Federal Reserve System, Division of Research and Statistics, National Income and Wages, Prices, and Productivity Sections, Washington, D.C.

Coen, Robert M. and Hickman, Bert G. (1984) Tax policy, federal deficits, and US growth in the 1980s. *National Tax Journal*, 37, March, 89–104.

Cohen, Darrel and Clark, Peter B. (1984) *The Effect of Fiscal Policy on the US Economy*, Staff Studies 136, Board of Governors of the Federal Reserve System, Washington, DC.

Enzler, Jared and Johnson, Lewis (1981) Cycles resulting from money stock targeting, in *New Monetary Control Procedures*, vol. 1, Board of Governors of the Federal Reserve System, Washington, DC.

Friend, Irwin and Hasbrouck, Joel (1983) Saving and after-tax rates of return. *Review of Economics and Statistics*, 65, November, 537–43.

Haberler, Gottfried (1985) International issues raised by criticisms of the US budget deficits, in *Essays in Contemporary Economic Problems, 1985: The Economic in Deficit* (eds Phillip Cagan and Eduardo Somensatto), American Enterprise Institute, Washington DC, pp. 121–45.

Hooper, Peter (1985) International repercussions of the US budget deficit. *Aussenwirtschaft*, 40, May, 119–55 (first published as International Finance Discussion Paper 246, Board of Governors of the Federal Reserve System, September 1984).

Steindel, Charles (1981) The determinants of private saving, in *Public Policy and Capital Formation*, Board of Governors of the Federal Reserve System, Washington, DC, pp. 101–14.

Stevens, Guy V.G. et al. (1984) *The US Economy in an Interdependent World: A Multicountry Model*, Board of Governors of the Federal Reserve System, Washington, DC, pp. 135–84.

Summers, Lawrence H. (1984) The legacy of current macroeconomic policies, in *The Legacy of Reaganomics: Prospects for Long-Term Growth* (eds Charles R. Hulten and Isabel V. Sawhill), Urban Institute, Washington DC, pp. 179–98.

Symansky, Steven (1985) The US budget deficit, monetary policy and the world recovery: an MCM simulation analysis, in *Probleme und Perspectiven der Weltwirtschaftlichen Entwicklung*, Duncker and Humbolt, Berlin, pp. 549–72.

Throop, Adrian W. (1985) Current fiscal policy: is it stimulating investment or consumption? Federal Reserve Bank of San Francisco, *Economic Review* (Winter issue), pp. 19–44.

Tobin, James (1980) *Asset Accumulation and Economic Theory: Reflections on Contemporary Macroeconomic Theory*, Yrjö Jahnsson lectures, University of Chicago Press, pp. 31, 50, 54–57.

9

Use of anticipations data in the anticipations model

WALTER NAGGL

9.1 INTRODUCTION AND SUMMARY

Forecasting the future and simulation are the main goals of econometric modelling. Forecasts have not been totally successful in worldwide practice so far. Neither the improvements in estimation methods, nor the continued enlargement of econometric models, have abolished wrong forecasts. Such renowned observers as McNees (1979) or Zarnowitz (1979), for instance, conclude that even the large-scale commercial models of Chase, Data Resources, Inc. (DRI), and Wharton are not superior to other forecasting methods. The saying that econometricians forecast better than econometric models[1] is still true.

In this situation, econometricians have long used empirical plans and expectations, i.e. anticipations variables (A-variables), in order to improve the forecasting record of econometric models. Small models based on A-variables were estimated by Friend and Taubman (1964) or by Fair (1971) and large structural models were turned into anticipations versions, such as the Wharton model (Adams and Duggal, 1974) or the DRI model (Eckstein *et al.*, 1984). Eckstein concludes that the estimation of the DRI futures market model was a great success:

> 'If the method had been used in late 1981, it would have sent up some warning flags for the DRI forecasts, and the actual estimates would have been better than any of the major forecasts published at that time' (Eckstein, 1984, p. 190).

Nevertheless, with regard to the total research efforts carried out, these experiments can only be qualified as half hearted, if compared for example with the flood of articles on A-variables which appeared in connection with the rational expectations hypothesis. It is characteristic that Eckstein (1984, p. 183), out of a

---

[1] This means that subjective adjustments of model results by econometricians have often improved forecasts.

literature which is alledgedly so large, refers only to five prior publications, made between 1958 and 1972, on the incorporation of A-variables in econometric models. Why have these attempts, which have always been qualified as successful by their authors, met comparatively little general acceptance? Two points of criticism, brought forth by Evans (1969) can perhaps serve as an explanation. Evans criticizes:

(1) Small models with A-variables do not incorporate the structure of the economy. They can only explain GNP and some of its components, a purpose which in his opinion can be served just as well by a one-equation model for GNP.
(2) Since A-variables are usually[2] not explained in A-models, they are an obstacle to simulation and thus to an important goal of econometric modelling.

In the estimation of the anticipations model (A-model) presented here the advantages that are offered by A-variables were used as extensively as possible. In 80 equations the industrial production in OECD countries and – in detail – income generation and expenditures in West Germany are explained by a model with quarterly data. Out of the 300 variables of the model 60 are A-variables, considerably more than in any other existing model.[3] In estimating a comparatively large model in this way, the criticisms mentioned by Evans were mostly met. The model predicts a large part of the variables that are of interest with regard to the cyclical development of the economy and, partly, also allows simulations. Without major difficulties it is also possible to estimate a fully structural version of the A-model. The usual way of going from the structural to the anticipations version can be reversed here.

9.2 THE FORECASTING VALUE OF A-VARIABLES

A-variables are defined here as empirical plans with regard to own activity of firms or consumers, as empirical expectations with regard to the environment, and sometimes, as empirical judgements.[4] They can either be used directly to forecast actual activity or the actual state of the environment, or else they can be applied in realization functions[5] together with variables that explain modifications of plans or expectations.

A-variables are collected worldwide mainly by the method of business tests (compare Strigel (1981) or Nerb (1975)), which was originally introduced by the Ifo Institute in Munich in 1949. This method minimizes the effort on the part of those who are questioned by asking them only, whether a certain variable will

---

[2] Adams and Klein make use of A-variables as explanatory variables and explain them in turn, which makes it possible for them to carry out simulations. Their simulation results show that use of A-variables improves the forecasts of their model.
[3] The models of Friend and Taubman (1964) or Fair (1971) are based on only two A-variables, for example.
[4] The 'assessment of inventories' depends heavily on expected sales.
[5] Compare Modigliani and Cohen (1961), chapter C, for the realization function.

(a) increase, (b) stay the same, or (c) decrease. The difference between (a) and (c) is then used as the quantitative realization of the variable in question (compare Anderson et al., 1955). Theil (1952) proposed quantifying business test data under the assumption that rates of change of the variable that is tested are normally distributed over all firms. This has important practical disadvantages, however (Foster and Gregory, 1977). Business test data are therefore still published in the form of differences and consequently used in difference form here.

Hypotheses about the generation of plans and expectations (expectations hereafter) usually rival the collection and use of A-variables, because they easily suggest the conclusion that the explanatory mechanism for expectations could replace A-variables. In the simplest case, expectations X^* with regard to variable X are explained by distributed lags of X:

$$X_t^* = \Sigma \alpha_i X_{t-i} \qquad (9.1)$$

If one takes into account that expectations, just like any other variable, are also influenced by a whole range of further variables Z_j, one arrives at a causal model:

$$X_t^* = \Sigma \alpha_i X_{t-i} + \Sigma \beta_j Z_j \qquad (9.2)$$

The theory of rational expectations states that the explanatory variables X_{t-i} and Z_j have to be derived from a complete model.

The question whether the mechanisms of Equations 9.1 or 9.2 are superior to A-variables is empirical in nature and has been answered in different ways. In order to be able to assess the answers, it is useful to specify which circumstances are favourable for either result. The following are favourable for A-variables:

(a) a complicated process of expectations formation which is not easily cast into an equation

(b) the case where the variable to be forecast is a decision variable (plan) for the person questioned, since the informational value of decision variables is usually high.

For the use of the mechanisms in Equations 9.1 or 9.2 simple and stable processes of expectation-formation are favourable.

With regard to consumer A-variables Evans (1969, p. 466) found that they do not contribute to the explanation of automobile purchases in the United States between 1957 and 1962, when financial variables are included in the regression equation. Adams and Klein (1972, p. 191) stated that 'anticipatory variables in the consumer sector are the least firm of those used in economic forecasting'. For West Germany, Tewes (1984) investigated whether empirical consumer buying plans and expectations could explain the residuals from a regression of consumer durables on income and short-term interest rates. He found no influence of the consumer A-variables. These results can be explained by the hypothesis that plans of consumers are simple in structure, as already proposed by Katona (1951). Their effect on consumption is therefore frequently contained in other explanatory variables.

In the case of firm-specific A-variables, which are more likely to stand for complicated processes of expectations formation (compare Katona, 1951), we get completely different results. Wolters (1984), who investigated empirical price expectations of West German firms in the investment goods industry and corresponding actual prices, finds that, while these empirical expectations influence actual prices, there is no influence from past prices to expectations as postulated by the model of Equation 9.1.

With regard to the model of Equation 9.2, it is frequently argued that it is possible to explain a large part of the variance of expectational data by other variables. However, these equations are usually unstable and thus useless for forecasting purposes. Knöbl for example writes about an equation of type 9.2

'First, the lag between demand pressure and price expectations is not constant.... Second, the functional relationship between dp [demand pressure] and p^e [price expectations] seems to be unstable.'

On the other hand, he arrives at the opposite conclusions with regard to the explanation of actual prices by empirical price expectations (1974, p. 93):

'It is clear that price expectations have influenced actual price developments and that price expectations give a better explanation of price behavior in the period 1965–72 than any of the demand pressure variables that have been tested.'

Even more than for econometric purposes, A-variables are used as coincident and leading cyclical indicators. In a detailed investigation of leading macroeconomic indicators for West Germany, Roberts (1975, p. 70f) finds that the business climate from the Ifo business test is the best leading indicator, both with regard to the length of the lead and correspondence to the amplitude of the reference series. Furthermore, the business climate is a very smooth curve, allowing quick conclusions to be drawn from changes in its direction. In spite of these high qualities, the business climate does not totally exclude false turning-points, though. Roberts therefore comes to the same conclusion as the NBER before him, namely that a large number of cyclical indicators under consideration for their theoretical justification should be used for business cycle forecasting.

The construction of an econometric model based on A-variables turns up as a logical consequence here. Evans (1969, p. 491), for example, writes:

'the combining of investment anticipations and other exogenous variables into a consistent framework is a method that many forecasters could well emulate'.

Poser and Hecheltjen (1973, p. 33) motivate their model based on A-variables in the following way:

'anticipatory data are widely used individually as single indicators of economic activity, and it seemed necessary to look at them in the context of a complete model'.

9.3 CONSTRUCTION OF A FORECASTING MODEL BASED ON A-VARIABLES

Estimation of a model based on A-variables is not only a logical consequence from the point of view of leading indicators, it is also desirable with regard to forecasting from the perspective of modelling. This has been investigated especially by Haitovsky and Treyz (1971). They point out that the inclusion of A-variables in an equation of a model has consequences which go beyond the improvement in the R-squared of the equation considered. Unlike unlagged exogenous variables, A-variables need not be forecast to the extent of their lead when model forecasts are made, so that they do not contribute to exogenous forecast errors. This reduces error cumulation within the model to the extent that the endogenous variable of the equation considered is itself an explanatory variable in other equations. This effect is still increased by the fact that inclusion of A-variables in an equation usually reduces the parameter values of the other explanatory variables, thereby reducing the possibility of error cumulation even more.

How should an econometric model based on A-variables be constructed then?

(1) Since even good leading indicators occasionally give false signals, it is wise to rely on a large number of A-variables instead of just a few.
(2) Not all A-variables are equally good. It was shown above that consumer A-variables are rather weak and that the explanation of private consumption by income and other traditional variables is therefore preferable.
(3) Empirical plans of firms (forecasts of own activity) are preferable to empirical expectations (forecasts of the environment) since plans have a higher forecasting quality.
(4) The model ought to be large enough in order to be able to forecast many variables that are of interest and in order to be as close as possible to a structural model. This requirement partly meets with the first.
(5) Anticipations should be applied in the relevant behavioural functions, as by Adams and Klein (1972), and not simply as broad indicators of economy activity.

The model by Friend and Taubman (1964) is not followed here, since it consists only of four structural equations, one identity, and two A-variables.[6] Neither is the Fair model (1971) which is also based on two A-variables and to a large extent upon the SRC index of consumer sentiment. The model by Poser and Hecheltjen (1973) is more promising under the above considerations, but it was not maintained by its authors.

In constructing the A-model (see Fig. 9.1), the above requirements (1) to (5) were met in the following way: private consumption was not explained by consumer plans and sentiments, but instead mainly by income. To this end the

---

[6]Biart and Praet (1985) have estimated a model for the large EC countries similar to the Friend and Taubman model.

170 Use of anticipations data

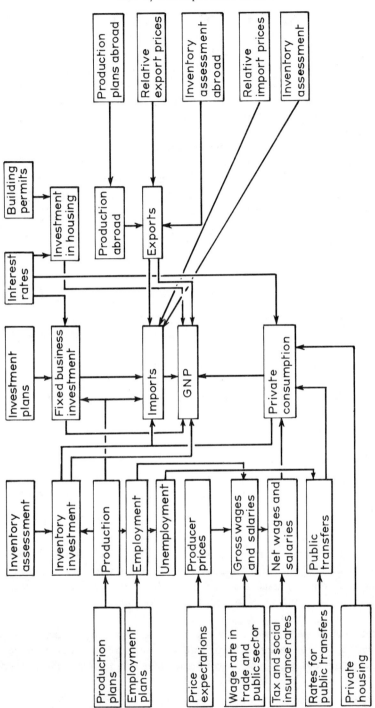

Fig. 9.1 Flow chart for the anticipations model

process of income generation was modelled in detail. Empirical plans about firms' own production or their planned employment from business tests are available for this purpose. Production in industry and construction is forecast by planned production in these sectors. Expectations with regard to future business from wholesale and retail trade serve to forecast sales in trade. In the short run, production and sales respectively are in turn the major determinants of employment. In industry and construction, employment plans are also an additional explanatory variable for employment. Employment in the services sector is explained by a traditional employment function which incorporates several activity variables and the wage rate. Government employment is set exogenously in forecasts.

Producer prices are forecast with the help of price expectations from the business survey. Producer prices, production, and employment serve to forecast gross wages and salaries per employee in the industrial sector and in construction. Wage changes in the other sectors are partly determined by wages in industry and partly set exogenously. Deduction of taxes and social insurance payments from gross wages and salaries leads to net wages and salaries. A large number of institutional requirements is specified in the calculation of these deductions and of transfer payments, since the information about changes in tax rates is usually given in advance and should not be wasted.

Net wages and salaries, transfer payments, and a lagged interest rate variable are the major determinants of private consumption. The imputed rent of private home owners contained in official consumption data is forecast separately, since it is independent of current income.

Fixed business investment is mainly influenced by lagged interest rates and lagged production variables. Current production, which also enters as an explanatory variable, is already explained at a prior stage. It is thus possible to explain fixed business investment largely without A-variables. The estimated regression contains only one A-variable, which is planned investment in industry. However, in order to make use of the wide range of plan data on investment, such as the investment plans of Bundespost and Bundesbahn, of the energy sector, or of construction, fixed business investment was explained alternatively for eight sectors by realization functions containing investment plans and unlagged production or sales variables. Investment in housing was determined by lagged housing permits and interest rate variables. Investment in inventories is explained to a large extent by the A-variable 'assessment of inventories'.

Exports could be explained by 'expectations on exports' from the business test with a single function. Since these expectations are less reliable than firms' production plans[7] and since a variable as important as exports should not depend so heavily on just one A-variable, this way of forecasting exports was not chosen. Instead, forecasts of exports are based on the information about future activity contained in the production plans of industry in other OECD countries. To this

---

[7] Compare the ranking given to French business test data by Devilliers (1983).

end, production in ten OECD countries is forecast with the help of planned production from business tests in these countries. Together with relative export prices, these production variables explain West German exports by countries. Exports to OPEC countries are explained by the receipts of these countries which in turn are a function of the price of oil and production in industrial countries.

Imports are disaggregated to four components and explained by all the other demand components in the model, by relative prices, and by the A-variable 'Assessment of inventories' in West German industry and trade. Government consumption is explained by a traditional equation. GNP finally results as an identity. At present, the A-model does not include a monetary sector, but there are no major problems in making the necessary adjustments. The model is mostly recursive. It has been estimated by ordinary least squares (OLS) and by the Hildreth-Lu method for correction of autocorrelation in the residuals.

Simulations can be performed with regard to various issues. For instance, the short-run effects of changes in tax rates can be investigated for income, consumption, imports and GNP. Exchange rate changes can be investigated for their influence upon exports to the country directly involved, upon exports to other countries through substitution effects, and upon total exports and imports. Furthermore, important intermediate results, such as industrial production in the USA, can be replaced by assumptions or, for instance, by the median of other forecasts and a new solution for the model can be calculated then. Finally, the model can be made 'more structural' by substituting endogenous variables for A-variables.

9.4 FORECASTING WITH THE A-MODEL

Since 1983 forecasts were calculated with this model two quarters into the future at the end of every quarter. (Since 1986 the model has been changed to forecast four quarters into the future.) At every forecast the model is re-estimated with the most recent data available. The process of forecasting is completely automated through computer programs. It requires two days to store the latest data and one day to execute the calculations.

In some cases, the maintenance of the model between forecasting dates requires respecification of equations which exhibit signs of instability after the inclusion of the most recent data. After the first two forecasts in July and October 1983 it was necessary to revise a large number of equations, such as the export functions and the consumption function. Since then, however, the number of equations that need respecification has decreased to about one equation after every forecast. The necessary revisons are now of minor importance, so that the model is practically stable in the parameters.

A special feature of the A-model is the fact that forecasts are not subjectively adjusted. Although the adjustment of model forecasts is a common practice, at least it has no theoretical basis (McNees, 1982, p. 39). By renouncing on adjustments, on the other hand, the forecasting ability of the model can be isolated from the forecasting ability of the econometrician. Specific characteristics in the A-model reduce the need to carry out subjective adjustments. Due to the use of a large number of leading variables, model forecasts are much less conditional on the values of exogenous variables, than is the case for other models. A recalculation of the forecasts with different values for the exogenous variables would therefore not result in great changes in the forecast values. Furthermore, the effect of such variables as entrepreneurs' judgement in a forecasting equation can be interpreted in the same way as the effect of the so-called 'constant-term adjustment'. In both cases the level of the forecasts is changed. However, when business test data are used, this change comes about by the judgement of the participants of economic decisions and not by the subjective judgement of the econometrician. Finally, since business test data catch cyclical movements of the economy very well, the inclusion of business test variables reduces the risk of structural breaks and therefore the need for subjective adjustments in forecasting equations for the business cycle.

9.5 EVALUATION OF FORECASTS

When forecasts with the A-model were first published in 1983, the model was still in its construction phase. It turned out then, that a large number of equations, especially export functions, had to be revised. In addition, economic activity itself was subject to several heavy singular influences. In the second quarter of 1984 the economy deviated from its cyclical path due to the dispute in the metal industry. In the first quarter of 1985 growth weakened due to the extremely cold weather. Thus, the database for an evaluation of the model's forecasts is not only short, but also characterized by strong singular events in a period of modest growth. Nevertheless the available forecasts of industrial production and the main components of aggregate demand will be compared with actual data here, in order to allow an early assessment of results. The evaluation is performed with percentage changes over the corresponding quarter of the previous year for forecast and actual values in order to eliminate seasonal influences (Table 9.1).

Industrial production (actual values), just as with all the other variables investigated here, clearly exhibited a turning-point during the forecast periods. This turning-point was already correctly anticipated by the first forecast dating from July of 1983 and nearly all the forecasts are close to the actual values. Even the largest error-of-forecast that occurred in October 1983 for 1984.1 does not exceed 3 percentage points.

Private consumption reached peak growth rates of 4% in the time period investigated during the fourth quarter of 1983 and the first quarter of 1984, after 2.5% during the fourth quarter of 1982. This increase in the growth rates of private consumption, as well as the decrease that followed, was anticipated by the forecasts, although with an upward bias.

Table 9.1 Forecast and actual values

	831	832	833	834	841	842	843	844	851	852	853
Industrial production: percentage change over previous year											
Actual values	−3.6	−1.2	1.7	5.4	6.7	−1.7	3.9	4.5	1.4	9.1	5.8
forecast in											
July 1983		−0.8	2.4	4.8	3.9	5.5	5.4	3.7	2.4	9.2	4.6
Oct 1983			2.3	4.1	5.2	4.6	4.8	4.9	3.6	9.4	5.5
Jan 1984				3.4	5.7	4.3	4.9	4.8	3.3	9.4	
Mar 1984											
June 1984											
Oct 1984											
Dec 1984											
Mar 1985											
June 1985											
Private consumption, nominal: percentage change over previous year											
Actual values	3.4	3.6	4.3	4.1	4.0	3.4	3.0	2.8	2.1	3.3	4.4
forecast in											
July 1983		4.4	5.3	5.5	5.4	4.1	4.3	3.9	2.9	3.0	3.4
Oct 1983			2.3	4.4	5.5	4.0	3.9	2.8	3.1	3.5	4.3
Jan 1984				4.7	4.6	4.3	3.2	2.8	3.1	4.1	
Mar 1984											
June 1984											
Oct 1984											
Dec 1984											
Mar 1985											
June 1985											

(Quarters forecast)

Evaluation of forecasts 175

Inventory investment, real: absolute change over previous year

Actual values	−1.5	0.6	2.7	6.2	7.1	2.0	0.0	−1.5	0.7	−0.8	−0.9
forecast in											
July 1983		−1.3	1.5	4.6	4.4	3.1					
Oct 1983			1.1	5.8	5.6	1.0					
Jan 1984				5.3	5.0	−0.1					
Mar 1984							0.2	0.6	−2.4	−0.1	
June 1984							−1.4	2.5	−2.0	1.5	0.0
Oct 1984							2.1	2.0	−1.8	0.3	0.1
Dec 1984											
Mar 1985											
June 1985											

Fixed investment, real: percentage change over previous year

Actual values	0.2	3.4	3.3	5.1	5.2	−1.6	0.8	1.7	−5.5	1.7	0.8
forecast in											
July 1983		−1.0	−1.0	0.6	5.9						
Oct 1983			2.8	1.8	6.2						
Jan 1984				2.8	6.7	6.3					
Mar 1984						8.0	4.6				
June 1984						8.5	5.1	0.7			
Oct 1984							6.5	−0.1	2.8	7.9	2.5
Dec 1984								1.6	1.0	8.2	0.1
Mar 1985									0.0	6.8	
June 1985											

Table 9.1 Continued

	831	832	833	834	841	842	843	844	851	852	853
						Quarters forecast					

Exports of goods, real: percentage change over previous year

	831	832	833	834	841	842	843	844	851	852	853
Actual values	−4.9	−2.2	2.6	6.1	11.1	6.2	8.7	10.2	5.8	9.7	7.0
forecast in											
July 1983											
Oct 1983			−1.0	0.6	1.3						
Jan 1984				1.5	6.1	8.9					
Mar 1984						11.6	11.3				
June 1984							14.7	13.2			
Oct 1984								13.7	9.1		
Dec 1984									8.1	15.3	
Mar 1985										12.5	9.6
June 1985											9.3

Imports of goods, real: percentage change over previous year

	831	832	833	834	841	842	843	844	851	852	853
Actual values	−0.4	3.8	7.4	10.2	10.1	3.4	4.5	1.6	2.9	2.5	6.8
forecast in											
July 1983											
Oct 1983			5.0	10.2	5.7						
Jan 1984				6.2	8.9	5.8					
Mar 1984						6.2	7.2				
June 1984							8.6	6.9			
Oct 1984								5.6	3.1		
Dec 1984									1.2	4.2	
Mar 1985										5.6	4.0
June 1985											3.9

Evaluation of forecasts

Investment in inventories has increased by large amounts at the end of 1983 and beginning of 1984, an increase which was captured well by the forecasts, in its intensity as well as its timing.[8] The forecasts for the first quarter of 1984 are relatively far off from the actual value, due to additional inventory investment in anticipation of the labour dispute in the metal industry. In 1984.4 and 1985.1, the offsetting errors-of-forecast are due to speculative movements in the stocks of oil products.

Real fixed business investment was forecast worst. But here, in addition to the special influences already mentioned, the termination of investment subsides also distorted actual investment, which leaves hardly any evidence from which to draw conclusions.

Exports of goods, just as for fixed investment, were underestimated by forcecasts during the starting phase of the model and export functions were revised thereupon. From January of 1984 on, the strong rise in exports was then anticipated by the forecasts. As will be noticed, growth rates of actual exports varied extremely during the period investigated, namely from −4.9% in 1983.1 up to 11.1% in 1984.1.

Imports of goods first rose considerably during the end of 1983 and then slowed down again at the end of 1984. These turning-points in imports were largely captured by the forecasts.

The evaluation of the forecasts by the mean error ME, the mean average error MAE, the root mean squared error RMSE, and by Theil's U in Table 9.2 conforms to the direct investigation of the data.

Table 9.2 Coefficients of forecasting accuracy

Variable	Quarters ahead	ME	MAE	RMSE	U
Industrial production	1	−0.3	1.1	1.2	0.22
Industrial production	2	0.4	1.1	1.4	0.25
Private consumption	1	−0.4	0.6	0.8	0.22
Private consumption	2	−0.7	0.9	1.1	0.29
Inventory investment	1	−0.4	1.8	2.1	0.58
Inventory investment	2	0.5	1.6	1.9	0.51
Fixed investment	1	−1.0	3.5	4.1	1.20
Fixed investment	2	−2.2	3.8	4.6	1.33
Imports of goods	1	−0.3	2.9	3.3	0.50
Imports of goods	2	0.1	2.5	2.8	0.44
Exports of goods	1	−0.4	3.7	3.9	0.48
Exports of goods	2	−0.3	4.7	5.3	0.62

Sample period 1983.4–1984.1 and 1984.3–1985.3. For one quarter ahead forecast 1983.3 is also included.

[8] This interpretation needs to be measured by what is possible. Investment in inventories fluctuates heavily and contains large errors in the data. Revisions of official statistics of up to 2 billion DM are not uncommon.

In summary, the model forecast the modest improvement in economic conditions in West Germany correctly, without indication of false turning-points. The above forecasting record can be considered as a minimum capability, since the model has been improved continually since the first forecasts were made.

REFERENCES

Adams, F. G. and Duggal, V. G. (1974) Anticipations variables in an econometric model: performance of the anticipations version of Wharton Mark III. *International Economic Review*, 15:2, 267–84.

Adams, F. G. and Klein, L. R. (1972) Anticipations variables in macro-econometric models, in *Human Behavior in Economic Affairs* (eds B. Strümpel, J. Morgan and E. Zahn), Elsevier, Amsterdam, London, New York.

Anderson, O., Bauer, R. K. and Giehl, R. (1955) Zur Theorie des Konjunkturtests Modelltheoretische Betrachtungen. *Ifo-Studien*, 1.

Biart, M. and Praet, P. (1985) Forecasting aggregate demand components with business and consumer surveys in the four main EC countries, CIRET Conference, Vienna.

Devilliers, M. (1983) The use of business surveys in short-term forecasts. The French experience, CIRET Conference, Washington.

Eckstein, O., Mosser, P and Cebry, M. (1984) The DRI market expectations model. *The Review of Economics and Statistics*, 66, 181–191.

Evans, M. K. (1969) *Macroeconomic Activity. Theory, Forecasting and Control*, Evanston, New York, London.

Fair, R. C. (1971) *A Short-Run Forecasting Model of the United States Economy*, D. C. Heath and Co. Lexington.

Friend, J. and Taubman, P. (1964) A short-term forecasting model. *The Review of Economics and Statistics*, 46, 229–36.

Foster, J. and Gregory, M. (1977) Inflation expectations: the use of qualitative survey data. *Applied Economics*, 9, 319–29.

Haitovsky, Y. and Treyz, G. I. (1971) The informational value of anticipatory data in macroeconomic model forecast, CIRET Conference, Brussels.

Katona, G. (1951) *Psychological Analysis of Economic Behavior*, McGraw-Hill, New York, Toronto, London.

Knöbl, A. (1974) Price expectations and actual price behavior in Germany. *IMF Staff Papers*, 21, 83–100.

McNees, S. K. (1979) The forecasting record for the 1970s. *New England Economic Review*, pp. 33–53.

McNees, S. K. (1982) The role of macroeconomic models in forecasting and policy analysis in the United States. *Journal of Forecasting*, 1, 37–48.

Modigliani, F. and Cohen, K. (1961) The role of anticipations and plans in economic analysis and forecasting, Studies in Business Expectations and Planning No. 4, University of Illinois, Urbana.

Nerb, G. (1975) *Konjunktruprognose mit Hilfe von Urteilen und Erwartungen der Konsumenten und der Unternehmer*, Duncker and Humbolt, Berlin.

Poser, G. and Hecheltjen, P. (1973) The use of anticipation data in a quarterly econometric model of the Federal Republic of Germany's economy, CIRET Conference, London.

Roberts, Ch. C. (1985) Makroökonomische Konjunkturindikatoren für die Bundesrepublik Deutschland, *Centre for International Research on Economic Tendency Surveys*, Munich.

Strigel, W. H. (1981) *Business Cycle Surveys – Economic Data Based on Business and Consumer Evaluations and Expectations*, Campus–Verlag, Frankfurt, New York.

Tewes, T. (1984) Konsumentenvertrauen und privater Verbrauch in der Bundesrepublik Deutschland. *Die Weltwirtschaft*, pp. 47–61.
Theil, H. (1952) On the time shape of economic microvariables and the Munich Business Test, *Review of the International Statistical Institute*, 20, 105–20.
Wolters, J. (1984) Preiserwartungen des Ifo-Konjunkturtestes und die tatsächliche Preisentwicklung. *Ifo-Studien*, 30, 29–61.
Zarnowitz, V. (1979) An analysis of annual and multiperiod quarterly forecasts of aggregate income, output, and the price level. *Journal of Business*, 52, 1–33.

APPENDIX A A-MODEL FORECASTS FROM 30 DECEMBER 1985

Steady growth

Prospects haven't been so good in a long time, is the conclusion drawn from nearly all national forecasts for 1986. In addition, these positive messages are brought forth with a unanimity which is rare for forecasters. The most recent forecasts for real GNP range from a growth of 2.8% as published by the research institutes to 4% issued by Commerzbank. Calculations with the A-model gave a growth rate of 3.5% (Fig. 9.A1 and Table 9.A1).

Since economic growth is presently dominated by private consumption, forecasters apparently profit from the stability in the propensity to consume, already mentioned by Keynes. With the rate of increase of consumer prices assumed to be 1.8%, the A-model forecasts an increase of 3.7% in real private consumption over 1985, which is close to most other forecasts. Other such 'constants' in forecasts published presently are a slight decrease in investment in housing of 1% (figures from the A-model) and an increase of 4% in real fixed investment. The rate of growth in real imports forecast at 6% exceeds the rate of growth of exports at 5.4% for the first time after two years. Due to improved terms of trade the surplus in the balance of trade will nevertheless increase by 13 billion DM over 1985 and will hit a new record of 90 billion DM.

Contrary to the above forecasts, the evolution of the US economy is much more uncertain at present. Although a continued stagnation[9] in the United States

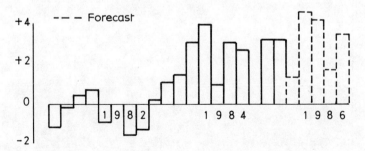

Fig. 9.A1 GNP at constant prices: percentage over previous year

---

[9] Judging the present situation as stagnation is based on the rate of growth in industrial production, which was 2% in October of 1985.

Table 9.A1

	Level	Percentage change over previous year						
						1986		
	1986	1984	1985	1986	1985 IV	I	II	III	IV

	Level 1986	1984	1985	1986	1985 IV	1986 I	1986 II	1986 III	1986 IV
Gross national product at constant prices									
Gross national product	1620	2.7	2.0	3.5	1.4	4.7	4.3	1.8	3.5
Fixed investment	323	0.8	−1.5	4.4	−2.2	11.0	6.0	−0.4	3.1
thereof: Housing	83	2.0	−13.9	−1.3	−14.0	7.5	−2.0	−5.2	−3.2
Inventory investment[a]	13	11.7	1.7	2.9	2.7	−1.3	1.4	0.7	2.0
Private consumption	882	0.6	1.4	3.7	2.1	4.0	4.0	2.7	4.0
Public consumption	320	2.4	1.9	1.8	2.2	2.7	1.7	1.7	1.1
Exports	577	8.0	7.6	5.4	3.4	4.1	6.9	5.1	5.7
Imports	496	5.5	5.0	6.0	5.8	3.0	7.5	6.0	7.3
Income generation									
Industrial production	107.5	3.4	4.5	3.5	1.7	2.5	5.1	2.3	4.2
Employment in industry	7.0	−1.1	1.2	1.4	1.5	1.4	1.4	1.1	1.8
Employment in construction	1.0	−1.4	−9.1	−0.5	−8.9	−1.7	0.6	−0.5	−0.3
Total employment	22.5	0.0	0.8	1.4	0.8	1.1	1.4	1.4	1.7
Unemployment[a]	2.2	0.0	0.0	−0.1	0.0	−0.1	−0.1	−0.1	−0.1
Gross wages and salaries	865	3.0	3.3	5.6	2.7	5.9	5.6	5.4	5.4
Redistribution of income									
Social insurance payments	123	5.3	4.5	7.5	4.1	7.9	7.4	7.6	7.3
Income tax wages and salaries	149	6.4	6.2	4.6	4.3	4.6	3.4	5.6	4.6
Net wages and salaries	592	1.7	2.4	5.4	2.0	5.7	5.8	4.9	5.3
Transfer payments	266	0.7	2.1	2.1	2.6	2.9	1.8	1.8	2.0
Nonprofit income	858	1.4	2.3	4.4	2.2	4.8	4.5	3.9	4.3
Trade in goods									
Exports of goods, real	477	9.2	7.9	5.4	4.7	3.1	7.4	6.2	4.9
Imports of goods, real	401	5.1	4.9	7.9	7.3	4.4	11.4	7.6	8.3
Balance of trade in goods[a]	88	12.0	21.4	13.1	3.8	6.5	5.0	3.4	−1.8

[a] Absolute change.

would not invalidate the 1986 forecasts for West Germany, it would decisively change the dynamics of future developments. When industrial production increased with rates of growth around 15% in 1984, European countries followed with increased growth rates. At present, growth in industrial production in Europe has decreased to an average[10] of around 2%, also under the influence of the slowdown in the United States.

Growth in the United States is at present mostly forecast to be modest.[11] Calculations with the A-model, however,[12] result in a strong increase in American industrial production in 1986 of 8% over the previous year, compared with 2.5% in 1985 (Fig. 9.A2). This will cause increased production in other countries too. Industrial production of West German export countries is forecast to grow by 4.5% on average in 1986, after 2.5% in 1985. From this it follows that there will be no[13] major changes in the expectations of firms and consumers and that there will be steady growth in 1986.

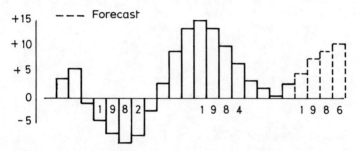

Fig. 9.A2 Industrial production, USA: percentage over previous year

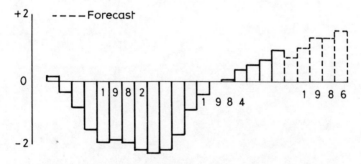

Fig. 9.A3 Total employment: percentage over previous year

[10] The given rate of growth is based on a weighted index of industrial production for these countries. The weights are derived from shares in West German exports.

[11] The German economic research institutes forecast an increase in real GNP of 3% for the USA in 1986, compared with 2.5% in 1985. Compare Wirtschaftskonjunktur, 1985, No. 10.

[12] The results from the A-model are supported by the evolution of several leading indicators. For instance, the leading indicators of the OECD and of Business Week (McGraw-Hill) have increased considerably more than actual production during the last year (1985).

[13] This is based of course on the assumption that such events as an international debt crisis will not occur.

Exports to the USA will – at an exchange rate of DM 2.50 per dollar – grow by 8% in 1986, compared with 20% in 1985. Total West German exports in goods will grow by 5.5%. The decline in the growth rates of exports in 1985.4 and 1986.1 to 4% is therefore not to be interpreted as a turning-point. The same is true of course for the weak rate of growth of GNP of 1.4% in 1985.4.

From the forecasts (Fig. 9.A3) there results an increase in employment of 300 000 persons in 1986 and a decrease in unemployment of 100 000 persons if unemployment is defined as until the end of 1985. Since unemployed persons older than 58 years who are not looking for a job no longer enter the official unemployment figures, the latter might decrease around 200 000.

10

An endogenously time-varying parameter (TVP) model of investment behaviour: theory and application to Belgian data

MARCEL GÉRARD
and CARINE VANDEN BERGHE

10.1 INTRODUCTION

There has long been a debate on the relative roles of the cost of capital on the one hand and the expected level of sales or output on the other as determinants of the investment behaviour of firms. Linked to the role of the cost of capital is the question of the efficiency of cost of capital government policies, including changes in tax parameters, as well as the question of whether or not investors behave as in a walrasian or notional world. Otherwise the role of expected sales opportunities refers to the so-called accelerator principle, reinterpreted in a more modern way as the illustration that investors are facing quantitative constraints on the market for the goods they produce.

Chirinko and Eisner (1983) have illustrated this debate quite well since they write, after comparing most major US macroeconomic econometric models, that the models generally show only modest effects of tax parameter changes on investment. They add that, essentially, this is because investment is determined by expectation of the future as well as by the current situation. Our contribution goes a bit further in that debate. It is based upon an analysis of investment behaviour at industry and aggregate level in Belgium. Section 10.2 presents a general framework which helps to locate that paper in an agenda of research in the econometric analysis of investment behaviour. Sections 10.3 and 10.4 are devoted, respectively, to the theory and application of an endogenously time-varying parameter model of investment behaviour.

This model is in the same spirit as our 1984 article (Gérard and Vanden Berghe, 1984), which is well summarized by Sneessens and Drèze (1986a, b). These authors write that we recognize that at any point in time some firms operate on competitive product markets and gear investment to a desired capital stock reflecting relative prices; whereas other firms operate on imperfectly competitive product markets and gear investment to a desired capital stock reflecting effective demand. An aggregation procedure comparable to that of Lambert (1984) leads again to an approximate expression for the desired stock of capital as a CES function of two expressions, one of which involves relative prices and the other effective demand. Estimates of the parameters of that expression imply estimates of the elasticity of desired capital with respect to relative prices and with respect to effective demand. These two elasticities vary over time. The former is positively related to the proportion of firms constrained by sales expectations, the latter is negatively related. Estimates suggest a rapidly growing influence of effective demand after 1974. Tentative as it may still be, this finding is worth keeping in mind when speculating about the determinants of investment.

The present text differs in some aspects from that commented on by Sneessens and Drèze (1986) since its organization is different, its scope is somewhat broader, interesting properties, including encompassing properties from the point of view of the economic theory of investment, are put forward and the estimation period is longer. However, the findings are confirmed.

Let us add that the observed declining importance of the cost of capital, as a determinant of the *level* of investment when the economic conditions deteriorate, doesn't mean that such a variable has nothing to do with investment. Actually, tax policy operating via the cost of capital appears to be an efficient tool for acting on the *financing* of investment. However, the latter aspect, which we investigated in Gérard (1985a, b), is beyond the scope of the present chapter. The consequences of the tentative findings of that research for the conduct of economic policy are straightforward enough not to be commented on further here. Gordon (1976), commenting on Lucas (1976), says that we could write that mechanical extrapolation of fixed parameter econometric models might provide (he says has provided) policymakers with misleading advice.

Finally, the excercise might also be viewed as an attempt to answer the question put by Chirinko and Eisner, i.e. how expectations of the future change with changes in observed data, as well as an attempt to answer closely related questions. More generally, the question to be answered is: how does responsiveness to determinants change with changes in observed data?

10.2 A GENERAL FRAMEWORK

Assume the following investment function

$$I_t = \sum_{i=0}^{m} \beta_{it} x_{it} + u_t \tag{10.1}$$

where I is investment, x_i's are determinants of investment (you can imagine some

$x_{i+j,t}$ being $x_{i,t-j}$), β_{it} coefficients are allowed to vary over time, and u_t is a residual. Then suppose that the β_{it} coefficients are permitted to vary according to the following expression:

$$\beta_{it} = \bar{\gamma}_0^i + \gamma_0^i \beta_{it-1} + \sum_{j=1}^{m} \gamma_j^i x_{jt} + \sum_{h=1}^{k} \lambda_h^i z_{ht} + \varepsilon_t^i \qquad (10.2)$$

where γ and λ are coefficients, ε is a residual and z are indicators, i.e. variables other than x_j's but with an economic content apparently relevant for explaining actual investment behaviour at a given time and place, despite the fact that they are not determinants if the investment behaves according to economic theory.

From Equation 10.2 there are three reasons for the β_j's to vary over time.

First, the responsiveness to determinants might depend on a series of economic and perhaps noneconomic factors such as the degree of credibility of the government and the economic climate domestically and/or abroad. In that case, new information z is incorporated in the model. We speak then of exogenous variation and of an *exogenously time-varying model of investment behaviour*. We have already investigated along those lines and plan to do it again in the future – see Cattier and Gérard (1981a, b) in French and Gérard (1982) in English.

Second, the responsiveness to determinants might depend on the size of the stimulus or on the relative values of the stimuli. Let us be more explicit. The responsiveness to changes in the cost of capital due for example to tax changes may depend on the level of expected sales opportunities; it is not counterfactual to imagine that investors will be less responsive to decreases in the cost of capital when they feel rationed on the market for the goods that they are producing, then when they don't feel so. Or to imagine that they will be very sensitive to an increased demand for their product when the cost of capital is desperately high. That is what we mean by endogenous variation and an *endogenously time-varying parameter model of investment behaviour*. We started exploiting that approach in our article (Gérard and Vanden Berghe, 1984), reproduced in Weiserbs (1985), and in that respect we owe an intellectual debt to a seminal contribution by Artus and Muet (1984) – also reproduced in Weiserbs (1985). The present chapter continues in that vein.

Third, the responsiveness to determinants might also depend on its own history because investors are learning by experimenting. Therefore the state of the β_j's at time t is a function of that state at time $t-1$ and we could identify the past-based variation as an *own past-based TVP model of investment behaviour*, with TVP denoting time varying parameters. Kalman filtering might be a way to cope will this kind of variation, which we plan to investigate in the coming months.

As stated above, we concentrate on the second type of variation in the rest of the chapter. Otherwise the reader may recognize some link between these approaches and the so-called Lucas critique, (Lucas, 1976).

10.3 AN ENDOGENOUSLY TVP MODEL OF INVESTMENT BEHAVIOUR: THEORY

Starting with the assumption that the actual industrial investor is rarely a pure walrasian agent nor a fully expected sales dominated agent but, at any given moment of time, exhibits behaviour which is some mixture of those extremes, we first try to modelize the behaviour of that individual agent. This enables us to derive *effective demand* equations as combinations of both *notional* (or walrasian) *demand* on the one side and *sales dominated* (or constrained) *demand* on the other; Taylor expansion and the minimization process appear to be two ways of coping with this.

An aggregation over individuals is then proposed, following an avenue opened by Lambert (1984) and assuming the minimization process at individual level. Aggregation leads to *aggregate functions*. After specifying the production process, we then examine some properties of the model. The model indeed offers endogenously TVP opportunities on the one hand, and exhibits some encompassing properties on the other, the standard model of investment behaviour being a particular case.

10.3.1 A model of investment behaviour at individual level

Assume an individual and industrial investor and adopt the following notation: r for both the rate of interest and the discount rate for future costs and earnings, p for the price of produced goods, k for the capital factor, δ for the rate of capital deterioration and replacement, q for the unit price of capital goods, Q for the level of production. The production function is assumed to be $f(k)$, a strictly quasiconcave function; other output, if any, is proportional to k. Moreover, for simplicity, we suppose here that there is no tax and no inflation and that expectations about future prices are myopic.

The investor then considers his or her NPV function which is written down as

$$\sum_{t=1}^{\infty} (1+r)^{-t}(p_t Q_t - c_t k_t) \tag{10.3}$$

subject to

$$Q_t = f(k_t) \tag{10.4}$$

Again, if there are inputs other than the capital they are to be added in the above function; c_t is the cost of capital defined by

$$c = (r + \delta)q \tag{10.5}$$

Now our investor may be either a walrasian investor or an expected sales minded investor. Or most often a mixture of those two extremes.

The walrasian investor believes in the properties of a competitive world and maximizes the NPV with respect to k after substituting $f(k)$ for Q. The outcome of the process is very well known since it is the standard microeconomic equilibrium

$$\frac{\partial f}{\partial k} = \frac{(r+\delta)q}{p} \equiv \frac{c}{p} \quad (10.6)$$

from which we draw the *notional demand function* for capital goods

$$\tilde{k} = \tilde{k}\left(\frac{c}{p}\right) \quad (10.7)$$

whose first derivative is known to be negative.

The expected sales minded investor decides on his or her necessary level of capital goods in the function of the sales he or she expects for the future, Q. For the investor we build up a *sales dominated demand function* for capital goods,

$$\bar{k} = \bar{k}(\bar{Q}, p) \quad (10.8)$$
$$\phantom{\bar{k} = \bar{k}(\bar{Q}, p)}{+ \ -}$$

where $\bar{Q}$ is some exogenous demand for his or her output and p is again the price of the goods produced. The actual investor in fact is some mixture of these extremes. But which mixture? We consider two cases.

First assume that the actual behaviour is simply some unknown function F of both extreme behaviours, so that *effective demand* is

$$k = F\left[\tilde{k}\left(\frac{c}{p}\right), \bar{k}(\bar{Q}, p)\right] \quad (10.9)$$

Exploiting the Taylor expansion formula, this equation might be approximated by

$$k = F_0 + x'F'_0 + x'F''_0 x \quad (10.10)$$

with F_0 the value of the function at point 0, F'_0 and F''_0 the corresponding first and second derivatives, F'_0 being a vector and F''_0 a matrix; finally $x' = [\tilde{k}(c/p), \bar{k}(\bar{Q}, p)]$. In other words, we have, for instance,

$$\log k = \alpha + (\beta_0 + \beta_1 \log \tilde{k} + \beta_2 \log \bar{k}) \log \tilde{k} + (\lambda_0 + \lambda_1 \log \tilde{k} + \lambda_2 \log \bar{k}) \log \bar{k}$$
$$(10.11)$$

Assuming then that $\tilde{k}$ is proportional to $(c/p)^{-1}$ and that $\bar{k}$ is proportional to $\bar{Q}$, and also that F''_0 is symmetric in the sense that $\lambda_1 = \beta_2$, we have

$$E_{k,(c/p)} = \beta_0 + 2\beta_1 \log \tilde{k} + 2\beta_2 \log \bar{k} \quad (10.12)$$

and

$$E_{k,Q} = \lambda_0 + 2\lambda_1 \log \tilde{k} + 2\lambda_2 \log \bar{k} \quad (10.13)$$

where E stands for elasticity. These elasticities are *endogenously time-varying parameters* as they vary over time in line with the explanatory variables of the behavioural model, i.e. c/p and Q. Note that if we impose $F''_0 = 0$, Equation 10.11 reduces to

$$\log k = \alpha + \beta_0 \log \frac{p}{c} + \lambda_0 \log \bar{Q} \quad (10.14)$$

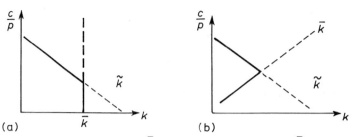

Fig. 10.1 The minimization process (a) $\bar{k}$ independent of p, or fixed; (b) $\bar{k}$ a negative function of p

which actually is the Eisner–Nadiri model precisely, but derived differently.

Instead of using an approximation in logarithms we can also use the alternative equation,

$$k^{-\rho} = \alpha + (\beta_0 + \beta_1 \tilde{k}^{-\rho} + \beta_2 \bar{k}^{-\rho})\tilde{k}^{-\rho} + (\lambda_0 + \lambda_1 \tilde{k}^{-\rho} + \lambda_2 \bar{k}^{-\rho})\bar{k}^{-\rho} \quad (10.15)$$

or if we assume $F_0'' = 0$,

$$k^{-\rho} = \alpha + \beta_0 \tilde{k}^{-\rho} + \lambda_0 \bar{k}^{-\rho} \quad (10.16)$$

i.e.

$$k = \left[\alpha + \beta_0 \left(\frac{p}{c}\right)^{-\rho} + \lambda_0 \bar{Q}^{-\rho} \right]^{-(1/\rho)} \quad (10.17)$$

In all this material the economist will recognize well-known specifications but with a renewed interpretation here, for example, that the last equation above is a typical CES form. However, the interpretation is different; the CES here appears to be a mathematical representation of a mixture of typical behaviours towards investment; for that reason, assuming that those typical behaviours are expressed logarithmically, it is not impossible to obtain a form like, for instance,

$$\log k = \left[\alpha + \beta_0 \left(\log \frac{p}{c}\right)^{-\rho} + \lambda_0 (\log \bar{Q})^{-\rho} \right]^{-(1/\rho)} \quad (10.18)$$

Another mixture takes its roots in the economics of rationing and consists of assuming that *effective demand* is ruled by

$$k = \min(\tilde{k}, \bar{k}) \quad (10.19)$$

Unlike previous assumptions one assumes here a priori a particular form for $F(.)$. Figure 10.1 illustrates the minimization process; the solid line corresponding to effective demand.

10.3.2 Aggregation over individuals

Since statistics are available at industrial, sectoral or more aggregate level, we have to aggregate over individual investors. The problem is then to proceed from

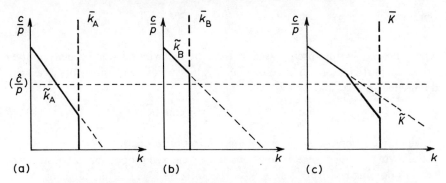

Fig. 10.2 (a) Individual A; (b) individual B; (c) aggregate over A and B

individual functions

$$k = F(\tilde{k}, \bar{k}) \tag{10.20}$$

to aggregate functions

$$K = K(\ldots) \tag{10.21}$$

K being the *aggregate effective demand* for capital goods.

If individuals are perfectly identical the problem is very simple and in some senses trivial. In that case we compute aggregate $\tilde{K}$ and aggregate $\bar{K}$ and $K(\tilde{K}, \bar{K})$ is formally identical to $F(\tilde{k},\bar{k})$. However, that is a kind of degenerate case. If we admit that different investors form different mixtures the problem becomes less simple but also more relevant. To make the point clear consider the two individual graphical examples in Fig. 10.2. The figure shows that if $(\hat{c}/p)$ prevails individual A will exhibit walrasian behaviour while individual B will be dominated by expected sales. At aggregate level, both $\tilde{K}$ and $\bar{K}$ come out from an aggregation over individual $\tilde{k}$ and $\bar{k}$ but clearly, effective demand, in case it is ruled by a minimization process, at aggregate level is not the aggregation over individual effective demands.

In other words, and generalizing, we have

$$\tilde{K} = \sum_{i=1}^{N} \tilde{k}_i \tag{10.22}$$

$$\bar{K} = \sum_{i=1}^{N} \bar{k}_i \tag{10.23}$$

but

$$K \leq \min[\tilde{K}, \bar{K}] \tag{10.24}$$

The recognition of this led Lambert to propose an aggregation process which, assuming as above that $\tilde{k}_i$ and $\bar{k}_i$'s are jointly and lognormally distributed, ends up with an attractive CES functional form for the aggregate effective demand. Transposing from his framework to our problem of investment behaviour, we obtain

$$K = (\tilde{K}^{-\rho} + \bar{K}^{-\rho})^{-(1/\rho)} \tag{10.25}$$

where ρ is to be interpreted as a statistic of behavioural homogeneity among firms. In particular if all investors are walrasian or if all investors are sales dominated, $\rho \to \infty$ and

$$\lim_{\rho \to \infty} (\tilde{K}^{-\rho} + \bar{K}^{-\rho})^{-(1/\rho)} = \min(\tilde{K}, \bar{K}) \tag{10.26}$$

in which we recognize the aggregate minimization process.

Following Lambert we can now compute a weighted proportion of constrained, or sales dominated, firms

$$p_W^c = 1 \bigg/ \left[1 + \left(\frac{\tilde{K}}{\bar{K}}\right)^{-\rho} \right] \tag{10.27}$$

Given the aggregate function K, this proportion is a decreasing function of $\bar{K}$ since

$$\frac{\partial p_W^c}{\partial \bar{K}} = \left\{ -1 \bigg/ \left[1 + \left(\frac{\tilde{K}}{\bar{K}}\right)^{-\rho} \right]^2 \right\} \left(\frac{\tilde{K}}{\bar{K}}\right)^{-\rho-1} \rho \frac{\tilde{K}}{\bar{K}^2} < 0 \tag{10.28}$$

illustrating that p_W^c increases when $\bar{K}$ becomes smaller.

10.3.3 Specification of $\tilde{k}$ and $\bar{k}$, $\tilde{K}$ and $\bar{K}$

Suppose that, at individual level for firm i, the production process obeys a function

$$f(k) = n_i k^v, \quad 0 < v < 1 \tag{10.29}$$

where k is the stock of capital goods, while n_i and v are parameters. Other inputs are assumed to be in a strict complementarity relation with capital.

The *notional demand for capital goods at individual level* may then be written, with p and c already defined

$$\tilde{k}_i = (n_i v)^{1/(1-v)} \left(\frac{p}{c}\right)^{1/(1-v)} \tag{10.30}$$

or, in a more compact way

$$\tilde{k}_i = \alpha_i \left(\frac{p}{c}\right)^\beta \tag{10.31}$$

The *constrained demand for capital goods at individual level* may also be obtained from Equation 10.29 and is

$$\bar{k}_i = \left(\frac{\bar{q}_i}{n_i}\right)^{1/v} \tag{10.32}$$

where $\bar{q}$ stands for the level of sales or demand expected by the firm.

The *aggregate notional demand function for capital goods* using Equation 10.31 now becomes

$$\tilde{K} = \sum_{i=1}^{N} \alpha_i \left(\frac{p}{c}\right)^{\beta} \qquad (10.33)$$

where N stands for the number of firms or, in logarithms, setting

$$\gamma = \log \sum_{i=1}^{N} \alpha_i, \qquad (10.34)$$

$$\log \tilde{K} = \gamma + \beta \log \left(\frac{p}{c}\right) \qquad (10.35)$$

On the other hand, considering that for any firm i, $\bar{q}_i$ is a proportion s_i of a given quantity $\bar{Q}$ at aggregate level, we have the following expression for the *aggregate constrained demand function for capital goods*, using Equation 10.32

$$\bar{K} = \bar{Q}^{1/v} \sum_{i=1}^{N} \left(\frac{s_i}{n_i}\right)^{1/v} \qquad (10.36)$$

or, again in logarithms, setting

$$\varepsilon = \sum_{i=1}^{N} \left(\frac{s_i}{n_i}\right)^{1/v}, \qquad (10.37)$$

$$\log \bar{K} = \varepsilon + \frac{1}{v} \log \bar{Q} \qquad (10.38)$$

Then, assuming that $\bar{Q}$ is determined by the actual produced quantity Q, we may rewrite Equation 10.38 as

$$\log \bar{K} = \varepsilon + \lambda \log Q \qquad (10.39)$$

Alternatively, we may assume that $\bar{Q}$ depends on past values of Q.

It turns out that the *aggregate effective demand function for capital goods* is finally

$$\log K = \left\{ \left[\gamma + \beta \log \left(\frac{p}{c}\right)\right]^{-\rho} + [\varepsilon + \lambda \log Q]^{-\rho} \right\}^{-(1/\rho)} \qquad (10.40)$$

and we will estimate this function in the empirical section of the chapter.

10.3.4 Some time-varying and encompasing properties of the functional form of the aggregate effective demand function for capital goods

The functional form of Equation 10.40 exhibits two kinds of properties. First, it enables us to measure *elasticities* of capital goods demand which are *allowed to vary endogenously* over a length of time, i.e. whose values vary in line with changes in the components of aggregate notional and constrained demand for capital goods. Second, it also exhibits *encompassing properties* since many investment functions may be viewed as particular cases of Equation 10.40, including the Eisner–Nadiri function.

Indeed, from Equation 10.40 we can derive the elasticities of capital goods demand with respect to p/c and to Q, i.e.

$$E_{K,p/c} = \beta \left[1 + \left(\frac{\varepsilon + \lambda \log Q}{\gamma + \beta \log p/c} \right)^{-\rho} \right]^{-(1+\rho)/\rho} \qquad (10.41)$$

$$E_{K,Q} = \lambda \left[1 + \left(\frac{\gamma + \beta \log p/c}{\varepsilon + \lambda \log Q} \right)^{-\rho} \right]^{-(1+\rho)/\rho} \qquad (10.42)$$

We observe that the elasticity of capital goods demand with respect to Q is going up when Q is going down, suggesting that, when economic conditions deteriorate, $\bar{K}$ becomes smaller and a larger number of individual investors regard their investment behaviour as constrained by limited expected sales opportunities; investors then become more sensitive to the level of expected demand for the goods they are producing and consequently less sensitive to the notional determinant of investment. That latter point is illustrated by the fact that the elasticity of capital goods demand with respect to p/c – the notional determinant of investment – is an increasing function of Q.

Conversely, when Q is going up, investors are less constrained by sales opportunities and become more optimistic regarding expected demand for their product; therefore they are less sensitive to the constrained determinant of investment behaviour – Q – and consequently more sensitive to the notional determinant – p/c.

Finally, the weighted proportion of constrained firms defined by Equation 10.27 becomes, from Equation 10.40

$$p_w^c = 1 \bigg/ \left[1 + \left(\frac{\gamma + \beta \log p/c}{\varepsilon + \lambda \log Q} \right)^{-\rho} \right] \qquad (10.43)$$

and appears to be negatively related to the output, i.e. the proportion of constrained firms goes up when economic activity declines.

Regarding *encompassing properties* we will limit the discussion here to two particular values of ρ, i.e. $\rho \to \infty$ and $\rho = -1$

When $\rho \to \infty$, Equation 10.40 is to be rewritten

$$\log K = \min \left[(\gamma + \beta \log p/c), (\varepsilon + \lambda \log Q) \right] \qquad (10.44)$$

This is the 'aggregate minimum,' already mentioned above. We immediately see that we have then either a *purely walrasian* demand function for capital goods or a *pure accelerator* effect. In the former case we have

$$\log K = \gamma + \beta \log \frac{p}{c} \qquad (10.45)$$

and consequently
$$E_{K,p/c} = \beta, E_{K,Q} = 0 \tag{10.46}$$
both fixed over time. In the latter case,
$$\log K = \varepsilon + \lambda \log Q \tag{10.47}$$
and
$$E_{K,p/c} = 0, \quad E_{K,Q} = \lambda \tag{10.48}$$
also fixed over time.

When $\rho = -1$, Equation 10.40 becomes
$$\log K = \alpha + \beta \log \frac{p}{c} + \lambda \log Q \tag{10.49}$$
where α stands for $\gamma + \varepsilon$. Equation 10.49 is the function of capital goods demand suggested by Eisner and Nadiri (1968); again it is a function with fixed coefficients. Then,
$$E_{K,p/c} = \beta, E_{K,Q} = \lambda \tag{10.50}$$
also not allowed to vary endogenously over a length of time. Moreover, if for any reason β is equal to λ, Equation 10.29 becomes
$$\log K = \alpha + \beta \log \frac{pQ}{c} \tag{10.51}$$
which is precisely the so-called *standard neoclassical* function suggested by Jorgenson – see for example Jorgenson (1963) or Hall and Jorgenson (1971).

10.4 AN ENDOGENOUSLY TVP MODEL OF INVESTMENT BEHAVIOUR: APPLICATION

We are now able to estimate Equation 10.40 for a series of Belgian industrial sectors and aggregates. Using the estimated values of the parameters of Equation 10.40 we shall then compute the statistics defined by Equations 10.41–10.43 for the period 1956–1984 and comment on the observed movements in the series obtained.

10.4.1 Estimation

The data used for the estimation cover the years 1953 (1956 for the dependent variable) to 1983 for a series of Belgian industrial sectors and aggregates. They come from the National Account, except the rate of interest, which is that charged by the National Company for Industrial Credit (Société Nationale de Crédit à l'Industrie or SNCI in French, Nationaal Maatschappij van Krediet aan de Nijverheid or NMKN in Dutch) on its 5–10 year loans.

Table 10.1 An endogenously TVP model of investment behaviour: application to Belgian data

Sectors	γ	β	ε	λ	$\bar{R}^2$	SSR
03 Textile industries	13.73^a (2.11)	-5.22 (-1.61)	2.06^a (3.35)	0.45 (2.69)	0.637	0.126
05 Timber and furniture[b]	0.00 (0.00)	0.00 (0.00)	0.44^a (9.10)	0.81^a (47.80)	0.987	—
07 Chemicals	-4.91^a (-31.25)	-0.17 (-1.27)	-1.18^a (-4.68)	1.74^a (19.30)	0.985	0.088
09 Iron, steel, nonferrous[b]	-4.70^a (-12.80)	0.81 (0.50)	-6.10^a (-5.20)	3.34^a (8.20)	0.928	—
10 Metal products	-5.28^a (-18.96)	0.11 (0.52)	-1.46^a (-2.95)	1.37^a (10.82)	0.982	0.064
11 Total of manufacturing industry	6.19^a (33.90)	0.39^a (2.56)	-3.45^a (-6.42)	1.65^a (16.75)	0.993	0.032
12 Total of GAFA	8.49^a (43.08)	0.13 (0.69)	-5.97^a (-8.19)	1.90^a (17.84)	0.997	0.021

$(\ldots) = t$-values.
[a] Significantly different from zero at the 5% level.
[b] From Gérard and Vanden Berghe (1984).
GAFA = gross accumulation of fixed assets.

Unlike previous exercises we concentrate the estimation on the specification provided by Equation 10.40 this time, using the appropriate routine of RATS on a micro-computer. However, on the basis of past experiment we have fixed ρ equal to 10. The results are reported in Table 10.1 and suggest that output is the stronger determinant of investment behaviour. Using the estimates reported in Table 10.1 we can now compute statistics 10.41–10.43 for the 1956–1984 period and comment on the movements observed in the series obtained.

10.4.2 The $E_{K,p/c}$, $E_{K,Q}$ and p_w^c statistics

In computing these statistics we have to cope with the effect of the general upward trend in Q over the period investigated. Therefore after a first attempt, the series of statistics have been recomputed in differential with respect to their trend and the average values in the initial series have been used as benchmarks. The results have been translated into a series of figures.

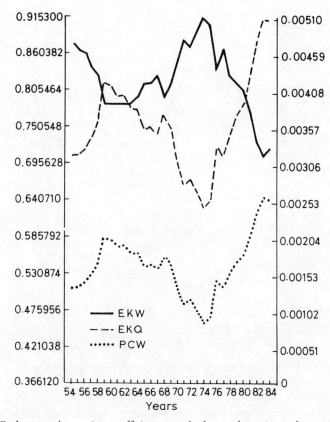

Fig. 10.3 Endogenously varying coefficients: total of manufacturing industry (sector 11)

In the figures, the letters EKW, EKQ and PCW have been used instead of $E_{K,p/c}$, $E_{K,Q}$ and p_w^c respectively. A special scale on the right of the figure has usually been introduced for the EKW statistics; this statistic has been omitted in Fig. 10.4 since the data were irrelevant in this instance.

An inspection of the figures seems to support the hypothesis that elasticities of investment or, more precisely, capital goods demand with respect to its determinants, can be regarded as varying in line with the economic situation: a recession means more pessimistic expectations amongst investors so that they become more demand minded while a recovery seems to have the opposite effect. In this respect it should be noted that since the beginning of the present economic crisis, say 1974, a general upward movement is observed in the EKQ elasticity – investors become more pessimistic and demand minded – but a peak seems to be passed in 1984. Indeed, when that year is introduced, a decrease in EKQ – or a slightly more moderate increase, in the case of metal products – is observed, illustrating that some economic recovery or a change in the economic climate is bringing a touch of optimism to the spirit of investors.

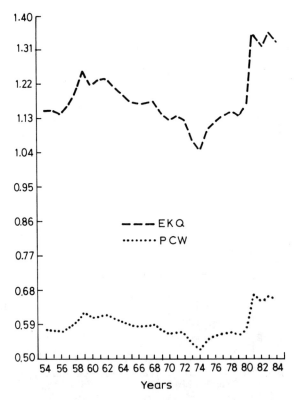

Fig. 10.4 Endogenously varying coefficients: total of the gross accumulation of fixed assets (GAFA) (sector 12)

TVP model of investment behaviour: application

We can now turn to an inspection of each figure separately and first consider the two aggregates (Figs 10.3 and 10.4), i.e. the total of manufacturing industry and the total of gross accumulation of fixed assets. Both exhibit a peak in expected demand elasticity – the EKQ curve – in 1959 and 1968 and a trough in 1974, immediately followed by a sharp rise of the curve, briefly interrupted in 1977 for the first aggregate and interrupted two years later for the other aggregate. The latter aggregate also presents an intermission in the rising movement of EKQ in 1982 while for both a new peak seems to be passed in 1984. The movement of the PCW curve follows that of the EKQ curve and an inverse movement is observed for EKW in Fig. 10.3.

Metal products (Fig. 10.5) also peak in 1959 and 1968 for EKQ and PCW while troughs appear in 1966 and 1971. The latter trough is immediately followed by a continuous rise for EKQ and PCW whose slope, however, becomes more moderate after 1982. An inverse movement is again observed for EKW. For the iron, steel and nonferrous sector (Fig. 10.6), the present upward movement in

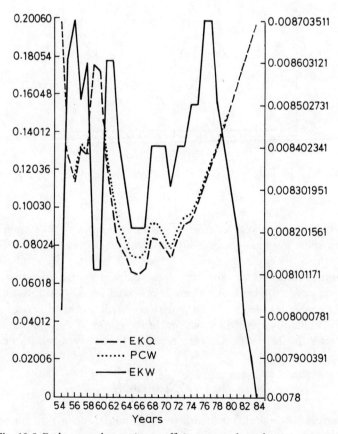

Fig. 10.5 Endogenously varying coefficients: metal products (sector 10)

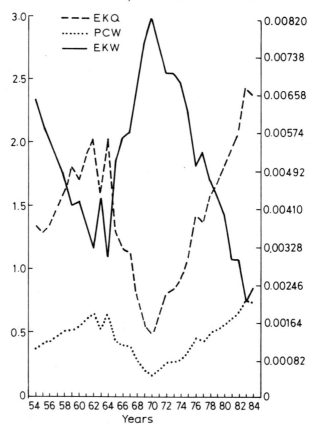

Fig. 10.6 Endogenously varying coefficients: iron, steel and nonferrous (sector 09)

EKQ starts in 1970 – a year of trough – with, as for the total of manufacturing industry, a short intermission in 1977 and a change of direction in 1984.

Two peaks in EKQ clearly appear in 1968 and 1976 for chemicals (Fig. 10.7) and as for most other sectors and aggregates under investigation here a new peak seems to be passed in 1984. A trough is observed in 1974 and in addition the 1977–1979 subperiod seems to be a temporary break in the growing pessimism. Finally, a continuous upward movement, broken in 1977 and apparently in 1984 too, characterizes the recent movements in EKQ for timber and furniture.

10.5 CONCLUSION

After describing a general framework for TVP models of investment behaviour, we have concentrated on what we call an endogenously time-varying parameter

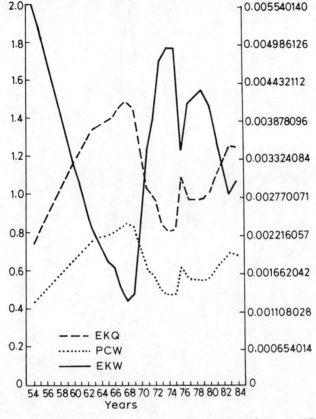

Fig. 10.7 Endogenously varying coefficients: chemicals (sector 07)

model of investment behaviour, first presenting the theory supporting that model and second applying the model to a series of Belgian industries and aggregates.

The theory supporting the model owes much to the developing literature on rationing, the aggregation process and TVP models. The empirical exercise first confirms that investors are more responsive to sales opportunities than to the cost of capital. But it also demonstrates that the responsiveness itself varies over time in line with changes in the intensity of economic stimuli. It appears therefore that investors become less responsive to the cost of capital and thus to tax changes when the level of expected sales opportunities, which illustrates the economic climate, deteriorates. Alternatively, they become more responsive to this cost when the economic climate improves. This asymmetry seems to have important consequences for the efficiency of economic policies and thus should not be ignored by policy-makers.

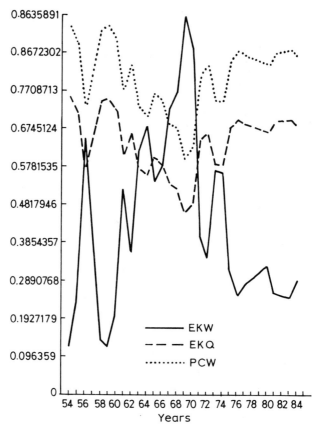

Fig. 10.8 Endogenously varying coefficients: timber and furniture (sector 05)

ACKNOWLEDGEMENTS

The present chapter is part of a research programme on the behaviour of investors; it is a revision and re-estimation of a paper that we presented on a number of occasions in 1985, including seminars in Antwerp and Namur, and conferences in Venice, Cambridge (Mass.) (the World Congress of the Econometric Society), and Vienna (CIRET); among the many people who provided us with comments and suggestions we should like to mention P. Artus, R. Faini, D. Kendrick, M. Miller and P.A. Muet. We are also indebted to the editorial department at Chapman and Hall whose revision of the text has improved it substantially, although of course any errors remain our own.

REFERENCES

Artus, P. and Muet, P. A. (1984) Investment, output and labour constraints, and financial constraints: estimation of a model with several regimes. *Recherches Economiques de Louvain*, 50, Nos 1–2; also in Weiserbs (1985).

Cattier, J. and Gérard, M. (1981a) L'incidence des mouvements économiques sur les coefficients économétriques d'une fonction d'investissement: les retards échelonnés variables. *Prévision et Analyse Économique*, 2, No. 2.

Cattier, J. and Gérard, M. (1981b) La dynamisation des coefficients économétriques: application à des données sectorielles belges. *Cahiers Economiques de Bruxelles*, 97.

Chirinko, R. and Eisner, R. (1983) Tax policy and investment in major US macroeconomic econometric models. *Journal of Public Economics*, 20 No. 2.

Eisner, R. and Nadiri, M. I. (1968) Investment behaviour and the neoclassical theory. *Review of Economics and Statistics*, 50, August, 369–82.

Gérard, M. (1982) Sectoral investment, business cycle and economic policy: an econometric study for Belgium (1953–1976), in *Selected Papers on Contemporary Econometric Problems* (presented at the Econometric Society European Meeting, Athens, 1979, and dedicated to the memory of Stephan Valavanis) (ed. G. Charatsis), The Athens School of Business Administration, Athens.

Gérard, M. (1985a) Investment and the cost of capital in the EEC countries: general formalization and measurement (1975–1982), paper presented at the Econometric Society World Congress, Cambridge, Mass., 19 August.

Gérard, M. (1985b) *La Fiscalité et l'Investissement dans la Communauté Européenne*, rapport de recherche.

Gérard, M. and Vanden Berghe, C. (1984) Econometric analysis of sectoral investment in Belgium (1956–1982). *Recherches Economiques de Louvain*, 50 Nos 1–2; also in Weiserbs (1985).

Gordon, R. (1976) Can econometric policy evaluations be salvaged? A comment, in *The Phillips Curve and Labour Market* (eds K. Brunner and A. Meltzer) Carnegie–Rochester Conferences Series on Public Policy, Vol. 1, North Holland, Amsterdam.

Hall, R. and Jorgenson, D. W. (1971) Application of the theory of optimal capital accumulation, in *Tax Incentives and Capital Spending*, (ed. G. Fromm). The Brookings Institution, Washington

Jorgenson, D. W. (1963) Capital theory and investment behaviour. *American Economic Review*, 53, 247–59.

Lambert, J. P. (1984) *Disequilibrium macro models based on business survey data: theory and estimation for the Belgian manufacturing sector*, Thèse de Doctorat, University of Louvain. Cambridge University Press (to be published).

Lucas, R. (1976) Econometric policy evaluation: a critique, in *The Phillips Curve and Labour Market*, (eds K. Brunner and A. Meltzer), Carnegie–Rochester Conferences Series on Public Policy, Vol. 1. North-Holland, Amsterdam.

Sneessens, H. and Drèze, J. (1986a) A discussion of Belgian unemployment: combining traditional concepts and disequilibrium econometrics, *Economica*, 53, S89–S119.

Sneessens, H. and Drèze, J. (1986b) What, if anything, have we learned from the rise of unemployment in Belgium, 1974–1983? *Cahiers Economiques de Bruxelles*, 110–11.

Weiserbs, D. (ed.) (1985) *Industrial Investment in Europe: Theory and Measurement*. Martinus Nijhoff.

11

Budget consolidation, effective demand and employment

WULFHEINRICH VON NATZMER

The current West German economic policy can be characterized as a budget consolidation policy augmented by several supply-side measures. This policy has been quite successful in consolidation but it could not avoid unemployment figures exceeding 2 million. This situation triggered a discussion whether the government should pursue an anticyclical deficit policy which supports the strengthening of the supply side by an increase in demand. This chapter evaluates the effects of a postponement of the first budget consolidation attempt which was run from 1976 to 1979. The first simulation only shifts the dates when borrowing takes place. In contrast, the second increases total government debt. Both evaluations are carried through in two different theoretical frameworks. Case 1 uses a Keynesian-type consumption function whereas case 2 incorporates a negative impact of public deficits on private consumption to consider the Barro–Ricardo equivalence proposition.

11.1 INTRODUCTION

The last ten years of West Germany's economic performance have been accompanied by considerable government deficits. They are at the centre not only of economic but also of political discussion. They were an additional factor in stimulating the 1982 change in the party coalition. The government in power since then has made it its proclaimed goal to bring down federal budget deficits within the next few years. Keeping in mind this political framework, we will confine ourselves here to an economic analysis of government debt with regard to its effects on economic performance.

The pros and cons of government deficits are the subject of wide public discussion. This sophisticated debate becomes more complex due to the distinction between particular components of the deficits. They are often seen as a combination of cyclical and noncyclical elements which demand different policy reactions, e.g. Wille (1983). Arguments range from the middle position of 'fiscal neutrality', e.g. Barro (1974), Miller (1983, p. 3), Evans (1985, p. 85), Dwyer (1985, p. 656), to the hypothesis of a self-consolidating 'debt paradox', e.g. Oberhauser (1985), Gandenberger (1983), on the one hand, and to the hypothesis of counterproductive effects due to the crowding out of private investment on the other, e.g. Woll (1984, p. 174).

The following analysis discusses the debt paradox hypothesis and considers the proposition of fiscal neutrality or the Barro–Ricardo equivalence proposition. A discussion of the relevance of a crowding out of private investment is beyond the scope of the chapter. In Section 11.2 the central features of the arguments are discussed. One way of testing the theories would require the specification and estimation of econometric systems which model the main lines of reasoning of the two arguments. If this were successful one would be able to weigh one theory against the other. A second approach would be to use an existing model, for example the Freiburger and Tübinger quarterly econometric model, version 81, as was done in this study. Consequently, the results can hardly be interpreted as a detailed test of the two countrary theories. But they do tell us something about the character of the model used. They may supply some points and answers useful to a discussion of the validity of the different arguments. But the limitations of the analysis differ for each of the two propositions. The Freiburger and Tübinger econometric model is demand orientated; it incorporates adaptive expectations and takes capacity into account by way of investment effects on potential output. Therefore it belongs to the class of models that include a supply sector although the supply side is '...less developed than the demand side' (Lambelet, 1985, p. 88). Because of this characteristic one can expect it to be more suitable for questions emphasizing the demand side of the economy.

Section 11.3 presents the developments of several central economic variables in West Germany and discusses the different experiments which were carried through. I use the term 'experiments' because the main objective of the simulations is to obtain an idea of the multiplier effects. Furthermore, the implementation of the debt paradox policy requires that several conditions be fulfilled if the experiments are to serve as a serious test. The importance of the economic situation is particularly emphasized in this discussion of demand policy, e.g. Wille and Kronenberger (1984, p. 611). But views may diverge as to whether the estimation period of the model and the requirements of the debt paradox policy are compatible. Their correspondence is vital not only for this particular type of policy, but also for every demand policy, because otherwise detrimental side effects may erode supply-side conditions, e.g. Sievert (1984, pp. 383–4). These conditions are assumed given in the experiments. Their influences do not differ from those involved in the specification and estimation process. Not considered in the experiments are effects based on changes in the willingness to work, and in the claims to labour compensation; or those effects resulting from the effort to improve economic life, i.e. to detect new and cheaper ways to satisfy needs; also disregarded are effects resulting from the willingness to postpone the satisfaction of wants and to bear greater risks in the prospect of greater opportunity, which are essential to supply-side policy (Sievert, 1984, pp. 382–3).

Section 11.4 discusses important features and causal relationships in the Freiburger and Tübinger econometric model. Section 11.5 presents the results of the policy experiments.

11.2 DEBT PARADOX AND FISCAL NEUTRALITY

The central proposition of the debt paradox is that production and employment depend on demand; in other words, the multiplier theory is empirically valid, e.g. Oberhauser (1985), Scherf (1985). Public debt which increases real effective demand induces positive production and employment effects.[1] Therefore one of the main economic policy objectives could be attained in this way. A byproduct of the higher level of economic activity is the capacity to painlessly service the debt out of increased government revenues and reduced unemployment payments. The crucial question is whether rising public debts trigger higher real demand. In a situation of cyclical underemployment anticyclical deficits are claimed to have this effect. In the course of a business cycle a nonrestrictive policy such as this promises both increased employment and consolidation of the additional government debt. It is unimportant whether the increase in total real demand stems from additional government spending or from reduced taxes and social security contributions.[2] But it is crucial that private investment varies parallel with the capacity utilization rate (Oberhauser, 1985, p. 335). Consequently, an increase in economic activity has to accelerate investment. Such reinforcements of economic activity lead to sustained circular-flow effects. Within this theory rational agents are aware that in the future they can pay taxes out of the higher income induced by this economic policy to service the additional government debt. There is no need to increase savings.

In contrast, the fiscal neutrality proposition argues the other way around. Because private agents do not believe in the expansionary effects of additional government debt, they reduce their demand by an amount necessary to pay future taxes. Neutrality is given when the additional public demand, i.e. public investment, or the private demand, resulting from a reduction in fiscal charges, is compensated by an increase in savings to pay off the additional debt in the future. In terms of the consumption function, in the debt paradox case (case 1)

---

[1] The proposition is not equivalent to the balanced budget view as discussed in Stein (1983, p. 26).
[2] In von Natzmer (1987) a reduction in the social security contribution rate, i.e. a policy parameter, is analysed within the Freiburger and Tübinger quarterly econometric model (Lüdeke et al., 1984). The experiment in which the lower government revenues are financed by domestic and foreign public borrowing leads to an increase in consumption and investment. When the disburdening of private agents ceases, the positive effects on production fade. If the disburdening of the private sector is financed by a reduction in transfer payments to private households the expansive effect on demand is nearly compensated, because the income is mainly shifted within the aggregated private disposable income from those who receive transfer payments to those who work and invest (von Natzmer, 1985b). Supply-side effects which stimulate the economic climate can hardly be detected in this form of modelling.

no dampening influences on private demand result from the rise in government debt. A simple but empirically valid consumption function can be based on a Keynesian approach including a habit persistence effect augmented by an interest rate impact, e.g. Lüdeke et al. (1984).[3]

The Barro–Ricardo equivalence proposition operates through a change in the parameters of the consumption function. An easy way to consider this hypothesis is to enclose a shift term which depends on public debt, i.e., except for the constant term the parameters are unaffected by changes in government debt. Blinder (1985, p. 686) formulates the proposition that '...Under certain circumstances, an outward shift of the supply curve of government bonds will call forth an exactly equal outward shift of the demand curve because people will want to save to pay their future taxes'. In the approach mentioned above this increase in savings induced by additional government debt shifts the consumption function outward as well. Miller (1983, p. 7) argues against an approach like this. In his opinion a regression of consumption on government deficit '...has little to say about the proposition'. He develops his argument in a Rational Expectations framework with government behaviour which can be expressed in policy rules. A change in policy is equivalent to a change of a parameter in the policy rule. Consequently the Rational Expectations formation transfers the policy parameter alteration to all or most of the reduced form parameters which determine the evolution of economic variables. It is crucial for this objection that government debt policy follows a detectable rule, whereas the Rational Expectations proposition is of minor importance for this result, e.g. von Natzmer (1985a). Whether agents were given an opportunity to detect such a deficit policy rule or not is discussed in Section 11.3.

Changing the specification of the consumption function in the model by adding a deficit term, case 2, shows a significant influence.[4] In 1980 prices the regression

---

[3] The empirical validity is proposed in respect of using such a common approach in an aggregated model. Hansen (1984) shows the importance of a distinction between durables and nondurables for an explanation of private consumption for West Germany. Blümle (1985) argues that savings decisions are far too complicated to be viewed as a residual of such a simple consumption function.

[4] Aschauer (1985, p. 122) runs a regression on real private consumption with lagged consumption and lagged per capita net deficit of the total government sector. In contrast, the consumption function in this chapter includes current total government deficit and private disposable income. These two variables can be considered as endogenous. Therefore ordinary least squares may be inappropriate. An instrumental variable approach, using the corresponding nominal variables lagged four quarters as instruments, leads to the following results:

$$C = 9.61 + 0.466C\,(t-4) + 0.484Y/PC - 1.12r - 0.220D/PC$$
$$(1.10)\ (3.22)\qquad\qquad (3.10)\qquad\quad (4.56)\ (1.86)$$

and

$$C = \ 18 \times 0.53 + (1-0.53)C(t-4) + 0.91 \times 0.53Y/PC - 2.1 \times 0.53r$$
$$- 0.41 \times 0.53D/PC$$

$R^{**}2 = 0.988, \quad DW = 1.984;$

in the second equation the adjustment parameter is explicitly considered. The results of a more detailed analysis of the influence of the government deficit as an explanatory variable in the consumption function are forthcoming.

with 62 observations leads to the following results:

$$C = 4.94 + 0.384C(t-4) + 0.560Y/PC - 1.07r$$
$$(0.923)\,(5.41) \qquad\qquad (7.28) \qquad\quad (-6.87)$$
$$- 0.119D/PC + \text{seasonal dummies}$$
$$(-2.78)$$

or, with the adjustment parameter explicitly considered,

$$C = 8 \times 0.62 + (1 - 0.62)C(t-4) + 0.91 \times 0.62Y/PC - 1.7 \times 0.62r$$
$$(0.84) \qquad (8.69) \qquad\qquad\qquad (33.1) \qquad\qquad (-4.5)$$
$$- 0.19 \times 0.62D/PC + \text{seasonal dummies}$$
$$(-2.89)$$

regression period: first quarter 1970 – second quarter 1985

$R^{**}2 = 0.995$
DW = 1.87
 t-values in brackets
 C = private consumption, 1980 prices
 Y = disposable income, current prices
 PC = deflator private consumption, 1980 = 1
 r = nominal interest rate
 D = government deficit (national account), current prices.

Using 1970 prices, which is consistent with the version of the model used for the experiments, the regressions lead to different parameter values. Table 11.1 shows the two consumption functions. To maintain consistency with the model both equations are estimated by ordinary least squares. An additional government deficit of DM 1 billion in 1970 prices decreases private consumption by DM 0.186 billion in the long run, all other things being equal. On the other hand an increase

Table 11.1 Consumption functions and government deficits

	Case 1	Case 2
Adjustment parameter	0.35862	0.37998
Long run:		
Marginal propensity to consume (real)	0.73451	0.79666
Interest rate, nominal	−2.1142	−1.9745
Budget deficit, real	—	−0.18587
Short run:		
Marginal propensity to consume (real)	0.26341	0.30272
Interest rate, nominal	−0.75820	−0.75029
Budget deficit, real	—	−0.07626

in disposable income by DM 1 billion in 1970 prices increases private consumption by DM 0.797 billion in the long run. The total effect depends on the influence of additional government debt on disposable income, on consumption prices and on interest rates. Parts of the negative effects of government debt on consumption are compensated by higher marginal propensities to consume, long run and short run, compared with the consumption function without a debt term. This result is due to the method of regression. In contrast the effects of the interest rate are approximately the same in the two equations. The significant parameter of the deficit variable can only be seen as a rough approximation of the influences on private consumption. But none the less there is an immediate negative, though small, impact of government deficit on private consumption. Empirical analysis of this relationship which investigates a longer time period will tell us whether this impact has a solid basis or whether it is a proxy for other variables, e.g. unemployment, influencing private consumption demand.

11.3 THE ECONOMIC SITUATION AND THE DESIGN OF EXPERIMENTS

In the first part of the eighties the steady growth of private consumption (Fig. 11.1) ended in stagnation combined with a rise in the average propensity to consume. Unemployment rose to numbers which are reminiscent of the figures of the late forties. Private and public real investment developed cyclically, with a downward tendency which started in 1980 and lasted for several years. The increase in the

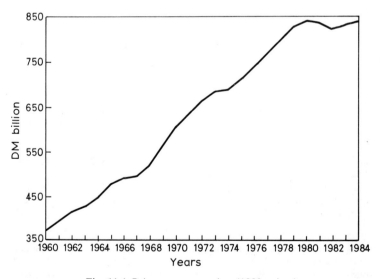

Fig. 11.1 Private consumption (1980 prices)

The economic situation and the design of experiments

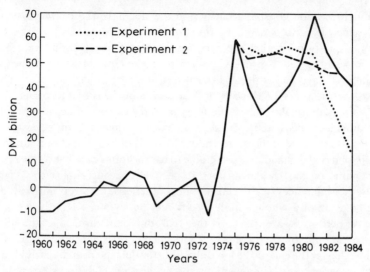

Fig. 11.2 Government deficit (national account, current prices)

gross national product in 1980 prices, which began in 1983, was mainly due to increasing exports. Most forecasters agree that the growth in exports spilled over to consumption and private investment and triggered thereby a boom which will continue in 1986 and 1987. The stagnation in private consumption was accompanied by a reduction in government investment even in current prices. One major reason for this restrained behaviour was the explosive increase of government debt starting in 1975. Figure 11.2 shows an M-shaped course of the deficit. During the four years from 1976 to 1979 the government tried to bring the deficit down; however, in the post-election year 1981 the deficit had reached a record height, DM 69 billion.

Economic agents who analyse the public deficit policy to detect a rule must distinguish between the period up to 1973 on the one hand and the period starting in 1974 on the other. The first period is characterized by modest deficits and surpluses, i.e. the stock of government debt declined in those subperiods. Even the Keynesian demand management policy pursued in the years 1967 to 1970 did not involve deficits exceeding DM 7 billion per year. Economic agents acting on the assumption of a balanced government budget policy rule could be fairly certain. This judgement of government behaviour proved invalid in 1975. Confronted with a sharp jump in unemployment figures, partly due to the first oil price crisis and partly for demographic reasons well known for several years, fiscal policy tolerated a deficit of nearly DM 60 billion in that year. This injection of government money was accompanied by slightly declining unemployment and rising private consumption and investment.

Although the first consolidation period did not have the characteristics of a slump, which is a necessary condition for the debt paradox to work, it is an interesting question whether a postponement of the first consolidation attempt would have triggered a process similar to a debt paradox. In the empirical experiments fiscal policy is altered in such a way that the deficit peak present in 1981 is shifted to the years 1976 to 1979. In an economy sensitive to public demand (case 1), the additional demand stemming from the deficit would raise production and employment and thereby would have made the stagnation of the early eighties less severe. In the case of 'fiscal neutrality' (case 2), one cannot expect any sizeable effect on total real demand. It is preferable to use the terms case 1 and case 2, instead of 'debt paradox' and 'fiscal neutrality' respectively, because these terms characterize only the differences in the consumption functions. They do not describe the implementation of a whole bundle of different policy measures.

Two experiments are run for the two different cases, i.e. the different consumption functions in the model. Experiment 1 does not increase the stock of the public debt at the end of 1984. It only shifts the deficits from the early eighties to the late seventies, thereby intensifying the budget consolidation starting in 1982. Experiment 2 increases the stock of public debt by DM 40 billion at the end of 1984. The altered course of the deficit declines smoothly from its peak in 1975 and coincides with the actual values in 1983 and 1984.

The government deficits in the period starting in 1975 have extraordinary values, e.g. in the fourth quarter 1985 government net borrowing was DM 16 billion, (Deutsche Bundesbank, 1986, p. 21). Therefore small changes in the time series of government deficits are not sufficient in an experiment which analyses the impact of a change in deficit policy. The high variability of this policy variable (see Table 11.2) restrains the validity of the Lucas critique (Lucas, 1976). Agents observe two peaks in the government deficit followed by more or less successful attempts to slow the growth of the public debt burden. Sievert (1984, p. 397) sees spectacular progress in economic policy in regard to this objective in the eighties. Leibfritz (1986) points out that a decrease in the deficit must be accompanied by a decrease in private savings, which will induce investment through lower interest rates, or by an increase either in private consumption or in the current account surplus. The latter may be seen as the main stimulus both for the current economic upswing and the progress in the consolidation of the budget.

The public should have no problems comprehending the current fiscal policy because the government does not attempt to conceal its policy goals. But it is rather difficult to include this policy in an equation with parameters which determine the policy chosen by the government. Because of the lack of such a policy equation policy changes have to alter the instrument variables, i.e. government deficit and government medium-term and long-term borrowing. As there is no stable policy rule for which the parameters could be altered, it is not clear in which way the experimental changes will or will not alter the parameters of other equations of the model. Although it would be preferable to change policy parameters, e.g. von Natzmer (1985a), it is beyond the range of this problem.

Table 11.2 Actual deficits and deficits of the experiments (quarterly data aggregated for expository purposes)

	Actual		Experiment 1			Experiment 2		
		% of GNP		Difference	% of GNP		Difference	% of GNP
1970	−2.19	0.32						
1971	1.15	0.15						
1972	4.02	0.49						
1973	−10.93	1.19						
1974	13.56	1.38						
1975	59.79	5.81		Unchanged			Unchanged	
1976	40.15	3.56	56.15	16.00	4.99	52.15	12.00	4.63
1977	29.22	2.44	53.22	24.00	4.44	53.22	24.00	4.44
1978	34.46	2.67	54.46	20.00	4.22	54.46	20.00	4.22
1979	40.95	2.93	56.95	16.00	4.08	52.95	12.00	3.79
1980	51.14	3.44	55.14	4.00	3.71	51.14	0.00	3.44
1981	69.87	4.52	53.87	−16.00	3.49	49.87	−20.00	3.23
1982	54.76	3.43	38.76	−16.00	2.43	46.76	−8.00	2.93
1983	46.29	2.76	26.29	−20.00	1.59	46.29	0.00	2.76
1984	40.92	2.33	12.92	−28.00	0.74	40.92	0.00	2.33
Sum	407.73		407.73	0.0		447.73	40.00	

11.4 THE MODEL

The model used for this investigation is a modified version of the Freiburger and Tübinger quarterly econometric model (Lüdeke et al., 1984). The estimation period is first quarter 1970 to fourth quarter 1981. Policy experiments are implemented in the first quarter 1976 and last for 36 quarters, i.e. up to 1984. The impacts of the first 24 quarters refer to *ex post* forecasts, whereas in the last 12 quarters the outcomes of the experiments are compared with *ex ante* forecasts of the model.

Government deficits financed by public borrowing enter the causal chains of the model at two points. Point one is a direct effect of the deficit in public investment in current prices. Policy changes in this aggregate were assumed to leave the corresponding deflator unaffected. This assumption is based on the results of the experiment, which exhibited only minor influences on the price index of private consumption and of gross national product.[5] Therefore, in this constellation price effects are not important for real government demand. Intensifications and dampenings stem from influences on public consumption on the one hand and on government revenues, including social security, on the other. The endogenized public investment is determined by the following definition:

> government investment
> = government revenues
> − government spending, excluding investment
> + government deficit.

Endogenous influences on government spending, excluding investment, arise from government consumption. But the impacts on government revenues are more important. Indirect taxes, wage taxes and taxes on firms as well as employers' and employees' contributions to social security depend on economic activity and therefore on the influences of government debt policy. The implication of this specification is that government investment is controlled by the deficit and amplified or dampened by the influences arising from taxes and social security contributions. In other words, in case 1 (no deficit variable in the consumption function), an increase in revenues is spent as additional investment and not used to reduce the exogenously determined deficits of the experiments. Of course, the same mechanism operates in the case of a deficit variable in the consumption function, i.e. in the so-called 'fiscal neutrality' case.

The second point of impact runs through the changes in government demand for medium-term and long-term credit. This flow variable enters the equation of quarterly changes of government current accounts as a quarterly difference, but does not enter the equation of total demand for medium and long-term credits.[6] The latter is affected by government borrowing following certain intermediate

---

[5] The model showed this characteristic in the studies of a disburdening of private households as well (von Natzmer, 1987). For a contrary view see, for example, Stein (1983, p. 27).
[6] The financial sector of the Freiburger and Tübinger econometric model is specified in flows, not in stocks. Therefore only changes in changes matter.

steps. Both the quarterly difference and the indirect influence on total demand for credit weakens the impact of additional government debt. It results in a high flexibility of the specified financial system with respect to government borrowing which appears to be a special characteristic of the model. It can hardly be interpreted as a feature of the Barro–Ricardo equivalence proposition or, as Dwyer (1985, p. 656) puts it, that this '... theorem can account for the tenuousness of any relationship between government debt and the interest rate'. If this reasoning is accepted the demand sector of the model should incorporate a mechanism which is compatible with the proposition.

If additional public deficits do not considerably increase interest rates, it is not surprising that fiscal stringency is also insufficient to lower interest rates and unable to keep them down. Britton (1985, p. 23) observes this insufficiency for the UK. Heilemann (1983) comes to the conclusion that no substantial evidence of a crowding out can be found for West Germany; where crowding out subsumes all effects that compensate the impacts of an expansive fiscal policy not financed by money creation. As far as the Freiburger and Tübinger model, version 81, is concerned, we can summarize that many causal relationships within the financial sector are affected but that the amounts of the impacts on interest rates, price deflators and variables of the real sector are small or nearly negligible. Figure 11.3

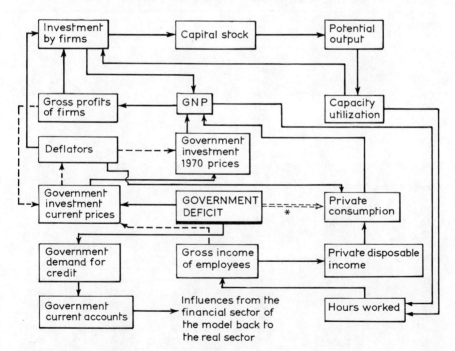

Fig. 11.3 Essential causal relationships affected by government deficits. *This impact is only specified in case 2. In case 1 government deficits do not influence private consumption directly

214 *Budget consolidation, effective demand and employment*

shows essential causal relationships, mainly within the real sector of the model, affected by government deficits.

11.5 RESULTS AND CONCLUDING REMARKS

The results of experiments show similar patterns in the way that government deficits influence the endogenous variables, although they differ in numerical values. The details are presented in Table 11.3 and in Figs 11.4 and 11.5.

1. Large changes in the government deficit induce parallel changes in production and employment. As long as the deficit is in excess of the actual values, the experiments lead to positive production and employment effects. But when the deficits in experiment 1 are lower than in reality, not only do gross national product and employment decline as well, but they are also lower than without a policy change. What is gained in the first subperiod is lost, at least in part, in the second. The modifications in gross national product

Table 11.3 Effects on gross national product and employment

Percentage of actual GNP:		Case 1		Case 2	
		Experiment 1	*Experiment 2*	*Experiment 1*	*Experiment 2*
1976		2.973	2.245	2.856	2.159
1977		4.437	4.319	4.291	4.159
1978		3.796	3.676	3.657	3.619
1979		3.132	2.609	2.843	2.451
1980		1.834	1.300	1.691	1.209
1981		−0.237	−0.845	−0.371	−0.875
1982		−0.650	−0.036	−0.645	0.028
1983		−1.867	0.135	−1.762	0.147
1984		−3.186	−0.033	−2.933	−0.108
Unemployment, yearly average (million):					
	Actual				
1970	0.1481				
1971	0.1890				
1972	0.2468				
1973	0.2821				
1974	0.6017		*Unchanged*		
1975	1.086				
1976	1.055	0.4606	0.6015	0.4844	0.6309
1977	1.030	0.2296	0.2326	0.2538	0.2601
1978	0.9893	0.3277	0.3544	0.3623	0.3786
1979	0.8702	0.3097	0.4153	0.3583	0.4434
1980	0.8995	0.5970	0.7059	0.6020	0.6802
1981	1.296	1.387	1.503	1.399	1.493
1982	2.006	2.101	1.959	2.096	1.930
1983	2.263	2.573	2.165	2.551	2.175
1984	2.265	2.806	2.196	2.757	2.203

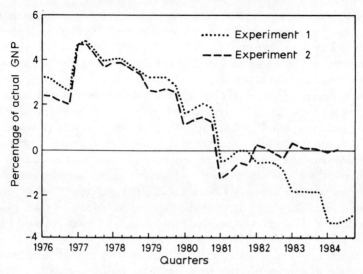

Fig. 11.4 Effects on GNP: case 1, no deficit variable in the C-function

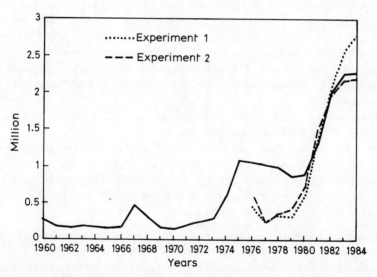

Fig. 11.5 Unemployment: case 1, no deficit variable in the C-function

induced by the alterations in the deficits range from 4.4% of the corresponding actual values to −3.1%. Concerning experiment 2, the conclusion is that the additional deficit spending from 1976 to 1979 induces positive output and employment effects which vanish when the deficit returns to the actual values in 1983. The lower deficits in 1981 and 1982 are accompanied by slightly lower

GNP values than without a policy change. Due to the dynamics of the system the loss in output caused by a deficit cut of DM 20 billion in 1981 is much smaller than the expansionary effect of an additional deficit of the same amount in 1978. Up to 1980 the experiments attain nearly full employment, i.e. the figures for additional employed workers correspond to the unemployment figures. This is only an approximation, because additional hours worked are translated by average working hours per day and number of days worked per period into additional employed persons. Changes in the composition of full-time and part-time workers are neglected, just as one can expect overtime work to increase with higher capacity utilization. These numbers are subtracted from and added to the unemployment series, respectively (see Table 11.3 and Fig. 11.5). In experiment 1, which keeps the total deficit constant but shifts the dates when borrowing takes place, the symmetry of the effects is realized in an increase of unemployment by half a million in 1984 to an average of 2.8 million.

2. The attempt to take the Barro–Ricardo equivalence proposition into account in the consumption function by a parameter which depends on government deficit was not successful, although the deficit variable has a significant influence even when estimated by an instrumental variable approach.[7] There is a dampening of demand effects, induced by a shift in private consumption, but this effect is far removed from fiscal neutrality. This result holds both ways, i.e. in the case of consolidation, too, the expectation of future lower tax payments as specified in the model is not able to prevent an eventual decline in private consumption. The main characteristic of parallel movements of deficits and endogenous variables predominates over the outcomes of the experiments.

3. The influences of government borrowing are small because of the specification of the model discussed in Section 11.4.

4. Prices and interest rates do not react sensitively to the experimental changes. This result suggests that there is little crowding out due to interest rate fluctuations. The rigidity of prices is based on the exogenous nominal wage rate, which is important for price movements and on the only small influences of interest rates specified in the model.

Although the model accentuates demand effects induced by government behaviour these demand effects vary almost parallel to public deficits. The higher economic activity at the beginning of the policy alteration is not sustained and does not avoid a decline beneath the base line values in experiment 1 when budget consolidation sharply reduces the deficits. In experiment 2 economic activity returns to the base line level in the last years of the investigation period. The experiments emphasize the importance of demand but they can hardly be interpreted as supporting a debt paradox. Considering this conclusion, one has to

---

[7] See Section 11.2.

keep in mind that an existing model has been used and that the experiments were not designed particularly to analyse a debt paradox. Instead they examined the effects of smoothing out the course of the deficit in the period 1976 to 1984.

REFERENCES

Aschauer, D. A. (1985) Fiscal policy and aggregate demand. *American Economic Review*, 75, 117–27.
Barro, R. J. (1974) Are government bonds net wealth? *Journal of Political Economy*, 82, 1092–117.
Blinder, A. S. (1985) Comment on federal deficits, interest rates and monetary policy. *Journal of Money, Credit, and Banking*, 17, 685–9.
Blümle, G. (1985) Zum heutigen Stand der Theorie des Sparens. Supplements to *Kredit und Kapital*, 9, 27–52.
Britton, A. (1985) The budget and its critics. *Fiscal Studies* 6, 23–30.
Deutsche Bundesbank (1986) *Öffentliche Finanzen Monatsbericht* No. 2, February.
Dwyer, G. P., Jr. (1985) Federal deficits, interest rates, and monetary policy. *Journal of Money, Credit, and Banking*, 17, 655–85.
Evans, P. (1985) Do large deficits produce high interest rates? *American Economic Review*, 75, 68–87.
Gandenberger, O. (1983) Thesen zur Staatsverschuldung, in *Staatsfinanzierung im Wandel* (ed. K.H. Hansmeyer), Duncker and Humblodt, Berlin, pp. 843–65.
Hansen, G. (1984) Der Einfluß von Zinsen und Preisen auf die die Ersparnisse und die Nachfrage nach dauerhaften Gütern in der Bundesrepublik 1961–1981. *Zeitschrift für Wirtschafts- und Sozialwissenschaften*, 104, 227–49.
Heilemann, U. (1983) Zur empirischen Relevanz des crowding-out Effektes für die Bundesrepublik Deutschland. *Beihefte der Konjunkturpolitik*, 30, 111–45.
Lambelet, J-C. (1985) Should systems like LINK be used for long-range forecasts and simulations? *Economic Modelling*, 2, 83–92.
Leibfritz, W. (1986) Konjunkturreport: Öffentliche Finanzen. *Wirtschaftskonjunktur*, 1/86, R1–R4.
Lucas, R. E. (1976) Econometric policy evaluation: a critique, in *The Phillips Curve and Labor Markets* (eds K. Brunner and A. H. Meltzer), North-Holland, Amsterdam, pp. 19–45.
Lüdeke, D., Friedrich, D., Hummel, W., von Natzmer, W., Röger, W., Röhling, W. and Termin, J. (1984) Freiburger and Tübinger quarterly econometric model for the Federal Republic of Germany: an overview. *Economic Modelling*, 1, 139–232.
Miller, J. (1983) Examining the proposition that federal budget deficits matter, in *The Economic Consequences of Government Deficits* (ed. L. H. Meyer), Kluwer-Nijhoff, Boston, pp. 3–24.
von Natzmer, W. (1987) Social security contributions, economic activity, and distributions, *Empirical Economics*, 12, 29–49.
von Natzmer, W. (1985a) Econometric policy evaluation and expectations, *Economic Modelling*, 2, 52–8.
von Natzmer, W. (1985b) Public sector shares and economic performance, in *Policy Adjustments to Changes in Economic Structure in Japan and in the Federal Republic of Germany* (eds T. Dams *et al.*) (Contributions to the 9th Joint Seminar Nagoya/Freiburg i. Br. in Nagoya, Japan, 18–20 March 1985) (to be published).
Oberhauser, A. (1985) Das Schuldenparadox. *Jahrbücher für Nationalökonomie und Statistik*, 200, 333–48.
Scherf, W. (1985) Budgetmultiplikatoren – Eine Analyse der fiskalischen Wirkungen

konjunkturbedingter und antizyklischer Defizite. *Jahrbücher für Nationalökonomie und Statistik*, **200**, 349–63.

Sievert, O. (1984) Angebotsorientierte Wirtschaftspolitik in der Bundesrepublik Deutschland. *List Forum*, **12**, 382–411.

Stein, J. L. (1983) Discussion on Miller (1983), *The Economic Consequences of Government Deficits* (ed. L. H. Meyer), Kluwer-Nijhoff, Boston, pp. 25–37.

Wille, E. (1983) Zum Konsolidierungsbedarf der öffentlichen Haushalte, in *Perspektiven der deutschen Wirtschaftspolitik* (ed. H. Siebert) Kohlhammer, Stuttgart, pp. 97–111.

Wille, E. and Kronenberger, S. (1984) Zielkonflikte im Kontext der Staatsverschuldung. Einige Anmerkungen zum empirischen Bezug, in *Intertemporale Allokation* (ed. H. Siebert), Lang, Frankfurt, pp. 607–47.

Woll, A. (1984) *Wirtschaftspolitik*, Vahlen, München.

12

Interaction between economic growth and financial flows: presentation of a model analysing the impact of short-term financial disturbances on economic growth

HASSE EKSTEDT and LARS WESTBERG

12.1 THE AIM OF THE STUDY AND THE THEORETICAL APPROACH

It is our purpose to describe the medium-term economic development in the Swedish economy during two periods, one from the beginning of the 1960s up to the middle of the 1970s, characterized by a substantial growth of production, and a second period from the middle of the 1970s up to the beginning of the 1980s, when production either stagnated or had a low growth rate in the medium term.

This study is part of a more ambitious work to construct a compact macro model to be used to study the interrelation between real and financial development both in the medium term and in the short run.

In the study that we present here we are specially interested to test our hypothesis concerning this interaction in the market economy sector between the real and the financial development. This is done in a model which analyses flows on real and financial markets and behaviour equations for these flows. In our following studies the model will also include stocks and stock adjustments.

It appears that during the period from the beginning of the sixties to the middle of the seventies the medium-term growth rate for the actual production in industry was higher than the potential production. After the middle of the seventies potential production stagnated and actual production was decreasing up to the beginning of the eighties. This is shown in Fig. 12.1.

Such a development could be explained by the fact that the long period of relatively stable growth that was interrupted by the oil price crisis had been

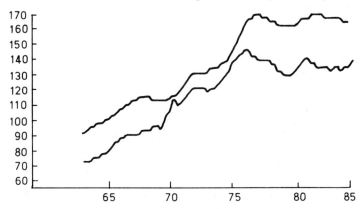

Fig. 12.1 Actual and potential production in industry in Sweden (four-quarter moving average)

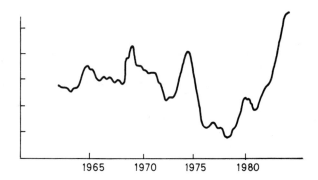

Fig. 12.2 The production value of industry in relation to its labour costs

accompanied by a continuous change in the financial conditions, which brought with it decreasing cash balances, a rising ratio of debt to equity finance and a rising ratio of debt payments to income (Davidson, 1978; Arrow and Hahn, 1971).

Figure 12.2 shows the production value of industry in relation to its labour costs. From this can be deduced that the prolonged growth process from the sixties up to 1973 was accompanied by a decline in the profit share of total production. Between 1969 and 1973 this decline was much deeper than the normal cyclical variation. This development led to a higher dependence on external finance.

Figure 12.3 shows a constant price index for equity prices in industry. It indicates that the cyclical fall in equity prices was interrupted in 1972. The big decline in industries' share prices did not start before 1974. This development can be interpreted as depending upon the prolonged high production growth rate. The

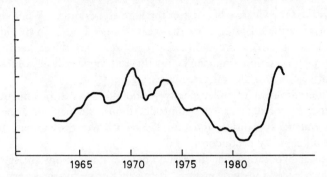

Fig. 12.3 Equity prices in industry, constant prices

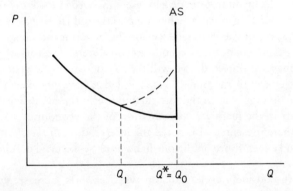

Fig. 12.4 Hypothetical analysis of the cost situation in industry before and after the oil price crisis: AS = aggregate supply; Q^* = full capacity output; Q_0 = actual output before the oil price crisis; Q_1 = actual output after the oil price crisis

decline in the production growth rate after 1974 led to an accelerated decrease in share prices and a weakening of the financial position in industry. The explosive rise in costs after the oil price crisis made necessary a reduction of actual production from the earlier full capacity production level.

Figure 12.4 shows in principle how actual production and full capacity production had been on the same level before the oil price crisis. $Q_0 = Q^*$. This production level is before the oil price crisis also the optimal production level. The oil price crisis leads to rising costs and the supply curve moves upwards to the left. Actual production and optimal production come down to a lower level at Q_1.

The decline of production leads to sinking profit expectations and therefore to falling equity prices. During earlier years the financial development had already been characterized by growing debts in relation to production because of falling profits in relation to production. The relation of debts to equity had, however, not risen in the same degree, because of the high equity prices. When these prices went down after the oil price crisis the weak financial position of industry was disclosed. Actual production could therefore not return to the earlier full capacity level. Optimal production remained below this full capacity level, which had been the optimal level before the oil price crisis.

But the development of full capacity production was also influenced in the medium term. By forcing down actual production the financial deterioration kept down investment expenditure and thus led to a lower growth rate for capital stock and full capacity production.

It is thus our hypothesis that short-run financial fluctuations create changes in the medium-term real growth process. If the real growth process is interrupted it will involve more profound modifications in real output and expenditure depending upon how far the financial position of the economy has deteriorated.[1]

The fact that the medium-term development in Sweden since 1983 has witnessed an increase in actual production while potential production has been unchanged raises the question of the consequences of the government's financial policy on the financial development and expenditures in different sectors of the economy. In the next phase of our study we will try to find out if the general idea of the crowding-out effects of the budget balance deficit have not been overestimated. We are of the opinion that in the intermedium term the big deficit has played an important part in the financial consolidation of the economy.

Our study here presented also noted the interrelation in *the short run* between production and expenditure on the one hand and the financial development on the other.

Normally the business cycle turns up when exports increase, which leads to increases in productivity and in the utilization of production capacity. In this phase of the cycle the relation of profits to production is increasing, which results in higher equity prices. The growth rate of prices for capital goods is normally moderate in this phase, while the yield of capital in relation to its cost is increasing (Kaldor, 1940; 1956–57). Except for stocks, investment will, however, be postponed until the economy has reached a higher utilization of productive capacity and the production has reached what can be assumed to be the medium-term growth rate.

In about the same phase of the cycle the increasing rate of growth in wages will make the profit share decline. The share of internal finance will go down and equity prices will decrease, although investment expenditure may still increase for a while because of the lag structure. External finance will increase in relation to production and be more expensive, which will ultimately lead to decreasing expenditure in final goods and a decline in the business cycle. This description of the interrelations between real and financial changes during the different phases of the Swedish business cycle is confirmed by our statistical material.

---

[1] Our theoretical approach has specially been influenced by the work of Minsky and Tobin. See, for example, Minsky (1964) and Tobin (1982).

The aim of the theoretical approach

Variables

- Q^* = full capacity product
- L^* = effective labour force
- K = real capital stock
- Q = actual production
- C = private consumption
- I = private investment
- G = government expenditure
- T = taxes
- t = tax rate
- X = exports
- Z = imports
- Y^F = profits after taxes
- Y^H = household disposable income
- L = employment
- w = wage rate
- VX = exchange rate
- r = discount rate
- P = prices of goods
- P^A = prices of equities
- P^B = prices of bonds
- P^X = international prices
- WT = world trade
- r^X = international interest rate

(Some symbols are not used in the flow chart but enter the following model.)

The main connections between real and financial variables are summarized in Fig. 12.5. The flow chart above the dotted line indicates the linkages between the variables that are of main importance for the short-term fluctuations of the economy. The arrows indicate not simply quantitative flows but also causal links. Shaded rectangles symbolize exogenous variables.

Following this part of the chart from left to right we see how the four types of expenditure of final goods, gross investment, exports, government expenditure, and private consumption influence the gross production of the market economy. Gross production, besides influencing in its turn expenditure of investment and consumption, has effects on imports and taxes. The relation between actual gross production (Q), and production at full capacity (Q^*), which is located in the lower part of the chart below the dotted line, influences the development of productivity. The utilization of full capacity production together with changes in the productivity lead to changes in wage rates and thus to changes in the distribution of factor income, which by its influence on supply and demand on the markets for different kinds of assets lead to changes in the asset prices. These prices, which are related to the official discount rate – in its turn depending upon the exchange

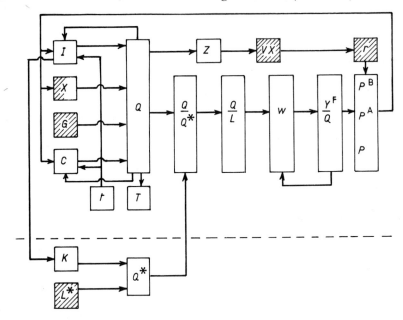

Fig. 12.5 Flow chart of the economy

rate and thus on exports and imports – will in the short run influence the private expenditures for investment and consumption.

In order to clarify the short-run interrelations between the real and the financial development, these relations have been simplified as much as possible in the chart, and some important connections have been left out, as, for example, the dependence of exports on the development of production capacity. Prices influence exports, however, which are only partly exogenous. This is indicated by the rectangle for exports, which is only partly shaded.

The complex financial sector is only included in an indirect way by the prices of the two types of financial assets, equities and bonds. The bond prices are dependent upon the official discount rate. The latter is controlled by the government in order to stabilize the exchange rates.

The variables and linkages which are of importance in the medium term are indicated below the dotted line. The exogenous growth rate of the effective labour force together with the capital stock influence full capacity production and thus the relation between the actual total product and the full capacity product. This relation influences the development of productivity in the short run. The chart also shows that the full capacity product is on the other hand dependent on the short-term development by the links between capital stock growth and investment expenditure. As investment expenditure in its turn is dependent upon the financial development and the asset prices, these financial conditions will be of decisive significance for full capacity production in the medium term, which has been argued earlier in the text.

12.2 THE ANALYTICAL MODEL

The casual links indicated in the flow chart can be developed into a model for the analysis of short-term and medium-term relations between real and financial development. The model is similar to Tobin (1980, Chapter 4) and Tobin (1982). This model will later in the chapter be transformed into an empirical model. The definitions of the symbols are given above.

$$Q^* = f_{Q^*}(L^*, K); \qquad Q_1^* \geqslant 0; Q_2^* \geqslant 0 \qquad (12.1)$$

$$Q = f_Q(C, I, G, X); \qquad Q_1 > 0; Q_2 > 0; Q_3 > 0; Q_4 > 0 \qquad (12.2)$$

$$C = f_C(Y^H, L/L^*); \qquad C_1 \geqslant 0; C_2 \geqslant 0 \qquad (12.3)$$

$$Y^H = f_{Y^H}(Q, T); \qquad Y_1^H > 0; Y_2^H < 0 \qquad (12.4)$$

$$I = f_I(Q/Q^*, P^A/K, r); \qquad I_1 > 0; I_2 > 0; I_3 < 0 \qquad (12.5)$$

$$X = f_X(WT, VX, P/P^X); \qquad X_1 > 0; X_2 > 0; X_3 < 0 \qquad (12.6)$$

$$Z = f_Z(C, I, VX, P/P^X); \qquad Z_1 > 0; Z_2 > 0; Z_3 < 0; Z_4 > 0 \qquad (12.7)$$

$$K = f_K(I); \qquad K_1 = 0 \qquad (12.8)$$

$$L = f_L(Q/Q^*); \qquad L_1 > 0 \qquad (12.9)$$

$$w = f_w(Q/L, (Y^F/Q)_{\text{LAG}}); \qquad w_1 > 0; w_2 > 0 \qquad (12.10)$$

$$P = f_P(Q/Q^*, w, P^X); \qquad P_1 \lessgtr 0; P_2 > 0; P_3 > 0 \qquad (12.11)$$

$$P^A = f_{P^A}(Q/Q^*, Y^F/Q); \qquad P_1^A > 0; P_2^A > 0 \qquad (12.12)$$

$$P^B = f_{P^B}(P^A, X/Z, G/T, r, r^X); \qquad P_1^B < 0; P_2^B > 0; P_3^B < 0; P_4^B < 0; P_5^B < 0 \qquad (12.13)$$

$$Y^F = f_{Y^F}(Q/Q^*, t); \qquad Y_1^F > 0; Y_2^F < 0 \qquad (12.14)$$

$$T = f_T(Q, t); \qquad T_1 > 0; T_2 > 0 \qquad (12.15)$$

The signs of the relations in the equations are indicated to the right of the equations. The indices refer to the arguments in each equation in successive order. The model includes 23 variables, of which 15 are endogenous, and has 15 equations.

The model is both supply and demand orientated, as can be seen from the first two equations, the first making the full capacity output of the market economy a function of the exogenous effective labour force and of the real capital stock. The second equation makes demand for the total product of the market economy a function of expenditures for private consumption, private investment, exports and government expenditures for goods and services.

The relation of the actual total production to the full capacity production enters all the functions of importance for short-term cyclical movements of the economy, and will thus become the dominant factor explaining the interrelated real and financial movements included in these fluctuations. The short-term financial fluctuations will, however, have medium-term consequences for the growth of production capacity, which in its turn will influence asset prices, liquidity and the medium-term financial development (Gurley and Shaw, 1960; Kragh, 1955 and 1967). The model can be said in the first place to aim at a description of the economy in the short and in the medium term. It is a dynamic model and it has nonequilibrium properties, both in the short and in the medium term.

In the short run the gap between the actual production and the production at full capacity will be perpetually changing and by these changes create the never-ending fluctuations of real and financial nature which interact upon each other. Simulations will in a later stage be used to find out if evidence is consistent with our theory of the short-run fluctuations, which undisturbed would display only damped cyclical characteristics, but which are perpetuated by external shocks, especially from the export sector.

The financial fluctuations which will be a byproduct of these short term cycles will, by their influence on the medium-term development of the growth of production capacity, have more long-term consequences for the stability of the economy.

The stability qualities of the short-term and the medium-term model may also be deduced from the lagstructure and the values of the coefficients of the behaviouristic equations of the model. This will be discussed in the empirical part of our study.

12.3 EMPIRICAL FINDINGS: BASED ON THE ANALYTICAL MODEL

As mentioned above, we have not built a model for the entire economy, but for the most market-orientated part of the economy. Thus we have chosen the industrial sector where we have the most complete data. As Sweden is a small open economy much of the dependency of the outside world will go through the industrial sector and this will create cross-border financial flows. In this section we will account for the most important structural equations of the theoretical model. Some estimations have induced us to consider other theoretical arguments and some have led us to improve proxies or to discover inconsistencies in the database. However, we believe that it will be of some interest to discuss empirical findings based on a data set containing two periods, one characterized by growth and the other by zero or negative growth.

In the empirical analysis we work with quarterly data. This will of course imply severe restrictions on the existence of data, especially on the financial side, but in the first run it is important to clear out the short-run variations of different variables. Another problem in models based on quarterly data is the time lags. Some lags are internal to the model, for example, investments as dependent upon

Potential and actual production

the development of the profit and capacity utilization, while other time lags are dependent upon structural rigidities changing from time to time.

We have by no means finished the analysis of the lag structure. As a matter of fact we believe that lags, especially of the latter kind, can cause important unexpected financial flows which during certain conditions may have important effects on medium and long-term real growth. An analysis of the lag structure is therefore one of the main issues in our further work.

All variables are taken from the National Account Statistics except international prices and interests and development of trade, which are taken from International Financial Statistics (IFS). The variables are transformed into indices where the third quarter 1972 is set to 100. The time series run from the first quarter 1961 to the fourth quarter 1984. Seasonal variations are adjusted by a four quarter moving average, but we also used dummy techniques. All the statistics are performed at Göteborgs Datacentral.

In this chapter we present the structural equations estimated by ordinary least squares (OLS), and where autocorrelation is prevalent, the Yule–Walker autoregressive method. As mentioned above, we have not been able to include data for stocks on the financial side. Because of revisions they were only available on a yearly basis since 1970. In our model of the entire economy based on yearly data these stocks and successive revaluations of the stocks are fundamental. With respect to this the system presented in this chapter is misspecified. However, our purpose is that the flows estimated here and the argument used can give some important information on the underlying model.

12.4 POTENTIAL AND ACTUAL PRODUCTION

A central variable in our model is capacity utilization. Many different methods of estimating this variable have been used. In Sweden official estimations are based either on unemployment statistics or on questionnaires to a sample of firms. The latter method has been used since the beginning of the seventies by Konjunkturinstitutet. For our purpose both of these methods have the disadvantage that they measure the short and medium-term variations in capacity utilization but they do not relate the variations to the long-term development of capacity.

We have tried to handle this problem by constructing a 'potential production function'. This function must be based on an estimation of the actual production. To estimate actual production we have chosen labour (man-hours) and energy (kWh) as arguments. The reason for using energy instead of capital is that we mean that energy is a good proxy for the actual utilization of the capital stock. Energy for heating is subtracted. For the long-term development we estimate energy consumption per man-hour as a function of the capital stock.

In Fig. 12.6 the dotted curve measures output per man-hour while the other curve is output per kWh. In Fig. 12.7 the curve measures energy consumption per

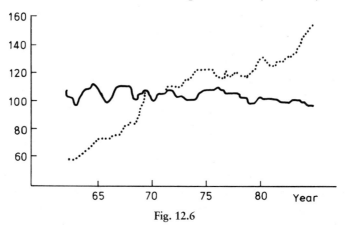

Fig. 12.6

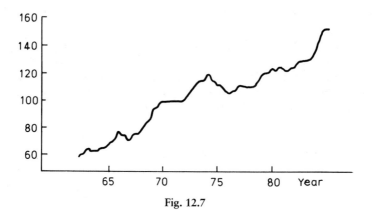

Fig. 12.7

man-hour. We use the following notation:

Q_t^* = potential production at time t
L_t^* = full employment in the industrial sector
L_t = actual employment in the industrial sector
E_t^* = maximal use of energy at a given level of employment
E_t = actual use of energy
K_t = capital stock valued at replacement costs at time t
Q_t = actual production at time t.

We define $L_t^* = L_t (1 + U_t)$ where U_t is the rate of unemployment in the industrial sector.

The variable E_t^* is more difficult to define. We have based the construction on the residuals of the long-run trend of the energy consumption per man-hour. We estimate the maximum consumption of energy per man-hour on the local maxima of the residuals to the linear trend. Of course it would have been more appropriate to base the construction on the relation of energy to capital. In that case we would evidently have restriction on output imposed by the capital stock. However, the estimations based on energy to capital stock were bad in the sense that they gave a systematic disturbance of the cycles compared with Konjunkturinstitutet's investigations while the method used was very precise both according to the phases and to the relative strength of the cycles during periods possible to compare. Equations to give the main steps of the method are:

$$\ln Q_t = -1.397 + 0.365 \ln L_t + 0.963 \ln E_t \quad \begin{array}{l} R^2 = 0.87 \\ F = 307 \end{array} \quad (12.16)$$
$$ (-1.4) \quad (1.2) \quad (22.5)$$

$$\left\{\frac{E_t}{L_t}\right\} = 0.574 + 0.0087\,T \quad \begin{array}{l} R^2 = 0.83 \\ F = 472 \end{array} \quad (12.17)$$
$$\phantom{\left\{\frac{E_t}{L_t}\right\} = } (25.2) \quad (21.7)$$

$$\left\{\frac{E_t}{L_t}\right\}_{\max} = 0.668 + 0.01\,T \quad (12.18)$$

The method used can be defended by the fact that enquiries have shown that installed electrical effect is far beyond the actual need. This can explain the bad results of the estimation based on the relation energy to capital. Figure 12.1 shows the constructed potential production function in relation to actual production.

12.4.1 Consumption

We have tested three types of arguments, income, unemployment and financial instruments. The sharp increase of bonds in the end of the seventies and the beginning of the eighties should have a negative effect on consumption. We also believe that the rise in prices of shares in the beginning of the eighties could have the same effect, but in this case we must add a reminder that shares are mostly traded by institutions and not so much by households.

From the life-cycle hypothesis and inquiries based on this we would expect the unemployment variable to have a positive effect on consumption. But empirical results in this direction are often based on yearly data. Quarterly data may weaken the positive effect because we measure the different phases of the cycle more carefully. Therefore we have added the change of unemployment between two successive periods as an argument (Table 12.1).

The parameters show a high degree of robustness which speaks in favour of no major misspecification. The increase in the intercept reflects a lagged wealth effect which is strengthened when we bring in the financial instruments. When we compare C2 and C3 we notice that the unemployment effect divides into one positive 'level' effect and one negative 'change' effect. The parameters of C_{t-1} and Y_t^h are similar to other studies made on Swedish consumption data (i.e. Berg, 1983).

Table 12.1 Dependent variable – consumption, C_t (mean value = 104.5)

Eqn	Intercept	Y^h_{t-1}	Y^h_t	B_t/r_t	P^A_t	U_t	ΔU_t	DW	R^2	F
12.19	15.87 (5.4)	0.227 (7.3)	0.607 (18.6)					1.65	0.91	480
12.20	15.98 (5.9)	0.196 (6.6)	0.581 (18.7)			0.094 (3.9)		1.85	0.93	377
12.21	15.33 (5.9)	0.209 (7.3)	0.565 (18.8)			0.094 (4.3)	−0.149 (−3.2)	1.71	0.93	315
12.22	43.33 (7.8)	0.176 (6.3)	0.567 (19.7)	−0.559 (−5.7)	−0.008 (−0.2)			1.91	0.94	337
12.23	37.32 (7.1)	0.180 (6.8)	0.545 (20.1)	−0.457 (−4.7)		0.057 (3.0)	−0.138 (−3.3)	1.95	0.95	318

(Estimation method: OLS).

12.4.2 Investments

In the empirical analysis of the theoretical equations we separate total investments in the industrial sector into investments in buildings and investments in machinery and transport equipments. Investments in buildings merely increase the long-run capacity while investments in machinery increase the short-term capacity and change the technology. Fundamentally investments in buildings reflect an improvement in the long-term expectations while investments in machinery do not necessarily do so (Hicks, 1977, chapter 1).

Three fundamental arguments govern investments in buildings: the ratio of profit to value added, π_t, the financial position to the capital stock, TQ_t, and finally the interest on bonds, r_t, and foreign assets, r_t^X. Except for the interests we must use proxies.

As a proxy for the ratio of profit to value added we use the ratio of gross revenues to total labour costs. The implicit assumption in this proxy is that all other input prices except for labour costs immediately affect the output prices.

The second argument is known as Tobin's q. A proper definition is the sum of stock market value of equity and debt divided by the existing stock of capital at current replacement cost (Tobin, 1969). While we have no data for the debts we have to use the index for equity prices as a proxy for the numerator. $TQ = P_t^A / K_t$.

The four arguments have different time horizons in the moment of decision. Profit rate is a medium-term variable over the cycle. Compared with profit rate financial position is a slowly moving variable while the interests are variables of momentary effect.

We also add capacity utilization to see whether it has any effect. It is probable that the effect is not very important, because investments in buildings are founded on long-term expectations and furthermore the time lag between the decision and the actual outlay displaces the outlays relative to short-run variations in capacity utilization.

A serious problem in estimating investments and especially investments in buildings is that data are build upon outlays, which means that there could be a considerable timelag between a decision and the actual outlay. To handle this we use the average value during four preceding quarters for the variables TQ_t and π_t. We also use an autoregressive model to estimate the functions.

Table 12.2(a) shows the estimations results for the periods 1961–1984 and (b) for 1972–1984. The estimations are, at best, a base for further inquiries. The difference between parts (a) and (b) in Table 12.2 displays the difficulties of long time series. The negative sign and the weak significance of variable π in (a) depend most of all on the changes of growth in the middle of the seventies. The same can be said about the capacity utilization, Q_t/Q_t^*. The proxy for Tobin's q, $\overline{TQ}$, has the wrong sign and a low significance. As it is a proxy we have to study this variable properly defined on another database. There is one interesting variable, r_t, effective rate of interest on bonds. It is positive and increases its strength in

Table 12.2 Dependent variable – investments in buildings

Eqn	Intercept	$\bar{\pi}$	$T\bar{Q}$	r_t	r_t^X	Q_t/Q_t^*	Regression R^2	Total R^2
(a) Period 1961–1984								
12.24	30.779 (0.72)	−0.141 (−0.58)	−0.240 (−1.5)	1.082 (4.2)	−1.559 (−0.44)		0.25	0.75
12.25	54.55 (1.2)	−0.031 (−0.1)	−0.206 (−1.3)	1.131 (4.3)	−0.847 (0.2)	−0.515 (−1.1)	0.26	0.75
(b) Period 1972–1984								
12.26	−148.47 (−2.9)	0.799 (2.6)	−0.164 (−0.8)	2.056 (5.7)	−0.035 (−0.9)		0.48	0.82
12.27	−173.20 (−2.8)	0.724 (2.3)	−0.288 (−1.2)	1.892 (4.4)	−0.051 (−1.1)	−0.610 (0.7)	0.50	0.82

Estimation method: Yule–Walker. Autoregressive model. Lags: $t-1$ to $t-4$.

Table 12.3 Dependent variable – investments in machinery and transport equipment

Eqn	Intercept	$\bar{\alpha}$	$\bar{\pi}$	r_t	r_t^f	d4	Regression R^2	Total R^2
(a) Period 1961–1984								
12.28	33.95 (2.3)	0.581 (3.6)				41.80 (17.7)	0.80	0.77
12.29	37.92 (1.2)	0.568 (2.9)		−0.037 (−0.2)	−0.050 (−0.01)	41.78 (17.3)	0.80	0.77
12.30	95.30 (2.3)	0.602 (3.5)	−0.453 (−1.5)	−0.085 (−0.5)	−0.724 (−0.2)	41.75 (16.8)	0.79	0.78
(b) Period 1972–1984								
12.31	94.24 (1.8)	0.091 (0.2)				56.15 (13.7)	0.82	0.76
12.32	−70.38 (−1.3)	1.879 (2.4)		1.416 (4.4)	−0.064 (−1.8)	54.6 (13.3)	0.84	0.82
12.33	−70.30 (−1.3)	0.410 (0.5)	0.363 (0.6)	1.455 (4.5)	−0.051 (−1.3)	54.41 (13.0)	0.85	0.83

Estimation method: Yule–Walker. Autoregressive model.
Lags: $t - 1$.

(b). This could be a sign of a wealth effect of the government debt. We will return to this point when we discuss the estimations of equity prices.

Investments in machinery and transport equipment are in general based on judgements on the short-run need for capacity and the need for technological improvements. Empirically this will result in two arguments, a long-term trend, where we use the ratio of energy consumption to labour, and the momentary capacity utilization. We also test the influence of the effective rate of interest on bonds and foreign assets and the total revenue on the labour costs, $\bar{\pi}$, the average for the four preceding quarters. We use a multiplicative form of a long-term trend and a short-term capacity utilization. We define: $\alpha_t = (E_t/L_t)\,(Q_t/Q_t^*)$. To adjust for lags between decision and outlays we take the average of four preceding quarters, $\bar{\alpha}$. As we use the relation between energy and labour as a proxy for the relative capital intensity, we might see the variable α_t as a variant of the accelerator.

For these estimations also we have used an autoregressive model. Table 12.3(a) accounts for the estimations for the period 1961–1984 while (b) accounts for the period 1972–1984. We use a dummy for the fourth quarter.

The most interesting result of our estimations is the remarkable difference in the effect of market rate of interest on bonds. Estimations for the total period from 1961 show an insignificant negative effect while in the estimations for the period 1972 to 1984 we receive a significant positive effect. Furthermore the effect of $\bar{\alpha}$ diminishes during the latter period. We interpret this as an indication of a financial consolidation caused by the public debt.

The weaker relation between investment in machinery and the variable $\bar{\alpha}$, during the period 1972–1984, can possibly be explained by a structural change within the industrial sector. There is a higher differentiation between growth rates in different parts of the industrial sector. This leads to a lower aggregate capacity utilization.

12.4.3 Foreign trade

As in the case of investments the trade development has changed dramatically during the seventies. We therefore estimate our functions on the whole period and on the second half, 1972–1984. The latter period covers approximately the time since the failure of the Bretton–Woods system.

12.4.4 Export

The three main variables for export are international trade development, the relation of internal to external inflation and the exchange rate. We expect the first and the third variables to have a positive sign, while the second has a negative sign. Many inquiries have shown that changes in relative inflation have a delayed effect. In this study we measure only a short-term effect. We have also tested a fourth variable, capacity utilization. A reasonable a priori attitude could be that

Table 12.4 Dependent variable – exports

Eqn	Intercept	WT_t	$\Delta P_t - \Delta P_t^x$	Vx_t	Q_t/Q_t^*	Regression R^2	Total R^2
(a) Period 1961–1984							
12.34	−65.47 (−3.2)	0.348 (5.9)	−0.071 (−0.3)	0.214 (7.1)		0.44	0.86
12.35	−144.0 (−9.2)	0.381 (10.6)	0.006 (0.02)	0.235 (10.7)	0.739 (7.9)	0.77	0.89
(b) Period 1972–1984							
12.36	−2.352 (−0.1)	0.110 (2.0)	−0.404 (−1.3)	0.189 (7.8)		0.73	0.66
12.37	−144.79 (−5.4)	0.239 (4.2)	−0.116 (−0.4)	0.239 (10.1)	1.034 (7.4)	0.81	0.80

Estimation method: Yule–Walker. Autoregressive model.
Lags: $t-1$ to $t-4$.

it would have a negative effect on export, both through an increase in prices and through an increased internal demand when the capacity utilization increases.

On the other hand there is some correlation between external and internal trade cycles. We also have the effect of capacity utilization on prices which we said earlier would be positive. This will be shown to be a crucial assumption, as it will have an opposite sign in our estimations.

As can be seen from Table 12.4(a) the parameters of Equation 12.34 are what they were expected to be according to sign and strength. The two most important variables are the development of international trade and the exchange rate. The difference between internal and external inflation has a negative sign but is insignificant.

We have entered capacity utilization into Equation 12.35. On the one hand capacity utilization is in a certain degree correlated to the international trade cycles and has thus a positive effect on export. On the other hand capacity utilization affects exports through its influence on prices. As we will discuss later, according to the estimations of output prices, the capacity utilization has a negative effect upon internal price changes. This gains support in Equation 12.35. The entrance of capacity utilization in Equation 12.35 leaves the parameter for international trade development unchanged but changes the price parameter. In Table 12.4(b) we estimate the same functions as in Table 12.4(a) for the period 1972–1984.

In Sweden two parts of this period are of particular interest, namely 1974–1978 and 1982–1984. In 1974 we faced a heavy boom combined with low labour costs as a consequence of an agreement between the trade unions and the employer organizations in the shadow of the first oil crisis. During the boom the industrial (especially the export industry) profits were the best since the end of the fifties. The agreement extended over two years. This meant that the wage increases

1975–1976 were severely displaced relative to the trade cycle. At the same time the Swedish exchange rate was bound to some other European currencies, of which the German mark was the most important, which implied a continuous revaluation to currencies outside the currency co-operation. In 1977 Sweden left the currency co-operation and devalued the crown.

The latest part of the period has been dominated by the devaluation 1982 and relatively modest wage increases. All this would make us believe that the difference in inflation rates and exchange rate should increase in importance during the second half of the seventies. However, we also have to account for structural changes in demand for export goods caused partly by the first oil crisis and partly by the increased competition in traditional exporting sectors of Swedish industry from newly industrialized countries.

The comparison of the two sets of estimations in Table 12.4(a) and (b) clearly indicates that on the *aggregate* level there has been no increase in significance of prices and exchange rate. This does not mean that the prices have not increased in importance but they have to be measured on a more disaggregated level and considered in relation to specific structural variables. The decrease in Swedish exports during the late seventies has been explained by the sharp increase in wages for 1975–76 which in its turn increased export prices. This has been the ruling explanation and very little attention has been paid to the structural problems. Our estimation shows that this one-sided explanation has exaggerated the importance of prices as an isolated explanation factor and introduces capacity utilization as an important explaining factor.

12.4.5 Imports

In Table 12.5 we account for the estimations of import function. As demand arguments we use consumption, investments in machinery and transport equipments and export. Also here we notice an insignificant momentary effect of relative inflation but contrary to export estimations the effect of exchange rate is also insignificant. A reasonable interpretation is that import goods have on the average a low price elasticity. On the other hand the capacity utilization has a significant negative effect on imports which combined with the negative effect of capacity utilization on prices can indicate an indirect price effect. The estimations for the period after 1972 indicate the unstable development of imports, especially for the end of the seventies, but the results are in the same directions as the estimations for the whole period.

12.5 WAGES, PRICES OF GOODS AND EQUITIES

12.5.1 Wages

In principle there are two kinds of wage changes, those agreed upon in central negotiations and those produced by wage drift. During the sixties the centrally

Table 12.5 Dependent variable – imports

Eqn	Intercept	C_t	I_t^m	X_t	$\Delta P_t - \Delta P_t^x$	Vx_t	Q_t/Q_t^*	Regression R^2	Total R^2
(a) Period 1961–1984									
12.38	15.92 (1.3)	0.125 (1.5)	0.143 (2.8)	0.462 (6.0)	−0.206 (−0.9)	1.482 (0.6)		0.73	0.86
12.39	35.3 (2.2)	0.116 (1.4)	0.137 (2.8)	0.507 (6.3)	−0.244 (−1.1)	0.186 (0.1)	−0.168 (−1.8)	0.74	0.87
(b) Period 1972–1984									
12.40	36.77 (1.5)	0.383 (1.3)	0.152 (2.6)	0.119 (1.2)	0.133 (0.7)	0.018 (0.5)		0.58	0.60
12.41	92.53 (3.1)	0.020 (0.1)	0.216 (3.7)	0.357 (2.8)	0.124 (0.7)	−0.024 (−0.9)	−0.360 (−2.6)	0.66	0.67

Estimation method: Yule–Walker. Autoregressive model.

set wages were dominating. From the vigorous boom of 69–70 until the middle of the seventies the two factors have been of equal importance on the average. In later years both trade unions and employers' organizations try to control the wage drift but the success is not overwhelming. Our fundamental hypothesis is that workers do not accept decreasing nominal wages.

As explanatory variables we use labour productivity, Q_t/L_t, and the average ratio of total revenue to labour costs during the preceding four quarters, $\bar{\pi}$. The latter variable indicates whether a low share of labour cost will lead to later wage increases. To test the *wage changes* we use the *changes* in the variables mentioned but we also enter the change in unemployment. The Swedish labour market is highly centralized and one of the goals of the Confederation of Trade Unions is to equalize wage increases between different sectors. This is one of the reasons why the labour unions want to control the wage drift. This goal makes it difficult to estimate wage changes without institutional variables. As will be seen from the estimations in Table 12.7 the explaining power of our estimations of wage changes is low. We use an autoregressive model lagged four quarters to estimate wages. In our estimations of *wage changes* we use OLS.

Table 12.6 shows that the labour productivity has decreased in importance during the second half of the period. There are many reasons for this, but the two most important are:

(1) During the sixties and in the beginning of the seventies the wages in the industrial sector, particularly the exporting sector, were a common 'norm' to the rest of the economy, while during the late seventies and eighties this has not been the case, which has implied that different groups try to compensate for wage increases in other sectors.
(2) Structural problems within the industrial sector have caused the overall productivity to grow slowly.

These two factors in combination are an important explanation for the decreased importance of labour productivity.

Table 12.6 Dependent variable – wages

Eqn	Intercept	$\bar{\pi}$	Q_t/L_t	Regression R^2	Total R^2
(a) Period 1961–1984					
12.42	46.68 (0.9)	0.427 (1.4)	0.846 (3.7)	0.19	0.97
(b) Period 1972–1984					
12.43	71.55 (0.9)	0.983 (1.6)	0.383 (3.3)	0.16	0.95

Estimation method: Yule–Walker. Autoregressive model.
Lags: $t-1$ to $t-4$.

Table 12.7 Dependent variable – change in wages

Eqn	Intercept	$\Delta \pi$	$\Delta(Q_t/Q_t^*)$	Δu	DW	R^2	F
(a) Period 1961–1984							
12.44	3.72	−0.574	0.648	−0.021	1.8	0.34	16.0
	(6.2)	(−5.4)	(5.9)	(−0.5)			
(b) Period 1961–1984							
12.45	6.51	−1.205	1.133	−0.097	2.1	0.69	34.6
	(9.1)	(−8.4)	(8.6)	(−1.8)			

Estimation method: OLS.

This explanation is supported by the reversed relationship between the two periods when it comes to the estimations of wage changes which are displayed in Table 12.7. The negative effect of the changes in $\bar{\pi}$ is due to the fact that wage increases occur at rather a late stage in the trade cycle.

12.5.2 Prices

We use the index for manufactured goods, domestic supply, as price index. The two most important explaining factors are labour costs and prices of inputs. About 40–50% of the inputs are imported, we therefore use the import prices as the price of inputs. We also enter capacity utilization and labour productivity as explaining factors. In an economy with a high capital–labour ratio one would expect capacity utilization to have a small or negative effect on price changes. This depends on the close relationship between capacity utilization and labour productivity along the trade cycle, and the fact that capital costs do not disappear when capacity utilization decreases. Among Keynesian economists this is hardly any surprise. Kaldor (1982, pp. 64–8), for example, discusses it when he criticizes the policy, 'strength through misery'. Here we estimate both the prices and the price changes (Tables 12.8 and 12.9).

There are no major differences in the sizes of the parameters between the two periods. The big difference is in the estimation of price changes where the significance and the correlation decrease in the estimations for the period 1972–1984; perhaps an effect of the structural problems, it could also reflect a growing tendency of the government to incur different types of price control. Furthermore, employers' taxes and fees have been increasing during this period.

The effects of capacity utilization and labour productivity on price changes are negative and significant. This result can be seen as supporting Kaldor.

The effect of import prices has increased slightly during the later part of the observation period, which could be a result of an increasing share of import goods in the total inputs.

Table 12.8 Dependent variable – prices

Eqn	Intercept	w_t	P_t^x	Q_t/Q_t^*	Q_t/L_t	Regression R^2	Total R^2
(a) Period 1961–1984							
12.46	25.08	0.203	0.583	−0.062		0.997	0.999
	(12.2)	(9.8)	(30.4)	(−3.3)			
12.47	23.34	0.217	0.579		−0.046	0.997	0.999
	(14.3)	(10.2)	(30.4)		(−3.2)		
(b) Period 1972–1984							
12.48	30.17	0.168	0.606	−0.092		0.995	0.999
	(7.8)	(4.8)	(21.6)	(−3.0)			
12.49	27.55	0.176	0.609		−0.092	0.995	0.999
	(8.2)	(5.0)	(21.7)		(−2.9)		

Estimation method: Yule–Walker. Autoregressive model.
Lags: $t-1$ to $t-4$.

Table 12.9 Dependent variable – changes in prices, ΔP_t

Eqn	Intercept	Δw_t	ΔP_t^x	$\Delta(Q_t/Q_t^*)$	$\Delta Q_t/L_t$	DW	R^2	F
(a) Period 1961–1984								
12.50	0.944	0.084	0.500	−0.056		2.2	0.77	99.7
	(3.6)	(2.6)	(16.6)	(−4.1)				
12.51	0.923	0.097	0.504		−0.048	2.1	0.77	109.2
(b) Period 1972–1984								
12.52	2.092	0.055	0.447	−0.076		2.2	0.70	38.2
	(3.9)	(1.3)	(10.6)	(−3.7)				
12.53	2.068	0.063	0.452		−0.052	2.2	0.71	39.6
	(3.9)	(1.5)	(10.8)		(−3.9)			

Estimation method: OLS.

12.5.3 Equity prices

Our main interest is to study the effects of government deficit and total government debt. We believe that a short-run negative effect of a government deficit can be exceeded by a long-run positive wealth effect of the total debt (Enthoven, 1960 and Werin, 1983). To study this hypothesis to its full extent requires not only financial flows but also stocks. Some indications, however, are possible to detect in our present estimations. As explaining factors we use the relation of total revenue to labour costs, foreign short-term interest and interest on bonds.

To these fundamental variables we add the net supply of new bonds and the total government debt. There is a substantial degree of autocorrelation in equity prices. We have therefore based our estimations on an autoregressive model, three

Table 12.10 Dependent variable – market prices of equities.

Eqn	Intercept	$\bar{\pi}$	r_t	r_t^x	ΔB_t	B_t	Regression R^2	Total R^2
(a) Period 1961–1984								
12.54	−69.61 (−2.7)	1.193 (5.8)	0.243 (2.1)	−0.100 (−4.2)			0.44	0.87
12.55	−68.73 (−2.4)	1.192 (5.8)	0.237 (2.0)	−0.102 (−4.2)	0.0003 (0.03)		0.43	0.87
12.56	−81.47 (−3.1)	1.094 (5.3)	0.465 (2.9)	−0.118 (−5.1)		0.084 (1.9)	0.51	0.88
12.57	−80.51 (−3.1)	1.086 (5.3)	0.469 (3.0)	−0.119 (−5.1)	−0.005 (−0.6)	0.089 (2.0)	0.51	0.88
(b) Period 1972–1984								
12.58	−18.80 (−0.5)	1.261 (5.5)	−0.633 (−2.5)	−0.082 (−2.6)			0.70	0.89
12.59	−15.81 (−0.4)	1.250 (5.4)	−0.630 (−2.5)	−0.086 (−2.8)	−0.006 (−0.6)		0.70	0.89
12.60	−92.28 (−1.8)	1.154 (5.4)	0.329 (0.6)	−0.113 (−3.7)		0.151 (2.1)	0.76	0.90
12.61	−99.42 (−2.0)	1.137 (5.4)	0.469 (0.9)	−0.121 (−3.9)	−0.014 (−1.4)	0.171 (2.4)	0.78	0.91

Estimation method: Yule–Walker. Autoregressive model.
Lags: $t-1$ to $t-4$.

quarter lagged. In Table 12.10(a) we estimate the equations for the period 1961–1984 and in (b) we estimate them for the period 1972–1984.

Equations 12.54 and 12.55 in Table 12.10(a) display the expected sign for the foreign interest rates and the relation of revenue to labour costs, where the latter is the most important variable. The appearance of net supply of new bonds, ΔB, has a negative effect on equity prices. Looking at Equations 12.56 and 12.57 we can see that government debt, B_t, entirely reduces the effect on market rates of interest of bonds and that it has a positive effect on equity prices. This pattern is even more clear if we look at the estimations in Table 12.10(b). The effect of the revenue variable $\bar{\pi}$ is here very weak and the effect of government debt becomes more important. The increase in equity prices at the beginning of the eighties was not preceded by an increase in profits. In our opinion these estimations clearly indicate a wealth effect of the government debt which creates a financial consolidation.

ACKNOWLEDGEMENT

The authors gratefully acknowledge financial support from the Department of Economics, Göteborg University and from Göteborgs Handelshögskolefonder.

REFERENCES

Arrow, K. J. and Hahn, F. H. (1971) *General Competitive Analysis*, Holden-Day Inc. Amsterdam.
Berg, L. (1983) *Consumption and savings – a study of household behaviour*. Alinquist & Wiksell, Uppsala.
Davidson, P. (1978) *Money and the Real World*, 2nd edn, Macmillan Press Ltd, London.
Enthoven, A. C. A. A neoclassical model of money, debt and economic growth, Appendix to: *Money in a Theory of Finance*, The Brookings Institution, Washington.
Gurley, J. and Shaw, E. (1960) *Money in a Theory of Finance*, The Brookings Institution, Washington.
Hicks, J. (1977) *Economic Perspectives*, Clarendon Press, Oxford.
Kaldor, N. (1940) A model of the trade cycle. *Economic Journal*, 50.
Kaldor, N. (1956–57) Alternative theories of distribution. *Review of Economic Studies*, 23/24.
Kaldor, N. (1982) *The Scourge of Monetarism*, Oxford University Press, Oxford.
Keynes, J. M. (1930) *Treatise on Money*, Macmillan & Co. Ltd, London.
Keynes. J. M. (1936) *The General Theory of Employment, Interest and Money*, Macmillan & Co. Ltd, London.
Kragh, B. (1955) The meaning and use of liquidity curves in Keynesian interest theory. *International Economic Papers*, No. 5.
Kragh, B. (1967) Finansiella långtidsperspektiv. *Statens Offenttiga Utredmingar*, 6, Stockholm.
Kindleberger, Ch. P. and Lafforgue, J. P. (1982) ed. *Financial Crisis*, Cambridge University Press, Cambridge/Paris.
Minsky, H. P. (1964) *Financial Crisis, Financial Systems, and the Performance of the*

Economy, Private Capital Markets, The Commission of Money and Credit. Prentice-Hall Inc, Englewood Cliffs NJ.

Tobin, J. (1969) A general equilibrium approach to monetary theory. *Journal of Money, Credit and Banking*, **1**, February.

Tobin, J. (1980) *Asset Accumulation and Economic Activity*, Basil Blackwell, Oxford.

Tobin, J. (1982) *Money and Finance in the Macro-Economic Process*, The Nobel Foundation, Stockholm.

Werin, L. (1983) Budgetunderskott, portföljval och tillgångsmarknader. *Finansdepartementet DsFi*, **29**, Stockholm.

13

Asymmetry in conservation: a capital stock analysis

JONATHAN V. GREENMAN

13.1 INTRODUCTION

The world is now experiencing a third energy shock as the price of oil plummets towards its natural market level, taking the prices of other fuels with it. In the first two shocks, in the 1970s, the oil price rose substantially as a consequence of political events in the Middle East. In the United States, for example, the price of gasoline doubled in real terms over the decade and the aggregate price of energy to industry by a factor of three (Doblin, 1982). The demand response to these price rises has been impressive with the OECD energy/GDP ratio falling almost 20% since 1973, with most of the fall occurring since 1979 (Fig. 13.1). This sharp fall finally put to rest the belief that energy and GDP were inseparably linked, a belief that was reinforced by the apparent lack of response of the ratio to the first price shock.

This conservation has been achieved in three distinct ways:

(1) Behavioural conservation. Energy has been saved by using the existing energy-consuming capital stock less intensively. For households, leisure driving has been curtailed and thermostats have been set at a lower level.
(2) Technical conservation. The energy required to operate a unit of a given capital stock at a given intensity has been reduced by retrofitting and redesign. More radically, energy has been saved by changing to a new technology rather than by continuing to improve the old. A good example of this is the introduction of continuous casting in the iron and steel industry.
(3) Structural conservation. The energy shocks have contributed to the trend in the OECD economies from high to lower energy-intensive activities, both across sectors – from manufacturing to services – and within sectors – from steel to plastics, for example. Accompanying these shifts there has been an 'export' of energy demand as the location of energy-intensive industries moves to less developed countries (LDCs) with lower factor costs.

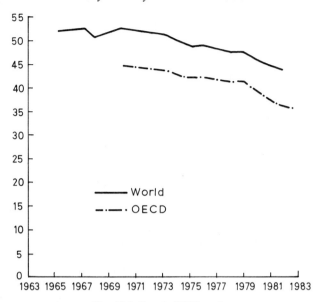

Fig. 13.1 Energy/GDP ratios

The question we address in this chapter is whether the conservation gains that have been achieved since 1973 will be reversed as energy prices fall. Clearly the behavioural component is likely to unravel as thermostats are turned up and leisure driving resumes. However, it is unlikely that there will be a substantial reversal in technical gains – insulation will not be torn out of buildings and there will not be a reversion to 'soaking pits' in steel production. Hence with a fall in real energy prices to their pre-crises levels there should be a net gain in the efficiency of the economy.

The asymmetry in price response is not captured by many of the available energy–economy models, in particular those that are linear or log linear in prices, and hence their forecasts would tend to exaggerate the demand response in falling price regimes. To model the asymmetry we have to distinguish between behavioural and technical conservation. One such model is presented in Section 3 and tested for robustness against four energy end-uses: residential heating, private use of motor vehicles, scheduled airline services and iron and steel.

To prepare the ground we provide an overview of the development of energy–economy modelling, from the early contribution of Hotelling to the large-scale models used by government, universities and consultancy groups.

13.2 ENERGY–ECONOMY MODELS

The simplest model to describe the demand for energy in a macro-economy takes the demand (D) to be a function of GDP (Y) and energy price (p_E):

$$D = AY^\alpha p_E^\beta \tag{13.1}$$

Such a model was used in the World Energy Conference study (WEC, 1978) set up immediately after the first energy crisis to examine the likely paths of energy demand under different GDP and price scenarios. The paths were tested for feasibility against what were thought to be reasonable supply projections (Ulph, 1980).

A market supply/demand model can be constructed by adding to Equation 13.1 a cumulative supply function with cost rising without bound as the cumulative supply approaches the resource limit. A monopoly situation was considered by Hotelling (1931) who took the rate of resource depletion to be the result of profit maximizing behaviour on the part of the producer, constrained by the size of the resource base and the demand response (13.1).

A more elaborate model can be obtained by adding resources as an input factor to the production function of a macro growth model. Stiglitz (1975), for example, used a simple Cobb–Douglas form:

$$Y = AL^\alpha K^\beta R^\gamma$$

where L, K, R denote, respectively, the services of labour and capital and the rate of resource consumption. The rate of capital accumulation and resource depletion is determined by optimizing a standard social welfare function. The transaction prices are taken to be the shadow prices obtained in the process of solving the optimization problem.

A model incorporating both capital stock and resources was also used by the *Limits to Growth* team (Meadows, 1962) to explore the ramifications of resource depletion. The rather pessimistic forecasts that were obtained from their model, subsequently well publicized, followed inexorably from certain restrictive features of their model structure. They assumed constant savings and capital–output ratios and an increasing fraction of capital devoted to resource extraction (Kay and Mirlees, 1974). This model illustrates the possible dangers of using empirically untested structures for forecasting purposes.

A relatively simple model that has been used for practical policy analysis is ETA-MACRO constructed by Manne (1979a, b) at Stanford. On the macro side it is similar in spirit to that of Stiglitz, optimizing a social welfare function with capital, labour and energy (electric and nonelectric) as inputs to an aggregate production function (of constant elasticity of substitution – CES – form). The key to the macro side is a single parameter, the elasticity of substitution; it measures the ability of the economy to adsorb energy shocks. This parameter is fixed judgementally. The energy side is more elaborate with sufficient structure to enable issues in the 'great nuclear power debate' to be addressed. The energy submodel determines the best way of satisfying final energy needs by minimizing costs over a set of production technologies.

The Hudson–Jorgenson model (1974; 1977) is much more ambitious, overcoming the weakness of ETA-MACRO by providing a complete and disaggregated macro model which is fully estimated (Fig. 13.2). Optimization is decentralized to the participating economic agents. Consumer behaviour is described by an indirect utility function and producers by minimum cost functions. Industry is

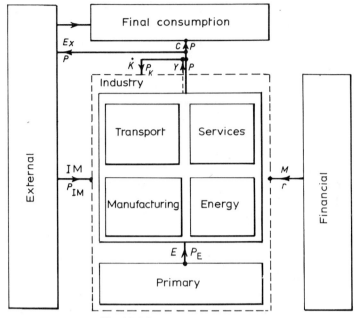

Fig. 13.2 Energy–economy models

Model A: Demand $E = Ap_E^\alpha Y^\beta$ [WEC]

Model B: Supply/demand
(a) Cumulative supply function (b) Hotelling

Model C: growth model
$$K = SY - \delta K$$
$$Y = F(K, \bullet, E)$$
$$Y = C + \dot{K}$$
(a) Stiglitz (C, E controls) (b) Meadows

Model D: +energy sector
ETA-MACRO (energy LP + welfare optimization)

Model E: macro + I/O (Input/Output)
(a) DoE (US Department of Energy)
(b) DRI (Data Resources Inc.)
(c) Hudson–Jorgenson
$$X_i = \sum_j a_{ij} X_j + Y_i$$
$$a_{ij} = a_{ij}(P)$$

disaggregated into four nonenergy and five energy sectors, the first four matching the normal categories into which final energy use is disaggregated – transport, manufacturing, services and agriculture. The interactions between these sectors are described by an input–output model with the added feature that the coefficients are price dependent. Precisely, if final demand for the *i*th good is denoted by y_i and its production by x_i then:
$$x_i = \sum_j a_{ij} x_j + y_i$$

where $a_{ij}(p)$ denote the input–output coefficients. Their evaluation follows from the analysis of Berndt–Wood (1975) using Shephard quality. Let p_i denote the price of good i and x_{ij} be the input of good j in the production of x_i of good i. Minimizing the total cost of factor inputs:

$$\sum_j p_j x_{ij}$$

for the ith good subject to the production constraint:

$$F_i(x_{ij}) = x_i$$

yields the minimum cost function $G_i(p_j, x_i)$ with optimal factor inputs given by:

$$x_{ij} = \partial G_i / \partial p_j.$$

When constant returns to scale operate the input–output, coefficients are given by $a_{ij} = x_{ij}/x_i$.

Berndt–Wood and Hudson–Jorgenson both used translog functions for cost functions G to provide a quadratic generalization to the Cobb–Douglas linear case. The original Berndt–Wood analysis was restricted to the US manufacturing sector with factors: capital (K), labour (L), energy (E) and materials (M). Their immediate objective was a rapprochement between the engineering and econometric approaches to energy conservation. This distinction is made clearer in the nested CES approach of Prywes (1986) and others. The production function for a given sector is taken to be:

$$F = F^{(3)}(M, F^{(2)}(L, F^{(1)}(K, E)))$$

where

$$F^{(i)}(X, Y) = (a_i X^{-\rho} + b_i Y^{-\rho})^{-1/\rho} f_i(t).$$

The dual function has the same structure with prices substituted for volumes. The elasticities of substitution are given by:

$$\sigma = 1/(1 + \rho).$$

In the 'engineering' way of seeing things, composite capital, $F^{(1)}$, is fixed and hence K and E are necessarily substitutes under relative price changes. In the 'economic' perspective output, $F^{(3)}$, is fixed and composite capital becomes substitutable with labour and materials. With these extra degrees of freedom K and E could be either substitutes or complements depending on the magnitude of the parameters in the component functions $F^{(i)}$. Berndt–Wood found that they were in fact complements with an energy price rise reducing capital at fixed output. This has recently been confirmed by Prywes who looked at individual sectors within manufacturing for the crisis period 1971–1976. This conclusion however has been questioned by Griffin–Gregory (1976) and Pindyck (1979) in their cross-sectional studies. They argue that the essentially static Berndt–Wood model will only pick up short-term effects.

The issue has been further examined with the dynamic continuous time model of Drollas–Greenman (1983). This model is similar in size and structure to the

Bergstrom–Wymer model (1976) of the UK economy. The new features are the inclusion of energy in the (nested CES) production function and the determination of factor inputs by Shephard duality. The engineering elasticity of substitution was found to be essentially zero, i.e. capital and energy are, modulo a time trend, close to being in fixed proportions; further the elasticity of 'composite capital' and labour was close to unity generating between them a Cobb–Douglas production function. Again this model probably underestimates the longer term responses since it was estimated only up to 1977 when energy efficiency gains had not fully worked through to the capital stock as a whole. It also does not discriminate between the different forms of conservation and hence does not address the issue of asymmetry of responses.

Our objective is to improve this model in these respects. As a first stage we analyse in this chapter energy use in certain key sectors where homogeneity permits a clearer understanding of the dynamic processes at work. From this analysis we propose a generalized model of energy use at macro level.

13.3 CONSERVATION

To see how to separate out technical from behavioural conservation we will examine first a simplified version of a model used in the analysis of gasoline consumption in the US (Pindyck, 1979; Dahl, 1986; Drollas, 1984).

13.3.1 The demand for gasoline in the US

Gasoline consumption, E, can be factored into the following terms:

$$E = URK \tag{13.2}$$

where K denotes the size of the car fleet, U its utilization (i.e. the average miles driven per year per car) and R the gallons consumed on average per mile (see Fig. 13.3). Broadly speaking behavioural conservation corresponds to utilization U (with leisure driving sacrificed as the gasoline price rises) and R to technical conservation (with consumption decreasing with greater efficiency).

The time series for U, shown in Fig. 13.4, exhibits great volatility with price, with falls of up to 10% in response to the price shocks. The consequent sharp rebounds are driven by rising per capita income and decreasing costs per mile as efficiency gains feed through to the fleet. Reduction in consumption through technical improvement is therefore offset by an increase in utilization.

The time series for R^{-1} (i.e. miles per gallon) shows the impressive gains that have been achieved in the efficiency of new cars under the influence of the Corporate Average Fund Economy (CAFE) mandatory standards set by the US government (Fig. 13.5). Over the last decade there has been a 90% improvement in new car performance due in part to a reduction in weight and the power-to-weight ratio. This has not yet worked through to the fleet as a whole – capital rotation taking roughly ten years in the case of cars. The full 90% however will not be achieved

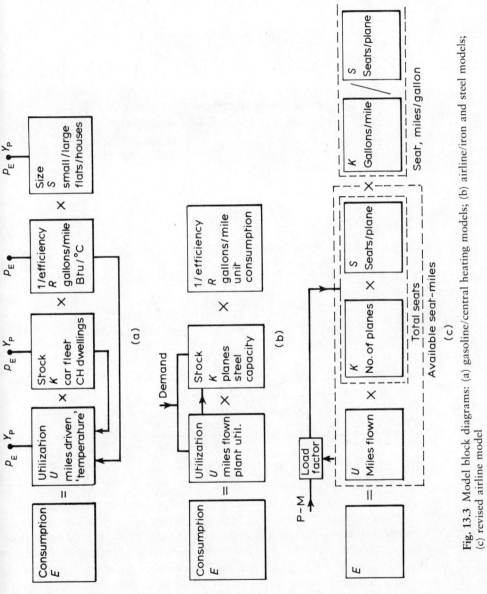

Fig. 13.3 Model block diagrams: (a) gasoline/central heating models; (b) airline/iron and steel models; (c) revised airline model

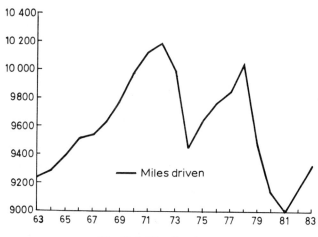

Fig. 13.4 US miles driven

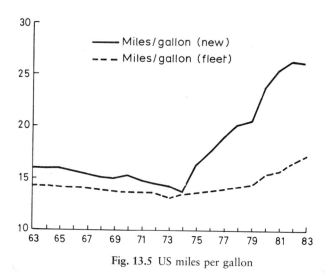

Fig. 13.5 US miles per gallon

since there will be an inevitable deterioration in performance with age and indifferent maintenance.

The number of cars per capita, K_p, has shown continued growth throughout the 1960s and 1970s, rising to a level of 800 cars per thousand persons of driving age (Fig. 13.6). Since the second oil crisis however this growth has stopped, momentarily if one believes the cause to be recession, permanently if the explanation is saturating demand. The former is more likely since there has been a hiatus in growth in most previous recessions. This has been confirmed by the 1984 and 1985 per capita stock figures which show a resumption in growth.

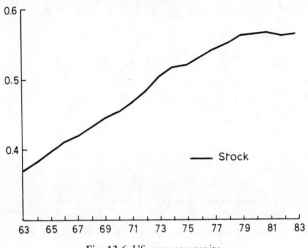

Fig. 13.6 US cars per capita

To convert factorization (Equation 13.2) into a model we have to specify equations for the component variables. The simplest model we have used has the following structure:

$$U = \alpha_0 + \alpha_1 \ln Rp_{E_-} + \alpha_2 \ln Y_p \qquad (13.3)$$

$$D(\ln K_p) = \gamma_1 (\ln K_p^* - \ln K_p) \quad (D = d/dt) \qquad (13.4)$$

where

$$\ln K_p^* = \beta_0 + \beta_1 \ln p_E + \beta_2 \ln Y_p$$

$$D(\ln R^{-1}) = \gamma_2(\theta \ln R_N^{-1} - \ln R^{-1}) \qquad (13.5)$$

where Y_p indicates per capita income. The parameter values and their associated t-values are listed in Table A1 of Appendix A. The crucial feature of the first equation, for utilization, is the dependence on cost per mile, Rp_E, rather than p_E, the cost per gallon. To capture accurately the timing of the utilization response the cost variable is lagged one period.

The per capita stock equation is driven primarily by per capita income with an elasticity close to one and hence over the estimation period there is no evidence of saturation. This is the experience of other analysis including the OECD. To forecast for the longer term one has to include saturation. Unfortunately the standard method, of using a simple logistic curve, has proved unsuccessful. Our best fit has been achieved with the functional form

$$\ln K_p^* = a + b \tan^{-1}(c \ln Y_p - d) \qquad (13.6)$$

We are however analysing the issue in more depth. The efficiency equation, R^{-1}, describes rotation of the car stock with efficiency gains in new cars R_N^{-1}, taking some years to have an impact on the car fleet as a whole. Parameter θ describes

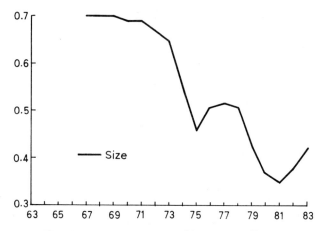

Fig. 13.7 US car size (ratio of large to small cars)

the deterioration of performance with age. Over the last decade R_N^{-1} was not market determined and hence the market relationship between final price and efficiency cannot easily be determined. In forecasting with the model therefore we have taken it to be exogenous.

In running the model forward in a regime with falling prices conservation gains through lower utilization will be reversed but efficiency gains will be maintained. Hence if prices return to their pre-crises levels (in real terms) consumption will not rejoin its previous growth path.

The positioning of the new lower growth path is determined by the relative strengths of behavioural and technical conservation and the interaction between them evidenced by the presence of an R factor in the U equation. There is a second mechanism by which these two modes of conservation interact, arising from the heterogeneity of the car stock. With falling prices there will be a shift back to large cars, running costs becoming less important when compared with other attributes. This trend is already in evidence (Fig. 13.7). To model this effect a 'size factor' S should be included in the factorization (Equation 13.2):

$$E = URSK \qquad (13.7)$$

to boost consumption with an increasing share of large cars in the fleet. R is redefined as the gallons per mile of the 'average' sized car. Behavioural conservation therefore has two components – a short-term effect embodied in U and a longer term effect through factor S.

13.3.2 Residential central heating

To apply the factorization (Equation 13.2) to fuel consumption in residential central heating we take K to be the number of households with central heating

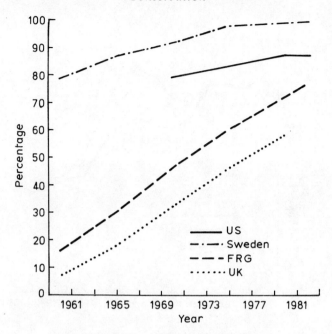

Fig. 13.8 OECD central heating penetration

(CH), R to denote the fuel required to achieve a given base environment and U the heating services required (in terms of temperature level, duration of the heating cycle and area heated) relative to that of the 'base' environment (see Fig. 13.3). Unlike US gasoline consumption, comprehensive data to identify components U, R, K are not available. From the work of Schipper et al. (1985) at the Lawrence Berkeley Laboratory and others there are however data available for stock K and unit consumption UR for several European countries.

Series for CH penetration are shown in Fig. 13.8. Penetration is reaching saturation in Sweden and the United States but is still in the rapid growth phase in EEC countries. For West Germany (FRG) and the UK the process has been satisfactorily explained using Equation 13.4 with saturation form (Equation 13.6).

Series for unit consumption, UR, are plotted in Fig. 13.9; again they show the volatility of consumer behaviour in the face of rapid price changes. More detail on changes in UR and its components have been provided by snapshot surveys which have been carried out in several countries. From these surveys the contributions of behavioural and technical conservation to the fall in unit consumption can be roughly calculated. Schipper's judgement, for example, is that roughly 65% of conservation in the FRG was behavioural but only 50% in the United States. The use of reduced forms in separating U and R, the standard procedure in this situation, is difficult for many countries because of the monotonicity of price changes in the estimation period and the similarity of lag

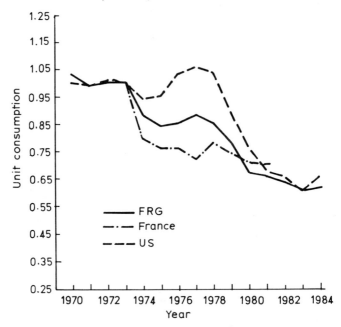

Fig. 13.9 Oil central heating utilization

structures for factors U, R. In the case of US gasoline, efficiency gains take some considerable time to work through since we have to wait for stock rotation but for residential heating increased efficiency can be achieved rapidly by retrofitting.

The equation we estimated for unit consumption has the form:

$$D(\ln UR) = \gamma_3(\ln UR^* - \ln UR)$$

where:

$$\ln UR^* = \alpha_0 + \alpha_1 \ln p_E + \alpha_3 W(\ln p_E) + \alpha_2 \ln (Y_p)$$

W denotes the Wolffram operator (Wolffram, 1971) defined by:

$$W(\ln p_E(t)) = \max_{t_1 \leq t} (\ln p_E(t_1))$$

and p_E is the aggregate fuel price. The Wolffram term models asymmetry in the consumption response and proxies technical gains. For the FRG we are unable to separately identify α_1, α_3 because of the unavailability of data for a period in which the fuel price falls. For the United States we were able to identify them; we found their values to be -0.341, -0.301 respectively (t-values: 1.84, 2.82) agreeing with the view of Schipper on a 50%–50% split between behavioural and technical conservation. The values of the other parameters for the FRG are listed in Table A1 of Appendix A.

Conservation

A size effect also operates in residential heating. New dwellings increase in size with per capita income and with a shift from apartments to single-family dwellings. Unlike US gasoline these trends are little affected by energy price perturbations.

Factorization (Equation 13.2) and its variant (Equation 13.3) have proved useful in analysing the two major components of domestic fuel consumption. It suggests a simple causal structure; capital stock is driven by income and the utilization of that stock adjusts to adsorb price shocks as an initial response. Subsequent retrofitting and capital rotation underpin the response and lead to permanent gains in energy saving.

To test the universality of this factorization procedure we have applied it to two radically different situations outside the domestic sector – the first being 'US airlines' as an example from the service sector, and the second, 'iron and steel' from the manufacturing sector.

13.3.3 US airlines

There have been three phases in the history of airline travel since the Second World War:

(1) The transition from turboprop to jet planes in the 1960s
(2) The progressive increase in the size of planes in response to growth in demand
(3) Gains in engine efficiency following the distortions in the energy price.

Since size is a crucial factor in the analysis of this sector we rewrite the basic equation:

$$E = UKR$$

(where U denotes average miles flown per plane, K the number of planes and R the average number of gallons consumed per mile) in the form (see Fig. 13.14):

$$E = U(KS)/(S/R) \qquad (13.8)$$

where S is the average number of seats per plane, KS is the total number of seats available and S/R denotes the 'seat-miles' per gallon (Fig. 13.10).

Unlike the two cases already considered, utilization here does not play as a central role in initially absorbing the price shocks. U is in the short term constrained by schedule constraints (Fig. 13.11). (The rise in U during the 1960s was due to the greater speed of jet planes permitting greater distances to be flown in a year.)

Rises in the airline price index together with the subsequent recessions depressed the demand for travel (passenger-miles) resulting in a lower load factor, LF (the proportion of seats occupied) and hence a lower growth in the fleet (Fig. 13.12). Costs were subsequently reduced through progressive increases in size (by replacement of old planes) and in fuel efficiency, generating in conjunction with increased growth in the economy, a new cycle of demand and fleet growth.

The model we used to describe these interactions has the following form:

$$D(\ln U) = \gamma_1 (\ln U^* - \ln U)$$

258 *Asymmetry in conservation*

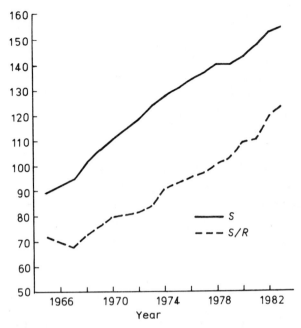

Fig. 13.10 US airlines

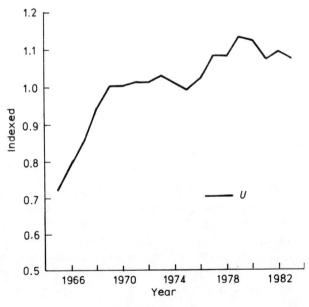

Fig. 13.11 US airlines

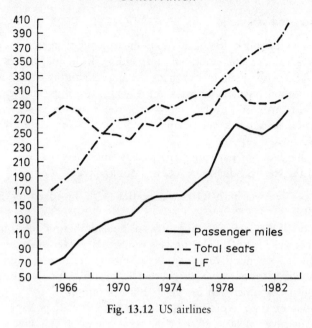

Fig. 13.12 US airlines

where

$$\ln U^* = \alpha_0 + \alpha_1 \ln (\text{speed}) + \alpha_2 \ln (\text{P-M}) + \alpha_3 t$$

$$D(\ln KS) = \gamma_2 (\ln KS^* - \ln KS)$$

where

$$\ln KS^* = \theta_2 (\ln (\text{P-M})^* - \ln U)$$

and

$$\ln (\text{P-M})^* = \ln (\text{P-M}) + \theta_1 D \ln (\text{P-M})$$

$$D^2 \ln (S/R) + aD \ln (S/R) + b \ln (S/R) = b \ln (S/R)^*$$

where

$$\ln (S/R)^* = \beta_0 + \beta_1 \ln S + \beta_2 W(\ln P_E)$$

Utilization responds to the services required (P-M; passenger-miles) and to technology (increasing average speed). (See Table A2 of Appendix A for the parameter values.) Total seating capacity, KS, is driven by expected demand for seats evaluated as the expected passenger-miles (P-M)* required, divided by current utilization. Seat-miles per gallon was found to be a second order process driven by two forces: the economics of scale generated by the increasing size of planes (the β_1 term) and increased engine efficiency induced by price rises (the β_2 term).

13.3.4 US iron and steel

In the short term the utilization of capital in the iron and steel sector is determined by demand with the capital stock only responding in the longer term. The level of demand reflects the general health of the manufacturing sector as it passes through the normal business cycles and resource price induced recessions. The vulnerability of this sector to the vagaries of the economy is shown vividly in Fig. 13.13; it is clear that swings of up to 15% in production are the norm in this industry.

Gains in efficiency, in particular energy efficiency, have occurred in each of the stages in the steel production process. For example:

(1) Increases of scale in blast-furnace operation with improved recycling of the waste gases.
(2) Penetration of the basic oxygen process using endothermic reactions as a heat source (Poznanskl, 1983). This process replaces the older open-hearth process which is twice as energy intensive. Because of the restricted ability to handle scrap, growth in the basic oxygen process has been accompanied by a growth in electric furnaces which can use up to 100% scrap.
(3) Continuous casting, avoiding the intermediate reheating stage (Poznanskl, 1983). This saves at least 20% of the total energy in this final stage of the steelmaking process.

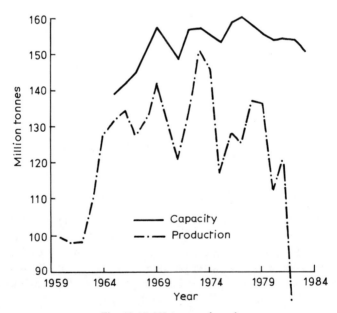

Fig. 13.13 US iron and steel

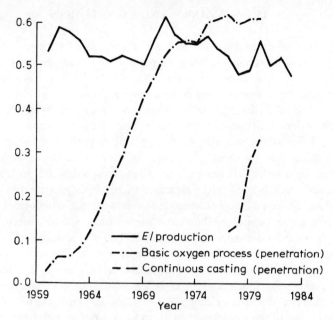

Fig. 13.14 US iron and steel

Such improvements have been implemented even when energy prices remained low – as part of the continual exercise to reduce costs and compete effectively. This is evident from the graph in Fig. 13.14 which shows energy per unit of production. The oscillations in the graph are caused by the efficiency/inefficiency of energy use in periods of high/low utilization. The general trend of the graph is clearly downwards with much of the gain occurring in the 1960s. We have been able to satisfactorily explain this series by regressing energy use per unit of capital against utilization and share of the basic oxygen process.

The equations defining our model for the iron and steel sector are based on factorization (Equation 13.2) and have the form:

$$\ln U = \theta_2 (\ln Q - \ln K)$$

$$\ln R = \ln R_0 = \ln (A + B(\text{OH}))$$

$$c^2 D^2 \ln K + 2cD \ln K + \ln K = \theta_1 \ln Q$$

where Q denotes output, K steelmaking capacity and OH the share of open hearth furnaces in the steelmaking process. The capital stock adjustment equation is second order, to model the long lead times of investment in this sector. The values of the parameters are listed in Appendix A.

13.4 SUMMARY AND CONCLUSION

The issue that introduced this chapter was the distinction between behavioural and technical conservation and the problem of how to capture this distinction at the macro level. At a sector level there are models that achieve the required differentiation – gasoline demand being a case in point.

We have looked in detail at one of these models, defined by a factorization of consumption into utilization, capital stock and efficiency terms – the first term modelling behavioural and the third technical responses. We have tested its general applicability by using it to describe a broad range of activities. First we looked at residential heating and then two industrial sectors – airlines and iron and steel. Naturally the structure of the equations for the component terms differs between the domestic and industrial sectors. In the former, utilization is under the direct control of the sector agents and responds directly to energy price. In the industrial sector, energy price affects both product/service prices and GNP and hence the demand for that product or service. The sector response acts through capital stock adjustment and increased fuel efficiency. It is clear however from the analysis that efficiency is an ongoing process and continues even if energy prices do not rise (because of economies of scale and attempts to reduce costs).

The extension of the model to the macro level is not straightforward since the applications we have looked at have been relatively simple due to homogeneity of the capital stock and its measurement in volume terms. At the macro level this is not possible nor is its decoupling from other factor inputs feasible. It is also not possible to embody energy efficiency in particular processes as we did to the iron and steel sector. The simplest model that takes into account the insights gained in the analysis has the form:

$$D(\ln L) = \gamma_1(\ln L^* - \ln L)$$
$$D(\ln K) = \gamma_2(\ln K^* - \ln K)$$
$$aD^2 \ln R + bD \ln R + \ln R = \ln R^*$$

where

$$\ln L^* = \partial G/\partial p_L; \ln K^* = \partial G/\partial p_K$$
$$G = YA_0^{-1} e^{-\lambda_1 t}(p_L/p_K)$$
$$p_K = p(a_1 + a_2 r - a_3 D(\ln p)) + Rp_E$$
$$\ln R^* = \alpha_0 + \alpha_1 t + \alpha_2 W(\ln p_E)$$

Here Y, L, p_L, p denote GNP, labour, wage rate and price level respectively and G the minimum cost function derived by Shephard duality from a nested CES production function for K, L, E of the type discussed in Section 2. This is a simple extension of the production submodel used by Drollas–Greenman (1983). It is currently being estimated and the results will be reported on in due course.

Appendix A

ACKNOWLEDGEMENTS

I would like to thank Dr L.P. Drollas for continued support in developing the capital stock model. Indeed he has been a leading advocate of this type of model for some time. I would also like to thank Professor Wymer of the IMF for the use of his suite of estimation/simulation programs. The views expressed in this chapter are not necessarily those of the British Petroleum Company.

APPENDIX A

The models presented in this chapter were estimated by full information maximum likelihood techniques taking into account cross-equation parameter constraints.

Table 13.A1

Parameter	US cars		FRG houses	
	Estimate	t-value	Estimate	t-value
α_0	8.934	86.85	7.216	43.35
α_1	−0.264	8.48	−0.475	14.75
α_2	0.119	7.41	0.0*	
a	−1.006	3.62	−1.514	15.33
b	0.495	1.87	1.044	8.51
c	2.922	1.54	4.037	7.32
d	−15.706	1.62	11.727	6.93
γ_1^{-1}	3.646	3.34	0.0*	
γ_2^{-1}	8.091	7.89	—	

Table 13.A2

Parameter	Airlines		Parameter	Iron and steel	
	Estimate	t-value		Estimate	t-value
α_0	−4.983	10.40	c	4.071	2.16
α_1	0.526	5.00	θ_1	1.04	35.24
α_2	0.392	7.60	A	0.451	13.60
α_3	−0.017	5.43	B	0.132	2.55
θ_1	2.223	2.40	θ_2	0.721	3.58
θ_2	1.136	56.16			
β_0	−0.854	2.20			
β_1	0.783	7.33			
β_2	0.146	3.98			
γ_1^{-1}	0.223	6.42			
γ_2^{-1}	10.076	2.79			
b	1.351	2.51			
a	1.973	3.29			

The parameter values and associated *t*-values obtained are listed in Table A1, firstly for the gasoline and heating models and secondly for the industrial models – airlines and iron and steel.

REFERENCES

Bergstrom, A. R. (ed) (1976) *Statistical Inference in Continuous Time Economic Models*, North-Holland, Amsterdam.

Berndt, E. R. and Wood, D. O. (1975) Technology, prices and the derived demand for energy. *Review of Economics and Statistics*, **57**, 259–68.

Dahl, C. (1986) Gasoline demand survey. *The Energy Journal*, **7**, 67–82.

Doblin, C. P. (1982) The Growth of Energy Consumption and Prices in the USA, FRG, France and the UK: 1950–1980, Report RR-82-18, May, International Institute for Applied Systems Analysis, Luxenburg, Austria.

Drollas, L. P. (1984) The demand for gasoline. *Energy Eonomics*, January, 71–82.

Drollas, L. P. and Greenman, J. V. (1987) The price of energy and factor substitution in the US economy (paper presented at the Econometric Society Conference, Pisa). *Energy Economics* July, pp. 159–66.

Griffin, J. M. and Gregory, P. R. (1976) An intercountry translog model of energy substitution responses. *American Economic Review*, December, 845–57.

Hotelling, H. (1931) The economics of exhaustible resources. *Journal of Political Economy*, April.

Hudson, E. A. and Jorgenson, D. W. (1974) US energy policy and economic growth 1975–2000. *Bell Journal of Economics*, **5**, 461–514.

Hoffman, K. C. and Jorgenson, D. W. (1977) Economic and technological models for evaluation of energy policy. *Bell Journal of Economics*, **8**, 444–66.

Kay, J. A. and Mirlees, J. (1974) The desirability of natural resource depletion, paper presented to the EESG/IES Conference on Natural Resources, London.

Manne, A. S. (1979a) Energy policy modelling: a survey. *Operations Research*, **27**, 1–36.

Manne, A. S. (1979b) ETA-MACRO: A model of energy-economy interactions'. *Advances in the Economics of Energy and Resources*, **2**, 205–33.

Meadows, D. H. *et al.* (1972) *The Limits to Growth*, Pan Books, London.

Pindyck, R. S. (1979) *The Structure of World Energy Demand*, MIT Press, Cambridge, Mass.

Poznanskl, K. Z. (1983) The international diffusion of steel technologies. *Technological Forecasting and Social Change*, **23**, 305–23.

Prywes, M. (1986) A nested CES approach to capital-energy substitution. *Energy Economics*, January 22–8.

Schipper, L. *et al.* (1985) Explaining residential energy use by international bottom-up comparisons. *Annual Review of Energy*, **10**, 341–405.

Stiglitz, J. (1975) Growth with exhaustible natural resources: efficient and optimal growth paths. *Review of Economic Studies*, pp. 123–152.

Ulph, A. M. (1980) World energy models – a survey and critique. *Energy Economics*, January, 46–59.

Wolffram, R. (1971) Positivistic measures of aggregate supply elasticities: some new approaches. *American Journal of Agricultural Economics*, **53**, No. 2, 356–9.

World Energy Conference (1978) *Study Group Report on World Energy Demand*, IPC Press, Kingston upon Thames.

14

Adjustment options for the US economy

JAN C. SIEBRAND and JOB SWANK

14.1 INTRODUCTION

After some years of staggering expansion, the US economy is faced with large deficits, both in the government budget and on the current account. These deficits make its future rather uncertain, in particular because the likely policy reactions are difficult to predict. One way to reduce this uncertainty is to evaluate different policy options, both with respect to their potential effectivity and with respect to possible side effects. Such an evaluation almost certainly narrows the set of *ex ante* feasible options. It is worthwhile, therefore, to undertake such an exercise.

This chapter evaluates some options for adjustment. This evaluation is based on an annual macroeconometric model of the US economy, model RASMUS 2b, which is discussed in Section 14.2. Section 14.3 deals with the simulation results of a number of policy measures aimed at reducing the above-mentioned deficits. Conclusions are drawn in a final section.

14.2 THE MODEL

14.2.1 Overall structure

Model RASMUS 2b is an annual macroeconometric model of the US economy, based on an integrated set of national and financial accounts. It contains 37 stochastic equations and 168 endogenous variables. The overall structure of the model is described in Swank and Siebrand (1986). This model and a similar one covering a block of six members of the European Community (EC) were designed as parts of a three-block world model, to be used for the evaluation of the options for national economic policy in an international setting and in particular the gains from international co-ordination or co-operation. The RASMUS models are intended to cover a wide range of theoretical notions with respect to the effectiveness of economic policy. Therefore, *r*ival *a*rguments are *s*imultaneously *m*odelled in (virtually) *u*niform *s*ystems. The RASMUS models contain the determination of real variables, like product demand and supply, labour demand and supply, as well as a rather detailed monetary sector describing the main financial relations between the central bank, commercial banks, government and the private and foreign sector. The traditional Keynesian type of argument is represented in the present model by elements like the multiplier/accelerator mechanism, mark-up pricing, the Phillips curve, expansion-induced short-term productivity gains, dependence of international trade flows on relative prices, as well as on pressure of demand in different countries. Relevant for supply conditions are the CES production structure, factor demand based on relative prices, capital costs based on endogenous interest rates, and the encouraging effect of shorter labour time on labour supply. These elements are reinforced by financial feedbacks, such as the impact of the financial position of government and of firms on their borrowing behaviour and therefore on interest rates and on debt service. As a result of this combination of Keynesian, neoclassical and financial elements the multipliers of RASMUS 2b show features of quite different lines of reasoning.

The model contains seven more or less interdependent blocks. The first block represents the supply side of the US economy. It is based on a CES-type production function with an estimated elasticity of substitution of about 0.5, which constrains the cost-minimization problem from which the factor-demand equations, as well as the determination of the output deflator, are derived. These equations have been estimated simultaneously in order to assure their mutual consistency. Capacity utilization follows from the confrontation of aggregate demand with the short-run CES function. The utilization rate of labour is captured by an index of average working hours, which depends on both secular and cyclical influences. Imports of goods and services are the foreign contributors to aggregate supply. Their main determinants are total sales, relative prices and utilization rates.

The main components of aggregate demand are determined in the second block. Private consumption depends on both past and actual income streams, real interest rates and on private liquidity. Disposable income and interest rates also play a part in the determination of residential construction, next to prices and capacity utilization. The equation explaining nonresidential business investment describes the gradual adjustment of the actual capital stock to its planned level, which follows from cost minimization, as mentioned above. The speed of adjustment depends on both the rate of capacity utilization and on private liquidity. Capital costs are determined by the price of investment goods, interest rates, the rate of physical decay and a number of parameters representing the influence of corporate taxes, the investment tax credit and depreciation deduction. Changes in inventories are related to sales and the lagged stock in a flexible-accelerator approach. Furthermore, allowance is made for the existence of unplanned inventories. Besides, the unemployment rate represents the precautionary motive for stock-building. Exports follow from world trade, relative prices – with a substitution elasticity of about −1.2 − and from pressure of demand, represented by both domestic and foreign utilization rates.

Wages and prices are determined in the third block. Wages are discerned in those paid by companies and those paid by the government and social security funds. Next to compensation of consumer-price increases and changes in labour productivity, the hourly wage rate paid by companies is very sensitive with respect to the state of the labour market, represented by the conventional Phillips–Lipsey effects. Forward shifting of taxes and social security contributions − potentially a major crowding-out mechanism − was not found to be significant in combination with the explanatory variables included in the present equation. The wage rate in the public sector follows that of the private sector, apart from an additional Phillips curve effect.

The fourth block contains private labour demand and labour supply. Labour demand refers to man-hours and is related to planned capacity and real labour costs. The adjustment of actual towards desired labour inputs is assumed to vary with the rate of capacity utilization. Labour supply is based on an endogenous participation rate, which is negatively affected by real wages and average hours and positively by the tension in the labour market as represented by the rate of employment.

Block five describes the social security sector. Social benefits are explained by the private wage rate, unemployment and the excess savings of social security funds, next to a trend term representing the secular growth of the number of beneficiaries. The government share in social benefits mainly depends on the level of unemployment. Total contributions are linked to total benefits. The shares of the contributions paid by employers and employees vary with the unemployment rate.

The government sector (block six) is relatively simple. Taxes are endogenous. The expenditures in constant prices and government employment are exogenous. The government wage rate follows the private one. Interest payments are endogenous and split up in payments to residents and payments to foreigners.

The financial surpluses resulting from the real transactions of the private sector, the foreign sector and the government, are the major inputs for the monetary sector (block seven). This sector is divided in four subsectors: the central bank, commercial banks, the private and foreign sector. For each of these the balance sheets are specified. Stochastic equations determine the demand for currency, demand deposits, net loans of the private sector and commercial banks to foreigners and bank lending to the private and the public sector. The other items are either explained by definitions or are exogenous. Interest rates are explained by the ratios of assets and liabilities of the private sector, the discount rate and the foreign interest rate.

The major spillovers between real and monetary sector are the financial surpluses and interest rates. Any action in one sector has an impact in both the real and the financial system. For example, an increase in government spending not only induces 'real' effects, such as production growth, an increase in utilization rates, price- and wage changes, etc., but also financial effects. All ways of financial additional spending induce monetary feedbacks which dampen the initial 'real' effects, at least after some time.

14.2.2 Forecasting performance

The performance of the complete model has been examined from a dynamic (policy-neutral) simulation covering the period 1967–1981. Such a test is rather stringent, since both current *and lagged* endogenous variables are generated by the model, which may result in an accumulation of prediction errors. On the other hand, the simulation is merely *ex post*, meaning that historical values are used for the exogenous variables.

Usually, the forecast accuracy of a model is judged from summary statistics such as the root mean square error, Theil's inequality coefficient, and so on. Since these error measures give only a very rough idea of the tracking ability of a model, preference has been given to a graphic evaluation. To that end, the simulation results for 18 important variables, together with the corresponding realized values, have been plotted against time in Fig. 14.1.

Evidently, the predictions for real GDP and its components are fairly good, except for inventory formation, which is a notoriously difficult variable to predict in an annual model, however. Especially the course of real private consumption is followed very closely, and the model performs surprisingly well with regard to the volume of residential construction. Larger errors occur in the volume of fixed business investment, which are mainly due to errors in the growth rate of aggregate demand, sometimes reinforced by errors in private liquidity. The recovery from the oil-price shock, predicted for 1975, came only in 1976; this is the main cause of the error accumulation arising after 1975. Moreover, the recession of 1980 is not sharply foreseen by the model, but is projected one year later. The prediction errors in commodity exports and commodity imports are partly caused by single-equation residuals and partly by inaccurate estimations of economic activity.

The course of wages and prices is followed rather closely, although the acceleration in inflation rates that developed at the end of the sample period is not fully captured by the model. This can partly be traced back to the increasing underestimation of the long-term interest rate from 1978 onwards, which affects prices (and hence wages) through the user cost of capital with a rather long lag. As the errors in wages and prices are largely offsetting in the determination of labour demand, labour supply and average hours, and since single-equation residuals in the latter three variables are relatively minor, the errors arising in dependent employment of companies and in the unemployment rate are primarily attributable to errors in private GDP and the rate of capacity utilization.

The model's performance with respect to the monetary sector is not completely satisfactory. Private liquidity is continuously underestimated, which is largely due to residuals in the estimated equation for the stock of demand deposits, occurring at the beginning of the simulation period. The fitful course of private borrowing

The model

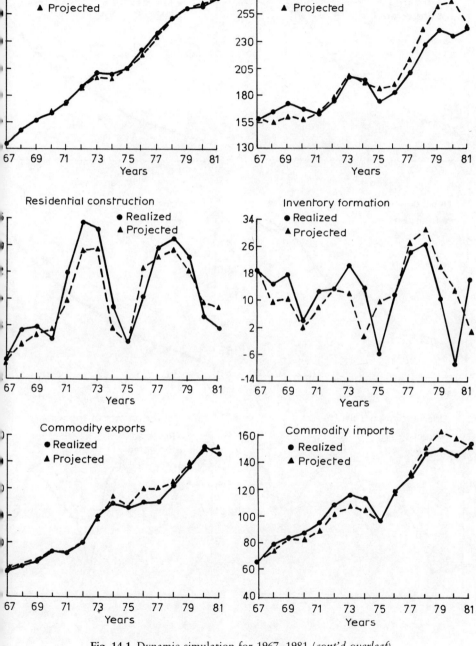

Fig. 14.1 Dynamic simulation for 1967–1981 (*cont'd overleaf*)

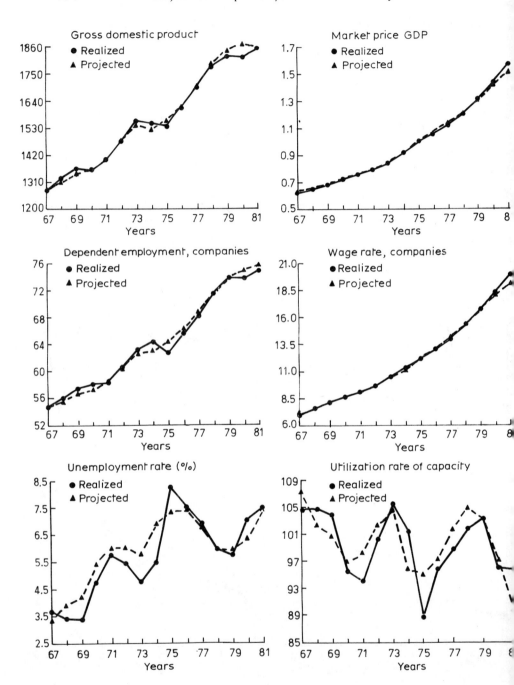

The model

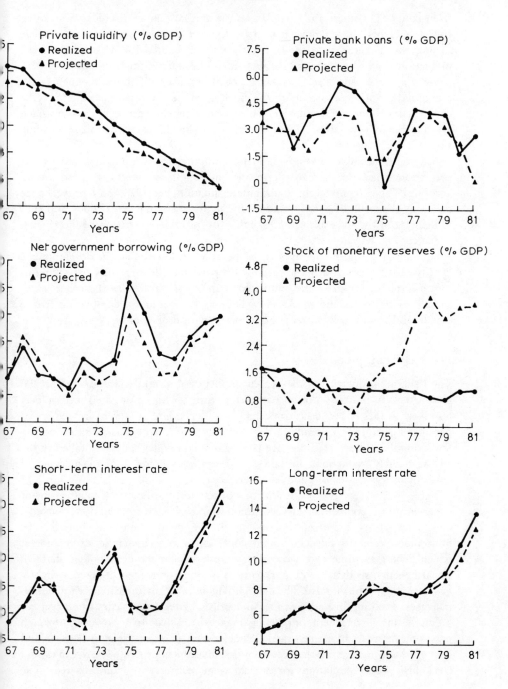

from banks is captured only partly. Also in this case most discrepancies can be traced as errors already present in the relevant estimated equation. The change in domestic government debt is slightly underestimated for most years of the prediction span, which is mainly brought about by underestimations in government payments to the private sector (wages, social benefits and interest payments). An accumulation of errors occurs in the stock of monetary reserves, which may be connected with the residual determination of this item in the model. This approach is somewhat questionable in view of the system of flexible exchange rates prevailing since the early seventies.

In spite of the sometimes large and accumulating errors in financial stocks, the predictions for both the short-term and the long-term interest rate are rather accurate. The comparatively small underestimations for 1978 and later years are primarily the result of overestimations in the private banks' reserve position and, in the case of the long-term interest rate, also of underestimation of the tension in the capital market.

Summarizing, the model appears to be able to describe actual developments over the focus period rather accurately. In particular the variables pertaining to the real sphere are generally simulated with small errors. On the other hand, some financial flows and stocks show large and sometimes accumulating errors, which tend to offset, however, in the determination of interest rates.

14.2.3 Dynamic properties

The dynamic properties of the present model depend on the value of some 200 parameters. In practice, not all parameters turn out to be of equal importance, however. In fact, a few crucial mechanisms, depending on a very limited set of parameters, appear to dominate the simulations. We will illustrate that with a few examples. These examples are typical of some policy measures taken in the 1980s, namely reductions in taxes paid by households and an increase in government purchases of goods and services.

Table 14.1(a) describes the results of a once-and-for-all increase in government expenditure on goods and services, amounting to 0.5% of net national income, and financed by increased borrowing in the capital market.[1] Initially the multiplier/accelerator effects dominate and a general rise in activity can be noted, which generates some extra tax income and reduces the government share in social benefits, so that the *ex post* burden on the government budget is slightly less than that of the impulse. The unemployment rate falls and capacity utilization increases, resulting in higher wages and prices. After a few years, the effects of accumulating government borrowing make themselves felt, however, through persistent increases in interest rates, which first affect residential construction and later gradually undermine the expansion-fed investment process. The rise in capital costs also boosts inflation for which wage earners are compensated. The

---

[1] All policy simulations discussed in this chapter cover the period 1981–1990.

The model

Table 14.1(a) Accumulated effects of an increase in government purchases of 0.5% of net national income, financed in the capital market (exogenous discount rate)

	Year		
	1	5	10
Percentage changes[a]			
Volume of:			
(1) Private consumption	0.1	0.6	1.2
(2) Business fixed investment	1.2	1.9	0.4
(3) Residential construction	0.0	−2.2	−1.2
(4) Exports of goods	−0.5	−0.3	−0.7
(5) Imports of goods	2.2	4.4	7.0
(6) Private GDP	0.7	0.8	0.7
Price of:			
(7) Private consumption	0.3	0.8	2.0
(8) Production	0.3	0.9	2.3
(9) Private wage rate	0.5	1.3	3.3
(10) Employment of companies	0.3	0.4	0.4
Levels[b]			
(11) Unemployment rate (%)	−0.2	−0.2	−0.2
(12) Utilization rate (%)	1.1	0.8	0.3
(13) Current account[c]	−0.2	−0.3	−0.5
(14) Government budget deficit[c]	0.4	0.6	1.3
(15) Government debt to private sector[c]	0.4	2.2	5.6
(16) Private debt to banks[c]	0.3	0.3	0.4
(17) Unborrowed reserves[c]	−0.1	−0.4	−1.1
(18) Official reserves[c]	−0.1	−0.4	−1.0
(19) Long-term interest rate	0.0	0.3	0.7
(20) Short-term interest rate	0.0	0.4	0.9

[a] Calculated according to $(X^s_i - X_i)/(0.01 X_i)$, where X^s_i and X_i are level variables, referring to the shocked solution and the control solution respectively $(i = 1, \ldots, 10)$.
[b] $X^s_i - X_i$.
[c] As a percentage of net national income.

deterioration of the current account resulting from increased activity and relative price increases is reflected in a substantial reduction in official reserves, which eventually would make the measure unmaintainable.

We feared that this simulation of fiscal policy not conditioned by monetary policy measures might be unrealistic, since some adjustment of the official discount rate would be plausible. Therefore, we also simulated the increase in government expenditure under the assumption that the official discount rate would adjust in a normal way. In order to define 'normal' policies we analysed discount rate adjustment in the sample period. In line with earlier analysis, and similar studies for the EC countries, US discount rate policy showed a rather stable pattern. The discount rate was found to react positively to the inflation rate and negatively to the external value of the US dollar as well as to the unemployment rate. The same determinants were significant for the subperiod 1972–1981, in which the dollar floated. For that period, an additional positive impact of the liquidity ratio proved significant. As floating exchange rates are relevant for the simulation period, we decided to adopt the latter equation. (Both equations are presented in Appendix A.)

The integration of 'normal' discount-rate policy in the model increases its interest flexibility. This is demonstrated by Table 14.1(b). The interest rates now react quicker to the increased level of activity, which diminishes its momentum. Discount rate policies appear to be immobilized by the conflict between the combat

Table 14.1(b) Accumulated effects of an increase in government purchases of 0.5% of net national income, financed in the capital market (endogenous discount rate)

	Year		
	1	5	10
Percentage changes[a]			
Volume of:			
(1) Private consumption	0.1	0.5	1.4
(2) Business fixed investment	1.2	1.8	1.0
(3) Residential construction	0.0	−2.6	−1.8
(4) Exports of goods	−0.5	−0.4	−0.6
(5) Imports of goods	2.1	4.0	7.2
(6) Private GDP	0.7	0.7	0.9
Price of:			
(7) Private consumption	0.3	0.8	1.9
(8) Production	0.3	0.8	2.2
(9) Private wage rate	0.4	1.2	3.1
(10) Employment of companies	0.3	0.4	0.5
Levels[b]			
(11) Unemployment rate (%)	−0.2	−0.2	−0.2
(12) Utilization rate (%)	1.1	0.7	0.5
(13) Current account[c]	−0.2	−0.3	−0.6
(14) Government budget deficit[c]	0.4	0.7	1.4
(15) Government debt to private sector[c]	0.4	2.3	6.0
(16) Private debt to banks[c]	0.2	0.2	0.3
(17) Unborrowed reserves[c]	−0.1	−0.3	−1.1
(18) Official reserves[c]	−0.1	−0.3	−1.0
(19) Long-term interest rate	0.1	0.3	0.7
(20) Short-term interest rate	0.1	0.5	0.9
(21) Discount rate	0.1	0.2	0.3

[a] Calculated according to $(X_i^s - X_i)/(0.01 X_i)$, where X_i^s and X_i are level variables, referring to the shocked solution and the control solution respectively ($i = 1, \ldots, 10$).
[b] $X_i^s - X_i$.
[c] As a percentage of net national income.

of overliquidity and inflation on the one hand and that of unemployment on the other. As a result conventional discount rate policy does not prevent a substantial loss in official reserves. Another consequence of rising interest rates is, however, that the investment expansion is reduced much earlier. Nevertheless, crowding-out is limited to residential construction. In fact, the occurrence of crowding out appears to hinge on the conditions in the period considered. Simulation of a similar impulse over the 1970s instead of the 1980s – not reported here – suggests that in periods with on average higher utilization rates and lower unemployment rates crowding out is much more likely.

Also in the case of a reduction in taxes paid by households we present two alternatives. Table 14.2(a) gives the effects when monetary policy is not adjusted,

Table 14.2(a) Accumulated effects of a reduction in direct taxes of households of 0.5% of net national income (exogenous discount rate)

	Year		
	1	5	10
Percentage changes[a]			
Volume of:			
(1) Private consumption	0.3	1.4	2.3
(2) Business fixed investment	0.5	1.6	1.4
(3) Residential construction	0.0	0.0	1.6
(4) Exports of goods	−0.2	−0.2	−0.6
(5) Imports of goods	0.9	4.4	9.0
(6) Private GDP	0.3	0.9	0.9
Price of:			
(7) Private consumption	0.1	0.7	2.0
(8) Production	0.1	0.8	2.3
(9) Private wage rate	0.2	1.2	3.6
(10) Employment of companies	0.1	0.5	0.5
Levels[b]			
(11) Unemployment rate (%)	−0.1	−0.2	−0.2
(12) Utilization rate (%)	0.4	0.9	0.2
(13) Current account[c]	−0.1	−0.3	−0.6
(14) Government budget deficit[c]	0.4	0.5	1.0
(15) Government debt to private sector[c]	0.3	1.5	3.7
(16) Private debt to banks[c]	0.1	0.5	1.3
(17) Unborrowed reserves[c]	−0.1	−0.5	−1.7
(18) Official reserves[c]	−0.0	−0.5	−1.5
(19) Long-term interest rate	0.0	0.2	0.5
(20) Short-term interest rate	−0.0	0.1	0.7

[a] Calculated according to $(X_i^s - X_i)/(0.01 X_i)$, where X_i^s and X_i are level variables, referring to the shocked solution and the control solution respectively $(i=1,\ldots,10)$.
[b] $X_i^s - X_i$.
[c] As a percentage of net national income.

Table 14.2(b) Accumulated effects of a reduction in direct taxes of households of 0.5% of net national income. Endogenous discount rate

	Year		
	1	5	10
Percentage changes[a]			
Volume of:			
(1) Private consumption	0.3	1.3	2.3
(2) Business fixed investment	0.4	1.2	1.4
(3) Residential construction	0.0	−0.3	0.7
(4) Exports of goods	−0.2	−0.2	−0.6
(5) Imports of goods	0.8	3.8	8.4
(6) Private GDP	0.3	0.8	1.0
Price of:			
(7) Private consumption	0.1	0.7	1.9
(8) Production	0.1	0.7	2.1
(9) Private wage rate	0.2	1.1	3.1
(10) Employment of companies	0.1	0.4	0.5
Levels[b]			
(11) Unemployment rate (%)	−0.1	−0.2	−0.2
(12) Utilization rate (%)	0.4	0.8	0.4
(13) Current account[c]	−0.1	−0.3	−0.6
(14) Government budget deficit[c]	0.4	0.5	1.2
(15) Government debt to private sector[c]	0.3	1.6	4.3
(16) Private debt to banks[c]	0.1	0.4	1.0
(17) Unborrowed reserves[c]	−0.0	−0.4	−1.4
(18) Official reserves[c]	−0.0	−0.3	−1.3
(19) Long-term interest rate	0.0	0.2	0.6
(20) Short-term interest rate	0.0	0.3	0.8
(21) Discount rate	0.0	0.2	0.4

[a] Calculated according to $(X_i^s - X_i)/(0.01 X_i)$, where X_i^s and X_i are level variables, referring to the shocked solution and the control solution respectively $(i = 1, \ldots, 10)$.
[b] $X_i^s - X_i$.
[c] As a percentage of net national income.

Table 14.2(b) the effects with an endogenous discount rate. In general, the effects are comparable with their counterparts in Tables 14.1(a) and (b), as was to be expected, but there are some remarkable differences. When the discount rate is not adjusted, tax reduction is less effective in the short run than extra government spending as a way to stimulate the economy, but after a few years the opposite is true. The relatively lower burden of the government sector induces lower increases in the interest rates, from which private investment benefits. With an endogenous discount rate, the longer-term effects on productive investment are hardly different, but residential construction suffers more, due to the higher short-term interest rate. Taken together, Tables 14.1 and 14.2 suggest that a demand-stimulating fiscal policy is effective in stimulating private activity in periods with modest factor utilization, but such policies would cause an ever-increasing deficit, both in the government budget and on the current account, and hence increases in interest rates.

As far as these policies are representative for what actually happened in the 1980s, they illustrate the background of the present problems of the US economy. The results also suggest that a reverse policy would take a long time to become effective. We will investigate some relevant options in the next section.

14.3 POLICY OPTIONS

Though also in the United States unemployment is still substantial, the imbalance in the current account and the related imbalance in the government budget are administratively more urgent problems. Some tentative projections we made, based on a moderate development of world demand, nearly stable international prices, and slightly falling external interest rates, and a freeze of average tax burdens and government expenditure after 1986 in the United States, do not suggest any significant spontaneous improvement in this situation. On the contrary, the ever-increasing government debt tends to worsen both deficits. It is therefore obvious that something has to be done.

As we do not expect monetary policy to be of much use on its own, we have analysed eight types of fiscal measures with respect to their potential effectiveness for the reduction of the imbalance mentioned before. A survey of the results is presented in Table 14.3.

All measures are standardized in the sense that the original impulse (roughly) amounts to 0.5% of net national income. Four different types of tax increases are simulated. Firms pay more if the investment tax credit is reduced, households pay more if their direct taxes are increased; the incidence of indirect taxes is in principle uncertain, but in the present model families ultimately pay the burden, as the wage equation does not allow for backward shifting of indirect taxes. A 10% increase in import duties obviously is not a very orthodox fiscal measure, but perhaps deserves consideration in view of the mounting deficit on the current account.

A reduction in the investment tax credit is by far the least attractive measure from an American point of view. In the medium term it relieves the government budget, but the implied increase in capital costs has inflationary effects and the urgent improvement in the current account comes hardly nearer. Similar increases in taxes of households are perhaps slightly less effective in terms of the budget, but they are much less harmful in terms of output and with this tax measure prices tend to fall with the decrease in interest rates. As families carry the burden of indirect taxes, the effects of this measure are largely similar to that of an increase in direct family taxes. The levying of import duties is a debatable measure, also from a US point of view. Indexation of prices and wages is responsible for

Table 14.3 Survey of 5- and 10-years accumulated effects of some fiscal policy measures and monetary conditions[a]

		Private GDP			Deficit			Official reserves	Long-term interest rate
	Year	Volume	Deflator		Current account	Government budget			

	Year	Volume	Deflator	Current account	Government budget	Official reserves	Long-term interest rate
Tax increases[b]							
Investment tax credit reduction	5	−1.6	1.9	−0.4	−0.7	0.9	−0.1
	10	−2.2	1.8	−0.4	−1.1	1.5	−0.4
Direct taxes households	5	−0.8	−0.7	−0.3	−0.5	0.3	−0.2
	10	−0.9	−2.0	−0.6	−1.2	1.3	−0.6
Indirect taxes	5	−0.8	−1.0	−0.3	−0.6	0.5	−0.3
	10	−1.1	−2.8	−0.7	−1.4	1.6	−0.7
Import duties[c]	5	−0.9	3.2	−0.7	−1.5	1.8	−0.3
	10	−1.8	2.9	−1.1	−2.1	3.3	−0.7
Government expenditure reductions[b]							
Purchases	5	−0.7	−0.8	−0.3	−0.7	0.3	−0.3
	10	−0.8	−2.1	−0.6	−1.4	1.0	−0.7
Social benefits	5	−0.6	−0.5	−0.2	−0.6	0.2	−0.1
	10	−0.9	−1.6	−0.5	−1.2	0.9	−0.4
Number of employees	5	−0.3	−1.7	−0.3	−0.7	0.1	−0.3
	10	−0.5	−4.4	−0.8	−1.6	1.2	−0.8
Government wage rate	5	−0.9	−0.7	−0.7	−3.0	0.8	−0.2
	10	−4.4	−5.5	−2.8	−8.1	3.9	−1.1
Monetary conditions							
Reduction discount rate by 1%	2	0.6	0.1	+0.1	−0.2	−0.6	−0.4
	5	0.5	0.2	+0.1	−0.2	−0.7	−0.2
	10	−0.1	0.2	−0.0	−0.1	−0.8	−0.1
Depreciation US dollar by 10%	2	0.6	0.6	−0.2	−0.3	−0.3	−0.4
	5	1.1	1.4	+0.0	−0.4	−0.2	−0.3
	10	0.4	2.2	−0.1	−0.2	0.1	−0.2

[a] Effects are based on endogenous discount rates and exogenous exchange rates. Units of variables correspond with those in preceding tables.
[b] Impulses are (roughly) equivalent to 0.5% NNI.

a substantial increase in inflation. The induced improvements in both the current account and the government budget reduce interest rates.

The second part of Table 14.3 shows the effects of a reduction in four different types of government expenditure, each initially amounting to about 0.5% of net national income. The categories discerned are government purchases of goods and services, the government share in social benefits, the number of government employees and the government wage rate. There are a few striking differences, but overall the similarities dominate, and the results are highly comparable to those of increased taxes paid by households, which could be expected, since the mechanisms involved do not differ much. A reduction of the government wage bill turns out to be a strongly deflationary measure. Especially a decrease in the government wage rate substantially relieves the government budget. In contrast to a reduction of the government share in social benefits and a fall in government employment, this measure does not directly lead to compensating income transfers.

The effects of the different fiscal measures presented in Table 14.3 are based on an endogenous discount rate and exogenous exchange rates. The realism of the last assumption is not warranted, but the results do not seem to suffer from some obvious bias, as the signals for exchange-rate adjustment do not indicate a certain direction of adjustment. The current account improves, but interest rates fall and the implied change in official reserves is never large.

In order to illustrate the sensitivity of our results to monetary conditions, we nevertheless present some simulations that involve manipulation of the official discount rate and a fall in the value of the US dollar, the latter under an endogenous discount rate. The stimulating effects of a reduced discount rate are relatively short lived and tend to produce inflation and a substantial loss in official reserves, without really contributing to the problems of the government budget or the current account. Due to domestic reactions to foreign prices and modest elasticities of substitution in foreign-trade equations, a depreciation of the US dollar appears not very effective in reducing the deficit on the US current account.

Reviewing our fiscal measures, and adding some political arguments to our economic discussion, some conclusions can be drawn.

Increase in direct taxes would be a U-turn in economic policy. Though it would be not without precedent, any administration will be inclined to avoid such a dramatic change. Import duties would at any rate have negative results for exports, because of their effects on world income and obviously also lead to retaliation. Therefore, indirect taxes, which still are relatively low in the United States, are the most likely tax instrument to be used to reduce the deficits.

On the expenditure side, a reduction in the number of government employees stands out as the most promising. This measure is rather effective and does not generate severe losses in production. A reduction in any spending category will face the opposition of pressure groups, but in the case of government officials their power may be relatively weak. The competing reduction in wage rates could cause problems in the selection of officials, as experiences in other countries demonstrate.

We are inclined to conclude that some combination of increased indirect taxes and a reduction in the number of government employees seems the most appealing package. In fact, the contribution of the last measure is bound to be much lower as the government wage bill is only about 10% of the net national income and the government is likely to encounter organizational problems if the number of officials is reduced too drastically. Our last exercise is based on this argument. It involves the simulation of an increase in the indirect-taxation tariffs from 13% to 15%, a reduction in the number of government employees by 2%, and an (additional) reduction in the discount rate by 2%. The results are shown in Table 14.4. As could be expected the implementation of this package will not

Table 14.4 Accumulated effects of an increase in indirect taxes from 13% to 15%, a reduction in government employment of 2% and a decrease in the official discount rate of 2%, all measures spread over two years

	Year		
	1	5	10
Percentage changes[a]			
Volume of:			
(1) Private consumption	−0.9	−3.3	−8.3
(2) Business fixed investment	−1.0	1.7	−8.3
(3) Residential construction	0.0	−2.5	9.7
(4) Exports of goods	0.5	1.2	2.4
(5) Imports of goods	−2.6	−9.7	−30.8
(6) Private GDP	−0.7	−1.4	−3.6
Price of:			
(7) Private consumption	0.9	0.2	−4.9
(8) Production	−0.4	−2.8	−8.8
(9) Private wage rate	−0.6	−4.0	−12.7
(10) Employment of companies	−0.3	−0.8	−1.8
Levels[b]			
(11) Unemployment rate (%)	0.3	0.7	1.1
(12) Utilization rate (%)	−1.2	−1.9	−3.3
(13) Current account[c]	0.3	0.8	2.5
(14) Government budget deficit[c]	−0.8	−2.5	−4.7
(15) Government debt to private sector[c]	−0.7	−6.9	−18.4
(16) Private debt to banks[c]	−0.2	−0.7	−3.4
(17) Unborrowed reserves[c]	0.0	0.5	4.3
(18) Official reserves[c]	−0.0	0.4	3.8
(19) Long-term interest rate	−0.4	−1.0	−2.4
(20) Short-term interest rate	−1.2	−1.2	−2.9

[a] Calculated according to $(X_i^s - X_i)/(0.01 X_i)$, where X_i^s and X_i are level variables, referring to the shocked solution and the control solution respectively $(i = 1, \ldots, 10)$.
[b] $X_i^s - X_i$.
[c] As a percentage of net national income.

lead to quick results, but in a medium-term perspective the problems of the US economy appear considerably reduced.

14.4 CONCLUSIONS

In this chapter we evaluated some policy options for the US economy on the basis of a macroeconometric model with an elaborated financial sector. Though this model generates large interest effects from increased government borrowing it turns out to yield very limited crowding-out effects, even if discount rates are endogenized. On the other hand, increased government spending does not generate enough tax revenues to compensate for their burden on the government budget. As a consequence, demand stimulation can offset a process of expansion that may persist for years without an effective check on the deterioration of the government budget and the current account. The last model property might be unrealistic since the exchange rate is not endogenous. On the other hand we have witnessed a rise in the dollar value in spite of a strong deterioration of the US current account. If this expansion can go on for many years, it is hardly surprising that the redress of such a situation also would take much time. Nevertheless, an increase in indirect taxes, accompanied by a fall in the number of government employees and some monetary relaxation could in the medium term contribute substantially to the solution of the present problems.

REFERENCE

Swank, J. and Siebrand, J. C. (1986) *RASMUS-2b, an annual model of the US economy* (2nd revised edition), Discussion Paper Series 8608/G, Institute for Economic Research, Rotterdam.

APPENDIX A

Equations explaining the US discount rate

(1) **Sample period 1958–1981:**

$$\Delta r_{CB} = -0.08 \, \mathring{E}R + 0.11 \, \mathring{p}y - 0.89 \, \Delta UR_{-\frac{1}{4}}$$
$$\quad (3.0) \quad\quad (3.5) \quad\quad (4.2)$$

Statistics: $\bar{S}_e = 0.82$; $DW = 1.86$
(OLS)

(2) **Sample period 1972–1981:**

$$\Delta r_{CB} = \underset{(5.7)}{-0.08\,\mathring{E}R} + \underset{(7.1)}{0.35\,\mathring{p}y} - \underset{(9.0)}{1.35\,\Delta UR_{-\frac{1}{4}}}$$

$$+ \underset{(4.6)}{2.48\,\Delta}\left(\frac{M1}{0.01VT_{-\frac{1}{2}}}\right)$$

Statistics: $\bar{S}_e = 0.40;\ DW = 2.27$
(OLS)

Description of variables

r_{CB} = discount rate
ER = external value US dollar
py = price GDP
UR = unemployment rate
M1 = stock currency and demand deposits
VT = value of total sales

15

Model building for decision aid in the agri-economic field

PATRICK ANGLARD, FRANCOISE GENDREAU

and A. RAULT

For the past ten years ADERSA has been involved in modelling agri-economic phenomena at the French level as well as at the European level, with a view to decision-aiding. The approach taken is original, due to the control engineering background of the team. The object of this chapter is to present the methodology through various examples (French cattle and milk production, European hog livestock, egg production and price formation). The various procedural steps, in some ways analogous to those used in control engineering, will be presented, with emphasis on the methods and tools (Kalman filtering, nonlinear optimization and identification) which, to the authors' knowledge, have not perfused yet into the econometric field. Beyond the methodological aspects, results interesting at both the French and European levels are presented.

15.1 INTRODUCTION

Applications of control engineering to economic modelling have often been studied from a theoretical point of view (Aoki, 1976). But practical applications are not very often found. This chapter is based on a 10-year-long experience in the agri-economic field. It describes briefly what are the control engineering techniques that can be used to cope with some of the problems well known to model designers. These techniques are:

(1) Kalman filtering for data coherence of stochastic processes
(2) extended Kalman filtering for bias correction
(3) nonlinear identification of model parameters
(4) optimal control for decision-aiding.

Numerous examples are provided. These examples all come from three main demographic models which all involve animal livestock. These models are:

(1) a model of a French cattle herd
(2) a model of EEC hog livestock
(3) a model of EEC egg production and prices.

The chapter ends with a description of the problems met when transferring the models to their final users and implementing them on microcomputers.

15.2 MODELS AND METHODS

15.2.1 The main features of the models

The models are basically demographic models, dealing with populations in a wide sense: animal, biological, or even stock. Such systems evolve according to an aging process, and birth and death laws. They have common characteristics:

(1) Available observations are heterogeneous and are often obtained through a polling system (sample survey) therefore noise corrupted.
(2) Physical phenomena are rather well known.
(3) The system is unstable: without control, a free population increases indefinitely.
(4) Only limited data are generally available and new experiments cannot be performed so that identification is always a passive process.

These differents points and their methodological implications will be developed (Rault and Leibundgut, 1981).

(a) Dealing with heterogeneous data

Though it may seem a minor concern, one of the first motivations for modelling has been to check the validity and coherence of statistical data. Available data are indeed hardly comparable, because they are heterogeneous from several points of view:

(1) Their origins are varied: we have national or European statistics; others come from private institutes. For example, for the egg model, data come from:

   (i) SCEES (French agricultural statistics)
   (ii) OSCE (EEC statistics)
   (iii) ITAVI (a French technical institute)
   (iv) customs for import and export data.

(2) Different time series have different sampling periods. As an example, for the hog livestock, production data (slaughtering, exports, imports) are available

monthly, but information on the population is provided every four months. For the beef cattle herd, only one sample survey is made each year.
(3) The field of study is not homogeneous: data are often collected from different samples. Moreover, a change in the sampling may induce a discontinuity in the corresponding time series. This happened in 1975 in the sample surveys made for the hog and beef cattle herds in France.
(4) Methods of classification in surveys can also create difficulties. For instance, slaughtered cows are classified according to their number of teeth, as we need age classes.

The model has to deal with these different data and should enable a check to be made on their coherence. We have to consider another point: the statistical validity of sample surveys. Their results have a theoretically known uncertainty; but they may also be biased if the sample is not well representative of the whole population. We shall see how to take into account this uncertainty and possible biases.

(b) Population models

Besides these difficulties with data, fortunately we deal with well-known physical phenomena that can easily be modelled. The three examples are all cohort population models. The general form of this type of model is given in Appendix A. These models are written under state space form, with a linear state equation of the form:

$$X(n+1) = AX(n) - U(n)$$

The state vector $X(n)$ consists of age groups of equal age span (see Appendix A for the remaining notation). This type of model can easily be divided into submodels which correspond to well-defined subsystems:

(1) male/female distinction for the beef cattle herd
(2) fattening/breeding system distinction for the hog livestock
(3) discrimination between industrial and traditional breeding of chickens (see Appendix B for some details of these models).

These models belong to the range of comprehensive models, as opposed to 'black box' or representation models. (A comprehensive model is based on a physical analysis of the phenomena, and integrates internal variables.) It is generally decomposed into elementary phenomena whose evolution is easy to describe. On the other hand, the only criterion of a representation model is an external behaviour identity, or goodness of fit – but the parameters of such models (as macroeconomic models) have no physical meaning. By contrast, parameters of population models have a zootechnical meaning (e.g. fertility rate, milk productivity, mortality rate) and identification of these parameters can improve our knowledge in this area.

(c) Price model

The above-mentioned population system is in fact part of a closed-loop system formed of a demographic block and an economic block, as shown in Fig. 15.1. Demographic phenomena are functions of imports and exports and of cattles-raisers' decisions, such as slaughtering or mating. They are also subject to some perturbations, such as climatic variations, complementary artificial feed and improvements on the breeds. All this determines a certain state of the livestock, which leads to a certain production (of milk, beef, pork, eggs, etc.)

Economic phenomena, based on the level of this production with external economic influence (consumption; standard of living, EEC policies, etc.) lead to producer prices. In return, these prices have a strong influence on cattle-raiser decisions. For the first two models (beef and porcine livestock) the economic point of view has been little studied and not in a conclusive way. However, for the egg problem, both demographic and economic phenomena have been modelled with the price–production loop.

The price model developed takes the classical form of a supply and demand equilibrium taking into account the elasticity of prices with respect to quantities:

(1) The supply appears as the production given by the population model, plus the imports–exports balance (exogenous data)
(2) The demand is represented as a decomposition into three components: a constant level, a trend and a multiplicative seasonal effect.

This model is closer to representation models. Some parameters do have a physical meaning (e.g. initial level or trend of demand) but others, as price elasticities, are more conceptual.

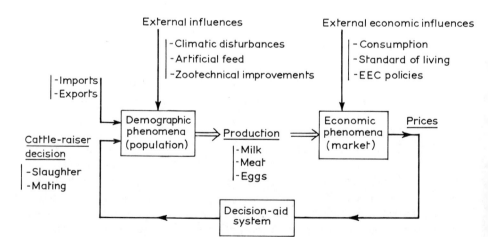

Fig. 15.1 General structure of the livestock system

(d) Stochastic aspects of the models

In the previous sections, we have described deterministic models – which is correct in so far as the aging process is concerned. But birth and death phenomena are indeed stochastic processes – let us see what are the implications for the models. We have stated that the evolution of the population was represented by Equation 15.1.

$$X(n+1) = AX(n) - U(n) \tag{15.1}$$

Matrices A and C contain parameters such as mortality or fertility rates, which are in fact stochastic variables – and we use their mean values. Random variations but also approximations made in modelling (linerization for example) can be represented by a *process noise*. Moreover, there is an *input noise* due to the origin of input data in the input vector $U(n)$: some errors may appear in slaughtering or import–export data. These different noises are globally represented by an additive noise vector $v(n)$; thus the state equation becomes:

$$X(n+1) = AX(n) - U(n) + v(n) \tag{15.2}$$

where $v(n)$ is a zero mean random vector.

Furthermore the statistical nature of observations generates uncertainty, called the *output noise*. The observation equation, which relates observations to internal variables, includes this noise vector $w(n)$:

$$S(n) = CX(n) + w(n) \tag{15.3}$$

where $w(n)$ is a zero-mean random vector, whose covariance can be derived from the confidence intervals of statistics.

15.2.2 The Kalman filter

The relevant references are Anderson and Moore (1979); Aoki (1976); Jazwinski (1970) and Maybeck (1979). Kalman filtering is a classical technique in aeronautics. It takes into account the stochastic aspect of a system and enables its internal variables to be estimated in an optimal way. It is a recursive data processing algorithm. The Kalman filter incorporates all information that can be provided to it as:

(1) the dynamics characteristics of the system, through a state space model
(2) the statistical description of process and output noises and uncertainty in the dynamic model
(3) any available information about initial conditions of the internal variables.

Some assumptions are necessary. At first, this technique deals with a dynamic stochastic model, written under state space form as:

$$\left. \begin{array}{l} X(n+1) = AX(n) - U(n) + v(n) \\ S(n) = CX(n) + w(n) \end{array} \right\} \tag{15.4}$$

where:

$X(n)$ is the state vector, with initial value $X(0)$
$S(n)$ is the output vector or measurement vector
$U(n)$ is the input vector
A, C are known matrices (they may be time varying)
$v(n)$ is the process noise
$w(n)$ is the output noise (see Section 15.2.1(d))

(a) Assumptions concerning the noises

These are as follows:

(1) $v(n)$ and $w(n)$ are gaussian zero-mean white noises (i.e. uncorrelated in time) with known *covariances* represented as Q and R matrices
(2) $v(n)$ and $w(n)$ are independent
(3) the matrices Q and R are positive definite.

Let us describe the steps of the filter

(1) Let an optimal estimate be given of state vector $X(n)$ at time n, noted as $X_{n+1/n}$. A covariance matrix is also computed
(2) The measurement vector $S(n+1)$ is then collected at time $n+1$, and compared with its predicted value $CX_{n+1/n}$; the residual is called 'innovations'
(3) The filtered or a posteriori estimate $X_{n+1/n+1}$ is then derived: the a priori estimate is updated by using the innovations balanced by a matrix called Kalman gain.

The Kalman filter estimate is therefore a linear combination of the a priori estimate and measurements. The relative weight of observations is given by the Kalman gain. This is schematized in Fig. 15.2, where the different estimates are represented with their *confidence interval*.

The computing of Kalman gain involves the covariance matrices Q and R of process and output noise; their relative weight is important. If the process noise is greater, the filtered estimate will be close to measurements (Fig. 15.2(a)). On the other hand, if the output noise is greater, the estimate will be closer to the

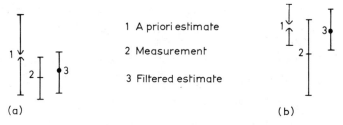

Fig. 15.2 Illustration of Kalman filter behaviour according to relative weight of process noise against measurement noise

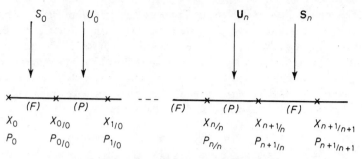

Fig. 15.3 Diagram of the Kalman filter

a priori estimate (Fig. 15.2(b)). Data for each measurement reduce the variance of the estimate and then improve the accuracy. Under the above assumptions, it can be shown that the Kalman filter yields the best estimate of the state vector, while taking into account the relative uncertainty of the system dynamics on the one hand, and measurements on the other. This estimate is optimal, in the sense of an unbiased minimum variance estimate. The steps of the algorithm are schematized in Fig. 15.3.

(F)   = filtering
(P)   = prediction or forecast
$X_{n+1/n}$ = a priori estimate with covariance matrix $P_{n+1/n}$
$X_{n+1/n+1}$ = a posteriori or filtered estimate with covariance matrix $P_{n+1/n+1}$
S(n)   = observation vector
U(n)   = input vector

(b) Parameter identification

Identification consists in finding the parameters minimizing a criterion which represents the error between the physical system (seen through observations) and the model (through its outputs). The calculation of this criterion depends on the problem and will be not be detailed here. The analytical form of this criterion is generally complex, and nonlinear with respect to the parameters. That is why we have chosen an heuristic method: the flexible polyhedron search (Himmelblau, 1972). It is also a geometric method, which progresses by drawing successive polyhedrons with $(n + 1)$ vertices, in an n-dimensional parametric space. At each step, a polyhedron is distorted in a direction opposite to the vertex to which corresponds the highest criterion.

(c) Identification of an unstable system

The relevant references are Jazwinski (1970); and Mehra (1971a, b). Besides the stochastic nature of the system there is another reason which justifies the use of a Kalman filter: the system is unstable as any free population is naturally unstable. Some parameters of this system are unknown, especially for beef and porcine livestock models. Thus we have to identify an unstable dynamic stochastic system, modelled under state space form by Equations 15.4. The A and C matrices depend on a parameter vector $\mathbf{p}$; its components are zootechnical parameters such as mortality, fertility rate and sex ratio; these matrices may also depend on time. Consequently, Equations 15.4 can be written as:

$$\left.\begin{aligned} X(n+1) &= A(\mathbf{p}, n) X(n) - U(n) + v(n) \\ S(n) &= C(\mathbf{p}, n) X(n) + w(n) \end{aligned}\right\} \quad (15.5)$$

The identification problem is then to find the optimal parameter vector $\mathbf{p}$. Formulation of the maximum likelihood estimation problem yields the following criterion D to be minimized with respect to the parameter vector $\mathbf{p}$, the input sequence U^* and the initial state $X(0)$:

$$\begin{aligned} D(\mathbf{p}, U^*, X(0)) = &(X(0) - X_0)^T P_0^{-1}(X(0) - X_0) \\ &+ \Sigma_n (S(n) - S(n)^*)^T R^{-1}(S(n) - S(n)^*) \\ &+ \Sigma_n (U(n) - U(n)^*) Q^{-1}(U(n) - U(n)^*) \end{aligned}$$

where $X(0)$, P_0 are the given initial state and its covariance matrix:

$$S(n)^* = S(n) - v(n) \quad U(n)^* = U(n) - w(n) \quad U^* = \{U(0)^*, U(1)^*, \ldots, U(n)^*\}$$

For a given parameter vector $\mathbf{p}$, the minimization of $(\mathbf{p}, U^*, X(0))$ with respect to U^* and $X(0)$ is a problem of optimal control with quadratic cost. This can be classically solved via dynamic programming. And Jazwinski (1970) has shown the equivalence between dynamic programming and Kalman filtering for solving such a problem.

Thus the identification problem is solved in two phases:

(1) a Kalman filter solves the minimization problem with respect to U^* and $U(0)$ for a given $\mathbf{p}$:

$$D(\mathbf{p}) = \min_{U^*, X(0)} D(\mathbf{p}, U^*, X(0)) = \Sigma_n \varepsilon(n)^T (R + C(\mathbf{p}, n) X_{n/n-1} C(\mathbf{p}, n)^T)^{-1} \varepsilon(n)$$

with $\varepsilon(n)$ the innovation process: $\varepsilon(n) = S(n) - C(\mathbf{p}, n) X_{n/n-1}$
$P_{n/n-1}$ the covariance matrix of the predicted value $x_{n/n-1}$

(2) then a classical nonlinear optimization procedure yields the optimum with respect to $\mathbf{p}$:

$$D(\mathbf{p}) = \min_{\mathbf{p}} D(\mathbf{p})$$

The flexible polyhedron search is used to perform the identification of $\mathbf{p}$ (Leibundgut et al., 1983).

(d) Optimal control methods (Aoki, 1976)

This type of method has been used for the egg production model. The problem is to determine the input to the production model, i.e. chick placements, that are best suited to induce a desired level of output (i.e. production of eggs). Both input and output should span a 12 to 24 month time horizon. Different methods have been tested (linear quadratic control, gradient methods, heuristic methods) and a gradient procedure was finally chosen.

The control problem is equivalent to minimizing $J(U,S)$, where $J(U,S)$ denotes:

$$J(U,S) = \alpha \frac{1}{2} \sum_{i=0}^{N-1} (S(i) - S(i)^*)^2 + \beta \frac{1}{2} \sum_{i=0}^{N-1} (U(i) - U(i)^0)^2 + \gamma \frac{1}{2}(S(N) - S(N)^*)^2 \quad (15.6)$$

where the output S and the input U must satisfy the production equations:

$$X(n+1) = AX(n) - U(n)$$
$$S(n) = CX(n)$$

S^* denotes the 'desired' output
S denotes the 'current' output
U^0 denotes the initial input
U denotes the 'current' input
α, β, γ are real constants

The minimization problem is also equivalent to the minimization of the Lagrangian $L(n,S)$ with respect to U and S where:

$$L(U,S) = J(U,S) + \Sigma_i \lambda^T(i+1)(AX(i) + U(i) - X(i+1)) \quad (15.7)$$

where $\lambda(i)$ denotes Lagrange's multiplicator's vector.
Minimization of $L(U,S)$ yields Equations 15.8a and b:

$$\frac{\partial L}{\partial S} = 0 \quad (15.8a)$$

$$\frac{\partial L}{\partial U} = 0 \quad (15.8b)$$

Equation 15.8a yields

$$\left. \begin{array}{ll} \lambda(i) = A^T \lambda(i+1) + \alpha(S(i) - S(i)^*) & i = 0, N-1 \\ \lambda(N) = \gamma(S(N) - S(N)^*) & i = N \end{array} \right\} \quad (15.9)$$

Note that the coefficients $\lambda(i)$ are computed through a backward scheme. Equation 15.8b yields:

$$\beta(U(i) - U(i)^0) + \lambda(i+1) = \text{Grad}_U J(U,S)_i = 0 \quad i = 0, N-1 \quad (15.10)$$

To ensure Equation 15.10 an iterative scheme is used.

$$U_{(i)}^{n+1} = U_{(i)}^n - k \, \text{Grad}_U J(U,S)_i \quad (15.11)$$

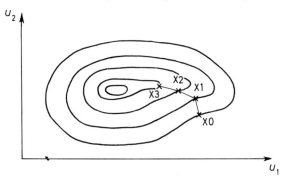

Fig. 15.4 ISO cost function $J(X)$ where $X = (u_1, u_2)$

From one iteration to the other, a change on inputs is made according to the steepest descent direction.

Figure 15.4 is a geometrical two-dimensional example. This example also suggests that a final convergence can be slow to achieve.

(e) Correction of biased observations (Friedland, 1969; Jazwinski, 1970) (extended Kalman filtering).

The assumptions for Kalman filtering have been discussed in Section 15.2.2(a). One of the assumptions is that the input noise process must have a zero mean. Unfortunately, this is not always true. The porcine livestock model is a counter example. There are different causes for bias:

(1) A bias occurs between small and large-scale sampling surveys. Three surveys are made each year but they are not equally reliable. A large-scale survey is made in December: it is said to be rather accurate. Two small-scale surveys are made in April and August. They have larger confidence intervals. Moreover, statistical tests have proved that biases occur on the small-scale surveys.
(2) Sample representativity: national data are derived from sampling surveys in a number of stock farms. The sample basis remains unchanged for ten years or more and thus it might lose its representativity. For instance, the number of large stock farms tends to increase. Thus the bias is time dependent.
(3) Changes of sample: to correct the above default, the sample is sometimes changed, which induces a sudden discontinuity in data.

To cope with this problem, a bias correction algorithm (Friedland, 1969) was implemented. This method uses the theory of extended Kalman filtering (commonly used for nonlinear filtering in control engineering). To implement this method, several assumptions must be made on the bias:

(1) Is it constant or time dependent?
(2) Does it affect the state variables or the observed variables?

Assuming that biases affect only the measured variables, Equations 15.4 yield:

$$\left. \begin{array}{l} X(n+1) = AX(n) - U(n) + v(n) \\ S(n) = CX(n) + Lb(n) + w(n) \end{array} \right\} \quad (15.12)$$

$b(n)$ denotes the bias vector at time n
L is a matrix that represents the effect of the biases on the different observations
$v(n)$ and $w(n)$ are white noise processes with zero mean

Other symbols are unchanged. Here the main idea is to include biases in the set of state variables and then to estimate them as well as the state variables.

Let $\mathscr{X}(n) = \begin{pmatrix} X(n) \\ b(n) \end{pmatrix}$ be the augmented state vector

(i) ASSUMING THAT THE BIASES ARE CONSTANT

In this case:

$$b(n+1) = b(n) \quad (15.13)$$

and Equations 15.12 and 15.13 become:

$$\left. \begin{array}{l} \begin{pmatrix} X(n+1) \\ b(n+1) \end{pmatrix} = \begin{pmatrix} A \\ 0 \end{pmatrix} \begin{pmatrix} X(n) \\ b(n) \end{pmatrix} - U(n) + v(n) \\ \\ S(n) = (CL) \begin{pmatrix} X(n) \\ b(n) \end{pmatrix} + w(n) \end{array} \right\} \quad (15.14)$$

They can be rewritten as:

$$\left. \begin{array}{l} \mathscr{X}(n+1) = \mathscr{A}\mathscr{X}(n) - U(n) + v(n) \\ S(n) = \mathscr{C}\mathscr{X}(n) + w(n) \end{array} \right\} \quad (15.15)$$

Conventional Kalman filtering techniques can now be applied to this new system, providing estimates of X_n and b_n simultaneously. But one drawback of this method is that the order of the system is increased and so are the sizes of the different matrices.

Friedland has developed an algorithm which consists of an iterative uncoupled estimation of $X(n)$ and $b(n)$, thus reducing the complexity of the computation.

(ii) ASSUMING THAT THE BIASES ARE NOT CONSTANT

The biases dynamic can then be represented by

$$b(n+1) = Z(n)b(n) \quad (Z(n) \text{ is a square matrix})$$

It can be assumed that there is a nonsingular matrix, W_n, defined by:

$$W(n + 1) = Z(n)\mathbf{b}(n)$$
$$W(0) = I \qquad \text{(I identity matrix)}$$

Then defining $b'(n)$ as $b'(n) = W_{(n)}^{-1}$, the problem becomes similar to the constant biases estimation.

15.3 THE USE OF OUR MODELS

15.3.1 Generalities

(a) Estimation of unknown parameters

The above sections stated that population models are comprehensive models. Thus their parameters have a true physical meaning. For instance, these parameters are: mortality rates, fertility, and growth rates. Their actual values are not always precisely known. Thanks to a model with its observed input and output and to an identification scheme, their accuracy can be improved. However, it should be noted that the identified parameters apply only to the global system, i.e. the national livestock. Small-scale studies on specific herds can show different results, which, in turn, can be compared with their national counterparts.

EXAMPLE 1

Parameters identified in our model of the French cattle herd are:

(1) mortality rates of young animals, adults, males and females
(2) fertility rates of females for each age class
(3) average milk production per cow
(4) technical gain on milk production due to artificial feed
(5) sex ratio at birth (percentage of females among new-born calves).

Our estimates for these parameters proved to be close to those commonly known by specialists, except for the sex ratio at birth. We identified a ratio of 47% female calves and the commonly known ratio was 48%. The sex ratio is now a key parameter. A 1% difference can deeply affect the long-term male-to-female structure of the herd.

EXAMPLE 2

Similar parameters have been estimated for the porcine livestock model. But here, with a model for each EEC country, estimates can be compared between countries, leading to an understanding of difference in production and livestock structures.

(b) Statistics filtering

Statistical data are always subject to uncertainty (i.e. additional noise) and sometimes bias. The above sections have described the Kalman filtering techniques

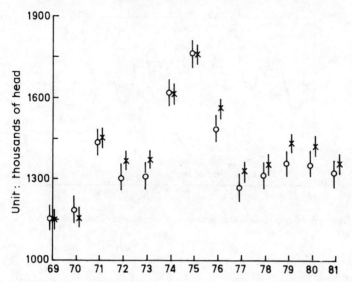

Fig. 15.5 Number of males 1–2 years old (example of an observed state variable). Comparison between the observed value (◇) and the a posteriori estimate (×)

as well suited for filtering and bias correction of observations and input to a system. This indirectly provides means of filtering statistical data.

EXAMPLE 1

See Fig. 15.5

EXAMPLE 2

Model of the French cattle herd

It has been mentioned previously that, in this model, the slaughtering input is subject to a rather important noise process. Assuming that this noise process $v(n)$ is additive to the state:

$$X(n+1) = AX(n) - U(n) + v(n)$$

yields

$$U^*(n) = AX_{n/n} - X_{n-1/n+1}$$

where:

$U^*(n)$ is the filtered input

$X_{n/n}$ and $X_{n+1/n+1}$ denote state estimates after Kalman filtering of observations

$S(n)$ and $S(n+1)$ are yearly data

Using seasonal coefficients, monthly estimates of slaughtering can then be derived from the yearly estimates $U^*(n)$.

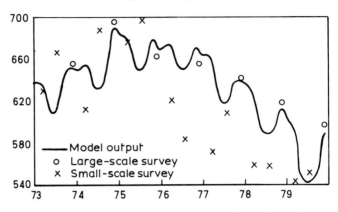

Fig. 15.6 Example of biased hog statistics in France: mated sows (unit: thousands of head)

EXAMPLE 3

The model of porcine livestock
The use of extended Kalman filtering leads to estimates of statistical bias on observations, and then provides means of bias correction on various time series such as age class sizes (Section 15.2.2e). This method has been used for the model of hog livestock on the time horizon 1974–1981. Two biases were to be estimated:

(1) a time-dependent bias corresponding to the loss accuracy of the sample
(2) a constant bias according for small and large-scale sample discrepancies (Fig. 15.6).

The results showed nonzero biases, proving that the small-scale surveys have biases. Some age groups were underestimated, others were overestimated, and the total was under estimated. The time increasing loss of representativity of the samples was made certain. It led to an underestimation of the porcine livestock of 1.5% in 1974 to 13% in 1981. This result was confirmed by the general census of agriculture of 1980. The new data turned out to be exactly the same as the estimates provided by the extended Kalman filter (see Fig. 15.7).

(c) Understanding reality

When building the model of eggs production, we noted (see Fig. 15.8) that once converted to ECU the wholesale prices of eggs of all EEC countries were almost identical. The EEC markets for wholesale eggs are all connected. The question was then to find out whether a country was dominating this market or, on the other hand, if all the productions of the different countries were to be aggregated in order to model prices.

The price model gave an answer which can be understood by using the results

The use of our models

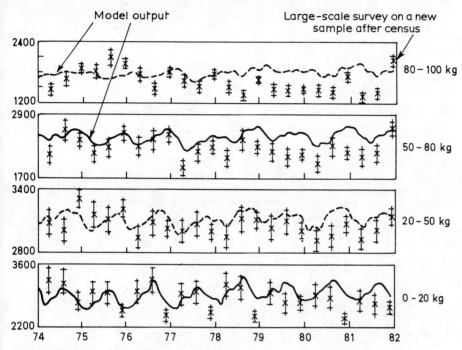

Fig. 15.7 Comparison of data and model output after bias estimation for fattening pigs: ✶ statistical data with their confidence interval

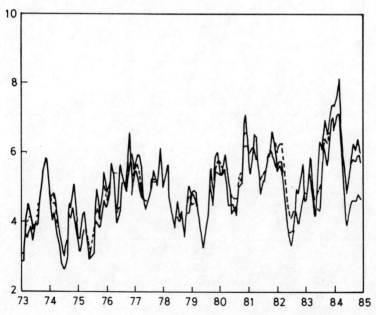

Fig. 15.8 Prices of eggs for France, the Netherlands and EEC9 (unit: ECU for 100 eggs)

of the price simulation (see Fig. 15.9) corresponding to the respective production of:

(1) France
(2) EEC5 (West Germany, France, Holland, Belgium, Luxembourg)
(3) EEC9 (EEC5 + Italy, UK, Ireland, Denmark).

The time horizon (77–86) is divided into three subperiods:

(1) 77–82: identification period
(2) 83–84: validation period (real prices are observed)

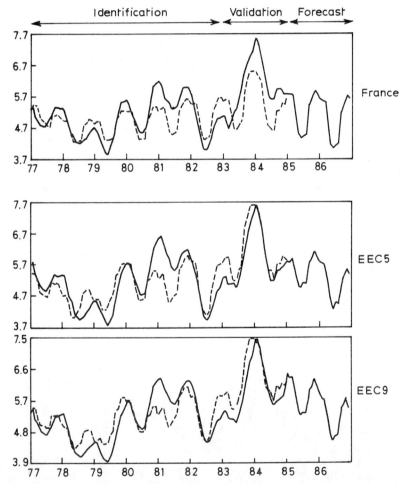

Fig. 15.9 Comparison of market and model prices: continuous lines, the price (ECU per 100 eggs) as observed on the main wholesale market; broken lines, the price (ECU per 100 eggs) as a result of a simulation using the production of the corresponding country (plus the import/export balance); the production is a result of our production model (there being no such measurement in any EEC country)

The use of our models 299

(3) 85–86: forcast period (observations are not available)

It is easy to see that the model for France (see Fig. 15.9, top chart) is not very accurate, especially over the validation period. But the EEC9 model is much more satisfactory (see Fig. 15.9, lower chart). By contrast, the EEC5 and EEC9 models (see Fig. 15.9, middle and lower charts) are very similar. This suggests that the wholesale price of eggs is a result of the EEC5 production since the EEC9 model is not more accurate then the EEC5 model. It must also be noted that all models fail to represent the high level of prices observed at the end of 1980 and beginning of 1981. However specialists know that French poultry livestock suffered from a disease at that time and that the actual production was lower than that predicted by the production model.

(d) Forecasting

Our models are indeed used for forecasting purposes. Some assumptions are required for each scenario: according to the nature of these assumptions, distinctions are made between:

(1) *Conjunctural scenarios*, where assumptions are made on the future input levels.
(2) *Structural scenarios*, where structural parameters change, inducing changes in output.

(e) Optimal control

The need for optimal control schemes first appeared in France when we tried to understand how to control the price of eggs. Examination of the recent past shows that the price of eggs is subject to cyclical overproduction crises, leading to a drop in price below production costs. To overcome these crises, producers sometimes have to use massive slaughtering of poultry which is indeed costly and irrational.

By using a production model, future production can be forecast. Using a price model and estimates of production costs, a crisis can then be foreseen. But then, instead of running 'what if' types of simulation, one might want a direct evaluation of the future monthly input levels to the system (i.e. one-day-old chicks) that are required to avoid overproduction. This can be achieved through an optimal control scheme using the theory of Section 15.2.2(e). See results below in Section 15.3.2.

15.3.2 A detailed example: the model of egg production and price EEC-EGG

(a) The structure of the software

With a few simplifications, the EEC-EGG software consists mainly of two submodels:

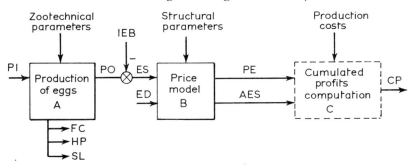

Fig. 15.10 Structure of the software for the egg model. PI = production input (one-day-old chicks); IEB = import–export balance; AES = quantity of eggs 'sold' at equilibrium; CP = cumulated profits; FC = feed consumption of laying hens; HP = laying hen population; SL = slaughtering of 72-week-old laying hen population; PO = production output; ES = egg supply; ED = egg demand; PE = price of eggs at equilibrium

(1) the production model
(2) the price model.

These two models (see below) are linked (Fig. 15.10).

All input and output figures are monthly time series, but the production model has, internally, a weekly structure which is more suited to the way chickens are bred. The demand for eggs, ED, is not an observed item of data. It is simulated by:

$$ED(M, Y) = ID \times (1 + TREND)^Y + COEF(M)$$

Y : year
TREND : annual growth of demand
COEF(M): 12 seasonal coefficients
M : month number within a year
ID : initial demand

$$\sum_{M=1}^{12} COEF(M) = 1$$

These seasonal coefficients help to explain the seasonal fluctuations in prices that could be observed in the past. ID, TREND and COEF are, together with the structural parameters, results of a nonlinear identification scheme (see Section 15.2.2(b)). Finally, we have a very simple third model to compute the estimated cumulated profits of producers.

A simulation usually operates from the production model (A) through the price model (B) and gives estimated profits (C). But this path can be reversed for decision-aiding purposes. The user can set a desired level of profit over a period of time. He (or she) also has to set a subjective forecast for production costs. C used 'backward' then, provides a 'desired' price PE* at market equilibrium. Using the price model B 'backward' one then obtains the 'desired' egg supply (ES*) which, together with the egg demand (ED*), would induce the 'desired' price PE* at market equilibrium. With the 'desired' egg supply ES* and its prediction for the import–export balance IEB, the user can then use the optimal control procedure described in Section 15.2.2(d). The result is an 'optimal' monthly input to the production system (PI*) whose corresponding production PO$^+$ together with the predicted import–export balance IEB is such that:

$$PO^+ + IEB = ES^*$$

(b) Results

Figures 15.11 to 15.15 display graphical results of simulations performed in 1984 over the time horizon 1977/1986.

Our first assumption for the years 1985 and 1986 was that producers would increase their input to the system. We assumed that:

$$1986\ \text{input} = 1985\ \text{input} = 1984\ \text{input} + 3\%$$

As a result, the production model (see Fig. 15.11) shows an increase in output which matches the unprecedented record of 1982 (the monthly average is 2500×10^6 eggs!!). Also note that, in return, the 1983 production is low. Reacting to low prices, producers have reduced their input.

The price model (broken curves in Fig. 15.12) reflects the fluctuations of

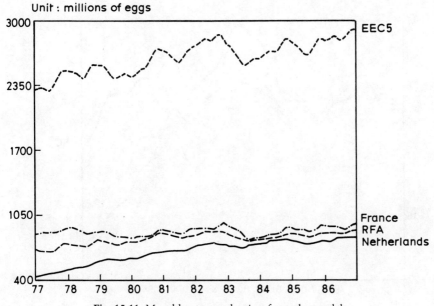

Fig. 15.11 Monthly egg production from the model

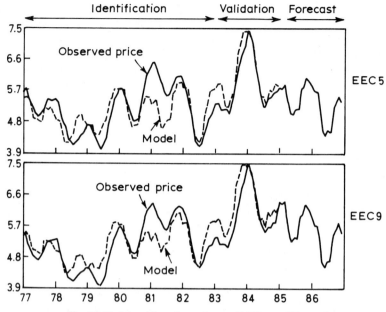

Fig. 15.12 Monthly prices of eggs (ECU per 100 eggs)

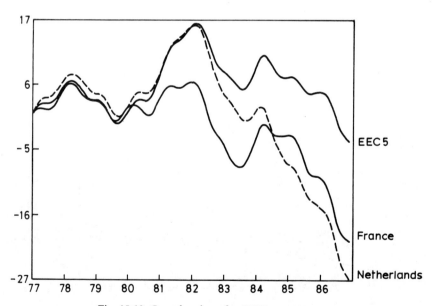

Fig. 15.13 Cumulated profits: ECU per 100 eggs

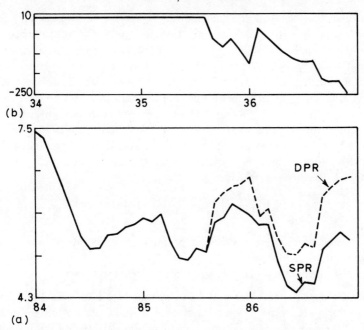

Fig. 15.14 (a) Simulated price (SPR) and desired price (DPR) (unit: ECU per 100 eggs); (b) production change required to change prices (unit: millions of eggs)

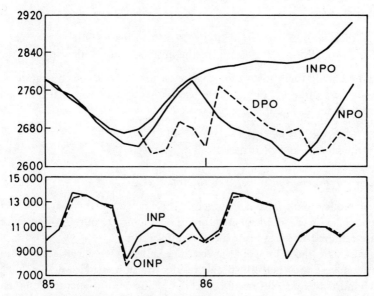

Fig. 15.15 Top chart: INPO = initially simulated production; DPO = 'desired' production; NPO = production after regulation (millions of eggs). Lower chart: INP = initial input to the production system; OINP = input after regulation

production. The 1982 current prices are very low but as a result of an oversized producer's reaction (i.e. massive reduction in input) 1983/1984 prices reach unprecedented heights. The lack of insight (at that time, our model did not exist!) leads to 'stop and go' behaviour which induces a three to four year cycle. Therefore, our assumption for the 1985/1986 input results in a steady price decrease over 1985 and 1986. This, together with our second assumption on production costs: a 1.5 to 2% increase over 1985 and 1986 in constant currencies, induces a quick drop in cumulated profits (see Fig. 15.13). To keep the cumulated profits close to zero, which is regarded as a good result by specialists, prices must change. Fig. 15.14(b) (top), shows the 'desired' drop. The optimal control algorithm is then applied. Results are shown in Fig. 15.15.

TOP CHART

INPO: initial production – i.e. *before* optimal control strategy
DPO: 'desired' production
NPO: new production – i.e. *after* optimal control strategy

It should be noted that NPO and DPO differ. This is due to the optimized criterion which (see Section 15.2.2(d)) also takes input and not only output into account. Therefore 'desired' output is not exactly met but 'optimal' input still has seasonal variations, as it usually does.

LOWER CHART

INP: initial input to the production system (continuous line)
OINP: 'optimal' input as a result of the optimal control algorithm

It should be noted that most of the change occurs in 1985, which is quite normal since the system has a 5 month time lag, as it takes 20 weeks for a new-born chick to become an adult laying hen.

15.3.3 Our models and their users

(a) Microcomputers and transfer to the final user

The first models built (e.g. the model of the French cattle herd) took years of research. They were implemented on minicomputers and the corresponding software required rather a large memory allocation.

At that stage, the user would specify the different scenarios to be run. The simulations were then performed and results transmitted. Nowadays things can be much more flexible. Our clients quickly became used to microcomputers. And these machines are now as powerful, if not more powerful, than the previous generation. Therefore we were naturally led to transferring our models on to

The use of our models

microcomputers and then to our clients. They can now specify their own simulations, and try them at no cost. However, we soon discovered that we were entering a different era; we had to acquire new skills.

(b) Hints for a successful transfer

We first learned that we could not 'get by' with 'put-up' programs. We had to understand that the models users did *not* have the designer's knowledge of the software. We also had to understand that the final user is usually not a computer 'whiz'. Therefore, the software must be carefully designed with methods and concepts commonly used in computer science:

THE SOFTWARE MUST BE USER FRIENDLY

The questions and their possible answers must be clear. The required assumptions to run a simulation must be clearly understood, as well as the way to specify them.

THE SOFTWARE MUST BE FAULTPROOF

The user's answers to questions must be checked. Nonambiguous messages must prompt the user to correct mistakes. Files must be protected to avoid frustrating loss of data.

USER MANUAL

A detailed user's manual must be provided and upgraded until it contains all the information the user might wish for.

MAINTENANCE

The code itself must contain extensive comment lines. Thanks to the use of explicit parameters, changes will be easy and safe. It should not be forgotten that one day the model's first designers will move or simply forget about some odd details that will suddenly become crucial points.

TRAINING

The user must be trained. His (or her) learning about the model will be greatly eased by well-designed software.

Most of these requirements can be met if the model's software is built using top–down structured methods. The programming language must stick to a standard, thereby enhancing portability.

ACKOWLEDGEMENTS

The research set forth in this chapter was supported by:

ONILAIT (beef herd model)
OFIVAL (hog livestock model)
UNIGRAINS (egg production and prices model)

The authors would like to thank Ms L. Bel and F. Berger who have also been part of the team involved with these models.

REFERENCES

Anderson, B. D. O. and Moore, J. B. (1979) *Optimal filtering*, Prentice Hall, Englewood Cliffs.
Aoki, M. (1976) *Optimal Control and System Theory in Dynamic Economic Analysis*, North-Holland Amsterdam.
Friedland, B. (1969) Treatment of bias in recursive filtering. *IEEE Transactions on Automatic Control*, 14, August, 359.
Himmelblau, D. (1972) *Applied Non-linear programming*, McGraw Hill, New York.
Jazwinski, A. H. (1970) *Stochastic Processes and Filtering Theory*, Academic Press, New York.
Leibundgut, B., Rault, A. and Gendreau, F. (1983) Applications of Kalman filtering to demographic models. *IEEE Transaction on Automatic Control*, 28, March.
Luenberger, D. G. (1979) *Introduction to Dynamic Systems: Theory, Models and Applications*, Jonn Wiley, New York.
Maybeck, P. S. (1979) *Stochastic Models, Estimation and Control*, vol. 1, Academic Press, New York.
Mehra, R. K. (1971a) Identification of stochastic linear dynamic systems using Kalman filter representation. *AIAA Journal*, 9, No 1, January.
Mehra, R. K. (1971b) On-line identification of linear dynamic systems with applications to Kalman filtering. *IEEE Transactions on Automatic Control*, 16, No. 1, February.
Rault, A. and Leibundgut, B. (1981) Demographic models of French livestock, *Proceedings 1st Working Conference On Computer Applications in Food Production and Agriculture Engineering*, IFIP, Cuba.
Rouhani, R. and Tse, E. (1981) Structural design for classes of positive linear systems. *IEEE Transactions on Systems, Man and Cybernetics*, 11, No. 2, February.
Smith, D. and Keyfitz, M. (1977) *Mathematical Demography*, Springer-Verlag.

APPENDIX A

GENERAL FORM OF A COHORT POPULATION MODEL

The relevant references are Luenberger (1979); Rouhani and Tse (1981); and Smith and Keyfitz (1977).

Free population

The population is divided into age groups (or cohorts) of equal age span, say one year for the beef cattle herd. That is, the first group consists of all those members of the population between the ages of zero and one year, the second consists of those between one and two years, and so on. The cohort model itself is a discrete-time dynamic system with the duration of a single time period (sampling period) corresponding to the basic cohort span (e.g. one year).

As an example, let us describe the equations of the model by considering only the female population. We consider $(i+1)$ groups with number $0, 1, \ldots i$. Let $x_i(n)$ be the population of the ith age group at time period n. We can write:

$$x_{i+1}(n+1) = \beta_i x_i(n) \qquad i = 0, 1, \ldots, i-2$$

where β_i is the survival rate of the ith age group during one period. The ith age group aggregates all those who are $(i-1)$ years old or more. So we have:

$$x_i(n+1) = \beta_{i-1} x_{i-1}(n) + \beta_i x_i(n)$$

The first age group x_0 consists of individuals born during the last time period. The size of this group depends on the birth rate of the other age groups:

$$x_0(n+1) = \alpha_1 x_1(n) + \alpha_2 x_2(n) + \cdots + \alpha_i x_i(n)$$

where α_i is the birth rate of the ith cohort group, expressed in numbers of births per time period per female.

In matrix form this general cohort model becomes:

$$\begin{bmatrix} x_0(n+1) \\ x_1(n+1) \\ \vdots \\ x_i(n+1) \end{bmatrix} = \begin{bmatrix} 0 & \alpha_1 & \alpha_2 & \cdots & \alpha_i \\ \beta_0 & 0 & & \cdots & 0 \\ 0 & \beta_1 & 0 & \cdots & \\ \vdots & & & & \\ 0 & & \beta_{i-1} & & \beta_i \end{bmatrix} \begin{bmatrix} x_0(n) \\ \vdots \\ x_i(n) \end{bmatrix} \qquad (A1)$$

or:

$$\mathbf{X}(n+1) = A\mathbf{X}(n) \qquad (A2)$$

where $\mathbf{X}(n)$ is called the state vector and A the transition matrix.

Controlled population

The above model represents only the natural evolution of a population. However, in domestic cattle, some external factors are to be considered: slaughtering, import and export. Thus the state equation (A2) becomes

$$\mathbf{X}(n+1) = A\mathbf{X}(n) - \mathbf{U}(n) \qquad (A3)$$

with:

$$\mathbf{U}(n) = [u_0(n) u_1(n) \ldots u_n(n)]^T \text{ the input vector}$$

where: $u_i(n)$ is the balance of slaughtering, exports and imports over the ith class during period n.

APPENDIX B

MAIN CHARACTERISTICS OF THE EXAMPLE MODELS

The beef cattle model

The livestock is divided into males and females, with five age groups for the female system and four age groups for the male. The sample period for the model is one year. An annual survey provides the following data: male and female populations aged 0–1 year, 1–2 years and over 2 years; and the total number of cows. We also have data about natural deaths. Other data are available monthly: milk production, slaughtering, import–export balance. Slaughtering and export-imports are input data. A problem has been met when collecting the slaughtering data: they are not given according to the age of the animals but according to their number of teeth. Milk production is an output of the model. It depends on the number of cows, and on the average production of the females of each age class. This productivity is time increasing due to technical improvements such as artifical feed and cross-breeding. This feature is taken into account via a technical gain-parameter.

To sum up, the model can be described by the state, control, observation and parameter vectors:

State vector **y**:	females 0–1 year, 1–2 years, 2–3 years, 3–4 years, over 4 years
	males 0–1 year, 1–2 years, 2–3 years, over 3 years
Control vector **u**:	slaughtering + exports–imports, for males and females: calves, 0 teeth, 1–4 teeth, over 4 teeth
Observation vector **s**:	females 0–1 year, 1–2 years, over 2 years
	males 0–1 year, 1–2 years, over 2 years
	number of cows, milk production, dead animals
Parameter vector **p**:	sex ratio
	mortality rates of calves and adult animals
	fertility rates for each age class
	percentage of cows per age class
	average production of milk and technical gain per year

The hog livestock model

The particular model structure used in this case was induced both by the structure of the porcine livestock and by the type of data available. Hog livestock is characterized by the following items:

Appendix B

(1) It is divided into two distinct parts: fattening pigs (which are slaughtered when they have reached a given weight) and breeding sows (which are assigned to reproduction).
(2) The reproduction phenomena are rather fast: the gestation is less than four months long and a sow gives birth to about ten piglets each time; this implies a quick rotation in the livestock.
(3) Input data (slaughtering, exports–imports) are available monthly; but observation data (pig populations) are provided every four months: fattening pigs are observed in weight classes while sows are classified according to their physiological state (for instance, breeding sows unmated, or sows mated for the first time).

To take these points into consideration, the general methodology has been modified in the following way:

(1) Two systems are considered: a 'fattening system' and a 'breeding system' only constituted by sows as the breeding boars are very few and so neglected. The state variables of the breeding system are three physiological states of sows because it would have been too difficult to link these observed states to age classes. As for the fattening system, it is necessary to draw up a relation between the age classes (state variables) and the weight classes (observations): it forms the observation equation.
(2) For both systems, the sampling period is one month; this value allows all available data and reproduction phenomena, to be taken into account. But as filtering requires data about pig populations, it can be realized only every four sampling periods; between two filtering times, the evolution of the model is computed every month through prediction equations.
(3) Although the two systems have their own evolution, they are not quite independent and this is modelled by introducing input which is not only slaughter. For instance, a certain amount of fattening pigs provides input for the breeding system; moreover, birth modelling is no longer carried out in the state transition matrix, but births are computed as a function of the different classes of sows (using fertility rates) and at the next period are introduced as input in the first class of the fattening system. This structure is illustrated in Fig. 15.B1.

The different vectors used in modelling will be defined for each system (FS = fattening system, BS = breeding system).

State vector **y**:   FS: 0–1 month, 1–2, 2–3, 3–4, 4–5, 5–6 months, over 6 months
                                     BS: sows mated for the first time, mated sows (not for the first time), breeding sows not mated
Control vector **u**:   FS: slaughtering on the last class, birth on the first class
                                     BS: part issued of FS (input on sows mated for the first time), slaughtering on unmated sows.

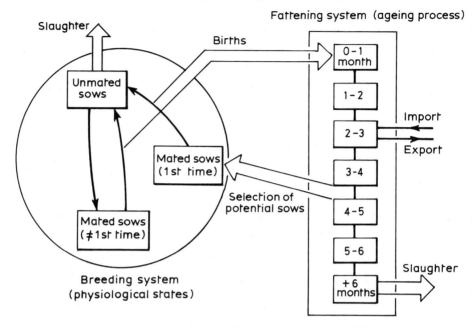

Fig. 15.B1 Structure of the porcine livestock model

Observation vector **s**: FS: observations are weight classes: 0–20 kg, 20–50 kg, 50–80 kg, 80–110 kg, over 110 kg
BS: state variables are also observations (the observation equation is thus identity).

Parameter vector **p**: mortality rates
sow performance (transition probability between physiological states), fertility rates and number of piglets per farrow, growth rates (to obtain the age–weight relations).

The egg production model

The model has a general structure and could be applied to any poultry livestock, with any system of housing. But the actual system we refer to, corresponds to industrial production, with battery housing of layers.

This model takes into account the weekly egg yield per layer, its improvements due to artificial feeding and the natural mortality of layers. Hens begin to lay when they are 20 weeks old, and they are slaughtered at 72 weeks when their productivity is decreasing. Input for the system consists of chicks, at the age of one day. The internal sample period of the model is one week, for the state equation. But the output is computed monthly. Very few data are available: the egg production and

Appendix B

the population size are known only yearly. Therefore this model cannot be identified. We use zootechnical knowledge for the parameters, such as egg yield per layer and per week according to the age, mortality and feedingstuff consumption. Chick placement data are available for periods of 4 or 5 weeks, and must be converted to monthly data.

The vectors for this model are the following:

State vector: 0–1 week, 1–2 weeks, ..., 71–72 week layers
Input vector: number of chicks placed each month
Output vector: egg production (weekly output aggregated per month)
total number of layers
number of slaughtered layers
consumption of feedingstuff per month

16

Estimated optimal lags for the optimization of models: a method for estimating the optimal lag between economic variables: Part IV

KAORU ICHIKAWA

When we wish to build an optimal econometric model, it is possible to do it in various ways. Our approach uses optimized lags in a lagged model. We use the coefficient of multiple correlation as measure for the accuracy of the model. Using optimized lags results, I think, in a relatively high value for the multiple correlation coefficient. This is why I began researching a method for estimating the optimal lag between economic variables in 1982, which I now verify with data for the first time in the present chapter. In the concluding section, moreover, I give a systematic procedure for building an optimally lagged model.

Since 1982, I have published several articles on methods and their applications in this field. These are introduced in Section 16.3.

16.1 THE PRINCIPLE OF ANALYSIS

Suppose there is a model in which the coefficient of multiple correlation is not high enough to be used for economic prediction or policy. In such a case, we can usually substitute a better economic variable, an explanatory variable, for the economic variable. However, suppose the model seems to be 'good' for both theoretical and empirical reasons. We must then make the coefficient high without substituting economic variables. In addition, let us assume that the model is not a lagged model, or, where it is a lagged model, that the lag is not optimal.

In such a case, we can make the coefficient of multiple correlation high by using optimal lags, not by guesswork as before (e.g. 'I *think* this is a one-year lag'), but on the basis of measurement and calculation. In the next section, we verify with data that the model is optimized by means of this principle.

16.2 VERIFICATION

Reading the economic white papers published by the Economic Planning Agency, we are surprised to find that very good models have been built having multiple correlations of 0.95 or 0.99.

However, a kind of econometric model with relatively low multiple correlation (of about 0.8) is described on page 225 of the 1980 issue. We denote rising rate of wages by W, rising rate of consumer prices by C, and unemployment rates by U. The Japanese model for 1961–1979 is

$$W = 17.248 + 0.757C - 6.632U$$
$$R = 0.797$$

where R represents the coefficient of multiple correlation. The white paper says that the model for the United Kingdom has $R = 0.801$, for Italy $R = 0.865$, for the USA $R = 0.898$ and for Germany $R = 0.681$. The coefficient of multiple correlation for Germany is specially low. This prompted us to analyse this kind of model. The white paper says 'data from 1961–1979 was used for estimating these models', and 'the source of this data is *Main Economic Indicators* by OECD'. Although we could not obtain the same data, we took quarterly and monthly data of the same kind for 1980–1983 from recent issues of *Main Economic Indicators*. By estimating the Japanese model from the quarterly data using the method of least squares,

$$W = 156.50 + 38.67C - 75.92U$$
$$R = 0.886$$

is obtained. This value of R approximates to the previous value of 0.797. Thus, we conjecture that the Japanese model in the white paper of the Economic Planning Agency is built upon quarterly data. Using monthly data, a different type of model

$$W = 332.15 - 1.496C - 123.35U$$
$$R = 0.508$$
(16.1)

is formed. The value of R is very small. I think the reasons for this are as follows. In the analysis using quarterly data, the influence of systematic movement appears, causing R to become high. In the case of monthly data, the influence of random movement is strong, causing R to become low. I think it is a good idea to use the model with a low value for R to show clearly the effect of optimizing a lag.

We thus compare Equation 16.1 with the lagged model obtained from the same data. The series of W and C values used for building our model were made by substituting the values of wages and consumer prices from the *Main Economic*

Verification

Table 16.1 Data used for obtaining Equation 16.1

Year	Month	Rising rate of wages (%)	Rising rate of consumer prices (%)	Unemployment rates (%)
1982	10	0.118	0.274	2.4
	11	4.00	−1.002	2.3
	12	178.98	−0.092	2.3
1983	1	−64.68	0.092	2.8
	2	−2.75	−0.276	2.9
	3	1.89	0.554	3.0
	4	1.04	0.459	2.9
	5	0.69	1.005	2.6
	6	58.13	−0.633	2.5
	7	31.44	−0.364	2.4
	8	−45.65	−0.274	2.7
	9	−10.67	1.190	2.6
	10	0.113	0.814	2.5
	11	5.63	−0.539	2.5

Indicators into the formula

$$\frac{\text{Current value} - \text{Last value}}{\text{Last value}},$$

respectively. The series of U values are then the values of unemployment (percentage of total labour force) from the book. These are shown in Table 16.1.

Let us examine the relation between W and C, and the relation between W and U, respectively. Suppose there is a space of three dimensions consisting of a W-axis, a C-axis, and a U-axis, as in Fig. 16.1. We plot any point P with co-ordinates (c_i, u_i, w_i) in the space to describe a scatter diagram. For example, thinking of

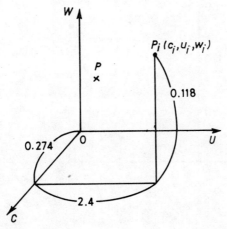

Fig. 16.1 Point $P_i(c_i, u_i, w_i)$ in three dimensions

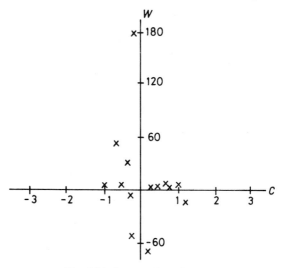

Fig. 16.2 Scatter plot of (C, W)

$c_i = 0.274$, $u_i = 2.4$, and $w_i = 0.118$, this point represents the situation of October 1982. In the same way, 14 points are plotted. Project the scatter to the W–C plane (Fig. 16.2) and to the W–U plane (Fig. 16.3) and to the C–U plane (Fig. 16.4). Looking at these scatter diagrams, we can conclude that there are almost no relationships between any two variables. It is bad that there is no relation between W and C, because it is the relation between the explanatory variable and the explained variable. For the same reason, it is bad that there is no relation between W and U. It is good that there exists no relation between U and C due to the problem of multicollinearity, because this is the relation between the explanatory variables. I

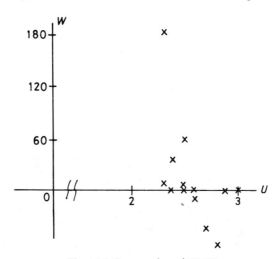

Fig. 16.3 Scatter plot of (U, W)

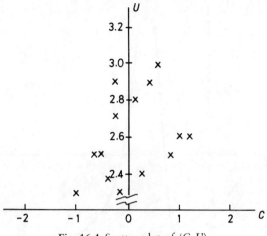

Fig. 16.4 Scatter plot of (C, U)

am afraid that this example is unsuitable, because of simple correlation coefficients $r_{wc} = -0.184$, $r_{wu} = -0.508$. However, it may be a good thing to use this model only for showing that the coefficient of multiple correlation becomes higher when optimal lags are used.

We will now estimate the optimal lag between W and C, and between W and U, respectively. Using the lags we will then examine whether the multiple correlation of the model becomes high or not. Looking at the relation between W and C in the diagram of the time series (Fig. 16.5), C's wave goes ahead of W's wave by about two months. Next, let us estimate the optimal lag between W and C by calculation. We

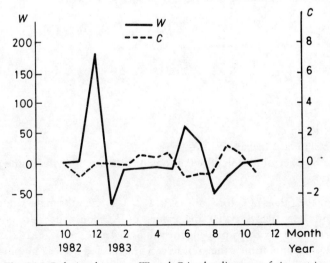

Fig. 16.5 Relation between W and C in the diagram of time series

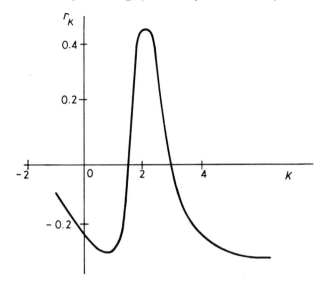

Fig. 16.6 Movement of the r_K between W and C

use here the method of Part II of the study (Ichikawa, 1984), in which we use only lag correlation (denoted by r_K) and fit a curve by the freehand method. We denote a time series by $\{X_t\}$, the time series of its result by $\{Y_t\}$, a time lag by K, and an optimal lag by κ. The freehand method is used because of the asymmetrical nature of the curve in Fig. 16.6. The data are unsuitable for fitting a quadratic curve of shape ∩ by the method of least squares. For estimating the value of κ a suitable approach is to trace the movement of points by the freehand method. Figure 16.6 was described as follows. Denoting k by $k = |K|$, we write:

$$r_{-k} = \frac{1}{\tau - k} \sum_{t=k+1}^{\tau} \frac{X_t - \bar{X}_t}{S_{X_t}} \frac{Y_{t-k} - \bar{Y}_{t-k}}{S_{Y_{t-k}}}, \quad k = 1, 2, \ldots \quad (16.2)$$

$$r_k = \frac{1}{\tau - k} \sum_{t=1}^{\tau - k} \frac{X_t - \bar{X}_t}{S_{X_t}} \frac{Y_{t+k} - \bar{Y}_{t+k}}{S_{Y_{t+k}}}, \quad k = 0, 1, 2, \ldots \quad (16.3)$$

where

$\bar{X}$ = the mean of X
S_X = the standard deviation of X
τ = the number of sample observations or number of terms of the given time series

Computing the values of r_K for $K = -1, 0, 1, 2, 3, 4, 5$, results in the values in Table 16.2. Figure 16.5 suggests that the values of K should be selected from the neighbourhood of $K = 2$. By plotting the values in Table 16.2, Fig. 16.6 is obtained. The result of computation is the same as can be expected from examining Fig. 16.5, so we can conclude that a higher correlation can be obtained by combining W_t with

Verification

Table 16.2 The values of Fig. 16.6

K	r_K
−1	−0.1468
0	−0.1849
1	−0.2957
2	0.4591
3	−0.0525
4	−0.2435
5	−0.2801

C_{t-2}. In fact, W and C influence each other. Of course, there is then a factor to infer that W's wave goes ahead of C's wave. However, we treat only the problem connected with Equation 16.1.

Let us next examine the relation between W and U. Describing the time series of W and U gives Fig. 16.7. It suggests that U's movement goes ahead of W's movement by two or three months, although it is also possible to read from the same graph that W goes ahead of U by about three months. Applying Equations 16.2 and 16.3 to sample observations generates a series of lag correlations (see Table 16.3 and Fig. 16.8). The series of r_K reaches maxima at $K = 3$ and $K = -3$. The former is large ($r_3 = 0.576$). However, the latter is a relatively small maximum ($r_{-3} = 0.335$), which is the reason why W's wave seems to go ahead of U's wave by about three months. The minimum is reached at $K = 0$, that is, $r_0 = -0.508$. The negative value means a retrocycle. From Fig. 16.7 we can see that two time series retroact, at December 1982, at August 1983, and so on. Although the series of r_K has some peaks, we deal with only $r_3 = 0.576$ for two reasons: (1) because

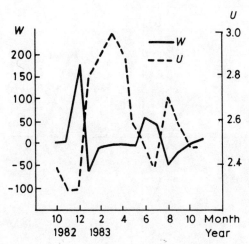

Fig. 16.7 Relation between W and U in the diagram of times series

Estimated optimal lags for the optimization of models

Table 16.3 The values of Fig. 16.8

K	r_K
−4	0.2868
−3	0.3351
−2	0.2217
−1	0.0296
0	−0.5078
1	−0.1892
2	0.0426
3	0.5760
4	0.2255
5	−0.0530

it has the largest absolute value, and (2) because it is connected with Equation 16.1. Therefore, the optimal lag results in $\kappa = 3$ from computation. We thus obtain

$$W_t = \alpha + \beta C_{t-2} + \gamma U_{t-3} \tag{16.4}$$

as the model with optimal lags.

We next estimate the parameters of Equation 16.4 to establish that the multiple correlation of the estimated Equation 16.4 is of a higher value than that obtained from Equation 16.1. Rewriting Table 16.1 into Table 16.4 reduces eleven pairs of sample observations. Although I am afraid that the sample size is too small, estimating the parameters from these data results in

$$W_t = -80.139 + 29.422 C_{t-2} + 29.756 U_{t-3}$$
$$R = 0.763 \tag{16.5}$$

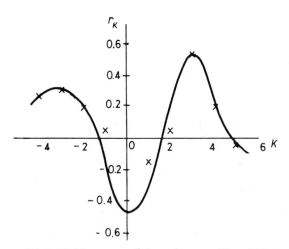

Fig. 16.8 Movement of the r_K between W and U

The origin of the theory for estimating the optimal lag

Table 16.4 Data used for obtaining Equation 16.5

Data number	W_t	C_{t-2}	U_{t-3}
	0.118		
	4.00		
	178.98	0.274	
1	−64.68	−1.002	2.4
2	−2.75	−0.092	2.3
3	1.89	0.092	2.3
4	1.04	−0.276	2.8
5	0.69	0.554	2.9
6	58.13	0.459	3.0
7	31.44	1.005	2.9
8	−45.65	−0.633	2.6
9	−10.67	−0.364	2.5
10	0.113	−0.274	2.4
11	5.63	1.190	2.7
		0.814	2.6
		−0.539	2.5
			2.5

The R value of Equation 16.5 is 0.763, higher than the value 0.508 obtained from Equation 16.1. It is thus verified that a lagged model with optimal lags obtained by fitting a curve is the best model.

16.3 THE ORIGIN OF THE THEORY FOR ESTIMATING THE OPTIMAL LAG

In this section, results published in Japanese are introduced in English. I reported on Part I of the study at the congress of the Japan Statistical Society in 1982 (Ichikawa, 1982). Parts II and III were published in Japanese in 1984 in *The Journal of Law and Economics* (Ichikawa, 1984). The applications were reported at the congress of the Society for the Economic Study of Securities in 1984 (Ichikawa, 1985).

The Method given in Part I of the study (Ichikawa, 1982) is outlined below. In the first part of that report, I state 'Since both for economic prediction and for the building of lagged models it is probably necessary to invent a method for estimating the optimal lag between two time series I gave consideration to the problem'. I conclude as follows: 'We denote the lag correlation (named by Kazuo Mizutani) by r_K, and the lag regression (named by Kaoru Ichikawa) by B_K,

322   Estimated optimal lags for the optimization of models

where $\kappa = -(n-1)/2, \ldots, -1, 0, 1, \ldots, (n-1)/2$. The value of n represents the number of terms calculated about the $r_K + B_K$ series. In this case,

$$\kappa = -\frac{1}{2}\left\{\frac{\sum_K K(r_K + B_K)}{\sum_K K^2}\right\}\left\{\frac{n\sum_K K^2(r_K + B_K) - \sum_K K^2 \sum_K (r_K + B_K)}{n\sum_K K^4 - \left(\sum_K K^2\right)^2}\right\}^{-1} \quad (16.6)$$

gives the optimal lag.' To avoid a misapplication, we explain this statement in detail. The 'optimal lag' means the time lag giving the highest of n values about absolute values of $r_K + B_K$ (or r_K alternatively). Here, we use a lag correlation r_K and a lag regression B_K to estimate an optimal lag κ. Suppose the notation defined in the previous sections has the same meaning as that in this section. We have

$$r_{-k} = \frac{1}{\tau - k}\sum_{t=k+1}^{\tau}\frac{X_t - \bar{X}_t}{S_{X_t}}\frac{Y_{t-k} - \bar{Y}_{t-k}}{S_{Y_{t-k}}}, \qquad k = 1, 2, \ldots, \frac{n-1}{2}$$

$$r_k = \frac{1}{\tau - k}\sum_{t=1}^{\tau-\kappa}\frac{X_t - \bar{X}_t}{S_{X_t}}\frac{Y_{t+k} - \bar{Y}_{t+k}}{S_{Y_{t+k}}}, \qquad k = 0, 1, 2, \ldots, \frac{n-1}{2}$$

$$B_{-k} = \frac{\frac{1}{\tau - k}\sum_{t=k+1}^{\tau}(X_t - \bar{X}_t)(Y_{t-k} - \bar{Y}_{t-k})}{\frac{1}{\tau - k}\sum_{t=k+1}^{\tau}(X_t - \bar{X}_t)^2}, \qquad k = 1, 2, \ldots, \frac{n-1}{2}$$

$$B_k = \frac{\frac{1}{\tau - k}\sum_{t=1}^{\tau-\kappa}(X_t - \bar{X}_t)(Y_{t+k} - \bar{Y}_{t+k})}{\frac{1}{\tau - k}\sum_{t=1}^{\tau-\kappa}(X_t - \bar{X}_t)^2}, \qquad k = 0, 1, 2, \ldots, \frac{n-1}{2}$$

The measure for estimating the strength of a relationship between two variables is not only the correlation coefficient but also the regression coefficient. However, we cannot necessarily say that the optimal lag resulting from the lag-correlation coefficient is equal to that resulting from the lag-regression coefficient. So, we use both coefficients in the form of $r_K + B_K$. After calculating the values of $r_K + B_K$ for $K = -(n-1)/2, \ldots, -1, 0, 1, \ldots, (n-1)/2$ using four equations, we plot them

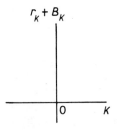

Fig. 16.9 Grid for plotting the values of $r_K + B_K$.

The origin of the theory for estimating the optimal lag

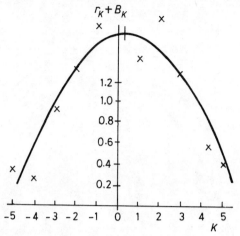

Fig. 16.10 $r_K + B_K$ between the volume of shipment of Chinese cabbage + princemelons and the income of the Ibaraki Prefecture

on a graph as in Fig. 16.9. After that, we pass a quadratic curve of shape ∩ through the points by the method of least squares. We then examine the maximum point on the curve, that is, we differentiate its estimated equation with respect to K, equate to 0, and solve the equation. This solution is Equation 16.6 or an optimal lag κ. At $K = \kappa$, we read that the movements of two time series show the strongest degree of analogy. If the value of κ is negative, we interpret it as meaning that the economic time series, that is thought to go ahead of another time series, in fact follows it. Next, what necessitates drawing a curve? The answer is as follows: to keep from missing the maximum between the observations since the observations are discrete. In the case of Fig. 16.10, the method of Ichikawa (1982), i.e. of Part I of the study, is suitable to use, but in the case of Fig. 16.11, it is unsuitable. Figure 16.10 shows the relationship between the volume of shipment of Chinese cabbage + princemelons and the income of the Ibaraki Prefecture. The result calculated by Equation 16.6 is $\kappa = 0.316$ (using yearly data); in other words, 3.8 months; this is because

$$12 \times 0.316 = 3.78 \text{ (months)}.$$

It coincides with the result shown in Fig. 16.10. Hence, we see that variation in the volume of shipment influences the movement of income about 3.8 months later. Figure 16.11 shows the relationship between the Tokyo stock price index and foreign exchange reserves. Draw the quadratic curve l estimated by the method of least squares on the figure. Also, superimpose a curve l' by the freehand method for comparison. As the points arrange themselves in a curve pattern, the freehand method is suitable to use. The point Q showing a maximum on the curve l is significantly distinct from the point P showing a maximum on the l' curve. Which would you prefer, P or Q, for our purposes? You probably prefer P to Q, as do I.

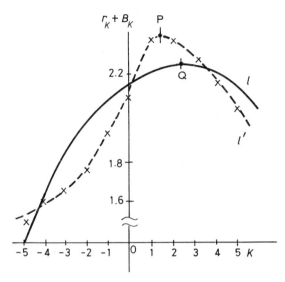

Fig. 16.11 $r_K + B_K$ between Tokyo stock price index and foreign exchange reserves

The method of Part II of the study (Ichikawa, 1984) is a method for using only r_K exclusive of B_K, and for drawing curves freehand. The absolute value of r_K does not rise above 1, and we can then assume that the movement of r_K has a peak. However, the value of B_K does not have such a limitation, and in fact sometimes has a very large value. Also, the B_K value does not necessarily reach a peak. In such a case we cannot treat B_K with the same weight as r_K. Additionally, the movement of B_K is various. Figure 16.12 shows the relationship between the mining and manufacturing production index and the employment index. In this case – the top of the curve is flat – the place where the curve reaches a peak is not clear, so that there is a large margin of error for estimation of κ. To avoid the troubles involved in using B_K, we deal only with r_K in this method.

The method of Part III of the study (Ichikawa, 1984) is a method using only r_K, and a quadratic curve is fitted to the data. The formula for estimating the optimal lag in this case is obtained by substituting $B_K = 0$ in Equation 16.6, that

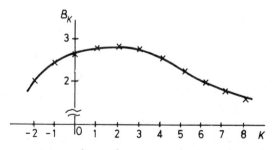

Fig. 16.12 B_K between mining and manufacturing production index and employment index

is, we obtain

$$\kappa = -\frac{1}{2}\left(\frac{\sum\limits_{K} Kr_K}{\sum\limits_{K} K^2}\right)\left\{\frac{n\sum\limits_{K} K^2 r_K - \sum\limits_{K} K^2 \sum\limits_{K} r_K}{n\sum\limits_{K} K^4 - (\sum\limits_{K} K^2)^2}\right\}^{-1} \quad (16.7)$$

16.4 THE METHOD OF PART IV (THE PRESENT CHAPTER)

In this chapter, we propose the necessity of the method of using $r_K + B_K$ and of drawing freehand curves. This is because there is also a case in which the curve of B_K has a peak, the nature of the curve is sharp, the order of the value of B_K is of the same order as r_K, and the arrangement of observed values of $r_K + B_K$ is in a curve pattern and the form of the curve is not symmetrical. For an example, refer to Fig. 16.11. The movement of B_K only is shown in Fig. 16.13. We thus name this method: A method for estimating the optimal lag between economic variables: Part IV (see Fig. 16.14).

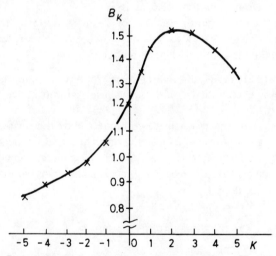

Fig. 16.13 B_K between Tokyo stock price index and foreign exchange reserves

	$r_k + B_k$	r_k
Quadratic curve	I	III
Curve by freehand method	IV	II

Fig. 16.14 Relative position of Part IV

We therefore have four methods – those of Parts I, II, III and IV – for the one purpose of estimating optimal lags. Although it seems to be anomalous, it is not. To represent the location of distribution, we have many measures – mean, mode, median, etc. – as representative. Our problem is similar to this, I think.

16.5 PROCEDURE FOR BUILDING AN OPTIMALLY LAGGED MODEL

Instead of a conclusion, we give here a series of steps for building an optimally lagged model.

(1) Denoting $(m+1)$ economic variables by $Y\ X_1\ X_2 \ldots X_m$, suppose $X_1\ X_2 \ldots X_m$ are Y's explanatory variables. Make m pairs of $(Y, X_1), (Y, X_2), \ldots, (Y, X_m)$ from them.
(2) For each pair, respectively, describe the movements of r_K and B_K on a graph. By means of this, choose one of the methods of Parts I, II, III or IV.
(3) Estimate the optimal lag from pair to pair: When drawing a curve freehand, estimate this from the graph. When fitting a quadratic curve to the data, obtain this from Equations 16.6 or 16.7.
(4) Denoting the estimated optimal lag by κ, the subscript of the explanatory variable is written as $t-\kappa$ for Y's subscript t in each pair. Pooling m pairs, we can make an optimally lagged model.

The method of Part IV (the present chapter), moreover, is another proposition in this study. However, as it has already been introduced above (Section 16.4), it is omitted here.

Many optimal lags appear when we continue sliding the phases of two kinds of wave. In such a case, we can choose the optimal lag in the neighbourhood of $K=0$. The reason why I say this is as follows. We generally choose a principal value, since an inverse trigonometrical function has many solutions – and our problem bears an analogy to this.

REFERENCES

Ichikawa, K. (1982) The estimation of time lag (or optimal lag) between economic variables – the industrial production index of mining and manufacturing and the international balance of payments (A method for estimating the optimal lag between economic variables: Part I), Congress of the Japan Statistical Society, *Journal of the Japanese Statistical Society*, **12**(2) (in Japanese).

Ichikawa, K. (1984) The analysis of econometric models connected with Chinese cabbage and princemelons (Methods for estimating the optimal lag between economic variables: Parts II and III). *The Journal of Law and Economics*, **16**, July, Chiba University (in Japanese).

Ichikawa, K. (1985) The time lags between fluctuation in stock price and some time series associated with it, Congress of the Society for the Economic Study of Securities, June 1984. *Annals of the Society for the Economic Study of Securities* **20**, (in Japanese).

Shimizugawa, S. and Ichikawa, K. (1974) Econometrics, Seirinshoin (in Japanese).

17

Macroeconomic policy and aggregate supply in the UK

MICHAEL BEENSTOCK
and
PAUL LEWINGTON

17.1 INTRODUCTION

Traditionally the theory of macroeconomic policy has been concerned with the effects of various fiscal and monetary measures on aggregate demand. While the new-classical approach to macroeconomics has perhaps offered new insights into our understanding of the effects of policy on demand its central focus has still been with the reworking of the IS–LM framework under alternative hypotheses about expectations and price formation. In doing so it has given more prominence than was previously the case to the so-called 'supply side' of the economy. However, there has been relatively little consideration given to the effects of macroeconomic policy on aggregate supply.

The central purpose in this chapter is therefore to explore the effects of policy on the supply-side of the economy. The exercise is carried out with reference to the City University Business School (CUBS) econometric model of the UK economy which attempts to specify the determinants of aggregate supply as well as aggregate demand. We consider the effects of fiscal policy in terms of public expenditure (capital and current) as well as direct and indirect taxes. We also consider the effects of social security policies in terms of benefit rates and National Insurance contributions. Such policies are likely to have nonneutral effects on aggregate supply and the supply-side of the economy is likely to be influenced by the way in which unbalanced budgets are financed. Money-financed deficits theoretically have different supply-side effects from bond-financed deficits. Therefore, the supply-side effects of macroeconomic policy in general reflect the structure of taxes and public expenditure, the nature of the social security system and the way in which the Exchequer finances itself.

In the next section we set out our theoretical premises. There are two basic aspects to this. First we highlight the main theoretical structure of the CUBS model in so far as it concerns our present subject matter. Space prevents a complete exposition of the model which in any event is available elsewhere in Beenstock et al. (1986). Secondly we consider some of the main theoretical dimensions to the debate about the effects of macroeconomic policy on the supply-side.

In Section 17.3 we simulate the CUBS model in order to illustrate the empirical effects of various macroeconomic policies on demand as well as supply. These effects are contrasted with the policy multipliers generated by conventional models of the economy.

Glossary of symbols

A	proportionate present value of capital allowances
$\bar{B}$	corporate bond stock
B^D	bonds held by the nonbank private sector
B_g	government bond stock
C	user cost of capital
e	employers' rate of National Insurance contributions
K	capital stock used to produce Q
K_s	social capital stock
L	labour force used to produce Q
$\bar{L}$	total employment
L_g	general government employment
L^s	working population
P	price of gross output
M	raw materials used to produce Q
P_c	consumer price index
P_m	price of raw materials (domestic currency)
POP	population of working age
Q	gross output excluding oil and general government output
R	real rate of interest
RES	gold and foreign exchange reserves
R_w	world (real) rate of interest
t	standard rate of personal income tax and NI contributions
t_c	corporate tax rate
u^*	equilibrium rate of unemployment
U	unemployment
V	net output excluding oil and general government output
$\bar{V}$	autonomous net output (oil and general government)
W	wage rate
Z	supplementary benefit rate
π	profits
δ	rate of depreciation

17.2 THEORY

17.2.1 Aggregate supply

We postulate an economy in which profit-maximizing firms sell their gross output in markets which do not necessarily clear instantaneously. This short-term nonmarket clearing may reflect institutional rigidities in markets for goods and services or it may reflect contracting behaviour which prevents prices from adjusting rapidly and implies the involuntary buildup or depletion of stocks. In the present version of the model firms are price takers; we have been unable to apply successfully the imperfectly competitive framework used for example by Layard and Nickell (1985) in which firms are price makers and in which the level of aggregate demand affects the demand for factors of production by firms. Our specification is therefore quite general and permits different forms of competition and market behaviour according to the data on which the model is estimated. The basic nature of the model thus reflects empirical considerations rather than a priori prejudices.

Firms produce their gross output (Q) with labour (L), capital (K) and inputs of raw materials and energy (M). In the CUBS model we disaggregate these inputs (M) into their component parts, but here they are aggregated together for expositional purposes. To produce their output firms also use the (exogenously determined) social capital stock (K_s) which consists for example of roads whose services are consumed free of charge. Thus the gross output production function is

$$Q = F_1(K, L, M, K_s) \qquad (17.1)$$

where $F'_{1j} > 0$ and $F''_{1j} < 0$. The net output of these firms is

$$V = Q - \frac{MP_m}{P}$$

where

P = gross output price
P_m = price of raw materials

The gross domestic product is defined as

$$GDP = V + \bar{V}$$

where $\bar{V}$ includes autonomous output (as far as the model is concerned) such as the output of general government and North Sea Oil. Firms' profits are equal to

$$\pi = QP - LW(1 + e) - KC - MP_m$$

where

W = wage rate
e = employment taxes
C = user cost of capital

If there are diminishing returns to scale it may be shown that profit maximization implies the following set of factor demand relationships:

$$L^D = F_2(\overset{+}{K}, \overset{-}{W(1+e)/P}, \overset{-}{P_m/P}, \overset{+}{K_s}) \tag{17.2a}$$

$$M^D = F_3(\overset{+}{K}, \overset{-}{W(1+e)/P}, \overset{-}{P_m/P}, \overset{+}{K_s}) \tag{17.2b}$$

$$K^D = F_4(\overset{-}{C/P}, \overset{-}{W(1+e)/P}, \overset{-}{P_m/P}, \overset{+}{K_s}) \tag{17.2c}$$

The signs of partial derivatives are indicated over the variables to which they refer. Some of these are robust with respect to most classes of production function (e.g. that own product wages reduce the demand for labour) while others are less robust (e.g. that raw materials are gross complements to labour). The pattern of partial derivatives, however, reflects the empirical structure of the CUBS model.

From Equations 17.1 and 17.2 it can be inferred that e and K_s directly affect aggregate supply. These are variables that are under the control of the authorities. If employers' National Insurance contributions are raised, employment, the capital stock and the demand for raw materials decline in which case aggregate supply contracts. If the social capital stock is raised the marginal products of labour, capital and raw material are augmented, their factor demands increase and aggregate supply expands.

However, the influence of the public sector upon aggregate supply does not end here. These are merely the direct effects. In the next section we discuss how the public sector indirectly influences the level of wages. Thereafter we consider the influences the public sector has in the capital and product markets. In so far as policy affects W, C and P Equations 17.1 and 17.2 imply that aggregate supply will be affected.

17.2.2 The labour market

A homogeneous labour market is postulated in which firms and the public sector compete for labour. The demand for labour by firms has already been considered in Equation 17.2a. The demand for labour by the public sector (L_g) is assumed to be exogenous. To some extent the cash limits policy implies that L_g is elastic with respect to real wages. However, we overlook this complication in what follows. In the short term public sector pay differs from private sector pay, but in the longer term there is evidence that relative rates of pay between these sectors are stable suggesting that in the longer term the homogeneity assumption is reasonable enough.

In the CUBS model employment is demand determined in the short run while in the long run it is determined by both demand and supply. Aggregate employment is therefore equal to

$$\tilde{L} = L + L_g$$

while unemployment is equal to

$$U = L^s - \tilde{L}$$

where L^s is the working population. In the CUBS model the participation ratio depends upon the disposable real wage, i.e.

$$\frac{L^s}{\text{POP}} = F_5(W(1\overset{+}{-}t)/P_c) \tag{17.3}$$

where

POP = population of working age
t = standard rate of tax plus employee NI contributions
P_c = consumer price index

The equilibrium or 'natural' rate of unemployment depends on the real level of social security benefits (rather than the replacement ratio which was not found to be statistically significant), i.e.

$$u^* = F_6(Z\overset{+}{/}P_c) \tag{17.4}$$

where Z is the level of social security benefits. The equilibrium condition in the labour market is that the effective supply of labour equals the aggregate demand for labour.

$$L^s(1-u^*) = \tilde{L} \tag{17.5}$$

which implies that the equilibrium own product real wage is equal to

$$\frac{W}{P} = F_7(\overset{-}{e}, \overset{+}{t}, \overset{+}{L_g}, \overset{-}{\text{POP}}, Z\overset{+}{/}P_c, \overset{-}{P_c}/P, \overset{-}{K}, P_m\overset{-}{/}P, \overset{+}{K_s}) \tag{17.6}$$

Equation (17.6) implies that aggregate supply is indirectly affected by further policy variables.

(1) Higher tax and employee NI contributions (t) lower the participation ratio. The reduced supply of labour raises the equilibrium real wage which lowers employment via Equation 17.2a. This in turn adversely affects other factor demands which lowers aggregate output through the production function. Whatever the effects of higher tax rates may be upon aggregate demand (in the CUBS model fiscal policy does not affect aggregate demand) the effects on aggregate supply are adverse.

(2) An increase in general government employment (L_g) raises the equilibrium real wage because it raises the demand for labour. The higher own product real wage reduces the demand for labour which contracts firms' output (V). However, GDP as a whole is unlikely to fall because general govenment output rises in proportion to the increase in general government employment. This reflects the way the Central Statistical Office measures the contribution of general government to the output measure of GDP. The output of an economist in HM Treasury (or City University) is assumed, at a given rate of pay, to be the same as his or her output would have been under alternative employment in the private sector. We do not wish to engage here in the debate (e.g. Bacon and Eltis, 1980) about the real contribution of general government employees

to the welfare of the nation and whether the output of general government is output proper. Instead we note that if marketed output falls market prices are likely to be raised for a given aggregate demand schedule.

(3) An increase in the real level of social security benefits (Z/P_c) increases the equilibrium unemployment rate which reduces the supply of labour. This in turn raises the equilibrium real wage which reduces the demand for labour via Equation 17.2a thereby contracting aggregate supply. As in the previous cases, the higher real wage adversely affects firms' demand for capital and raw materials.

(4) Indirect taxes and subsidies will affect the ratio of market prices to factor cost prices. Higher VAT rates, for example, will raise P_c/P either because market prices rise or because factor cost prices fall according to the distribution of tax incidence. Thus higher indirect taxes will lower the real disposable wage thereby exerting upward pressure on wages. At the same time the downward pressure on factor cost prices will raise own product real wages; the demand for labour will fall and aggregate supply will decline.

Thus there are, several macroeconomic policy measures (whose influence on aggregate demand has been the traditional focus of applied macroeconomics) which theory suggests will affect the supply-side of the economy in terms of their implications for the labour market. Another set of policy measures affects the supply-side via the capital market.

17.2.3 The capital market

Firms are assumed to finance their fixed capital on a homogeneous bond market, i.e. debt, debentures and equity are aggregated together. Similarly the public sector finances itself in the homogeneous bond or capital market, i.e. interest-bearing public sector debt is aggregated together. Corporate and government bonds are assumed to be substitutes in the portfolios of wealth holders although no attempt is made to explain the risk premium on corporate capital. In equilibrium the post tax return on fixed assets is hypothesized to be equal to the user cost of capital, i.e.

$$\frac{\pi(1-t_c)}{K} = C$$

This assumes that corporate income is effectively taxed at the mainstream corporate tax rate and that it is not double-taxed, once at the corporate tax rate and again at the personal tax rate when dividends are distributed. The user cost of capital is defined (abstracting from capital gains) as

$$C = (1-A)(R(1-t_c) + \delta)$$

where

A = proportionate present value of capital allowances

R = rate of interest (real)
t_c = corporate tax rate
δ = rate of depreciation

The net demand for bonds by wealth holders is hypothesized to depend upon the post tax real return on savings:

$$B^D = F_8(R(\overset{+}{1}-t), \overset{+}{Y_p}, \overset{-}{R_w}) \qquad (17.7)$$

where Y_p denotes permanent income (proxied by the trend in disposable income) and R_w is the world interest rate. The stock of bonds issued by firms is the counterpart of their fixed assets, hence

$$\tilde{B} = K,$$

while the stock of government bonds (B_g) outstanding is the counterpart of the public sector budget constraint, i.e.

$$\Delta B_g = \text{PSBR} - \Delta M_0 + \Delta \text{RES}$$

where

PSBR = public sector borrowing requirement
M_0 = monetary base
RES = gold and foreign exchange reserves

The equilibrium condition in the bond market is

$$B^D = \tilde{B} + B_g$$

which implies the following equilibrium solution for the user cost of capital

$$C = F_9(\overset{-}{A}, \overset{+}{t}, \overset{+}{K}, \overset{+}{B_g}, \overset{+}{R_w}, \overset{-}{t_c}) \qquad (17.8)$$

According to Equations 17.8, 17.2 and 17.1 macroeconomic policy will affect aggregate supply in several respects.

(1) An increase in capital allowances (A) reduces the cost of capital which in Equation 17.2c raises the capital stock. Other factor demands increase and a higher level of output is implied by the production function. For example the reduction in capital allowances announced in the 1984 Budget is likely to reduce aggregate output while its net effect on employment will be adverse (contrary to the Chancellor's claims). This is because the economy moves onto a lower isoquant and reduces the demand for labour despite a fall in the capital–labour ratio. Just as higher real wages can be bad for investment although they raise the capital–labour ratio (i.e. they are worse for employment than investment) so higher capital costs can be bad for employment (although they are worse for investment than for employment).
(2) Higher personal tax rates reduce the demand to hold wealth. This raises the cost of capital because the resulting excess demand for loanable funds raises

interest rates. The higher cost of capital adversely affects aggregate supply along the lines already considered. Thus higher tax rates have adverse supply-side effects through both labour and capital markets.

(3) An increase in the stock of public sector debt outstanding (B_g) raises interest rates and 'crowds out' the supply of bonds by firms because capital costs are increased. The degree to which this occurs varies inversely with the elasticity of substitution between domestic and foreign bonds. In the limit B_g has no effect on C if domestic and foreign bonds are perfect substitutes. In this case $R = R_w$. Nor will 'crowding out' occur in the capital market if, as Barro (1974) suggests, government bonds do not count as net wealth because they are not backed by real assets. Instead they should be regarded as deferred taxation. This 'superneutrality' hypothesis is not, however, explored here.

17.2.4 Aggregate demand and fiscal policy

Despite numerous attempts we have been unable to detect any effects of fiscal policy upon aggregate demand. This is the case no matter how fiscal policy is measured in terms of tax rates, public expenditure or stocks and flows of public sector debt. Instead the primary domestic determinant of aggregate demand is the real money stock. This appears to suggest that the LM schedule is vertical and that there is complete crowding out in the product markets. On the other hand, we find that aggregate demand varies directly with world trade and inversely with the real exchange rate which is inconsistent with a vertical LM schedule. Alternatively, it might be argued (although we do not) from Barro that fiscal policy is 'superneutral'.

Our model building philosophy is highly empirical. Starting out with a highly general theoretical framework we include empirical effects if we can estimate them and exclude them otherwise even if this implies that the model is not completely consistent. The embarrassment of inconsistencies is a healthier basis for research than papering over the cracks to remove them. Therefore, here as elsewhere in the model, we retain empirical inconsistencies in the hope that we will be able to resolve them with the progress of research.

The efficacy of fiscal policy in other models of the economy may reflect specification errors since on the whole they ignore the supply side. The irrelevance of fiscal policy for aggregate demand in the CUBS model may reflect the highly aggregated nature of the model. Clearly we cannot explain away, at this stage, why fiscal policy is effective in other models but not in our own. In the meanwhile we note that in the simulations reported in Section 17.3 the traditional (but not necessarily valid) effects of fiscal policy on aggregate demand are absent.

17.2.5 General equilibrium

Before proceeding to the empirical analysis in Section 17.3 a simplified overview of the interaction between macroeconomic policy and aggregate supply is presented

Theory 335

in Fig. 17.1. Since there are several arguments in the production function (Equation 17.1), two-dimensional diagrams cannot do full justice to the model that has been outlined. Thus panel IV illustrates the production function in terms of capital and labour only and V_1 is a two-dimensional cross section of a four-dimensional isoquant. The labour market is depicted in panel II. The schedule

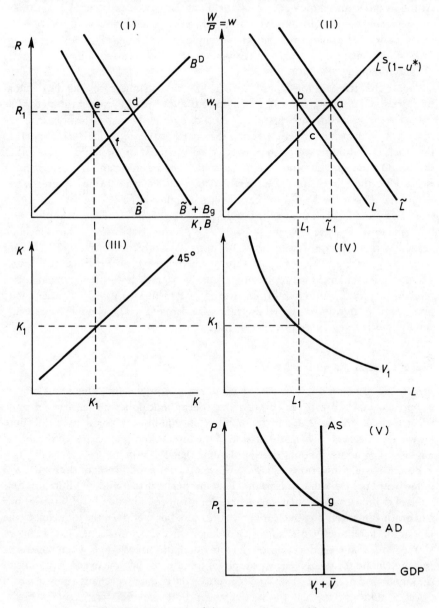

Fig. 17.1 Overview of the economy in equilibrium

marked $L^s(1 - u^*)$ is the aggregate supply schedule for labour while the $\tilde{L}$ schedule represents the aggregate demand for labour. Firms' demand for labour is represented by the L schedule, thus the horizontal distance between the two schedules equals L_g, employment by general government.

The labour market is in equilibrium at a and the equilibrium real wage is w_1. At this wage firms employ L_1, of labour. Had there been no general government employment, the equilibrium would have been at c and firms would have hired more labour. Thus general government employment has partially 'crowded out' firms' employment. The degree of 'crowding out' varies inversely with the elasticity of labour supply.

Panel I characterizes the capital market. The B^D schedule represents the demand for bonds and is drawn for given assumptions about world interest rates etc. The $\tilde{B}$ schedule is derived from Equation 17.2a and represents the stock supply schedule of corporate debt while the horizontal distance between $\tilde{B}$ and $\tilde{B} + B_g$ is equal to the stock of public sector debt. Supply equals demand in the bond market at d, the equilibrium rate of interest is R_1 at which the desired capital stock of firms is K_1. Had there been no public sector debt to hold in portfolios interest rates would have been lower and the capital stock higher at f.

Thus K_1 and L_1 are the equilibrium levels of the capital stock and employment for firms in which case in terms of panel IV V_1 is their net output. The associated vertical aggregate supply schedule is AS in panel V which depicts the market for goods and services. The aggregate supply schedule is vertical because real wages and the real exchange rate are homogeneous to degree zero in the aggregate price level. From panel IV the equilibrium level of GDP is equal to $V_1 + \bar{V}$. The aggregate price level is determined at g where aggregate demand (AD) is equal to aggregate supply and the aggregate price level is P_1.

17.2.6 General disequilibrium

Figure 17.1 gave an overview of the economy in equilibrium. This focus reflects our present concern with the effects of macroeconomic policy on aggregate supply, effects which are more likely to manifest in equilibrium than disequilibrium. However, a necessarily brief discussion of the disequilibrium features of the model may assist in understanding the simulations that follow.

Because of lags in various equations the temporary equilibrium of the model will differ from its long-run equilibrium. This implies that the model will in any case tend, over time, towards its permanent or long-run equilibrium. In addition labour and product markets do not clear in the short run, but they do in the long run. In the labour market money wages adjust according to an expectations augmented Phillips Curve while in the product market the rate of inflation accelerates when there is an excess demand for goods (excess supply of money). Over time excess demands in the labour and product markets tend to zero as the cumulative change in wages and prices is sufficient to clear the labour and product markets respectively. In the simulations that we report the dynamic multipliers reflect the equilibration processes that have been described.

17.3 EMPIRICAL ANALYSIS

To illustrate the empirical order of magnitude of the effects of macroeconomic policy on aggregate supply the CUBS model is simulated. As far as possible the discussion is related to the theoretical effects considered in Section 17.2. However, we also consider several related policy issues that have some political currency in so far as the supply-side of the economy significantly affects the argument.

17.3.1 Model properties

The properties of the CUBS model used in the present exercise have already been described by Beenstock *et al.* (1986). Here we can do no more than summarize the main characteristics of the model. $\bar{V}$ (exogenous output) consists of the output of general government and oil production. In the product markets the long-run level of output is determined through a production function in which M is disaggregated into energy and other raw materials.

Not discussed in Fig. 17.1 is the foreign exchange market in which the exchange rate adjusts through time to balance the supply and demand for foreign exchange. In the long run the exchange rate is determined by a variable purchasing power parity (PPP) rate which rises with the level of oil production and the price of oil (when the UK is a net exporter of oil) and reflects the ratio of the price of traded and nontraded goods at home relative to the same ratio abroad. In the short run the exchange rate deviates from its PPP level but these deviations are not accounted for by interest rate movements and changes in exchange rate expectations. We have therefore been unable to identify any systematic influences of capital account shocks upon the exchange rate. In the long run a higher real exchange rate raises output because it lowers input costs of energy and raw materials. In the short run, however, it lowers output because it temporarily reduces aggregate demand as competitiveness is reduced.

The model was estimated from annual data over the period 1950–1983. Validation exercises and tracking tests are reported by Beenstock *et al.* (1986).

17.3.2 National Insurance contributions

As discussed theoretically in Section 17.2 an increase in employers' rate of National Insurance contributions (e) is the equivalent of a tax on jobs. In Table 17.1 we simulate the effects of a permanent increase in the rate of contribution by 5 percentage points. The simulation is carried out under the assumption that the authorities adhere to an M_0 target and that the higher National Insurance receipts

Table 17.1 Rise in employer National Insurance contributions by 5 percentage points

Year	Employment (%)	Real wage (%)	GDP (%)	Capital stock (%)	Profit rate (%)	Working population (%)
1	−0.9	−0.4	−1.5	−0.2	−3.4	0
2	−2.0	−2.8	−0.9	−0.5	−0.6	−0.5
3	−2.5	−1.7	−1.4	−1.1	−1.3	−0.9
4	−2.6	−1.7	−1.1	−1.1	−0.9	−1.4
5	−2.6	−1.8	−1.4	−1.2	−1.1	−1.6
6	−2.6	−2.1	−1.2	−0.9	−0.8	−1.5
7	−2.6	−2.3	−0.9	−0.4	−0.4	−1.4
8	−2.4	−2.6	−0.6	0.1	−0.2	−1.4
9	−2.1	−2.8	−0.5	0.6	−0.2	−1.3
10	−1.7	−3.1	−0.5	0.8	−0.3	−1.2
15	−0.6	−3.3	−0.6	−0.1	−0.8	−0.4
20	−1.1	−3.7	−0.4	0	0	−0.8

are spent on 'neutral' public expenditure, e.g. current output and transfers such as pensions. These expenditures are neutral in terms of the model because, unlike public expenditure on employment, capital goods and social security benefits, they do not affect the supply-side of the economy. Since fiscal policy does not exert an independent influence upon aggregate demand the assumption about monetary policy insulates the economy for demand-side effects in which case the results in Table 17.1 are essentially supply-side disturbances.

Table 17.1 indicates that the level of employment falls by about 2.6% in the first decade and that real wage rates fall by about 2% over the same period. Allowing for the increase in contribution rates real wage cost to firms will have risen by 5 − 2 = 3% which implies that the general equilibrium real wage elasticity of demand for labour is − 0.9. The incidence of the 'employment tax' is borne by both employers and employees although it is slightly greater in the former case. The adverse effects of increased employment costs on profits initially reduce net investment so that the capital stock declines as does the long-run profit rate. The lower profit rate in turn implies that interest rates fall by about one percentage point in line with the lower demand for loanable funds. However, as might be expected, output becomes less labour intensive and the capital–labour ratio rises since the fall in employment is proportionately greater than the fall in the capital stock. The oscillatory behaviour of the capital stock is induced by the oscillatory behaviour of the real exchange rate (not shown).

The fall in the equilibrium level of employment and capital adversely affects aggregate supply though the production function and GDP falls by about 1.5%. The fall in marketed output (V_0) is greater than this but some of this may by attributed to changes in the equilibrium demand for energy and raw materials by firms. The final column in Table 17.1 implies that the participation ratio falls; the reduction in real wages undermines the propensity to participate.

Table 17.2 Employment effects of the National Insurance reforms of the 1985 Budget (1000s)

	Employer contributions	Employee contributions	Total
1986	2.2	0.4	2.6
1987	3.8	0.6	4.4
1988	4.4	0.9	5.3
1989	4.7	1.5	6.2
1990	5.5	2.1	7.6

A conventional analysis of an increase in National Insurance contribution might imply a temporary deflation in aggregate demand. In contrast our analysis suggest a permanent loss of output with each percentage point of contribution reducing output by 0.2%, which is a relatively large order of magnitude.

We treat employee contributions as an addition to personal taxation the effects of which are considered below. In Table 17.2 we simulate the effects of the National Insurance reforms announced in the Budget of 1985. These reforms had two elements. First, employer contribution rates were lowered for people on wages less than £4860 per annum but the upper earnings limit of £13780 was abolished making it cheaper to hire low paid workers and more expensive to employ higher paid workers. However, the reform was not revenue neutral and the average rate of contribution was lowered slightly, saving employers about £80 million per year.

At the same time employee contributions were lowered for people on low pay such that the average rate of contribution was reduced by $2\frac{1}{2}\%$. The simulated net effects of these reforms on jobs are reported in Table 17.2. The effects are very small with fewer than 10 000 jobs being created after five years. These effects reflect the assumption in the model that the labour market for the high paid operates on the same basis as the labour market for the low paid in which case there is nothing to be gained by altering the way in which the burden of contributions is spread between high and low paid workers. The demand for high paid workers falls by almost as much as the demand for low paid workers increases and the wages of low paid workers rise in about the same proportion in which the wages of high paid workers decline. The effects of the reforms of employee contributions (column 2) are small because the model does not disaggregate labour supply by high and low paid workers. In so far as the 'poverty' and 'unemployment' traps are important in this context the model will underestimate the likely effects of the reforms on labour supply.

17.3.3 Social Security benefits

It is the real level of Social Security benefits rather than the replacement ratio that effects the equilibrium rate of unemployment in the model. In conventional models an increase in benefits can be expected to raise the level of aggregate demand and

Table 17.3 Effects of a 5% rise in real supplementary benefit

Year	Unemployment rate (%)	Real wage (%)	GDP (%)	Profit rate (%)	Price level (%)
1	0.07	0.3	−0.1	−0.2	0.3
2	0.14	0.3	−0.1	−0.2	0.4
3	0.2	0.4	−0.2	−0.2	0.4
4	0.24	0.5	−0.2	−0.2	0.3
5	0.29	0.6	−0.3	−0.3	0.3
6	0.34	0.7	−0.3	−0.3	0.2
7	0.39	0.8	−0.3	−0.3	0.2
8	0.42	0.9	−0.3	−0.3	0.2
9	0.45	0.9	−0.3	−0.3	0.2
10	0.47	1.0	−0.4	−0.3	0.2
20	0.6	1.4	−0.5	−0.4	0.3

output. In contrast the opposite is expected in the context of the present model; the higher real benefits out of employment lower the effective supply of labour which raises real wages and reduces employment and output. These results are reported in Table 17.3 where the simulation is conducted on the same fiscal and monetary assumptions as for Table 17.1 in order to isolate the supply-side effects of a 5% rise in the real level of supplementary and unemployment benefits.

The simulated increase in unemployment (which after 10 years is equal to half a percentage point on the unemployment rate) entirely reflects an increase in equilibrium unemployment rather than involuntary unemployment. Thus by for example year 10 the effective supply of labour has fallen by about 100 000. The cut in the effective supply of labour has the effect of raising the equilibrium real wage by approximately 1% over the same period. These increased costs reduce the demand for labour and adversely effect the profit rate. The capital stock and other factor demands decline but, as theory suggests, output becomes less labour intensive (not shown). The lower employment of capital, labour, energy and raw materials by entrepreneurs adversely affects aggregate supply through the production function and GDP falls permanently by about 0.4%. For a given monetary policy the lower equilibrium level of output implies a higher aggregate price level.

The simulation suggests that the elasticity of GDP with respect to the real benefit rate is −0.1. Increased benefits do not stimulate output through demand-side effects, they reduce it through supply-side effects.

17.3.4 Personal taxation

Unfortunately the model does not distinguish the effects of personal tax allowances, therefore we can only analyse the effects of changing the standard rate of tax. In Table 17.4 we simulate the effects of an increase in the standard rate of tax by 5p

Empirical analysis

Table 17.4 Effects of a 5p increase in the standard rate of income tax

Year	Real wage (%)	Working population (%)	Unemployment rate (%)	GDP (%)	Price level (%)
1	0.2	−0.5	−0.52	−0.1	0.3
2	0	0	0.04	0	0.1
3	0.2	−0.2	−0.09	−0.1	0.1
4	0.3	−0.5	−0.32	−0.1	0.2
5	0.3	−0.6	−0.37	−0.2	0.2
6	0.4	−0.6	−0.29	−0.2	0.2
7	0.5	−0.6	−0.24	−0.2	0.2
8	0.5	−0.6	−0.19	−0.2	0.1
9	0.6	−0.6	−0.15	−0.2	0.1
10	0.6	−0.7	−0.13	−0.2	0.2
20	0.9	−0.7	−0.05	−0.3	0.2

using the same fiscal and monetary assumptions as before.

Higher tax rates reduce the real disposable wage which adversely affects the effective supply of labour. This in turn puts upward pressure on wage rates which reduces the demand for labour. After ten years real wages are 0.6% higher and the working population is 0.7% lower. Output is permanently reduced by 0.2% which suggests that output falls by about 0.04% for every 1p rise in tax rates. This is not a large effect and reflects the fact that the burden of the tax is largely borne by employees whose supply of labour is relatively inelastic.

Higher tax rates reduce unemployment because they lower the number of people looking for jobs. This contrasts with the conventional analysis that higher tax rates depress aggregate demand and increase unemployment. A further contrast relates to the negative balanced budged multiplier that is implied by Table 17.4. The extra tax revenue is spent on public expenditure so that the budget is balanced. The balanced budget multiplier is positive in demand-side models because the marginal propensity to consume is less than one. However, the conventional multiplier ignores the supply-side effects of higher tax rates. When the appropriate corrections are made it turns out that the balanced budget multiplier is negative as suggested for example by Eltis (1980).

As far as the CUBS model is concerned the effects of higher employee rates of National Insurance contributions are similar to increases in direct tax rates. Therefore we do not report any independent simulations of changes in employee contributions.

17.3.5 The Laffer curve

Table 17.4 indicated that higher tax rates adversely affect aggregate supply and the equilibrium level of employment. It might be inferred from this that tax revenue does not increase in proportion with the increase in the standard rate of tax because the real level of economic activity has been reduced by the tax increase. This, of course, is the insight that lies behind the Laffer curve. But in the context of the present simulation exercise this inference is wrong. First of all, although higher marginal personal tax rates reduce the level of employment they increase equilibrium real wages and in terms of the CUBS model the latter effect dominates the former. The personal tax base may therefore rise rather than fall and the Laffer curve for personal taxes may slope the wrong way (i.e. it is concave rather than convex). On the other hand reduced rates of economic activity and profits will reduce corporate and indirect tax revenues which will introduce an element of convexity into the Laffer curve.

Figure 17.2 plots the simulated Laffer curve that is implied by the model when the standard rate of tax is successively increased. The vertical axis measures the percentage change in real tax revenue as a whole (i.e. not just personal taxes). The horizontal axis measures the percentage change in the standard rate of tax (i.e. not tax rates across the economy as a whole). The dotted line indicates the percentage change in tax revenue that would be expected if the real economy was neutral with

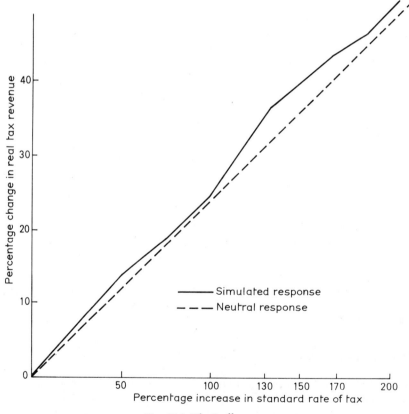

Fig. 17.2 The Laffer curve

respect to the tax rate. Its slope reflects the share of personal taxes in total tax revenue. The schedule indicates the simulated long-run (after 10 years) effects on tax revenues for different percentage increases in the standard rate of tax.

Laffer's thesis implies that the solid schedule should be below the dotted line and that it should be convex. The chart reveals that the opposite applies; the simulated effects lie above the dotted line – a given percentage increase in the standard rate of tax brings about a disproportionate increase in real tax revenue. This reflects the fact that the real wage bill rises when tax rates are raised and that personal income is more tax intensive than corporate income and expenditure. It should be recalled that the model is on the whole loglinear in behavioural equations so that the incentive effects of the very large tax changes in the chart may be understated. Nevertheless the chart serves to emphasize that disincentive effects and tax revenue effects are not as interchangable as the simple Laffer curve suggests.

17.3.6 Lowering the age of retirement

In 1983 the government lowered the effective retirement age for certain men aged between 60 and 65 years. Inevitably this lowered the level of unemployment. More generally it has been argued that lowering the age of retirement might alleviate the unemployment problem by taking people out of the labour market altogether and by making jobs available to the remaining unemployed. In a similar way the extension of the Youth Training Scheme announced in the 1985 Budget has the effect of removing (young) people from the workforce.

Here we simulate the effects of reducing the male retirement age to 60. In 1985 this is equivalent to a reduction in the population of working age of 1.492 millions and in 1995 it is equivalent to a reduction of 1.326 millions. For a given participation ratio it is inevitable that the working population will fall proportionately and for a given volume of employment it is inevitable that those currently unemployed are more likely to obtain a vacated job. In the longer run, however, unemployment will revert to its 'natural' rate. The same does not apply to the equilibrium level of employment since the reduction in labour supply raises the equilibrium real wage which adversely affects labour demand.

But the story does not end here because the increased pension bill has to be financed through higher taxes or National Insurance contributions. Indeed, the model suggests that by the fourth year the standard rate of tax would have to be raised by 4p in order to preserve revenue neutrality. This increase takes into account all the indirect effects on tax revenue induced by the reduction in labour supply. The higher tax rates adversely affect the equilibrium level of employment (see Table 17.5) in which case lowering the age of retirement may reduce employment rather than increase it.

The results of the simulation are reported in Table 17.5. They show, as expected, that initially there is a large drop in the rate of unemployment as the labour force is reduced and as people under 60 move into the vacated jobs left by people over 60. However, real wages rise because of the contraction in the supply of labour and

Table 17.5 Effects of lowering retirement age of men to 60

Year	Unemployment rate (%)	Employment (%)	Real wage (%)	GDP (%)
1	−5.3	−0.4	2.2	−0.9
2	−3.1	−1.0	1.8	−0.8
3	−1.1	−1.4	2.1	−0.9
4	−0.5	−1.6	2.6	−0.9
5	−0.5	−1.9	2.9	−1.2
6	−0.6	−2.3	3.2	−1.4
7	−0.6	−2.7	3.3	−1.2
8	−0.5	−2.9	3.4	−1.2
9	−0.6	−3.0	3.6	−1.3
10	−0.8	−2.8	3.7	−1.5
16	−0.2	−3.3	4.8	−1.7

because the higher tax rates needed to finance the higher pension bill adversely affect labour supply incentives. Because real wages rise the demand for labour falls so that eventually employment falls by about 3%. With fewer people employed (as well as other factors of production) output falls via the production function. The rate of unemployment eventually reverts to its 'natural' rate in which case the incidence of unemployment amongst the under-60s will be unchanged unless equilibrium unemployment rates depend on age.

17.3.7 Simulation of a cut in indirect taxes

In this simulation the rate of VAT is cut from 15% to 10%. The lost revenue is offset as before by reducing 'neutral' public expenditure thus leaving the PSBR unchanged.

In its present form the CUBS model does not take account of the effects of VAT changes on consumer behaviour. It might be expected that consumers would make a switch from savings toward consumption in response to a cut in VAT. This effect is absent in the following simulation.

In the CUBS model there are two aggregate price indices. One is the deflator for GDP at market prices; the other is the deflator for GDP at factor cost. The 'price' equation is set in terms of market prices; i.e. it is market prices which enter the demand for money and hence market prices are the proper determinant in the price equation.

When VAT is cut there is no direct long-term effect on market prices. What happens instead is that the gap between market prices and factor cost prices narrows, so that factor cost prices rise. The reason is as follows: GDP at factor cost at constant 1980 prices (i.e. 'real' output) is not directly affected by VAT changes. But part of the incidence of VAT on prices is absorbed by the producer, so when

VAT falls the full benefit is not passed on to the consumer. Thus current price GDP at factor cost rises and the deflator rises to offset this by a similar amount. Since GDP at current market prices does not change (when GDP at factor cost at current prices rises) then the factor cost adjustment at current prices (effectively taxes on expenditure less subsidies) falls.

The increase in factor cost prices lowers own product real wages as well as the relative user cost of other factors of production. The increased profitability that results triggers an expansion in employment and the demand for factors of production as entrepreneurs take advantage of the higher profit margins. Through the production function aggregate supply is increased which lowers the equilibrium price level expressed in terms of market prices. The results of the simulation are reported in Table 17.6

The impact effect of lower VAT is a fall in market prices by about 1/3%. This effect occurs through a delta term in the price equation. When this influence ends (in the second year) the contributions to changes in market prices come only from higher output and from change in unit wage costs (the effect of which are again only transitory). In the long run only the higher output should influence market prices in a downward direction for a given quantity of money. It is clear, however, that market prices have not settled down to be negative after 16 years although the average over the entire period is negative.

Factor cost prices rise immediately though not by as much as might be expected had market prices not fallen. In the long run factor cost prices rise by around 1%, i.e. the effect of a 5% cut in VAT is shared in the proportions 4% to the consumer

Table 17.6 Effects of a 5% cut in the rate of VAT

Year	Deflator for GDP at factor cost (%)	Deflator for GDP at market prices (%)	GDP at factor cost (%)	Real wage (%)
1	0.83	−0.15	0.04	−0.03
2	0.66	−0.33	0.09	−0.29
3	1.07	0.08	−0.03	−0.11
4	1.1	0.11	0.05	−0.21
5	1.16	0.17	0	−0.18
6	1.03	0.05	0.04	−0.16
7	0.9	−0.09	0.05	−0.13
8	0.77	−0.22	0.07	−0.11
9	0.74	−0.25	0.06	−0.11
10	0.8	−0.19	0.05	−0.13
11	0.91	−0.08	0.04	−0.16
12	1.02	0.03	0.04	−0.18
13	1.1	0.11	0.05	−0.18
14	1.12	0.13	0.05	−0.17
15	1.07	0.08	0.06	−0.15
16	0.99	0	0.06	−0.13

346   *Macroeconomic policy and aggregate supply in the UK*

and 1% to the producer. Since money wages are influenced initially by market prices, the rise in factor cost prices relative to market prices lowers the equilibrium own product real wage which induces a small expansion in aggregate supply.

17.3.8 The effects of SERPS

The Government Actuary estimated that the State Earnings-Related Pension Scheme (SERPS) would raise the rate of National Insurance contributions by 3–4% (over its 1975 level) by 2008–9. More recent studies have questioned the basis of this estimate and have gone on to look at the cost of SERPS over a much longer period. Hemming and Kay (1982) estimate that in the 'long term' (where 'long term' is in excess of 50 years) SERPS will raise the rate of National Insurance contributions by 13.4%.

We have taken the Government Actuary's figure for the period until 2008 and then from this date increased it in steps by the same amount as Hemming and Kay. Thus, by the time SERPS reaches maturity (some time after 2033) the rate of National Insurance contributions is some 10.2% higher (than its 1975 level). By this means we are able to examine the disincentive effects SERPS would have had if it were allowed to continue. These are shown in Table 17.7.

Since National Insurance rates rise gradually over time their effect also rises gradually over time. In the first ten years the level of employment falls by 1/5%. Gradually the effect becomes stronger until by the time SERPS has matured (after 2033) employment is down by 1.6%. The real wage rises correspondingly – by 2033 it is some 1.7% higher. The reduced employment adversely affects factor demands

Table 17.7 Effects of SERPS

Year	Private sector employment (%)	Real wage (%)	GDP (%)
1	−0.01	0.04	−0.014
2	−0.02	0	−0.0
3	−0.02	0.03	−0.01
4	−0.04	0.1	−0.04
5	−0.07	0.08	−0.03
6	−0.09	0.11	−0.05
7	−0.12	0.14	−0.06
8	−0.15	0.16	−0.07
9	−0.18	0.23	−0.09
10	−0.21	0.22	−0.09
15	−0.38	0.39	−0.15
20	−0.57	0.64	−0.24
25	−0.83	0.9	−0.31
30	−1.04	1.09	−0.37
35	−1.15	1.25	−0.45
40	−1.38	1.5	−0.56
45	−1.6	1.72	−0.66

and output falls in terms of the production function. At maturity GDP is 0.66% lower than otherwise would have been the case on account of SERPS.

Most probably the Government Actuary as well as Hemming and Kay have understated the higher National Insurance contributions that are required because they did not allow for any adverse supply-side effects in their calculations. Correspondingly, the results in Table 17.7 most probably are an understatement of the effects of SERPS on the economy.

The simulation abstracts from any effects that SERPS might have upon savings behaviour and capital markets. In so far as SERPS reduces the need to save for pensions the private sector's demand for wealth will be lower and the equilibrium real rate of interest is likely to be higher. Therefore the simulation will tend to understate the adverse effects of SERPS on output.

17.3.9 Raising gas prices by 5%

Gas prices have two direct effects in the CUBS model. Firstly, they affect industrial energy costs, and secondly they affect the revenue of the British Gas Corporation. We estimate that a 5% rise in gas prices would raise overall energy costs by 1.7%. This lowers private sector output via the production function and lowers the demand for capital and labour through the derived factor demand equations. The lower demand for labour lowers the real wage. Unemployment remains unchanged in the long run.

A 5% rise in gas prices would, in 1983/84, have raised the British Gas

Table 17.8 Effects of a 5% rise in gas prices changes

Year	Private sector output (%)	Private sector employment (%)	Private sector capital stock (%)	PSBR (£ million)
1	−0.18	−0.31	0.07	−67
2	−0.43	−0.41	−0.05	132
3	−0.41	−0.38	−0.13	78
4	−0.29	−0.29	−0.13	−88
5	−0.24	−0.19	−0.1	−171
6	−0.23	−0.13	−0.07	−181
7	−0.23	−0.11	−0.05	−181
8	−0.21	−0.12	−0.04	−195
9	−0.21	−0.14	−0.05	−203
10	−0.23	−0.16	−0.06	−181
11	−0.24	−0.18	−0.08	−165
12	−0.25	−0.2	−0.08	−166
13	−0.25	−0.22	−0.08	−181
14	−0.24	−0.22	−0.08	−211
15	−0.25	−0.21	−0.07	−194
16	−0.25	−0.2	−0.06	−239

Corporation's profits by £320 million. We assume that this increase in profits is indexed to the inflation rate. In the very near future the Gas Corporation is to be privatized. We nevertheless assume that this increase in profits finds its way into the Exchequer's coffers. There are several more or less implausible scenarios which could account for these profits accruing to the government. For example, the government might impose a tax on gas sales, the incidence of which fell entirely on the consumer. Alternatively the government might impose a 'windfall' tax. Whatever the story we assume that the PSBR is reduced by £320 million (indexed to the rate of inflation).

The net result of these changes is shown in Table 17.8. Private sector output falls by 0.43% in the short run but settles down to a level 0.25% lower in the long run. Private sector employment follows a similar path ending up at 0.2% lower in the long run. The beneficial effects of the higher gas prices on the PSBR are, to a large extent, offset by lower tax revenues. Therefore when the supply-side is taken into consideration the PSBR is weakened rather than strengthened when British Gas raises its prices.

17.3.10 Social capital stock (raised by 1%)

Social capital in the context of the CUBS model is exclusively the capital stock of roads. The implications for channelling investment into the construction or improvement of the road system is illustrative of the effect of improving the country's infrastructure in general. The magnitude of the effects would, however, be different if the investment were channelled into, say, hospitals or schools.

In this simulation the 'social capital stock' is raised by 1%. The investment this entails is financed through cuts in 'neutral' public expenditure as previously defined. The effect of this policy is simple and straightforward. The 'social capital stock' enters the production function and raises the level of private sector output. Thus in the long run (after about 3-4 years) private sector output increases by about 0.5%. This low elasticity reflects the relatively low share of social capital in the aggregate stock of productive capital.

Notice that we have excluded schools, hospitals etc. from the social capital stock because these items, while of self-evident importance, cannot (directly at least) be considered as arguments in an aggregate production function. We have therefore concentrated our attention on productive social capital. We have also excluded the capital stock of for example British Rail and the GPO. This is because the services of these nationalized companies are paid for at the point of consumption and are not consumed collectively. Elsewhere, Beenstock *et al.* (1984) have attempted to include the user price of the services of public utility companies as an argument in the production function, i.e. higher user costs depress aggregate supply in the same way as higher wage costs depress aggregate supply. Certain estimates suggested that if the relative user cost of public service and utility companies rises by 1% aggregate supply tends to fall by about 0.06%.

17.3.11 Increasing general government employment

In this simulation general government employment is raised by 3% in year 1. Thereafter the level remains unchanged. The extra expenditure this entails is, as before, offset through cuts in 'neutral' public expenditure.

The rise in general government employment shifts the aggregate demand curve for labour to the right and raises the own product real wage. In the public sector this is unimportant (in terms of the simulation but not in general) because the government would simply be more draconian in its other spending cuts. In the private sector however the higher wage rate causes firms to cut back on their demand for labour. Since capital, energy and labour are gross complements, the cutback in private sector employment is accompanied by a cut in the demand for energy and employment. This results in a fall in private sector output (Table 17.9).

By the above effects the increase in general government employment crowds out private sector output and employment. The net effect, however, must be a rise in total employment. The only way in which public sector employment could crowd out private sector employment completely would be if the supply curve for labour were vertical. In the CUBS model the supply of labour is inelastic but not vertical – thus there is significant, but not complete, crowding out.

In terms of total output (GDP) the net effect of the policy is to produce a net increase in GDP as calculated. Since general government output (by definition in the CUBS model) is not sold on any market it is not priced. Thus the 'value added' in the public sector is simply the wage bill. For example, if the government decided to

Table 17.9 Increase of 3% in general government employment

Year	Private sector employment (%)	Total employment (%)	Own product real wage (%)	Private sector output (%)	GDP (%)
1	−0.06	0.67	0.24	−0.1	0.44
2	−0.15	0.58	0.2	−0.08	0.43
3	−0.22	0.51	0.31	−0.1	0.41
4	−0.29	0.45	0.41	−0.13	0.37
5	−0.38	0.39	0.45	−0.2	0.3
6	−0.49	0.3	0.47	−0.23	0.27
7	−0.58	0.23	0.48	−0.22	0.28
8	−0.61	0.2	0.5	−0.21	0.28
9	−0.58	0.21	0.51	−0.23	0.26
10	−0.52	0.25	0.53	−0.26	0.22
11	−0.47	0.29	0.54	−0.27	0.2
12	−0.45	0.31	0.57	−0.27	0.2
13	−0.47	0.29	0.61	−0.26	0.2
14	−0.53	0.26	0.65	−0.27	0.19
15	−0.6	0.21	0.67	−0.28	0.17
16	−0.66	0.16	0.68	−0.3	0.15

double the number of park attendants then public sector output would rise by the amount of the wages paid to those new attendants irrespective of whether the value of the output of parks increases by a like amount. In the private sector, by contrast, the level of output is derived from a production function. The level of GDP rises for the following reason. The fall in private sector employment is less than the rise in general government employment. This would tend, other things being equal, to raise GDP. Furthermore, the level of output in the general government sector rises proportionately with the rise in general government employment, whereas in the private sector the level of output rises less than proportionately with employment. Despite the fact that general government output is a much smaller fraction of GDP than private sector output it is the case that GDP rises.

The simulation results are shown in Table 17.9. In the long run private sector employment falls by around $\frac{1}{2}$% as the result of a 0.6% rise in the own product real wage. Total employment rises by around $\frac{1}{4}$% showing that the crowding out in the private sector is insufficient to offset the rise in general government employment. Private sector output also falls by about $\frac{1}{4}$% whereas GDP rises by around 1/5%. Thus the 3% rise in general government output offsets the 0.25% fall in private sector output.

17.3.12 Changing the balance of public sector finance

In this simulation we consider the effects of changing the balance of financing the PSBR from money to bond finance. Thus the growth rate of M_0 is reduced by 1% per annum and the government's budget constraint is satisfied by selling more interest bearing government debt. The lower rate of growth of M_0 lowers the rate of inflation by a slightly disproportionate amount (homogeneity has not been imposed upon the model) while the increased funding of the PSBR raises real interest rates by about half a percentage point per annum. Initially private investment is crowded

Table 17.10 Effects of open market sales leading to 1% p.a. lower M_0 growth

Year	Inflation (% p.a.)	Real interest rate (% p.a.)	Capital stock (%)	Real exchange rate (%)	GDP (%)
1	−0.5	0.1	0	0	−0.09
2	−0.4	0	−0.02	0.2	−0.15
3	−1.1	0.3	−0.02	0.3	−0.07
4	−1.4	0.3	−0.05	0.6	−0.06
5	−1.6	0.4	−0.06	1.0	−0.03
6	−1.7	0.6	−0.05	1.4	−0.04
7	−1.5	0.5	−0.01	1.5	−0.08
8	−1.3	0.4	0.06	1.3	−0.13
9	−1.2	0.5	0.15	1.0	−0.17
10	−1.0	0.4	0.23	0.7	−0.22
15	−1.5	0.6	0.12	1.1	−0.39

out but the reduced relative rate of inflation raises the real exchange rate through a portfolio balance effect as confidence in sterling is strengthened. The higher real exchange rate lowers input costs of energy and raw materials whose prices are set internationally. These cost reductions boost profits which lead to a stimulus in investment. This effect begins to predominate by year 8. Indeed, in the longer run this latter effect dominates the former effect so that increased funding does not adversely affect investment.

Despite this long-run increase in the capital stock, GDP falls. The short-run fall reflects the rise in the real exchange rate and the fall in real money balances which are demand-side influences. The long-run decline reflects supply-side influences which occur despite the rise in the capital stock. In terms of the model the 'natural' rate of unemployment varies inversely with the rate of inflation. This effect, while incompatible with Equation 17.6, was clearly suggested by the data. It implies that workers are more risk averse in their wage demands when inflation is high than when inflation is low. Therefore, lower inflation raises the natural rate of unemployment (by $\frac{1}{2}\%$), thereby reducing the supply of labour and raising real wages. So although the capital stock rises, the level of employment falls; the latter effect dominates the former and GDP declines.

17.4 CONCLUSIONS

We make no attempt to summarize all the simulations that we have carried out. Instead we limit ourselves to points of a more general nature. First and foremost we wish to emphasize the importance of supply alongside demand and models which abstract from the supply-side are surely misspecified, especially in the longer term. If in the longer term output is constrained by the production function it should go without saying that it is crucial to estimate the effects of macroeconomic policy on aggregate supply.

Secondly, our estimates reflect the novelty of this line of research. In particular, estimates of numerical orders of magnitude of policy changes may alter when in future work we disaggregate marketed output into manufactures and services and when expectations are made consistent or rational.

Finally, our approach suggests that economic policy affects the macroeconomy in more ways than is perhaps traditionally supposed. In addition to fiscal and monetary policies it turns out that energy and social security policies are likely to affect the state of the economy. In due course it will most probably be necessary to add housing policy and other policy areas where the supply-side is affected.

REFERENCES

Bacon, R. and Eltis, W. (1980) *Britain's Economic Problem: Too Few Producers*, 2nd edn, Macmillan, London.

Barro, R. (1974) Are government bonds net wealth? *Journal of Political Economy*, **82** (6), 1095–117.

Beenstock, M., Dalziel, A. Lewington, P. and Warburton, P. (1986) A macroeconometric model of aggregate supply and demand for the UK. *Economic Modelling*, **3** (October), 242–68.

Beenstock, M., Dalziel, A. and Warburton, P. (1984) Aggregate investment and output in the UK. *Recherches Economiques de Louvain*, **50**.

Eltis, W., (1980) The need to cut public expenditure and taxation, in *Is Monetarism Enough?*, Institute for Economic Affairs, Reading, p. 24.

Hemming, R. and Kay, J. (1982) The cost of the state earnings related pension scheme. *Economic Journal*, **92** (June), 300–19.

Layard, P. R. and Nickell, S. J. (1985) The causes of UK unemployment. *National Institute Economic Review*, **111** (February), 62–85.

18

Two recent trends combined in an econometric model for the Netherlands: the supply-side and sectoral approach

JOHAN P. VERBRUGGEN

18.1 INTRODUCTION

Distinctive features of the Dutch economy are openness and a relatively large public sector. Compared with an average total government outlay for the OECD countries of just over 40% of GDP in 1983, the Dutch ratio was considerably more than 60%. The openness of the Dutch economy is clear from the fact that exports and imports accounted for 44% and 40% respectively of total sales in 1985. These features are closely related to a growing interest in the modelling of supply-side elements and sectoral aspects of the economy: two recent trends in economic modelling in the Netherlands, which are central to this chapter.

The supply-side trend stems from a growing interest in supply-side economics once it had become fashionable in the United States. Unfortunately, there is no consensus on what this economic trend stands for, which makes empirical testing difficult.[1] To some, supply-side economics means renewed interest in the supply or production side of the economy, which in the past decades had been upstaged by Keynesian influences. This broad definition does not warrant referring to a recent trend in economic modelling in the Netherlands, since the supply block based on the clay–clay vintage approach has for over ten years been a vital element in the models used for policy-making in the Netherlands. In our view, the recent interest in the modelling of supply-side elements must be connected with a narrower interpretation of supply-side economics, the main point of which is that aggregate output, employment and unemployment depend predominantly on microeconomic factors determining aggregate supply. On this interpretation the supply-

[1] Van Duijn (1982) distinguishes eight different views of what supply-side economics stands for.

determination of the economy is rather the essence of the (neo) classical theory and for the supply-siders 'in a narrow sense' it is one of the main axioms.

The supply-side elements to be modelled are linked with a number of constantly recurrent themes, such as the influence of the public sector's size on economic growth, the public sector burden and incentives, (de)regulation, trade unions, the black/informal economy and innovation. In this chapter, an attempt will be made to model some of these elements within an existing sector model,[2] starting from the view that, if the often microeconomically-orientated supply-side elements operate in the economy systematically and to an appreciable extent, they should be capable of being made visible in some form in econometric models. It is emphasized that ours is a tentative effort which still needs theoretical and empirical elaboration.

The second trend is the development of sector models (see, for example, den Hartog and Tjan (1979); Driehuis and van den Noord (1980); Kuipers *et al.* (1980); Muller *et al.* (1980); van Schaik (1980); van Zon (1985); Kuipers *et al.* (1985)). Its underlying reason is that analysis of medium- and long-term economic developments is hardly feasible unless differences and interrelations between sectors are taken into account. The relative openness of the Dutch economy plays a major part in this. Generally speaking, this openness makes it possible to divide the private sector into two subsectors, a capital-intensive tradable one and a labour-intensive nontradable one. An example of a recent sector model for the Netherlands is the VICTOR model (see Verbruggen, 1985a), which serves as a basis for our empirical research into the modelling of some supply-side elements. This gives rise to a combination of the two recent trends in econometric modelling in the Netherlands.

The organization of this chapter is as follows. Before discussing the modelling of supply-side elements in general in Section 18.3, a general outline of the VICTOR model is presented in Section 18.2. Section 18.4 presents the empirical results, while Section 18.5 examines – with the aid of some simulations – the significance of the modelled supply-side elements for economic policy. The final Section 18.6 is a summary.

18.2 THE VICTOR MODEL

There may be various reasons for selecting an aggregation level below that of the private sector as a whole. One reason is that there are great structural differences between sectors on the demand and supply sides. Another major reason is that the composition of the macroeconomic aggregates does not remain the same over time (see Kmenta and Ramsey, 1981). The two reasons may be said to be mutually

---

[2] Various efforts to quantify certain supply-side elements have been made at a micro level in the past; see Brown and Jackson (1980) and OECD (1985) for a survey. For some recent attempts, see Blomquist (1983), Hansson (1984) and Stuart (1984). The larger part of the empirical research to date has been into the relationship between taxation and labour supply.

reinforcing. Therefore it is the more desirable to disaggregate if sectors, besides being structurally different, also have different shares of the aggregates over time. As both kinds of difference are found in the Dutch economy, a 'one-sector' macroeconomic model is not in principle the most suitable for analysing longer-term economic developments.

The VICTOR model divides the private sector into four: the capital-intensive sector, the service sector, the building industry and a rest-of-the-economy sector. The first can be regarded as a tradable sector: about 50% of its total sales are exported. It is made up of manufacturing, transport, storage and communication firms. The building industry and the various industries making up the service sector are labour intensive and serve mainly the home market (the nontradable sector). The service and building industries export about 12% and 3% of their total sales respectively. The rest-of-the-economy sector comprises farming, public utilities, mining (including gas and oil production) and home-ownership.

This subdivision strikes a balance between what is theoretically desirable and practically feasible. We have opted for an arrangement in which the industries in each of the first three sectors are reasonably congruent as to production and marketing structures. The fourth sector just comprises industries whose specific structures disqualify them for inclusion in any of the other three. The behavioural equations for this sector have not been estimated but are included in the model for the time being with the aid of simple rules of thumb.

It goes without saying that, constructing the VICTOR model, a fair amount of time and attention has been devoted to sectoral equations. The equations in the macro part follow those of existing macro models for the Netherlands.[3] In conformity with these models, the VICTOR model contains an exogenous exchange rate. The VICTOR model has four blocks:

(1) The first block is made up of the sectoral behavioural equations. It contains equations for each sector for employment, wage rates, investment, exports, imports, sales of consumer goods and services, domestic sales prices, export prices and production capacities.
(2) The sectoral equations of the second block are definitions, concerning production, domestic sales, total sales, labour productivity, production prices, total sales prices, nonlabour income and utilization rates.
(3) The third block contains the equations of the monetary sector.
(4) The fourth and final block covers the public sector and contains some remaining macro equations.

Compared with other sector models for the Netherlands (see, for example, den Hartog and Tjan (1979); Driehuis and van den Noord (1980); Kuipers et al. (1980); Muller et al. (1980); van Schaik (1980); van Zon (1985); Kuipers et al. (1985)), the number of variables for which sectoral behavioural equations are specified in the

---

[3] The monetary part has been taken over from Buitelaar (1985). For the public sector and some other macro equations, such as those for labour supply and depreciation by firms, we have followed the Central Planning Bureau's (1983a) most recent medium-term yearly model, called FREIA.

18.3 MODELLING SUPPLY-SIDE ELEMENTS

When modelling supply-side elements, one must distinguish between modelling the supply or production side of the economy and modelling the predominantly microeconomic factors determining aggregate supply, which supply-siders stress. As for the former (concerning production functions, equations for investment, capital stock and capacity) Dutch econometricians have lots of experience, based so far in the main on the clay–clay vintage approach. See, for example, den Hartog *et al.* (1975); Central Planning Bureau (1977), (1983a), (1983b) and (1985). For an overview of developments of the macroeconomic models of the Central Planning Bureau, see den Hartog (1983). This chapter, however, deals with the latter factors, with which there is hardly any experience, whether in the Netherlands or in any other country.[4] Comparatively little research has been done on the macroeconomic effects of the measures proposed by supply-side economists. There are some reasons for this, which, incidentally, make the results of our empirical research less robust than they seem.

First, the measures proposed by supply-siders often seek to reverse trends, given the factors that in most Western economies the burdens of taxes and social security contributions have increased, regulation has been intensified and labour markets have become more rigid over the past decades (see van Duijn, 1982).

Second, the proposed measures have their effects mainly in the microeconomic sphere, so that they by definition hardly lend themselves to macroeconomic analysis. In empirically verifying some supply-side elements, we have experimented with variables from the macroeconomic sphere (especially the public sector burden) which function as proxies for some underlying, predominantly microeconomic factors.[5] This approach has the advantage of making it possible to endogenize these proxy variables in the model, which is helpful for simulations. A drawback is, however, that the relationships between the underlying factors and the proxy variables are not distinct, so that, in interpreting the results, the underlying assumptions must not be neglected.[6]

Third, the factors stressed by supply-siders, such as labour market flexibility, regulation and the unobserved sector, are as a rule hard to quantify and consequently hard to analyse.

---

[4] This does not hold true for all supply-side elements. Experience has been built up in the Netherlands with the shifting forward of employees' taxes and social security contributions in wage costs. See Central Planning Bureau (1971; 1977; 1983a); Driehuis (1972) and Knoester and van der Windt (1985).
[5] The same approach has been adopted by Rutten (1985) and Buitelaar (1985).
[6] One of the assumptions, for instance, says that increases in the public sector burden are accompanied by more regulation. Although this has generally been so in the past, the relationship between the two is neither unambiguous nor causal.

This chapter nevertheless attempts to model some supply-side elements, although it will be clear from the foregoing that it is doing so tentatively and merely to elicit further study. Besides, this chapter does not seek to be exhaustive; it has just experimented with some supply-side elements in the sectoral behavioural equations for investment, production capacity, exports, wage rates and consumption.[7]

18.4 EMPIRICAL RESULTS

18.4.1 Investment

One of the supply-side issues is the relationship between the size of the public sector and structural economic growth. As the latter is determined, in the supply-siders' view, by the growth of aggregate supply in which investment plays a crucial part, the relationship between the public sector and investment is a major point of the issue in question.

A large public sector means as a rule a great deal of government intervention in the economy. Regulations and statutes on minimum wages, dismissal, environmental pollution, physical planning, etc. limit the scope of action for entrepreneurs. As a result, industry sees its badly needed flexibility curbed and may decide to omit or reduce investment or invest abroad. Besides, a country with a large public sector will as a rule have high taxes, social security contributions and compensatory benefits, which may have bad effects on the intensity and quality of work, on people's readiness to accept jobs and on human capital investment. These circumstances too, may make firms less inclined to invest.

From this angle, the relationship between the size of the public sector and corporate investment is a negative one, although two remarks should be made. First, some public expenditure, e.g. to eliminate constraints on economic growth concerning infrastructure, education and technology, might boost business investment. Hence it is not only the size but also the composition of the public expenditure that is important. Second, regulation may imply extra investment by firms (especially capital-intensive ones) to comply with imposed criteria as to, say, environmental protection and the safety of workers and consumers.[8] All in all, the relationships between the size of the public sector and investment are not exclusively negative.

As proxy for the size of the public sector we have opted for the public sector burden variable 'dr'.[9] Other proxy variables are possible, though. We have assumed

---

[7] International literature pays a great deal of attention to the impact of taxes and social security contributions on labour supply (see footnote 2). In the Netherlands with its relatively high labour force growth this effect is less prominent at the moment.

[8] On the figures supplied by Evans (1981) it is calculated that US investment to meet pollution, health and safety standards averaged about 5% of total fixed business investment in the 1970s.

[9] This variable has been compiled on the basis of the burden of employees' wage taxes and social security contributions on disposable earned incomes and the burden of the employers' social security contributions on gross wages.

that the level of the public sector burden affects the investment level, or changes in capital stock. Although the percentage changes in investment are explained, changes in the public sector burden appear as an explanatory variable. Moreover, it seems likely that the effect of an increase in the burden is more unfavourable when the level of the public sector burden is already high. This hypothesis is not disproved by empirical testing, since the change in the public sector burden yields more significant results if level-weighted than it does without weighting.

A detailed description of the specification of the investment equation would not fit the scope of this chapter, so we confine ourselves to the basic specification:[10]

$$\dot{i} = \left(\frac{k}{i}\right)_{-1} [\lambda(\dot{k}^* - \dot{k}_{-1}) + \delta \dot{k}_{-1}].$$

Desirable capital stock growth ($\dot{k}^*$) is determined not only by the burden variable but also by the following factors: expected production growth (represented by the production growth trend $\dot{y}_t$); changes in the utilization rate (Δq) and profit development deflated by the price of investment goods ($\dot{Z} - \dot{p}_{iet}$). In addition, two monetary variables have been included, namely, changes in real interest rates ($r_1 - \dot{p}_y$) and a monetary tension variable ($\dot{Bm} - \dot{V}$). The latter term is based on the assumption that economic subjects, in this case entrepreneurs, respond to changes in monetary tension not only by price adjustments but also by nonprice volume adjustments.[11] A monetary alleviation ($\dot{Bm} > \dot{V}$) or tightening ($\dot{Bm} < \dot{V}$) will therefore lead to extra or less spending, in this case investment.

Since for the capital-intensive sector we have based ourselves on the clay–clay vintage approach, Equation 18.1 includes a real wage cost variable ($\dot{p}_1 - \dot{p}_y$), which represents the investment growth resulting from a (partial) replacement of the capital goods of the oldest vintage(s) scrapped for economic reasons. See, for example, den Hartog, van de Klundert and Tjan (1975) and de Ridder (1977). Since the capital-related approach is felt to be inappropriate for the service sector, Equation 18.2 contains no real wage cost variable. Finally, both equations contain a variable (is) in connection with the government's policy to stimulate investment through subsidization.[12] The found effectiveness of investment facilities can be termed fairly high in comparison with other research for the Netherlands. The investment equations are as follows.[13]

---

[10] For a detailed theoretical underpinning of the investment equation, see van Riet (1987).
[11] In the literature this is called the buffer mechanism. See, for example, Knoester (1984).
[12] The investment subsidization scheme was launched in 1978 as a substitute for tax breaks consisting of accelerated depreciation and investment allowances; see van Sinderen (1985). For an overview of the findings of various studies on the effectiveness of investment subsidization in the Netherlands, see van Sinderen and Verbruggen (1986).
[13] The estimation equations in this chapter have been estimated by the ordinary least squares (OLS) method. For each equation the unadjusted squared correlation coefficient (R^2), adjusted standard error of estimation (Se), the Durbin–Watson statistic (DW) and the estimation period (EP) are given. Underneath the estimation coefficients the t-values are shown. For the meanings of the letters and symbols used, see the list of symbols. It has proved impossible to estimate a reasonable equation for the building industry.

$$\dot{i}^c_{et} = 4.47\Delta\dot{y}^c_t + 1.0\dot{y}^c_{t-1} + 1.61\Delta q^c_{-5/4} + 0.82(\dot{Z}^c_{-1} - \dot{p}_{iet}) +$$
$$\phantom{\dot{i}^c_{et} = } (3.04) \quad\quad (-) \quad\quad (1.55) \quad\quad\quad (5.73)$$
$$+ 0.91[(Bm\dot{/}V) - (Bm\dot{/}V)_{111}]_{-1} - 2.27\Delta(r_1 - \dot{p}^c_{y532})_{-2} +$$
$$ (2.89) \quad\quad\quad\quad\quad\quad\quad\quad (-2.67)$$
$$+ 3.09\Delta is_{-1\frac{1}{2}} + 3.75\Delta(\dot{p}^c_1 - \dot{p}^c_y)_{s-1} - 3.99[dr(\Delta dr)]_{-\frac{1}{2}} \quad\quad (18.1)$$
$$ (3.68) \quad\quad\quad (2.04) \quad\quad\quad\quad\quad (-2.73)$$

$R^2 = 0.82$ $Se = 6.18$ $DW = 2.27$ $EP = 54\text{–}82$

$$\dot{i}^s_{et} = 11.79\Delta\dot{y}^s_t + 1.0\dot{y}^s_{t-1} + 1.35\Delta q^s_{-1} + 0.46(\dot{Z}^s_{-1} - \dot{p}_{iet}) +$$
$$\phantom{\dot{i}^s_{et} = } (5.84) \quad\quad (-) \quad\quad (1.64) \quad\quad\quad (2.32)$$
$$+ 1.18[(Bm\dot{/}V) - (Bm\dot{/}V)_{111}]_{-3/4} - 1.99\Delta(r_1 - \dot{p}^s_y)_{-2/3} +$$
$$ (3.30) \quad\quad\quad\quad\quad\quad\quad\quad (-2.06)$$
$$+ 2.53\Delta is_{-1\frac{1}{2}} - 6.18[dr(\Delta r)]_{1111} + 6.60 \quad\quad (18.2)$$
$$ (2.50) \quad\quad (-2.16) \quad\quad\quad (2.40)$$

$R^2 = 0.81$ $Se = 7.78$ $DW = 2.50$ $EP = 54\text{–}82$

Given the said reservations as to the empirical results, it will be clear that the significant contributions made by the supply-side variables to the explanations of investment illustrate and suggest, but do not prove an unfavourable effect of an increase in the public sector size on investment. It is noted that, on its own, the coefficient of the burden variable is smaller for the capital-intensive sector than for the service sector. In the aforementioned basic specification, however, it is found that the variables determining desirable capital-stock growth are weighted with the factor $\lambda(k/i)_{-1}$, which is much higher for the service sector than for the capital-intensive sector. If this is taken into account, the effect of the public sector size on desirable capital stock growth ($\dot{k}^*$) can be calculated and is then found to be smaller for the service sector. Attribution of Equations 18.1 and 18.2 shows that the unfavourable effects of the growing public sector on investment have been great. It may serve as an indication that, on the estimates presented here, investments in the capital-intensive and service sectors lost 2.7% and 4.2% respectively per annum on average over the estimation period, owing to the growth of the public sector. An outcome which looks rather high at first sight.

18.4.2 Production capacity

As mentioned before, supply-siders hold that the relationship between public sector and structural economic growth runs via aggregate supply or corporate production capacity.[14] The latter is determined, generally speaking, by the size of the capital

---

[14] The term 'corporate production capacity' stands for the maximum output achievable with a normal utilization of the means of production available to firms. Since directly observable production capacity figures (per sector) are lacking, the yearly size of this variable has been fixed by the peak-to-peak line method.

stock (see Section 18.4.1) and employment on the one hand and by capital and labour productivity on the other.

According to the clay–clay vintage approach, which is deemed to be applicable to the capital-intensive sector, production capacity is determined in principle by the size and composition of the capital stock, which is considered to be built up of vintages with embodied labour-augmenting technological progress, increasing from year to year. Capital productivity is in many cases assumed to be constant, whereas structural labour productivity is determined for each vintage by the (embodied and disembodied) labour-saving technological development and effective working time. See, for example, Central Planning Bureau (1983a). Conceivably, labour productivity may be affected also by the size of the public sector, since a large public sector means as a rule high (marginal) tax/transfer rates, which may have a bad effect on the intensity and quality of the labour performances. See, for example, Lindbeck (1981). From this point of view labour productivity and therefore production capacity is determined not only by the embodied technology and effective working time but also by the motivation and attitudes of the employees. In conformity with the disembodied technological progress, the effects of this supply-side phenomenon do not differ from vintage to vintage. As proxy for this supply-side element we have opted for the public sector burden on employees (dr^{wk}), since we are here concerned in particular with worker application. There remains the question of the dimension of this burden variable or whether labour productivity is influenced by the level of, or changes in, dr^{wk}. Since neither effect can be ruled out beforehand, we have estimated with both dr^{wk} and Δdr^{wk}.

Besides the burden variable(s) the linearized capacity equation for the capital-intensive sector includes the standard terms of the clay–clay vintage approach.[15] These concern the investment ratio (i/y), the real wage cost variable in connection with the economic scrapping of vintages $(\dot{p}_1 - \dot{p}_y)$ and working time $(\dot{w}t)$.

$$\begin{aligned}\dot{cap}^c = & \; 0.31\,(i/y)^c_{1111} - \; 0.21\,(\dot{p}_1 - \dot{p}_y)^c_{111} + 0.18\dot{w}t^c_{-1/2} + \\ & (10.61) \qquad\qquad (-2.19) \qquad\qquad\qquad (2.58) \\ & -0.23\Delta dr^{wk}_{532} - \; 0.12 \\ & (-2.17) \qquad\quad (-0.22)\end{aligned} \qquad (18.3)$$

$$R^2 = 0.88 \qquad Se = 0.37 \qquad DW = 1.61 \qquad EP = 56\text{–}82$$

In the estimates only Δdr^{wk} got a significant coefficient. The relationship between public sector size and production capacity in the capital-intensive sector runs all in all via three channels. Apart from the direct effect (Δdr^{wk}) there is an indirect effect through investment (see Section 18.4.1) and wage costs (see Section 18.4.4).

The clay–clay vintage approach is deemed inappropriate to the labour-intensive service sector and building industry, which are assumed to have rather more 'putty–putty-like' production structures. Production capacity growth depends on the volume of the available means of production via the investment ratio (i/y) and

---

[15] For the derivation of the linear capacity equation on the basis of the clay–clay vintage approach, see Knoester and van Sinderen (1984).

employment growth ($\dot{ab}$) on the one hand, and labour productivity ($\dot{h}$) on the other. This last factor is determined by the structural development of productivity ($\dot{h}_t$), represented by the average real productivity growth in the last four years, and the aforedescribed incentive effect. We have again estimated with the level of, and changes in, the public sector burden on employees (dr^{wk}).

The estimates show that, as contrasted with the capital-intensive sector, in the service and building industries the public sector burden level is found to be the main influence. If the constant term is seen in full as an adjustment to this level variable, it is found that the public sector burden has depressed production capacity in the building industry since the early 1960s and in the service sector only since the first oil shock.

$$\dot{cap}^s = \underset{(5.07)}{0.20(i/y)^s_{-1}} + \underset{(2.48)}{0.28\dot{ab}^s_{-1/4}} + \underset{(8.71)}{0.72\dot{h}^s_t} +$$
$$+ \underset{(1.91)}{0.16\dot{wt}_{-t}} - \underset{(-3.93)}{0.07dr^{wk}_{1111}} + \underset{(3.26)}{2.30} \qquad (18.4)$$

$$R^2 = 0.93 \quad Se = 0.32 \quad DW = 1.74 \quad EP = 60\text{–}82$$

$$\dot{cap}^b = \underset{(3.76)}{0.73(i/y)^b} + \underset{(3.11)}{0.16\dot{ab}^b} + \underset{(3.71)}{0.25\dot{h}^b_t} +$$
$$- \underset{(-6.00)}{0.20dr^{wk}_{1111}} + \underset{(4.21)}{4.23} \qquad (18.5)$$

$$R^2 = 0.90 \quad Se = 0.68 \quad DW = 1.10 \quad EP = 60\text{–}82$$

The different dimensions of the supply-side variables in the capacity equations indicate that further research is needed on these equations in particular.

18.4.3 Exports of goods and services

The capital-intensive sector, accounting for more than 70% of the Netherlands' total exports, is by far the most important sector for the country's export trade. Export performance in the service sector (mainly trade) depends strongly on that of the capital-intensive sector, therefore in the model it is assumed that both export developments (in constant prices) are the same. Exports by the building industry are negligible and exogenous in the model. Hence only the export performance of the capital-intensive sector will be discussed in the following.

So far, modelling of exports has leaned heavily on the notion that exports are mainly determined by demand. Exports often depend on foreign demand for goods and services (world imports) and relative prices. In addition, models for the Netherlands have from the beginning of the 1960s included sometimes the so-called 'home-pressure-of-demand effect', presented by changes in utilization rate (Δq). As the name makes clear, this cyclical effect was considered mainly from the demand side. Yet, since both production and production capacity play a role in determining utilization rates, the effect can be interpreted likewise from the supply-side angle.

Evans (1981) clarified the supply-side character of the term as follows: '... an increase in the gap between actual and maximum potential GNP raises exports, since the greater capacity of the US economy permits, the production of more goods and services for export markets as well'.

Besides this more or less cyclical supply-side element production capacity, or supply, can also have a more structural effect on exports, as two aspects may show. First, a large portion of the exports of the capital-intensive sector is accounted for by multinationals. When in a small, open economy like the Netherlands' a multinational expands production capacity it is largely assured of having an international outlet for the extra output. Second, capacity expansion in the capital-intensive sector will as a rule mean innovative investment which enhances the quality and technological sophistication of the products and also their international marketability.[16] To represent this supply-side effect, capacity growth in the capital-intensive sector ($c\dot{a}p^c$) has been included.

$$\dot{e}^c = \underset{(8.52)}{0.95\dot{m}_w} - \underset{(-3.39)}{1.42(\dot{p}_e - \dot{p}'_e)_{532}} - \underset{(-2.41)}{0.57\Delta q^c_{-1\frac{1}{2}}} + \underset{(2.10)}{0.55 c\dot{a}p^c_{-1/2}} - \underset{(-2.78)}{3.55} \quad (18.6)$$

$R^2 = 0.87 \quad Se = 2.24 \quad DW = 2.06 \quad EP = 56-82$

Exports by the capital-intensive sector are on the whole explained by a specific demand factor ($\dot{m}_w$), a specific supply factor ($c\dot{a}p^c$) and by two terms depending on both demand and supply factors.

18.4.4 Wage rate

A more-or-less familiar supply-side element is the public sector burden in the wage equation, concerning the so-called forward shifting of higher taxes and social security contributions from employees to employers in order to protect disposable earned income.

Various studies have shown that direct taxes and social security contributions have been shifted forward on a relatively large scale in the Netherlands in the past. See Knoester (1983) and Knoester and van der Windt (1985). Another aspect of wage determination in the Netherlands is that the outcome of bargaining in some capital-intensive industries forms a major guideline for wage negotiations in other branches. Hence the 'wage-leader hypothesis' has been tested and was confirmed for the commercial service sector, that is to say that explanation of gross wages in this sector by those in the capital-intensive sector gave some better results than a complete explanation with the relevant variables for this part of the service sector.[17]

---

[16] Similar supply-side-orientated arguments can be found in Draper (1985) and Rutten (1986). The exports equation in the most recent macroeconomic model of the Central Planning Bureau (1985) also includes a supply-side variable, namely, the investment ratio in the Netherlands in respect of the average investment ratio in the OECD.

[17] Although one may also argue on statistical grounds that the commercial service sector is leading instead of following the capital-intensive sector or that pay settlements are determined on the macro

Empirical results

In the building industry, though, there was no question of wages following those paid in the capital-intensive sector.

Besides the forward-shifting term (Δdr^{wk}), the wage equations include variables for price compensation ($\dot{p}_c$), labour productivity ($\dot{h}$), Phillips curve mechanism ($1/u$) and social security contributions paid by the employers (Δspw). Where employers manage to shift part of an increase in their social security contributions on to the workers this is expressed in a coefficient which is less than one for the variable in question. Where employees manage to shift part of the increase in their direct taxes and social security contributions on to the employers the result is a coefficient larger than zero for the forward-shifting variable. In view of the specific definition of this term forward-shifting is complete when the coefficient is about 1.2.

$$\dot{p}_1^c = 1.0 \dot{p}_{c-1/2} + 1.05 \dot{h}_{111} + 3.18(1/u) + 0.41 \Delta dr^{wk} +$$
$$(—) \quad\quad (6.19) \quad\quad (3.79) \quad\quad (2.10)$$

$$+ 0.81 \left(\frac{\Delta spw}{1 + 0.01\, spw_{-1}} \right)^c - 1.54 \tag{18.7}$$
$$(2.35) \quad\quad\quad\quad\quad\quad (-2.46)$$

$$R^2 = 0.90 \quad Se = 1.52 \quad DW = 2.09 \quad EP = 56\text{--}82$$

$$\dot{p}_1^{cs} - \left(\frac{\Delta spw}{1 + 0.01 spw_{-1}} \right)^s = 1.01 \left[\dot{p}_1^c - \left(\frac{\Delta spw}{1 + 0.01 spw_{-1}} \right)^c \right] - 0.54 \tag{18.8}$$
$$(15.30) \quad\quad\quad\quad\quad\quad\quad\quad (-0.82)$$

$$R^2 = 0.90 \quad Se = 1.17 \quad DW = 1.74 \quad EP = 56\text{--}82$$

$$\dot{p}_1^b = 1.0 \dot{p}_{c-1/2} + 0.58 \dot{h}_{532} + 2.01(1/u)_{-1}^b + 1.09 \Delta dr^{wk}$$
$$(—) \quad\quad (3.43) \quad\quad (2.34) \quad\quad (2.88)$$

$$+ 0.70 \left(\frac{\Delta spw}{1 + 0.01 spw_{-1}} \right)^b \tag{18.9}$$
$$(2.13)$$

$$R^2 = 0.80 \quad Se = 2.16 \quad DW = 2.72 \quad EP = 56\text{--}82$$

These equations make it clear that the forward shifting from workers to employers is nearly complete in the building industry. In the capital-intensive sector, and consequently in the (commercial) service sector too, about 35% of an increase in direct taxes and social security contributions is shifted forward. In this way public sector growth, accompanied by an increase in taxes and social security contributions, can lead to a wage-cost increase that may in turn result in a decline in production capacity (in the capital-intensive sector) through economic scrapping of

---

level between the employers' and trade union federations (the determining factors in Equation 18.7 being macro variables with the exception of ($\Delta spw/1 - spw_{-1}$)), these claims are untenable, considering experience with wage bargaining in the past. See van Drimmelen and van Hulst (1981).

capital goods and in a growing black/informal economy (see Section 18.4.5). Moreover, such a wage-cost rise may also curtail production capacity through falling investment as a result of shrinking profitability.

18.4.5 Private consumption

The last supply-side issue considered in this chapter in its tentative modelling of supply-side elements is the black economy or, on a broader definition, the untaxed/informal sector. The high Dutch taxes and social security contributions on earned income are making it attractive for people to enter the black economy as buyers or sellers or resort to do-it-yourself activities. For the main part this goes for labour-intensive consumer sales by the service and building industries. Part of these effects are already included in the price-substitution variables of the sectoral consumer sales equations. Higher wage taxes and social security contributions raise wage costs (see Section 18.4.4), which eventually results in price rises that are comparatively large for labour-intensive consumption. This makes such products relatively expensive and induces consumers to switch to more labour-extensive consumption. They can partially replace these labour-intensive products by making them themselves. This is not officially registered as consumption except for the initial expense of the necessary materials.

Besides, it seems likely that the high Dutch taxes and social security contributions have changed the moral standards of citizens where the black economy is concerned. As a result, the high public sector burden will also have direct consequences, i.e. not through the relative price variables, for the consumption of labour-intensive products in particular (black consumption is not expressed in the official statistics). To make this effect and the consequences of do-it-yourself activities visible, we have experimented in the consumer sales equations with the same proxy variable as is included in the investment equations. Here, too, changes in the public sector burden (Δdr) have been weighted with the level of the public sector burden.

Unsurprisingly, the burden variable was found to make no significant contribution to explaining consumer sales of the capital-intensive sector. The same holds true for the service sector. The building industry, however, did show a significant negative effect, which corresponds with the notion of a relatively large black and do-it-yourself economy in this industry.[18] (Consumption from this sector mainly consists of operations such as home repairs and maintenance.) Because of the small volume of consumer sales in this sector this supply-side element is of limited importance on the macro level.

Consumer sales by the building industry are explained not only by the burden variable but also by macro consumption ($\dot{c}$) as a scale factor, the relative prices of consumer sales by this industry ($\dot{p}_c - \dot{p}_c^{nb}$) and an income distribution variable

---

[18] An insignificant contribution need not mean that there are no black or do-it-yourself circuits in the sectors in question. The drop in consumption may conceivably be compensated by consumption paid with black money.

(LD − ZD). The last factor represents the phenomenon that the proportion of labour and transfer incomes spent on consumption supplied by the building industry is smaller than the proportion of nonlabour incomes so spent. The same presents itself for consumer sales by the capital-intensive sector, whereas the reverse is true for consumption supplied by the service industry.

$$\dot{c}^b = \underset{(2.86)}{0.95\dot{c}} - \underset{(-5.35)}{0.91(\dot{p}_c^b - \dot{p}_c^{nb})} - \underset{(-3.48)}{0.44(\dot{L}D - \dot{Z}D)_{4321}} +$$

$$- \underset{(-2.21)}{3.24\,[dr(\Delta r)]} + \underset{(2.67)}{5.32} \tag{18.10}$$

$$R^2 = 0.81 \quad Se = 3.61 \quad DW = 2.44 \quad EP = 56\text{–}80$$

Compared with the equation without public-sector-burden variable, the coefficient of the relative price variable in Equation 18.10 has dropped a little, while the explanation rate rises.[19] This confirms the hypothesis that − in an equation without a burden variable − part of the effect in question is included in the coefficient of the relative price variable. Compared with the relevant coefficients of the other consumer sales equations, the cross price elasticity remains high, though. For the capital-intensive sector and the service sector we obtain cross price elasticities of − 0.66 and − 0.38 respectively.

18.4.6 Conclusion

This section has reported on our attempt to model some supply-side elements with the aid of proxy variables from the macroeconomic sphere. In nine equations significant coefficients of such proxy variables have been found. Most of the modelled supply-side elements have (indirectly) something to do with the negative relationship mentioned by supply-siders, between public-sector size and structural economic growth. Although we feel that in the state of the art our method is acceptable, the results should be interpreted with caution. Modelling supply-side elements in this manner should be seen as an illustration of a way of taking such elements into account in a macro model rather than as empirical evidence of the presence of such elements in economic reality. Much theoretical and empirical research will be needed before the supply-side elements can be included in the range of 'standard elements' embodied in nearly every econometric model.

18.5 SIMULATIONS

18.5.1 Introduction

To obtain some insight into the macroeconomic and sectoral consequences of supply-side policies and the influence of the supply-side elements discussed in the

---

[19] In the original equation the cross price elasticity is − 1.03 and the R^2 statistic 0.765. See Verbruggen (1985a).

foregoing on these consequences, we shall in this section present some simulations as calculated with the VICTOR model with and without these supply-side elements.[20] Some preliminary remarks to begin with.

First, it is emphasized that the quantitative results of simulations have as a rule wide margins of uncertainty; this holds true in particular for the results obtained by the VICTOR model in which various supply-side elements have been modelled tentatively.

Second, if the consequences of some typical supply-side measures can be calculated with a macroeconomic model at all, it can be done so only tentatively. Cases in point are deregulation and policy measures to increase income differentials, stimulate innovation and get rid of labour market rigidities, which can in principle be better analysed with a partial, more microeconomically-orientated (labour market) model.

To begin with, we will deal with the policy to 'scale down the public sector' by tax cuts with and without accompanying public sector expenditure cuts. Next we shall see what the consequences are of an increase in production capacity and wage-cost moderation, focusing on the consequences for economic growth, employment and the budget deficit, since the Netherlands' present economic problems are centred on these points.

To trace the effects of the supply-side elements, the simulation results will be calculated likewise with a version of the model in which the influence of these factors is eliminated. This has been done in a very simple way, by reducing all coefficients of the supply-side terms in question to zero.

18.5.2 Tax reduction

Whenever in the initial years of supply-side economics discussions touched on tax cuts, the immediate response was a reference to the Laffer curve and a stock of famous 'success stories'.[21] In recent practice, however, advocates of tax cuts have not been invoking the Laffer curve nearly as much as the finding that large public sectors are becoming an increasing constraint on economic growth. In addition to supply-side effects, part of which are expressed in the model via public sector burden variables, Keynesian expenditure effects also play a role in tax reduction. On the basis of the following results, it will be endeavoured to make some distinction between expenditure effects and supply-side effects.

Table 18.1 shows the consequences of a cut in direct taxes amounting to 1% of net national income (NNI), as computed with the VICTOR model with and without the supply-side elements discussed in Section 18.4. We shall discuss first of all the results (in brackets) without supply-side elements.

It follows from Table 18.1 that private consumption, investment and imports

---

[20] In accordance with Klein (1983), we consider supply-side policies to be structural policies, aiming at specific issues, specific economic activities, and specific groups.
[21] Three well-known examples are the Kennedy–Johnson tax cuts in 1964, the Mellon tax cuts of 1922–1925 and the rescue of the City of New York after 1975.

Table 18.1 Simulated effects of a once-and-for-all reduction of direct taxes by 1% of NNI (in 1980)[a]

Cumulated effects after year	1		5		10	
Employment in:						
Capital-intensive sector (%)	0.3	(0.1)	2.1	(0.3)	4.4	(0.7)
(Commercial) service sector (%)	0.4	(0.3)	0.8	(0.2)	1.1	(0.2)
Building industry (%)	1.1	(0.7)	1.5	(0.4)	1.8	(0.5)
Total private sector (%)	0.4	(0.2)	1.1	(0.2)	2.0	(0.4)
Volume of production in:						
Capital-intensive sector (%)	0.6	(0.4)	1.6	(0.3)	3.3	(0.4)
(Commercial) service sector (%)	0.9	(0.9)	1.4	(0.6)	2.9	(0.6)
Building industry (%)	1.1	(0.7)	1.5	(0.4)	2.1	(0.5)
Total private sector (%)	0.8	(0.7)	1.5	(0.5)	2.9	(0.6)
Other volumes						
Production capacity (%)	0.3	(0.1)	1.7	(0.6)	4.2	(1.0)
Private consumption (%)	0.8	(1.1)	0.9	(1.1)	1.9	(1.0)
Corporate investment in equipment (%)	7.9	(5.1)	9.1	(2.3)	12.4	(2.8)
Exports (%)	0.3	(0.1)	1.6	(0.0)	3.8	(0.0)
Imports (%)	1.4	(1.2)	2.1	(0.8)	3.8	(0.8)
Wage rate in market sector (%)	−0.8	(0.1)	−1.4	(0.3)	−1.4	(0.1)
Production price (%)	−0.2	(0.1)	−1.0	(0.1)	−1.6	(−0.1)
Public sector deficit (% pt NNI)	0.8	(0.9)	0.6	(1.0)	0.3	(1.3)

[a] Figures in brackets are found when supply-side elements are excluded.

will all rise in the short term, owing to the favourable expenditure effects. As a result, production in companies will rise by 0.7% in the first year, to sag slightly afterwards, mainly due to crowding out of investment in the second and third years. It is striking to see that inflation will not or hardly pick up. The reasons for this are the high rate of unemployment in the base year (1980), so that the Phillips curve mechanism hardly operates (low cost–push inflation), and the openness of the Dutch economy, making much of the extra demand leak abroad (low demand–pull inflation).

Inclusion of the supply-side elements makes relatively little real change in the first year, but does produce a rather different picture in the medium and long term. Instead of remaining almost stable, corporate production and employment rise sharply when supply-side elements are included. The economy moves on to a higher growth track. A major contribution to the brisk economic growth is made by a rise in exports, caused entirely by the inclusion of the supply-side elements. Production capacity growth plays a crucial role in this. In the first few years most of the export growth is caused by improved price competitiveness as a result of the lower wage cost following the tax cut. After a few years the 'quality improvement' of supply resulting from higher (innovative) investments becomes important. In the longer term the utilization rate falls because production capacity continues to grow strongly as a result of the tax-cut-induced retrenchment of the public sector to the benefit of the private sector. This enables firms to step up production for export. Moreover, competitiveness improves even further because of the price fall through the inverted demand–pull effect.

Also when supply-side elements are included investment drops in the second and third years, owing to crowding out. This effect is offset, however, by favourable consequences of the decline in the public sector burden among other things. It is assumed implicitly, though, that this decline owing to the cut in direct taxes will also reduce the underlying factors of this proxy variable, such as regulation, rigidities and work disincentives. When interpreting the results one must keep in mind this assumption, which follows from our way of modelling supply-side elements.

The cut in direct taxes will increase somewhat the structural macro labour productivity, confirming the supply-siders' views on this point. It is striking to see how this variable differs for each sector, which is mainly due to the different developments of employment in the various sectors. Employment in the capital-intensive sector is relatively wage sensitive whereas, moreover, real wage costs in this sector fall sharply as compared with other sectors, owing to its specific marketing and cost structures. As a result, employment in the capital-intensive sector rises substantially more than in the labour-intensive sectors.

The supply-side elements also have a major long-term impact on the public sector deficit. Without these elements it rises by more than the initial one percentage point of NNI whereas inclusion of the supply-side elements keeps the growth in the public sector deficit below one percentage point of NNI. It does not, however, make for zero growth, so that in countries where these deficits are (too) high already tax reduction involves financial consequences.

All in all, the conclusion is that a tax cut has favourable economic consequences, but not for the public sector deficit.[22] In the short term, favourable expenditure effects play a major part in this, but in the longer run these effects are cancelled out in part by crowding-out effects. Yet the latter are in turn overcompensated by the favourable effects of the modelled supply-side elements, so that the economy moves on to a higher path of growth and the public sector-deficit growth is nearly halved in the long term.

18.5.3 Balanced-budget policy

Supply-siders like to depict the economy 'as a coiled spring held down by the weight of government'. Remove the weight, they say, and the spring will reveal its inherent force (see Heilbronner, 1982). We shall now test this view by a simulation in which both public sector outlays and income are reduced once and for all by 1% NNI in a

---

[22] The sizes of the effects depend also on the selected base year 1980, marked by a slump and a high rate of unemployment. When the economy is booming and the labour market stringent, inflation will be pushed up more, resulting in worse real effects.

Table 18.2 Simulated effects of a once-and-for-all reduction in public-sector income and outlays, amounting to 1% NNI (in 1980)[a]

Cumulated effects after year	1	5	10
Employment in:			
Capital-intensive sector (%)	0.1 (0.1)	1.8 (0.7)	3.6 (1.2)
(Commercial) service sector (%)	0.0 (−0.0)	0.3 (−0.0)	0.6 (−0.0)
Building industry (%)	−0.3 (−0.5)	0.3 (−0.4)	0.5 (−0.4)
Total private sector (%)	0.0 (−0.0)	0.7 (0.2)	1.4 (0.3)
Volume of production in:			
Capital-intensive sector (%)	0.1 (−0.0)	1.0 (0.1)	2.2 (0.2)
(Commercial) service sector (%)	0.0 (0.0)	0.8 (0.2)	1.9 (0.4)
Building industry (%)	−0.5 (−0.6)	0.2 (−0.6)	0.6 (−0.5)
Total private sector (%)	−0.0 (−0.1)	0.7 (0.1)	1.7 (0.2)
Other volumes			
Production capacity (%)	0.1 (0.0)	0.9 (0.3)	2.6 (0.6)
Private consumption (%)	−0.1 (0.0)	−0.1 (0.0)	0.7 (0.1)
Corporate investment in equipment (%)	2.5 (1.5)	6.0 (1.3)	7.9 (1.2)
Exports (%)	0.1 (0.0)	1.2 (0.2)	2.8 (0.3)
Imports (%)	0.1 (0.0)	0.8 (−0.1)	2.0 (−0.1)
Wage rate in market sector (%)	−0.8 (−0.2)	−2.0 (−0.9)	−2.3 (−1.2)
Production price (%)	−0.2 (−0.1)	−1.2 (−0.6)	−1.8 (−0.9)
Public sector deficit (% pt NNI)	0.0 (0.1)	−0.1 (0.1)	−0.5 (0.1)

[a] Figures in brackets are found when supply-side elements are excluded.

way that will leave the public sector deficit *ex ante* unchanged.[23]

Taking a look at the results shown in Table 18.2 (in brackets), excluding supply-side elements, we see hardly any changes in production and employment for the private sector either in the short or in the longer term. The same holds true for the various expenditure categories, apart from investment, which rises especially in the short run as a result mainly of the increased disposable real nonlabour incomes. The wage and price falls are linked with the productivity decrease in the capital-intensive sector and the utilization rate drops.

Inclusion of the supply-side elements makes hardly any changes in the results for the first year. In the medium term, however, production and employment rise clearly, which confirms the supply-siders' idea about a negative balanced-budget multiplier. As in the case of tax reduction, the economy moves on to a higher path of growth, even though less high than appears from Table 18.1. Exports and investment again play crucial parts. Employment growth is concentrated in the capital-intensive sector for the same reasons as under the tax-cut simulation. A

---

[23] The spending and income package of the public sector, in which the negative impulses have been given, is comprehensive and representative in its composition.

comparison of sector employment between Tables 18.1 and 18.2 shows that the reduction in public sector spending is mainly at the expense of sectors orientated relatively strongly towards the domestic market. The same picture emerges as to the short-term sectoral production development.

Table 18.2 shows that, given the corporate production capacity, public sector retrenchment strengthens the supply side of the economy considerably. On the demand side there is a diverging development, especially in the short term. Unlike exports and investment, private consumption stagnates. In the longer run, though, a negative balanced-budget policy also benefits consumption, as a result of the improved economic growth and employment. The same reasons underlie the decrease in the public sector deficit. The government could use this financial elbow-room to cut direct taxes even more, with extra favourable effects as a result (see Table 18.1). In that case the initial sacrifice in terms of public sector expenditures is rather small compared with the resulting favourable effects in terms of economic growth, employment and consumption.

All in all, it is found that inclusion of supply-side elements in the model makes the balanced-budget multiplier negative.[24] Knoester (1983) called this result the 'inverted Haavelmo effect'. Pen (1985) said: 'These results are more or less in line with what supply-siders assert. They give no indication that we are in the right-hand part of the Laffer curve, but nevertheless they are not encouraging for those who want to overcome the stagnation by increasing the size of the public sector, and certainly not for those who levy additional taxes for that purpose'. Our analysis indicates that in the latter case there is a fair chance of the economy switching structurally to a lower path of growth.

18.5.4 Capacity increase

Supply-siders point to the importance of supply-reinforcing measures for economic growth. Many of such measures, like tax cuts to curb the public sector for the benefit of the private sector, have expenditure effects in addition to supply effects. However, one can think of other measures which in the first instance have only supply effects, such as deregulation and removal of market rigidities. Another possibility is adjusting the structure of public sector outlays to stimulate economic growth. An example would be a shift in government investment from buildings to infrastructure to remove bottlenecks. Although the spending effects of either sort of investment will not differ much – and are even completely identical in most models – such a shift would very probably increase the supply potential of the private sector. An interesting question is whether such supply-reinforcing measures are capable of stimulating economic growth on their own, i.e. without direct expenditure effects.[25]

---

[24] A similar result has been found by Knoester (1983) and Rutten (1985).

[25] It would be more correct to speak of 'supply-reinforcing measures without direct *domestic* spending effects', since production capacity plays a major part in exports.

Table 18.3 Simulated effects of a once-and-for-all 1% increase in production capacity (in 1980)[a]

Cumulated effect after year	1	5	10
Employment in:			
Capital-intensive sector (%)	0.0 (−0.0)	0.1 (−0.3)	−0.2 (−0.5)
(Commercial) service sector (%)	0.0 (−0.0)	0.1 (−0.1)	−0.0 (−0.2)
Building industry (%)	0.1 (−0.0)	0.0 (−0.2)	−0.1 (−0.2)
Total private sector (%)	0.0 (−0.0)	0.1 (−0.1)	−0.1 (−0.3)
Volume of production in:			
Capital-intensive sector (%)	0.3 (0.0)	0.7 (0.2)	0.8 (0.2)
(Commercial) service sector (%)	0.2 (0.1)	0.7 (0.3)	0.8 (0.3)
Building industry (%)	0.1 (−0.0)	0.2 (0.0)	0.1 (0.0)
Total private sector (%)	0.2 (0.0)	0.7 (0.2)	0.7 (0.2)
Other volumes			
Production capacity (%)	1.0 (1.0)	1.3 (1.0)	1.4 (1.0)
Private consumption (%)	0.1 (−0.0)	0.6 (0.2)	0.8 (0.3)
Corporate investment in equipment (%)	0.5 (−0.2)	1.1 (−0.3)	0.0 (−0.3)
Exports (%)	0.3 (0.0)	0.9 (0.2)	1.0 (0.2)
Imports (%)	0.2 (−0.1)	0.8 (0.0)	0.8 (0.1)
Wage rate in market sector (%)	−0.0 (−0.0)	0.3 (−0.0)	0.6 (0.1)
Production price (%)	−0.1 (−0.1)	−0.3 (−0.5)	−0.3 (−0.5)
Public sector deficit (% pt NNI)	−0.0 (0.0)	−0.2 (0.0)	−0.2 (0.0)

[a] Figures in brackets are found when supply-side elements are excluded.

To answer this question we have simulated a once-and-for-all 1% increase in production capacity via an autonomous variable. Then the increase in supply is not a result of an increase in capital stock but is caused by a general improvement of market-sector efficiency, boosting the structural labour and capital productivity.

Again we look first at the results without supply-side elements (Table 18.3 in brackets) and find that there is hardly any economic growth. The small production rise in combination with the (implicitly assumed) increase in labour and capital productivity leads on balance to a drop in employment and investment. In the model the drop in employment is caused by a decline in the utilization rate, the effect of which causes a production-price drop and hence a rise in real wage costs.

Inclusion of the supply-side elements gives a much higher economic growth rate, so that the question whether supply-reinforcing measures without direct expenditure effects can on their own generate economic growth can be answered more or less affirmatively. In the medium and long run, private production rises by 0.7%. As production capacity grows endogenously in the long term likewise, the utilization rate continues to be lower. At first sight this outcome seems to clash with Say's law, in which most adherents of supply-side economics seem to believe. But on closer inspection this simulation, in which capacity expansion does not coincide with positive spending effects, cannot in our view be regarded as a theoretically correct test of Say's law, since Say assumed that, in order to produce goods, it was necessary to purchase materials and capital goods and to hire labour, while the money spent on these factor inputs would in turn be spent on goods and services by its recipients. See, for example, Evans (1983).

By far the larger part of the extra output will be exported in the first few years. Hence the fact that so much of the extra supply generates extra demand on its own is connected with the openness of the Dutch economy. In the longer run, however, private consumption too, rises substantially. But investment hardly grows in the long term, which is not surprising, since the existing capital stock is large enough to turn out the extra production.

Notwithstanding the inclusion of the supply-side elements employment results remain disappointing, owing to the rise in real wage costs. The inverted demand–pull effect depresses production prices, while nominal wages rise with growing labour productivity. It is questionable, though, whether in times of widespread unemployment productivity growth can be fully passed on in wages. For this reason the capacity simulation was calculated again, but this time with an adjusted wage equation, without such passing-on in wages. Then employment results (see Table 18.4) are much better, especially in the capital-intensive sector.

Table 18.4 also indicates that the increase in capacity causes a decline in the public sector deficit. Should government use this financial elbow-room to cut direct taxes, then even better results emerge. A tax reduction which would on balance leave the deficit the same after five years, would in the long run add 1% employment, whereas the autonomous increase in production potential goes hand in hand with a similar increase in the economic growth rate already in four years' time.

Table 18.4 Simulated effects on a once-and-for-all 1% increase in production capacity, with adjusted wage equations (in 1980)[a]

Cumulated effects after year	1	5	10
Employment in:			
Capital-intensive sector (%)	0.0	0.5	1.0
(Commercial) service sector (%)	0.0	0.2	0.3
Building industry (%)	0.1	0.2	0.2
Total private sector (%)	0.0	0.3	0.5
Volume of production in:			
Capital-intensive sector (%)	0.3	0.9	1.2
(Commercial) service sector (%)	0.2	0.8	1.1
Building industry (%)	0.1	0.4	0.4
Total private sector (%)	0.2	0.7	1.0
Public sector deficit (% pt NNI)	−0.0	−0.2	−0.4

[a] Model including supply-side elements.

18.5.5 Wage moderation

Wages are not only crucially important for the demand side but also for the supply side of the economy. Generally speaking, a wage rise has negative consequences for the supply side if real wages grow faster than labour productivity, so that labour's share of national income (or more specifically the ratio of wage bills to net added value in companies) increases and firms become less inclined to invest. In addition, a real wage rise affects the supply side of the economy through an economic scrapping of capital goods.

In a small open economy like the Netherlands' a wage rise will have unfavourable consequences for the supply side sooner than in a closed economy. This is because a wage-cost rise often cannot be passed on in product prices for reasons of competitiveness, with declining profitability as a result. All in all, wage moderation, especially in the Dutch situation, can be regarded as a supply-side policy. Where the taxation and capacity simulations were accompanied in the first instance by favourable and neutral expenditure effects respectively, this supply-side policy is in the short term even accompanied by unfavourable spending effects.

Table 18.5 (in brackets) sets out the results of a once-and-for-all 2% wage moderation in the market sector as calculated with the model without supply-side

Table 18.5 Simulated effects of a once-and-for-all 2% wage moderation in the market sector (in 1980)[a]

Cumulated effects after year	1	5	10
Employment in:			
Capital-intensive sector (%)	0.3 (0.3)	3.2 (2.6)	5.5 (3.6)
(Commercial) service sector (%)	0.2 (0.2)	1.1 (0.8)	1.5 (1.0)
Building industry (%)	0.4 (0.3)	1.1 (0.5)	1.3 (0.4)
Total private sector (%)	0.2 (0.2)	1.5 (1.2)	2.5 (1.6)
Volume of production in:			
Capital-intensive sector (%)	0.1 (0.1)	1.3 (0.6)	2.4 (0.6)
(Commercial) service sector (%)	−0.3 (−0.3)	0.7 (0.2)	1.8 (0.5)
Building industry (%)	0.2 (0.2)	0.8 (0.3)	1.2 (0.3)
Total private sector (%)	−0.2 (−0.2)	0.7 (0.2)	1.7 (0.3)
Other volumes			
Production capacity (%)	0.0 (0.0)	0.6 (0.2)	2.1 (0.5)
Private consumption (%)	−0.9 (−0.9)	−1.1 (−1.2)	−0.3 (−0.9)
Corporate investment in equipment (%)	1.6 (1.6)	5.3 (1.7)	7.2 (1.2)
Exports (%)	0.2 (0.2)	1.7 (0.8)	3.1 (0.8)
Imports (%)	−0.2 (−0.2)	0.4 (−0.4)	1.6 (−0.5)
Wage rate in market sector (%)	−2.3 (−2.2)	−4.9 (−4.4)	−5.3 (−4.9)
Production price (%)	−0.7 (−0.7)	−2.4 (−2.1)	−2.9 (−2.4)
Public sector deficit (% pt NNI)	−0.1 (−0.1)	−0.5 (−0.2)	−0.1 (−0.4)
Public sector burden (% pt NNI)	−0.1 (−0.1)	−0.5 (−0.4)	−0.6 (−0.4)
Labour's share of NI (% pt)	−1.0 (−1.0)	−1.3 (−1.0)	−1.3 (−1.0)

[a] Figures in brackets are found when supply-side elements are excluded.

elements. At first, production falls off as a result of a decline in private consumption. As a relatively large portion of the service sector's production is sold to private consumers, the consequences of a drop in domestic consumption hit this sector in particular. In the medium and long term, however, the macro production drop that occurred in the first year is compensated. Production and employment increases relatively fast and sharply in the capital-intensive sector. The first because exports, which constitute the main outlet of this sector, are not affected by the decline in domestic demand and even go up as a result of the improved competitiveness. The latter because employment in the capital-intensive sector is relatively wage sensitive whereas real wage costs in this sector fall sharply as compared with other sectors. Investment also benefits, mainly under the influence of enhanced profitability. The favourable effect of a wage moderation on the supply side of the economy is reflected in growing production capacity and a drop in labour's share of national income. Finally, the public sector deficit shrinks as a result of lower public sector spending which in turn is a result of the indexation of much of these expenditures to the market sector wage rate while the overall economic pick-up increases tax revenues and decreases outlays even more.

Inclusion of supply-side elements again proves to have little effect in the short term, but in the longer run the favourable economic consequences are greatly strengthened. Unlike the other simulations, this one does not show fundamental differences between the results with and without supply-side elements.

Notwithstanding the relatively strong production rise in the capital-intensive sector, the service sector makes the biggest contribution to economic growth in the long run. This may seem surprising since private consumption, the service sector's most important sales category, drops in the long run too. But that does not mean a long-term decline likewise in all the five components of private consumption, namely consumption supplied by our four domestic sectors and foreign countries. Table 18.6 shows that consumption of service products, unlike imported consumption, performs relatively well, because the consumer sales equations contain a

Table 18.6 Simulated effects of a once-and-for-all 2% wage moderation in the market sector (in 1980) for the composition of the consumption package (including supply-side elements)[a]

Cumulated effects after year	1	5	10
Volume of consumption supplied by:			
Capital-intensive sector (%)	−0.7 (−0.5)	−0.9 (−0.4)	−0.6 (0.1)
Service sector (%)	−0.7 (−0.9)	−0.1 (−1.3)	1.1 (−0.7)
Building industry (%)	0.0 (−0.2)	2.9 (1.9)	4.8 (3.5)
Rest-of-the-economy sector (%)	−1.0 (−1.0)	−1.4 (−1.2)	−0.7 (−0.6)
Foreign countries (%)	−1.8 (−1.5)	−3.6 (−1.8)	−3.0 (−0.9)
Volume of private consumption:			
Total (%)	−0.9 (−1.0)	−1.1 (−1.2)	−0.3 (−0.6)

[a] Figures in brackets are the results if price substitution by the consumers is excluded.

relative price variable representing the consequences of price substitution by the consumers.

The wage moderation has the result that the prices of the consumption supplied by the domestic sectors drop, while the prices of imported consumption remain unaffected. In consequence imported consumption declines. Despite the relatively small cross price elasticity of consumer sales by the service industry, this large consumption category grows, owing to the relatively hefty drop in the prices of the goods and services supplied by this sector. The reasons for this drop are the relatively large wage-cost component in this labour-intensive industry and the fact that changes in costs in this 'nontradable sector' are fully passed on in prices. Without price elasticities in the consumer sales equations (Table 18.6, in brackets) consumption of services declines more than the average. Consumption supplied by the building industry rises sharply due to the shift in the distribution of incomes from labour to nonlabour incomes.[26]

All in all, it is found that the positive consequences of wage moderation do not depend on the modelled supply-side elements, although they are strengthened by them. This holds true especially for economic growth, which improves by an extra 1% in the long run when supply-side elements are included. However, even without supply-side elements wage moderation is favourable for the economy in general and for employment in particular.

18.6 SUMMARY

This chapter has attempted to model some supply-side elements in a four-sector model for the Netherlands' economy, called the VICTOR model, and to analyse the influence of these elements' inclusion on the simulation results of some supply-side policies.

The starting point has been the thought that, if the often microeconomically-orientated supply-side elements are systematically and to an appreciable extent operative in the economy, it should be possible to make them visible in macroeconometric models. This chapter has made an initial step in such an endeavour, using some proxy variables from the macroeconomic sphere. These variables (especially the public sector burden) function as proxies for some underlying, predominantly microeconomic factors. This approach has the advantage of making it possible to endogenize these variables in the model, which is helpful for simulations. The main drawback is that the relationships between the proxy variables and their underlying factors are not distinct, which makes the empirical estimation results tentative. These results indicate that by this method it is possible to model significantly some supply-side elements in nine sectoral equations (on investment, production capacity, exports, wages and consumption).

Next four policy options have been analysed by means of the VICTOR model

---

[26] For a detailed description of the sectoral consequences of wage moderation, see Verbruggen (1985b).

including and excluding the modelled supply-side elements. It goes without saying that the results have wide margins of uncertainty. Their value, as we see it, is illustrative rather than predictive.

First we have simulated the effects of a direct tax cut. One of the most important findings is that, through incorporation of the supply-side elements, retrenchment of the public sector by means of a direct tax cut may move the economy on to a higher path of growth. Exports and business investment developments play a crucial role in this. There will be an increase in the public sector deficit, but it will be considerably smaller than the initial amount with which the taxes are reduced. Exclusion of the modelled supply-side elements makes the results far less good. In the long run the public sector deficit increases by more than the initial tax cut.

In the second simulated policy option the tax cuts are accompanied by public sector expenditure cuts. Again the economy moves on to a higher path of growth through incorporation of the modelled supply-side elements, albeit less high than with tax cuts on their own. The balanced-budget multiplier is clearly negative in our calculations. Without the supply-side elements the results are hardly good if at all. In that case the balanced-budget multiplier is approximately zero.

The detrimental influence of a large public sector can also be mitigated through measures aiming at deregulation and removal of market rigidities. Assuming that these measures can increase private sector's production potential through improved efficiency, we have simulated these measures by an autonomous increase in production capacity. The question whether such supply-reinforcing measures, unaccompanied by direct expenditure effects, can on their own generate economic growth can be answered in the affirmative, according to our calculations. Without supply-side elements, such a policy seems hardly if at all to promise favourable results.

Finally, the consequences of a wage-moderation policy have been analysed. Although such a policy seems to yield good economic consequences even without the supply-side elements, their inclusion gives much better results. Via price substitution by consumers, not only the export-orientated capital-intensive sector stands to benefit by wage moderation but also the home-market-orientated service sector, if the benefits for the latter are less great and take longer to make themselves felt.

All in all, the incorporation of the modelled supply-side elements might be rather important for the simulation results of policy options in general and for supply-side policies in particular. Therefore, we believe that further micro- and macro-economic research into the significance of these elements merits high priority. The more so, because modelling of supply-side elements is still in its infancy.

ACKNOWLEDGEMENTS

There are many people to whom I owe thanks for their advice and comments on various drafts. Above all, I am grateful to Ab van Ravestein, Jarig van Sinderen and Gerrit Zalm.

REFERENCES

Blomquist, N. S. (1983) The effect of income taxation on the labour supply of married men in Sweden. *Journal of Public Economics*, November.

Brown, C. V. and Jackson, P. M. (1980) *Public Sector Economics*, 2nd edn, M. Robertson, Oxford.

Buitelaar, P. (1985) De geschatte gedragsvergelijkingen van het model REMON en REMONA, onderzoeksmemorandum 8501, Ministry of Economic Affairs, The Hauge.

Central Planning Bureau (1971) Centraal Economisch Plan 1971, The Hague.

Central Planning Bureau (1977) Een macro model voor de Nederlandse economie op middellange termijn, occasional paper 12, The Hague.

Central Planning Bureau (1983a) FREIA, Een macro-economisch model voor de middellange termijn, monografie 25, The Hague.

Central Planning Bureau (1983b), KOMPAS, Kwartaalmodel voor prognoses, analyse en simulatie, monografie 26, The Hague.

Central Planning Bureau (1985) FREIA–KOMPAS '85, Een kwartaalmodel voor Nederland voor de korte en middellange termijn, monografie 28, The Hague.

Cross, R. (1982) *Economic Theory and Policy in the UK*, M. Robertson, Oxford.

Driehuis, W. (1972) *Fluctuations and Growth in a Near Full Employment Economy*, Rotterdam University Press, Rotterdam.

Driehuis, W. and Noord, P. J. van den (1980) Produktie, werkgelegenheid sectorstructuur en betalingsbalans in Nederland, 1960–1985, WRR, The Hague.

Duijn, J. J. van (1982) Aanbodeconomie in Nederland, in *De economie van het aanbod, Preadviezen van de Vereniging voor de Staathuishoudkunde*, Stenfert Kroese, Leiden.

Evans, M. K. (1981) An econometric model incorporating the supply-side effects of economic policy, in *The Supply-Side Effects of Economic Policy*, (ed. L. H. Meyer), CSAB, Federal Reserve Bank of St Louis, Kluwer-Nijhoff, Boston.

Evans, M. K. (1983) *The Truth About Supply-Side Economics*, Basic Books, New York.

Draper, D. A. G. (1985) Exports of the manufacturing industry: an econometric analysis of the significance of capacity. *The Economist*, 133, No. 3.

Drimmelèn, W. van and Hulst, N. van (1981) *Loonvorming en loonpolitiek in Nederland*, Wolters-Noordhof, Groningen.

Hansson, I. (1984) Marginal cost of public funds for different tax instruments and government expenditures, *Scandinavian Journal of Economics*, 86, No. 2.

Hartog, H. den, Klundert, Th. van de and Tjan, H. S. (1975) De structurele ontwikkeling van de werkgelegenheid in macro-economisch perspectief, in *Werkloosheid, Preadviezen van de Vereniging voor de Staathuishoudkunde*, Stenfert Kroese, The Hague.

Hartog, H. den and Tjan, H. S. (1979) A clay–clay vintage model approach for sectors of industry in the Netherlands, occasional paper 17, Central Planning Bureau, The Hague.

Hartog, H. den (1983) Employment in the Netherlands: the analysis by the Central Planning Bureau on the macro level, in *Unemployment: the Dutch Perspective*, Ministry of Social Affairs and Employment, The Hague.

Heilbronner, R. L. (1982) The demand for the supply-side, in *Supply-Side Economics: A Critical Appraisal* (ed. R. H. Fink), University Publications of America, Maryland.

Klein, L. R. (1983), *The Economics of Supply and Demand*, Basil Blackwell, Oxford.

Kmenta, J. and Ramsey, J. B. (1981) Model size, quality of forecast accuracy and economic theory, in *Large-Scale Macro-Econometric Models*, (eds J. Kmenta and J. B. Ramsey), North-Holland, Amsterdam.

Knoester, A. (1983) Stagnation and the inverted Haavelmo effect: some international evidence, discussion paper 8301, Ministry of Economic Affairs, The Hague.

Knoester, A. (1984) Theoretical principles of the buffer mechanism, monetary quasi-equilibrium and its spillover effects. *Kredit und Kapital*, 17, No. 2.

Knoester, A. and Sinderen, J. van (1984) A simple way of determining the supply-side in macro-economic models. *Economics Letters*, 16, Nos. 1–2.

Knoester, A. and Windt, N. van der (1985) Real wages and taxation in ten OECD countries, discussion paper 8501 G/M, Institute for Economic Research, Erasmus University, Rotterdam.

Kuipers, S. K., Muysken, J., Berg, D. J. van den and Zon, A. H. van (1980) Sectorstructuur en economische groei: een eenvoudig groeimodel met zes sectoren van de Nederlandse economie in de periode na de tweede wereldoorlog, WRR, The Hague.

Kuipers, S. K., Andriesen, C. M., Jacobs, J. P. A. M. and Kuper, S. H. (1985) A putty–clay vintage model for sectors of industry in the Netherlands. *The Economist*, 133, No. 2.

Lindbeck, A. (1981) Work disincentives in the welfare state, reprint 176, Institute for International Economic Studies, University of Stockholm.

Muller, F., Lesuis, P. J. J. and Boxhoorn, N. M. (1980) Een multisectormodel voor de Nederlandse economie in 23 bedrijfstakken, WRR, The Hague.

OECD (1985) The role of the public sector, causes and consequences of the growth of government. *Economic Studies*, No. 4, OECD, Paris.

Olson, M. (1982) *The Rise and Decline of Nations*, Yale University Press, New Haven.

Pen J. (1985) *Among Economists, Reflections of a Neo-Classical Post Keynesian*, North-Holland, Amsterdam.

Ridder, P. B. de (1977) Een jaargangenmodel met vaste technische coëfficiënten en in kapitaal geïncorporeerde arbeidsbesparende technische vooruitgang, occasional paper 14, Central Planning Bureau, The Hague.

Riet, A. van (1987) Substitutie en complementariteit: een onderzoek naar de invloed van faktorkosten op het investeringsgedrag, Discussion Paper 8702, Ministry of Economic Affairs, The Hague.

Rutten, F. W. (1985) Berekeningen over economische groei en werkgelegenheid, in *Voor praktijk of wetenschap* (eds W. Begeer, C. A. Oomens and W. F. M. de Vries), Central Bureau of Statistics, The Hague.

Rutten, F. W. (1986) Naar een hogere economische groei. *Economisch Statistische Berichten*, No. 3537.

Schaik, A. B. T. M. van (1980) Arbeidsplaatsen, bezettingsgraad en werkgelegenheid in dertien bedrijfstakken, WRR, The Hague.

Sinderen, J. van (1985) Some major causes of developments in Dutch business investment, discussion paper 8503, Ministry of Economic Affairs, The Hague.

Sinderen, J. van and Verbruggen, J. P. (1986) Over de effectiviteit van premiëring van investeringen, Discussion Paper 8601, Ministry of Economic Affairs, The Hague.

Stuart, C. (1984) Welfare costs per dollar of additional tax revenue in the United States. *American Economic Review*, 74, No. 3.

Verbruggen, J. P. (1985a) VICTOR, Een vier-sectorenmodel voor de Nederlandse economie, discussienota 8502, Ministry of Economic Affairs, The Hague.

Verbruggen, J. P. (1985b) Sectorale gevolgen van loonmatiging. *Maandschrift Economie*, No. 6.

Zon, A. H. van (1985) A simple multisector model with six sectors of production. *The Economist*, 133, No. 3.

APPENDIX

LIST OF SYMBOLS

With respect to the symbols used the following principles are valid:

(1) Symbols in capital letters are in current prices

Appendix

(2) symbols in small letters are in constant prices
(3) first differences are indicated with a 'Δ'
(4) percentage changes are indicated with a dot.

When a variable has one of the following suffixes, this indicates sectoral application:

- b   building industry
- c   capital-intensive sector
- cs   commercial services
- s   service sector.

Besides common lag-indicators in years (e.g. $-1/4$, $-1/2$ and -1) the next indicators are used:

$$x_{111} = 0.333(x + x_{-1} + x_{-2})$$
$$x_{532} = 0.5x + 0.3x_{-1} + 0.2x_{-2}$$
$$x_{1111} = 0.25(x + x_{-1} + x_{-2} + x_{-3})$$
$$x_{4321} = 0.4x + 0.3x_{-1} + 0.2x_{-2} + 0.1x_{-3}$$
$$x_{11111} = 0.2(x + x_{-1} + x_{-2} + x_{-3} + x_{-4})$$

ab	= employment
Bm	= base money (redefined)
c	= private consumption
cap	= production capacity
dr	= $dr^{wk} + spw$
dr^{wk}	= burden of employees' wage taxes and social security contributions on disposable earned income
e	= exports of goods and services
h	= labour productivity
h_t	= structural (lagged) value of h
is	= investment incentives premiums
i	= gross private fixed investment
i_{et}	= gross private investment in equipment and means of transport
k	= capital stock
k^*	= desirable capital stock
LD	= disposable labour income
m_w	= world trade (twice reweighted)
p_c	= consumption price deflator
p_c^{nb}	= price deflator of consumption not supplied by the building industry
p_e	= exports price deflator
p_e'	= exports price deflator of competitors in foreign markets (twice reweighted)
p_{iet}	= price deflator of i_{et}
p_1	= wage rate

p_y	= gross added value price deflator
q	= utilization rate
r_1	= long-term interest rate
spw	= burden of employers' and social security contributions on gross wages
u	= unemployment rate
V	= total sales
wt	= working time
y	= gross added value
y_t	= structural (lagged) value of y
Z	= disposable nonlabour income (gross)
ZD	= disposable nonlabour income (net)

19

The supply side of RIKMOD: Short-run producer behaviour in a model of monopolistic competition

MICHAEL HOEL and RAGNAR NYMOEN

The supply side of RIKMOD – which is a quarterly model of the Norwegian economy – is derived from the short-run decisions of a profit-maximizing firm in a monopolistically competitive market. We first derive the levels of price, output and employment which would maximize profit if employment could be adjusted costlessly. This gives actual price, and the optimal level of employment as functions of factor and foreign prices, capacity utilization, and demand shocks. However, the actual level of employment may deviate from the optimal level, due to costs associated with adjusting employment. The sum of these costs and various costs associated with having a nonoptimal level of employment are minimized. This procedure gives employment, working hours and inventories as functions of employment and inventory levels in the previous period and the corresponding levels which would have been optimal if employment adjustment were costless.

This chapter reports empirical results for the manufacturing sectors of the RIKMOD model. The preferred estimated equations – which are made subject to misspecification tests – are in concord with the theory. Typically, the results reveal a 'Keynesian' response pattern to demand shocks. 'Classical' results only appear for relatively high degrees of capacity utilization.

19.1 INTRODUCTION

Producer behaviour comprises decisions about prices and the level of output, changes in inventories and employment as well as investment in machinery and buildings. The level of investment depends primarily on long-term considerations such as expectations about profitability, whereas a shorter time horizon is probably relevant for the other decisions. This is why investment behaviour may be modelled somewhat separated from the other supply-side decisions. On the other hand, short-term decisions are closely connected. This chapter describes main features of the modelling of short-term producer behaviour in the RIKMOD model of the Norwegian economy. In this model output prices, production levels, investment in inventories and the level of employment are determined jointly. By producer behaviour in the short run we mean the behaviour of firms within a time span short enough to allow us to consider the level of capacity as fixed.

There are three important aspects of the model of short-term producer behaviour in RIKMOD. Firstly, the analysis is based on the principles of monopolistic competition. Each firm is assumed to face a downward sloping demand curve. The level of each of these demand schedules depends on the prices of other firms within the same sector, i.e. the sectoral price level. Each individual firm considers the sectoral price level as given. Moreoever, the individual demand schedules, and the sector demand schedule, depend on factors exogenous to the sector as a whole, such as prices of competing imports and the level of gross domestic demand.

The second important aspect of the model of short-term producer behaviour, is the explicit allowance for costs of adjusting the level of employment. One way to avoid abrupt changes in employment is to allow changes in demand to be partly offset through changes in inventories. Employment and inventories of finished goods are thus determined simultaneously.

The third aspect of our modelling of short-term producer behaviour is the simplifying assumption that the producers' decisions may be modelled as a two-step process: first the optimal price, production and level of employment are calculated, assuming no changes in inventories, and no costs of adjusting employment. Next, these optimal values of production and employment are modified when changes in inventories are allowed for and we introduce costs of changing the level of employment. This model of two-step optimizing behaviour is not completely satisfactory from a theoretical point of view. However, some simplification of a general intertemporal optimizing framework seems indispensable in the context of a numerical model.

In Section 19.2 of the chapter, the theoretical model is spelled out in some detail. Section 19.3 contains the empirical results and some simulation properties.

19.2 THEORETICAL CONSIDERATIONS

This section is organized according to the dichotomy of the producers' optimalization. In Section 19.2.1 we consider the model of producer decisions when changes in inventories are ignored together with costs of adjusting employment, i.e. step one of the two-step optimizing procedure. In particular, we focus on the

Theoretical considerations

specification of technology and the price equation. In Section 19.2.2 we turn to the joint determination of employment and inventories when costs of adjusting employment are introduced.

19.2.1 Producer behaviour without inventories and with costless adjustments of employment

In order to simplify notation, we assume that a production sector is made up of identical firms. The notation is:

Lower case letters	= denote variables concerning a representative firm
Capital letters	= denote variables concerning the sector as a whole
n	= the number of firms comprising the sector
p, P	= price of output (firm, sector)
x, X	= level of output (firm, sector)
y, Y	= level of demand (firm, sector)
*	= denotes a variable's expected value
$d(\cdot), D(\cdot)$	= demand function (firm, sector)
$c(\cdot), C(\cdot)$	= cost function (firm, sector)
$\mathbf{Z}$	= Vector of exogenous variables.

Each firm faces a specific demand schedule:

$$y = d(\underset{-}{p}, \underset{+}{P}, \mathbf{Z}) \tag{19.1}$$

Demand is higher the lower the firm's own price, and the higher the level of the sector price. Demand is also dependent on exogenous factors ($\mathbf{Z}$), to which we will return below.

Assuming a symmetric equilibrium, defined by $p = P$, sectoral demand is given by:

$$Y = D(P, \mathbf{Z}) = nd(p, P, \mathbf{Z}) \tag{19.2}$$

The own-price elasticity ε and sectoral price elasticity E are given by:

$$-\varepsilon = \frac{\partial d(p, P, \mathbf{Z})}{\partial p} \frac{p}{d(p, P, \mathbf{Z})} > 0 \tag{19.3}$$

$$E = \frac{\partial D}{\partial P} \frac{P}{D} = \varepsilon + \left(\frac{\partial d}{\partial P} \frac{P}{d}\right) \text{i.e. for } p = P \tag{19.4}$$

Since the last term in Equation 19.4 is positive, $-\varepsilon$ is greater than $-E$.

We assume that ε is a constant. Hence Equation 19.1 may be written:

$$y = p^\varepsilon \tilde{d}(P, \mathbf{Z}) \tag{19.5}$$

The firm's short-term cost function is $c(X, \mathbf{Z})$. Profits are:

$$\pi = pd(p, P, \mathbf{Z}) - c(d(p, P, \mathbf{Z}), \mathbf{Z}) \tag{19.6}$$

We assume that the firms choose p subject to uncertainty about the values of P and Z. As a simplification we assume that the firms decide their own price subject to the expected values P^* and Z^*.

Substituting P^* and Z^* for P and Z in Equation 19.6, maximization of π gives:

$$p\left(1 + \frac{1}{\varepsilon}\right) = c_x(y^*, Z^*) \tag{19.7}$$

where c_x is marginal costs and

$$y^* = d(p, P^*, Z^*) \tag{19.8}$$

is expected demand when price is given by Equation 19.7. Equation 19.7 may be written as

$$p = (1 + m)c_x(y^*, Z^*) \tag{19.9}$$

where the mark-up factor m is:

$$m = \frac{1}{(-\varepsilon) - 1} \tag{19.10}$$

Since $-\varepsilon > 1$ at the optimum, m is strictly positive.

In a symmetric equilibrium, all prices are equal, i.e. $p = P$. We also assume that firms have perfect foresight concerning the sectoral price level. If $p = P = P^*$ is substituted in Equation 19.8 we obtain:

$$Y^* = nd^* = nd(P, P, Z^*) = D(P, Z^*) \tag{19.11}$$

The sector cost function is given by

$$C(X; Z) = nc\left(\frac{X}{n}, Z\right) \tag{19.12}$$

so that

$$C_X(X; Z) = c_X\left(\frac{X}{n}, Z\right) \tag{19.13}$$

Together with $p = P$ and Equation 19.9 this gives:

$$P = (1 + m)C_X(Y^*, Z^*) \tag{19.14}$$

where Y^* is given by Equation 19.11.

Equations 19.11 and 19.14 together determine the sectoral price level and planned output. This plan is based on the producers' expectations concerning the exogenous variables in Z. *Ex post* these expectations may prove wrong, i.e. $Z \neq Z^*$. However, if Z satisfies:

$$p = (1 + m)c_x(d(p, P, Z^*), Z^*) \geq c_x(d(p, P, Z), Z) \tag{19.15}$$

the firm will choose to produce $y = d(p, P, Z)$ at the *ex ante* fixed price. We assume this response pattern to be predominant in industrial sectors, thus we obtain:

$$Y = D(P, Z) \tag{19.16}$$

where P is determined by Equations 19.11 and 19.14. In a multisectoral model, the analytical counterpart to Equation 19.16 is the entire input–output and expenditure block. Very often such models contain mark-up price equations, while production is typically demand determined. Equation 19.16 may be seen as a necessary condition for the overall theoretical consistency of these models.

In most models of small open economies, prices of domestically produced goods are affected by world prices of similar products, here denoted by Q^K. This relationship is only indirectly represented by Equation 19.14, in that Q^K may be an element in the Z-vector of exogenous variables affecting Y through Equation 19.11. However, at least two arguments might be advanced for a more direct influence from world prices on domestic prices. Firstly, the demand elasticity ε (and hence the mark-up coefficient m) is a constant in our case. However, the demand elasticity may depend on Q^K. Secondly, in some industries, price determination may be characterized by collusive behaviour rather than monopolistic competition. Within a collusive framework firms might be expected to settle for the price of competing imports (Eastman and Stykolt (1966), Harris (1984)).

Against this background the price Equation 19.14 is modified in order to allow for direct influence from competing prices (Q^K) on P:

$$P = \alpha C_X(Y, Z) + \beta Q^K \qquad (19.17)$$

α, β are nonnegative constants. The special case $\alpha = 1 + m, \beta = 0$ corresponds to Equation 19.14. In Equation 19.17 differences between expected and actual values of the variables are ignored. More generally, therefore, Y, Z and Q^K are replaced by their expectations Y^*, Z^* and Q^{K*}.

We now turn to the specification of technology. Variable costs are made up of cost of materials and the wage bill. We write variable costs $C(\cdot)$ as:

$$C(X, Z) = Q^V v X + W G(X, Z) \qquad (19.18)$$

where Q^V is price per unit of material input, v is a coefficient of material input per unit of output and W is nominal wage costs per man-hour. Note that there are no *ex post* substitution possibilities between materials and labour. The function $G(\cdot)$ is the inverse of the production function:

$$X = F(L, Z), \qquad (19.19)$$

where L is man-hours and the Z-vector includes the amount of capital.

A sectoral production function based on aggregation from micro units with fixed capacity and man-hours per unit of production, is shown in Fig. 19.1. Production units are ranked according to their requirements of man-hours per unit of production $(l_i = L_i/X_i)$, such that $l_1 \leq l_2 \cdots < l_N$ if there are N production units. Production capacity of unit number i is denoted by K_i.

In Fig. 19.1 there are initially three production units. The unbroken curve represents the sectoral step production function. Over time new production units will enter, and old ones will be scrapped. Scrapping may be due to physical or economic depreciation. In the latter case units with high l_i are scrapped because

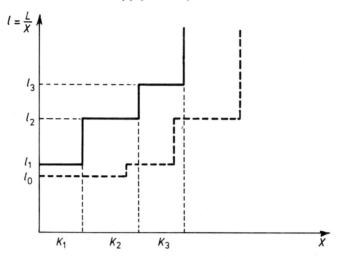

Fig. 19.1 Sectoral production function based on aggregation of micro units with fixed input coefficients

sales do not cover variable costs. New units will be based on the latest technology available, and will also be well adapted to the level and trend in wage costs. When wages are rising, new units will be characterized by low l_i's, i.e. labour productivity is high. In Fig. 19.1 the broken curve represents the situation after the scrapping of unit 3, and the entry of unit 0 ($l_0 < l_1$ and $K_0 > K_3$). The sectoral curve will typically shift downwards over time as a result of investment in new units. In Fig. 19.1 we have assumed that sector capacity is growing. Of course this will only be the case if gross investment exceeds the capacity of scrapped units.

Aggregation of sector production functions as illustrated by Fig. 19.1 is of course only straightforward when production units are price takers. Within a framework of monopolistic competition, the sectoral production function is not to be interpreted as a technological relationship as indicated by Fig. 19.1, but rather as an economic relationship (see Hoel and Nymoen (1986)). However, we feel that even if the exact conditions of aggregation are quite complicated to sort out in the case of monopolistic competition, this should not keep us from applying the basic idea to macro economic modelling.

In Fig. 19.2 the solid $L'(X)$ curve represents an approximation to the production function in Fig. 19.1, giving the *marginal* labour requirement as a function of output. The broken curve $l(X)$ gives the corresponding *average* labour requirement. By convention, this kind of approximation is justifiable by the fact that the number of units in each sector is quite large. In Fig. 19.2, K represents the aggregate capacity of the sector, while T represents the technique of the most efficient unit.

The sectoral production function in Fig. 19.1 assumes that all labour is associated with the level of production. More realistically, a part of the labour input is related to capacity, not to production. Thus L_i and ($l_i = L_i/X_i$) should be

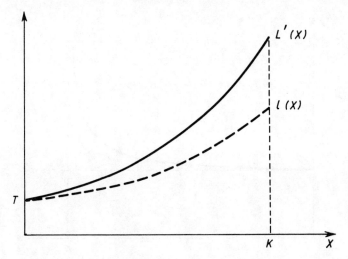

Fig. 19.2 Continuous sectoral production function

reinterpreted in terms of that part of the man-hours which is proportional to production.

Following the distinction above we define:

$$L = L^K + L^X = G(X, Z), \qquad (19.20)$$

where L is total man-hours. L^X is employment associated with production, L^K is capacity-dependent employment. We may further write:

$$G(X, Z) = L^K + \int_0^X G_X(s, Z)\, ds \qquad (19.21)$$

G_X is man-hours requirements associated with a small increase in production. The solid curve $L'(X)$ in Fig. 19.2 corresponds to $G_X(X, Z)$ with $G_{XX}(X, Z) > 0$. Total use of labour per unit of production is given by:

$$\frac{L}{X} = \frac{G(X, Z)}{X} = \frac{L^K}{X} + \frac{1}{X} \int_0^X G_X(s, Z)\, ds \qquad (19.22)$$

Due to the first term on the right-hand side of the equation, $G(X, Z)/X$ may or may not rise with the scale of production. The broken curve $l(X)$ in Fig. 19.2 corresponds to the second term on the right-hand side of Equation 19.22.

The following two equations complete our specification:

$$L^K = kTK \qquad (19.23)$$

$$L^X = TX + g\left[X + (K - X) \log\left(\frac{K - X}{K}\right) \right] \qquad (19.24)$$

g and k are non-negative constants.

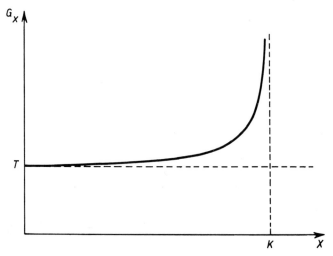

Fig. 19.3 The shape of the G_X curve

The properties of the $G_X(X, Z)$ function are of course crucial to the price equation. From Equation 19.24 we obtain:

$$G_X(X, Z) = T - g \log\left(\frac{K - X}{K}\right) \geq T \tag{19.25}$$

$$G_{XX}(X, Z) = \frac{g}{K - X} > 0 \tag{19.26}$$

$$\lim_{X \to K} G_X(X, Z) = \infty \tag{19.27}$$

$$\lim_{X \to 0} G_X(X, Z) = T \tag{19.28}$$

Figure 19.3 shows $G_X(X, Z)$ for fixed values of T, K and g. Figures 19.4 and 19.5 illustrate how an increase in K and a reduction in T affect the G_X curve.

The technology parameter, T, is assumed to follow an exogenous trend:

$$T = T_0(1 - ht) \approx T_0 e^{-ht} \tag{19.29}$$

h is a non-negative parameter, and t represents time. Substituting Equations 19.18, 19.25 and 19.29 into the price Equation 19.17 gives

$$P = \alpha\left\{vQ^V + W\left[T_0(1 - ht) - g \log\left(\frac{K - X}{K}\right)\right]\right\} + \beta Q^K \tag{19.30}$$

Note that this analysis easily extends to the case where individual firms face two separate demand curves, one representing domestic demand, and the other

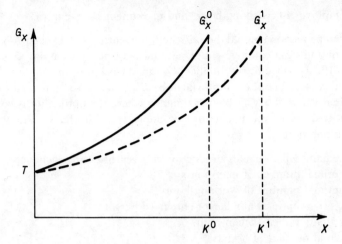

Fig. 19.4 Shifting the G_x curve: increased capacity

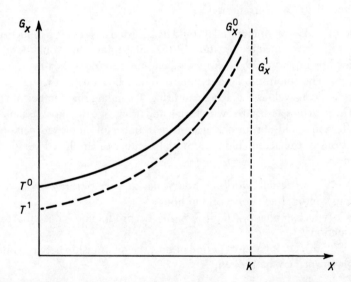

Fig. 19.5 Shifting the G_x curve: change in technology of most efficient unit

representing foreign demand (exports). In this case two separate price equations are derived, which we may write:

$$P_j = \alpha_j \left\{ vQ^v + W \left[T_0(1-ht) - g \log\left(\frac{K-X}{K}\right) \right] \right\} + \beta_j Q_j^K \quad (19.31)$$

where $j = H$ indicates variables and parameters with reference to the domestic market, and $j = A$ refers to the equation for export prices.

19.2.2 Employment, working hours and investment in inventories

The preceding section ignored the existence of inventories of finished goods, and of the cost of adjusting employment. The number of persons employed is given by Equation 19.20 divided by the number of optimal working hours. In this section we denote this optimal level of employment by L^o. Lower case 'o' is used to indicate the optimal value of a variable throughout, where the optimality is conditional upon the absence of adjustment costs of employment. In this section, we use the following notation:

Y_t = demand which equals production if inventories are unchanged
L_t^o = optimal number of man-hours
H_t^o = optimal number of working hours
$N_t^o = L_t^o/H_t^o$ = optimal number of employed persons
L_t = actual number of man-hours
H_t = actual number of hours
$N_t = L_t/H_t$ = actual number of persons employed
S_t^o = optimal stock of inventories
S_t = actual stock of inventories.

If labour (and input of materials) could be adjusted costlessly, production would be equal to Y determined by Equation 19.16, and employment would be equal to L^o determined by Equation 19.20. Adjusting employment does, however, incur costs on the firm. The amount of labour employed may be adjusted through changes in working hours as well as by hiring and firing. Changing the number of employed persons incurs costs associated with hiring and firing. We therefore assume that the firm prefers to avoid abrupt changes in the number of employed persons.

Deviations of the actual and the optimal number of employed persons may be obtained by:

(1) $H \neq H^o$, i.e. actual working hours deviate temporarily from the (exogenous) optimal level of working hours
(2) $X \neq Y$ (which implies $L \neq L^o = G(Y, Z)$), i.e. production is not equal to optimal production
(3) $L \neq G(X, Z)$, i.e. for a short period of time the relationship between output and employment may be broken.

However, none of these alternatives are costless: $H > H^o$ results in overtime compensation. $H < H^o$ accompanied by reduced wages is often not feasible. Furthermore, wage costs are partly independent of working hours; thus $H < H^o$ will also incur extra costs. If $X \neq X^o$, the firm must either allow price to deviate from its *ex ante* level (Equation 19.7), or production must deviate from actual (*ex post*) demand. As discussed in Section 19.2.1 we assume that firms prefer to avoid abrupt price changes, and prices are thus fixed at their *ex ante* level. $X < Y$ either implies rationing, or a rundown of stocks to below their optimal level. Both of these possibilities are adverse to the firm. Similarly, $X > Y$ implies an adverse increase in inventories.

Theoretical considerations

Deviations from the production function at 'low' production ($L > G(X,Z)$) are often referred to as labour hoarding. $L < G(X,Z)$ is also possible if workers are (temporarily) working harder than normal. Within an implicit contract framework this may be feasible for a short period of time (Strand, 1983). Formally, this possibility may be represented by writing the production function $X = F(EL, Z)$ where E is 'work intensity'. The normal level of E is 1, but for a short period of time E may deviate from 1.

From the above discussion we conclude that firms will prefer small numerical values of $(N_t - N_{t-1})$, $(S_t - S_t^o)$, $(H_t - H_t^o)$ and $(E_t - 1)$. Formally, we assume that firms minimize the loss function:

$$B = B(N_t - N_{t-1}, S_t - S_t^o, H_t - H_t^o, E_t - 1) \tag{19.32}$$

subject to the constraints:

$$X_t = Y_t + (S_t - S_{t-1}) \tag{19.33}$$

$$X_t = F(E_t, H_t, N_t, Z_t) \tag{19.34}$$

and for given values of $Y_t, H_t^o, S_t^o, S_{t-1}$ and N_{t-1}. In general, the solution to this problem will define:

$$I_t = \Phi^I(N_{t-1}, S_{t-1}, Y_t, S_t^o, H_t^o), \quad I = N, S, H \tag{19.35}$$

The three equations in Equation 19.35, together with Equations 19.33 and 19.34, will determine the five endogenous variables; X_t, N_t, S_t, H_t and E_t. The Φ^I functions may of course be rather complicated. In the following we will approximate the system by linear functions. We also make the system subject to the following stationarity conditions:

$$N_t = N_t^o, \ S_t = S_t^o, H_t = H_t^o, X_t = Y_t \text{ and } E_t = 1.$$

Then we obtain the following linear system:

$$N_t = a_{NN} N_{t-1} + (1 - a_{NN})N_t^o + a_{NS}(S_{t-1} - S_t^o) \tag{19.36}$$

$$S_t = a_{SS} S_{t-1} + (1 - a_{SS})S_t^o + a_{SN}(N_{t-1} - N_t^o) \tag{19.37}$$

$$H_t = H_t^o + a_{HN}(N_{t-1} - N_t^o) + a_{HS}(S_{t-1} - S_t^o) \tag{19.38}$$

In order to comment on the signs of the parameters, we assume that the system is initially in a condition of stationarity. Assume first that Y_t and N_t^o increase, while S_t^o is unchanged. Presumably actual employment N_t will increase, but less than N_t^o. At the same time, H_t increase while S_t is reduced. That is to say

$$0 < a_{NN} < 1$$
$$a_{SN} > 0 \tag{19.39}$$
$$a_{HN} < 0$$

Assume next that Y_t and N_t^o are unchanged, while S_t^o increases (for example due to lower costs of investment in inventories). We assume that S_t will increase, but less than S_t^o. Furthermore N_t and H_t increase, as production must rise in order to meet increased demand for inventories. This implies the following signs for the relevant coefficients:

$$0 < a_{SS} < 1$$
$$a_{NS} < 0 \qquad (19.40)$$
$$a_{HS} < 0$$

The following supplementary condition (given conditions 19.39 and 19.40) is necessary and sufficient to ensure overall stability of the system:

$$a_{NN}a_{SS} - a_{SN}a_{NS} < 1 \qquad (19.41)$$

Rosanna (1984) obtains a system similar to the above within an explicit intertemporal framework. The restrictions on the parameters in Rosanna's model are quite similar to ours, with the exception of a_{HN} which may be positive in Rosanna's model. This is due to his specification of the production function, where working time enters as a separate argument.

By specifying relations for optimal employment and optimal stock of inventories, we obtain a simultaneous model of price and employment determination. The optimal number of employed persons is given by:

$$N^o = \frac{G(X, Z)}{H^o} \qquad (19.42)$$

As for optimal stocks of finished goods, we simply assume:

$$S^o = sX \qquad (19.43)$$

Equations 19.16, 19.30, 19.33, 19.36–19.38, 19.42 and 19.43 make up a simultaneous model which determines the eight endogenous variables Y, X, N^o, S^o, P, N, S and H.

19.3 EMPIRICAL PROPERTIES

This section describes some salient features of supply-side response in the RIKMOD model of the Norwegian economy. The examples concentrate on the three manufacturing sectors of the model: export industries excluding oil (1), manufacturing of investment goods (2) and manufacturing of food and beverages (3).

Over the past decade or so, one of the most influential contributions to the econometric analysis of time series, has been the methodological proposals of Sargan, Hendry and Mizon among others, frequently referred to as the 'LSE methodology'. A recent contribution to this literature is McAleer et al. (1985). When specifying the price equations, we proceed in much the same way as proposed in McAleer et al. The price equations are first estimated as general autoregressive distributed lag (AD) equations. Due to the relatively short period of estimation (1967(1)–1982(4)) the most general models include four quarter lags of all variables. Sequential testing of simplified specifications is subsequently carried out. The simplified equations have been made subject to standard misspecification tests: Lagrange multiplier tests of autocorrelation and heteroscedasticity up to eight quarters and Chow-tests of parameter stability.

Received theory suggests explanatory variables which were omitted from our maintained hypothesis. Misspecification tests also include significance tests of these variables. As an example we ran tests of variables representing the level or change in aggregate demand. These tests did not give significance to *direct* influence from aggregated demand on prices. However, we shall see that demand affects prices through the level of capacity utilization in the production sectors.

The specified dynamic response pattern of the model of employment, inventories and working hours, precluded use of the 'general to specific' methology when estimating this model. However, misspecification tests were employed to the same extent as is the case with the price equations. Details of the estimated equations are reported in Appendix A.

19.3.1 Determination of prices

The data used for the estimation of the price equations are mainly from the Norwegian quarterly national accounts. The exception is the capacity measure, where we have analysed several approaches, namely the modified Wharton index and indices based on estimation of frontier production functions (Cappelen and Jansen, 1984). Our investigations as to which method is best suited for short-term modelling remain, however, inconclusive.[1]

In Table 19.1 we report elasticities of price of goods delivered to the Norwegian market, with respect to wage costs per hour, price of materials and price of

Table 19.1 Elasticities of price of goods delivered to the Norwegian market, with respect to sectoral wage cost per hour (W), price of material input (PM) and price of imports (PB)

	Sectors		
	Export (1)	Capital goods (2)	Food (3)
W	0.23 (0.25)	0.43 (0.45)	0.22 (0.23)
PB	0.20 (0.34)	— (0.22)	0.21 (0.26)
PM	0.57	0.45	0.57
Extra-sectoral prices	(0.41)	(0.41)	(0.51)

Input–output corrected elasticities in round brackets

[1] The Econometric Modelling Group at the Norwegian Bureau of Statistics have had some success using indices based on trend through capital–output peaks on Norwegian data, output being adjusted for variations in the number of working days (Cappelen and von der Fehr, 1985).

imports. The elasticities refer to the stationary solutions of the dynamic price equations.

The elasticity with respect to wage cost is highest in the capital producing industries. This reflects the high labour intensity of this industry, in which construction of residential and business buildings is an important subsector. The direct elasticities of sector 2 do not sum to unity, due to a constant term which is included in the estimated equation. (Thus the elasticity attributed to external prices includes the effect of this constant term.)

Being a multisectoral model, an input–output accounting structure is an integral part of the RIKMOD model. The price of a sector's material inputs is thus endogenous in the model (Jansen, 1984) and is dependent on own-sector wage costs and import price, as well as wage costs and prices of other sectors. The numbers in round brackets in Table 19.1 show the reduced form elasticities with respect to wage costs and import price of each sector. We find that import prices is attributed the highest elasticity both in export industries (sector 1), and in manufacturing of food (sector 2). Also for production of capital goods (sector 2), we find that import prices of capital goods have a substantial impact on domestic prices. This effect, however, stems exclusively from the input–output accounting structure, since no *direct* effect of import prices on prices of domestic production is picked up in the estimated equation.

When appreciating the total effect of import prices on domestic price formation, one should also include the effect on wages. Recent empirical work on wage determination in Norwegian manufacturing suggests that this effect is substantial (Hoel and Nymoen, 1986; Stølen, 1985). Thus, recent research on price and wage formation in Norway seemingly lends support to the main hypothesis of the Scandinavian model of inflation (Aukrust, 1977); i.e. in the long run the price development of a small open economy is determined by competitors' prices and productivity growth. Often this kind of inflationary response is taken as evidence of price taking firms. Our results do not lend support to this kind of inference. Instead we suggest that a behavioural basis for the hypothesis of the Scandinavian model of inflation is to be found in the interplay between a model of monopolistically competitive firms and a bargaining model of wage formation (Nymoen, 1986).

Finally, Table 19.1 illustrates the importance of costs of cross-sectoral deliveries of intermediate goods. We take sector 3 as an example; consider a 1% increase in wage costs and import prices on input from other sectors. (The level of output is assumed to be unaffected by this change.) Due to homogeneity, domestic prices of food and beverages will be up by 1%. However, Table 19.1 suggests that only one-half of the price increase may be attributed to intrasectoral cost–push elements, the other half is explained by increased costs of inputs from other sectors. Therefore important cost–push elements of sellers prices from sector 3 will be prices of deliveries from sheltered industries such as fisheries and farming. As for export industries, the price of hydroelectric power is an important extrasectoral cost–push element.

Table 19.2 Elasticity of domestic price with respect to wage costs, as function of capacity utilization

	Capacity utilization						
Sector	0.50	0.60	0.70	0.80	0.90	0.95	0.99
Export (1)	0.23	0.23	0.24	0.25	0.26	0.28	0.31
Capital (2)	0.41	0.41	0.41	0.42	0.42	0.42	0.43
Food (3)	0.21	0.21	0.22	0.22	0.23	0.24	0.25

In the above discussion, we have assumed a constant level of capacity utilization. However, changes in demand will influence prices through changes in capacity utilization. Table 19.2 shows the elasticity of price with respect to wage costs, for different values of capacity utilization. The effect of capacity utilization is most pronounced for export industries. According to Table 19.2 the technology in capital-producing industries has approximately constant returns to scale. This result is likely to be due to the problem of measuring capacity. The high labour intensity in this industry suggests a synthetic concept of capacity (Johansen, 1968). In order to measure a synthetic concept of capacity, one may start by estimating a frontier production function (Cappelen and Jansen, 1984). When constructing series on capacity from the estimated production function, one should use data on available man-hours rather than actual man-hours. So far, however, we have used actual man-hours, which may greatly diminish the relevance of our capacity measure.

In the present version of the model only export prices for the sector labelled export industries is modelled, export prices for goods produced by other sectors being determined on the world market. The estimated equation determines export price as a mark-up on marginal costs and also allows for direct influence of competing prices. The result suggests an elasticity of 0.77 with respect to competing prices and 0.23 with respect to marginal costs. This confirms expectations about behavioural differences between export and domestic prices, as cost shifting is less typical for export prices. (See Frantzen (1985) for evidence on Belgian data).

19.3.2 Employment, inventories and working hours

The three-equation model of employment, inventories and working hours has been estimated for the three manufacturing sectors for which reliable data on inventories are available on a yearly basis from the statistics of Norwegian manufacturing industry. These data have been transformed to quarterly data by utilizing quarterly indices of stocks of inventories.[2] Thus our data on inventories

---

[2] The indices of stocks of inventories have kindly been made available by the Econometric Modelling Group at the Central Bureau of Statistics.

should be better suited for econometric work than the National Accounts data on changes in inventories.

Data on employment and working hours are extracted from the employment accounts developed at the Central Bureau of Statistics (Harildstad, 1986). The number of employees is measured as if all work is carried out by full-time wage earners.

We estimate the models using several dummy variables. In addition to seasonal dummies, we introduce dummies in order to capture the policy measures directed towards the manufacturing industry during the seventies. These measures include export credits and cheap loans to finance investment in inventories. Although primarily directed towards export of manufactures, export credits – along with an official guarantee for the debt of shipowners and the high investment level in the North Sea – also alleviated the crises in Norwegian shipyards in the wake of the OPEC 1 oil price shock in 1973. Details on the estimation results are available in Appendix A.

Inference on the basis of these results is relatively more tentative than is the case for the price equations. This is particularly the case for the inventory equations, where standard residuals are relatively large, and where parameter constancy seems to be a problem in sectors 1 and 3. Thus large prediction errors on inventories remain probable. However, post sample dynamic tracking experiments do not suggest that prediction errors on inventories are large enough to push employment predictions off beam.

19.3.3 Simulation properties

This section illustrates the dynamic response pattern implied by the estimated behavioural equations. To this end we run simulations on three separate models, one for each manufacturing sector of the RIKMOD model. The simulated models may be seen as empirical versions of the theoretical models discussed in Sections 19.2.1 and 19.2.2. In order to capture the endogenous nature of costs of input of materials, we have added equations describing material input prices as linear functions of own and extrasectoral import prices. The coefficients of the equations are extracted from the input–output accounting structure of the RIKMOD model. (The base-year at present is 1983.)

In the three sectoral models, demand-side response is rather simple and is represented by estimated sectoral demand functions (e.g. Equation 19.16). Sectoral demand is described as (log linear) functions of relative prices and indicators of aggregate demand. As for the export sector, output of manufactures in OECD countries is used as an indicator of aggregate demand; for production of capital goods we use domestic investment, and for production of food and beverage we use private consumption. Domestic investment and consumption (at current prices) is kept exogenous in the simulations. Endogeneity with respect to these variables would involve larger parts of the RIKMOD model, and this is not feasible at present. Moreover, we feel that our representation of the demand side

Table 19.3 Effect on prices, inventories and employment resulting from a 1% increase in production in OECD countries (sector 1), a 1% increase in domestic investment (sector 2), and a 1% increase in domestic private consumption (sector 3). Percentage spread between control- and experiment simulation

		Export				Capital			Food		
		P_H	P_A	S	N	P_H	S	N	P_H	S	N
Year:	1	0.18	0.01	−0.15	0.14	0.04	0.05	0.05	0.02	0.08	0.12
	2	0.38	0.04	−0.04	0.52	0.06	0.17	0.20	0.02	0.21	0.33
	3	0.33	0.02	0.34	0.72	0.06	0.19	0.28	0.05	0.26	0.43
	4	0.22	0.01	0.55	0.84	0.06	0.18	0.30	0.02	0.28	0.51
	5	0.22	0.01	0.68	0.87	0.06	0.21	0.32	0.02	0.28	0.51
	6	0.25	0.02	0.70	0.88	0.06	0.21	0.33	0.02	0.28	0.53

is sufficient for the purpose of demonstrating salient features of supply-side response.

Our first experiment is related to a sustained 1% increase in the nonprice exogenous variables in the demand equations (i.e. production of manufactures in OECD countries (sector 1), aggregate domestic investment demand (sector 2), and private consumption (sector 3)). Control and experimental simulations were run for each of the three sectoral models over the time period 1978(1) to 1983(4). The results of the experiments are shown in Table 19.3.

Firstly, note that increased marginal costs in export industries, resulting from increased production in order to meet foreign demand, results in a significant rise in prices of goods sold on the *domestic market*, while export prices only increase modestly. This reflects the fact that firms in this industry face steeper demand curves at home than abroad. Consequently shifting of costs on to export prices is modest.

Price responses in the two other sectors are much weaker than in the export industries. In the case of capital-producing industries this largely reflects an overall constant-returns-to scale technology (cf. the above discussion). In the case of sector 3, however, the results are partly due to considerable excess capacity implied by the control simulation. Thus, a 5% increase in consumer expenditure results in a mean percentage price increase ten times larger than the increase implied by Table 19.3. This result is of course due to the shape of the marginal cost curve, which implies 'Keynesian' price response at low levels of capacity utilization, while a sharp price increase may be the result if demand is increased at high levels of capacity utilization.

We now turn to quantity response. In our model of monopolistic competition, increased demand is met by increased production since by assumption *instantaneous* price adjustment does not take place. Accordingly, both the optimal level of employment and the optimal level of inventories rise when demand is increased. However, increased optimal employment affects stocks of finished goods negati-

Table 19.4 Effects of a 1% increase in own-sector wage cost per hour

		Export				Capital			Food		
		P_H	P_A	N	S	P_H	N	S	P_H	N	S
Year:	1	0.26	0.03	−0.02	0.03	0.41	−0.06	−0.08	0.22	−0.02	−0.01
	2	0.20	0.02	−0.11	0.01	0.34	−0.28	−0.29	0.29	−0.02	−0.05
	3	0.15	0.02	−0.14	−0.07	0.32	−0.62	−0.35	0.29	−0.12	−0.07
	4	0.16	0.02	−0.13	−0.11	0.32	−0.73	−0.46	0.29	−0.14	−0.08
	5	0.10	0.02	−0.11	−0.10	0.32	−0.78	−0.51	0.29	−0.15	−0.08
	6	0.08	0.02	−0.10	−0.09	0.32	−0.79	−0.50	0.29	−0.16	−0.08

vely. Accordingly, when both the optimal levels of stocks and employment are increased, actual stocks of finished goods may or may not rise, dependent on the magnitude of these two effects. From Table 19.3 we find that the level of inventories in export industries is *run down* as a first response to increased demand. Thus the negative effects from increased optimal employment initially outweigh the positive effects resulting from the increased level of optimal stocks of finished goods. Eventually, actual stocks of finished goods must rise to equal optimal stocks. From Table 19.3 we find that stocks are up by 0.34% compared with the base run after 3 years, and continue to rise until the end of the simulation period.

As for the two other sectors, the positive effects from increased optimal stocks of finished goods outweigh the negative effect from increased employment. The percentage increase in employment is greater than the percentage increase in production. In sector one, production is up by 0.78% after 6 years; employment is up by 0.88%. This effect reflects the decreasing returns technology.

Table 19.4 shows the effects of a 1% sustained increase in sectoral wage costs. As regards the export industries, most of the cost–push effect appears on sales to the domestic market, export prices showing only a modest increase. Note also that the initial price increase is reduced somewhat as marginal costs go down due to reduced production. Both price and employment effects are most pronounced in the capital-producing sectors. This is due to the combined effect of relatively low productivity and high own-sector deliveries of material input.

The simulations reported in Table 19.4 exclude effects from increased earnings on sectoral demand. On the other hand, induced price increases on input from other sectors are also neglected. The first of these effects would of course contribute to employment stabilization, while the second mechanism would affect employment adversely. Clearly, in order to appreciate the net effect, the price and employment equations have to be incorporated in the complete input–out structure of RIKMOD.

In Table 19.5 we show the results of simulation experiments where competing prices on goods are increased permanently by 1%. Note that export industry prices are more elastic with respect to increased import and world market prices than with respect to wage costs. As for capital-producing industries, employment

Table 19.5 Effects of a 1% increase in competing prices

		Export				Capital			Food		
		P_H	P_A	N	S	P_H	N	S	P_H	N	S
Year:	1	0.23	0.68	0.08	−0.09	0.11	0.10	0.13	0.23	−0.01	−0.01
	2	0.56	0.78	0.24	0.00	0.30	0.55	0.58	0.29	−0.04	−0.02
	3	0.52	0.80	0.37	0.20	0.31	1.06	0.81	0.29	−0.06	−0.03
	4	0.50	0.81	0.41	0.29	0.32	1.25	0.77	0.27	−0.07	−0.04
	5	0.49	0.81	0.42	0.33	0.32	1.33	0.86	0.27	−0.06	−0.03
	6	0.49	0.81	0.42	0.34	0.32	1.36	0.86	0.26	−0.06	−0.03

effects are very strong, corresponding to the high elasticities with respect to import and exprice implied by the sectoral demand equation. Finally, increased competitors' prices have an adverse effect on employment in manufacturing of food and beverages, due to the combined effect of a relatively small substitution effect in demand, and relatively high direct cost–push effect in the estimated price equation.

19.4 CONCLUSIONS

In this chapter we have derived relations concerning price, employment and investment of inventories within an explicit model of monopolistic competition.

We find the overall results of our work so far rather encouraging. Admitted, the estimated equations have their fair share of inadequacies. As an example, lack of parameter constancy may indicate that the inventory equations are misspecified. We feel, however, that changes in the specification, which are compatible with the general approach of this chapter, may remedy this shortcoming. As an example our present specification describing optimal stock of inventories as a linear function of production may be far too simple.

The approach taken in this chapter has exciting prospects for future work. For one thing, a better understanding of the transmission of world inflation to domestic inflation, so typical of small open economies, seems to be within reach by combining an explicit model of monopolistic competition, a bargaining model of wage formation and the input–output structure of a multisectoral model. Taken together this gives us three analytically distinctive and *logically consistent* channels of transmission of world inflation to domestic prices. Firstly, the model of monopolistic competition spelled out in Section 19.2 gives a rationale to include competitors' prices in the domestic price equations. Evidence on the significance of this *direct* effect is reported in Section 19.3 and in Appendix A. Secondly, import prices affect domestic prices through shifting of costs, since material input prices are dependent on import prices through the input–output structure. Thirdly, a bargaining model of wage formation, assuming that negotiators take account of a labour demand schedule consistent with firm behaviour under monopolistic competition, identifies competing prices as an important explanatory variable for wage formation. In this way the short-run producer behaviour under monopolistic competition becomes an integral part of a model which explains price and wage formation in open economies.

REFERENCES

Aukrust, O. (1977) Inflation in open economy: A Norwegian Model, in *Worldwide Inflation: Theory and Recent Experience* (eds L. B. Kraine and W. S. Sâlant), Brookings Inst., Washington, DC.
Cappelen, Å. and Jansen, E. S. (1984) The measurement of potential output: a frontier production function approach, Working Paper 1984/1, Bank of Norway, Oslo.
Cappelen, Å. and von der Fehr, N. H. M. (1985) Kapasitetsutnyttelse i KVARTS-sektorer, Central Bureau of Statistics, Oslo.
Eastman, H. C. and Stykolt, S. (1966) *The Tariff and Competition in Canada*, University of Toronto Press, Toronto:
Frantzen, D. J. (1985) The pricing of manufactures in an open economy: a study of Belgium. *Cambridge Journal of Economics*, 9, 371–82.
Harildstad, A. (1986) Det norske arbeidskraftsregnskapet. *Økonomiske Analyser*, No. 2, 25–34.
Harris, R. (1984) Applied general equilibrium analysis of small open economics with scale economics and imperfect competition. *American Economic Review*, 74, 1016–32.
Hoel, M. and Nymoen, R. (1986a) Kortsiktig produsentatferd i RIKMOD, Bank of Norway, Oslo.
Hoel, M. and Nymoen, R. (1986b) Wage determination in Norwegian manufacturing: an empirical application of a theoretical bargaining model, Bank of Norway, Oslo.
Jansen, E. S. (1984) Model building in Norges Bank. *Economic Bulletin*, No. 3, Bank of Norway, Oslo.
Johansen, L. (1968) Production functions and the concept of capacity. *Recherches Recente sur la Fonction de Production*, No. 2, Cerena Namur.
McAleer, M., Pagan, A. R. and Volker, P. A. (1985) What will take the con out of econometrics? *American Economic Review*, 75, 293–307.
Nymoen, R. (1986) Lønnsdannelse i industrien. En teoretisk forhandlingsmodell og empiriske lønnsrelasjoner, arbeidsnotat 1, Bank of Norway, Oslo.
Rosanna, R. J. (1984) A model of the demand for investment in inventories of finished goods and employment. *International Economic Review*, 25, 737–47.
Strand, J. (1983) Layoff, labor productivity and worker seniority rules, Memorandum 20, University of Oslo, Oslo.
Stolen, N. M. (1985) Faktorer bak lønnsveksten. *Økonomiske Analyser*, No. 9, 29–49.

APPENDIX A DEFINITION OF VARIABLES AND ESTIMATION RESULTS

Definition of variables

(a) Price equations

PH_j = domestic price index; production sector j (1983 = 1)
PA_j = export price index; production sector j (1983 = 1)
PB_j = price index of competing imports; production sector j (1983 = 1)

PM$_j$ = price index of *external* material input; production sector j (1983 = 1)
WPH$_j$ = labour cost per hour; nominal NOK
XC$_j$ = capacity utilization rate, sector j
t = trend variable
V$_j$ = material input per unit of production
MC$_j$ = marginal cost of production; nominal NOK.

PM$_j$ is used as an instrumental variable for the price index on total material input.

(b) Equations for employment and stocks of finished goods

N$_j$ = number of full-time equivalent employed wage earners (1000 persons)
$\bar{L}_{xj}$ = the optimal number of employed persons associated with the level of production
S_j, S_j^0 = Actual and optimal level stocks, sector j
D1, D2, D3 = dummy for 1, 2, and 3 quarter
LAG$_j$ = economic policy dummy in equations for stocks of finished goods
SLK$_j$ = economic policy dummy used in employment equation.

Statistics

$\eta(j, N-k)$ = modified Lagrange multiplier test of autocorrelation of order j (N = number of observations, k = number of regressors)
ARCH(j) = Lagrange multiplier test of j-order autoregressive conditional heteroscedasticity
CH($k, N-2k$) = two-sample Chow-test
The two subsamples are 1967/1 to 1976/4 and 1977/1 to 1982/4
CHPS($N_2, N_1 - k$) = post sample Chow-test
N_1 is the number of observations 1967/1–1979/4; N_2 is the number of observations 1980/1–1982/4.
Single coefficient t-values are in round brackets.
SSR = sum of squared residuals
SER = standard error of regression

Domestic prices: export sector

1967(1)–1982(4)

Method: 1. Order Cochrane Orcutt; RHO(1) = 0.616
$\qquad\qquad\qquad\qquad\qquad\qquad\qquad\quad$ (6.1)

SER = 0.0136   LHS-MEAN = 0.509
$\eta(2, 45) = 0.71$ $\eta(4, 43) = 0.63$ $\eta(8, 39) = 0.42$
ARCH(1) = 0.25   ARCH(4) = 1.69
CH(8, 48) = 1.56   CHPS(12, 43) = 1.45

$$PH_1 = 0.896\,(V_1 PM_1) + 0.005\,WPH_1 + 0.192\,PB_1(-1)$$
$$(9.19)(5.37)(2.71)$$
$$-5.73 \times 10^{-5}\,WPH_1 \times t - 5.31 \times 10^{-5}\,WPH_1\,LOG\,(1-XC_1)$$
$$(5.77)(3.41)$$
$$-6.99 \times 10^{-5}\,WPH_1(-1)\,LOG\,(1-XC_1(-1))$$
$$(4.25)$$
$$-5.6101 \times 10^{-5}\,WPH_1(-2)\,LOG\,(1-XC_1(-2))$$
$$(2.99)$$

Export prices: export sector

1967(1)–1982(4)

Method: ordinary least squares (OLS)

SER = 0.017   LHS-MEAN = 0.009
$\eta(1, 50) = 0.8$ $\eta(4, 47) = 0.92$ $\eta(8, 43) = 0.87$
ARCH(1) = 0.001   ARCH(4) = 4.32   CH(3, 58) = 0.4   CHPS(12, 49) = 1.35

$$\Delta PA_1 = 0.07\Delta MC_1 + 0.54\Delta PB_1 + 0.29\Delta PA_1(-1)$$
$$(1.4)(6.75)(3.22)$$

Domestic price: production of capital goods

1967(1)–1984(4)

Methods: OLS

SER = 0.0068   LHS-MEAN = 0.556
$\eta(1, 47) = 0.07$ $\eta(4, 44) = 0.37$ $\eta(8, 40) = 0.83$
ARCH(1) = 0.22   ARCH(4) = 0.08
CH(7, 50) = 2.24   CHPS(12, 45) = 1.31

$$PH_2 = 0.054 + 0.48(V_2 PM_2) + 0.33(V_2(-1)PM_2(-1))$$
$$(5.4)(4.36)(3.30)$$
$$+ 0.005 WPH_2 - 1.86 \times 10^{-5} WPH_2 t$$
$$(6.25)(3.64)$$
$$- 2.58 \times 10^{-5} WPH_2 LOG\,(1 - XC_2)$$
$$(2.96)$$

Domestic prices: production of food and beverages

1967(1)–1982(4)

Method: OLS

SER = 0.0707   LHS-MEAN = 0.014
$\eta(1, 49) = 0.004$ $\eta(4, 46) = 0.66$ $\eta(8, 46) = 0.71$
ARCH(1) = 0.68   ARCH(4) = 5.27

$CH(4, 56) = 3.2$ $CHPS(12, 48) = 1.34$

$$\Delta PH_3 = 0.88\Delta(V_3PM_3) + 0.004\Delta WPH_3(-1)$$
$$\quad\quad (10.1) \quad\quad\quad\quad (4.44)$$
$$+ 4.6 \times 10^{-5}\Delta(WPH_3 LOG(1 - XC_3))$$
$$(3.83)$$
$$+ 0.19\Delta PB_3(-1)$$
$$(2.37)$$

Stocks of finished goods: export sector

1970(2)–1982(4) $R^2 = 0.9319$ $SSR = 2.503 \times 16^6$
$SER = 241.306$ $\eta(1, 38) = 0.78$ $\eta(4, 35) = 2.12$
$ARCH(1) = 0.22$ $ARCH(2) = 6.37$

LHS-variable is S_1

RHS-variable	Estimate	t-value
$(S_1(-1) - S_1^0)$	0.3937	4.33
$(N_1(-1) - \bar{L}_{X1})$	48.8768	3.79
Constant	2.1012×10^3	3.04
LAG_1	719.936	6.10
$D1$	361.17	3.06
$D2$	-74.88	0.68
$D3$	-44.95	0.47
t	-28.961	2.70

Employment: export sector

1970(2)–1982(4)

Method: OLS

$R^2 = 0.9965$ $SSR\ 39.7$
$SER = 0.964$ $\eta(1, 39) = 2.76$ $\eta(3, 46) = 1.36$
$ARCH(1) = 0.86$ $ARCH(4) = 2.92$

LHS-variable is N_1

RHS-variable	Estimate	t-value
$N_1(-1) - \bar{L}_{X1}$	0.8645	21.5
Constant	-11.1639	3.85
SLK_1	2.4140	1.30
$S1$	1.543	3.97
$D2$	3.39	8.69
$D3$	2.22	6.0
t	0.143	2.8
$SLK_1 t$	-0.070	1.16

Employment: production of capital goods

1970(2)–1982(4)

Method: OLS

$R^2 = 0.98$ SSR = 935.86
SER = 4.72 $\eta(1, 37) = 3.51$ $\eta(4, 34) = 3.32$
ARCH(1) = 6.18 ARCH(4) = 5.37

LHS-variable is N_2

RHS-variable	Estimate	t-value
$N_2(-1) - \bar{L}_{X2}$	0.7656	9.69
$S_2(-1) - S_2^0$	−0.00395	0.43
Constant	11.53	1.99
SLK_2	−24.17	1.93
D1	−1.43	0.74
D2	5.13	2.37
D3	10.2	4.9
t	−1.11	3.11
$SLK_2 t$	0.67	2.27

Stocks of finished goods: production of capital goods

1970(2)–1982(4) $R^2 = 0.60$ SSR = 3.5×10^6

SER = 285.39 $\eta(1, 37) = 0.02$ $\eta(4, 34) = 0.32$
ARCH(1) = 0.39 ARCH(4) = 1.32

LHS-variable is S_2

RHS-variable	Estimate	t-value
Constant	−65.35	0.43
$(N_2(-1) - N_2)$	3.073	1.19
$(S_2(-1) - S_2^0)$	0.158	1.31
D1	331.8	2.74
D2	173.68	1.37
D3	−13.077	0.10
t	5.9363	0.98
LAG_2	722.244	5.03

Appendix A

Employment: production of food and beverages

1970(2)–1982(4) $R^2 = 0.993$ SSR = 119.185

SER = 1.685 $\eta(1, 37) = 0.62$ $\eta(4, 34) = 0.78$
ARCH(1) = 0.45   ARCH(4) = 1.97

LHS-variable is N_3

RHS-variable	Estimate	t-value
$(N_3(-1) - \bar{L}_3)$	0.71856	5.61
$(S_3(-1) - S_3^0)$	−0.002632	2.63
Constant	−0.825	0.28
SLK_3	11.825	2.23
D1	4.218	3.81
D2	4.947	5.4
D3	8.52	9.5
t	0.3638	2.01
$SLK_3 t$	0.1976	2.37

Stocks of finished goods: production of food and beverages

1970(2)–1982(4) $R^2 = 0.682$ SSR = 8.810×10^5

SER = 139.97 $\eta(1, 37) = 2.12$ $\eta(4, 34) = 1.05$
ARCH(1) = 2.62   ARCH(4) = 3.09

LHS-variable is S_3

RHS-variable	Estimate	t-value
$(S_3(-1) - S_3^0)$	0.71347	7.51
Constant	17.872	0.25
D1	243.852	4.33
D2	59.672	1.06
D3	−0.861	1.30
t	−1.5159	1.14

20

Direct interventions, interest rate shocks and monetary disturbances in the Canadian foreign exchange market: a simulation study

KANTA MARWAH and HALLDOR P. PALSSON

20.1 INTRODUCTION

Caught between the tactics of a 'pesky platoon' of 'Chicago bears' and the sentiments of New York's financial fundamentalists, the Canadian dollar encountered intense downward pressures in the exchange markets in early 1986. This pressure, which started building up towards the end of October 1985, began to intensify late in January 1986. Initially, the Bank of Canada sought to resist this pressure by gradually increasing the short-term interest rates and by direct interventions in the exchange markets. By the beginning of February 1986, the gradual increase in the interest rates had already amounted to a cumulative increase of $2\frac{1}{2}$ percentage points since the end of October. At the same time, the differentials above similar rates in the US had widened to over 3 percentage points from $\frac{1}{2}$ percentage point. However, in the judgement of the Bank of Canada, 'the extent of this resistance evidently was either not widely realized or was not convincing'[1] as the negative sentiment against the Canadian dollar continued to mount. So much so that by the first week of February the dollar had fallen to almost 69 cents in terms of the US dollar from its initial level of 73 cents of late October. The psychological barrier of 70 cents had been broken for the first time.

Any further decline had to be thwarted, 'the extreme and unwarranted sell-off'[1] of Canadian dollars had to be countered. Convinced of their view that the pessimistic perceptions in the financial markets about the state of the Canadian

---

[1] Introductory statement by Gerald K. Bouey, Governor of the Bank of Canada, in an appearance before the House of Commons Standing Committee on Finance, Trade and Economic Affairs, Tuesday, February 11, 1986.

economy were rather misinformed and misplaced, and did not reflect the true status of its economic and financial fundamentals, the government and the Bank of Canada decided to move fast, and to move more aggressively.[2] The government bolstered its exchange reserves by new borrowings of $2.4 billion US (or $3.4 billion Canadian) and there was a massive intervention in the currency market by the Bank of Canada. A simultaneous and a forceful move into the Canadian money market pushed the short-term interest rates up, and led to an increase in the Bank Rate from 10.80% to 11.41%.

The response of the exchange market was prompt and as desired. The dollar jumped smartly above 71 cents US and the authorities felt euphoric. The dollar front became relatively quiet. Soon after, on 26 February, the Finance Minister presented his budget with a message that trumpeted the soundness of the Canadian economy and the government's agenda for control over its deficits. The exchange markets were not impressed and the pounding of the dollar started all over again. Today, the market is still volatile, and it is too early to predict the final outcome. Nonetheless, the gains achieved by massive interventions seem to be on the verge of extinction.

One may tend to view this recent saga of the Canadian dollar exchange market as not an isolated case. For the government interventions are simply a matter of short-term strategy of buying time that provides an artificial strength to the currency.[3] It is generally believed that the aim of Canadian interventions has always been to 'maintain orderly markets' by limiting excessive fluctuations but still letting the dollar find its own value.[4]

The purpose of this chapter is to study the short-to-medium run dynamics of the Canadian dollar and the associated entities under a variety of policy scenarios

---

[2] In fact, in his statement (see footnote 1), Governor Bouey rationalized the Bank's assessment of the current situation by comparing it with previous but recent times of intense pressures.

> 'During 1984 and again in early 1985 our dollar was subject to periods of intense exchange market pressure arising from increases in US interest rates and from the overall strength of the US dollar against all currencies. On this most recent occasion the external situation has been different – US short-term interest rates have been generally stable, their long-term rates have declined and the US dollar has fallen against other major currencies.'

Nonetheless, the Canadian dollar has declined. On the misinformed perceptions about the state of the Canadian economy, he observes,

> 'the Canadian economy has been showing an underlying strength and our balance of payments situation is sound despite the recent fall in oil prices and the more general weakness in commodity prices. Although there is evident room for further progress, our economic performance has been improving markedly, and there is nothing in the recent news which justifies the extent of the negative sentiment which has dominated financial markets in the past number of weeks.'

[3] Often at great costs in terms of unemployment and growth, especially when pursued through interest rates.

[4] Longworth (1980), Rose (1983), and the Department of Finance (1982) found that the policy has been symmetric since the end of the Bretton Woods system. However, Rose was able to identify fourteen subperiods during which different intervention regimes were operational. Thus while the aim of the policy was always the same, the mechanics of it were not constant.

including direct interventions. These scenarios are generated through the application of a recently developed (Marwah, 1985) balance-of-payments structural model of the foreign exchange market of Canada. The capital flows, the spot and forward exchange rates, and the monetary sector of Canada have all been endogenized in the model. The policy dynamics generated by the model are quite reasonable, and in most cases accord well with a priori expectations. We shall briefly describe the model and reaffirm its assumptions before presenting the policy scenarios.

20.2 THE MODEL

In its broad structure, Marwah's (1985) prototype model may be characterized as a 'balance-of-payments components' model in which the foreign exchange market of Canada has been fully intergrated with its monetary sector. The model endogenizes net capital flows, the spot and forward exchange rates, the short and long-term interest rates, and the demand and supply quantities of both money and quasimoney. The model is particularly relevant for a system of managed float whose two polar regimes of fixed and fully flexible exchange rates become its special cases. Government interventions have been related to the market adjustment process; they are generated endogenously but are quantified implicitly and globally across various markets.

Net capital flows have been disaggregated into *ten* functional and ownership categories. These are: direct investment, portfolio investment, resident official long-term capital, deposit money banks long-term capital, other long-term capital, resident official short-term capital, deposit money banks short-term capital, other short-term capital, errors and omissions, and effective financing and counterpart items. The exchange rates of the Canadian dollar including the spot, forward and forward-parity rates have been analysed against *five* major currencies which currently constitute the valuation basket of a unit of special drawing right (SDR). These are namely the French franc, the West German mark, the Japanese yen, the pound sterling and the US dollar. The model is designed to track their nominal movements in the short-to-medium run.

The equation structure and the parameter estimates of the model used in the present study are listed in Appendix A. The model contains altogether 37 equations, of which 27 are structural and ten are identities. It has been estimated by quarterly data for 1971–81 and has been tested by in-sample simulations and *ex-post* forecasts. Some structural highlights of the model may be briefly noted as follows.

In the model, the exchange rate is viewed primarily as an instrument tending to equilibrate the external balances on both current and capital accounts. It is neither treated simply as a relative price of two assets nor of two monies or two goods but a little bit of each. The model is thus neither pure portfolio nor pure monetary but combines the key elements from both. It embodies a portfolio allocation process in which assets denominated in different currencies are *not* assumed to be perfect substitutes because risk and uncertainty factors can neither be entirely eliminated nor made invariant. Furthermore, there is a massive evidence to support the contentions that foreign exchange markets are not fully efficient and interest arbitrage conditions do not essentially hold.[5] Thus, without conforming strictly to the monetary tenets of interest rate parity and purchasing power parity, it does recognize both these relationships, and, at the same time, emphasizes the role of capital flows.

Recognizing an emerging consensus on the role of market fundamentals, the model incorporates current and disaggregated capital account balances, GNP, domestic and foreign interest rates jointly with other major entities of domestic money market and foreign exchange market, and the role of speculation and expectations involving risk and uncertainties. Expectations and risk enter through variables such as forward exchange rate, expected yield differential and the variance of expected yield differential. In doing so, the model simultaneously determines capital flows, exchange rates and interest rates.

The primary fundamental factors which explain the capital flows are: the level, the change in the level and the variance of yield differentials on assets denominated in different currencies, price parity deviations, the rate of growth of portfolio of foreign assets and the credit needs arising out of current account transactions. The difference in the sensitivities of different categories of capital flows to these factors is clearly revealed by the parameter estimates. For example, direct investment seems to be subject to the 'flow' decision of portfolio allocation, but portfolio investment is subject to both 'flow' and 'stock' decisions. Moreover, in the case of portfolio investment, it is the speculative interactive yield component which seems to be all important. There is also evidence that residence official capital flows tend to counterbalance private flows; this tendency is specially true for long-term capital flows. Some other components move in congruity with each other.

The stock of net foreign assets is expected to remain unchanged for the foreign exchange market to be in equilibrium under a fully flexible exchange rate system. For it is only then that the demand for home currency generated by the current account balance will be equal to its supply generated by the net outflow of capital. An equilibrium exchange rate is obtained in the model by stipulating this condition in the form that the sum of current and capital account balances must be equal to zero. A nonzero actual change in the stock of net foreign assets is treated as an implicit measure of government interventions. The difference between the actual spot rate and the equilibrium spot rate is thus made a function of these interventions. In the final analysis, fundamental factors such as growth of net foreign assets, GNP, domestic interest rate and lagged forward rate enter directly in the equations explaining the spot rate of the Canadian dollar in terms of each of the five currencies. The lagged forward rate is significant to the extent that it processes

---

[5] See Section 20.3. And also, to the references cited in Marwah (1985), one may add Baillie and McMahon (1985).

the expectations about the future spot rate and serves as its predictor.[6] Judging by the magnitudes of partial elasticities, the Canadian dollar/Deutschmark forward exchange rate seems to be most effective in predicting its future spot rate. Similarly, Canadian dollar/yen exchange rate seems to be most responsive to changes in the Canadian interest rate and the Canadian dollar/USA dollar is the least.

The forward rate itself is determined recursively by the interest parity forward rate and the moving standard deviation of yield differentials on competing assets. Thus, whereas the forward market does account statistically for elements of market efficiency, it also explains its lapses from full efficiency, and shows that adjustments of the exchange rates of the Canadian dollar in terms of the pound sterling and the US dollar are somewhat slow.

Finally, the interest rate which links the foreign exchange sector with the domestic money market is determined through the money market analysis in terms of M_1 and M_q, the quasimoney. The quasimoney includes savings plus foreign currency deposits. The statistical results reveal a significant degree of disparity between the behaviour of these two components, which subsequently affects the behaviour of short-term and long-term interest rates. The long-term interest rate primarily ensures equilibrium in the domestic money market and the short-term interest rate dominates the external balances. These features become specially noteworthy in the simulation results.

20.3 RETESTING THE ASSUMPTIONS

The two fundamental assumptions, interest parity and the perfect efficiency of the foreign exchange market, which were dropped in the construction of the model, need to be tested with updated data before the model is used for policy scenarios. These assumptions continue to attract a great deal of attention in the quantitative studies of foreign exchange markets in spite of the fact that enough evidence has been amassed against their general validity. For the sample period used in the model, these assumptions were clearly found not to be valid. However, there still remains a question of whether more experience with the free exchange market has led to any perceptible movement towards greater efficiency. These assumptions are therefore retested through the same series of null hypotheses as stated in Marwah (1984).

20.3.1 The interest parity assumption

Any test of the interest parity assumption entails a test of the joint hypothesis that international investors act rationally and the assets denominated in various currencies are perfect substitutes. For only then would the expected rates of return

---

[6] In the literature on the efficiency of the foreign exchange market, it is this rate which is assumed to be an unbiased predictor of the future spot rate.

on these interest-bearing assets be equal. This parity result may be related either to a covered arbitrage (with some risk) or an uncovered arbitrage (with nearly no risk). An uncovered parity is a much stronger assumption than a covered parity. We test the validity of this assumption for a covered arbitrage. The test is conducted in two forms.

(A) Assuming zero or constant risk across all markets, a simple parity test is employed in which the forward exchange rate (ε_{fcj}) is regressed on the forward interest parity (ε^P_{fcj}) rate under the null hypothesis that the intercept of this regression must be equal to zero and the slope coefficient equal to unity with the error term serially uncorrelated. Formally,

$$\varepsilon_{fcj} = \beta_0 + \beta_1 \varepsilon^P_{fcj} + \text{error}, \tag{20.1}$$
$$H_0: \beta_0 = 0, \quad \beta_1 = 1; \quad H_1: \beta_0 \neq 0, \quad \beta_1 \neq 1$$

(B) The simple parity test is augmented by adding the role of speculation in the forward exchange market. In this augmented test, the forward exchange rate is estimated as a weighted average of the forward interest parity rate and the expected spot rate (ε^e_{scj}). If the expectations about the spot rate are formulated rationally and perfect foresight does exist, then the expected spot rate can be approximated by one-period led actual spot rate ($\varepsilon^e_{scj} = \varepsilon_{scj+1}$). The null hypothesis can then be formulated as:

$$\varepsilon_{fcj} = \beta_0 + \beta_1 \varepsilon^P_{fcj} + \beta_2 \varepsilon_{scj+1} + \text{error}. \tag{20.2}$$
$$H_0: \beta_0, \beta_2 = 0, \quad \beta_1 = 1, \quad \text{no serial correlation};$$
$$H_1: \beta_0, \beta_2 \neq 0, \quad \beta_1 \neq 1, \quad \text{error serially correlated}.$$

Table 20.1 A test of interest-parity assumption

$$\varepsilon_{fcj} = \beta_0 + \beta_1 \varepsilon^P_{fcj} + \text{error},$$
$$H_0: \beta_0 = 0, \quad \beta_1 = 1; \quad H_1: \beta_0 \neq 0, \quad \beta_1 \neq 1$$

Explanatory variables	France C$/franc	West Germany C$/DM	Japan C$/yen	UK C$/£	US C$/US$
Intercept	0.0115	0.0327	0.00037	0.1457	0.0605
	(4.91)	(7.86)	(5.28)	(2.13)	(5.47)
ε^P_{fcj}	0.9524	0.9142	0.8984	0.9323	0.9278
	(86.79)	(110.65)	(63.74)	(30.63)	(82.81)
$\bar{R}^2$	0.992	0.995	0.989	0.940	0.991
d	0.542	0.495	0.724	1.77	0.444
SE	0.0036	0.0081	0.0001	0.0805	0.0110
t for $H_0: \beta_1 = 1$	(4.33)	(10.38)	(7.23)	(2.22)	(6.45)

Note: $\varepsilon^P_{fcj} = \left(\dfrac{100 + r}{100 + r_j}\right) \varepsilon_{scj}$; sample period 1970(1)–1985(2) except for Japan; for Japan: 1973(4)–1985(2). The numbers within the parentheses below the coefficients are t-ratios.

The test results of interest parity assumption for the Canadian dollar market in terms of each of the five currencies are presented in Tables 20.1 and 20.2. The definitions of the variables are listed in Appendix A. These results based on sample period 1970(1)–1985(2) confirm the evidence presented in Marwah (1984) for a period ended 1981(4): not even in one single case does the interest parity condition strictly hold. In Table 20.1 the estimated intercept in all cases is statistically different from zero (at the 0.01 level), the slope coefficients are highly significant, different from zero as well as unity. The t-ratio for the null hypothesis of slope coefficient being unity varies from 2.22 for the UK to 10.38 for West Germany. Apparently, the forward rate parity deviations are significant from a statistical point of view and must be explained by other factors.

The results of the augmented test presented in Table 20.2 in which the interest parity relation is jointly examined with the rate of speculation reconfirm the initial results: the null hypothesis that $\beta_1 = 1$ is conclusively rejected for each and every case, and, except for West Germany where evidence is somewhat less strong, for $\beta_2 = 0$ as well. Moreover, the statistical significance of β_2 estimates underlines the importance of the role of speculative activity in determining the forward exchange rate of the Canadian dollar. Transaction costs alone are unlikely to explain away successfuly all the deviations from the interest parity rate.

The rejection of the interest parity condition involves rejecting any or all the three subconditions related to the general issue of market efficiency and accepting alternatively that investors may not be rational, financial assets denominated in different currencies may not be perfect substitutes, or, there may exist larger risk

Table 20.2 A further test of interest parity and speculations

$$\varepsilon_{fcj} = \beta_0 + \beta_1 \varepsilon^{p}_{fcj} + \beta_2 \varepsilon^{e}_{scj} + \text{error}$$
$$H_0: \beta_0 = 0, \quad \beta_1 = 1, \quad \beta_2 = 0; \quad H_1: \beta_0 \text{ and/or } \beta_2 \neq 0, \quad \beta_1 \neq 1$$

Explanatory variables	France	West Germany	Japan	UK	US
	C$/franc	C$/DM	C$/yen	C$/£	C$/US$
Intercept	0.0094	0.0273	0.000228	0.0759	0.0616
	(4.28)	(6.13)	(3.32)	(1.28)	(5.02)
ε^{p}_{fcj}	0.8398	0.8553	0.7863	0.6972	0.7517
	(30.12)	(26.93)	(22.33)	(13.58)	(15.57)
ε^{e}_{scj}	0.1222	0.0692	0.1460	0.2656	0.1842
	(4.22)	(1.92)	(3.55)	(5.35)	(3.68)
$\bar{R}^2$	0.994	0.995	0.993	0.959	0.992
d	0.924	0.627	1.067	1.67	0.774
SE	0.0032	0.0080	0.00009	0.0669	0.0100
t for $H_0: \beta_1 = 1$	(5.75)	(4.56)	(6.07)	(5.89)	(5.14)

$\varepsilon^{p}_{fcj} = \left(\dfrac{100+r}{100+r_j}\right) \varepsilon_{scj}$, $\varepsilon^{e}_{scj} = \varepsilon_{scj+1}$; sample period: 1970(1)–1985(1) except for Japan; for Japan: 1973(4)–1985(1). The numbers within the parentheses below the coefficients are t-ratios.

premiums than what is accounted by the covered parity. This brings us to the test of the market efficiency.

20.3.2 Market efficiency and all that

Assuming that all available market information is used rationally by risk-neutral agents in their estimate of the forward rate, the efficiency question is examined by testing the unbiasedness hypothesis stated in two forms: (a) a 'weak-form' which states that the forward exchange rate is an unbiased predictor of the future spot rate; and (b) a 'strong-form' in which the forward premium is expected to be an unbiased predictor of the *change* in the future spot rate. Formally, in the regressions

$$\varepsilon_{scj} = \beta_0 + \beta_1 \varepsilon_{fcj-1} + \text{error, and} \tag{20.3}$$

$$\varepsilon_{scj} - \varepsilon_{scj-1} = \beta'_0 + \beta'_1 (\varepsilon_{fcj} - \varepsilon_{scj})_{-1} + \text{error}, \tag{20.4}$$

$H_0: \beta_0 = 0, \quad \beta_1 = 1; \quad H_1: \beta_0 \neq 0, \quad \beta_1 \neq 1 \quad \text{by 'weak-form'}$
$H_0: \beta'_0 = 0, \quad \beta'_1 = 1; \quad H_1: \beta'_0 \neq 0, \quad \beta'_1 \neq 1 \quad \text{by 'strong-form'}$

In addition, if H_0 is true, then error must not be serially correlated.

A priori, if the forward market can indeed predict the future spot rate without bias, and since the current spot rate is also known at the same time as the forward rate, the current forward premium should also predict the *change* in the future spot rate without bias. It can be easily seen that the 'strong-form' test is more stringent than the 'weak-form' which could possibly lead erroneously towards acceptance of the hypothesis. For example, if $\beta'_1 = 1$, β_1 would also be unity; however, if $\beta'_1 \neq 1$, β_1

Table 20.3 Exchange market efficiency: 'weak-form'

$$\varepsilon_{scj} = \beta_0 + \beta_1 \varepsilon_{fcj-1} + \text{error},$$
$$H_0: \beta_0 = 0, \quad \beta_1 = 1; \quad H_1: \beta_0 \neq 0, \quad \beta_1 \neq 1$$

Explanatory variables	France C$/franc	West Germany C$/DM	Japan C$/yen	UK C$/£	US C$/US$
Intercept	0.0084	0.0296	0.000495	0.0667	−0.0144
	(0.93)	(2.03)	(2.35)	(0.51)	(0.48)
ε_{fcj-1}	0.9564	0.9309	0.8960	0.9749	1.0165
	(22.68)	(29.97)	(20.51)	(16.83)	(38.22)
$\bar{R}^2$	0.897	0.938	0.905	0.827	0.961
d	1.50	1.789	1.532	1.523	1.915
SE	0.0130	0.0279	0.0003	0.1453	0.0235
t for $H_0: \beta_1 = 1$	(1.03)	(2.23)	(2.38)	(0.434)	(0.620)

Note: Sample period: 1970(2)–1985(2) except for Japan; for Japan: 1974(1)–1985(2). The numbers within the parentheses below the coefficients are *t*-ratios.

could still be equal to unity, provided the slope coefficient in the regression of lagged spot rate on the lagged forward rate is equal to one. This relationship has been identified in Marwah and Bodkin (1984).

The results of the efficiency test for all the five countries are presented in Tables 20.3 and 20.4. The null hypothesis in the 'weak-form' as judged by β's is accepted for C\$/franc, C\$/£ and C\$/US\$ markets; it is rejected for C\$/DM and C\$/yen. However, the 'strong-form' test results do not support this conclusion. In the case of the C\$/franc market, β'_1 is not only negative and statistically significant, it is also significantly different from unity. Similarly, in the case of the C\$/US\$ market, the 'strong-form' efficiency is rejected as β'_1 is neither statistically different from zero nor from unity. It is only in the case of the C\$/£ market that the value of β'_1 in the 'strong-form' test confirms the acceptance of the efficiency hypothesis, but then we have the problem of the error structure which evidently contains serial correlation. It is also noteworthy that, contrary to the expectations, there are negative forward premiums for three markets, and the negative one for France is even statistically significant. In short, not unexpectedly, the null hypothesis continues to be rejected in all cases.

In spite of all its limitations, the results of the unbiasedness test are indeed damaging to the hypothesis of speculative efficiency. If one is willing to reject the adequacy of the efficient market models, one must then either search for other economic and noneconomic (political) fundamental factors or/and assume that market adjustments to information flow are rather sluggish. In any case, one must examine both the spot and the forward exchange markets simultaneously. This is precisely what is done in the prototype model.

Table 20.4 Exchange market efficiency: 'strong-form'

$$\varepsilon_{scj} - \varepsilon_{scj-1} = \beta'_0 + \beta'_1 (\varepsilon_{fcj} - \varepsilon_{scj})_{-1} + \text{error},$$
$$H_0: \beta'_0 = 0, \quad \beta'_1 = 1; \quad H_1: \beta'_0 \neq 0, \quad \beta'_1 \neq 1$$

Explanatory variables	France C\$/franc	West Germany C\$/DM	Japan C\$/yen	UK C\$/£	US C\$/US\$
Intercept	−0.00074	0.00429	0.000055	0.0077	0.0043
	(0.47)	(0.71)	(1.16)	(0.40)	(1.42)
$(\varepsilon_{fcj} - \varepsilon_{scj})_{-1}$	−3.0035	−0.4121	−0.3700	0.8670	0.423
	(2.42)	(0.38)	(0.665)	(3.65)	(0.56)
$\bar{R}^2$	0.090	0.002	0.01	0.185	0.005
d	1.653	1.832	1.615	1.603	1.872
SE	0.012	0.0287	0.00029	0.1452	0.0235
t for $H_0: \beta_1 = 1$	(3.22)	(1.30)	(2.46)	(0.56)	(1.02)

Note: Sample period: 1970(2)–1985(2) except for Japan; for Japan: 1974(1)–1985(2). The numbers within the parentheses below the coefficients are *t*-ratios.

20.4 THE POLICY SIMULATIONS

In this section we simulate six scenarios by using the prototype model with the primary purpose of obtaining insight into some topical issues of the Canadian foreign exchange market. For example, (1) what is the impact of direct intervention by the government on the exchange rate of the Canadian dollar? The question is examined in scenario 1. (2) Are secondary interventions directed through the money market effective? These interventions may either be undertaken to counteract the undesired interest rate shocks originating from abroad or be adopted independently. Scenarios 2–4 examine the effects of interest rate shocks on the Canadian foreign exchange market. (3) Are there any gains to be made if the quantity uncertainty in money supply is eliminated? Scenarios 5 and 6 deal with the effects of a Friedman-type rule on the foreign exchange and money markets of Canada.

The quantitative assumptions underlying these scenarios are:

Scenario 1: The stock of net foreign assets is increased (ΔNFA) by 500 millions of SDRs in 1975: first quarter (1). It is a one-time shock.

Scenario 2: All foreign interest rates (YTB_j) are increased by 1 percentage point from 1975(1) to 1975(4).

Scenario 3: All foreign interest rates (YTB_j) are increased by 1 percentage point from 1975(1) to 1975(4), but the Canadian short-term rate (YTB_j) is held at the baseline.

Scenario 4: Canadian short-term interest rate (YTB) is increased by 1 percentage point in 1975(1).

Scenario 5: Money base is allowed to grow at an annual growth rate of 9.7% over the simulation period.

Scenario 6: Money base is increased by 5% in 1975(1).

The results of each scenario are presented in terms of deviations from the control-baseline solution. The control solution is the original status-quo solution of the prototype model (variant III).

20.4.1 Direct interventions in the foreign exchange market: Scenario 1

The model is simulated under the assumption that the stock of net foreign assets is raised by 500 millions of SDRs presumably through Exchange Fund Account (borrowing) operations.[7] The results are presented in Table 20.5 and Fig. 20.1.

As expected, the value of the Canadian dollar immediately appreciates against all currencies, its global exchange rate falls from the baseline by 1.33%. The largest

---

[7] The magnitude of this policy shock was considered reasonable as it fell within the realm of observed amounts of operation of the Exchange Fund Account. For example, throughout 1977 and 1978 the Bank of Canada carried out these operations to the tune of $50 million per week on the average. The US dollar holdings of the Exchange Fund fell by $1.2 billion US and rose by $4.9 billion US in 1978 as large sums were borrowed to support the Canadian dollar.

Table 20.5 Scenario 1: ΔNFA increased by SDRs 500 million in 1975(1) – deviations from control solution (Δ%)

	ε_{scC}	ε_{scFR}	ε_{scGR}	ε_{scJP}	ε_{scUK}	ε_{scUS}	YTB	YGB	M_1	$M_1 + M_q$
1975(1)	−1.33	−2.21	−1.57	−3.55	−1.03	−1.44	−0.69	0.29	0.12	−0.02
1975(4)	−0.53	−1.08	−1.06	−1.16	−0.41	−0.46	−0.06	0.19	0.03	−0.09
1976(4)	−0.38	−0.46	−0.68	−0.35	−0.28	−0.38	0.00	0.11	0.01	−0.07
1977(4)	−0.26	−0.17	−0.41	−0.10	−0.17	−0.28	0.00	0.07	0.01	−0.05
1978(4)	−0.19	−0.07	−0.28	−0.03	−0.11	−0.23	0.00	0.05	0.00	−0.04
1979(4)	−0.15	−0.03	−0.21	−0.01	−0.08	−0.17	0.00	0.03	0.00	−0.03
1980(4)	−0.11	−0.01	−0.15	−0.01	−0.05	−0.14	0.00	0.02	0.00	−0.02
1981(4)	−0.09	−0.01	−0.12	0.00	−0.04	−0.12	0.00	0.01	0.00	−0.01
1982(4)	−0.06	0.00	−0.08	0.00	−0.03	−0.09	0.00	0.01	0.00	−0.01
$\sum (\Delta\%)$	−3.81	−4.04	−4.56	−5.21	−2.20	−3.31	−0.75	0.78	0.17	−0.34
Mean Δ	−0.0023	−0.0007	−0.0023	−0.00001	−0.004	−0.0031	−0.0031	0.0068	0.0033	−0.0398

Note: $-\Delta\varepsilon_{scj}$ = appreciation of the Canadian dollar; $+\Delta\varepsilon_{scj}$ = depreciation of the Canadian dollar.

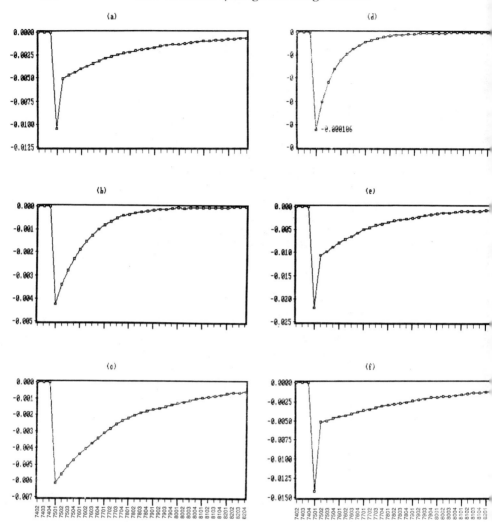

Fig. 20.1 Scenario 1: (a) ε_{scc}; (b) ε_{scFR}; (c) ε_{scGR}; (d) ε_{scJP}; (e) ε_{scUK}; (f) ε_{scUS}; (g) YTB; (h) YGB; (i) M_1; (j) $M_1 + M_q$

The policy simulations

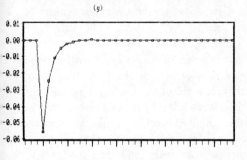
(g)

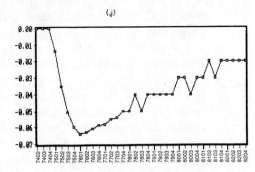
(j)

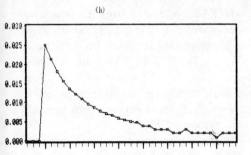
(h)

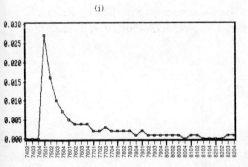
(i)

percentage gain is recorded in terms of the Japanese yen and the smallest against the pound sterling. It gains close to 1.4 cents against the US dollar. The differential impact on the cross exchange rates is not unexpected as it is known, and we have already seen, that the five markets do differ in terms of efficiency. However, these gains are not permanent. Although the exchange rate of the Canadian dollar against all the five currencies stays mostly below the control solution throughout the simulation period, more than half of the gains are lost by the end of 1975. All the rates tend to converge back to the baseline. The deviation of the global rate from the baseline solution was merely one-tenth of 1% by the end of 1980.

In the money market there is a small increase in M_1 and a fall in M_q; however, M_1 converges back to the baseline much faster than M_q. Correspondingly, there is an immediate fall in the short-term interest rate (YTB) and an increase in the long term (YGB). The short-term interest rate falls because of its role in the interest arbitrage relation; for the yield differential can now be maintained at a lower interest rate in the face of the strengthened Canadian dollar. By 1976(4), YTB returns to the original path as is shown in Fig. 20.1(g). An increase in YGB is caused by falling money supply ($M_1 + M_q$) which generates excess demand pressures in the domestic money market. However, the convergence of YGB to its original path is much slower compared with YTB; it follows the pattern of $M_1 + M_q$.

The cumulative effect of these interventions on aggregate net capital inflows is positive. However, their effect on different categories of capital inflows varies. For example, while direct investment in Canada increases, the portfolio investment falls for number of quarters after 1975(1) before converging to the original track. Similarly, the deposit money banks short-term capital and other short-term capital inflows register a small decline.

20.4.2 Interest rate disturbances

(a) Foreign interest rate shock: scenario 2

In the early 1980s, the interest rate in Canada climbed steeply in response to the high interest rate in the US. For example, the Canadian short-term interest rate (YTB) went up from 12.8% in 1980(3) to 17.7% in 1981(3). Scenario 2 was simulated to assess the impact of foreign interest rate shock on the foreign exchange market of Canada. This shock was quantitatively specified in terms of an increase in all foreign interest rates by 1 percentage point per quarter 1975(1) to 1975(4) without making any other adjustment to the model. The interactions between the forward exchange market and the domestic money market were left intact to absorb this shock. An overview of the results is provided in Table 20.6 and Fig. 20.2.

The impact of foreign interest rate shock on the exchange rate of the Canadian dollar filters through three basic channels: the response of the Canadian interest rate to the foreign interest rate and its direct impact on the exchange rate, the response of capital flows to changes in the relative interest rates and their indirect

Table 20.6 Scenario 2: all foreign interest rates (YTB$_j$) increased by 1 percentage point in 1975(1)–1975(4) – deviations from control solution (Δ%)

	ε_{scc}	ε_{scFR}	ε_{scGR}	ε_{scJP}	ε_{scUK}	ε_{scUS}	YTB	YGB	M_1	$M_1 + M_q$
1975(1)	−0.51	−0.76	−0.76	−2.47	−0.92	−0.17	5.73	0.43	−0.64	−0.01
1975(4)	−2.29	−3.78	−4.30	−9.31	−3.72	−0.85	10.70	1.77	−0.94	0.82
1976(4)	−1.50	−1.96	−3.15	−4.20	−2.47	−0.49	0.68	1.24	0.09	0.57
1977(4)	−0.92	−0.74	−1.92	−1.29	−1.49	−0.34	0.08	0.72	0.07	0.21
1978(4)	−0.63	−0.31	−1.27	−0.47	−0.95	−0.27	0.03	0.38	0.04	0.04
1979(4)	−0.46	−0.15	−0.95	−0.23	−0.69	−0.21	0.01	0.18	0.02	−0.02
1980(4)	−0.33	−0.07	−0.71	−0.09	−0.46	−0.17	0.00	0.09	0.01	−0.02
1981(4)	−0.22	−0.05	−0.50	−0.06	−0.30	−0.11	0.00	0.04	0.00	−0.02
1982(4)	−0.15	−0.03	−0.33	−0.02	−0.19	−0.08	0.00	0.03	0.00	−0.02
Σ (Δ%)	−7.01	−7.85	−13.89	−18.14	−11.00	−2.69	17.23	4.88	−1.35	1.55
Mean Δ	−0.0069	−0.0019	−0.0007	−0.00007	−0.0280	−0.0033	0.1092	0.0506	−0.0199	0.1437

Note: $-\Delta \varepsilon_{scj}$ = appreciation of the Canadian dollar; $+\Delta \varepsilon_{scj}$ = depreciation of the Canadian dollar.

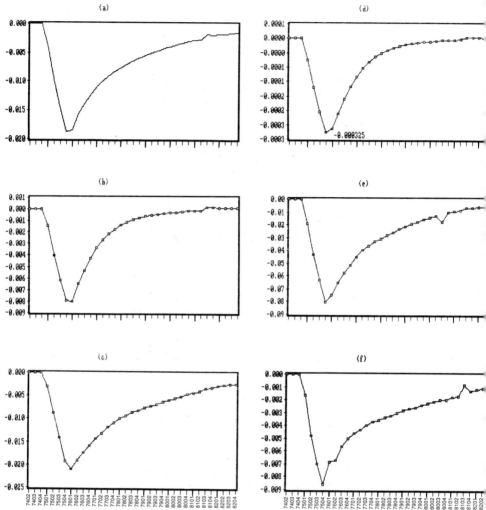

Fig. 20.2 Scenario 2: (a) ε_{scc}; (b) ε_{scFR}; (c) ε_{scGR}; (d) ε_{scJP}; (e) ε_{scUK}; (f) ε_{scUS}; (g) YTB; (h) YGB; (i) M_1; (j) $M_1 + M_q$

The policy simulations

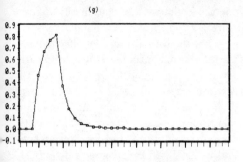

(g)

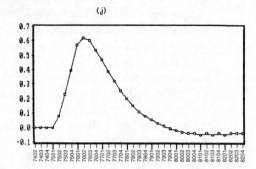

(j)

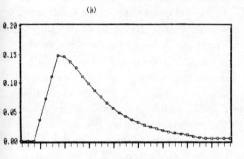

(h)

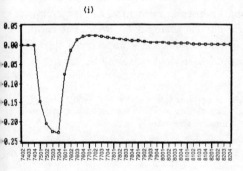

(i)

impact (via growth of net foreign assets) on the exchange rate, and finally via the interest arbitrage relationship operating through the forward exchange market.[8] The first two effects materialize immediately but the forward market effect enters with a one-period lag. However, as far as the relative magnitudes of these effects are concerned, the Beta coefficient for the forward rate is found to be largest for all currencies, followed in turn by interest rate and capital flows.[9]

From Table 20.6 and Fig. 20.2(g) and (h), it is apparent that the Canadian interest rates immediately start moving up in response to an increase in the interest rates abroad. However, the new gap between the domestic and foreign interest rates gets only partially closed. By the end of 1975, YTB had gone up by 0.81 percentage point (81 basis points) from the baseline solution. The change in the relative interest rates leads to a decline in the capital inflows. It was observed that capital inflows remained below the control solution by about one-tenth of 1% during the simulation. These two forces affected the exchange rate in the opposite direction; an increase in interest rate led to an appreciation of the Canadian dollar but a fall in the capital inflows led to its depreciation. Since the former effect is much stronger statistically than the latter, the net result was an appreciation of the Canadian dollar across all currencies. This seemingly anomalous net effect was then further reinforced by the forward market arbitrage. Given that any increase in the Canadian interest rate was smaller than the initial increase assumed in the foreign interest rates, the interest arbitrage relation could be maintained only at a higher future value of the Canadian dollar. In other words, the forward market tells us that the higher the foreign higher interest rate, the larger must be the premium against which domestic currency would be hedged if capital inflows are to be maintained. Thus the forward market compensates through higher value of the Canadian dollar for the lower yield on domestic securities.[10]

---

[8] These channels are represented respectively in the exchange rate equations by YTB, GNFA and ε_{fcj-1}.
[9] The partial β-coefficients measured approximately from the estimates of the reduced form equations derived by substituting the forward rate relations into the spot rates were as follows.

Explanatory variable	β coefficients				
	ε_{scFR}	ε_{scGR}	ε_{scJP}	ε_{scUK}	ε_{scUS}
GNFA	−0.1354	−0.0659	−0.1342	−0.0696	−0.1769
GNPR	0.3194	0.1689	0.6036	0.4773	0.2946
YTB	−0.4273	−0.2584	−0.5626	−0.5333	−0.2046
ε_{fcj-1}^{p}	0.9481	1.0384	0.8956	0.7812	0.5127
$\sigma_{s\mu d-1}$	−0.0819	−0.0085	−0.0732	−0.1968	−0.1122
ε_{scj-1}					0.3901

The direct examination of the 'beta' coefficients reveals that the lagged interest parity is by far the most important variable in all equations. Nonetheless, the contribution of other variables is by no means insignificant. For the estimates of the reduced form equations, see Marwah (1984, Table 5).

[10] The impact of foreign interest rate shock on the exchange rate of the Canadian dollar may appear seemingly anomalous. However, we must remember that the model is picking up short-to-medium term effects. In the very short run, given the forward parity rate, the impact would be distributed between domestic interest rates and the spot exchange rate.

The policy simulations

The Canadian dollar appreciates across all currencies and stays higher throughout the simulation period; the largest impact occurs in 1975(4). For example, in 1975(4), one US dollar could be exchanged for 1.008 Canadian dollar instead of 1.017 as recorded in the control solution. It may also be noted that the adjustment of the spot rate varies markedly across different currencies. Measured in terms of percentages, the Canadian dollar appreciates most against the Japanese yen and least against the US dollar. Nonetheless, the absolute magnitudes of the deviations from the baseline seem relatively small in comparison with the size of the shock.

(b) Foreign interest rate shock with Canadian YTB held at the baseline: scenario 3

The purpose of scenario 3 is to insulate the short-term domestic interest rate from shock of foreign interest rate changes and let the primary adjustments take place through the foreign exchange market. This experiment amounts basically to nothing more than the use of 'made in Canada' interest rate policy unsusceptible to secondary interventions, and it highlights the role of the forward exchange market.[11]

The quantitative nature of the shock is the same as in scenario 2 above, as all foreign interest rates are raised by 1 percentage point in 1975(1)–1975(4). The results are presented in Fig. 20.3 and Table 20.7. As expected, it was found that capital inflows were reduced sharply. The Canadian dollar still appreciates against all currencies; however, such appreciation takes place with a one-period lag, and, except against the US dollar, the amount of appreciation is somewhat less than what was recorded in scenario 2. The convergence to the baseline also proceeds at a slower pace. The brunt in the money market falls on YGB which increases as is shown in Fig. 20.3(h); $M_1 + M_q$ correspondingly falls.

The net cost of an absence of secondary interventions aimed at counteracting the foreign interests shocks appears to be in terms of a mild decline in the capital inflows.[12] Otherwise, as far as the exchange rate is concerned, the forward market appears to be strong enough by itself to absorb the vagaries of interest rates abroad.

(c) A Canadian interest-rate shock: scenario 4

The short-term interest rate was often used during the 1970s as an instrument to achieve money supply and exchange rate targets in Canada.[13] Scenario 4 was

---

[11] While this experiment of holding YTB to the baseline may appear somewhat unrealistic, especially since real GNP is treated exogenously, we believe that as far as the foreign exchange market is concerned, it is its impact on capital flows which is more relevant. And capital flows are treated endogenously.

[12] A decline in capital inflows, even when the exchange rate appreciates, underlines the role of other factors which control their movements. Moreover, it also reflects that an appreciation of the Canadian dollar is not sufficient to compensate for the indexed yield differential on securities denominated in Canadian dollars and securities denominated in foreign currencies.

[13] For the debate on the choice of the instruments, money supply or interest rate, as a part of the policy of monetary graduation, see Courchene (1978).

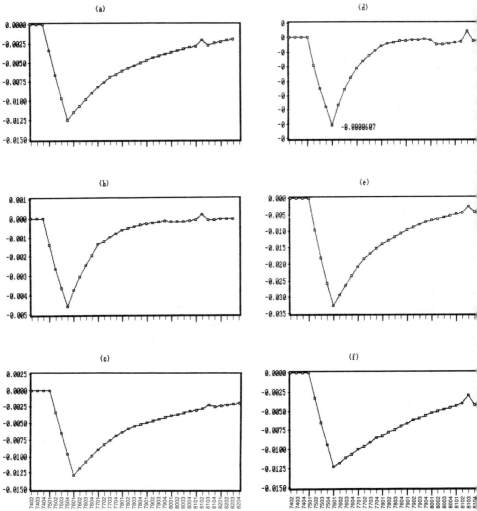

Fig. 20.3 Scenario 3: (a) ε_{scc}; (b) ε_{scFR} (c) ε_{scGR}; (d) ε_{scJP}; (e) ε_{scUK}; (f) ε_{scUS}; (g) YGB; (h) M_1; (i) $M_1 + M_q$

The policy simulations

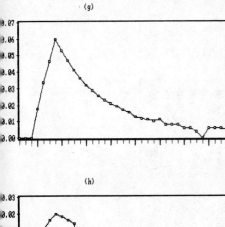

(g)

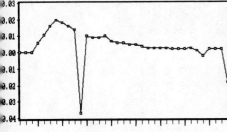

(h)

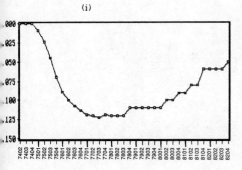

(i)

Table 20.7 Scenario 3: all foreign interest rates (YTB$_j$) increased by 1% percentage point in 1975(1)–1975(4) but the Canadian short-term rate (YTB) held constant at the baseline – deviations from control solution (Δ%)

	ε_{sc}	ε_{scFR}	ε_{scGR}	ε_{scJP}	ε_{scUK}	ε_{scUS}	YTB	YGB	M_1	$M_1 + M_q$
1975(2)	−0.43	−0.69	−0.81	−0.60	−0.45	−0.33	0.00	0.41	0.05	−0.04
1975(4)	−1.18	−1.73	−2.17	−1.37	−1.21	−0.92	0.00	0.72	0.08	−0.10
1976(4)	−1.18	−1.12	−1.98	−0.69	−1.14	−1.03	0.00	0.40	−0.14	−0.13
1977(4)	−0.78	−0.41	−1.20	−0.18	−0.68	−0.79	0.00	0.25	0.03	−0.12
1978(4)	−0.56	−0.17	−0.79	−0.04	−0.43	−0.63	0.00	0.16	0.02	−0.10
1979(4)	−0.43	−0.08	−0.60	−0.02	−0.31	−0.49	0.00	0.09	0.01	−0.08
1980(4)	−0.34	−0.07	−0.47	−0.07	−0.23	−0.39	0.00	0.06	0.01	−0.06
1981(4)	−0.30	−0.04	−0.37	−0.04	−0.19	−0.36	0.00	0.03	−0.01	−0.04
1982(4)	−0.19	−0.12	−0.26	−0.01	−0.11	−0.24	0.00	0.04	0.00	−0.03
∑ (Δ%)	−5.39	−4.33	−8.66	−3.01	−4.75	−4.18	0.00	2.16	0.05	−0.70
Mean Δ	−0.0053	−0.00099	−0.0053	−0.00001	−0.0116	−0.006	0.00	0.0199	0.0045	0.0903

Note: −Δε_{scj} = appreciation of the Canadian dollar; +Δε_{scj} = depreciation of the Canadian dollar.

generated to examine the effects of a discretionary increase in the short-term interest rate (YTB) intended to meet exchange rate and money supply objectives. This policy shock was quantitatively imposed through an increase in YTB by 1 percentage point in 1975(1).

The foreign exchange markets reacted to this shock as expected in four out of five cases. The results are presented in Fig. 20.4 and Table 20.8. The only foreign exchange market that comes close to passing 'weak' efficiency test, C$/US$, shows an overall depreciation of the Canadian dollar following an initial improvement. In its process of convergence back towards the baseline it seems to overshoot. In other words, the Canadian dollar appreciates against the US dollar in 1975(1), but starts depreciating in 1975(2) and continues on a lower path for the rest of the duration of the simulation period. However, the value of the Canadian dollar in terms of all other currencies does improve, but in varying degrees. The global exchange rate continues to show some appreciation until the end of 1979.

Initially, net capital inflows at an aggregate level do increase but then fall back below the baseline. Direct and portfolio investments follow almost the same pattern. Presumably, the movements in the C$/US$ market dominate the pattern.

In terms of money market effects, M_1 is reduced by $458 million in 1982(4), while the sum $M_1 + M_q$ goes up by $393 million. However, both aggregates converge to the baseline in 1980(3) and 1982(2) respectively. YTB exhausts the cumulative effect of the initial shock by 1979(4) and YGB in 1981(3). The overall dynamics of the money market are reasonable, as it takes about 3 to 4 years for an effect of monetary disturbances to peter out completely.

To sum up, a discretionary increase in the short-term interest rate does lead to an overall appreciation of the Canadian dollar, but much of these gains are neutralized by its failure to stay appreciated against the US dollar. In addition, its adverse effect on capital inflows makes the short-term interest rate all the less attractive as a policy instrument. The model predicts that there is very little to be gained by varying independently the short-term interest rate in Canada.

20.4.3 Monetary disturbances

The purpose of this section is to test whether a simple monetary rule which eliminates the quantity uncertainty with respect to the money supply does stabilize the foreign exchange market. The underlying hypothesis to be tested in fact is whether by adopting a rule for the growth of high-powered money (base) and adhering to it the government can lessen the variability of exchange rates. The uncertainty in money supply is presumed to enhance the uncertainty in foreign exchange market.

(a) Growth of high-powered money with a rule vis-à-vis observed growth rates: scenario 5

In this experiment the high powered money was allowed to grow smoothly at an annual rate of 9.72%. This rate was obtained by linear approximation of the low

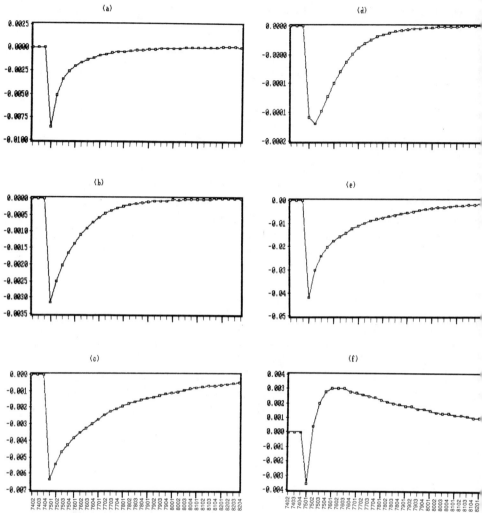

Fig. 20.4 Scenario 4: (a) ε_{scc}; (b) ε_{scFR}; (c) ε_{scGR}; (d) ε_{scJP}; (e) ε_{scUK}; (f) ε_{scUS}; (g) YTB; (h) YGB; (i) M_1; (j) $M_1 + M_q$; (k) CABR

The policy simulations

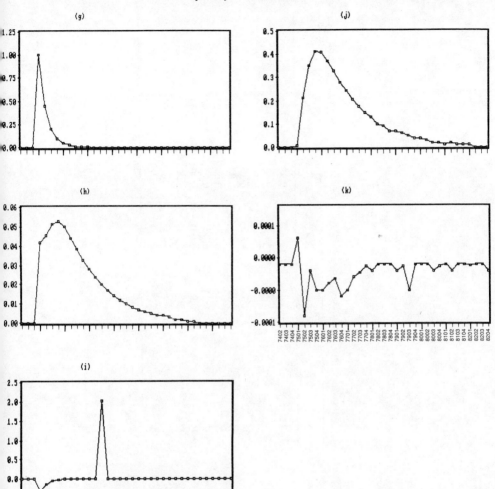

432   *The Canadian foreign exchange market*

Table 20.8 Scenario 4: Canadian short-term interest rate (YTB) increased by 1 percentage point in 1975(1) – deviations from control solution (Δ%)

	ε_{scc}	ε_{scFR}	ε_{scGR}	ε_{scJP}	ε_{scUK}	ε_{scUS}	YTB	YGB	M_1	$M_1 + M_q$
1975(1)	−1.09	−1.64	−1.64	−5.31	−1.99	−0.36	12.32	0.49	−1.44	−0.01
1975(4)	−0.31	−0.78	−0.95	−3.51	−0.94	−0.27	1.27	0.63	−0.06	0.59
1976(4)	−0.13	−0.34	−0.59	−1.24	−0.60	0.27	0.12	0.36	0.03	0.33
1977(4)	−0.05	−0.13	−0.36	−0.38	−0.37	0.22	0.02	0.18	0.02	0.15
1978(4)	−0.03	−0.05	−0.24	−0.14	−0.23	0.17	0.01	0.08	0.01	0.06
1979(4)	−0.01	−0.03	−0.18	−0.06	−0.17	0.14	0.00	0.03	0.00	0.03
1980(4)	0.00	−0.01	−0.13	−0.02	−0.11	0.11	0.00	0.01	0.00	0.01
1981(4)	0.00	0.00	−0.10	−0.01	−0.08	0.09	0.00	0.00	0.00	0.01
1982(4)	0.00	0.00	−0.07	0.00	−0.05	0.06	0.00	0.00	0.00	0.00
$\sum (\Delta \%)$	−1.62	−2.93	−7.19	−10.67	−4.54	0.97	13.74	1.78	−1.44	1.19
Mean Δ	−0.0009	−0.0005	−0.0020	−0.00003	−0.0089	−0.0016	0.0582	0.0163	0.0482	0.1128

Note: $-\Delta\varepsilon_{scj}$ = appreciation of the Canadian dollar; $+\Delta\varepsilon_{scj}$ = depreciation of the Canadian dollar.

Table 20.9 Scenario 5: money base growing as much by the rule as by the unknown process – deviations from control solution (Δ%)

	ε_{scc}	ε_{scFR}	ε_{scGR}	ε_{scJP}	ε_{scUK}	ε_{scUS}	YTB	YGB	M_1	$M_1 + M_q$
1975 (1)	−0.18	−0.35	−0.40	−1.49	−0.40	0.02	1.51	5.52	−5.42	−0.85
1975 (4)	−0.23	−0.45	−0.53	−1.80	−0.58	0.09	1.81	6.92	−6.56	−1.42
1976 (4)	−0.25	−0.48	−0.57	−1.83	−0.74	0.15	2.09	6.79	−6.68	−1.63
1977 (4)	−0.27	−0.54	−0.59	−1.85	−0.92	0.23	2.40	8.65	−8.21	−2.03
1978 (4)	−0.23	−0.50	−0.49	−1.77	−0.92	0.33	1.68	7.85	−8.21	−2.28
1979 (4)	−0.21	−0.50	−0.43	−1.91	−1.00	0.36	1.28	6.21	−7.96	−2.39
1980 (4)	−0.16	−0.48	−0.39	−1.76	−0.98	0.40	0.85	4.27	−6.80	−2.21
1981 (4)	−0.09	−0.37	−0.31	−1.31	−0.85	0.44	0.29	2.54	−4.66	−1.96
1982 (4)	−0.03	−0.18	−0.20	−0.59	−0.57	0.38	0.01	0.86	−2.38	−1.38
$\sum (\Delta\%)$	−1.65	−3.85	−3.91	−14.31	−6.96	2.40	11.92	49.61	−56.83	−16.15
Mean Δ	−0.0017	−0.001	−0.0025	−0.0001	−0.0178	0.0029	0.13	0.58	−1.90	−2.21

Note: $-\Delta\varepsilon_{scj}$ = appreciation of the Canadian dollar; $+\Delta\varepsilon_{scj}$ = depreciation of the Canadian dollar.

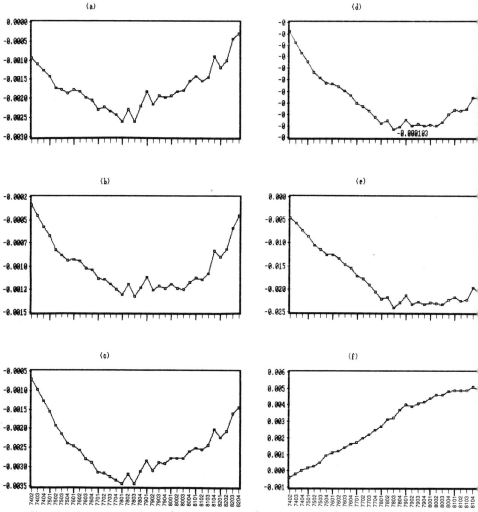

Fig. 20.5 Scenario 5: (a) ε_{scc}; (b) ε_{scFR}; (c) ε_{scGR}; (d) ε_{scJP}; (e) ε_{scUK}; (f) ε_{scUS}; (g) YTB; (h) YGB; (i) M_1; (j) $M_1 + M_q$; (k) CABR; (l) DINR

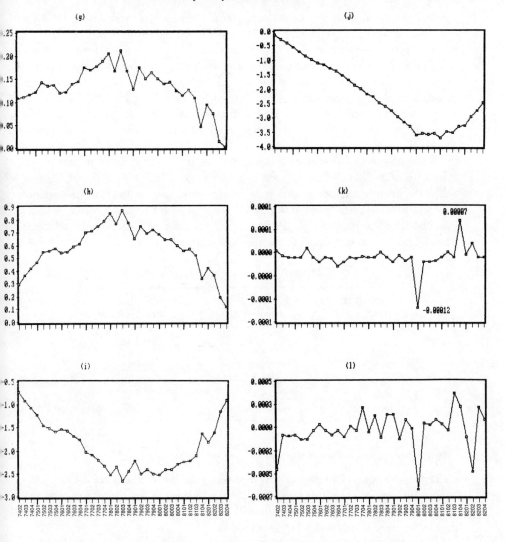

points of actual growth rates. Consequently, the money base assumed in this experiment was always equal to or less than its actual level.

The results generated by this scenario are summarized in Table 20.9 and Fig. 20.5. Apparently, because of the underestimate of money base, the money supply falls below the control solution and the interest rates are forced up. The Canadian dollar gains against all currencies except against US dollar. Its depreciation against the US dollar becomes increasingly pronounced after 1976. The mechanism of transmission of this shock seems to be the same as was noted when short-term interest rate was directly raised (see scenario 4).

If the dispersion of spot exchange rates can be accepted as an indicator or a measure of stability of exchange rates under the rule *vis-à-vis* no-rule regime, then the no-rule regime appears to be more stable. The coefficients of variation for various spot rates based on scenario 5 (rule) and on baseline (no-rule) solution are listed in Table 20.10. The difference between the two vectors of coefficients belonging to these regimes indeed appears negligible.

Table 20.10 Coefficients of variation (%)

Exchange rates	Rule (scenario 5)	No-rule (control solution)
ε_{scc}	8.38	8.35
ε_{scFR}	8.31	8.34
ε_{scGR}	18.95	18.92
ε_{scJP}	25.30	25.11
ε_{scUK}	6.49	6.57
ε_{scUS}	8.12	7.98

(b) Domestic monetary disturbance: scenario 6

The purpose of this experiment was to explore the effects of a domestic monetary shock on various exchange rates. The scenario generated by this experiment was similar to the one generated by Helliwell and Boothe (1983). It provides some basis for comparing the dynamics of two models.

We conducted this experiment by increasing the high-powered money by 5% in 1975(1). This increase in value by $491.8 million was once and for all, and no other adjustment was made to the model. The results are presented in Table 20.11 and Fig. 20.6.

In this scenario, the exchange rate of the Canadian dollar depreciates on the average against all currencies. The rates of depreciation, however, are quite small. For example, the global exchange rate depreciates by one-tenth of 1% immediately and then converges back fast towards the baseline. On the convergence path, there is some overshooting for exchanges rates in terms of the US dollar (ε_{scUS}) and the Japanese yen (ε_{scJP}). On the whole, the markets adjust very quickly to the monetary shock. The exchange rates are only slightly affected in value after 1975. Our results

Table 20.11 Scenario 6: money base increased by $491.8 million in 1975(1) – deviations from control solution (Δ%)

	ε_{scC}	ε_{scFR}	ε_{scGR}	ε_{scJP}	ε_{scUK}	ε_{scUS}	YTB	YGB	M_1	$M_1 + M_q$
1975(1)	0.118	0.177	0.179	0.580	0.217	0.038	−1.342	−3.44	3.199	0.219
1975(4)	0.001	0.009	0.020	0.049	0.014	−0.010	0.081	0.08	0.000	0.094
1976(4)	0.001	0.005	0.016	0.002	0.014	−0.010	0.011	0.04	0.004	0.042
1977(4)	0.001	0.000	0.009	−0.002	0.004	0.000	0.005	0.02	0.003	0.019
1978(4)	0.000	0.000	0.006	0.000	0.004	0.000	0.002	0.01	0.000	0.009
1979(4)	0.000	0.000	0.004	−0.002	0.004	0.000	0.000	0.01	0.000	0.000
1980(4)	0.000	0.000	0.003	0.000	0.000	0.000	0.000	0.00	0.000	0.000
1981(4)	0.000	0.000	0.001	0.000	0.000	0.000	0.000	0.00	0.000	0.000
1982(4)	0.000	0.000	0.001	0.000	0.000	0.000	0.000	0.00	0.000	0.000
$\sum$ (Δ%)	0.121	0.191	0.239	0.629	0.257	0.038	−1.243	−3.27	3.206	3.830
Mean Δ	0.00003	0.00001	0.00006	0.00001	0.0003	−0.00004	−0.0015	−0.0066	0.0225	0.0242

Note: $-\Delta\varepsilon_{scj}$ = appreciation of the Canadian dollar; $+\Delta\varepsilon_{scj}$ = depreciation of the Canadian dollar.

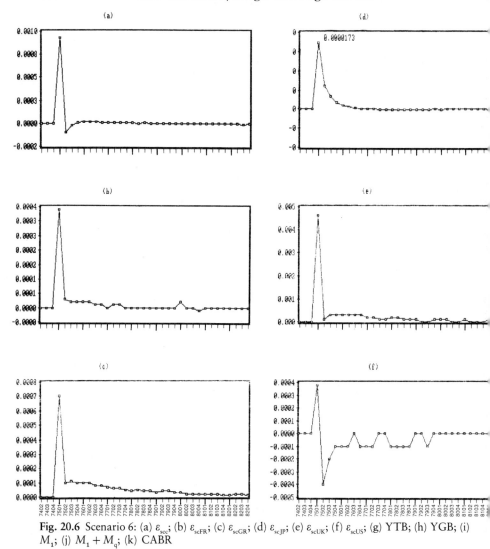

Fig. 20.6 Scenario 6: (a) ε_{scc}; (b) ε_{scFR}; (c) ε_{scGR}; (d) ε_{scJP}; (e) ε_{scUK}; (f) ε_{scUS}; (g) YTB; (h) YGB; (i) M_1; (j) $M_1 + M_q$; (k) CABR

The policy simulations

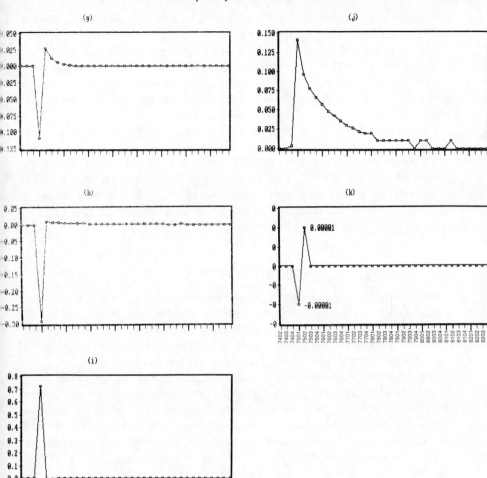

are consistent with what Helliwell and Boothe found using annual data for the 'portfolio' model which they simulated.

Consistent with the changes in the interest rates, the net capital inflows are adversely affected in 1975(1), improve in value in 1975(2) and then converge to the baseline solution. However, direct investment goes down but portfolio investment increases. Thus the monetary shocks seem to have very little effect on net capital inflows, but it does change their composition.

The money market effects appear more complex as the money supply was already growing at a high rate at the timing of the shock. The interest rates fall below the baseline immediately in 1975(1) but then move back to a permanently higher path. Most of the effect on M_1 peters out by 1975(4). However, this shock has a relatively more lasting impact on the level of $M_1 + M_q$; the convergence is achieved in 1979(4). Thus the adjustments to this shock seem to take rather long to work out fully through the money market.

20.5 CONCLUSIONS

The six scenarios generated with respect to the foreign exchange market of Canada and its domestic money market under a variety of policy assumptions and shocks are indeed revealing. They tell us what happens to the Canadian dollar and associated entities in a world where assets denominated in various currencies are not perfect substitutes and market operations are not fully efficient.

The dynamics generated in the short-to-medium run indicate that direct interventions by the government in the foreign exchange market do produce results in the desired direction, and the absence of secondary interventions in response to a change in foreign interest rates proves somewhat costly, especially in terms of capital flows. Nonetheless, anomalous as it may seem, the forward exchange markets, although not fully efficient, appear to be strong enough to absorb the vagaries of interest rates abroad. When the foreign interest rates increase, the Canadian dollar markets tend to overcompensate.

A domestic 'made in Canada' monetary shock in terms of a discretionary increase in the short-term interest rate does enhance the value of the Canadian dollar immediately, but most of these gains seem to get over-neutralized against the US dollar after a period of time. This delayed perverse reaction leads to the capital outflow. This result is specially noteworthy since the US dollar market is the only market among the five studied *vis-à-vis* the Canadian dollar which nearly passes the efficiency test. The simulation results point out that there is very little to be gained by using short-term interest rate as a policy instrument to achive an exchange rate target.

In fact, the policy simulations with our model do confirm that Canada has limited options when it comes to managing the exchange rate of its dollar. Direct interventions and interest rate policies do produce results in the desired directions but only temporarily, and at certain costs. Eventually, it is the fundamental economic factors which must determine the destiny of the Canadian dollar.

And finally, our results refute the hypothesis that the quantity uncertainty of the monetary authorities is a source of uncertainty in the foreign exchange market. It indeed matters how much the money supply grows over time but not by what process that growth takes place.

ACKNOWLEDGEMENT

We are grateful to Ronald G. Bodkin, Lawrence R. Klein and Lawrence Schembri for helpful comments.

REFERENCES

Baillie, R. T. and McMahon, P. C. (1985) Some joint tests of market efficiency: the case of the forward premium. *Journal of Macroeconomics*, 7, No. 2, Spring.

Canada, Department of Finance (1982) *Annual Report to Parliament on the operation of the Exchange Fund Account by the Minister of Finance*, Ottawa.

Courchene, T. (1978) *The Strategy of Gradualism: An Analysis of Bank of Canada Policy from mid-1975 to mid-1977*, C. D. Howe Research Institute, Montreal.

Helliwell, J. F. and Boothe, P. M. (1983) Macroeconomic implication of alternative exchange rate models, in *Exchange Rates in Multicountry Econometric Models* (eds P. DeGrauwe and T. Peeters), Macmillan, London.

Longworth, D. (1980) Canadian interventions in the foreign exchange market: a note. *Review of Economics and Statistics*, 62, No. 2, May.

Marwah, K. (1984) A structural model of the foreign exchange markets of the 1970s: a Canadian experience. *Carleton Economic Papers*, No. 84, 14.

Marwah, K. (1985) A prototype model of the foreign exchange market of Canada: forecasting capital flows and exchange rates. *Economic Modelling*, 2, No. 2, April.

Marwah, K. and Bodkin, R. G. (1984) A model of the Canadian global exchange rate: a test of the 1970s. *Journal of Policy Modeling*, 6, No. 4.

Rose, A. K. (1983) Intervention and the Canadian foreign exchanges: a microeconomic approach, unpublished M. Phil. thesis, Oxford University.

APPENDIX A

(Source: Marwah (1985) A prototype model of the foreign exchange market of Canada: forecasting capital flows and exchange rates)

The statistical estimates

The model is estimated for Canada by using 1971–1981 quarterly data. However, due to the incompleteness of some series, the sample period for the exchange rate equations of the Canadian dollar in terms of the Japanese yen starts with the first quarter of 1974. The primary data sources used are the *International Financial Statistics* and the *Balance of Payments* data tapes of the International Monetary Fund (IMF), *Bank of Canada Review* and the *Weekly Financial Statistics* of the Bank of Canada. The balance-of-payments identity is maintained in such a way that each capital flow component is exclusive of exceptional financing and liabilities constituting the foreign authorities' reserves. All capital flow components are measured as net inflows ($+$ = inflows, $-$ = outflows). The BOP and IFS codes, wherever relevant, have been indicated with the definitions of the variables.

The model has been estimated by three-stage least squares (3SLS) except for the cross exchange rate equations. These exchange rate equations could not be estimated by a system approach as their sample size was not uniform. They have, therefore, been estimated either by two-stage least squares corrected for serial correlation by Cochrane–Orcutt technique (2SCORC) or by the ordinary least squares (OLS) wherever it is justified. In presenting the estimates, the method of estimation has been noted with each equation. However, prior to estimating any parts of the model by 3SLS and 2SCORC, the entire model was estimated by ordinary least squares. For the sake of comparison and for assessing the goodness of fit of each individual equation, the complete set of OLS estimates has also been presented in this appendix.

Definitions of the variables and data sources

AF	= adjustment factor to reconcile the monetary aggregate of term deposits obtained from *Bank of Canada Review* and the *International Financial Statistics* (= term deposits which had not been included in M_q), millions of C$
BAL	= balance on capital account (CAB) less resident official long-term (ROL) and short-term (ROS) capital flow
BALR	= BAL normalized (BAL/NFA$_{-1}$)
CAB	= balance on capital account, $+$ = inflows and $-$ = outflows (D1.XA minus A..CA), millions of SDRs
CABR	= CAB normalized (CAB/NFA$_{-1}$)
CUB	= balance on current account, $+$ = surplus and $-$ = deficit, (A..CA), millions of SDRs
CUBR	= CUB normalized (CUB/NFA$_{-1}$)
D	= dummy variable taking unity up to 1973(Q1) and zero otherwise
DBL	= deposit money banks long-term capital inflow (5z1X4), millions of SDRs
DBLR	= DBL normalized (DBL/NFA$_{-1}$)
DBS	= deposit money banks short-term capital flow (5z2XA), millions of SDRs
DBSR	= DBS normalized (DBS/NFA$_{-1}$)
DIN	= direct investment (3..X4), millions of SDRs
DINR	= DIN normalized (DIN/NFA$_{-1}$)
EFCI	= exceptional financing, counterpart items and liabilities constituting foreign authorities' reserves (2..C4 plus .Y.XB plus 9W.X4), millions of SDRs
EFCIR	= EFCI normalized (EFCI/NFA$_{-1}$)
ERO	= errors and omissions capital flow (.A.X4), millions of SDRs

Appendix A

EROR	= ERO normalized (ERO/NFA$_{-1}$)
ε_{fcj}	= 3-months forward exchange rate of the Canadian dollar (C$) per unit of currency of country j; j = France (FR), West Germany (GR), Japan (JP), the UK and the US, end of the period
ε_{fcc}	= 3-months global forward exchange rate of the Canadian dollar computed as $\sum_j \omega_j \varepsilon_{fcj}$, $\sum \omega_j = 1$. The weights assigned are:

Currency	Weights
French franc	0.13
Deutschmark	0.19
Japanese yen	0.13
Pound sterling	0.13
US dollar	0.42

ε^p_{fcc}	= the global interest parity forward exchange rate of the Canadian dollar, based on the short-term interest rate and computed as

$$\sum_j \omega_j \left(\frac{100 + r^s}{100 + r^s_j}\right) \varepsilon_{scj}, \quad \sum \omega_j = 1$$

ε_{scj}	= the spot exchange rate of the Canadian dollar per unit of currency of country j, end of the period (IFS series ae)
ε_{scc}	= the global spot exchange rate of the Canadian dollar computed as $\sum_j \omega_j \varepsilon_{scj}$, $\sum \omega_j = 1$, as in ε_{fcc}
ε_{scSD}	= the exchange rate of the Canadian dollar per SDR, end of the period
GNFA	= the rate of change in the stock of net foreign assets, (NFA − NFA$_{-1}$)/NFA$_{-1}$
GNP	= gross national product (IFS series 99a), billions of C$
GNPR	= GNP normalized by foreign assets (GNP/NFA$_{-1}$)
H	= money base (reserve money, IFS series 14), billions of C$, end of the period
M_1	= currency plus demand deposits (IFS series 34), billions of C$, end of the period
M_q	= quasimoney (savings plus foreign currency deposits, IFS series 35), billion of C$, end of the period
NFA	= net foreign assets of Canada, millions of SDRs, end of the period, NFA$_t$ = NFA$_{1969}$ + $\sum_{i=1969}$ (CUB + CAB)$_{t-i}$, (NFA$_{1969}$ from IFS series 1..S). These international reserves include gold, holdings of SDRs, reserve position at the Fund, and foreign exchange reserves inclusive of such items as monetization, demonetization and valuation adjustment.
NFA$_{-1}$	= net foreign assets at the beginning of the period
p_e	= export prices of Canada, index of unit value of exports expressed in Canadian dollars, 1975 = 1.00 (IFS series 74)
p_{ej}	= index of unit value of exports of country j expressed in own currency unit, 1975 = 1.00 (IFS series 74)

p_{ec}	= weighted average of the export price level of j countries, 1975 = 1.00, computed as $\sum_j \omega_j p_{ej}$, $\sum \omega_j = 1$
PIN	= portfolio investment (6z1XA), millions of SDRs
PINR	= PIN normalized (PIN/NFA$_{-1}$)
Q	= quarter
RGNP	= real GNP in 1975 constant dollars (GNP/p_e)
ROL	= resident official long-term capital flow (4z1XA), millions of SDRs
ROLR	= ROL normalized (ROL/NFA$_{-1}$)
ROS	= resident official short-term capital flow (4z2XA), millions of SDRs
ROSR	= ROS normalized (ROS/NFA$_{-1}$)
RL	= other long-term capital flow (8z1X4), millions of SDRs
RLR	= RL normalized (RL/NFA$_{-1}$)
RS	= other short-term capital flow (8z2Z4), millions of SDRs
RSR	= RS normalized (RS/NFA$_{-1}$)
r^s	= short-term interest rate measured by the treasury bill rate (YTB), period average, percentage p.a. (IFS series 61C)
r^l	= long-term interest rate measured by the government bond yield (YGB), period average, percentage p.a. (IFS series 61A)
T	= time trend, 1970(Q1) = 1
μd_j	= the difference between the yield on Canadian domestic securities and the expected yield to Canadian investors on securities denominated in currency of country j, measured in terms of the Canadian dollar, percentage p.a., computed as

$$(100 + r) - (100 + r_j)\frac{\varepsilon_{fcj}}{\varepsilon_{scj}}.$$

μd	= weighted average of the yield differential μd_j, computed as $\sum_j \omega_j \mu d_j$, $\sum \omega_j = 1$; that is,

$$\mu d = (100 + r) - \sum_j \omega_j (100 + r_j)\frac{\varepsilon_{fcj}}{\varepsilon_{scj}}$$

$$\mu d = \mu d \text{HS} + \mu d \text{HFS}$$

$$\mu d \text{HS} = (100 + r)\sum_j \omega_j\left(1 - \frac{\varepsilon_{fcj}}{\varepsilon_{scj}}\right)$$

$$\mu d \text{HFS} = \sum_j \omega_j (r - r_j)\frac{\varepsilon_{fcj}}{\varepsilon_{scj}}$$

Note: An attached letter l denotes long term, computed by using long-term interest rate and s denotes short term computed by using short-term interest rates.

ω_j	= weight assigned to country j, the same as used in the construction of the global exchange rates.
YGB	= r^l

Appendix A

YTB	$= r^s$
z	= dummy variable taking value zero for the period prior to the adoption by the Bank of Canada of policy of monetary gradualism (up to 1975 (Q3)) and unity thereafter
$\sigma_{l\mu d}$	= 4-quarter moving standard deviation of the long-term yield differential ($l\mu d$) (moving quarters $t-0$ to $t-3$)
$\sigma_{s\mu d}$	= 4-quarter moving standard deviations of the short-term yield differential ($s\mu d$) (moving quarters $t-0$ to $t-3$)
j	= France (FR), West Germany (GR) Japan (JP), UK and the US
.	= dot above the variable indicates the rate of change
Δ	= change from the preceding period
OLS	= ordinary least squares
2SLS	= two-stage least squares
2SCORC	= two-stage least squares corrected for serial correlation by the Cochrane–Orcutt technique
3SLS	= three-stage least squares

The complete estimated model is presented below. The sample contains 44 quarterly observations for the years 1971–1981 unless specified otherwise. The numbers within parentheses below the coefficients are t-ratios.

The estimated model

Aggregate capital flow

$$\text{CABR} = 0.00004912 + 0.99957 \text{ GNFA} - 0.001395 \text{ GNPR} - 0.99998 \text{ CUBR}$$
$$(1.94) \quad\quad (10293.8) \quad\quad (3.98) \quad\quad\quad (15187.4)$$
$$+ 0.0000133 \, s\mu d - 0.000521 \, [\dot{\varepsilon}_{scc} - (\dot{p}_e - \dot{p}_{ec})] - 0.0000503 \, D \quad\quad \text{(A1)}$$
$$(1.69) \quad\quad\quad (2.41) \quad\quad\quad\quad\quad\quad\quad (0.17) \quad\quad\quad 3\text{SLS}$$

Direct investment

$$\text{DINR} = -1.31664 + 0.46499 \, (\text{GNFA} + \dot{\varepsilon}_{scSD}) - 7.34089 \text{ GNPR} + 0.03586) \, l\mu d$$
$$(2.82) \quad\quad (2.06) \quad\quad\quad\quad\quad\quad (5.32) \quad\quad\quad (1.26)$$
$$+ 5.27987 \text{ DBLR} + 1.60762 \, \frac{p_e}{p_{ec}} - 0.019482 \, D \quad\quad\quad\quad \text{(A2)}$$
$$(3.59) \quad\quad\quad (3.13) \quad\quad\quad\quad (0.33) \quad\quad\quad\quad\quad 3\text{SLS}$$

Portfolio investment

$$\text{PINR} = 0.17542 + 0.68803 \text{ GNFA} + 0.02585 \, s\mu d\text{HFS} - 0.051388 \, \Delta s\mu d\text{HFS}$$
$$(5.18) \quad\quad (3.74) \quad\quad\quad\quad (2.79) \quad\quad\quad\quad (3.70)$$
$$+ 2.26996 \text{ RLR} + 0.01049 \, \sigma_{s\mu d} - 0.013259 \, D \quad\quad\quad\quad \text{(A3)}$$
$$(8.29) \quad\quad\quad (1.98) \quad\quad\quad (2.61) \quad\quad\quad\quad\quad\quad 3\text{SLS}$$

Resident official long-term

$$\text{ROLR} = 0.164421 + 1.00234 \text{ GNFA} - 1.0032 \text{ BALR} - 1.03393 \text{ CUBR}$$
$$(3.10) \quad\quad (9.90) \quad\quad\quad\quad (10.49) \quad\quad\quad\quad (11.04)$$

$$- 0.17905 \frac{\dot{p}_e}{\dot{p}_{ec}} - 0.009914\, \sigma_{l\mu d} + 0.00581\, D \quad\quad (A4)$$
$$(3.46) \quad\quad\quad (1.51) \quad\quad\quad\quad (0.52) \quad\quad\quad \text{3SLS}$$

Deposit money banks long-term

$$\text{DBLR} = 0.00366 + 0.034979\, (\text{GNFA} + \dot{\varepsilon}_{scSD}) + 0.013501\, \text{CUBR} - 0.22246\, \text{RLR}$$
$$(1.35) \quad\quad (2.31) \quad\quad\quad\quad\quad\quad\quad\quad\quad (1.66) \quad\quad\quad\quad (10.72)$$
$$+ 0.073276\, [\dot{\varepsilon}_{scc} - (\dot{p}_e - \dot{p}_{ec})] + 0.008543\, \sigma_{l\mu d} - 0.00535\, D \quad\quad (A5)$$
$$(2.36) \quad\quad\quad\quad\quad\quad\quad\quad (7.22) \quad\quad\quad\quad (1.33) \quad\quad \text{3SLS}$$

Other long-term

$$\text{RLR} = 0.01777 + 0.244998\, (\text{GNFA} + \dot{\varepsilon}_{scSD}) - 0.010664\, l\mu d\text{HFS}$$
$$(2.15) \quad\quad (4.26) \quad\quad\quad\quad\quad\quad\quad\quad\quad (1.43)$$
$$- 0.51806\, \text{RSR} - 0.02953\, D \quad\quad (A6)$$
$$(8.57) \quad\quad\quad\quad (1.86) \quad\quad\quad \text{3SLS}$$

Resident official short-term

$$\text{ROSR} = -0.13664 - 0.07839\, \text{GNFA} + 0.06793\, \text{BAL} + 0.10663\, \text{CUBR}$$
$$(2.50) \quad\quad (0.74) \quad\quad\quad\quad (0.69) \quad\quad\quad\quad (1.098)$$
$$+ 1.47979 \frac{\dot{p}_e}{\dot{p}_{ec}} + 0.01383\, \sigma_{s\mu d} - 0.00888\, D \quad\quad (A7)$$
$$(2.74) \quad\quad\quad (2.45) \quad\quad\quad\quad (0.76) \quad\quad\quad \text{3SLS}$$

Deposits money banks short-term

$$\text{DBSR} = -0.14609 - 1.0325\, \text{CUBR} - 1.43705\, \text{DINR} - 0.08438\, \sigma_{s\mu d}$$
$$(2.19) \quad\quad (7.78) \quad\quad\quad\quad (8.27) \quad\quad\quad\quad (1.08)$$
$$+ 0.27017\, D \quad\quad (A8)$$
$$(2.60) \quad\quad\quad \text{3SLS}$$

Other short-term

$$\text{RSR} = 0.150086 + 0.17115\, \text{GNFA} - 2.05859\, \text{GNPR} + 0.03527\, s\mu d\text{HS}$$
$$(3.48) \quad\quad (1.06) \quad\quad\quad\quad (3.47) \quad\quad\quad\quad (1.69)$$
$$- 0.06562\, \sigma_{s\mu d} + 0.27501\, \text{RSR}_{-1} - 0.09854\, D \quad\quad (A9)$$
$$(1.60) \quad\quad\quad\quad (1.70) \quad\quad\quad\quad (2.33) \quad\quad \text{3SLS}$$

Appendix A

Errors and omissions

$$
\begin{aligned}
\text{EROR} = &-0.07245 - 0.54636\,\text{GNFA} + 0.07273\,\text{l}\mu d\text{HS} - 0.03880\,\Delta\text{l}\mu d\text{HS} \\
&\;\;(1.86)\quad\;(2.84)\qquad\qquad(2.38)\qquad\qquad\;\;(1.78) \\
&+ 0.53274\,\text{DINR} - 0.014625\,D \\
&\;\;(4.44)\qquad\qquad\;(0.20)
\end{aligned}
\qquad\text{(A10)}\;\;\text{3SLS}
$$

$$
\text{CABR} = \text{DINR} + \text{PINR} + \text{ROLR} + \text{DBLR} + \text{RLR} + \text{ROSR} + \text{DBSR} + \text{RSR} + \text{EROR} + \text{EFCIR} \qquad\text{(A11)}
$$

$$
\text{GNFA} = \text{CUBR} + \text{CABR} \qquad\text{(A12)}
$$

Spot exchange rates

$$
\begin{aligned}
\varepsilon_{\text{scFR}} = &\;0.02946 - 0.03866\,\text{GNFA} + 0.24462\,\text{GNPR} - 0.00313\,\text{YTB} \\
&(1.88)\quad\;(2.20)\qquad\qquad\;(1.73)\qquad\qquad\;(2.29) \\
&+ 0.9308\,\varepsilon_{\text{fcFR}-1} \\
&\;\;(12.07) \\
&\text{Mean} = 0.2269 \qquad \bar{R}^2 = 0.846 \\
&\text{SE} = 0.0131 \qquad\;\; d = 1.612
\end{aligned}
\qquad\begin{aligned}\text{(A13)}\\ \text{OLS}\end{aligned}
$$

$$
\begin{aligned}
\varepsilon_{\text{SCGR}} = &\;0.04170 - 0.05674\,\text{GNFA} + 0.46226\,\text{GNPR} - 0.00638\,\text{YTD} \\
&(2.04)\quad\;(1.44)\qquad\qquad\;(1.44)\qquad\qquad\;(2.13) \\
&+ 0.9756\,\varepsilon_{\text{fcGR}-1} \\
&\;\;(17.14) \\
&\text{Mean} = 0.4623 \qquad \bar{R}^2 = 0.949 \\
&\text{SE} = 0.0287 \qquad\;\; d = 1.836
\end{aligned}
\qquad\begin{aligned}\text{(A14)}\\ \text{OLS}\end{aligned}
$$

$$
\begin{aligned}
\varepsilon_{\text{scJP}} = &\;0.00127 - 0.0010\,\text{GNFA} + 0.01492\,\text{GNPR} - 0.00016\,\text{YTB} \\
&(6.16)\quad\;(3.74)\qquad\qquad(5.77)\qquad\qquad\;(6.48) \\
&+ 0.85709\,\varepsilon_{\text{fcJP}-1} \\
&\;\;(17.11) \\
&\text{Mean} = 0.00424 \qquad \bar{R}^2 = 0.969 \\
&\text{SE} = 0.00019 \qquad\;\; d = 1.770
\end{aligned}
\qquad\begin{aligned}\text{(A15)}\\ \text{OLS}\end{aligned}
$$

Sample: 1974(Q1)–1981(Q4)

$$
\begin{aligned}
\varepsilon_{\text{scUK}} = &\;0.3887 - 0.21133\,\text{GNFA} + 3.4844\,\text{GNPR} - 0.0420\,\text{YTB} \\
&(1.98)\quad(1.06)\qquad\qquad(2.10)\qquad\qquad(2.70) \\
&+ 0.91675\,\varepsilon_{\text{fcUK}-1} \\
&\;\;(11.28) \\
&\text{Mean} = 2.3213 \qquad \bar{R}^2 = 0.800 \\
&\text{SE} = 0.1483 \qquad\;\; d = 1.636
\end{aligned}
\qquad\begin{aligned}\text{(A16)}\\ \text{OLS}\end{aligned}
$$

$$
\begin{aligned}
\varepsilon_{\text{scUS}} = &\;0.07525 - 0.12538\,\text{GNFA} + 0.5210\,\text{GNPR} - 0.00354\,\text{YTB} \\
&(1.19)\quad\;(6.10)\qquad\qquad\;(2.34)\qquad\qquad\;(2.23) \\
&+ 0.93343\,\varepsilon_{\text{fcUS}-1} \\
&\;\;(14.49)
\end{aligned}
$$

$$\text{Mean} = 1.0764 \quad \bar{R}^2 = 0.972 \tag{A17}$$
$$\text{SE} = 0.0151 \quad d = 1.158 \quad \text{OLS}$$

$$\varepsilon_{scc} = 0.13\,\varepsilon_{scFR} + 0.19\,\varepsilon_{scGR} + 0.13\,\varepsilon_{scJP} + 0.13\,\varepsilon_{scUK} + 0.42\,\varepsilon_{scUS} \tag{A18}$$

Alternatively,

$$\varepsilon_{scc} = -0.08863 - 0.150467\,\text{GNFA} + 1.45029\,\text{GNPR} - 0.018143\,\text{YTB}$$
$$\quad\quad\;\;(0.86)\quad\quad\;(3.07)\quad\quad\quad\quad\;(4.26)\quad\quad\quad\quad\;(5.25)$$

$$+\,0.665092\,\frac{p_e}{p_{ec}} + 0.41038\,\varepsilon_{fcc-1} - 0.0362\,D \tag{A19}$$
$$\quad(4.68)\quad\quad\quad\quad\;(4.48)\quad\quad\quad\;(1.74) \quad\quad\quad \text{3SLS}$$

Forward exchange rates

$$\varepsilon_{fcFR} = 0.03234 + 0.87688\,\varepsilon^p_{fcFR} - 0.004866\,\sigma_{s\mu d}$$
$$\quad\quad\quad(4.32)\quad\;(28.60)\quad\quad\quad(2.72)$$

$$\text{Mean} = 0.2271 \quad \bar{R}^2 = 0.992$$
$$\text{SE} = 0.0028 \quad d = 1.780 \tag{A20}$$
$$\rho = 0.509 \quad\quad \text{2SCORC}$$

$$\varepsilon_{fcGR} = 0.03830 + 0.91073\,\varepsilon^p_{fcGR} - 0.006052\,\sigma_{s\mu d}$$
$$\quad\quad\quad(2.63)\quad\;(33.43)\quad\quad\quad(1.33)$$

$$\text{Mean} = 0.4623 \quad \bar{R}^2 = 0.998$$
$$\text{SE} = 0.0061 \quad d = 1.598 \tag{A21}$$
$$\rho = 0.704 \quad\quad \text{2SCORC}$$

$$\varepsilon_{fcJP} = 0.000487 + 0.89787\,\varepsilon^p_{fcJP} - 0.000166\,\sigma_{s\mu d}$$
$$\quad\quad\quad(4.54)\quad\quad(41.43)\quad\quad\quad(3.16)$$

$$\text{Mean} = 0.00456 \quad \bar{R}^2 = 0.995$$
$$\text{SE} = 0.000075 \quad d = 1.385 \tag{A22}$$
$$\text{Sample: 1974(Q1)–1981(Q4)} \quad \rho = 0.454 \quad \text{2SCORC}$$

$$\varepsilon_{fcUK} = 0.136309 + 0.55927\,\varepsilon^p_{fcUK} + 0.39858\,\varepsilon_{fcUK-1} - 0.07089\,\sigma_{s\mu d}$$
$$\quad\quad\quad(1.48)\quad\quad(6.63)\quad\quad\quad\quad(4.38)\quad\quad\quad\;(2.12)$$

$$\text{Mean} = 2.2891 \quad \bar{R}^2 = 0.938$$
$$\text{SE} = 0.0763 \quad d = 1.958 \tag{A23}$$
$$\quad\quad \text{OLS}$$

$$\varepsilon_{fcUS} = 0.04095 + 0.37518\,\varepsilon^p_{fcUS} + 0.59234\,\varepsilon^p_{fcUS-1} - 0.012893\,\sigma_{s\mu d}$$
$$\quad\quad\quad(1.52)\quad\;(2.19)\quad\quad\quad\;(3.10)\quad\quad\quad\quad(2.68)$$

$$\text{Mean} = 1.0773 \quad \bar{R}^2 = 0.981$$
$$\text{SE} = 0.01206 \quad d = 1.969 \tag{A24}$$
$$\rho = -0.106 \quad\quad \text{2SCORC}$$

$$\varepsilon_{fcc} = 0.13\,\varepsilon_{fcFR} + 0.19\,\varepsilon_{fcGR} + 0.13\,\varepsilon_{fcJP} + 0.13\,\varepsilon_{fcUK} + 0.42\,\varepsilon_{fcUS} \tag{A25}$$

Alternatively,

$$\varepsilon_{\text{fcc}} = 0.03848 + 0.83159\,\varepsilon^p_{\text{fcc}} + 0.12468\,\varepsilon_{\text{fcc}-1} - 0.01408\,\sigma_{s\mu d}$$
$$(1.81)\quad\ \ (17.18)\qquad\ \ (2.46)\qquad\quad\ \ (2.71)$$
$$+\ 0.11758\,D \tag{A26}$$
$$(15.16) \qquad\qquad\qquad\qquad\qquad\qquad \text{3SLS}$$

Forward parity exchange rates

$$\varepsilon^p_{\text{fcFR}} = \left(\frac{100 + \text{YTB}}{100 + \text{YTB}_{\text{FR}}}\right)\varepsilon_{\text{scFR}} \tag{A27}$$

$$\varepsilon^p_{\text{fcGR}} = \left(\frac{100 + \text{YTB}}{100 + \text{YTB}_{\text{GR}}}\right)\varepsilon_{\text{scGR}} \tag{A28}$$

$$\varepsilon^p_{\text{fcJP}} = \left(\frac{100 + \text{YTB}}{100 + \text{YTB}_{\text{JP}}}\right)\varepsilon_{\text{scJP}} \tag{A29}$$

$$\varepsilon^p_{\text{fcUK}} = \left(\frac{100 + \text{YTB}}{100 + \text{YTB}_{\text{UK}}}\right)\varepsilon_{\text{scUK}} \tag{A30}$$

$$\varepsilon^p_{\text{fcUS}} = \left(\frac{100 + \text{YTB}}{100 + \text{YTB}_{\text{US}}}\right)\varepsilon_{\text{scUS}} \tag{A31}$$

$$\varepsilon^p_{\text{fcc}} = 0.13\,\varepsilon^p_{\text{fcFR}} + 0.19\,\varepsilon^p_{\text{fcGR}} + 0.13\,\varepsilon^p_{\text{fcJP}} + 0.13\,\varepsilon^p_{\text{fcUK}} + 0.42\,\varepsilon^p_{\text{fcUS}} \tag{A32}$$

Monetary sector

$$\frac{M_1}{P_e} = 1.55371 + 0.102013\,\text{RGNP} - 0.47219\,\text{YTB} + 0.406224\left(\frac{M_1}{P_e}\right)_{-1}$$
$$\quad\ \ (6.32)\qquad\ (5.60)\qquad\qquad\ \ (7.43)\qquad\qquad\ (5.35)$$
$$-\ 1.72801\,z \tag{A33}$$
$$(3.25) \qquad\qquad\qquad\qquad\qquad\qquad \text{3SLS}$$

$$\frac{M_q}{P_e} = -9.74481 + 0.09744\,\text{RGNP} - 0.50183\,(\text{YGB} - \text{YTB})$$
$$\quad\ \ (2.02)\qquad\ (3.24)\qquad\qquad\ \ (2.21)$$

$$+\ 0.868638\left(\frac{M_q}{P_e}\right)_{-1} + 1.01990\,z \tag{A34}$$
$$(1.79)\qquad\qquad\qquad\ (9.48) \qquad\qquad \text{3SLS}$$

$$H = 4.4449 + 0.28993\,T - 0.000211\,\text{NFA} \tag{A35}$$
$$\quad (7.95)\quad\ (3.73)\qquad\ (2.38) \qquad\qquad\qquad \text{3SLS}$$

$$M_1 = 7.25132 + 1.600276\,H + 0.34025\,(\text{YGB} - \text{YTB}) \tag{A36}$$
$$\quad\ \ (9.14)\qquad\ (2.62)\qquad\ \ (2.78) \qquad\qquad\qquad \text{3SLS}$$

$$YGB = 12.86651 + 0.102469\ M_q - 0.472398\ H - 5.189428\ \varepsilon_{fcc}$$
$$(8.79)\quad\ \ (5.94)\qquad\quad (3.19)\qquad\quad (4.01)$$
$$+ 0.160738\ AF \qquad\qquad\qquad\qquad\qquad\qquad\qquad\quad (A37)$$
$$(5.62) \qquad\qquad\qquad\qquad\qquad\qquad\qquad\qquad\qquad 3SLS$$

$$YTB = 0.27912 + 0.494661\ YTB_{US} - 12.050972\ (\varepsilon_{fcc} - \varepsilon^p_{fcc})$$
$$(2.70)\quad\ \ (5.51)\qquad\qquad\quad (4.594)$$
$$+ 0.371338\ YGB - 0.245234\ YGB_{-1} + 0.440975\ YTB_{-1} \qquad (A38)$$
$$(1.24)\qquad\qquad (9.75)\qquad\qquad\quad (3.49) \qquad\qquad\qquad 3SLS$$

$$M_1 + M_q = \left(\frac{M_1}{p_e}\right)p_e + \left(\frac{M_q}{p_e}\right)p_e \qquad\qquad\qquad\qquad\qquad (A39)$$

(Ordinary least squares estimates)

Aggregate capital flow

$$CABR = 0.00004575 + 0.99964\ GNFA - 0.001489\ GNPR - 1.00001\ CUBR$$
$$(1.72)\qquad\qquad (11802.4)\qquad\ \ (3.84)\qquad\qquad (12720.9)$$
$$+ 0.000017\ s\mu d - 0.000441[\dot{\varepsilon}_{scc} - (\dot{p}_e - \dot{p}_{ec})] + 0.000004\ D$$
$$(1.75)\qquad\qquad (1.72)\qquad\qquad\qquad\qquad\quad (0.13)$$
$$\text{Mean} = 0.14124 \qquad \bar{R}^2 = 0.999$$
$$\text{SE} = 0.0000675 \qquad d = 2.30 \qquad\qquad\qquad\qquad\qquad (A40)$$

Direct investment

$$DINR = -1.45697 + 0.48787\ (GNFA + \dot{\varepsilon}_{scSD}) - 0.894389\ GNPR$$
$$(2.63)\qquad (2.56)\qquad\qquad\qquad\qquad\quad (5.93)$$
$$+ 0.026247\ l\mu d + 1.73082\ DBLR + 1.83055\ \frac{p_e}{p_{ec}} - 0.05365\ D$$
$$(0.75)\qquad\qquad (1.10)\qquad\qquad (3.03)\qquad\qquad (0.87)$$
$$\text{Mean} = -0.1105 \qquad \bar{R} = 0.786$$
$$\text{SE} = 0.13274 \qquad d = 2.049 \qquad\qquad\qquad\qquad\qquad (A41)$$

Portfolio investment

$$PINR = 0.15402 + 0.43529\ GNFA + 0.02272\ s\mu dHFS - 0.07806\ \Delta\mu dHFS$$
$$(4.17)\qquad (2.76)\qquad\qquad (1.96)\qquad\qquad\quad (3.82)$$
$$+ 2.14726\ RLR + 0.07067\sigma_{s\mu d} - 0.15703\ D$$
$$(7.56)\qquad\qquad (1.096)\qquad\quad (2.88)$$
$$\text{Mean} = 0.2250 \qquad \bar{R}^2 = 0.881$$
$$\text{SE} = 0.1178 \qquad d = 1.319 \qquad\qquad\qquad\qquad\qquad (A42)$$

Resident official long-term

$$ROLR = 0.18856 + 0.78177\ GNFA - 0.80282\ BAL - 0.82452\ CUBR$$
$$(3.24)\qquad (9.69)\qquad\qquad (10.21)\qquad\qquad (10.47)$$

$$-0.17905 \frac{p_e}{p_{ec}} - 0.0154\sigma_{l\mu d} + 0.00389 D$$
$$(3.59) \qquad (1.10) \qquad (0.32)$$

$$\text{Mean} = -0.0387 \qquad \bar{R}^2 = 0.821$$
$$\text{SE} = 0.02387 \qquad d = 1.954 \tag{A43}$$

Deposit money banks long-term

$$\text{DBLR} = 0.004565 + 0.032215 \, (\text{GNFA} + \dot{\varepsilon}_{scSD}) + 0.015569 \, \text{CUBR}$$
$$(1.53) \qquad (2.45) \qquad (1.63)$$
$$-0.19462 \, \text{RLR} + 0.052643[\dot{\varepsilon}_{scc} - (\dot{p}_e - \dot{p}_{ec})]$$
$$(10.05) \qquad (1.43)$$
$$-0.000909\sigma_{l\mu d} - 0.003714 D$$
$$(0.15) \qquad (0.87)$$
$$\text{Mean} = -0.0043 \qquad \bar{R}^2 = 0.787$$
$$\text{SE} = 0.00946 \qquad d = 2.499 \tag{A44}$$

Other long-term

$$\text{RLR} = 0.01127 + 0.27316 \, (\text{GNFA} + \dot{\varepsilon}_{scSD}) + 0.00606 \, l\mu d\text{HFS}$$
$$(1.32) \qquad (5.16) \qquad (0.67)$$
$$-0.42786 \, \text{RSR} - 0.018532 D$$
$$(7.35) \qquad (1.15) \tag{A45}$$
$$\text{Mean} = 0.0277 \qquad \bar{R}^2 = 0.793$$
$$\text{SE} = 0.04213 \qquad d = 2.152$$

Resident official short-term

$$\text{ROSR} = -0.16703 + 0.17630 \, \text{GNFA} - 0.16548 \, \text{BAL} - 0.13951 \, \text{CUBR}$$
$$(2.89) \qquad (2.17) \qquad (2.14) \qquad (1.77)$$
$$+0.17781 \frac{p_e}{p_{ec}} + 0.018507 \sigma_{s\mu d} - 0.00471 D$$
$$(3.08) \qquad (1.72) \qquad (0.42)$$
$$\text{Mean} = 0.01535 \qquad \bar{R}^2 = 0.399$$
$$\text{SE} = 0.02333 \qquad d = 1.998 \tag{A46}$$

Deposit money banks short-term

$$\text{DBSR} = -0.05631 - 1.0577 \, \text{CUBR} - 1.39575 \, \text{DINR}$$
$$(0.60) \qquad (3.71) \qquad (6.57)$$
$$-0.23612\sigma_{s\mu d} - 0.27909 D$$
$$(1.65) \qquad (2.40)$$
$$\text{Mean} = -0.17833 \qquad \bar{R}^2 = 0.697$$
$$\text{SE} = 0.2757 \qquad d = 2.33 \tag{A47}$$

Other short-term

$$\text{RSR} = 0.15952 + 0.03003 \text{ GNFA} - 2.0570 \text{ GNPR} + 0.03858\sigma_{s\mu d}$$
$$(3.46) \quad (2.24) \quad (3.22) \quad (1.58)$$
$$- 0.07981\sigma_{s\mu d} + 0.27245 \text{ RSR}_{-1} - 0.100277\, D$$
$$(1.71) \quad (1.53) \quad (2.34)$$

$$\text{Mean} = -0.0400 \qquad \bar{R}^2 = 0.694$$
$$\text{SE} = 0.0903 \qquad d = 2.155 \tag{A48}$$

Errors and Omissions

$$\text{EROR} = -0.09859 - 0.48066 \text{ GNFA} + 0.07810 \text{ l}\mu d\text{HS} - 0.06243\, \Delta\text{l}\mu d\text{HS}$$
$$(2.55) \quad (2.02) \quad (1.34) \quad (1.35)$$
$$+ 0.3598 \text{ DINR} - 0.01378\, D$$
$$(2.64) \quad (0.19)$$

$$\text{Mean} = -0.13843 \qquad \bar{R}^2 = 0.414$$
$$\text{SE} = 0.1950 \qquad d = 2.608 \tag{A49}$$

$$\text{CABR} = \text{DINR} + \text{PINR} + \text{ROLR} + \text{DBLR} + \text{RLR} + \text{ROSR} + \text{DBSR}$$
$$+ \text{RSR} + \text{EROR} + \text{EFCI} \tag{A50}$$

$$\text{GNFA} = \text{CUBR} + \text{CABR} \tag{A51}$$

Spot exchange rates

$$\varepsilon_{\text{scFR}} = 0.02946 - 0.03866 \text{ GNFA} + 0.24462 \text{ GNPR} - 0.00313 \text{ YTB}$$
$$(1.88) \quad (2.20) \quad (1.73) \quad (2.29)$$
$$+ 0.93088\, \varepsilon_{\text{fcFR}-1}$$
$$(12.07)$$

$$\text{Mean} = 0.2269 \qquad \bar{R}^2 = 0.846$$
$$\text{SE} = 0.0131 \qquad d = 1.612 \tag{A52}$$

$$\varepsilon_{\text{scGR}} = 0.04170 - 0.05674 \text{ GNFA} + 0.46226 \text{ GNPR} - 0.00638 \text{ YTB}$$
$$(2.04) \quad (1.44) \quad (1.44) \quad (2.13)$$
$$+ 0.9756\, \varepsilon_{\text{fcGR}-1}$$
$$(17.14)$$

$$\text{Mean} = 0.4623 \qquad \bar{R}^2 = 0.949$$
$$\text{SE} = 0.0287 \qquad d = 1.836 \tag{A53}$$

$$\varepsilon_{\text{scJP}} = 0.00127 - 0.0010 \text{ GNFA} + 0.01492 \text{ GNPR} - 0.00016 \text{ YTB}$$
$$(6.16) \quad (3.74) \quad (5.77) \quad (6.48)$$
$$+ 0.85709\, \varepsilon_{\text{fcJP}-1}$$
$$(17.11)$$

$$\text{Mean} = 0.00424 \qquad \bar{R}^2 = 0.969$$
$$\text{SE} = 0.00019 \qquad d = 1.770$$
Sample: 1974(Q1)–1984(Q4) \tag{A54}

Appendix A

$$\varepsilon_{scUK} = 0.3887 - 0.21133 \text{ GNFA} + 3.4844 \text{ GNPR} - 0.0420 \text{ YTB}$$
$$(1.98) \quad (1.06) \quad\quad (2.10) \quad\quad (2.70)$$
$$+ 0.91675 \; \varepsilon_{fcUK-1}$$
$$(11.28)$$

$\quad$ Mean $= 2.3213 \quad\quad \bar{R}^2 = 0.800$
$\quad$ SE $= 0.1483 \quad\quad d = 1.636$ $\quad\quad\quad\quad$ (A55)

$$\varepsilon_{scUS} = 0.07525 - 0.12538 \text{ GNFA} + 0.5210 \text{ GNPR} - 0.00354 \text{ YTB}$$
$$(1.19) \quad\quad (6.10) \quad\quad\quad (2.34) \quad\quad\quad (2.23)$$
$$+ 0.93343 \; \varepsilon_{fcUS-1}$$
$$(14.49)$$

$\quad$ Mean $= 1.0764 \quad\quad \bar{R}^2 = 0.972$
$\quad$ SE $= 0.0151 \quad\quad d = 1.158$ $\quad\quad\quad\quad$ (A56)

$$\varepsilon_{scc} = 0.13 \; \varepsilon_{scFR} + 0.19 \; \varepsilon_{scGR} + 0.13 \; \varepsilon_{scJP} + 0.13 \; \varepsilon_{scUK} + 0.42 \; \varepsilon_{scUS} \quad\quad\quad (A57)$$

Alternatively,

$$\varepsilon_{scc} = -0.06374 - 0.100426 \text{ GNFA} + 1.11807 \text{ GNPR} - 0.014479 \text{ YTB}$$
$$(0.54) \quad\quad (2.39) \quad\quad\quad (2.97) \quad\quad\quad (3.63)$$
$$+ 0.46678 \frac{p_e}{p_{ec}} + 0.60035 \; \varepsilon_{fcc-1} - 0.06529 \; D$$
$$(2.79) \quad\quad\quad (5.52) \quad\quad\quad (2.80)$$

$\quad$ Mean $= 0.8685 \quad\quad \bar{R}^2 = 0.880$
$\quad$ SE $= 0.03046 \quad\quad d = 1.404$ $\quad\quad\quad\quad$ (A58)

Forward exchange rates

$$\varepsilon_{fcFR} = 0.02592 + 0.9049 \; \varepsilon^p_{fcFR} - 0.0052 \; \sigma_{s\mu d}$$
$$(7.46) \quad\quad (64.25) \quad\quad (3.93)$$

$\quad$ Mean $= 0.2271 \quad\quad \bar{R}^2 = 0.991$
$\quad$ SE $= 0.0032 \quad\quad d = 1.006$ $\quad\quad\quad\quad$ (A59)

$$\varepsilon_{fcGR} = 0.03962 + 0.9110 \; \varepsilon^p_{fcGR} - 0.00908 \; \sigma_{s\mu d}$$
$$(7.50) \quad\quad (95.83) \quad\quad (2.62)$$

$\quad$ Mean $= 0.4667 \quad\quad \bar{R}^2 = 0.996$
$\quad$ SE $= 0.0084 \quad\quad d = 0.628$ $\quad\quad\quad\quad$ (A60)

$$\varepsilon_{fcJP} = 0.000446 + 0.90378 \; \varepsilon^p_{fcJP} - 0.00015 \; \sigma_{s\mu d}$$
$$(6.91) \quad\quad (68.4) \quad\quad (3.62)$$

$\quad$ Mean $= 0.00456 \quad\quad \bar{R}^2 = 0.994$
$\quad$ SE $= 0.00009 \quad\quad d = 0.992$
$\quad$ Sample: 1974(Q1)–1981(Q4) $\quad\quad\quad\quad$ (A61)

$$\varepsilon_{fcUK} = 0.13631 + 0.5593 \; \varepsilon^p_{fcUK} - 0.07089 \; \sigma_{s\mu d} + 0.39858 \; \varepsilon_{fcUK-1}$$
$$(1.48) \quad\quad (6.63) \quad\quad (2.12) \quad\quad\quad (4.38)$$

$\quad$ Mean $= 2.2891 \quad\quad \bar{R}^2 = 0.938$
$\quad$ SE $= 0.1483 \quad\quad d = 1.958$ $\quad\quad\quad\quad$ (A62)

$$\varepsilon_{fcUS} = 0.07231 + 0.6994\,\varepsilon^p_{fcUS} - 0.01078\,\sigma_{s\mu d} + 0.23281\,\varepsilon_{fcUS-1}$$
$$(3.97) \quad (12.17) \quad\quad (2.89) \quad\quad\quad (3.56)$$
$$\text{Mean} = 1.0773 \quad \bar{R}^2 = 0.99$$
$$\text{SE} = 0.00895 \quad d = 1.224 \tag{A63}$$

$$\varepsilon_{fcc} = 0.13\,\varepsilon_{fcFR} + 0.19\,\varepsilon_{fcGR} + 0.13\,\varepsilon_{fcJP} + 0.13\,\varepsilon_{fcUK} + 0.42\,\varepsilon_{fcUS} \tag{A64}$$

Alternatively,
$$\varepsilon_{fcc} = 0.03489 + 0.80108\,\varepsilon^p_{fcc} + 0.15901\,\varepsilon_{fcc-1} - 0.013242\,\sigma_{s\mu d}$$
$$(1.57) \quad\quad (15.18) \quad\quad (2.88) \quad\quad\quad (2.44)$$
$$+ 0.11324\,D$$
$$(13.75)$$
$$\text{Mean} = 0.89606 \quad \bar{R}^2 = 0.979$$
$$\text{SE} = 0.013 \quad d = 1.970 \tag{A65}$$

Forward parity exchange rates

$$\varepsilon^p_{fcFR} = \left(\frac{100 + \text{YTB}}{100 + \text{YTB}_{FR}}\right)\varepsilon_{scFR} \tag{A66}$$

$$\varepsilon^p_{fcGR} = \left(\frac{100 + \text{YTB}}{100 + \text{YTB}_{GR}}\right)\varepsilon_{scGR} \tag{A67}$$

$$\varepsilon^p_{fcJP} = \left(\frac{100 + \text{YTB}}{100 + \text{YTB}_{JP}}\right)\varepsilon_{scJP} \tag{A68}$$

$$\varepsilon^p_{fcUK} = \left(\frac{100 + \text{YTB}}{100 + \text{YTB}_{UK}}\right)\varepsilon_{scUK} \tag{A69}$$

$$\varepsilon^p_{fcUS} = \left(\frac{100 + \text{YTB}}{100 + \text{YTB}_{US}}\right)\varepsilon_{fcUS} \tag{A70}$$

$$\varepsilon^p_{fcc} = 0.13\,\varepsilon^p_{fcFR} + 0.19\,\varepsilon^p_{fcGR} + 0.13\,\varepsilon^p_{fcJP} + 0.13\,\varepsilon^p_{fcUK} + 0.42\,\varepsilon^p_{fcUS} \tag{A71}$$

Monetary sector

$$\frac{M_1}{p_e} = 0.69315 + 0.10393\,\text{RGNP} - 0.41437\,\text{YTB} + 0.4163331\left(\frac{M_1}{p_e}\right)_{-1}$$
$$(0.20) \quad\quad (4.11) \quad\quad\quad (4.99) \quad\quad\quad (3.99)$$
$$- 2.11994\,z$$
$$(2.88)$$
$$\text{Mean} = 24.748 \quad \bar{R}^2 = 0.932$$
$$\text{SE} = 1.0916 \quad d = 1.686 \tag{A72}$$

$$\frac{M_q}{p_e} = -6.42499 + 0.07661 \, \text{RGNP} - 0.49162 \, (\text{YGB} - \text{YTB})$$
$$\phantom{\frac{M_q}{p_e} =} (1.27) \quad\quad (2.43) \quad\quad\quad\quad (2.13)$$

$$+ 0.8754 \left(\frac{M_q}{p_e}\right)_{-1} + 1.07563 \, z$$
$$(17.458) \quad\quad\quad\quad (0.9516)$$

$$\text{Mean} = 52.274 \quad\quad \bar{R}^2 = 0.984$$
$$\text{SE} = 1.7209 \quad\quad d = 1.595 \quad\quad\quad\quad (A73)$$

$$H = 5.09596 + 0.28080 \, T - 0.000308 \, \text{NFA}$$
$$(7.36) \quad\quad (2.79) \quad\quad\quad (30.45)$$

$$\text{Mean} = 11.247 \quad\quad \bar{R}^2 = 0.993$$
$$\text{SE} = 0.346 \quad\quad d = 2.28 \quad\quad\quad\quad (A74)$$

$$M_1 = 7.68195 + 1.56706 \, H + 0.27719 \, (\text{YGB} - \text{YTB})$$
$$(9.48) \quad\quad (25.13) \quad\quad\quad (2.20)$$

$$\text{Mean} = 25.557 \quad\quad \bar{R}^2 = 0.980$$
$$\text{SE} = 0.8334 \quad\quad d = 0.794 \quad\quad\quad\quad (A75)$$

$$\text{YGB} = 13.0148 + 0.10102 \, M_q - 0.46763 \, H - 5.32518 \, \varepsilon_{\text{fcc}} + 0.165399 \, \text{AF}$$
$$(8.14) \quad\quad (2.99) \quad\quad\quad (3.71) \quad\quad\quad (5.51) \quad\quad\quad (5.25)$$

$$\text{Mean} = 9.5295 \quad\quad \bar{R}^2 = 0.924$$
$$\text{SE} = 0.696 \quad\quad d = 0.919 \quad\quad\quad\quad (A76)$$

$$\text{YTB} = -1.13186 + 0.39981 \, \text{YTB}_{\text{US}} - 10.5908 \, (\varepsilon_{\text{fcc}} - \varepsilon_{\text{fcc}}^p)$$
$$(1.163) \quad\quad (5.37) \quad\quad\quad\quad (3.99)$$

$$+ 0.78129 \, \text{YGB} - 0.35339 \, \text{YGB}_{-1} + 0.34419 \, \text{YTB}_{-1}$$
$$(3.96) \quad\quad\quad (1.60) \quad\quad\quad\quad (2.68)$$

$$\text{Mean} = 8.6255 \quad\quad \bar{R}^2 = 0.975$$
$$\text{SE} = 0.698 \quad\quad d = 1.50 \quad\quad\quad\quad (A77)$$

$$M_1 + M_q = \left(\frac{M_1}{p_e}\right) p_e + \left(\frac{M_q}{p_e}\right) p_e \quad\quad\quad\quad (A78)$$

21

Effects of a fall in the price of oil: the case of a small oil-exporting country

KJELL BERGER, ÅDNE CAPPELEN,
VIDAR KNUDSEN and KJELL ROLAND

21.1 INTRODUCTION

This chapter explores the macroeconomic effects on the Norwegian economy of a large fall in the price of oil. The chapter is a condensed version of a more voluminous report being published by the Central Bureau of Statistics (see Berger et al. 1986). Our analysis is divided into three parts. In Section 21.2, we discuss some scenarios for the price of crude oil, using a world oil market model developed at the Central Bureau of Statistics (CBS), Oslo. A high price and a low price scenario are derived together with a number of alternatives according to different sets of assumptions. In Section 21.3 we review some studies of international macroeconomic repercussions of falling oil prices. Included here are results from a simulation that we have carried out on the LINK model system. Finally, in Section 21.4, building on the projected oil price paths and assumptions about international repercussions, we analyse the effects on the Norwegian economy of a lower price of oil. For this part we utilize MODAG A, a macroeconomic model of the Norwegian economy also developed at the Central Bureau of Statistics.

21.2 THE CRUDE OIL MARKET

21.2.1 Price scenarios

In this section two main price scenarios are outlined; the high price and the low price scenario. The words 'high' and 'low' refer to the outcome of two possible patterns of OPEC behaviour. The model used in this chapter to analyse the crude oil price is the world oil market (WOM) model (see Lorentsen and Roland, 1985).

WOM is a very simple, open model and could probably be characterized as a framework facilitating consistent reasoning about the market given exogenous assumptions on key parameters.

The model has three demand areas: the USA, the rest of the OECD and developing countries. Demand from each region is defined as functions of GDP, the present and lagged real price of petroleum products and prices for other energy carriers. It contains three supplying areas: OPEC, the rest of the market economies and net supply from centrally planned economy (CPE) countries. The two scenarios are based on best-guess estimates on critical exogenous parameters such as economic growth, exchange rates, prices for competing fuels, cost of transportation, refining, storage and distribution of crude oil.

Regarding supply from producers outside OPEC and CPE countries, supplies are projected by an econometric resource base submodel in WOM (see Weyant and Kline, 1980). Three central hypotheses are made: first, a constant fraction of known, recoverable reserves is produced each year. Second, the fraction of undiscovered oil-in-place discovered in a particular year is price responsive. Third, the optimal recovery factor depends on the price of oil. Adjusting to the optimal recovery factor is time consuming. All parameters in the model are of course subject to a considerable degree of uncertainty. Most important is probably the estimate of the remaining, undiscovered resource base of crude oil. Net export from the CPE area is exogenous and we assume that net export of crude oil and petroleum products from the CPE area stays roughly constant at 2.4 million barrels per day (mbd) after 1985.

Of utmost importance for the future development of the crude oil market are assumptions about the behaviour of each OPEC member country and their collective decisions at ministerial meetings. This is what separates the two main scenarios of the crude oil market outlined in this section.

In the high price scenario, we assume that OPEC members, after adjusting the price level of crude oil to $20 per barrel, are able to agree on production quotas despite the extreme pressure against the cohesion within OPEC and the financial hardship some countries are faced with. This implies that Saudi Arabia is still willing to play the role of a swing producer. In the weak market expected over the next few years, OPEC, in accordance with earlier behaviour, will support nominal prices and be flexible on output. Over time, when capacity utilization is increasing to more tolerable levels, prices will start to increase. To sum up, in this scenario OPEC is assumed to have adjusted the price level once and for all in January of 1986 to reduce the supply pressure and stimulate demand. The quotas are again restored and enforced and OPEC again acts as the swing supplier in the market.

Alternatively, in the low price scenario, OPEC's market power is reduced for the following reason. The fundamental precondition for OPEC to maintain market control in a weak market is its ability to be a swing-producer. This ability hinges on Saudi Arabia's willingness to take the greatest swings in production. A change in the balance of power in Saudi Arabia, costs experienced over the last 2–3 years from extremely low production and long-term interests in a healthy market could

The crude oil market

Table 21.1 The crude oil price in the high and low price scenarios ($ per barrel)

	Real crude oil price (1984 prices)		Nominal crude oil price	
	High price	Low price	High price	Low price
1984	28.0	28.0	28.0	28.0
1985	26.5	26.5	27.5	27.5
1986	18.5	14.5	20.0	15.5
1987	19.0	13.5	21.0	15.0
1988	20.0	13.0	23.0	15.0
1989	22.0	12.5	26.0	15.0
1990	24.5	17.5	30.5	22.0
1991	27.5	23.0	35.5	29.5
1992	30.5	26.5	40.5	35.5
1993	33.0	29.5	46.0	41.5
1994	35.5	30.5	51.5	44.0
1995	37.5	32.0	57.0	48.0
1996	39.5	33.0	62.0	52.0
1997	41.0	34.5	67.0	56.0
1998	42.5	35.5	72.0	60.5
1999	43.5	37.0	77.0	65.5
2000	45.0	38.5	83.0	70.5

turn out to reduce Norway's ability to carry the bulk of the costs involved in OPEC controlling the market. Altogether, this implies that when OPEC decides on quotas reducing supply, costs of these measures have to be evenly distributed between all member countries. This severely limits OPEC's market power and forces the organization in a weak market to a larger degree to undercut prices to defend volumes.

The GDP growth rates in the three demand regions and the exchange rates against the dollar are in accordance with the macroeconomics section later in this chapter. Furthermore, both in the high price and the low price scenario we assume competing fuel prices to stay constant in real terms and costs of transportation, refining, storage and distribution to follow the general GDP deflator. Finally, indirect taxes levied on crude oil or petroleum products are constant. On the basis of the above-mentioned parameters and assumptions on OPEC behaviour, production, consumption and market clearing prices were projected by the WOM-model.

As indicated by Table 21.1 and Fig. 21.1, the crude oil price in real terms decreased from $28 per barrel in 1984 to $26.5 per barrel in 1985. In the high price scenario the real crude oil price falls to $18.5 in 1986 ($20 in nominal terms). As outlined above we assume that OPEC countries stick together. From 1986 the real crude oil price starts to rise and reaches the 1985 price level again by 1992. Average annual growth rate from 1991 to 2000 is 5.5%. OPEC regains a firm grip on the

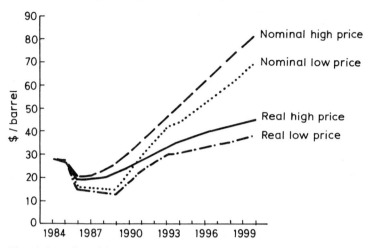

Fig. 21.1 High and low price scenarios: real and nominal crude oil prices

market in the late 80s. OPEC's income from oil production is more than 50% higher by 1990 than it was in 1985, and by the end of the century it will not be far from three times higher.

In the low price scenario the real crude oil price slides down to $12.5 per barrel and competes effectively against other sources of energy. By the end of the 1980s, OPEC's market power is again very strong. From 1989 to 1991 the real crude oil price increases by 84%. Thus, in this scenario a third oil price shock is embedded.

21.2.2 A low price on crude in the long run – is it possible?

The effects on the Norwegian economy of a fall in the crude oil price depends of course very much on whether the fall is temporary or not. The real crude oil price increases considerably both in the high and low price scenario in the next decade. In the oil industry, some people seem to favour arguments supporting the belief that the crude oil price could stay at a lower level than what prevails today over several decades. Arguments supporting this view are decoupling of demand for crude oil and economic growth, fierce competition between crude oil and other sources of energy (something close to a backstop technology in some end-uses) or discoveries of significant new hydrocarbon provinces.

To assess the probability of each of these events coming true is difficult. In the following, they are specified in such a way that in our judgement they are outliers if looked at separately, but not altogether unlikely. The following parameters have been changed:

(1) Decoupling of crude oil demand and economic growth: income elasticities are set at 0.25 instead of 0.8–0.9.

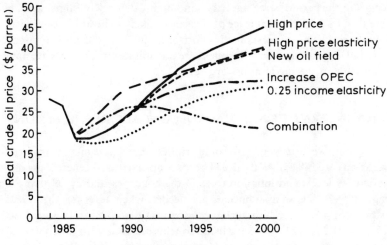

Fig. 21.2 Low price scenarios

(2) Fierce competition from other sources of energy: the price elasticity doubled in the long run.
(3) An optimistic resource base assumption: in the market economies outside OPEC, one new crude oil province with the approximate size of the North Sea is discovered. In addition, net exports from CPEs increase gradually by 2 mbd above today's level.
(4) OPEC production capacity increases gradually from 1991 to 2000 by 15 mbd.

In all other respects, these scenarios are similar to the high price scenario. Figure 21.2 shows the price paths calculated when the above-mentioned exogenous variables are varied one by one. The range of possible price trajectories widens considerably and prices down to $31 per barrel (1984 dollar) in 2000 now become feasible.

The next question obviously is: if each of these parameter changes are possible, what is the probability of a combination? We are not able to answer this question accurately. A priori, we see no reason to believe that one or the other outcome is strongly positively correlated. Thus, if they are judged possible but rather unlikely one by one, a world combining high price elasticities, decoupling of demand, an increase in OPEC capacity and an optimistic view of the resource base seem highly unlikely. Still, the possibility cannot be ruled out. In Fig. 21.2 this combination is labelled the combination scenario. In this world, the real crude oil price is falling from 1990 and is reduced to $22 per barrel by 2000. Because of the very high price elasticities the crude oil price trajectories in these cases are higher than the high price scenario over the next few years.

This leads us to conclude that a price for crude oil permanently at or slightly below $20 per barrel in constant terms is not viable in the long run. A slump in the real prices is likely to come over the next 5–8 years, the depth of the price slide and the length of the period being uncertain and highly dependent on personal beliefs. Despite this, by 1995 the real price of oil is most likely to be at or above last year's level. In the short run, as a response to short-term shocks in the market, prices could possibly fall through the floor. Despite these possibilities, such prices are not viable or sustainable in the medium or long term.

21.3 INTERNATIONAL MACROECONOMIC REPERCUSSIONS

In this section we briefly review some studies of international macroeconomic repercussions of a fall in the oil price. The main objective is to generate assumptions about activity levels and inflation rates in the countries that are Norway's main trading partners. These assumptions are needed when assessing the effects of a lower price of oil on the Norwegian economy as described in the next section.

In Tables 21.2 and 21.3 we have listed conclusions from four different studies in addition to results from our own computation on the LINK international macroeconomic model system. These LINK simulations are presented in more

Table 21.2 Impact on GNP of a 25% decrease in the price of oil (%)

	Source				
Area	OECD (1983) 1y 2y 3y	DRI (1984) 1y 2y 3y	Powell and Horton (1985) 1y 2y 3y	DRI (1986) 1y 2y 3y	LINK 1y 2y 3y
OECD total	0.6 0.9 1.2				0.7 0.9 1.1
USA		0.7 1.4 1.6			
Major Seven[a]			0.25 1.0 1.3		
Big Four[b]				0.7 1.4 1.3	

[a] USA, Canada, Japan, UK, France, West Germany and Italy.
[b] UK, France, West Germany and Italy.
1y: First year effects.
2y: Second year effects.
3y: Third year effects.

Table 21.3 Impact on GNP deflator of a 25% decrease in the price of oil (%)

	Source				
Area	OECD (1983) 1y 2y 3y	DRI (1984) 1y 2y 3y	Powell and Horton (1985) 1y 2y 3y	DRI (1986) 1y 2y 3y	LINK 1y 2y 3y
OECD total	−1.5 −2.5 −3.0				−0.9 −0.5 0.0
USA		−0.6 −1.1 −1.2			
Major Seven			−0.8 −1.5 −0.5		
Big Four				−0.7 −1.7 −1.7	

detail in Berger *et al.* (1986). We have scaled the figures so as to represent a 25% decrease in the price of oil, assuming that it is permissible to assume that the models used are log-linear.

We note from Table 21.3 that first-year effects on GNP of the oil price decrease are predicted to be in the range of 0.6–0.7%. The exception is Powell and Horton at only 0.25%. OECD (1983) assumes unchanged government expenditure in nominal terms. The effect on the general price level of a fall in the price of oil thus leads to an implicit change in the stance of fiscal policy; in this case towards more expansionary policy. Data Resources Inc. (DRI) have, in both studies, assumed an ease in monetary policy. However, Powell and Horton also make some policy assumptions; namely that the United States, France, Italy and Canada reduce their deficits. This means that automatic stabilizers are allowed to work in connection with the higher activity level following a decrease in the oil price. Japan, Germany and the UK, however, are assumed to ease fiscal policy so that budget deficits are held unchanged. The second and third-year effects on GNP are in all five studies estimated to be 1–1.5%, DRI being slightly higher than the others. If we compare these with the LINK simulations, we find similarities when looking at first-year effects. In the second and third year, the LINK results seem to predict a more modest impact on GNP. We have not included any policy responses in the LINK simulations, which may explain the differences.

In Table 21.3 we have listed the results for inflation. Note that while the two DRI studies and Powell and Horton show similar results, OECD (1983) reports a far stronger impact on inflation rates of a fall in the oil price (see Table 21.3). We have not been able to find an explanation for this difference. However, all four studies predict that second-year effects on inflation should be almost twice that of the first-year effect. The LINK results show comparable results for inflation in the first year, but later the effect is somewhat weaker. One reason for this may be that the LINK models contain a number of exogenous prices that have not been adjusted in the low price scenario.

As is evident from the tables, different studies may come to somewhat different conclusions regarding the impact on GNP and inflation in the world economy of a fall in the price of oil, depending among other things on the model being used and the assumptions being made. However, to evaluate impacts of lower oil prices on the Norwegian economy, we need an assessment of the impacts on Norwegian export markets and import prices. Any assumptions made on these issues contain great uncertainty. Consequently, results should be viewed more as a possible, rather than the most probable outcome. We have chosen something like 'middle of the road' impacts in the light of the results reported above. For GNP in the countries constituting Norway's main trading partners we have assessed the impact to be 0.6% in the first year, 1% in the second and 1.3% in the third year. Considering GNP deflators in the same countries the deflators are reduced by 0.7% in the first year and 1.5% from then on. As the real oil price is expected to increase rapidly around 1990 we assume that output is reduced and inflation is higher so that the positive development disappears in the 1990s.

21.4 IMPACTS OF LOWER OIL PRICES ON THE NORWEGIAN ECONOMY

In this section some possible effects on the Norwegian economy of a lower oil price are analysed. For the computations we utilized the macroeconomic model MODAG A (cf. Cappelen and Longva, 1986).

21.4.1 The high price scenario and the Norwegian economy

In our high oil price scenario (CBS-HOP), we utilize the GNP growth rates and inflation rates in the OECD area from the baseline LINK simulation mentioned in the previous section. With assumptions made on economic policy, we obtain a high oil price scenario for the Norwegian economy that is broadly in line with the government long-term programme for the period 1986 to 1989. For the following years we project lower growth rates, partly due to decreasing growth rates in production of oil and natural gas as well as investments in the oil sector. We do not go into detail on the high price scenario, except to say a few words about the development in the balance of payments, as this is one of the most important factors when discussing how economic policy might change as a reaction to lower oil prices.

Table 21.4 shows projections of components of the balance of payments in the CBS-HOP.

The current balance will be in deficit until 1990, and net foreign debt will at its highest constitute 12% of nominal GDP. In the 1990s, Norway is projected to become a net creditor nation, if our assumptions should prove correct. Although these figures must be interpreted with great care, we believe that they indicate that as long as oil and natural gas prices develop in accordance with the high price scenario, there is no reason to worry about the deficit in the balance of payments in the years ahead. Figure 21.3 shows some indicators of the CBS-HOP scenario, and the development of the government budget surplus follows in Fig. 21.4.

Table 21.4 Balance of payments in CBS-HOP, 1985–1993 (billion NOK)

	Trade balance	Net interest dividends	Net transfers	Current account	Net foreign debt % of nominal GDP
1985	41	−12	−4	25	8.1
1986	9	−12	−4	−7	9.5
1987	1	−10	−4	−13	11.5
1988	8	−10	−5	−7	12.0
1989	17	−11	−5	1	11.0
1990	40	−11	−6	23	6.5
1991	65	−13	−6	46	−0.5
1992	85	−12	−6	67	−8.7
1993	103	−8	−7	88	−17.5

NOK = Norwegian kroner.

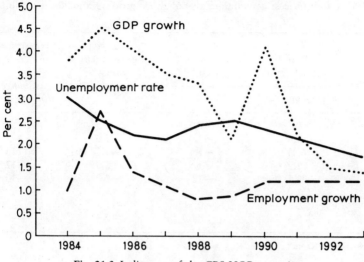

Fig. 21.3 Indicators of the CBS-HOP scenario

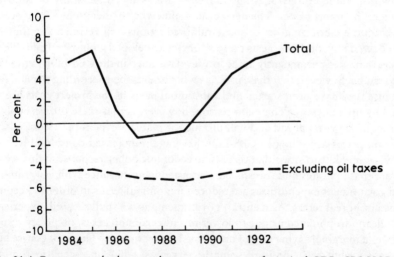

Fig. 21.4 Government budget surplus as a percentage of nominal GDP: CBS-HOP

21.4.2 The low price scenario and the Norwegian economy 1986–1993.

The low price scenario (CBS-LOP) is based on an assumption of a more 'aggressive' production policy by OPEC which will result in a lower oil price than in CBS-HOP, in particular from 1986 to 1989. Due to a lower oil price it is assumed that the dollar exchange rate depreciates somewhat more during the next years but appreciates more in 1990. It is worth noticing that although oil prices increase substantially around 1990 also in CBS-LOP, the oil price does not reach the level in the CBS-HOP scenario.

Table 21.5 Low oil prices: assumptions regarding exogenous variables: deviations from CBS-HOP

	GDP in the OECD area	Inflation in OECD	Wages and government transfers	Investment in the oil sector
	Changes (%)			Changes (billion NOK)
1986	0.6	−0.5	−0.5	0
1987	1.0	−1.0	−1.0	−7
1988	1.3	−1.5	−1.5	−13
1989	1.3	−1.5	−2.0	−12
1990	1.1	−1.0	−2.0	−10
1991	1.0	−0.5	−1.5	−10
1992	0.6	−0.5	−1.0	−9
1993	0.6	0.0	−0.5	−7

NOK = Norwegian kroner.

Some important assumptions regarding exogenous variables in the CBS-LOP are shown in Table 21.5. Wages and transfers are assumed generally to follow the changes in import prices. The deviation is somewhat higher from 1989 due to the effect on the labour market of substantial reductions in investments in the North Sea. Few, if any, of the projects presently being developed would be profitable if oil prices remained permanently at the low level assumed in the CBS-LOP in the 1980s. We assume, however, that these projects will be developed according to plan, partly because they have been started and substantial parts of the project costs are sunk costs by now anyway. The main reason, however, is that these projects come on stream in the years ahead and will produce at plateau level in the 1990s, at which time oil prices according to CBS-LOP have again increased considerably.

We would like to stress that there is no economic policy response in this scenario except that tax rates and transfers are adjusted in accordance with lower inflation and government expenditures are reduced in nominal terms in order to keep them constant in real terms. As (central) government, as we shall see, will experience a significant drop in income due to lower revenue from oil taxes, no policy response implies a marginal saving rate of unity for the government. This is of course highly unrealistic, but the CBS-LOP scenario serves as a baseline scenario for policy changes that will be studied in Section 21.4.3 below.

Table 21.6 shows the effect on some macroeconomic indicators of lower oil prices. The figures are shown as deviations in per cent from the CBS-HOP scenario. The real wage is actually higher in the CBS-LOP than in the CBS-HOP as lower oil prices reduce consumer prices by more than wages are assumed to be reduced. The reason for this result is that using a wage equation that mainly emphasises competitiveness in traditional manufacturing sectors in addition to labour market conditions, lower prices on refined oil products are not assumed to affect wages. Thus the effects of higher unemployment in the CBS-LOP (maximum 0.3

Table 21.6 Some main economic indicators: percentage deviations between CBS-HOP and CBS-LOP

	Consumer prices	Private consumption	Gross investment	Exports	Imports	GDP	Employment
1986	−1.1	0.5	0.1	0.3	0.5	0.2	0.1
1987	−1.6	0.4	−6.0	0.5	1.4	−0.5	−0.2
1988	−2.1	0.1	−11.5	0.8	−3.1	−1.0	−0.5
1989	−2.7	0.1	−10.5	1.0	−2.6	−0.8	−0.5
1990	−2.2	−0.1	−8.2	1.1	−2.0	−0.5	−0.4
1991	−1.6	−0.2	−8.2	1.1	−2.2	−0.4	−0.3
1992	−1.4	0.1	−7.5	0.9	−1.9	−0.3	−0.2
1993	−1.1	0.2	−5.9	0.9	−1.3	−0.2	−0.2

percentage points in 1988 and 1989) is not enough to offset the general effect of lower oil prices on consumer goods.

The decline in gross investment is mainly due to assumptions regarding oil investments, but the general reduction in output also has a negative effect. However, the negative effect on GDP is the result of mainly two opposite effects: that stemming from lower investments and the positive effect of higher exports. Higher world market demand will in particular stimulate those sectors of the economy that are export orientated. In Fig. 21.5 we show the effect on main aggregates of the manufacturing sector which in many ways reflects the macroeconomic figures in Table 21.5. The effect on the manufacturing sector producing consumer goods is small compared with the large negative impact on industries producing investment goods and the positive effect on export-orientated manufacturing sectors.

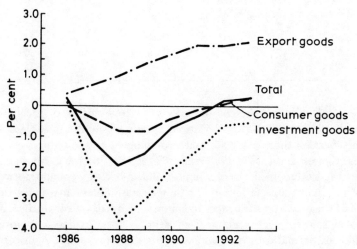

Fig. 21.5 Output in manufacturing: percentage deviations between CBS-HOP and CBS-LOP

Table 21.7 Deviations between CBS-LOP in balance of payments

	CBS-LOP			Changes in absolute levels between CBS-HOP and CBS-LOP		
	Trade balance (billion NOK)	Current balance	net foreign debt (in %) of nominal GDP	Trade balance (billion NOK)	Current balance	Net foreign debt (in %) of nominal GDP
1986	−5	−21	12.6	−14	−14	3.1
1987	−12	−26	17.6	−13	−13	6.1
1988	−5	−19	20.4	−13	−12	8.4
1989	−5	−27	23.1	−22	−23	12.1
1990	20	1	20.4	−20	−22	13.9
1991	55	32	13.5	−11	−14	14.0
1992	75	47	6.2	−10	−20	14.7
1993	91	63	−1.6	−12	−25	15.9

NOK = Norwegian kroner.

Table 21.8 Decomposition of changes in current balance between CBS-LOP and CBS-HOP

	Absolute changes due to:				
	Exports of oil and gas	+ Other exports	− Imports	+ Net interest and transfers	= Total
1986	−15	0	−2	−1	−14
1987	−19	−1	−7	0	−13
1988	−24	−2	−13	0	−12
1989	−32	−3	−13	−1	−23
1990	−22	−7	−9	−2	−22
1991	−19	0	−8	−4	−14
1992	−21	3	−8	−10	−20
1993	−19	2	−5	−13	−25

Tables 21.7 and 21.8 present the effects on components of the balance of payments both in levels and as deviations from CBS-HOP (cf. Table 21.4). The first-year effect on the current account is dominated by the reduction in export revenues from oil and gas. In 1987 and 1988 the effect on the current account of lower oil prices is modified by lower imports due to both lower import prices and a reduction in the volume of imports, mainly because of less investment in the oil sector. The reason why net transfers and interest do not deviate from the CBS-HOP scenario in spite of higher net foreign debt, is that dividends to foreign oil companies are dramatically reduced in the 1980s. It is only when oil prices start to pick up around 1990 that the net interest and transfers deteriorate significantly, mainly due to higher debt service on the foreign debt accumulated in the 1980s.

Impacts of lower oil prices on the Norwegian economy

The contribution to nominal export revenues from exports of nonoil goods is very small as price and quantity work in opposite directions. In the 1980s the price effect dominates. Lower export prices (which does not imply improved competitiveness as world market prices are lower as well) outweigh the positive volume effect. In the 1990s the opposite occurs.

Finally, the government budget balance is of course already in deficit in 1986 in the low oil price scenario and the deficit is increasing until 1989 when it is nearly 30 billion NOK or 5% of nominal GDP. Accumulated over the period 1986 to 1993 the deficit is approximately 10 billion NOK or close to zero. The government budget balance excluding tax revenue from oil production shows an almost identical development as in the CBS-HOP scenario. As a share of nominal GDP, the deficit is somewhat higher as nominal GDP is significantly lower in the CBS-LOP scenario. In 1989 the nominal GDP-difference between the two scenarios is almost 10%. In this respect the economic policy is even more expansionary in the low price scenario than in CBS-HOP. It is thus reasonable to assume a somewhat more restrictive fiscal stance in order for the (central) government to compensate for some of the loss in income due to lower oil prices.

21.4.3 Restrictive fiscal policy response (FR) to lower oil prices

Due to deficits on both the current account and the government budget balance, it is assumed that average rates are increased by one percentage point from mid-1986 and until 1990, after which the tax rate is reduced by half a percentage point. In addition the growth rate of government expenditures for consumption purposes is reduced from 3% annually to 1% each year from 1987 to 1990 and to 2% in 1991, after which the growth rate is as in the previous two scenarios. This reduction in public spending will increase unemployment quite substantially, and the growth in wage rates and government transfers to households are thus reduced as well. Otherwise this scenario, called CBS-LOP/FR, is the same as CBS-LOP. It turns out, as we shall see, that the unemployment rate increases beyond levels that have been experienced in Norway so far. What will happen to wage growth at these levels of unemployment is highly uncertain, and the results must be regarded only as tentative.

Reduced government employment and purchases of goods combined with higher taxes, reduce mainly private and public consumption and thus GDP during 1986 and 1987. Gross investments are reduced only gradually. In addition to the accelerator effects on investment, both higher user cost of capital relative to wages and lower profitability tend to reduce investments. Exports is gradually increasing compared with the CBS-LOP scenario as lower GDP reduces capacity utilization rates which again lead to lower export prices. In addition, lower wage rates improve price competitiveness. The main positive effect on the trade balance is, however, due to lower imports partly as a result of lower domestic prices compared with import prices but mainly because of lower domestic demand. The deficit on the current account is gradually reduced by some 10 billion NOK in 1990 and thereafter. The net foreign debt is reduced by 25 billion NOK in 1990 compared with the CBS-LOP scenario.

Net foreign debt as share of nominal GDP in the three scenarios is shown in Fig. 21.6. As is apparent from the figure, the government can achieve a substantial reduction in the debt burden. The price, however, is considerably lower employment and GDP, as shown in Fig. 21.7.

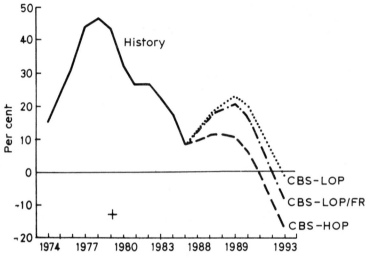

Fig. 21.6 Net foreign debt as a percentage of nominal GDP

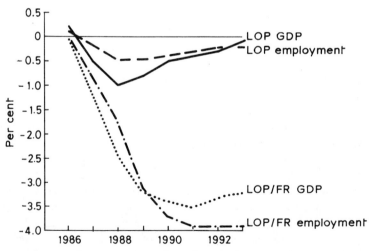

Fig. 21.7 Percentage change in GDP and employment between CBS-HOP, CBS-LOP and CBS-LOP/FR

The government budget balance is improved quite substantially due to lower expenditure and higher tax revenues. In 1990 the deficit is reduced by 9 billion NOK or nearly 2% of nominal GDP. In the light of the improvement on the current account this implies that net acquisition of financial assets by the private sector is virtually unchanged. The deficit excluding tax revenues on oil production is actually more or less constant in nominal terms in the CBS-LOP/FR scenario. Compared with the CBS-LOP where the deficit in per cent of nominal GDP was 5% in 1989, that share is now reduced to just over 3%.

21.5 CONCLUDING REMARKS

Based on our and other students' analyses both of the crude oil market and international repercussions we believe that it is highly unlikely that crude oil prices will stay at a low level for the rest of this century. If this conclusion is accepted the positive impact on GDP and inflation rates that the OECD area now experience following lower oil prices early in 1986, is only temporary. Higher oil prices in the 1990s will again change the world distribution of income and increase inflation rates and reduce GDP in most OECD countries. For Norway this implies that the deterioration in terms of trade is of a passing nature and that the balance of payments deficit that will persist during the rest of this decade will turn to perhaps a huge surplus in the early 1990s. The domestic macroeconomic policy response should take this into account. Judged by the huge deficits that Norway accumulated in the mid 1970s that resulted in a net foreign debt nearly half of nominal GDP in 1978, the debt ratio will increase only moderately even in the case of a low oil price. However, taking into account the (remote) possibility of a permanently low oil price, some adjustments in domestic policy already taking place, seem sensible.

REFERENCES

Berger, K, Cappelen, Å., Knudsen, V. and Roland, K. (1986) Effects on the Norwegian economy of lower oil prices, report, Central Bureau of Statistics, Oslo.

Cappelen, Å. and Longva, S. (1986) MODAG A: a medium-term annual Macroeconomic model of the Norwegian economy, in *Nordic Macroeconomic Models* (eds O. Bjerkholt and J. Rosted), Book Series On International Studies in Economic Modelling, Chapman and Hall, London.

DRI (1984) *US Review*, November, Data Resources, Inc. New York.

DRI (1986) *European Review*, January, Data Resources Inc., New York.

Lorentsen, L. and Roland, K. (1985) Markedet for råolje, reports 85/4, Central Bureau of Statistics, Oslo.

OECD (1983) Simulated macroeconomic effects of a large fall in oil prices, working papers 8, June, OECD, Economics and Statistics Department.

Powell, S. and Horton, G. (1985) The economic effects of lower oil prices, working paper 76, April, Government Economic Service, London.

Weyant, J. P. and Kline, D. M. (1982) OPEC and the oil glut: Outlook for oil export revenues during the 1980s and 1990s. *OPEC Review*, 6, No. 4.

22

Modelling the effects of investment subsidies

W. DRIEHUIS and P. J. VAN DEN NOORD

22.1 INTRODUCTION

From 1970 onwards unemployment in the Netherlands has risen continuously. In 1985 and 1986 a small decline can be observed, but the unemployment rate still amounts to 14.5%. The increase in unemployment was accompanied by a decline in the rate of inflation, which is almost nil in 1986, and a strong rise in the public deficit. Despite recent reductions in this deficit (as a percentage of national income) it is the view of the policy-maker that a further reduction is needed and this aim has first priority in economic policy.

It is against this background that discussion takes place on the abolition of investment subsidies. These subsidies were introduced in 1977 by the Law on Investment Account (Wet Investerings Rekening, WIR). The idea was that selective subsidizing of capital would stimulate investment decisions of firms and would therefore enlarge the number of jobs. In the Netherlands a shortage of jobs is held to be responsible for the greater part of unemployment, which is therefore called structural.[1]

This chapter studies the effects of the use of investment subsidies on employment. Is it true that the subsidies, which were given after 1977 and which have now increased to 1.3% of national income, have indeed increased employment? In Section 22.2 the policy-makers' view on the effects of investment subsidies is given. Section 22.3 is used for presenting our own modelling of investment subsidies, followed by empirical results in Section 22.4. Conclusions are given in Section 22.5.

---

[1] It should be pointed out that the use of the term structural unemployment by Dutch economists differs from the use of this term in the United States and the United Kingdom. In these countries structural unemployment means unemployment due to a malfunctioning of labour markets. The Dutch use of structural unemployment is identical to so-called classical unemployment. See Malinvaud (1980), who considers the term Marxian unemployment to be even more appropriate then the term classical unemployment.

22.2 THE POLICY-MAKERS' VIEW ON INVESTMENT SUBSIDIES

Private investment in the Netherlands has always been influenced by fiscal policy measures. Until 1977 accelerated depreciation and investment credits were generally accepted means of investment policy. However, they were mainly applied as instruments of business cycle policy rather than as a means of fostering economic growth in a selective way. Furthermore, these fiscal measures were only attractive for profit-making firms. With the introduction of investment subsidies it was possible to stimulate capital formation in loss-making firms as well and to aim at greater selectivity. The latter element is introduced by having basic premiums and supplementary premiums, the latter being differentiated, for instance, according to labour intensity and pollution intensity of investment outlays.

We deal here only with the possible macroeconomic consequences of investment subsidies on employment via the creation of jobs. The policy-makers' view in this respect was based on an analysis of the Central Planning Bureau (CPB). For this purpose the CPB used a macroeconomic model called VINTAF[2] at the time of introducing the investment subsidies, and uses FREIA[3] at present. Although these models differ in a number of aspects,[4] they have in common the modelling of the supply side of the goods market. This so-called supply block of FREIA (and VINTAF) can be summarized in the following linearized equations. The linearization is used in order to facilitate comparison with an alternative model of the Netherlands economy constructed by the authors, called SECMON (SEctoral MOdel for the Netherlands economy).[5] SECMON is a sectoral model with another, and, in our view, more adequate modelling of the supply side of the goods market.

The linearized version of the clay–clay vintage approach of the supply block for the *whole private sector* in the CPB models reads as follows (lags are omitted):[6]

$$\dot{y}^* = \frac{1}{\kappa_0} I + \theta(\Delta A - 1) + \delta_1 \dot{h} + \varepsilon - \pi \tag{22.1}$$

$$\dot{a}^* = \dot{y}^* + (\eta - \theta)\Delta A - \delta_2 \dot{h} - \mu - \chi \tag{22.2}$$

$$\chi = \rho \dot{y}^* + \chi_0 \tag{22.3}$$

$$\dot{a} = \dot{a}^* + \beta(\dot{y} - \dot{y}^*) \tag{22.4}$$

$$\Delta A = \frac{-(\dot{w} - \dot{p}) + \delta_2 \dot{h} + \chi}{\mu} + 1 \tag{22.5}$$

---

[2] VINTAF is published in CPB (1977).
[3] FREIA is published in CPB (1983).
[4] For a review See den Butter (1985).
[5] SECMON is published in Driehuis *et al.* (1983). A comparison of the supply blocks of SECMON and the CPB models is also found in den Hartog (1984).
[6] For the original version of this vintage model see den Hartog and Tjan (1976). The linearization is based on de Ridder (1977); see also Driehuis (1984).

The policy-makers' view on investment subsidies

where a dot over a variable means a relative rate of change, and:

y^* = productive capacity
κ_0 = capital coefficient in a base period
I = transformed investment ratio
A = economic lifetime of equipment
h = labour time = machine time
ε = disembodied capital augmenting technical progress
π = technical obsolescence of equipment
μ = embodied labour augmenting technical progress
χ = disembodied labour augmenting technical progress
a^* = number of jobs (potential employment)
a = employment
y = production of private sector
w = nominal wage costs per worker
p = output price of private sector
η = relative share of jobs on the eldest (marginal) vintage
θ = relative share of productive capacity on the eldest (marginal) vintage

The ratio of investment in equipment to output is defined as

$$I = \frac{i}{y^*} e^{\varepsilon t} h^{\delta_1}$$

where:

i = investment outlays on equipment (at constant prices).

Equation 22.1 describes the rate of growth of productive capacity as a function of the transformed investment ratio, economic obsolescence of equipment (represented by the term $\Delta A - 1$), machine time, capital augmenting technical progress and technical obsolescence of equipment. This capacity equation is assumed to hold for the whole business sector. In Equation 22.2 the employment under full capacity utilization, also called potential employment or jobs, is assumed to be a function of productive capacity, economic obsolescence of equipment, labour time and embodied and disembodied labour augmenting technical progress. Equation 22.3 is the so-called (second) Verdoorn's law implying that the rate of disembodied labour augmenting technical progress is positively related to potential production growth. See Verdoorn (1952, 1965) and Kaldor (1966). Equation 22.4 expresses actual employment as a function of the number of jobs and the rate of capacity utilization. At full capacity utilization employment equals potential employment, but not necessarily the labour force. If potential employment (a^*) is smaller than the labour force, structural unemployment due to a shortage of jobs is said to exist. In Equation 22.5 it is shown that, apart from changes in the hours worked, the economic obsolescence of equipment accelerates if real wage costs ($\dot{w} - \dot{p}$) are rising faster than the rate of embodied and disembodied labour augmenting technical progress ($\mu + \chi$).

Only for the sake of argument do we introduce the following equations, each describing relevant mechanisms modelled in FREIA in a simplified way (lags are omitted):

$$\dot{y} = -\alpha_1 \dot{p} + \alpha_2 \dot{y}_a - \alpha_3 \dot{p}_{ka} \tag{22.6}$$

$$\dot{p} = \lambda_1 \dot{w} - \lambda_2 (\dot{y} - \dot{a}) + \lambda_3 \dot{p}_k + \lambda_4 \dot{p}_a \tag{22.7}$$

$$\dot{w} = \xi_1 \dot{p} + \xi_2 (\dot{y} - \dot{a}) - \xi_3 (\dot{n} - \dot{a}) + \xi_4 \dot{w}_a \tag{22.8}$$

$$I = \gamma_1 \dot{y} - \gamma_2 (\Delta A - 1) - \gamma_3 \dot{p}_{ka} + \gamma_4 I_a \tag{22.9}$$

$$\dot{p}_k = \phi_1 \dot{p} + \phi_2 \dot{r} + \phi_3 \dot{p}_{ka} \tag{22.10}$$

where (new variables only):

y_a = all nonspecified influences on output
p_{ka} = the influence of investment subsidies on capital cost
p_a = all nonspecified influences on prices
p_k = user cost of capital
r = rate of interest
n = labour force
w_a = all nonspecified influences on wages
I_a = all nonspecified influences on investment.

Equation 22.6 is a kind of a reduced-form expenditure equation in which all exogenous foreign and policy variables are represented by y_a. Equation 22.7 is a price equation and Equation 22.8 is a wage equation in which the Phillips-curve effect is simply represented by the difference between labour force and employment growth. Equation 22.9 is an investment equation incorporating a simple acceleration mechanism, while economic obsolescence represents the influence of investment for replacement. Two autonomous terms are added, p_{ka} to have an income effect of subsidizing investment and I_a comprising all other effects on I. The impact of p_{ka} is *negative* since it is defined as a *cost* factor: subsidizing investment then implies a lowering of p_{ka}. According to Equation 22.10 the user cost of capital depends on the current price of capital goods (here: total output price), the interest rate and the impact of investment subsidies.[7]

We reduce the ten equations to three equations for output, inflation and employment, as follows:

$$\dot{y} = \omega_1 \dot{p} + \omega_2 \dot{p}_{ka} + (\cdots) \tag{22.11}$$

$$\dot{p} = \omega_3 \dot{y} + \omega_4 \dot{a} + \omega_5 \dot{p}_{ka} + (\cdots) \tag{22.12}$$

$$\dot{a} = \omega_6 \dot{y} + \omega_7 \dot{p} + \omega_8 \dot{p}_{ka} + (\cdots) \tag{22.13}$$

---

[7] For a review see Auerbach (1983).

where

$$\omega_1 = -\alpha_1$$
$$\omega_2 = -\alpha_3$$
$$\omega_3 = \frac{-(\lambda_2 - \lambda_1\xi_2)}{1 - \lambda_1\xi_1 - \lambda_3\phi_1}$$
$$\omega_4 = \frac{\lambda_1(\xi_3 - \xi_2) + \lambda_2}{1 - \lambda_1\xi_1 - \lambda_3\phi_1}$$
$$\omega_5 = \frac{\lambda_3\phi_3}{1 - \lambda_1\xi_1 - \lambda_3\phi_1}$$
$$\omega_6 = \frac{\beta + (1 - \beta - \rho)(1/\kappa_0)\gamma_1 + \omega_9\omega_{10}[\rho(1/\kappa_0)\gamma_1 - \xi_2]}{1 - \omega_9\omega_{10}(\xi_2 - \xi_3)}$$
$$\omega_7 = \frac{\omega_9\omega_{10}(1 - \xi_1)}{1 - \omega_9\omega_{10}(\xi_2 - \xi_3)}$$
$$\omega_8 = \frac{-[(1 - \beta - \rho) + \omega_9\omega_{10}\rho](1/\kappa_0)\gamma_3}{1 - \omega_9\omega_{10}(\xi_2 - \xi_3)}$$
$$\omega_9 = \eta - (\beta + \rho)\theta - (1 - \beta - \rho)(1/\kappa_0)\gamma_3$$
$$\omega_{10} = \frac{1}{\mu - \rho[\theta - (1/\kappa_0)\gamma_2]}$$

and where $(\cdots)$ denote all autonomous influences apart from p_{ka}.

This enables us to describe the most relevant effects of investment subsidies on three targets of economic policy. To start with output and inflation there are two effects. In the *first* place investment subsidies lower the amount of payments to the public sector, increase disposable profits and therefore increase investment expenditures and output. This is called the *direct output effect*. In the *second*

Table 22.1 Effects of the introduction of investment subsidies amounting to 1% of NNP in a base year according to the CPB

	Effect[a] after		
	1 year	3 years	6 years
Effect on production	0.1	1.1	1.5
Effect on inflation	−0.3	−1.3	−1.1
Effect on employment	0	0.3	0.8

[a] Accumulated percentage deviation from a central projection.
Source: CPB (1986).

place investment subsidies lower the user cost of capital and therefore inflation. This effect may be reinforced when investment subsidies stimulate the rate of labour productivity growth. Whether the final impact on inflation will be negative depends on the strength of the Phillips-curve effect. But when the initial level of unemployment is assumed to be high it may be safely assumed that inflation is ultimately reduced ($\xi_2 < \xi_3$). Its positive output effect, especially via exports, is called the *indirect output effect*. Turning now to employment there is a negative effect via an acceleration of the replacement of capital due to economic obsolescence. Provided that $\xi_1 < 1$, the effect of lower inflation on employment is negative. Clearly this must be the case since *real* wage costs are rising due to money illusion.[8] This is the negative *obsolescence effect*. On the other hand there are positive effects from higher output and from lower capital cost (via investment), the latter also being considered an output effect.

The effects just mentioned in general terms can also be found in a recent CPB study. Table 22.1 gives its results which are based on simulations with FREIA in which the impact of the monetary sector is eliminated as well as the direct effects of the investment share on output growth as an indication of possible supply constraints on output. This table gives an adequate representation of the policy-makers' view of the effects of investment subsidies. It will be clear that these effects will only materialize after a relatively long period of time, since considerable time lags, especially concerning obsolescence effects, are involved. It follows from Table 22.1 that in the CPB analysis the positive output effects on employment outweigh the negative obsolescence effect.

22.3 AN ALTERNATIVE WAY OF MODELLING INVESTMENT SUBSIDIES

The foregoing analysis is inadequate for at least two reasons:

(1) The economic obsolescence of equipment is not only influenced by real wage cost but also by real capital cost;[9]
(2) Since investment subsidies are mainly relevant for the capital-intensive sector it is necessary to distinguish this sector from the labour-intensive sector.

These mechanisms are incorporated in the large sectoral model called SECMON which is in use at SEO, Foundation for Economic Research of the University of Amsterdam. Details of SECMON are given in Appendix A. It may suffice to mention here that it distinguishes 12 capital-intensive sectors and 6 labour-intensive sectors. Economic obsolescence of equipment is only relevant in the capital-intensive sectors as well as Verdoorn's law. For the sake of argument we will

---

[8] It should be borne in mind that even when the rise in consumer price is fully passed on in wages, this need not to invalidate the assumption $\xi_1 < 1$ since producer prices usually rise less than consumer prices.
[9] For a theoretical foundation see Malcomson (1975), Nickell (1975) and van den Noord (1984, 1988).

An alternative way of modelling investment subsidies

only present the relevant mechanisms comparable to the set of equations in Section 22.3. Note that

(1) Real capital costs now influence economic obsolescence and hence influence the number of jobs, at least in the capital-intensive sectors (Equation 22.18)
(2) Real wage costs have an impact on employment in the labour-intensive sector and capital costs have not (Equation 22.19)
(3) The direct production effect of investment subsidies is only relevant in the capital-intensive sector (Equation 22.20)
(4) Verdoorn's law is only modelled for the capital-intensive sector (Equation 22.16)[10]
(5) Wage and price equations are now different per sector; the modelling of wage formation follows the lines of the so-called Scandinavian theory of inflation (see Edgren et al., 1973) (Equations 22.26 and 22.27)
(6) It is assumed that output of the labour-intensive sector is mainly dependent on output of the capital-intensive sector (Equation 22.21).

The condensed model reads (where 1 denotes the capital-intensive and 2 denotes the labour-intensive sector):

$$\dot{y}_1^* = \frac{1}{\kappa_1} I + \theta_1(\Delta A - 1) + \delta_3 \dot{h} + \varepsilon_1 - \pi_1 \tag{22.14}$$

$$\dot{a}_1^* = \dot{y}_1^* + (\eta_1 - \theta_1)\Delta A - \delta_4 \dot{h} - \mu_1 - \chi_1 \tag{22.15}$$

$$\chi_1 = \rho_1 \dot{y}_1^* + \chi_0 \tag{22.16}$$

$$\dot{a}_1 = \dot{a}_1^* + \beta_1(\dot{y}_1 - \dot{y}_1^*) \tag{22.17}$$

$$\Delta A = \sigma \left\{ \frac{-[(\dot{w}_1 - \dot{p}_1) - (\dot{p}_k - \dot{p}_1)] - (\delta_3 - \delta_4)\dot{h} + \chi_1}{\mu_1} + 1 \right\} \tag{22.18}$$

$$\dot{a}_2 = \beta_2 \dot{y}_2 - \beta_3(\dot{w}_2 - \dot{p}_2) - \chi_2 \tag{22.19}$$

$$\dot{y}_1 = -\alpha_4 \dot{p}_1 + \alpha_5 \dot{y}_{a_1} - \alpha_6 \dot{p}_{ka} \tag{22.20}$$

$$\dot{y}_2 = \alpha_7 \dot{y}_1 + \alpha_8 \dot{y}_{a_2} \tag{22.21}$$

$$I = \gamma_5 \dot{y}_1 - \gamma_6(\Delta A - 1) - \gamma_7 \dot{p}_{ka} + \gamma_8 I_a \tag{22.22}$$

$$\dot{p}_1 = \lambda_5 \dot{w}_1 - \lambda_6(\dot{y}_1 - \dot{a}_1) + \lambda_7 \dot{p}_k + \lambda_8 \dot{p}_{a_1} \tag{22.23}$$

$$\dot{p}_2 = \dot{w}_2 - (\dot{y}_2 - \dot{a}_2) \tag{22.24}$$

$$\dot{p}_k = \phi_1 \dot{p}_1 + \phi_2 \dot{r} + \phi_3 \dot{p}_{ka} \tag{22.25}$$

$$\dot{w}_1 = \xi_5 \dot{p}_1 + \xi_6(\dot{y}_1 - \dot{a}_1) - \xi_7(\dot{n} - \dot{a}_1) + \xi_8 \dot{w}_{a_1} \tag{22.26}$$

$$\dot{w}_2 = \dot{w}_1 \tag{22.27}$$

---

[10] The empirical evidence is found in Knibbe and van den Noord (1983).

where σ in Equation 22.18 is the relative share of capital cost in total cost on the newest vintage.

This model can be reduced to six equations, namely:

$$\dot{y}_1 = \omega_{11}\dot{p}_1 + \omega_{12}\dot{p}_{ka} + (\cdots) \tag{22.28}$$

$$\dot{p}_1 = \omega_{13}\dot{y}_1 + \omega_{14}\dot{a}_1 + \omega_{15}\dot{p}_{ka} + (\cdots) \tag{22.29}$$

$$\dot{a}_1 = \omega_{16}\dot{y}_1 + \omega_{17}\dot{p}_1 + \omega_{18}\dot{p}_{ka} + (\cdots) \tag{22.30}$$

$$\dot{y}_2 = \omega_{21}\dot{y}_1 + (\cdots) \tag{22.31}$$

$$\dot{p}_2 = \omega_{22}\dot{y}_1 + \omega_{23}\dot{p}_1 + \omega_{24}\dot{a}_1 - (\dot{y}_2 - \dot{a}_2) + (\cdots) \tag{22.32}$$

$$\dot{a}_2 = \omega_{25}\dot{y}_2 + (\cdots) \tag{22.33}$$

where

$$\omega_{11} = -\alpha_4$$

$$\omega_{12} = -\alpha_6$$

$$\omega_{13} = \frac{-(\lambda_6 - \lambda_5\xi_6)}{1 - \lambda_5\xi_5 - \lambda_7\phi_1}$$

$$\omega_{14} = \frac{\lambda_5(\xi_7 - \xi_6) + \lambda_6}{1 - \lambda_5\xi_5 - \lambda_7\phi_1}$$

$$\omega_{15} = \frac{\lambda_7\phi_3}{1 - \lambda_5\xi_5 - \lambda_7\phi_1}$$

$$\omega_{16} = \frac{\beta_1 + (1 - \beta_1 - \rho_1)(1/\kappa_1)\gamma_5 + \omega_{19}\omega_{20}[\rho_1(1/\kappa_1)\gamma_5 - \xi_6]}{1 - \omega_{19}\omega_{20}(\xi_6 - \xi_7)}$$

$$\omega_{17} = \frac{\omega_{19}\omega_{20}(\phi_1 - \xi_5)}{1 - \omega_{19}\omega_{20}(\xi_6 - \xi_7)}$$

$$\omega_{18} = \frac{\omega_{19}\omega_{20}\phi_3 - [\{1 - \beta_1 - \rho_1\} + \omega_{19}\omega_{20}\rho_1](1/\kappa_1)\gamma_7}{1 - \omega_{19}\omega_{20}(\xi_6 - \xi_7)}$$

$$\omega_{19} = \eta_1 - (\beta_1 + \rho_1)\theta_1 - (1 - \beta_1 - \rho_1)(1/\kappa_1)\gamma_6$$

$$\omega_{20} = \frac{\sigma}{\mu_1 - \sigma\rho_1(\theta_1 - (1/\kappa_1)\gamma_6)}$$

$$\omega_{21} = \alpha_7$$

$$\omega_{22} = \xi_6$$

$$\omega_{23} = \xi_7$$

$$\omega_{24} = -(\xi_6 - \xi_7)$$

$$\omega_{25} = \frac{(\beta_2 - \beta_3)}{(1 - \beta_3)}$$

while

$$\dot{p} = \zeta_1 \dot{p}_1 + (1 - \zeta_1)\dot{p}_2 \tag{22.34}$$

$$\dot{y} = \zeta_2 \dot{y}_1 + (1 - \zeta_2)\dot{y}_2 \tag{22.35}$$

$$\dot{a} = \zeta_3 \dot{a}_1 + (1 - \zeta_3)\dot{a}_2 \tag{22.36}$$

It is clear that the signs of the direct and indirect output effects in sector 1 are identical to those which were found in Section 22.2 for the whole economy. But the negative obsolescence effect is now stronger since not only does the increase in real labour cost in sector 1 stimulate the replacement of existing technology, but also falling real capital costs have such an effect. Any model, such as the one in Section 22.2, which neglects the impact of real capital costs on the economic obsolescence of equipment must therefore be considered as inadequate in analysing the effects of investment subsidies on employment. As concerns the effects in sector 2 we have on the one hand the spillover effect of the direct output effect in sector 1. On the other hand there is a positive indirect output effect via higher private consumption (modelled in SECMON, but omitted in this section).

In summary we must conclude that in SECMON the negative obsolescence effects are much stronger then in FREIA. It is therefore not impossible that these negative effects outweigh the positive effects of higher output. In the following section it is shown that this is indeed so.

22.4 EMPIRICAL RESULTS

With SECMON the development in the period 1978–1983 has been 'predicted' as if there had been no WIR investment subsidies at all. Next, the development is simulated starting from the *actual* WIR subsidies. The difference between the latter and the former may be attributed to the introduction of the WIR. The results of this exercise are given in Appendix B. In Table 22.2 these results are summarized in such a manner that they are compatible with Table 22.1. This means that the effects are shown of the *introduction* of investment subsidies amounting to 1% (instead of 1.3% as in Appendix B) of net national income.

Table 22.2 Effects of the introduction of investment subsidies amounting to 1% of NNP in a base year according to the SEO: production, inflation and employment

	Effect[a] after		
	1 year	3 years	6 years
Effect on production	0.4	0.6	1.1
Effect on inflation	− 1.0	− 0.9	− 1.3
Effect on employment	− 0.3	− 0.4	− 0.4

[a] Accumulated percentage deviation from a central projection.
Source: Appendix B, results divided by 1.3 to facilitate comparability with Table 22.1.

Table 22.3 Effects of the introduction of investment subsidies amounting to 1% of NNP in a base year according to the SEO: sectoral breakdown of employment

	Effect[a] after					
	1 year		3 years		6 years	
	Absolute	(%)	Absolute	(%)	Absolute	(%)
Capital intensive sector[b]	−7	(−0.4)	−12	(−0.7)	−16·	(−0.9)
Labour intensive sector[c]	−8	(−0.2)	−5	(−0.1)	−2	(−0.0)
Total	−15	(−0.3)	−17	(−0.4)	−18	(−0.4)

[a] Accumulated deviation from a central projection (× 1000 man-years).
[b] Industries 1, 2A–2I, 4 and 5C (see Appendix A).
[c] Industries 3, 5A–5F without 5C (see Appendix A).
Source: Appendix B, results divided by 1.3 to facilitate comparability with Table 22.1.

The negative *initial* effect of investment subsidies on *inflation* is now much stronger than in the analysis of the CPB. This is explained by the augmented obsolescence effect which forces labour productivity growth to accelerate rapidly. Because of the strong negative effect on inflation the initial positive effect on *production* is also stronger compared with the CPB analysis. The obsolescence effect furthermore effectively *decreases* employment which is *contrary* to the findings of the CPB. In the *medium term* (after six years) the effects on growth and inflation are rather similar to the CPB analysis, although the effects found by SEO are somewhat more moderate. The effects on employment, however, *remain contradictory*.

A *sectoral breakdown* of the effects of the introduction of investment subsidies on employment is given in Table 22.3. From these results it follows that the negative effects on employment are far stronger in the capital-intensive sector than in the labour-intensive sector, at least in the longer run. This is expected since the user cost of capital is not relevant in the labour-intensive sector as a determinant of employment growth.

An interesting aspect of the results in Table 22.3 is the *time shape* of the effects. It appears that the initial negative effect on employment in the labour-intensive sector tends to vanish after some time. Apparently in the long run the real wage cost and output effects outweigh each other. In the *capital-intensive sector*, however, the negative obsolescence effect becomes more predominant in the course of time. This must be attributed to the considerable lag of the impact of relative factor prices on investment and employment.

22.5 CONCLUSIONS

From the above analysis one is apt to conclude that subsidizing investment as a means of stimulating employment has failed. Without these subsidies employment growth in the Netherlands would have been less unfavourable then it actually has been. This is the authors' view, *not* the policy-makers' view. It should be stressed, however, that the policy-maker until now has neglected the impact of accelerating obsolescence of equipment in the capital-intensive sector due to investment subsidies.

One could take this conclusion as an argument for proposing the abolition of investment subsidies. However, these subsidies have also had *favourable* effects, especially with respect to production growth and inflation. It is not the authors' competence to judge whether the employment, production growth and inflation effects are on balance favourable or unfavourable. This is for the politician to decide. Evidently the politician's judgement must be based on an adequate analysis of the effects of investment subsidies. In the authors' view this has not been the case so far.

REFERENCES

Auerbach, A. J. (1983) Taxation, corporate financial policy and the cost of capital. *Journal of Economic Literature*, 21, 905–40.

Butter, F. A. G. den (1985) Freia and Kompas, the Central Planning Bureau's new generation of macroeconomic policy models: a review article. *De Economist*, 133, 43–63.

Central Planning Bureau (CPB) (1977) *Een macro model voor de Nederlandse economie op de middellange termijn (VINTAF-II)* (A macro model for the Netherlands economy in the medium term), occasional paper 12, CPB, The Hague.

Central Planning Bureau (CPB) (1983) *FREIA, een macro-economisch model voor de middellange termijn* (FREIA, a medium-term macroeconomic model), monografie 25, CPB, The Hague.

Central Planning Bureau (CPB) (1986) *Verlaging van WIR-premies, lastenverlichting en de prijsverhouding tussen kapitaal en arbeid: een macro-economische analyse* (Reduction of WIR-premiums, reduction of social security premiums and the ratio of the prices of capital and labour: a macroeconomic analysis), werkdocument 6, CPB, The Hague.

Driehuis, W. (1984) Macroeconomic investment and employment functions, in *Mathematical Methods in Economics* (ed. F. Van der Ploeg), John Wiley, New York, pp. 131–61.

Driehuis, W., van Ierland, E. C. and van den Noord, P. J. (1983) *A SECtoral MOdel for the Netherlands economy (SECMON-C)*, paper presented at the 7th International Conference on Models and Forecasts, Lódz, Poland, 10–14 October 1983.

Edgren, G., Faxèn, K. and Odhner, C. (1973) *Wage Formation and the Economy*, Allen and Unwin, London.

Hartog, H. den (1984) Empirical vintage models for the Netherlands: a review in outline. *De Economist*, 132, 326–49.

Hartog, H. den and Tjan, H. S. (1976) Investments, wages, prices and demand for labour. *De Economist*, 124, 32–55.

Kaldor, N. (1966) *Causes of the Slow Rate of Economic Growth of the United Kingdom*, Cambridge University Press, Cambridge.

Knibbe, A. and van den Noord, P. J. (1983) On the relevance and the stability of Verdoorns Law, research memorandum 8331, University of Amsterdam, Department of Economics.

Malcomson, J. (1975) Replacement and the rental value of capital equipment subject to obsolescence. *Journal of Economic Theory*, 10, 24–41.

Malinvaud, E. (1980) *Profitability and Unemployment*, Cambridge University Press, Cambridge.
Nickell, S. (1975) A closer look at replacement investment, *Journal of Economic Theory*, 10, 54–88.
Noord, P. J. van den (1984) *Microfoundations of a post-Keynesian vintage model*, research memorandum 8404, University of Amsterdam, Department of Economics.
Noord, P. J. van den (1988) *Kapitaalvorming, technische ontwikkeling en werkgelegenheid* (Capital formation, technical change and employment), PhD thesis (to be published).
Ridder, P. B. de (1977) *Een jaargangenmodel met vaste technische coëfficiënten en in kapitaal geïncorporeerde arbeidsbesparende technische vooruitgang* (A clay–clay vintage model with embodied labour saving technical progress), occasional paper 14, Central Planning Bureau, The Hague.
Verdoorn, P. J. (1952) *Welke zijn de achtergronden en vooruitzichten van de economische integratie in Europa en welke gevolgen zal deze integratie hebben met name voor de welvaart in Nederland?* (Backgrounds and prospects of the economic integration of Europe for the Netherlands), Prae-adviezen van de Vereniging voor Staathuishoudkunde 1952, Martinus Nijhoff, The Hague, pp. 48–132.
Verdoorn, P. J. (1965) Complementarity and long-range projections. *Econometrica*, 24, pp. 429–450.

APPENDIX A

GENERAL OUTLINE OF SECMON, SECTORAL MODEL FOR THE NETHERLANDS ECONOMY

SECMON incorporates demand and supply elements. Supply factors are not only taken into account by specifying (vintage type) production functions, but also via a sectorally differentiated explanation of sector outputs and sector prices, which in combination with sectoral production capacities add a dimension to the traditional demand model. In addition, SECMON has a detailed public sector incorporating expenditures, taxes and social premiums.

SECMON has six types of households: consumers, firms, government, social insurance, pension funds (including life insurance) and other countries. The category of firms has 18 sectors, namely:

(1) Agriculture
(2) Manufacturing
    2A Food, beverages and tobacco
    2B Textiles, clothing and leather
    2C Timber and stone
    2D Paper and printing
    2E Chemicals
    2F Basic metals
    2G Metal products and machinery
    2H Electrical products
    2I Transport equipment

(3) Construction
(4) Energy
(5) Services
    5A Housing services
    5B Trade
    5C Transport and communication
    5D Finance and insurance
    5E Health services
    5F Other services

The model has ten blocks, each of which describes related activities or processes, namely:

I   Gross output of firms
II   Final demand
III   Imports of goods and services
IV   Production capacity and utilization rate
V   Labour market
VI   Wages and prices
VII   Incomes
VIII Government, receipts and expenditures
IX   Social security and pension funds
X   Monetary variables

The *demand* for goods and services is explained in blocks I, II and III. *Final* demand of consumers, enterprises (investment), government and final demand from abroad (exports) as well as intermediate demand of enterprises are distinguished. Furthermore, concerning the final demand of consumers ten categories of goods and services are distinguished, namely:

A Food, beverages and tobacco
B Clothing and footwear
C Gross rent and water
D Fuel and power
E Furniture, furnishings and household equipment and operation
F Medical care and health expenses
G Private transport: purchases of means of transport
H Private transport: operation cost
I   Public transport
J Others

To determine the production per sector the imports of goods and services are subtracted from the domestic demand per sector in accordance with its domestic market share.

The *supply* of goods and services is described in block IV. The productive capacity is sectorally determined and confronted with actual production. In this way, the utilization rate of productive capacity is approximated. The *labour market* is described in block V. Both the actual and the potential demand for labour are calculated per sector. Labour supply, however, is not sectorally modelled. Blocks I–V mainly describe volume flows between households. Block VI, *wages and prices*, takes care of the connections between the volume flows and the money flows. This block contains wage equations per sector and price equations per sector with respect to gross output and all final deliveries. Since volumes and prices are determined in the above-mentioned blocks, national *income* and its *distribution* can be calculated. This is done in blocks VII–IX. The government plays an important role because of its influence on the secondary income distribution. The model calculates the primary (wages, profits) as well as the secondary income distribution (disposable incomes by households). Block X, finally, is aimed at modelling the monetary sector. At this stage, however, it is not completed.

APPENDIX B

DETAILED RESULTS

In this appendix the results are shown of an *ex post* simulation with SECMON for the period 1978–1983, assuming the absence of investment subsidies. This simulation is confronted with another simulation, starting from the actual investment subsidies. Tables 22.B1–B8 show the differences between the latter and the first simulation, with respect to:

Macroeconomic keyfigures	(Table B1)
Economic lifetime of capital equipment	(Table B2)
Gross investment in capital equipment	(Table B3)
Production capacity	(Table B4)
Rate of capacity utilization	(Table B5)
Potential employment	(Table B6)
Actual production	(Table B7)
Actual employment	(Table B8)

The investment subsidy rates in the period under consideration were (as percentages of total outlays):

Year	Equipment	Buildings
1978	11.3	27.5
1979	11.7	28.0
1980	10.2	18.0
1981	12.2	18.0
1982	12.0	14.0
1983	12.5	14.0

Appendix B

Table 22.B1 Macroeconomic keyfigures[a]

	1978	1979	1980	1981	1982	1983
Volumes						
Private consumption	0.9	1.1	1.0	1.3	1.4	1.5
Gross investment (enterprises excluding dwellings)	0.9	2.1	2.8	4.0	4.7	4.6
Exports of goods and services (excluding energy)	0.2	0.4	0.5	0.7	0.8	0.9
Imports of goods and services	0.6	0.7	0.8	1.2	1.4	1.4
Production of enterprises	0.6	0.9	0.9	1.2	1.4	1.5
Prices						
Private consumption	−1.3	−1.4	−1.2	−1.6	−1.7	−1.8
Exports of goods and services	−0.7	−0.7	−0.6	−0.9	−0.9	−0.9
Wage sum per employee (enterprises)	−0.0	−0.0	−0.0	−0.0	−0.1	−0.0
User cost of capital	−15.7	−16.3	−14.5	−17.3	−17.0	−17.6
Government						
Budget surplus (% of NNI)	−1.3	−1.4	−1.3	−1.7	−1.8	−2.0
Burden of social security premiums (ID)	0.3	0.2	0.2	0.2	0.2	0.3
Burden of direct taxes (ID)	0.2	0.1	0.1	0.1	0.1	0.2
Labour market						
Employment	−0.4	−0.5	−0.5	−0.6	−0.6	−0.5
Labour supply	−0.0	−0.0	−0.0	−0.0	−0.0	−0.0
Unemployment rate	0.4	0.5	0.5	0.6	0.6	0.5
Others						
Current account (% of NNI)	−0.6	−0.6	−0.5	−0.6	−0.6	−0.6
Share of labour income (% of NNI)	0.6	0.3	0.2	0.3	0.2	0.4

[a] Accumulated percentage deviation (percentage points in the case of ratios).

Table 22.B2 Economic lifetime of capital equipment[a]

	1978	1979	1980	1981	1982	1983
2A Food, beverages and tobacco	−0.3	−0.3	−0.2	−0.4	−0.4	−0.1
2B Textiles, clothing and leather	−0.2	−0.2	−0.2	−0.2	−0.2	−0.1
2C Timber and stone	−0.2	−0.1	−0.1	−0.2	−0.2	−0.0
2D Paper and printing	−0.6	−0.5	−0.4	−0.5	−0.5	−0.1
2E Chemicals	−10.4	−9.6	−9.2	−12.1	−11.6	−1.3
2F Basic metal	−0.3	−0.3	−0.3	−0.4	−0.4	−0.1
2G Metal products and machinery	−0.2	−0.1	−0.1	−0.1	−0.1	0.0
2H Electrical products	−0.1	−0.1	−0.1	−0.2	−0.2	−0.1
2I Transport equipment	−0.9	−0.7	−0.5	−0.8	−0.8	0.0
5C Transport and communication	−0.2	−0.2	−0.2	−0.2	−0.2	−0.0

[a] Deviation in number of vintages (nonaccumulated).

Table 22.B3 Gross investment in capital equipment[a]

	1978	1979	1980	1981	1982	1983
2A Food, beverages and tobacco	4.9	8.3	10.8	14.0	14.6	12.8
2B Textiles, clothing and leather	22.4	22.0	22.8	40.8	43.6	29.7
2C Timber and stone	13.8	13.8	9.8	15.3	23.3	9.2
2D Paper and printing	0.7	54.9	41.5	36.5	74.6	75.8
2E Chemicals	1.0	2.9	4.0	6.0	7.9	9.8
2F Basic metal	0.6	2.7	6.5	8.9	10.1	9.7
2G Metal products and machinery	5.3	15.1	15.3	16.6	20.6	20.1
2H Electrical products	9.5	8.9	11.0	16.0	14.6	16.2
2I Transport equipment	0.7	0.9	1.0	24.6	28.0	22.6

[a] Accumulated percentage deviation.

Table 22.B4 Production capacity[a]

	1978	1979	1980	1981	1982	1983
2A Food, beverages and tobacco	−1.1	−1.5	−1.3	−1.1	−0.9	0.3
2B Textiles, clothing and leather	−2.9	−5.0	−6.2	−7.8	−8.8	−6.8
2C Timber and stone	−1.0	−1.0	−0.7	−1.0	−1.2	−0.4
2D Paper and printing	−3.6	−5.8	−6.7	−8.8	−10.7	−9.3
2E Chemicals	0.1	0.6	0.9	1.7	2.4	2.9
2F Basic metal	−0.0	−0.0	0.0	0.2	0.6	1.1
2G Metal products and machinery	−3.1	−2.2	0.7	1.4	2.2	7.0
2H Electrical products	−0.7	−1.2	−1.3	−1.5	−1.5	−0.8
2I Transport equipment	−2.8	−4.3	−4.7	−6.7	−7.9	−5.5
5C Transport and communication	−0.0	−0.0	−0.1	−0.1	−0.1	−0.1

[a] Accumulated percentage deviation.

Table 22.B5 Rate of capacity utilization[a]

	1978	1979	1980	1981	1982	1983
2A Food, beverages and tobacco	1.5	1.9	1.6	1.4	1.2	0.4
2B Textiles, clothing and leather	3.7	6.9	9.0	11.8	13.6	11.3
2C Timber and stone	1.8	2.5	2.2	3.0	3.6	3.0
2D Paper and printing	3.8	6.5	7.9	10.4	12.9	12.0
2E Chemicals	0.4	0.6	0.4	0.1	−0.1	−0.4
2F Basic metal	0.4	1.0	1.0	1.3	1.3	1.0

(Contd.)

Appendix B

Table 22.B5 (Contd.)

	1978	1979	1980	1981	1982	1983
2G Metal products and machinery	3.4	2.9	0.2	−0.0	−0.4	−4.1
2H Electrical products	1.3	2.2	2.5	3.1	3.3	2.7
2I Transport equipment	2.7	4.5	5.2	7.4	8.7	6.3
5C Transport and communication	0.6	0.9	0.9	1.2	1.4	1.5

[a] Accumulated percentage-point deviation.

Table 22.B6 Potential employment[a]

	1978	1979	1980	1981	1982	1983
2A Food, beverages and tobacco	−3	−5	−5	−7	−8	−7
2B Textiles, clothing and leather	−3	−4	−5	−6	−6	−5
2C Timber and stone	−1	−1	−2	−2	−3	−3
2D Paper and printing	−3	−5	−6	−7	−9	−7
2E Chemicals	0	0	0	0	0	0
2F Basic metal	0	0	0	0	0	0
2G Metal products and machinery	−6	−7	−4	−5	−6	1
2H Electrical products	−1	−2	−3	−4	−5	−4
2I Transport equipment	−1	−2	−2	−3	−4	−3
2 Manufacturing	−19	−27	−27	−34	−40	−28
5C Transport and communication	0	0	0	0	0	0

[a] Accumulated deviation in man-years (× 1000).

Table 22.B7 Actual production[a]

	1978	1979	1980	1981	1982	1983
1 Agriculture	0.5	0.6	0.5	0.7	0.7	0.7
2A Food, beverages and tobacco	0.6	0.7	0.6	0.8	0.8	0.9
2B Textiles, clothing and leather	1.0	1.2	1.1	1.5	1.6	1.7
2C Timber and stone	0.7	1.2	1.2	1.7	2.0	2.1
2D Paper and printing	0.7	0.9	0.9	1.2	1.3	1.3
2E Chemicals	0.5	1.3	1.3	1.8	2.2	2.4
2F Basic metal	0.5	1.3	1.3	1.8	2.2	2.4
2G Metal products and machinery	0.7	0.9	1.0	1.4	1.7	1.8
2H Electrical products	0.7	0.9	1.0	1.4	1.7	1.8

(Contd.)

Table 22.B7 (Contd.)

		1978	1979	1980	1981	1982	1983
2I	Transport equipment	0.7	0.9	1.0	1.4	1.7	1.8
2	Manufacturing	0.6	1.0	1.0	1.4	1.6	1.7
3	Construction	0.3	0.8	1.0	1.4	1.7	1.7
4	Energy	0.2	0.3	0.4	0.5	0.6	0.6
5A	Housing services	0.6	0.8	0.7	0.9	1.0	1.1
5B	Trade	0.7	0.9	0.9	1.2	1.4	1.4
5C	Transport and communication	0.7	0.9	0.9	1.2	1.4	1.4
5D	Finance and insurance	0.7	1.0	1.0	1.3	1.5	1.5
5E	Health services	0.6	0.8	0.7	0.9	1.0	1.1
5F	Other services	0.6	0.9	0.8	1.1	1.2	1.2
5	Services	0.7	0.9	0.8	1.1	1.2	1.3
O	Total enterprises	0.6	0.9	0.9	1.2	1.4	1.5

[a] Accumulated percentage deviation.

Table 22.B8 Actual employment[a]

		1978	1979	1980	1981	1982	1983
1	Agriculture	1	2	2	2	2	2
2A	Food, beverages and tobacco	−2	−3	−3	−4	−5	−5
2B	Textiles, clothing and leather	−2	−3	−4	−5	−5	−4
2C	Timber and stone	−1	−1	−1	−2	−2	−2
2D	Paper and printing	−1	−2	−2	−2	−3	−3
2E	Chemicals	0	0	0	0	0	0
2F	Basic metal	0	0	0	0	0	0
2G	Metal products and machinery	−3	−4	−4	−5	−6	−4
2H	Electrical products	−1	−2	−2	−3	−4	−4
2I	Transport equipment	−1	−1	−1	−2	−3	−2
2	Manufacturing	−10	−16	−18	−24	−28	−23
3	Construction	0	1	1	2	3	3
4	Energy	0	0	0	0	0	0
5A	Housing services	0	0	0	0	0	0
5B	Trade	−9	−8	−4	−6	−5	−4
5C	Transport and communication	0	0	0	0	0	0
5D	Finance and insurance	0	0	1	1	1	1
5E	Health services	0	0	0	0	0	0
5F	Other services	−3	−3	−3	−4	−3	−3
5	Services	−12	−11	−7	−8	−7	−5
O	Total enterprises	−20	−24	−22	−28	−30	−23

[a] Accumulated deviation in man-years (× 1000).

23

Collective bargaining and macroeconomic performance: the case of West Germany

ULLRICH HEILEMANN

The present upswing of the West German economy has entered its fourth year (1986) and neither the GNP growth rate nor the inflation rate seem to show any alarming signs. The labour market is improving too, but rather slowly compared with previous cycles. The unemployment rate is persistently high. However, when seen from outside, in this respect the West German picture does not look very different from that of the rest of Europe (Blanchard *et al.*, 1985, p. 5). Nevertheless, West German public authorities and most observers are much concerned. Of course, a formal explanation for this continuing labour market imbalance can easily be found in the growth of the economy which relatively to the average growth of productivity of 2.5% is still too small to raise employment substantially. Especially so since the labour force is expanding and the influx from the 'Stille Reserve' is high. Causes and cures for the unsatisfying labour market developments can be grouped along several axes. The most important one in Germany is the demand–supply line, with the supply-side view centred on the role of wages.

The Council of Economic Experts (CEE) for example suggests in its recent report that negotiated wages in 1986 should rise only along the line of productivity increases and should do without any mark-up for the expected small rise of about 2% in the cost-of-living index (Sachverständigenrat, 1985, p. 192 *et seq.*). Such a policy – according to the Council – would not mean a dramatic and lasting shift towards profits since employment and mid-term investment prospects would be improved. Of course, not all labour market difficulties would be overcome immediately, especially not the 'structural' ones, but the path would be paved for further improvements. Unfortunately, the Council presents no numerical judgements on the effects of their proposal, so that one can only speculate about their magnitude and time path. This gave rise to the present chapter, which tries to examine some aspects of the working of such wage rules and presents some estimations of their outcome.

Table 23.1 Parameters and stability statistics of the gross wage income per employee function – moving regressions, 1961(3) to 1985(2)

Sample	A1 t1	A2 t2	A3 t3	RR F	DW1 DW4	RMSPE
1966 (2)	−0.242 0.5	0.909 14.2	0.452 5.7	0.893 154.2	1.411 2.490	13.6
1966 (3)	0.232 0.4	0.797 11.5	0.505 5.4	0.846 101.4	1.167 2.651	15.4
1966 (4)	0.389 0.6	0.741 11.0	0.547 5.7	0.833 92.3	1.087 2.853	15.4
1967 (1)	0.360 0.6	0.740 11.2	0.550 5.8	0.835 93.4	1.142 2.882	16.7
1967 (2)	0.370 0.6	0.745 11.3	0.543 5.8	0.837 94.9	1.154 2.903	15.3
1967 (3)	0.359 0.6	0.745 11.3	0.545 5.8	0.836 94.4	1.155 2.992	15.3
1967 (4)	0.351 0.5	0.744 11.3	0.550 5.8	0.836 94.6	1.140 3.031	15.3
1968 (1)	0.345 0.5	0.744 11.4	0.549 5.8	0.837 95.0	1.100 3.033	15.3
1968 (2)	0.506 0.8	0.739 11.4	0.531 5.6	0.834 93.1	1.043 2.981	13.7
1968 (3)	0.428 0.6	0.738 11.5	0.541 5.7	0.835 93.3	1.064 2.890	13.5
1968 (4)	0.386 0.6	0.739 11.5	0.544 5.7	0.835 93.5	1.068 2.895	13.5
1969 (1)	0.408 0.6	0.750 11.3	0.533 5.4	0.827 88.3	1.036 2.741	13.8
1969 (2)	0.408 0.6	0.735 10.9	0.537 5.2	0.815 81.4	1.126 2.625	14.4
1969 (3)	0.416 0.6	0.730 11.2	0.559 5.5	0.822 85.5	1.283 2.699	14.2
1969 (4)	0.459 0.6	0.742 11.4	0.536 5.4	0.820 84.2	1.226 2.616	14.2
1970 (1)	0.445 0.6	0.758 12.0	0.502 5.3	0.823 86.3	1.180 2.576	13.8
1970 (2)	0.493 0.7	0.758 12.1	0.497 5.7	0.824 86.9	1.174 2.721	13.8
1970 (3)	0.463 0.7	0.759 12.1	0.497 6.0	0.828 89.0	1.169 2.704	13.8
1970 (4)	0.390 0.6	0.762 12.3	0.498 6.4	0.836 94.1	1.142 2.747	13.6
1971 (1)	0.382 0.6	0.761 12.3	0.505 6.5	0.839 96.7	1.144 2.777	13.6
1971 (2)	0.359 0.5	0.762 12.4	0.507 6.5	0.842 98.5	1.148 2.774	13.7
1971 (3)	0.363 0.6	0.764 12.6	0.505 6.6	0.843 99.7	1.140 2.750	13.6
1971 (4)	0.378 0.6	0.763 12.6	0.503 6.6	0.843 99.6	1.139 2.752	13.5
1972 (1)	0.318 0.5	0.768 12.7	0.502 6.6	0.842 98.9	1.138 2.738	13.5
1972 (2)	0.201 0.3	0.774 12.7	0.509 6.5	0.838 96.0	1.140 2.739	13.5
1972 (3)	−0.003 0.0	0.786 12.7	0.527 6.5	0.832 91.9	1.159 2.789	12.8
1972 (4)	−0.078 0.1	0.791 12.3	0.534 6.3	0.817 82.7	1.160 2.802	12.7
1973 (1)	−0.079 0.1	0.791 11.7	0.534 .2	0.800 74.0	1.159 2.802	12.7
1973 (2)	−0.273 0.3	0.805 11.3	0.543 6.3	0.790 69.5	1.165 2.734	12.7

Table 23.1 (Contd.)

Sample	A1 t1	A2 t2	A3 t3	RR F	DW1 DW4	RMSPE
1973 (3)	−0.497 0.6	0.822 11.2	0.550 6.3	0.788 68.8	1.157 2.696	13.2
1973 (4)	−0.601 0.7	0.831 11.4	0.548 6.3	0.794 71.2	1.175 2.698	13.0
1974 (1)	−0.695 0.8	0.840 11.5	0.544 6.1	0.800 74.2	1.124 2.698	12.9
1974 (2)	−0.606 0.7	0.827 11.7	0.566 6.4	0.812 79.7	1.157 2.557	12.7
1974 (3)	−0.654 0.8	0.827 12.0	0.585 6.4	0.821 84.7	1.174 2.325	12.7
1974 (4)	−0.661 0.9	0.825 12.2	0.597 6.3	0.827 88.2	1.138 2.183	12.8
1975 (1)	−0.545 0.7	0.818 12.5	0.576 5.9	0.827 88.4	1.131 2.185	12.6
1975 (2)	−0.312 0.4	0.801 12.9	0.531 5.6	0.829 89.9	1.237 2.240	12.0
1975 (3)	0.150 0.2	0.767 13.9	0.453 5.2	0.845 100.9	1.580 1.776	11.1
1975 (4)	0.574 1.0	0.729 14.9	0.399 5.2	0.860 113.9	1.829 1.482	10.6
1976 (1)	0.804 1.4	0.711 13.9	0.375 5.0	0.842 98.8	1.839 1.502	10.8
1976 (2)	0.883 1.5	0.696 12.7	0.373 4.9	0.814 81.0	1.519 1.514	11.2
1976 (3)	0.278 0.5	0.767 14.9	0.411 6.4	0.860 113.5	2.091 1.668	10.1
1976 (4)	−0.154 0.3	0.827 16.8	0.424 7.5	0.887 145.2	2.400 1.862	10.0
1977 (1)	0.042 0.1	0.811 16.2	0.408 7.2	0.880 135.8	2.259 1.779	10.6
1977 (2)	0.437 0.9	0.778 14.6	0.373 6.2	0.857 110.7	1.938 1.854	12.2
1977 (3)	0.441 1.0	0.779 15.0	0.373 6.3	0.865 118.7	2.066 1.859	12.2
1977 (4)	0.503 1.2	0.775 15.2	0.368 6.4	0.871 124.4	2.041 1.853	12.3
1978 (1)	0.633 1.5	0.763 15.2	0.351 6.1	0.871 124.5	2.018 1.746	12.5
1978 (2)	0.684 1.7	0.755 15.4	0.335 5.8	0.873 127.0	2.072 1.480	12.3
1978 (3)	0.593 1.5	0.765 15.5	0.342 5.7	0.872 125.8	2.044 1.446	12.7
1978 (4)	0.626 1.6	0.760 15.3	0.336 5.5	0.867 120.7	1.902 1.462	12.7
1979 (1)	0.867 2.5	0.723 16.6	0.298 5.5	0.882 138.8	1.245 1.975	12.0
1979 (2)	0.671 2.1	0.760 18.0	0.336 6.7	0.898 162.4	1.141 2.075	12.4
1979 (3)	0.489 1.3	0.780 15.6	0.339 6.1	0.868 122.1	1.627 1.714	27.2
1979 (4)	0.452 1.1	0.776 13.4	0.335 5.7	0.830 90.6	1.467 1.546	30.9
1980 (1)	0.590 1.3	0.745 10.8	0.335 5.7	0.771 62.4	1.490 1.478	31.8
1980 (2)	0.182 0.4	0.803 9.4	0.351 5.2	0.740 52.7	1.259 1.871	85.5
1980 (3)	0.134 0.3	0.825 9.0	0.333 4.7	0.738 52.2	1.600 2.094	85.3

Source: author's computations.

23.1 DETERMINANTS OF NEGOTIATED WAGES IN THE FRG

Before examining the effects of the wage rules suggested, it seems appropriate to take at least a short look at the empirical determinants of German wages at the macroeconomic level (for a detailed analysis compare, for example, Flanagan et al. 1983, p. 208 et seq.).

As in most other industrialized countries, the explanations of the wage rate in West Germany follow an enlarged Phillips-curve approach:

$$w = f(U, p) \tag{23.1}$$

with

w = hourly wage rate, changes against previous year
U = number of unemployed
p = consumer price-index, changes against previous year

As will be illustrated later on, these factors explain up to 79% and on average about 65% of actual development. This leaves much room for the influence of other factors. Thus it cannot reasonably be denied that such factors as wage restraint or the aggressiveness of unions (Schmidt, 1972, p. 245 et seq.), strike activity (Fautz, 1979), union organization rate (Fautz, 1980) or political factors (Gärtner, 1980) have played important roles in the West German wage-setting process of the past. These influences, however, seem to have been of temporary or episodical importance only. To illustrate this point, we take a two-step approach suggested by Schmidt (1972) in which are explained, firstly, the negotiated wages and, secondly, the (actual) gross wage income per employee:

$$\text{TLGHJW}_t = A_1 + A_2 \text{PCPJW}_{t-1} + A_3 \frac{1}{4} \sum_{i=1}^{4} \text{AL}_{t-1} + u_t; \tag{23.2}$$

$$\text{BLGAJW}_t = A_1 + A_2 \text{TLGHJW}_t + A_3(0.6\, \text{BSP76JW}_t + 0.3 \tag{23.3}$$
$$\text{BSP76}_{t-1} + 0.1\, \text{BSP76JW}_{t-2}) + u_t.$$

with

TLGHJW = index of negotiated wages on an hourly basis, changes against previous year
PCPJW = consumer price-index, changes against previous year
AL = registered unemployed ($\times 1000$)
BLGAJW = gross wage income per employed person, changes against previous year
BSP76JW = GNP in 1976 prices, changes against previous year

Since we do not want to explain the complete wage-setting process in this chapter, we restrict our further analysis to Equation 23.2, keeping in mind that the effect of the negotiated wages is somewhat mitigated by Equation 23.3 according to the actual economic conditions as represented by the GNP growth rate (in Table 23.1, the regressions have been assigned to the centre of each sample period). The maximum of the 'wage-drift' as presented by maximum values of both A_2 and A_3 in Equation 23.3 seems to have happened around the mid-seventies when labour shortages were still very noticeable.

To examine the changing influence of the various factors of Equation 23.2, a number of moving regressions were run and several stability tests such as the CUSUM, CUSUM-square and the Quandt log-likelihood test were performed (Heilemann and Münch, 1984, p. 119 et seq.). The sample length of the moving regressions was restricted to 40 quarters. This period covers approximately two business cycles and should help to avoid any cyclical bias as far as possible. In the present context, the problem of cyclical bias is of special relevance, since the German wage-setting process used to be marked by substantial lags in the upswing as well as in the downswing phases (Giersch, 1970). The regression results of the negotiated wage function are presented in Table 23.2.

These results and the various stability statistics (not presented here) indicate that even the simple approach of Equation 23.2 shows a satisfying explanatory power. As already pointed out, the fit is not an overwhelmingly good one and could easily have been improved by including the variables mentioned above. The basic conclusions, however, would not have been very different from those drawn here. In addition, the well-known methodological difficulties of Phillips-curve-type equations (Santomero, and Seater, 1978, p. 513 et seq.) (serial correlation, multicollinearity, simultaneous equation bias) should be kept well in mind. Nevertheless, it can be stated that up to 1972 the two most important variables were the constant term and the price variable while later on the price variable was pushed aside by the unemployment variable. Although the constant term is difficult to interpret other than to represent missing variables, its values over a large part of the sample have induced some observers to regard it as representing the long-term trend in productivity (Schmidt, 1972, p. 251; Santomero and Seater, 1978, p. 511). Second, the coefficient of the price variable shows a wide range of variance within short periods (cf., for example, Woll et al., 1976; Zahn, 1973, p. 214 et seq.). Whether this reflects the omittance of important variables as mentioned above or is due to a varying adaptation rate of 'rational' behaving economic agents to inflation (Santomero and Seates, 1978, p. 528 et seq.; Flanagan et al., 1983, pp. 214, 300) must be left open. Third, the number of unemployed persons played an important role in determining negotiated wages during the first half of the sixties and began to do so again in the midseventies, while in the meantime it was not of significant influence. Interestingly, since 1975 the coefficient of the number of unemployed has been constantly diminishing, although the values of this variable nearly doubled during this period.

From the last observation – not forgetting our reservations – one might well conclude that clearing the labour market is not a prime goal of labour market agents.

Table 23.2 Parameters and statistics of the negotiated wages function – moving regressions, 1961–2 to 1986–1

Sample	A1 t1	A2 t2	A3 t3	RR F	DW1 DW4	RMSPE
1966 (1)	5.478 1.9	1.529 2.5	−0.010 1.7	0.461 15.8	0.243 0.712	29.7
1966 (2)	4.049 1.5	1.899 3.4	−0.008 1.3	0.509 19.2	0.276 0.731	31.1
1966 (3)	5.182 2.1	1.624 3.3	−0.010 1.7	0.505 18.8	0.349 0.924	30.6
1966 (4)	6.423 2.9	1.315 3.1	−0.012 2.1	0.489 17.7	0.310 1.044	29.8
1967 (1)	6.817 3.2	1.213 3.1	−0.013 2.3	0.484 17.4	0.286 1.113	29.5
1967 (2)	7.150 3.5	1.149 3.1	−0.013 2.5	0.485 17.4	0.279 1.205	29.5
1967 (3)	7.628 3.9	1.052 3.1	−0.014 2.8	0.481 17.2	0.252 1.148	29.6
1967 (4)	8.209 4.4	0.967 3.0	−0.015 3.1	0.492 17.9	0.255 1.155	29.1
1968 (1)	8.424 4.8	0.949 3.3	−0.016 3.3	0.508 19.1	0.262 1.178	28.8
1968 (2)	8.702 5.2	0.920 3.5	−0.016 3.5	0.523 20.3	0.268 1.205	28.4
1968 (3)	8.917 5.5	0.904 3.6	−0.017 3.6	0.534 21.2	0.277 1.234	27.9
1968 (4)	8.708 5.5	0.965 4.0	−0.016 3.6	0.547 22.3	0.300 1.280	27.7
1969 (1)	8.210 5.3	1.048 4.5	−0.015 3.3	0.541 21.8	0.301 1.318	28.2
1969 (2)	7.620 5.1	1.134 5.1	−0.013 3.1	0.538 21.5	0.289 1.372	28.8
1969 (3)	6.633 4.7	1.252 5.7	−0.010 2.6	0.532 21.0	0.286 1.367	30.4
1969 (4)	6.220 4.8	1.290 6.0	−0.009 2.6	0.530 20.9	0.294 1.334	31.2
1970 (1)	5.992 5.0	1.304 6.1	−0.008 2.7	0.525 20.5	0.294 1.303	31.7
1970 (2)	5.748 5.1	1.310 6.1	−0.007 2.9	0.521 20.1	0.274 1.262	32.7
1970 (3)	5.757 5.4	1.306 6.1	−0.007 3.3	0.529 20.8	0.279 1.296	33.4
1970 (4)	5.917 5.7	1.306 6.2	−0.007 4.0	0.548 22.4	0.293 1.321	33.4
1971 (1)	5.959 5.8	1.304 6.3	−0.007 4.4	0.566 24.1	0.312 1.363	34.3
1971 (2)	6.043 6.0	1.288 6.3	−0.007 4.7	0.580 25.5	0.320 1.412	35.7
1971 (3)	6.166 6.0	1.253 6.2	−0.006 4.8	0.582 25.8	0.315 1.426	37.7
1971 (4)	6.381 6.1	1.208 6.0	−0.006 4.8	0.582 25.7	0.316 1.508	39.8
1972 (1)	6.704 6.4	1.145 5.8	−0.006 4.9	0.579 25.5	0.321 1.571	40.6
1972 (2)	7.066 6.7	1.072 5.4	−0.006 5.0	0.572 24.7	0.328 1.613	40.6
1972 (3)	7.516 7.2	0.987 5.1	−0.006 5.2	0.570 24.6	0.345 1.721	35.0
1972 (4)	8.072 7.6	0.879 4.5	−0.005 5.5	0.565 24.1	0.360 1.829	26.2

Table 23.2 (Contd.)

Sample	A1 t1	A2 t2	A3 t3	RR F	DW1 DW4	RMSPE
1973 (1)	8.530 8.0	0.797 4.1	−0.005 5.8	0.573 24.9	0.373 1.937	23.6
1973 (2)	8.980 8.2	0.721 3.7	−0.005 6.1	0.585 26.1	0.385 2.033	22.6
1973 (3)	9.597 8.8	0.622 3.2	−0.006 6.6	0.609 28.9	0.405 2.133	20.3
1973 (4)	10.107 9.0	0.549 2.9	−0.006 7.0	0.639 32.7	0.428 2.108	19.4
1974 (1)	10.796 9.3	0.454 2.3	−0.006 7.6	0.671 37.7	0.452 2.140	18.7
1974 (2)	11.724 9.8	0.329 1.7	−0.007 8.5	0.712 45.8	0.437 2.193	17.4
1974 (3)	12.242 9.4	0.259 1.2	−0.007 8.5	0.720 47.6	0.389 1.999	18.9
1974 (4)	12.266 8.3	0.253 1.1	−0.007 8.1	0.715 46.5	0.375 1.862	19.6
1975 (1)	12.361 7.8	0.238 1.0	−0.007 7.8	0.712 45.7	0.383 1.772	19.7
1975 (2)	12.335 7.4	0.236 1.0	−0.007 7.5	0.704 44.1	0.338 1.723	19.6
1975 (3)	11.192 6.5	0.370 1.5	−0.006 6.8	0.702 43.5	0.372 1.725	18.7
1975 (4)	9.602 5.6	0.547 2.3	−0.006 6.0	0.713 45.9	0.454 1.678	17.6
1976 (1)	8.370 5.1	0.662 3.0	−0.005 5.6	0.727 49.3	0.572 1.639	17.2
1976 (2)	7.465 5.1	0.723 3.7	−0.004 5.4	0.747 54.7	0.351 1.583	17.2
1976 (3)	8.218 5.7	0.633 3.3	−0.005 6.0	0.736 51.5	0.311 1.469	17.5
1976 (4)	9.046 6.6	0.555 3.1	−0.005 6.9	0.740 52.7	0.298 1.433	17.7
1977 (1)	9.533 7.2	0.520 3.0	−0.006 7.7	0.756 57.2	0.306 1.455	17.4
1977 (2)	9.776 7.5	0.506 3.0	−0.006 8.1	0.772 62.7	0.353 1.594	17.9
1977 (3)	9.746 7.8	0.527 3.2	−0.006 8.7	0.787 68.3	0.370 1.765	19.2
1977 (4)	9.539 7.4	0.541 3.2	−0.006 8.4	0.779 65.3	0.361 1.779	21.7
1978 (1)	9.158 7.0	0.567 3.2	−0.005 8.1	0.766 60.5	0.340 1.681	24.1
1978 (2)	8.812 6.6	0.577 3.1	−0.005 7.5	0.745 54.0	0.317 1.461	26.7
1978 (3)	8.492 6.0	0.565 2.9	−0.004 6.7	0.715 46.5	0.251 1.187	28.4
1978 (4)	7.947 5.6	0.560 2.9	−0.004 6.1	0.689 40.9	0.242 1.176	27.8
1979 (1)	7.462 5.5	0.544 2.9	−0.003 5.8	0.676 38.7	0.263 1.221	25.6
1979 (2)	7.396 5.9	0.462 2.6	−0.003 5.9	0.664 36.5	0.283 1.299	22.7
1979 (3)	7.716 7.8	0.282 2.0	−0.003 6.9	0.695 42.2	0.314 1.528	18.5
1979 (4)	8.008 9.5	0.158 1.3	−0.003 7.8	0.724 48.5	0.365 1.538	15.9
1980 (1)	8.235 10.7	0.059 0.5	−0.002 8.4	0.731 50.3	0.454 1.694	14.8

$RR\phi = 0.621$
Source: author's computations.

23.2 NEGOTIATED WAGES AND MACROECONOMIC PERFORMANCE

The previous section should have shown that macroeconomic performance plays an important and varying role in the process of wage formation. In the following section, the question of the macroeconomic consequences of various wage policies will be examined. To obtain an overall impression, first the effects of a 'wage shock' will be analysed.

It is obvious that such a question can be answered appropriately only in a detailed econometric (or otherwise parameterized) model. Though even the most detailed models are usually open to a number of objections with regard to their picture of the economy. The econometric model to be used here is the RWI business-cycle model, a medium-sized, short-term model which has been used for forecasting and simulation purposes for more than 10 years. It consists of about 40 stochastic equations and 75 definitions (Heilemann and Münch, 1984; Heilemann, 1985). The demand block of the model contains 30 equations, the distribution part 21 equations, the production block 16 equations, while the price part explains 12 variables and the government sector is represented by 36 equations. Exogenous variables are the policy levers (social security contribution rate, government construction outlays, short-term and long-term interest rates) on the one side and mainly internationally determined variables (world trade, import prices) on the other side. The theoretical base of the single equations is – as in most applied econometric models – somewhat eclectic, comprising neoclassical elements as well as Keynesian or monetarist ones. The general scheme of the model is, of course, Keynesian, of the Keynes–Klein type. The reactions of the model may be labelled with respect to the role of demand and money and the stability of the private sector generally as post-Keynesian (Davidson, 1980). The formation of expectations is in most cases of the adaptive or 'weak rational' (Eckstein) type. Sample periods of the model are usually – as already mentioned above – the last 40 quarters of the data base available. For the present study, however, an older version of the model with the sample period 1973(1) to 1982(4) has been used, mainly because this version has already been widely examined (Heilemann, 1984). The parameters have been estimated with ordinary least squares (OLS). The *ex post* and *ex ante* performance of the model seems so far to be rather satisfying.

To give a general idea of the influence of the negotiated wage rates on the macroeconomic performance of the model, this variable has been exogenized. For the first four quarters of the sample period (1973(1) to 1973(4)) its values were reduced 1 percentage point below the actual development. Then a full dynamic simulation was run over the complete sample period. The effects of this policy (differences from the baseline simulation) are shown in Table 23.3.

The results of macroeconometric simulation exercises depend as with any scientific explanation on the assumptions made and on the model used. The model has already been described briefly, so that it is sufficient to recall that it is primarily a short-term model and any forecasts or simulations with a horizon of more than

Table 23.3 Effects of a 1 percentage point reduction in negotiated wages,[a] 1973–1982, billion DMark: differences from the baseline simulation

	1973(1)	1973(2)	1973(3)	1973(4)	1974(2)	1974(4)	1975(2)	1975(4)	1976(4)	1977(4)	1978(4)	1980(4)	1982(4)
GDP, origin													
Employed (× 1000)	−0.9	−3.3	2.8	8.5	9.8	5.6	4.4	3.9	14.8	23.5	29.2	39.0	45.5
Production per employee	−0.0	−0.1	−0.1	−0.1	−0.3	−0.3	−0.2	−0.2	−0.2	−0.2	−0.2	−0.3	−0.3
GNP, demand, real													
Private consumption	−0.1	−0.3	−0.4	−0.5	−0.7	−0.7	−0.7	−0.7	−0.6	−0.6	−0.6	−0.5	−0.5
Government consumption	0.0	0.0	0.0	0.0	0.0	0.0	−0.1	−0.1	−0.1	−0.1	−0.1	−0.1	−0.0
Investment	−0.0	−0.0	−0.1	−0.0	−0.2	−0.3	−0.2	−0.2	−0.1	−0.0	0.0	0.0	0.1
Investment in machinery	−0.0	−0.0	−0.1	−0.1	−0.2	−0.1	−0.1	−0.1	−0.0	−0.0	−0.0	0.0	0.0
Construction	−0.0	−0.0	0.0	0.0	−0.0	−0.1	−0.1	−0.1	−0.0	−0.1	−0.0	0.0	0.0
Investment, changes	−0.0	−0.0	−0.1	−0.1	−0.1	−0.1	−0.1	−0.0	0.0	0.0	−0.0	0.0	−0.0
Exports	0.0	0.1	0.1	0.2	0.2	0.2	0.2	0.2	0.2	0.2	0.2	0.3	0.3
Imports	−0.0	−0.1	−0.2	−0.2	−0.3	−0.3	−0.3	−0.2	−0.1	−0.1	−0.1	−0.1	−0.0
GNP	−0.0	−0.2	−0.3	−0.3	−0.6	−0.6	−0.6	−0.5	−0.4	−0.3	−0.3	−0.2	−0.2
Consumer prices, % change	−0.0	−0.0	−0.0	−0.0	−0.1	−0.1	−0.1	−0.1	−0.0	−0.0	−0.0	0.0	0.0
GNP prices, % change	−0.2	−0.2	−0.3	−0.4	−0.2	−0.1	−0.2	−0.1	−0.1	−0.0	−0.0	0.0	0.0
GNP, income distribution													
Gross wage income	−0.8	−0.9	−0.9	−1.0	−1.0	−1.1	−1.1	−1.2	−1.2	−1.2	−1.2	−1.2	−1.2
Gross profits	0.5	0.4	0.1	0.0	−0.2	−0.4	−0.5	−0.5	−0.8	−0.9	−1.0	−1.2	−1.2
National income	−0.4	−0.6	−0.8	−0.9	−1.2	−1.5	−1.6	−1.8	−2.0	−2.1	−2.2	−2.4	−2.4
Net wage income	−0.5	−0.5	−0.5	−0.5	−0.6	−0.6	−0.6	−0.7	−0.6	−0.6	−0.6	−0.6	−0.6
Net profits	0.5	0.4	0.1	−0.1	−0.2	−0.2	−0.4	−0.4	−0.7	−0.8	−0.9	−1.0	−1.0
Labour share	−0.3	−0.3	−0.2	−0.2	−0.1	−0.0	0.0	0.0	0.1	0.1	0.1	0.1	0.2
Government													
Income	−0.4	−0.5	−0.5	−0.5	−0.6	−1.0	−0.9	−1.0	−1.0	−1.1	−1.1	−1.2	−1.3
Expenditures	−0.3	−0.3	−0.3	−0.4	−0.4	−0.6	−0.6	−0.7	−0.9	−1.0	−1.1	−1.1	−1.1
Deficit	−0.1	−0.2	−0.2	−0.1	−0.2	−0.4	−0.3	−0.3	−0.1	−0.1	−0.1	−0.1	−0.1

Author's own computations. Errors in sums may be due to rounding effects.
[a] Reduction takes place 1973(1) to 1973(4). For details of the assumptions see text.

8–10 quarters have to be judged with some reserve. The essentially short-term character of the model results from a number of features. The most important in the present context seem to be, firstly, that demand is more or less always met; secondly, that the capacity effects of investments are neglected and, thirdly, that the short-term forecasting performance has always been an important selection criterion for the equations. The assumptions of a model or a simulation usually concern the exogenous variables. For the following investigation it seemed implausible to let the expected lower inflation rate make the real interest rate rise. Therefore an accommodating monetary policy (Eckstein, 1983, p. 35 et seq.) was assumed with the money interest rate varying with the inflation rate and the real interest rate being the one from the baseline simulation. The other exogenous variables remained unchanged. However, with the FRG having more than a 15% share in world trade it may be questioned whether these are very realistic assumptions.

It is not necessary to give here a detailed account of the various direct and indirect effects of a cut in money wages, nor is it necessary to assign them to classical or Keynesian arguments (Chick, 1983, p. 152). The most important result of the simulation is certainly the rise in employment. Although in the first two quarters there is a small decline, in the following periods there is a significant increase in employment. Whether these effects are regarded as large or small is in the end a matter of taste. But it should not be forgotten that the wage reductions are comparatively small (Table 23.4) and hold for one wage round only. The employment decline may be labelled according to Chick as Keynesian, since it is caused by the immediate response of private consumption to reduced purchasing

Table 23.4 Actual and cost-level neutral wages (CEE)[a] 1973 to 1982: changes against previous year

	Money wages		Real wages		
	Actual	CEE	Actual	CEE	
				A[b]	B[c]
1973	10.38	6.88	3.37	−0.13	−0.13
1974	13.27	9.47	5.99	2.19	2.49
1975	9.23	9.23	3.21	3.21	3.71
1976	5.93	8.23	1.69	3.99	4.49
1977	6.92	7.62	3.31	4.01	4.21
1978	5.75	6.15	2.94	3.34	3.34
1979	4.89	3.89	0.76	−0.24	−0.34
1980	6.75	2.85	1.22	−2.68	−2.78
1981	5.57	3.27	−0.47	−2.77	−2.67
1982	4.06	3.76	−0.96	−1.26	−1.06

Sources: Deutsche Bundesbank, Sachverständigenrat, 1983, p. 88, Table 19, and author's computations.
[a] Negotiated wages.
[b] Deflated with the consumption deflator.
[c] Deflated with simulated values of the consumption deflator (cf. Table 23.5).

power, whereas the rise in employment later on may be called classical, since reduced costs encourage firms to expand production and employment.

The production side of the model is represented by an implicit production function or labour-demand function with unit labour costs (gross labour cost per unit real GNP) playing a central role:

$$EWA_t = 2283.091 + 9.932 \ (CP76_t + IAN76_t + EX76_t)$$
$$\quad\quad\quad (4.7) \quad\quad (11.4)$$
$$\quad\quad - 3\,124\,436 \ LSTK_{t-1} + 0.835 \ EWA_{t-1}$$
$$\quad\quad\quad (10.9) \quad\quad\quad\quad (34.7)$$
$$\quad\quad + 27.029 \ DS1 + 366.0421 \ DS2 + 291.367 \ DS2$$
$$\quad\quad\quad (0.8) \quad\quad\quad (11.9) \quad\quad\quad (11.6)$$

$$R^2 = 0.987; \quad DW = 1.487; \quad RMSPE = 0.203; \quad MAE = 39.7$$

with:

EWA   = number of wage earners employed (× 1000)
CP76 = private consumption, in 1976 prices
IAN76 = investments, in 1976 prices
EX76 = exports, in 1976 prices
LSTK = unit labour costs
DS1, 2, 3 are seasonal dummy variables

Of course, the demand growth reduction acts as a counterweight to the cost effects mentioned. The reduction of the inflation rate is a consequence of the reduced labour costs as well as of the reduced capacity utilization. The growth reduction is mainly caused by diminished private consumption which is intensified for some time by the reductions in investment. However, this well-known interaction of multiplier and accelerator is increasingly disturbed by the reduction of wage costs. That it takes such a long time until the second effect dominates the first one is mainly a consequence of the slow reaction of private consumption, as indicated by a mean lag of about four quarters. However, the contractive effects of shrinking private consumption are offset by nearly 50% by expanding exports, caused by the reduced capacity utilization and the worsened terms of trade, and by reduced imports. A most interesting picture is offered by the (functional) income distribution. For more than three quarters the wage reduction causes a sharp rise in profits, but later on profits are affected by growth reductions, too. In the long run, the losses of entrepreneurs are in absolute terms as high as those of wage earners. Consequently, the wage share (corrected for employment effects) is diminished, mainly during a time of wage restraint, but already half a year later recaptures its old position. Government seems to be a loser, too, though the reduced income is nearly balanced by a reduction in expenditures, mainly a consequence of the reduction in transfer payments.

The previous simulation results have illustrated the great influence even of small reductions in money wages (cf. also Jahnke, 1985). (In fact, negotiated wages are the most influential variable (with regard to GNP, prices, employment) of the complete model, as parameter-sensitivity tests with the model according to the example of Kuh and Neese, 1981, have shown.) This effect and the well-known stability consciousness of the FRG (Flanagan et al., 1983, p. 250 et seq.) have since the early 60s caused much reasoning about the 'right' wage formula. The centre of this debate has been the CEE who developed such concepts as the productivity orientated wage policy or the cost-level neutral wage policy (Sachverständigenrat, 1964, p. 248; 1966, p. 302; 1967, p. 323). The main goals of these concepts have been the stability of the price level or constant labour share or both. A main precondition of all these concepts, however, has been a cleared labour market; that means more or less the absence of (involuntary) unemployment.

Although today the prerequisites of these concepts are not fulfilled, it might be interesting to look at the results of such policies over a longer period. This might hold even for a short-term model like the present one, since the comparatively low short-term employment sensitivity of this type of model with respect to wages is often mistaken for an empirical validation of the purchasing power theory. Unfortunately it is difficult to implement wage formulae not only in reality (Sachverständigenrat, 1967, p. 342; Molitor, 1973, p. 161 et seq.) but also in an econometric model such as the present one. The main difficulties arise, firstly, from the wide fluctuations for example of productivity and terms of trade changes on a quarterly base, and, secondly, from the absence of (direct) capital costs in the model employed. On the other hand, it should not be overlooked that the 'cost level neutral' wage development as suggested by the CEE completely neglects the various repercussions between the variables employed, especially wages and productivity; not to speak of the vagueness of concepts such as 'unavoidable price increases' or the difficulties of integrating labour time reductions. To overcome the modelling problems, we assumed that in the period 1973 to 1982 the observed negotiated wages were changed according to the differences between observed wages and cost level neutral wages (Sachverständigenrat, 1983/84, p. 88) (Table 23.4). Again, it was assumed that the real interest rate remained unchanged and the simulation was performed over 40 quarters.

The results in Table 23.5 make it evident that the FRG economy would in some respects have been considerably better off with a wage policy along the lines suggested by the CEE. With regard to employment and inflation, the gains are rather obvious. With regard to growth, there is in the first year a reduction of the GNP growth rate of about half a percentage point, but later on the old growth path is joined again. Inflation would also be lower – a main goal of the CEE concept. These reductions are greatest when measured by the GNP deflator, since consumption prices react comparatively slowly and with respect to wage cost also weakly. Though possible, we refrained from refeeding the lower inflation rate into the model via wages. With regard to the income distribution, the labour share would have been on average about 0.9 percentage point lower than observed. The government deficit is increased, mainly during the first three years. Primarily this is a consequence of the fact that government income such as taxes and social security

Table 23.5 Effects of a cost-level neutral wage policy (CEE),[a] 1973–1982, billion DMark: differences from the baseline simulation

	1973	1974	1975	1976	1977	1978	1979	1980	1981	1982
GDP, origin	8.2	65.0	118.3	128.6	142.7	179.4	213.6	289.9	419.7	509.0
Employed, (× 1000)	−0.4	−1.5	−2.2	−2.0	−1.6	−1.5	−1.6	−2.1	−3.2	−3.9
Production per employee										
GNP, demand, real										
Private consumption	−6.0	−17.7	−23.2	−20.4	−16.7	−14.4	−14.5	−20.7	−28.1	−31.0
Government consumption	0.0	−0.3	−1.1	−2.0	−2.3	−2.0	−1.7	−1.4	−1.5	−1.8
Investment	−0.9	−4.1	−6.7	−4.3	−0.7	1.5	1.7	1.0	−1.2	−2.2
Investment in machinery	−1.1	−3.3	−3.3	−1.2	0.1	0.8	0.7	0.0	−0.4	0.1
Construction	0.2	−0.8	−3.3	−3.1	−0.8	0.8	1.0	1.1	−0.7	−2.3
Investment, changes	−1.0	−2.5	−2.2	0.2	1.0	0.7	0.0	−0.7	−1.6	−1.3
Exports	2.0	5.7	6.4	5.8	5.2	4.9	6.1	10.3	14.7	15.7
Imports	−1.9	−6.1	−8.1	−5.8	−3.4	−2.5	−2.2	−3.0	−4.3	−4.5
GNP	−4.0	−12.7	−18.6	−15.0	−10.0	−6.9	−6.2	−8.6	−13.3	−16.1
Consumer prices, % change	−0.0	−0.3	−0.5	−0.5	−0.2	−0.0	0.1	0.1	−0.1	−0.2
GNP prices, % change	−1.1	−2.0	−1.4	−0.3	−0.0	−0.2	−0.2	−1.3	−1.4	−0.7
GNP, income distribution										
Gross wage income	−16.1	−35.8	−38.1	−30.0	−27.2	−24.8	−30.4	−52.9	−65.5	−66.6
Gross profits	4.5	1.6	−2.1	−21.3	−22.7	−23.8	−23.1	−21.2	−30.8	−40.9
National income	−11.6	−34.2	−50.2	−51.3	−49.9	−48.6	−53.5	−74.1	−96.2	−107.5
Net wage income	−9.1	−19.3	−21.0	−15.8	−14.3	−13.2	−16.0	−27.7	−34.0	−34.0
Net profits	4.1	0.8	−11.9	−18.7	−18.6	−19.1	−18.2	−16.9	−26.4	−34.8
Labour share	−1.2	−1.5	−0.3	0.7	0.9	0.9	0.7	−0.0	0.3	1.0
Government										
Income	−8.4	−21.1	−27.0	−27.2	−26.6	−25.3	−28.9	−42.4	−53.1	−58.8
Expenditures	−5.9	−15.2	−20.9	−23.0	−25.3	−24.9	−26.2	−35.0	−45.4	−51.8
Deficit	−2.5	−5.9	−6.2	−4.2	−1.3	−0.4	−2.7	−6.4	−7.7	−7.0

Own computations. Errors in sums may be due to rounding effects.
[a] For details of the simulated wage development see text and Table 23.4.

contributions react immediately on inflation while expenditures, especially transfer payments, lag behind up to two years.

Whether the reductions in money (and real) wages do pay off is difficult to judge on the evidence of simulations such as the present ones. Compared with the results of other German models for similar impulses, the outcome of the present study does not seem to be completely out of touch (Martiensen, 1984, p. 69 *et seq.*; Jahnke, 1985, p. 14 *et seq.*). Lüdeke (1980) has undertaken a similar simulation with an econometric model (Friedrich *et al.*, 1979) over the period 1969 to 1975. With regard to the employment and relative income effects, his results seem to be similar to those of the present study. With regard to the growth effects, however, there are wide differences for which there seem to be mainly two reasons. First, the different specifications of the investment in machinery functions: whereas in the model used by Lüdeke, wages enter with a positive sign, there is a negative relation in the RWI model. Secondly, real exports are exogenous in the Lüdeke study and endogenous, determined by capacity utilization (export-push hypothesis) and the terms of trade, in the model employed here. Of course, from the perspective of an integrative economic theory, the factor-demand approach used by Lüdeke is very convincing as far as those investments which are made to raise productivity are concerned.

Even if the results of the present simulations are accepted, a final judgement has to balance the gains in employment and the lower inflation rate against the losses in growth, the transitory reduction of labour's share in income and the enlarged government deficit. Against the results it may be said that there are some shortcomings in 'contemporary cliometrics' in general (Andreano, 1970) or in the model used. The last point especially should not be overlooked, since the model has been constructed primarily to make short-term analyses.

23.3 SUMMARY AND CONCLUSIONS

The present chapter has tried to examine the reciprocal effects of negotiated wages and macroeconomic performance. The empirical and methodological base were single equations as well as complete model simulations. Despite several methodological reservations, it could be shown that with regard to negotiated wages the inflation rate has been of varying, though mostly central influence. Classical Phillips-curve factors such as unemployment played an important role at the beginning of the 60s and from the end of the 70s onwards; however, in recent years of high and rising unemployment, labour-market conditions seem to have been of shrinking importance. In other words, Keynes' empirical founded statement, that nominal wages are 'sticky' (Keynes, 1936, p. 11; Chick, 1983, p. 152), seems to describe the West German wage scene in the period under study fairly well, in so far as wages do not seem to react very much on labour-market conditions.

The central role of negotiated wages for macroeconomic performance was examined in two simulations with a macroeconomic short-term model over the period 1973 to 1982. The economic evaluation of these simulations, however, is rather difficult to carry out since macroeconomic goals are affected in different and sometimes opposite ways. Positive effects are encountered on the employment side and with regard to inflation; negative effects, some of them transitory, are registered with respect to economic growth and to relative income distribution. In addition, for various methodological reasons, these findings have to be judged carefully. On the other side, the results seem to be not implausible and are to a certain degree backed by other simulations.

Results such as the present ones are usually addressed to a policy-maker, suggesting the use of this or that lever to improve the macroeconomic performance. With the present study things look different. The addressees are the employer associations and the unions; to be more specific: in 1983 about 2000 negotiated wage contracts were registered (Meyer and Vatthauer, 1985, p. 3). Both parties are not usually assumed to contribute to the common welfare but to improve the economic situation of their members. In the FRG things might look a little different, since the autonomy of labour-market parties in the collective bargaining process seems to be conditioned on their positive contribution to macroeconomic performance – at least in the medium term and in the long run. But even leaving these possible restrictions aside, perhaps the chapter gives cause for a rethink on West German wage policies.

REFERENCES

Andreano, R. L. (ed.) (1970) *The New Economic History. Recent Papers on Methodology*, John Wiley & Sons, New York.

Blanchard, O., Dornbusch, R., Drèze, J., Giersch, H., Layard, R. and Monti, M. (1985) Employment and growth in Europe: a two-handed approach, Report of the CEPS Macroeconomic Policy Group, Economic Papers, 36, EEC Brussels.

Chick, V. (1983) *Macro Economics after Keynes*, MIT Press, Cambridge MA.

Davidson, P. (1980) Post-Keynesian economics: Public interest (special issue), *The Crisis in Economic Theory*, New York, 20, 151–73.

Eckstein, O. (1983) *The DRI-Model of the US Economy*, McGraw-Hill, New York.

Fautz, W. (1979) Gewerkschaften, Streikaktivitäten und Lohninflation. *Zeitschrift für die gesamten Staatswissenschaften*, 135, 605–28.

Fautz, W. (1980) Sind Löhne und Preise wirklich inflexibel nach unten? *Zeitschrift für Wirtschafts- und Sozialwissenschaften*, 100, 111–39.

Flanagan, R. J., Soskice, D. W. and Ulman, L. (1983) Unionism, economic stabilization and income policies: European experience, *Brookings Studies in Wage–Price Policy*, Washington.

Friedrich, D., Kau, W., Lüdeke, D. and v. Natzmer, W. (1979) Ein ökonometrisches Vierteljahresmodell für den güterwirtschaftlichen und monetären Bereich der Bundesrepublik Deutschland. (Forschungsberichte aus dem Institut für Angewandte Wirtschaftsforschung, 3.) Tübingen.

Gärtner, M. (1980) Politisch-ökonomische Determinanten der Lohnentwicklung in Deutschland, Königstein/Ts.

Giersch, H. (1970) Growth, cycles, and exchange rates: the experience of West Germany, Wicksell Lectures, Almqvist & Wiksell, Stockholm.

Heilemann, U. (1985) The RWI econometric short term model: problems and practice, in

Macroeconometric Modelling of the West German Economy (eds B. Gahlen and M. Sailer), International Institute of Management/Industrial Policy, Berlin, pp. 189–202.

Heilemann, U. (1984) Zur Stabilität des RWI-Konjukturmodells. *Mitteilungen des Rheinisch-Westfälischen Instituts für Wirtschaftsforschung*, 35, 313–37.

Heilemann, U. and Münch, H. J. (1984) The great recession: a crisis in parameters? in *System-Modelling and Optimization*, Proceedings of the 11th Conference 1983 (ed. P. Thoft-Christensen). (*Lecture Notes in Control and Information Sciences*, 59), Berlin, pp. 117–29.

Heilemann, U. and Münch, H. J. (1984) Einige Bemerkungen zum RWI-Konjunkturmodell, in *Simulationsrechnungen mit ökonometrischen Makromodellen* (eds G. Langer, J. Martiensen and A. Quinke), München, pp. 355–85.

Jahnke, W. (1985) Simulation verschiedener Strategien zur Verringerung der Arbeitslosigkeit, paper presented to the Annual Meeting of the Statistische Gesellschaft, Bonn, September 27.

Keynes, J. M. (1936) *The General Theory of Employment, Interest and Money*, Macmillan, London.

Kuh, E. and Neese, J. (1981) Parameter sensitivity, dynamic behavior and model reliability: an initial exploration with the MQEM monetary sector, in *Proceedings of the Econometric Society European Meeting 1979*, Selected Econometric Papers in Memory of Stefan Valavanis (ed. E. G. Charatsis). (*Contributions to Economic Analysis*, 138.) Amsterdam, p. 121–68.

Lüdeke, D. (1980) Zur empirischen Relevanz der Lohn-Preis-Spirale und der Lohn-Investitionsaversion, in *Empirische Wirtschaftsforschung*, Festschrift für R. Krengel aus Anlass seines 60. Geburtstages (eds v. J. Frohn and R. Stäglin), Berlin, pp. 213–35.

Martiensen, J. (1984) Simulationsexperimente mit makroökonometrischen Modellen für die Bundesrepublik Deutschland, in *Simulationsexperimente mit ökonometrischen Makromodellen* (eds G. Langer, J. Martiensen and A. Quinke), München, pp. 9–75.

Meyer, W. and Vatthauer, M. (1985) Tarifbewegung in der Bundesrepublik Deutschland – Beschreibung der Datenbasis einer mikroökonometrischen Analyse für die Jahre 1960 bis 1983. Universität Hannover, Fachbereich Wirtschaftswissenschaften, Arbeitspapier 1985-1.

Molitor, B. (1973) Keine glückliche Hand – Der Sachverständigenrat und die Verteilungspolitik, in *Zehn Jahre Sachverständigenrat* (ed. R. Molitor) Fischer-Athenäum, Frankfurt/M., pp. 152–70.

Sachverständigenrat zur Begutachtung der gesamtwirtschaftlichen Entwicklung, various reports, 1964, Mainz 1965 et seq.

Santomero, A. and Seater, J. J. (1978) The inflation–unemployment trade-off: a critique of the literature. *Journal of Economic Literature*, 16, 499–544.

Schmidt, R. (1972) Kurzfristige Prognosefunktion für die Tarif- und Effektivlohnentwicklung in der Bundesrepublik Deutschland. *Die Weltwirtschaft*, pp. 237–64.

Woll, A., Faulwasser, B. and Ramb, B.-T. (1976) Zur Zeitstabilität des Phillips-Theorems. *Kredit und Kapital*, 9, 293–316.

Zahn, P. (1973) Die Phillips-Relation für Deutschland de Gruyter, Berlin.

24

A cost–push model of galloping inflation: the case of Yugoslavia

DAVORIN KRAČUN

This chapter introduces a model explaining the cost–push mechanisms present under Yugoslav economic policy and also considers the specifics of the Yugoslav economic system.

First it applies standard input–output analysis procedure, multiplying the horizontal vectors of value added by the matrix multiplier to find total coefficients. The rate of growth of producers' prices for the next year is then computed as a function of (1) the expected (= planned) rate of inflation, (2) the tendency of less profitable sectors to come up to average profitability through above-average price increases and (3) economic policy goals which are formulated as a system of linear functions. An iterative procedure is employed to provide an optional number of adjustments to the quantitative targets of economic policy depending on the actual growth of prices. The calculation is based on obtainable authentic data carefully selected and adapted for use in the model.

Inflation in the Yugoslav economy is considerably higher than in the economies of the OECD countries. While the OECD countries have substantially succeeded in lowering inflation during the 1980s, Yugoslavia experienced at the same time an escalation of inflation, unknown throughout Europe. This high rate of inflation in Yugoslavia is a result of the current difficulties that the country is encountering: high indebtedness (the payments on which account for 40% of the annual flow of foreign currency), a low level of development, a high unemployment rate (around 20%), huge differences in the levels of development of individual regions, poor competitiveness on the world market, etc. Aside from this the institutionalized economic system differs in many significant characteristics from the systems in OECD countries.

In the model illustrated in this chapter we analyse a cost–push mechanism which acts in circumstances of high inflation rates and economic stagnation. In the model we assume that realistic prices for the factors of production will be pursued (positive rates of interest, a realistic rate of exchange, etc.).

For the basic of the model, with which we intended to illustrate the laws of the cost–push inflation in Yugoslavia, we shall be applying an input–output model and the operation of the decomposition of material costs into the costs of primary production factors. The process of price growth in the individual sectors is mutually dependent on:

(1) the measures of economic policy in the field of factor prices or the distribution of income;
(2) the tendency of average and above-average sectors to defend their position in the primary distribution of national income by passing on increased costs through higher prices;
(3) the tendency of below-average sectors to improve their position in the primary distribution of national income;
(4) expected or planned inflation rates;
(5) the immediate adjustment of prices and incomes to compensate for deviations from expected or planned inflation.

To simulate the process mentioned above an iterative procedure is employed to provide an optional number of adjustments, following the decomposition procedure.

The model uses the available data for the Yugoslav economy. The principal sources of data are from the annual financial statement for the Yugoslav economy in 1984 collected and arranged by the Social Accounting and Auditing Service.[1] Additional sources of data on imports are obtained from customs declarations (processed by the National Bank) and the available statistical data for the growth of the physical volume of production, prices and the available input–output table for intermediate goods.

24.1 PRICE STRUCTURE AND DECOMPOSITION PROCEDURE

In the input–output model the production value in the economy is divided into n sectors. The production value of each sector in the domestic market can be divided into the value of intermediate goods, the value of imports and the value added:

$$P(j) = \text{SADA}(j) + \sum_{i=1}^{n} \text{YA}(i, j) \qquad (24.1)$$

$P(j)$ = the value of production in the domestic market in the j-sector
$\text{SADA}(j)$ = the value of intermediate goods used in the j-sector
$\text{YA}(i,j)$ = the elements of imports and value added in the j-sector

---

[1] The Social Accounting and Auditing Service of Yogoslavia is a state financial service which controls the business of all organizations of associated labour, i.e. the socially-owned enterprises. Each organization submits a complete financial report after every three month period. The data included in these reports enables us to compose the quantities that are useful for economic analysis.

Price structure and decomposition procedure

If Equation 24.1 is divided by $P(j)$, then the direct coefficients are obtained:

$$SAD(j) = SADA(j)/P(j) \qquad (24.2)$$

$$Y(i,j) = YA(i,j)/P(j) \qquad (24.3)$$

The direct coefficients of intermediate goods, imports and value added form unity.

$$SAD(j) + \sum_{i=1}^{n} Y(i,j) = 1 \qquad (24.4)$$

The difference, in more detail, between the coefficients of the intermediate goods and one (see Equation 24.4) is defined as follows:

$$\sum_{i=1}^{n} Y(i,j) = IM(j) - EX(j) + DF(j) + OC(j) + WS(j) + CT(j)$$
$$+ NI(j) + AC(j) \qquad (24.5)$$

Definition of symbols:

- $IM(j) - EX(j)$ = the net direct coefficient of imports; the difference between the direct coefficient for imports of intermediate goods and the direct coefficient for exports.
- $DF(j)$ = the direct coefficient for depreciation of fixed assets
- $OC(j)$ = the direct coefficient for other costs which are not included in the intermediate costs
- $WS(j)$ = the direct coefficient for wages and salaries
- $CT(j)$ = the direct coefficient for contributions and taxes
- $NI(j)$ = the direct coefficient for net interest; the difference between the interest paid and interest received by each sector
- $AC(j)$ = the direct coefficient for capital accumulation.

The production value of each sector is shown here as a net value. That is both the direct coefficient for imports and the direct coefficient for interest costs show only the net cost to production in each sector. Both coefficients may also have a negative value. This method of illustrating the value of production in each sector enables us to investigate exactly those costs that are relevant to price-formation on the domestic market.

Where the production of an economy is composed of n-sectors a part of the production values of other sectors is included in the intermediate domestic consumption of each sector. The production relations between the sectors are shown by a square matrix of domestic and technical coefficients **[AD]** where $AD(i,j)$ shows that part of the production value of the j-sector which originated in the i-sector.

The direct coefficient for the value of intermediate goods $SAD(j)$ is the vertical sum of coefficients seen in Equation 24.6:

$$SAD(j) = \sum_{i=1}^{n} AD(i,j) \qquad (24.6)$$

Through the matrix [AD] the decomposition (Babic, 1978; Sekulic, 1980) of intermediate goods to the primary sources of costs defined in Equation 24.5 is carried out. In matrix form this operation is written as:

$$[TY](i) = [Y](i)[I - AD]^{-1} \quad (24.7)$$

[TY](i) = the horizontal vector of total coefficients for imports or of values added

[Y](i) = the horizontal vector of direct coefficients for imports or of values added

$[I - AD]^{-1}$ = the inverse value of the difference between the matrix unit and domestic technological coefficients — the matrix multiplier

The sum of total coefficients for net imports and values added in each sector forms unity:

$$SAD(j) + \sum_{i=1}^{n} Y(i,j) = 1 \Rightarrow \sum_{i=1}^{n} TY(i,j) = 1 \quad (24.8)$$

In our model we operate with the following row vectors of total coefficients:

for imports

$$[TIM] = [IM][I - AD]^{-1} \quad (24.9)$$

for exports

$$[TEX] = [EX][I - AD]^{-1} \quad (24.10)$$

for depreciation of fixed assets

$$[TDF] = [DF][I - AD]^{-1} \quad (24.11)$$

for other costs

$$[TOC] = [OC][I - AD]^{-1} \quad (24.12)$$

for wages and salaries

$$[TWS] = [WS][I - AD]^{-1} \quad (24.13)$$

for taxes and contributions

$$[TCT] = [CT][I - AD]^{-1} \quad (24.14)$$

for net interest costs

$$[TNI] = [NI][I - AD]^{-1} \quad (24.15)$$

for capital accumulation

$$[TAC] = [AC][I - AD]^{-1} \quad (24.16)$$

Total coefficients in each sector form unity:

$$TIM(j) - TEX(j) + TDF(j) + TOC(j) + TWS(j)$$
$$+ TCT(j) + TNI(j) + TAC(j) = 1 \quad (24.17)$$

The aggregate of the sectors' total coefficients into total coefficients for more sectors or for the entire economy is possible by weighting based on the total value of production in each sector:

$$\text{TY}(i) = \left[\sum_{j=1}^{n} \text{TY}(i,j) P(j)\right] \bigg/ \sum_{j=1}^{n} P(j) \qquad (24.18)$$

$\text{TY}(i)$ = the total coefficient of i-primary costs for s-sectors
Total coefficients for the entire economy also form unity:

$$\sum_{i=1}^{n} \text{TY}(i) = 1 \qquad (24.19)$$

24.2 THE MECHANISM OF THE GROWTH OF COSTS AND PRICES

Total coefficients enable the investigation of the impact of price changes to primary production factors on the formation of costs and prices in each sector and on the total price level (Bole, 1982). The weightings by which the impact on price growth of the individual elements of the price structure on the entire sector are evaluated, play an important role in the evaluation of impacts on the growth of prices or the nominal value of each primary factor.

Although prices for most goods are determined on the market, in the Yugoslav economy it is impossible to expect a spontaneous establishment of balanced prices for production factors for at least two reasons (Kračun and Žižmond, 1982);

(1) the existing economic structure is not in harmony with available resources and therefore certain factors are in chronic shortage (foreign currency, capital), while others are in abundance (unemployment);
(2) the existing institutionalized economic system does not permit the normal functioning of the laws of supply and demand in the markets for certain production factors.

Thus, economic policy has a decisive effect on the prices for production factors and in turn on costs.

Economic policy in Yugoslavia includes such measures as the following: direct price and incomes controls, the determination of the official rate of exchange, customs duties, export bonuses,[2] the setting of obligatory depreciation charges and the level allowed for the revaluation of fixed assets, the setting of indices for wages and salaries, social accords on the distribution of income, the setting of contributions and taxes, inter-bank agreements on the rates of interest, etc. All have a decisive impact on costs.

These regulations have an important impact but their scope is objectively and relatively modest as it includes only a zone of a few percentage points above or below the inflation rate as, for example, the rate of exchange must correspond

---

[2] Among them the tax drawbacks have the most important role.

to the competitiveness of the economy on the world market; depreciation must in the long term preserve the substance of fixed assets, wages and salaries must not be under the existence minimum. These facts cause an interdependence between the actual inflation rate and the measures of economic policy.

The essence of the mechanism for the growth of costs and prices in the model lies in the reciprocal interaction of inflation and policy impacts on the nominal growth of elements of the costs structure. In practice the actual growth of prices has always substantially surpassed the expected, i.e. the planned growth, and forecasts of price growth must also take into account the fact that the increase in those costs regulated or effected by economic policy, will more or less follow the actual rather than the planned rate of inflation.

In economic practice there is a strong desire at the beginning of the year for the predicted levels of price growth not to be surpassed (if only the prediction had been realistic and policy measures based on this). It has always been the case that actual growth surpasses the framework of the planned growth and therefore it has been necessary to lift those restrictions which prevent the prices of production factors following inflation.

In recent years policy practice has turned away from fixed targets for the allowed increase in costs and prices. Increasingly such targets have been replaced by target ranges, i.e. percentage points above or below the actual rate of inflation. This mechanism of target ranges will be simulated in our model.

The problem of adjustment, where the actual growth rate of producers' prices is at the same time the exogenous and endogenous variable, is solved in the model by a procedure of several adjustable phases. In this model the planned growth rate of producers' prices is only one of the initial impulses. If the measures of economic policy are inconsistent and do not allow the planned rate to be achieved then a suitably increased growth rate of prices as the input variable is introduced into the following phase. The measures of economic policy are defined in the form of linear functions. In each subsequent phase the actual rate of growth of prices is modified and introduced into these linear functions as an independent variable.

24.2.1 The initial impulses of inflation

There are three initial impulses in the model: planned inflation, a disproportionate difference in wages, salaries and profits between various sectors and the burdening of the economy with additional costs.

(a) Planned inflation

Planned inflation or the planned average rate of growth of producers' prices (PP(0)) is the original and independent variable in the model. As regards the constellation of other variables and total coefficients, this variable through a sequence of adjustments will be increased while theoretically it could be decreased or left unchanged.

In view of the importance of this variable for the final result we can raise the following question: how exactly can it be determined?

In Yugoslav practice this variable can be simply determined as there exist numerous indications. Firstly, there is inflation in the current or previous year. It is probable that in the coming year inflation will be approximately the same as it was in the previous year and this is one of the fundamental sources of information.

The most important indicators for the determination of PP (0) are the annual plans of the socio-political community and, in particular, the Federal budget and the Resolution on Economic Policy. Enterprises and their annual plans and the policy of prices and distribution, as a rule, follow these indicators.

The psychological influence of expected inflation should also not be neglected. It is well known that if most of the subjects within an economy expect inflation then inflation will occur.

(b) Imbalance between sectors

Imbalance between sectors can be defined by a comparison of the wages and salaries per worker employed and the accumulation of capital as a percentage of capital employed. The starting point lies in the fact that the sectors where the wages and salaries per worker employed are below average, and contain loss-making enterprises, or where capital accumulation is below average, increase prices more rapidly than the general rate of price growth.

Herein we must consider that increases in wages and salaries will also cause higher taxes and contributions; this will be covered by the sector through corresponding increases in prices.

It is necessary to illustrate certain institutional particularities in the formation of wages and salaries (Bajt, 1983) and the accumulation of capital in Yugoslavia. The basic principle is that the workers dispose of the value of the total net product (= income) and they determine its distribution between wages and salaries and the accumulation of capital. In practice this means that wages and salaries for the same work in different enterprises can be different depending on the profitability of the business operation. Therefore, we assume in this model that in less successful sectors workers will try to correct the level of wages and salaries with an above-average increase in prices.

The absolute value for the correction of wages and salaries in the j-sector DWSA(j) would be:

$$\text{DWSA}(j) = (\text{WSA}/M)M(j)q(j) \times \text{LD} - \text{WSA}(j) \qquad (24.20)$$

DWSA(j) = the correction of wages and salaries in the j-sector
WSA = the value of total wages and salaries in the entire economy

M	= the number of employed in the entire economy
$M(j)$	= the number of employed in the j-sector
$q(j)$	= the coefficient of qualifications and working conditions in the j-sector. Under average conditions this value is 1, for sectors with above-average claims the value ranges from 1 to 1.1, for less demanding sectors it ranges from 0.9 to 1.
LD	= the desired level; this is included in the model as an exogenous variable
$WSA(j)$	= the wages and salaries paid in the j-sector

In the Yugoslav economic system the accumulation of capital is not only a residual item; it can also behave as the costs of the business operation. The enterprise is obliged, to a certain extent, to accumulate its own capital. In any case, less accumulative sectors will tend towards a faster increase in the accumulation of capital:

$$DACA(j) = (DACA/K)K(j) \times LD - ACA(j) \tag{24.21}$$

$DACA(j)$	= the additional accumulation of capital which the j-sector must achieve
DACA	= the total accumulation of capital
K	= the total capital employed
$K(j)$	= the capital employed in the j-sector
$ACA(j)$	= the actual accumulation of capital in the j-sector

From the absolute value of these additional corrections the coefficients are formed in such a way that they are divided by the corresponding $P(j)$. In the account below for the remainder of the model they are considered only if they have a positive value:

$$DWS(j) = \begin{cases} DWSA(j)/P(j), & \text{if } DWSA(j) > 0 \\ 0, & \text{if } DWSA(j) \leq 0 \end{cases} \tag{24.22}$$

$$DAC(j) = \begin{cases} DACA(j)/P(j), & \text{if } DACA(j) > 0 \\ 0, & \text{if } DACA(j) \leq 0 \end{cases} \tag{24.23}$$

The rationale behind the requirements in Equations 24.22 and 24.23 lies in the assumption that only the below averagely successful sectors will make an attempt to change their income situation. The average and the above average sectors will try to preserve their situation and therefore additional corrections to wages and salaries and the accumulation of capital are not undertaken.

The variable LD, which like PP(0) is an exogenous variable in the model, takes the value 1 when we expect that the below-average sectors will have as their objective a relative income position which is the average level of the previous year. By changing this variable we can intensify, reduce or even eliminate (with LD = 0) the impact of intersector disharmony.

The results (Equations 24.22 and 24.23) are the direct coefficients of additional wages and salaries and of the additional accumulation of capital. We obtain total

The mechanism of the growth of costs and prices

coefficients by post-multiplying with the matrix multiplier the vector of direct coefficients:

$$[TDWS] = [DWS][I - AD]^{-1} \qquad (24.24)$$

[TDWS] = the horizontal vector of total coefficients for the correction of wages and salaries; the individual elements are TDWS(j)

$$[TDAC] = [DAC][I - AD]^{-1} \qquad (24.25)$$

[TDAC] = the horizontal vector of total coefficients for the correction of capital accumulation; the individual elements are TDAC(j)

(c) Additional burdens

The additional burdens to the economy appear where economic policy causes additional costs either to the individual sectors or to the entire economy.

The most characteristic additional charges to the economy are higher taxes. Higher taxes in the j-sector can be either caused by higher rates of taxes or the result of relatively higher wages and salaries, which in the Yugoslav system are the principal basis of taxation.

$$DCT(j) = [CT(j)/WS(j)] \times DWS(j) + SCT \times [WS(j) + DWS(j)] \qquad (24.26)$$

DCT(j) = the direct coefficient of increased taxes and contribution in a (j) sector
CT(j) = the direct coefficient of taxes and contributions in the j-sector
WS(j) = the direct coefficient of wages and salaries in the j-sector
DWS(j) = the direct coefficient of the increased wages and salaries in the j-sector
SCT = the increase in the tax rate (exogenous variable)

In Yugoslavia the depreciation of fixed assets is made according to the prescribed depreciation rate. If the prescribed depreciation rates are changed then the economy feels this change as an additional cost. A faster rate of depreciation leads to an increase in cost and vice versa.

$$DDF(j) = SDF \times FA(j)/P(j) \qquad (24.27)$$

DDF(j) = the direct coefficient for the change to depreciation in the j-sector
SDF = the change to the prescribed depreciation rate
FA(j) = the purchase value of fixed assets in the j-sector

The vectors for the total coefficionts of additional charges are as follows:

$$[TDCT] = [DCT][I - AD]^{-1} \qquad (24.28)$$

for taxes and contributions and

$$[TDDF] = [DDF][I - AD]^{-1} \qquad (24.29)$$

for depreciation of fixed assets

In a similar way we can also consider the additional costs in other fields, e.g. increased customs duties which increase the cost of imports, etc.

24.2.2 The starting point for the adjustment stages of the model

As the starting point for the adjustment stages the total coefficients and values added, with corrections included, are applied. The general expression for the element in the starting vectors is shown by $NY(i, j, 0)$ where

$$NY(i, j, 0) = TY(i, j) + TDY(i, j) \tag{24.30}$$

and where $TDY(i, j)$ represents the element in the vector for the multiplied value of coefficients for the correction to primary costs:

$$[\mathbf{TDY}](i) = [\mathbf{DY}](i)[\mathbf{I} - \mathbf{AD}]^{-1} \tag{24.31}$$

If we add up the corresponding elements of all horizontal vectors $[\mathbf{NY}](i, 0)$ we get the horizontal vector $[\mathbf{NP}](0)$ whose elements $NP(j, 0)$ represent the initial coefficients for the growth of producers' prices in each sector.

In general, this is

$$NP(j, 0) = \sum_{i=1}^{m} NY(i, j, 0) \tag{24.32}$$

where $NY(i, j, 0)$ is the initial coefficient of the i-primary costs (net imports $NIM(j, 0) - NEX(j, 0)$, depreciation $NDF(j, 0)$, other costs $NOC(j, 0)$, wages and salaries $NWS(j, 0)$, taxes and contributions $NCT(j, 0)$, net interest costs $NNI(j, 0)$, capital accumulation $NAC(j, 0)$ in the j-sector before input into the first stage of adjustment. The value $NP(j, 0)$ is, as a rule, slightly higher than 1 due to the inclusion of corrections to primary costs. The difference $NP(j, 0) - 1$ in each sector represents that initial impulse to inflation which comes about as a result of the differences in the profitability of sectors or through additional charges.

24.2.3 The procedure for adjustment stages

Let us consider that in the observed period (e.g. a year) there will be N adjustments to prices and incomes corresponding to the actual inflation rate. In the k stage the adjustment to the i-primary costs in the j-sector will be as represented in Equation 24.33.

$$NY(i, j, k) = NY(i, j, k-1)[1 + F(i) \times PP(k-1) + G(i)]^{1/N} \tag{24.33}$$

The following symbols mean:

k = subsequent number of stages
N = total number of stages
$F(i), G(i)$ = parameter of economic policy measures of the i-element in primary costs

The mechanism of the growth of costs and prices

$PP(k-1)$ = average annual growth rate of producers' prices in the period before the k-stage.

The effect of economic policy measures or their impact on the individual measures of primary costs is illustrated in the form of a linear function where the independent variable is the actual inflation rate (this is the increase in producers' prices) before the k-stage. The growth of prices is, of course, in the following stage dependent on the measures followed in the previous stage so that the independent variable in the k-stage depends, in fact, on the parameters in the $k-1$ stage.

The linear form of the measures, which depend on the actual inflation rate, mostly corresponds to the practice of planning in Yugoslavia. The nominal movements of aggregates contained in the economic policy documents are, as a rule, illustrated in two possible forms. Firstly as a percentage of the achieved (and in the period of planning still unknown) inflation rate – in this case the coefficient $F(i)$ is suitably reduced or increased by the envisaged percentage. Secondly in the form of percentage points above or below the average growth of prices; in this case the parameter $G(i)$ is applied, with either a positive or a negative value.

With Equation 24.33 the change in value to the vector $[\mathbf{NY}](i,k)$ is shown through the adjustment stages.

The average annual growth of prices $PP(k)$, which all the subjects of the economy are adapting to, is changed from stage to stage. $PP(0)$ enters into the first stage ($k=1$) as it is an exogenous variable which corresponds to the expected average rate of price growth. The change to this variable from one stage to the next can be performed according to Equations 24.34 or 24.35. Therefore:

$$PP(k) = \left\{ \left[\sum_{j=1}^{n} NP(j,k)w(j) \right] \bigg/ \left[\sum_{j=1}^{n} NP(j,0)w(j) \right] \right\}^{N/k} - 1 \quad (24.34)$$

or

$$PP(k) = \left[\sum_{j=1}^{n} NP(j,k)w(j) \right]^{N/k} - 1 \quad (24.35)$$

In both cases $NP(j,k)$ is the sum of the coefficients of the primary costs after the k-stage in the sector; $w(j)$ the weighted value of the sector; while n = the number of sectors.

$$NP(j,k) = \sum_{i=1}^{n} NY(i,j,k) \quad (24.36)$$

$$w(j) = P(j) \bigg/ \sum_{j=1}^{n} P(j) \quad (24.37)$$

The difference between Equations 24.34 and 24.35 lies in the importance attributed to the initial cost impulses before the first stage (i.e. the tendency of less profitable sectors to improve their profitability by increasing prices faster than the average rate and the influence of additional charges on costs and prices). In

Equation 24.34 these impacts are considered directly without any compound effect – the presumption is that these corrections to prices will be implemented only in the last stage, i.e. at the end of the observed period (year). In Equation 24.35 we presume that the initial cost impulses are already included in prices at the beginning of the year, i.e. during the first stage. In this case this impact on prices is, of course, compounded through the further stages.

Solution 24.34 represents the lower limit while solution 24.35 the upper limit for the solution of the model. Therefore the most representative result is the geometric median of the two, unless we are able to predict when further significant cost pressures are expected. In this case we attach a larger or smaller significance to one of the solutions to the model.

As to Equation 24.33, we should explain that it expresses the fundamental principle by which the coefficients of primary costs are increased in the iterative stages. With certain coefficients this can be expressed more precisely in the model.

Exchange rate policy has a decisive impact on import costs. Yugoslavia has a floating rate of exchange which may either over or under-compensate for the growth in domestic prices. This model envisages that there will be one devaluation in each stage while the surpassing of or the lag in devaluation behind the growth of prices are determined by the exogenous variables FIM, FEX, GIM and GEX. The movement of the coefficient for imports will therefore be:

$$\mathrm{NIM}(j,k) = \mathrm{NIM}(j,k-1)[1 + \mathrm{FIM} \times \mathrm{PP}(k-1) + \mathrm{GIM}]^{1/N} \qquad (24.38)$$

while those for exports will be as follows:

$$\mathrm{NEX}(j,k) = \mathrm{NEX}(j,k-1)[1 + \mathrm{FEX} \times \mathrm{PP}(k-1) + \mathrm{GEX}]^{1/N} \qquad (24.39)$$

where the relevant coefficient for net imports is $\mathrm{NIM}(j,k) - \mathrm{NEX}(j,k)$.

Where there is no divergence in the movements of import costs and export revenue then the equations FIM = FEX and GIM = GEX are valid. We must also consider that an aggressive exchange rate policy will reduce export prices. Should we assume that devaluations are considerably greater than the growth of domestic prices, we can set the exogenous variable as:

$$\mathrm{FIM} > \mathrm{FEX} \quad \text{and/or} \quad \mathrm{GIM} > \mathrm{GEX}.$$

The influence of a possible increase in prescribed depreciation rates on costs was included in Equations 24.27 and 24.29 by the exogenous variable SDF. It is also possible to influence depreciation costs by a policy of revaluing fixed assets. The revaluation of fixed assets can lag behind or surpass the average growth of prices, which is illustrated by the exogenous variables FDF and GDF. The accounting costs for depreciation are proportional to the official defined value of fixed assets.

$$\mathrm{NDF}(j,k) = \mathrm{NDF}(j,k-1)[1 + \mathrm{FDF} \times \mathrm{PP}(k-1) + \mathrm{GDF}]^{1/N} \qquad (24.40)$$

Other costs are defined by the variables FOC and GOC:

$$\mathrm{NOC}(j,k) = \mathrm{NOC}(j,k-1)[1 + \mathrm{FOC} \times \mathrm{PP}(k-1) + \mathrm{GOC}]^{1/N} \qquad (24.41)$$

The mechanism of the growth of costs and prices

The policy for the nominal growth of wages and salaries in conjunction with the growth of prices is expressed by the exogenous variables FWS and GWS:

$$\text{NWS}(j,k) = \text{NWS}(j, k-1)[1 + \text{FWS} \times \text{PP}(k-1) + \text{GWS}]^{1/N} \quad (24.42)$$

Possible changes to tax rates can be expressed by the exogenous variable SCT in Equations 27.26 and 27.28. As the movement of taxes and contributions is proportionate to wages and salaries the same exogenous variables as used for wages and salaries (FWS and GWS) are relevant for this coefficient:

$$\text{NCT}(j,k) = \text{NCT}(j, k-1)[1 + \text{FWS} \times \text{PP}(k-1) + \text{GWS}]^{1/N} \quad (24.43)$$

For a number of years the rate of interest in Yugoslavia has been lower than the rate of inflation. This negative rate of interest considerably stimulated borrowing so that today the Yugoslav economy as a whole shows a general overindebtedness. The consequence of moving towards positive rates of interest has recently caused net interest as a cost item to increase much more rapidly than the growth of prices.

In our model we are interested in the net interest costs that individual sectors will bear if economic policy attempts to maintain a definite real rate of interest by immediately adjusting nominal rates of interest to the actual rate of the growth of prices.

The movement of the coefficient of net interest costs is based on a regressive analysis of the interdependance of the index of growth of net interest costs, the real discount rate (National Bank of Slovenia, 1985b) and the index of growth of producers' prices. This regressive analysis is based on the four years' data, where the rate of interest was gradually becoming positive and gave the following result:

$$y = -0.44991 + 1.84969 x_{(1)} + 1.71455 x_{(2)} \quad (24.44)$$

y = the index of net interest costs
$x_{(1)}$ = the real discount rate
$x_{(2)}$ = the index for the growth of producers' prices

The regressive Equation 24.44 has a determining coefficient of 0.999; the level of significance is measured at 99% (Anon, 1983).

On the basis of the regressive Equation 24.44 the procedure for changing the coefficient of the net rate of interest in the j-sector due to the maintenance of a particular real rate of interest can be written as:

$$\text{NNI}(j,k) = \text{NNI}(j, k-1)\{[1 + \text{PP}(k-1)] \times 1.71455 + \\ + \text{RES} \times 1.84969 - 0.44991\}^{1/N} \quad (24.45)$$

The exogenous variable RES represents the real discount rate that the central bank will try to maintain by adjusting the nominal discount rate to the actual growth of prices.

For Yugoslav enterprises capital accumulation is seen as a kind of duty. A large number of economic policy measures are designed to ensure capital accumulation since the system allows workers to decide whether they will accumulate income or use it for wages and salaries. Due to these measures requiring a definite level of capital accumulation from enterprises, capital accumulation becomes a kind of cost which is passed on through higher prices. Usually capital accumulation is required to increase more rapidly than income. For this reason the nominal change in the coefficient of capital accumulation is dealt with in the model in the same way as other types of costs. With the exogenous variables FAC and GAC the required growth of capital accumulation as regards the general growth of prices is defined as:

$$\text{NAC}(j,k) = \text{NAC}(j,k-1)[1 + \text{FAC} \times \text{PP}(k-1) + \text{GAC}]^{1/N} \quad (24.46)$$

N number of stages is an exogenous variable in the model. The purpose of the iterative procedure lies in the fact that the participants in the economy and within the context of inflation increasingly need to check their plans, and thus they have to adjust their nominal parameters of business operation for inflation frequently during the year in order to maintain their real value. In Yugoslavia such adjustments usually happen on a quarterly basis: quarterly periodical accounts enable the comparison of each enterprise with the average for that activity and with the general economy; approximately every three months the banks also check the rate of interest. The procedure of four stages corresponds to these practices. Certain other events do not have quarterly cycles: for example, there are monthly devaluations of the exchange rate − 12 adjustments would be justified here. Considering different requirements and arguments it is possible to define a definite number of adjustments in the model as an exogenous variable. In current Yugoslav circumstances a procedure of four stages is the most appropriate as it best duplicates the actual course of adjustments in the economy.

24.3 INPUT–OUTPUT TABLE OF THE YUGOSLAV ECONOMY

It is understandable that in circumstances of such high inflation as are found in Yugoslavia the input–output coefficients are not stable (Dolenc and Pfajfar, 1976; Vukman, 1982). As the compiling of complete original input–output tables is a comparatively demanding statistical project it is impossible to expect that the current data for each year will be available. The Yugoslav Federal Institute of Planning has so far only published complete input–output tables for 1982 (Federal Planning Office, 1984). As our intention is to apply the presented model to the economic structure of 1984 on the basis of which we could establish the laws of cost inflation for 1985 it is necessary to reconstruct the second and third quadrant of the input–output tables for 1984 on the basis of the available sources of data.

The quickest estimate for the value of the elements of value added is obtained by means of data collected by the Social Accounting and Auditing Service (Social Accounting and Auditing Service, 1985). All enterprises in social ownership submit financial statements composed of a profit and loss account, a balance sheet and many other types of data to the Social Accounting and Auditing Service. The financial statements contain around 600 different types of data on material and

Input–output table of the Yugoslav economy

financial flows including all costs, incomes, funds, employment, etc. The sector figures which are the source of data for value added, capital and employment in this application of the model can be aggregated from these data.

In Yugoslavia the import and export of goods is recorded through customs declarations. These data are promptly collected and processed by the National Bank of Yugoslavia. The data can also be aggregated for individual sectors (National Bank of Slovenia, 1985a).

The above-mentioned sources allow tables to be quickly constructed for imports and for value added. But they are not suitable as a basis for the reconstruction of the entire table for domestic intermediate consumption. Since we possess the data for total production value, value added, and imports and exports, this can be used to modify the technological coefficients compiled in 1982. By applying a bi-proportional method based on the knowledge of the sum of coefficients for intermediate consumption in each sector we obtain:

$$AD(i, j) = A(i, j) \times C(i) \times SAD(j) \bigg/ \sum_{j=1}^{n} [A(i, j) \times C(i)] \qquad (24.47)$$

In Equation 24.47 $AD(i, j)$ is a modified technological coefficient for the i-sector 'giver' and j-sector (receiver). $A(i, j)$ represent the primary technological coefficients and their source is the published input–output table for 1982 (Federal Planning Office, 1984).

$C(i)$ are the indices for the nominal growth of domestic product value of the j-sector between the year in which the primary matrix is derived (in our case 1982) and the year in which our matrix is modified (in our case 1984). The source of these data can either be the Social Accounting and Auditing Service or the official statistical data (Federal Statistical Office, 1985 and 1986–7) for price movements and quantities sold on the domestic market.

$SAD(j)$ is the sum of current technological coefficients. This can be obtained by subtracting from 1 the direct coefficients of value added and imports (4).

n is the number of sectors, in our case 37. The sectors correspond to the official Yugoslav classification of activities. Less important activities are collected under the sectors entitled sundry activities – numbers 32 and 37.

As usual for the bi-proportional method, the procedure in Equation 24.47 also encompasses two different effects: the effect of rows and the effect of columns (RAS). The effect of rows is provided by the indices of growth for domestic product values, while the effect of columns is based on current data where the whole value of domestic intermediate consumption is an established part.

The method of Equation 24.47 can be applied either to the initial data for the absolute values (the flow matrices) or to the coefficients (the matrix of coefficients). In both cases identical results are obtained.

It should be noted that Table 24.1 comprises only the socially-owned sector of the economy. There are not sufficiently reliable data available for the private sector; furthermore the private sector in Yugoslavia is relatively unimportant (14% of the social product) and is only present in agriculture and in service industries.

A cost–push model of galloping inflation

Table 24.1 The direct coefficients for intermediate consumption, imports, and values added for 37 sectors of the economy in 1984

(j)	Sector	j =	1	2	3
1	Electricity generation		0.1749	0.0474	0.0285
2	Coal mining		0.1991		0.1893
3	Coal processing		0.0009		
4	Production of oil and natural gas		0.0058	0.0004	
5	Oil refining		0.0901	0.0203	0.0104
6	Extraction of iron ore				
7	Iron and steel production		0.0034	0.0304	0.0006
8	Extraction of nonferrous metal ores				
9	Production of nonferrous metals		0.0011	0.0004	
10	Processing of nonferrous metals		0.0003	0.0007	
11	Production of nonmetals		0.0004	0.0002	
12	Processing of nonmetals		0.0037	0.0002	
13	Metal processing activities		0.0153	0.0410	0.0024
14	Production of machines		0.0010	0.0046	0.0021
15	Vehicle production (all types)		0.0014	0.0017	
16	Shipbuilding				
17	Electrical machines and appliances		0.0267	0.0055	0.0002
18	Production of chemical products		0.0034	0.0008	0.0032
19	Processing of chemical products		0.0002	0.0317	0.0009
20	Production of building materials		0.0030	0.0024	
21	Timber processing		0.0005	0.0022	
22	Wood final products		0.0007	0.0002	
23	Production and processing of paper		0.0017	0.0007	0.0001
24	Spinning material and textiles		0.0001	0.0001	
25	Final textile products		0.0008	0.0014	0.0002
26	Hide and furs		0.0001	0.0002	
27	Leatherwear and leather accessories		0.0010	0.0015	0.0003
28	Rubber processing		0.0015	0.0079	0.0004
29	Production of foodstuff products				
30	Production of beverages				
31	Production of fodder				
32	Other manufacturing		0.0029	0.0047	0.0007
	Total manufacturing and mining		0.5403	0.2067	0.2393
33	Agriculture and fishing				
34	Building industry		0.0138	0.0096	0.0001
35	Transport and communications		0.0309	0.0148	0.1475
36	Distributive trades		0.0329	0.0157	0.0283
37	Sundry economic activities		0.0538	0.0969	0.0200
	Total economy SAD(j)		0.6718	0.3437	0.4352
	Imports IM(j)		0.0225	0.0380	0.5606
	Exports EX(j)		0.0055	0.0067	0.1619
	Net imports IM(j) − EX(j)		0.0170	0.0313	0.3987
	Other costs OC(j)		0.0287	0.0566	0.0184
	Depreciation of fixed assets DF(j)		0.1226	0.0991	0.0362
	Wages and salaries WS(j)		0.0488	0.1933	0.0212
	Taxes and contributions CT(j)		0.0402	0.1390	0.0198
	Net interest NI(j)		0.0586	0.0332	0.0463
	Capital accumulation AC(j)		0.0123	0.1039	0.0243
	Production value P(j)		520 125	119 662	60.049

(Contd.)

Table 24.1 (Contd.)

4	5	6	7	8	9
0.0065	0.0053	0.0650	0.0421	0.0730	0.1200
		0.0009	0.0021	0.0007	0.0041
		0.0210	0.0929	0.0005	0.0254
0.0048	0.0544		0.0032	0.0001	
0.0034	0.0797	0.0894	0.0177	0.0289	0.0294
			0.0227		
0.0014	0.0005		0.4417	0.0192	0.0006
			0.0025	0.2806	0.3305
		0.0005	0.0039	0.0138	0.5493
	0.0001		0.0005	0.0001	0.0011
0.0001			0.0012		0.0007
0.0007	0.0001	0.0029	0.0070	0.0006	0.0191
0.0004	0.0075	0.0193	0.0200	0.0081	0.0057
0.0011	0.0002	0.0045	0.0064	0.0076	0.0017
0.0009	0.0003		0.0036	0.0012	0.0017
0.0007	0.0006	0.0047	0.0006	0.0037	0.0005
0.0001	0.0070		0.0070	0.0093	0.0061
0.0019	0.0036	0.0765	0.0019	0.0167	0.0023
0.0002	0.0007	0.0064	0.0012	0.0057	0.0029
		0.0002	0.0002	0.0009	0.0001
	0.0001	0.0004	0.0004	0.0002	0.0006
0.0001	0.0003	0.0014	0.0003	0.0002	0.0003
0.0002				0.0001	0.0002
0.0004	0.0002	0.0010	0.0006	0.0009	0.0003
			0.0001	0.0002	
0.0003	0.0002	0.0057	0.0003	0.0009	0.0004
		0.0122	0.0004	0.0097	0.0002
				0.0001	
0.0008	0.0006	0.0219	0.0129	0.0034	0.0076
0.0240	0.1615	0.3337	0.6935	0.4864	1.1111
0.0006	0.0002	0.0047	0.0019	0.0032	0.0008
0.0041	0.0310	0.0327	0.0370	0.0377	0.0568
0.0023	0.0141	0.0369	0.0255	0.0182	0.0252
0.0363	0.0098	0.2560	0.0910	0.0472	0.0346
0.0675	0.2167	0.6639	0.8489	0.5927	1.2286
0.2990	0.7659	0.0252	0.1420	0.1167	0.1938
0.0003	0.0536		0.1418	0.1753	0.7261
0.2987	0.7123	0.0252	0.0002	−0.0586	−0.5323
0.0450	0.0135	0.0309	0.0224	0.0383	0.0239
0.0821	0.0109	0.0923	0.0437	0.1257	0.0749
0.0252	0.0089	0.0814	0.0341	0.1253	0.0482
0.2497	0.0124	0.0602	0.0268	0.0941	0.0477
−0.0120	0.0282	0.0314	0.0314	0.0651	0.0848
0.2437	−0.0029	0.0148	−0.0076	0.0173	0.0242
74 487	484 425	19 354	542 180	75 790	120 009

(Contd.)

524 A cost–push model of galloping inflation

Table 24.1 (Contd.)

(j)	Sector j =	10	11	12
1	Electricity generation	0.0150	0.0721	0.0470
2	Coal mining	0.0005		0.0050
3	Coal processing	0.0009	0.0001	0.0001
4	Production of oil and natural gas	0.0005	0.0314	0.0230
5	Oil refining	0.0082	0.0518	0.0716
6	Extraction of iron ore			
7	Iron and steel production	0.0008	0.0218	0.0087
8	Extraction of nonferrous metal ores			0.0110
9	Production of nonferrous metals	0.6003	0.0003	0.0097
10	Processing of nonferrous metals	0.0281	0.0003	0.0045
11	Production of nonmetals	0.0002	0.0522	0.0860
12	Processing of nonmetals	0.0010	0.0114	0.0840
13	Metal processing activities	0.0120	0.0397	0.0254
14	Production of machines		0.0076	0.0029
15	Vehicle production (all types)	0.0007	0.0083	0.0036
16	Shipbuilding			0.0001
17	Electrical machines and appliances	0.0023	0.0081	0.0058
18	Production of chemical products	0.0016	0.0480	0.0226
19	Processing of chemical products	0.0043	0.0278	0.0268
20	Production of building materials	0.0017	0.0076	0.0067
21	Timber processing	0.0008	0.0017	0.0017
22	Wood final products	0.0035	0.0015	0.0096
23	Production and processing of paper	0.0028	0.0106	0.0191
24	Spinning material and textiles	0.0015		0.0037
25	Final textile products	0.0006	0.0062	0.0033
26	Hide and furs	0.0001	0.0011	0.0002
27	Leatherwear and leather accessories	0.0007	0.0028	0.0020
28	Rubber processing	0.0014	0.0159	0.0032
29	Production of foodstuff products			
30	Production of beverages			
31	Production of fodder			
32	Other manufacturing	0.0209	0.0117	0.0131
	Total manufacturing and mining	0.7102	0.4401	0.5004
33	Agriculture and fishing			
34	Building industry	0.0007	0.0024	0.0061
35	Transport and communications	0.0262	0.0741	0.0844
36	Distributive trades	0.0254	0.0352	0.0504
37	Sundry economic activities	0.0192	0.0629	0.0857
	Total economy SAD(j)	0.7817	0.6148	0.7269
	Imports IM(j)	0.2413	0.0280	0.1577
	Exports EX(j)	0.3617	0.0394	0.2425
	Net imports IM(j) − EX(j)	−0.1204	−0.0114	−0.0848
	Other costs OC(j)	0.0360	0.0397	0.0351
	Depreciation of fixed assets DF(j)	0.0564	0.0695	0.0605
	Wages and salaries WS(j)	0.0766	0.1473	0.1360
	Taxes and contributions CT(j)	0.0729	0.1000	0.0899
	Net interest NI(j)	0.0340	0.0297	0.0460
	Capital accumulation AC(j)	0.0628	0.0105	−0.0097
	Production value P(j)	70 867	22 898	96 765

(Contd.)

Input–output table of the Yugoslav economy

Table 24.1 (Contd.)

13	14	15	16	17	18
0.0231	0.0190	0.0115	0.0257	0.0215	0.0435
0.0008	0.0012	0.0006		0.0007	0.0018
0.0147	0.0031	0.0015	0.0064	0.0010	0.0078
0.0018	0.0012	0.0011	0.0022	0.0002	0.0077
0.0106	0.0122	0.0089	0.0065	0.0089	0.0414
0.1838	0.1608	0.0932	0.0465	0.0510	0.0006
0.0014					0.0061
0.0352	0.0131	0.0142	0.0986	0.1263	0.0126
0.0337	0.0216	0.0189	0.0271	0.0408	0.0043
0.0029	0.0001	0.0003	0.0003	0.0002	0.0001
0.0053	0.0059	0.0047	0.0003	0.0144	0.0006
0.1818	0.1499	0.1359	0.0967	0.0750	0.0076
0.0110	0.0740	0.0036	0.0255	0.0120	0.0019
0.0164	0.0243	0.2097	0.0164	0.0015	0.0014
0.0002	0.0036	0.0006	0.1356	0.0010	
0.0098	0.0443	0.0182	0.0826	0.1987	0.0024
0.0077	0.0070	0.0040	0.0340	0.0261	0.2659
0.0150	0.0068	0.0135	0.0225	0.0089	0.0287
0.0007	0.0017	0.0007	0.0019	0.0013	0.0021
0.0033	0.0021	0.0047	0.0078	0.0011	0.0001
0.0055	0.0020	0.0031	0.0060	0.0092	0.0014
0.0032	0.0024	0.0023	0.0013	0.0063	0.0169
0.0022	0.0003	0.0038	0.0022	0.0021	0.0055
0.0010	0.0006	0.0034	0.0008	0.0006	0.0018
	0.0004	0.0011		0.0003	0.0004
0.0010	0.0029	0.0007	0.0028	0.0014	0.0011
0.0024	0.0105	0.0397	0.0002	0.0063	0.0042
0.0002					0.0003
0.0026	0.0066	0.0026	0.0103	0.0068	0.0051
0.5774	0.5776	0.6026	0.6602	0.6235	0.4734
					0.0018
0.0140	0.0186	0.0044	0.0059	0.0058	0.0035
0.0361	0.0349	0.0244	0.0567	0.0321	0.0398
0.0603	0.0513	0.0356	0.0878	0.0361	0.0362
0.0104	0.0530	0.0413	0.0964	0.0427	0.0622
0.6982	0.7354	0.7083	0.9070	0.7403	0.6168
0.0968	0.1094	0.2365	0.6654	0.2517	0.3421
0.2038	0.3345	0.2611	1.4562	0.3630	0.1977
−0.1070	−0.2251	−0.0247	−0.7909	−0.1113	0.1445
0.0440	0.0603	0.0283	0.1242	0.0518	0.0264
0.0447	0.0408	0.0436	0.0374	0.0366	0.0741
0.1421	0.1581	0.1042	0.2670	0.1200	0.0460
0.1032	0.1255	0.0793	0.2681	0.0893	0.0379
0.0437	0.0615	0.0588	−0.0318	0.0738	0.0837
0.0311	0.0435	0.0022	0.2190	−0.0004	−0.0295
549 756	287 447	365 411	38 504	394 932	344 366

(Contd.)

526 A cost–push model of galloping inflation

Table 24.1 (Contd.)

(j)	Sector	j = 19	20	21
1	Electricity generation	0.0129	0.0564	0.0348
2	Coal mining	0.0006	0.0093	0.0010
3	Coal processing	0.0006	0.0018	0.0010
4	Production of oil and natural gas	0.0017	0.0134	0.0015
5	Oil refining	0.0241	0.1281	0.0284
6	Extraction of iron ore			
7	Iron and steel production	0.0019	0.0281	0.0026
8	Extraction of nonferrous metal ores	0.0019	0.0010	
9	Production of nonferrous metals	0.0068	0.0015	0.0001
10	Processing of nonferrous metals	0.0006	0.0006	0.0012
11	Production of nonmetals	0.0014	0.0153	0.0001
12	Processing of nonmetals	0.0073	0.0139	0.0003
13	Metal processing activities	0.0287	0.0378	0.0201
14	Production of machines	0.0007	0.0037	0.0026
15	Vehicle production (all types)	0.0002	0.0021	0.0004
16	Shipbuilding			
17	Electrical machines and appliances	0.0032	0.0043	0.0007
18	Production of chemical products	0.2056	0.0079	0.0159
19	Processing of chemical products	0.1617	0.0082	0.0180
20	Production of building materials	0.0015	0.0766	0.0030
21	Timber processing	0.0008	0.0027	0.0728
22	Wood final products	0.0018	0.0015	0.0256
23	Production and processing of paper	0.0312	0.0109	0.0075
24	Spinning material and textiles	0.0059	0.0011	0.0001
25	Final textile products	0.0034	0.0019	0.0005
26	Hide and furs	0.0004		
27	Leatherwear and leather accessories	0.0003	0.0010	0.0004
28	Rubber processing	0.0007	0.0067	0.0008
29	Production of foodstuff products	0.0061		
30	Production of beverages	0.0002		
31	Production of fodder			
32	Other manufacturing	0.0120	0.0371	0.0036
	Total manufacturing and mining	0.5240	0.4729	0.2431
33	Agriculture and fishing	0.0047		
34	Building industry	0.0030	0.0178	0.0037
35	Transport and communications	0.0371	0.1019	0.1083
36	Distributive trades	0.0400	0.0548	0.0221
37	Sundry economic activities	0.0449	0.0516	0.3914
	Total economy SAD(j)	0.6537	0.6990	0.7686
	Imports IM(j)	0.2509	0.0621	0.0718
	Exports EX(j)	0.2099	0.0362	0.2488
	Net imports IM(j) − EX(j)	0.0410	0.0260	−0.1770
	Other costs OC(j)	0.0395	0.0315	0.0459
	Depreciation of fixed assets DF(j)	0.0364	0.0612	0.0608
	Wages and salaries WS(j)	0.0746	0.1046	0.1339
	Taxes and contributions CT(j)	0.0634	0.0653	0.0889
	Net interest NI(j)	0.0453	0.0454	0.0502
	Capital accumulation AC(j)	0.0462	−0.0331	0.0287
	Production value P(j)	332 847	165 213	103 048

(Contd.)

Input–output table of the Yugoslav economy

Table 24.1 (Contd.)

22	23	24	25	26	27
0.0294	0.0742	0.0293	0.0112	0.0119	0.0074
0.0012	0.0056	0.0019	0.0007	0.0031	0.0012
0.0003				0.0006	0.0002
0.0009	0.0040	0.0010	0.0002		0.0003
0.0131	0.0171	0.0106	0.0077	0.0186	0.0043
0.0132	0.0007	0.0021	0.0003	0.0004	0.0056
	0.0005				
0.0001	0.0030	0.0002			
0.0026	0.0017	0.0002	0.0005	0.0001	0.0002
	0.0010			0.0014	
0.0110	0.0011	0.0001	0.0009	0.0012	0.0002
0.0938	0.0115	0.0072	0.0060	0.0051	0.0141
0.0005		0.0011	0.0004	0.0009	0.0013
0.0073	0.0002	0.0001	0.0001	0.0002	0.0006
0.0042	0.0070	0.0026	0.0013	0.0016	0.0009
0.0087	0.0556	0.0525	0.0069	0.0598	
0.0522	0.0367	0.0132	0.0140	0.0371	0.0468
0.0098	0.0017	0.0007	0.0004	0.0008	0.0002
0.2042	0.0005	0.0004	0.0004	0.0004	0.0005
0.0954	0.0039	0.0019	0.0024	0.0014	0.0092
0.0127	0.2380	0.0062	0.0052	0.0008	0.0036
0.0658	0.0012	0.2602	0.3941	0.0124	0.0284
0.0135	0.0012	0.0033	0.1216	0.0005	0.0040
0.0028	0.0001	0.0002	0.0063	0.0724	0.5675
0.0003	0.0006	0.0003	0.0011	0.0083	0.0946
0.0018	0.0012	0.0005	0.0005		0.0181
	0.0040	0.0012		0.3280	
0.0117	0.0180	0.0040	0.0098	0.0076	0.0129
0.6565	0.4903	0.4013	0.5922	0.5747	0.8221
0.0002	0.0043	0.0084	0.0041		
0.0020	0.0024	0.0014	0.0017	0.0012	0.0022
0.0939	0.0513	0.0236	0.0349	0.0292	0.0328
0.0305	0.0252	0.0176	0.0148	0.0486	0.0304
0.0564	0.1081	0.0290	0.0231	0.0353	0.0217
0.8394	0.6816	0.4813	0.6708	0.6891	0.9092
0.0787	0.1918	0.3122	0.1792	0.2018	0.1865
0.3031	0.1514	0.1771	0.3619	0.1189	0.5168
−0.2244	0.0404	0.1350	−0.1826	0.0829	−0.3303
0.0423	0.0282	0.0380	0.0435	0.0263	0.0397
0.0434	0.0661	0.0453	0.0316	0.0173	0.0169
0.1585	0.0665	0.1279	0.1902	0.0591	0.1566
0.0964	0.0511	0.0883	0.1243	0.0499	0.1062
0.0625	0.0546	0.0393	0.0491	0.0142	0.0479
−0.0181	0.0116	0.0449	0.0732	0.0612	0.0537
196 909	176 435	263 969	304 437	85 675	156 113

(Contd.)

528 A cost–push model of galloping inflation

Table 24.1 (*Contd.*)

(j)	Sector	j =	28	29	30
1	Electricity generation		0.0274	0.0101	0.0184
2	Coal mining		0.0008	0.0014	0.0009
3	Coal processing		0.0001	0.0012	
4	Production of oil and natural gas		0.0039	0.0003	0.0023
5	Oil refining		0.0158	0.0086	0.0196
6	Extraction of iron ore				
7	Iron and steel production		0.0248		0.0014
8	Extraction of nonferrous metal ores				
9	Production of nonferrous metals		0.0024	0.0001	0.0013
10	Processing of nonferrous metals			0.0013	
11	Production of nonmetals		0.0019	0.0013	0.0006
12	Processing of nonmetals		0.0007	0.0025	0.0308
13	Metal processing activities		0.0223	0.0095	0.0210
14	Production of machines		0.0041	0.0025	0.0019
15	Vehicle production (all types)		0.0003	0.0007	0.0011
16	Shipbuilding				
17	Electrical machines and appliances		0.0033	0.0016	0.0011
18	Production of chemical products		0.0642	0.0023	0.0035
19	Processing of chemical products		0.0425	0.0064	0.0202
20	Production of building materials		0.0011	0.0008	0.0012
21	Timber processing		0.0016	0.0002	0.0005
22	Wood final products		0.0014	0.0012	0.0058
23	Production and processing of paper		0.0032	0.0124	0.0078
24	Spinning material and textiles		0.0256	0.0004	0.0001
25	Final textile products		0.0137	0.0011	0.0007
26	Hide and furs		0.0036		0.0002
27	Leatherwear and leather accessories		0.0017	0.0003	0.0006
28	Rubber processing		0.2438	0.0024	0.0042
29	Production of foodstuff products		0.0020	0.2513	0.1537
30	Production of beverages			0.0004	0.1560
31	Production of fodder				
32	Other manufacturing		0.0051	0.0057	0.0147
	Total manufacturing and mining		0.5172	0.3258	0.4692
33	Agriculture and fishing		0.0037	0.4479	0.2479
34	Building industry		0.0032	0.0014	0.0032
35	Transport and communications		0.0352	0.0249	0.0265
36	Distributive trades		0.0338	0.0165	0.0301
37	Sundry economic activities		0.1166	0.0200	0.0325
	Total economy SAD(j)		0.7097	0.8366	0.8094
	Imports IM(j)		0.2518	0.0566	0.0343
	Exports EX(j)		0.2232	0.0530	0.0745
	Net imports IM(j) − EX(j)		0.0287	0.0036	−0.0402
	Other costs OC(j)		0.0233	0.0166	0.0302
	Depreciation of fixed assets DF(j)		0.0451	0.0254	0.0418
	Wages and salaries WS(j)		0.0795	0.0528	0.0752
	Taxes and contributions CT(j)		0.0549	0.0352	0.0553
	Net interest NI(j)		0.0363	0.0479	0.0707
	Capital accumulation AC(j)		0.0224	−0.0182	−0.0423
	Production value P(j)		109 752	948 490	128 553

(*Contd.*)

Input–output table of the Yugoslav economy

Table 24.1 (Contd.)

31	32	Total manufacturing	33	34	35
0.0063	0.0147	0.0357	0.0123	0.0091	0.0196
0.0004	0.0005	0.0167	0.0004	0.0005	0.0030
0.0002	0.0002	0.0092	0.0001	0.0002	0.0007
0.0004	0.0014	0.0060	0.0009	0.0013	
0.0040	0.0154	0.0275	0.0460	0.0212	0.1386
		0.0016			
0.0002	0.0036	0.0621	0.0003	0.0464	0.0070
		0.0090			
	0.0263	0.0279		0.0023	0.0004
0.0001	0.0035	0.0076		0.0060	0.0010
0.0023	0.0001	0.0023	0.0005	0.0004	0.0003
	0.0006	0.0057	0.0005	0.0163	0.0034
0.0010	0.0112	0.0432	0.0084	0.0441	0.0167
	0.0028	0.0062	0.0012	0.0024	0.0025
0.0001	0.0019	0.0136	0.0027	0.0034	0.0635
		0.0009		0.0006	0.0160
0.0002	0.0024	0.0176	0.0007	0.0218	0.0050
0.0231	0.0059	0.0324	0.0619	0.0030	0.0047
0.0023	0.0226	0.0205	0.0036	0.0090	0.0052
	0.0106	0.0036	0.0011	0.1024	0.0019
	0.0020	0.0074	0.0002	0.0117	0.0021
	0.0004	0.0052	0.0016	0.0361	0.0008
0.0084	0.1280	0.0158	0.0009	0.0010	0.0014
0.0006	0.0025	0.0293	0.0007	0.0006	0.0007
0.0002	0.0004	0.0067	0.0009	0.0017	0.0076
	0.0011	0.0132	0.0001	0.0001	0.0001
0.0001	0.0003	0.0028	0.0002	0.0009	0.0016
0.0015	0.0034	0.0083	0.0027	0.0035	0.0329
0.0012	0.0010	0.0387	0.0084		0.0033
0.0013		0.0028	0.0001		0.0007
0.0465		0.0008	0.1041		
0.0029	0.2026	0.0130	0.0018	0.0267	0.0112
0.1035	0.4653	0.4932	0.2623	0.3728	0.3518
0.6630	0.0759	0.0747	0.4387		0.0005
0.0010	0.0037	0.0051	0.0017	0.1059	0.0103
0.0284	0.0258	0.0379	0.0179	0.0720	0.0987
0.0400	0.0227	0.0309	0.0326	0.0804	0.0701
0.0164	0.1267	0.0506	0.0335	0.0480	0.1124
0.8524	0.7201	0.6924	0.7867	0.6792	0.6437
0.0592	0.0587	0.1920	0.0156	0.0159	0.0575
0.0179	0.1208	0.1849	0.0143	0.0458	0.2784
0.0413	−0.0621	0.0071	0.0013	−0.0299	−0.2209
0.0103	0.0338	0.0324	0.0367	0.0491	0.0623
0.0106	0.0377	0.0476	0.0237	0.0296	0.1244
0.0190	0.1231	0.0888	0.0605	0.1380	0.1793
0.0178	0.0882	0.0677	0.0467	0.0903	0.1207
0.0331	0.0367	0.0486	0.0358	0.0044	0.0361
0.0156	0.0225	0.0139	0.0086	0.0392	0.0543
120 839	214 180	7 493 485	1 120 787	1 025 853	699 341

(Contd.)

Table 24.1 (Contd.)

(j)	Sector	j =	36	37	Total economy
1	Electricity generation		0.0872	0.0204	0.0514
2	Coal mining		0.0045	0.0118	0.0093
3	Coal processing		0.0018	0.0018	0.0045
4	Production of oil and natural gas		0.0029	0.0080	0.0041
5	Oil refining		0.0477	0.0107	0.0390
6	Extraction of iron ore			0.0022	0.0008
7	Iron and steel production		0.0054	0.0504	0.0327
8	Extraction of nonferrous metal ores			0.0050	0.0038
9	Production of nonferrous metals		0.0006	0.0069	0.0118
10	Processing of nonferrous metals			0.0025	0.0035
11	Production of nonmetals			0.0012	0.0011
12	Processing of nonmetals		0.0033	0.0069	0.0050
13	Metal processing activities		0.0284	0.0062	0.0319
14	Production of machines		0.0037	0.0113	0.0050
15	Vehicle production (all types)		0.0102	0.0161	0.0131
16	Shipbuilding			0.0029	0.0012
17	Electrical machines and appliances		0.0120	0.0133	0.0139
18	Production of chemical products		0.0089	0.0309	0.0222
19	Processing of chemical products		0.0122	0.0153	0.0147
20	Production of building materials		0.0090	0.0093	0.0112
21	Timber processing		0.0023	0.0328	0.0068
22	Wood final products		0.0097	0.0088	0.0085
23	Production and processing of paper		0.0262	0.0178	0.0178
24	Spinning material and textiles		0.0041	0.0095	0.0138
25	Final textile products		0.0067	0.0067	0.0061
26	Hide and furs			0.0026	0.0054
27	Leatherwear and leather accessories		0.0027	0.0037	0.0025
28	Rubber processing		0.0139	0.0128	0.0111
29	Production of foodstuff products		0.0870	0.0211	0.0512
30	Production of beverages		0.0043	0.0047	0.0031
31	Production of fodder			0.0019	0.0065
32	Other manufacturing		0.0377	0.0263	0.0236
	Total manufacturing and mining		0.4325	0.3816	0.4366
33	Agriculture and fishing		0.0918	0.0469	0.0941
34	Building industry		0.0503	0.0569	0.0318
35	Transport and communications		0.1410	0.0686	0.0831
36	Distributive trades		0.0628	0.0373	0.0480
37	Sundry economic activities		0.1732	0.0848	0.1020
	Total economy SAD(j)		0.9516	0.6761	0.7955
	Imports IM(j)		0.0104	0.0173	0.0843
	Exports EX(j)		0.0395	0.1089	0.1089
	Net imports IM(j) − EX(j)		−0.0291	−0.0917	−0.0245
	Other costs OC(j)		0.0095	0.0485	0.0268
	Depreciation of fixed assets DF(j)		0.0031	0.0413	0.0302
	Wages and salaries WS(j)		0.0225	0.1526	0.0716
	Taxes and contributions CT(j)		0.0182	0.1075	0.0531
	Net interest NI(j)		0.0085	0.0071	0.0265
	Capital accumulation AC(j)		0.0158	0.0587	0.0205
	Production value $P(j)$		7 468 527	1 325 861	19 133 854

Results

The classification comprises all officially classified sectors in the economy. For this reason education, the health service, and government administration are not included under intersector relations. The flow of resources from the economy into the above-mentioned activities and into the private sector, which cannot be classified either by interest costs or by taxes, contributions, wages and salaries comes under other costs (OC).

The coefficients in Table 24.1 are calculated according to the net principle, the value 1 being the output for the domestic markets. Therefore it is possible for the sum of coefficients of intermediate domestic consumption to be greater than 1 ($SAD(j) > 1$) if the value of exports is greater than the sum of imports and value added.

Besides the direct coefficients, Table 24.1 also contains the absolute value of output for the domestic market for each sector in millions of dinars.

24.4 RESULTS

The input–output table for the Yugoslav economy in 1984 enables us, with the model, to illustrate cost pressures which are of key importance in price formation for the following year. The value of exogenous variables will be set to correspond to the economic policy that Yugoslavia sought to pursue in 1985 (Table 24.2).

Table 24.2 Exogenous variables for 1985

Variable	Value for 1985	Refers to
PP(0)	0.5	The planned rate of inflation
LD	0.9	The desired level of wages, salaries and capital accumulation
SCT	0.0	The change to the tax-rate
SDF	0.0	The change to the prescribed rate of depreciation
FIM	1.0	Imports
GIM	0.05	Imports
FEX	1.0	Exports
GEX	0.05	Exports
FDF	1.0	Depreciation of fixed assets
GDF	0.0	Depreciation of fixed assets
FOC	1.0	Other costs
GOC	0.0	Other costs
FWS	1.0	Wages and salaries
GWS	−0.05	Wages and salaries
RES	0.02	The net rate of interest
FAC	1.0	Capital accumulation
GAC	0.05	Capital accumulation
N	4	Number of stages

The inflation rate PP(0) of 50% will be taken as the planned rate. Such a rate of inflation was also envisaged in the planning documents for 1985.

Imbalances between sectors due to differences in wages and salaries and in the accumulation of capital will be dealt with by the variable LD = 0.9. The effect of this exogenous variable can be seen in Equations 24.20 and 24.21 and is the criterion by which we shall predict the sectors that should improve their position in the primary distribution of national income by using accelerated price growth. This includes all the sectors where the wages and salaries per employee or capital accumulation per capital employed are below 90% of the average in the Yugoslav

Table 24.3 Direct coefficients for additional increases to wages and salaries, taxes and contributions as well as to capital accumulation

(j)	Sector	Wages and salaries	Contributions and taxes	Capital accumulation
1	Electricity generation	—	—	0.0554
5	Oil refining	—	—	0.0074
6	Extraction of iron ore	0.0035	0.0026	0.0110
7	Iron and steel production	—	—	0.0258
8	Extraction of nonferrous metal ores	—	—	0.0042
9	Production of nonferrous metals	—	—	0.0043
11	Production of nonmetals	—	—	0.0252
12	Processing of nonmetals	—	—	0.0367
15	Vehicle production (all types)	—	—	0.0126
17	Electrical machines and appliances	—	—	0.0173
18	Production of chemical products	—	—	0.0444
20	Production of building materials	—	—	0.0633
22	Wood final products	—	—	0.0410
23	Production and processing of paper	—	—	0.0110
29	Production of foodstuff products	—	—	0.0325
30	Production of beverages	—	—	0.0686
32	Other manufacturing	—	—	0.0019
	Total manufacturing and mining	0.0000	0.0000	0.0181
33	Agriculture and fishing	—	—	0.0210
34	Building industry	0.0030	0.0019	—
35	Transport and communications	—	—	0.0009
	Total economy	0.0002	0.0001	0.0083

Note: The value of the coefficients for the activities not included in Table 24.3 is given as 0.

economy. It is envisaged that these sectors will surpass average price growth to enable wages and salaries and capital accumulation to reach the level of 90% of the average in the economy (Table 24.3).

Among the 37 sectors discussed above there were 19 in 1984 whose profitability did not allow sufficient capital accumulation as to reach 90% of the average. In one case (sector 34 – the building industry) the capital accumulation was satisfactory but wages and salaries were below the desired level.

The transformation of the direct coefficients from the input–output table into the coefficients illustrating the starting points for the adjustment stages is presented in Table 24.4.

Table 24.4 only presents coefficients for the entire economy. As with the exogenous variables we did not envisage additional costs burdens from government policy (SCT = 0, SDF = 0). The coefficients at the starting point of the adjustment stages are greater than one by exactly that amount which is the multiplied value of the coefficients of additional costs in Table 24.3. These price corrections which would enable the worst sectors to approach the level of 90% directly stimulate a price growth of 4.06%.

In the procedure of adjustment (in Table 24.5) we have – by using exogenous variables – predicted that certain costs are increasing faster and other slower than the general growth of prices. A faster increase is anticipated for those costs which are connected with production factors whose price, in Yugoslavia, is unrealistically low. For this reason we have predicted the growth of net interests costs that result from the maintenance of the nominal interest rate 2% points higher than average price growth (in Equation 24.45 RES = 0.02). We have also anticipated that the official rate of exchange will be devalued faster than the general rate of price-growth. By GIM = GEX = 0.05 it was determined that import costs as well as income from exports were increasing 5 percentage points faster. A faster growth is also envisaged for capital accumulation, since in the 1985 plan of economic policy, measures for the strengthening of the reproduction capacity of the economy were predicted.

Table 24.4 The transition of direct coefficients to obtain the coefficients at the starting point of the adjustment stages

	$Y(i)$	$TY(i)$	$DY(i)$	$TDY(i)$	$NY(i,0)$
Imports (IM)	0.0843	0.4500	—	—	0.4500
Exports (EX)	−0.1089	−0.4888	—	—	−0.4888
Depreciation (DF)	0.0302	0.1684	—	—	0.1684
Other costs (OC)	0.0268	0.1201	—	—	0.1201
Wages and salaries (WS)	0.0716	0.3090	0.0002	0.0003	0.3093
Contributions and taxes (CT)	0.0531	0.2311	0.0001	0.0002	0.2313
Net interest costs (NI)	0.0265	0.1276	—	—	0.1276
Capital accumulation (AC)	0.0205	0.0826	0.0083	0.0401	0.1227
Total	0.2041	1.0000	0.0086	0.0406	1.0406

Table 24.5 The adjustment of coefficients in four stages

	After the 1st phase NY(i,1)		After the 2nd phase NY(i,2)		After the 3rd phase NY(i,3)		After the 4th phase NY(i,4)	
	Min.	Max.	Min.	Max.	Min.	Max.	Min.	Max.
IM	0.5021	0.5021	0.5656	0.5880	0.6406	0.6928	0.7284	0.8198
EX	−0.5431	−0.5431	−0.6091	−0.6324	−0.6867	−0.7412	−0.7777	−0.8725
DF	0.1864	0.1864	0.2080	0.2167	0.2340	0.2535	0.2646	0.2986
OC	0.1326	0.1326	0.1486	0.1544	0.1670	0.1808	0.1885	0.2128
WS	0.3392	0.3392	0.3761	0.3918	0.4196	0.4560	0.4700	0.5328
CT	0.2539	0.2539	0.2813	0.2931	0.3138	0.3408	0.3513	0.3986
NI	0.1549	0.1549	0.1897	0.1989	0.2344	0.2569	0.2904	0.3349
AC	0.1371	0.1371	0.1543	0.1604	0.1746	0.1889	0.1988	0.2238
NP	1.1628	1.1628	1.3149	1.3708	1.4970	1.6293	1.7144	1.9497

IM = imports; EX = exports; DF = depreciation; OC = other costs; WS = wages and salaries; CT = contributions and taxes; NI = net interest costs; AC = capital accumulation; NP = total.
Note: The lower values of the coefficients are based on the procedure in Equation 24.34, the upper values on the procedure in Equation 24.35.

On the other hand wages and salaries together with taxes and contributions should increase more slowly than the general rate of price growth (GWS = − 0.05). A bigger drop should not be required as the real purchasing power of the population in Yugoslavia has recently been decreased and is now among the lowest in Europe.

The adjustment of incomes as well as prices in the economy is within the proportions determined by the exogenous variables. The movement of coefficients shows these proportions. Table 24.5 illustrates the changing of coefficients in each stage of adjustment.

The coefficients obtained after the fourth stage show the final result, i.e. the model forecast for the annual rate of price growth based on the mechanism of cost–push inflation. For 1985 the general rate of price growth in Yugoslavia was predicted at between 71% and 91%. The individual sectors increase prices as shown in Table 24.6.

As well as lower and higher variants of the forecast growth of prices Table 24.6 also gives the geometric median of both results. The most likely rate of price growth forecast by the model is, along with the already mentioned exogenous variables, the median result: 82.8%. By comparison we may state that the official statistical data (Federal Statistical Office, 1985 and 1986b) for the growth of prices in Yugoslavia in 1985 approach these calculations. In 1985 the production prices of industrial products were, in comparison with 1984, 81.5% higher while retail prices were in general 75.7% higher.

The last column of Table 24.6 shows the data for the recorded growth of prices of the producers in individual sectors. We can see that the growth of prices in the majority of sectors was within the limits calculated by the model. The majority of the index prices, that do not correspond to the results of the model, can be explained if we examine the Yugoslav economic situation more closely.

Results

Table 24.6 The model's forecast and the statistically recorded growth of prices for 37 sectors of the Yugoslav economy in 1985

	NP(j) Min.	NP(j) Max.	NP(j) Median	Statist. recorded
1 Electricity generation	1.7640	2.0017	1.8790	1.554
2 Coal mining	1.6166	1.8333	1.7215	1.629
3 Coal processing	1.6530	1.8728	1.7594	2.095
4 Production of oil and natural gas	1.5776	1.7806	1.6760	1.827
5 Oil refining	1.6503	1.8630	1.7534	1.830
6 Extraction of iron ore	1.6709	1.8969	1.7803	1.979
7 Iron and steel production	1.7568	1.9973	1.8731	1.871
8 Extraction of nonferrous metal ores	1.6871	1.9198	1.7996	1.839
9 Production of nonferrous metals	1.8788	2.1683	2.0183	1.964
10 Processing of nonferrous metals	1.8024	2.0723	1.9326	1.887
11 Production of nonmetals	1.6888	1.9177	1.7996	1.736
12 Processing of nonmetals	1.7358	1.9746	1.8513	1.967
13 Metal processing activities	1.6956	1.9324	1.8101	1.864
14 Production of machines	1.7018	1.9426	1.8182	1.640
15 Vehicle production (all types)	1.7296	1.9712	1.8464	1.692
16 Shipbuilding	1.6616	1.9114	1.7821	—
17 Electrical machines and appliances	1.7833	2.0397	1.9071	2.374
18 Production of chemical products	1.8148	2.0625	1.9346	1.888
19 Processing of chemical products	1.7110	1.9454	1.8244	1.634
20 Production of building materials	1.7818	2.0228	1.8984	1.862
21 Timber processing	1.6671	1.8996	1.7795	2.358
22 Wood final products	1.7681	2.0173	1.8885	1.706
23 Production and processing of paper	1.7287	1.9654	1.8432	1.835
24 Spinning material and textiles	1.6557	1.8786	1.7636	1.908
25 Final textile products	1.6552	1.8842	1.7659	1.808
26 Hide and furs	1.6772	1.9041	1.7870	1.674
27 Leatherwear and leather accessories	1.6941	1.9330	1.8096	1.700
28 Rubber processing	1.6899	1.9217	1.8020	1.974
29 Production of foodstuff products	1.7773	2.0210	1.8952	1.693
30 Production of beverages	1.8655	2.1227	1.9899	1.851
31 Production of fodder	1.7018	1.9335	1.8139	1.425
32 Other manufacturing	1.6840	1.9161	1.7963	—
Total manufacturing and mining	1.7296	1.9678	1.8448	1.815
33 Agriculture and fishing	1.7668	2.0064	1.8827	1.580
34 Building industry	1.6715	1.8988	1.7815	1.850
35 Transport and communications	1.6578	1.8865	1.7684	1.820
36 Distributive trades	1.7123	1.9467	1.8257	1.800
37 Sundry economic activities	1.6629	1.8904	1.7730	—
Total economy	1.7144	1.9497	1.8282	1.796

For a lower growth in prices than the one calculated by the model two reasons can, in particular, be suggested: stricter state control over prices, and the crisis of demand connected with the fall in purchasing power of the population.

Despite the principle that prices should be freely formed in a market economy prices in certain sectors, as a rule, attract greater attention. This is certainly true in the cases of food production and energy. For this reason electricity generation (sector 1), production of foodstuff products (sector 29), production of beverages (sector 30), production of fodder (sector 31) and agriculture (sector 33) record a growth in prices which is below the growth of prices given by the model. On the other hand the producers of durable and more expensive goods are facing the decreasing purchasing power of the population due to the lowering of wages and salaries and to the restrictive credit policy now operating in Yugoslavia. As a result of this manufacturers of furniture have experienced special difficulties (sector 22 – the production of final wood products), and partly the manufacturers of cars (sector 15 – vehicle production).

The explanation for a higher rate of growth than the result given in the model lies in the rent charged for natural resources and in supply shortages. The production of oil and of natural gas (sector 4), the extraction of iron ore (sector 6) and timber processing (sector 21) are based on rare natural resources. As the scarcity of natural resources has not been fully reflected in the Yugoslav price system, the faster growth of prices than the costs in these sectors is the result of this type of rent increasingly being charged.

The markets for electrical machines and appliances and the market for rubber products have recently been recording an excess of demand over supply. Thus we can explain the higher recorded price indices in sectors 17 and 28.

24.5 CONCLUDING REMARKS

The model illustrated shows the possibility of calculating price movements based on the dynamics of costs in circumstances of galloping inflation and economic stagnation. Its essence is the adjustment process where both prices and incomes are the subject of economic policy.

The basic purpose in the construction of the model was to analyse the operation of the inflation mechanism in the Yugoslav economy. The equations are based on the behaviour of the Yugoslav economy and therefore also include the specifics of the economic system and the management of economic policy. The input data, required by the model, are formulated on the basis of the available Yugoslav statistical sources. Despite these particularities we take into consideration all the rules of scientific analysis and we operate, as far as possible, with the concepts normally used in the OECD countries.

This chapter does not extensively examine inflation theory; it does not seem to explain the initial causes of inflation. The model illustrated is limited to the analysis of the cost–push mechanism in which inflation operates and where the inflation expectations of all economic subjects, the structural disproportions and economic policy measures in the spheres of prices and incomes, all play an important role.

Despite all these limits this model offers reliable results. It clearly shows why the Yugoslav economy in 1985 could not meet the 50% target for price growth and that in these circumstances a rate of approximately 80% could be expected. As this was in fact the outcome in 1985 we may conclude that the cost–push mechanism presented largely correspondends to the behaviour of the Yugoslav economy. The majority of sectors determine prices according to costs and only in certain sectors are there other factors whose impact on price formation is stronger (e.g. the world price, a clear disproportion between demand and supply, scarcity of natural resources, etc.).

ACKNOWLEDGEMENTS

Chapter 24 is based on the author's research work with the Institute of Economic Diagnosis and Prognosis, The School of Economics and Commerce, University of Maribor. The author is grateful to Drago Gajšt and Pravin G. Mirchandani for their help with the English version of this chapter and to Leo Ciglenečki for his help with the computer work.

REFERENCES

Anon (1983) *Statistical Package for Social Sciences – extended. Version 2.1, User's Guide*, McGraw-Hill Book Company and SPSS Inc.

Babić, M. (1978) *Osnove input–output analize*, Narodne novine, Zagreb.

Bajt, A. (1983) Mehanizmi inflacije u Jugoslaviji in *Savremeni problemi ekonomske stabilizacije*, Savet akademija nauka i umetnosti SFRJ, Titograd, pp. 445–61 (English summary available).

Bole, V. (1982) *Privreda u 1983 godini (modelska procena)*, Ekonomski institut Pravne fakultete, Ljubljana.

Dolenc, M. and Pfajfar, L. (1976) Primerjava nekaterih metod za ocenjevanje tehničnih koeficientov v medsektorskih tabelah. *Economic Analysis*, 10, No. 1–2, 88–102, Belgrade (English summary available).

Federal Planning Office (1984) *Input–Output Table of Yugoslavia for 1982*, Federal Planning Office, Belgrade.

Federal Statistical Office (1985 and 1986a) *Statistical Yearbooks of Yugoslavia 1984 and 1985*, Federal Statistical Office, Belgrade.

Federal Statistical Office (1985 and 1986b) *Saopštenje (Communique)*, Federal Statistical Office, Belgrade, Vol. 29 (1985), No. 500 and Vol. 30 (1986), No. 2.

Fomby, T. B., Hill, R. C. and Johnson, S. R. (1984) *Advanced Econometric Methods*, Springer-Verlag, New York.

Institute for Economic Research (1985) *Makroekonomsko modeliranje*, Institute for Economic Research, Ljubljana, June 1985.

Kračun, D. and Žižmond, E. (1982) Das jugoslavische Preissystem, in *Schriftenreihe der Artbeitsgemainschaft fuer Wirtschafts – und Sozialgeschichte*, Graz, No. 3, pp. 19–33.

National Bank of Slovenia (1985a) *Imports and Exports of Goods in Yugoslavia in 1984*. Provided by the National Bank of Slovenia, Ljubljana.

National Bank of Slovenia (1985b) *Denarna in devizna gibanja*, Vol. 14, No. 5, Narodna banka Slovenije, Analitsko raziskovalni center, Ljubljana, May 1985.

Sekulić, M. *Medjusektorski modeli i strukturna analiza*, Informator Zagreb.

Social Accounting and Auditing Service (1985) *Financial Statements of the Organisations of the Associated Labour for 1984*. Provided by the Social Accounting and Auditing Service of Yugoslavia, Belgrade.

Vukman, J. (1982) Pogojenost sistema enačb pri input–output analizi. *Naše gospodarstvo*, **28**, No. 2, 152–5, Maribor. (English synopsis available).

25

Short-term forecasting of wages, employment and output in Barbados

DANIEL O. BOAMAH

25.1 INTRODUCTION

The policy-maker's understanding of the process of real output determination in an economy is greatly enhanced by an analysis of the labour force and of the determinants of wages and employment. In Barbados, much of the work on the labour force takes the form of one-period employment surveys. A noted example of this is the seminal work by Cumper (1959). There is a noticeable dearth of studies on what determines employment and wages in Barbados. The exception is the study by Downes and McClean (1982). However, in our opinion, their interesting theoretical discourse provides an insight into the effect of the bargaining process on money wage determination but still requires broader empirical support.

In the Caribbean region as a whole, interesting studies have been carried out on the determinants of wages and employment. The works by Farrell (1980), Iyoha (1978), St Cyr (1981) and Brown (1980) merit special mention. However, apart from St Cyr, the others discuss the issues of wages and employment in general terms. The small contribution expected of this chapter is the empirical support it attempts to lend to some of the issues discussed.

This chapter attempts an analysis of the labour and commodity markets in Barbados. It explores some of the main forces that seem to influence changes in wages, employment and real output in the economy, in a simultaneous equation context. The organization of the chapter follows the usual pattern. We outline the framework of the model in the next section, and in Section 25.3 we discuss the empirical results encompassing the parametric estimates, simulation and forecasting as well as a review of data definitions, measurements and sources. Section 25.4 concludes the chapter with a summary of the main findings.

25.2 A FRAMEWORK FOR THE ANALYSIS

A two-sector model is adopted for the analysis. We distinguish between traded and nontraded goods and services in the economy. The latter are basically home goods and services that do not enter into international trade. They encompass nonsugar agriculture, utilities, and all services other than those arising from tourism. The tradable goods and services are traded in the international markets but they also compete with nontradables for domestic resources in production and, to some extent, consumption.

Figure 25.1 illustrates the behaviour of the relative price of nontradables to tradables over the period 1958 to 1980. The diagram clearly shows a sufficiently large upward movement in the ratio over the reference period. This variability of the relative price ratio provides enough justification for the separate treatment of tradables and nontradables in this chapter.

25.2.1 The wages and employment functions

Given the importance of trade unions in wage setting in Barbados, a wage model that takes explicit account of union power in addition to traditional variables such as expectation of future price movements, productivity and the state of the labour market would be the ideal choice. Unfortunately, we have been unable to obtain a series on trade union membership to serve as a useful proxy for union assertiveness (Irfan, 1982). In addition, the alternative proxies such as the average real earnings or real wages that have been employed by such authors as Ashenfelter *et al.* (1972), Ormerod (1982) and Brooks and Henry (1983) are not supported by formal

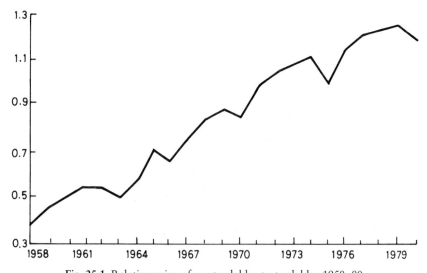

Fig. 25.1 Relative price of nontradables to tradables 1958–80

theoretical justification to encourage their use. Trade union influence in the determination of wages is thus only implicit in this chapter.

We take the view that during negotiations trade unions push to recoup losses in real wages brought about by inflationary pressures. Changes in nominal wages in the traded sector are assumed to be influenced by expected price changes and labour market conditions. The higher the unions perceive future price movements the harder they are likely to push for higher wage demands. Firms are also encouraged to raise wages the higher the price they expect to get for their products and the higher the level of workers' productivity. The latter is an important determinant of firms' profitability and a good indicator to workers of their contribution to the progress of the firms (Komiya and Yasui, 1984). However, wage demands by workers are likely to be moderated by any slackness in the labour market brought about by recessionary conditions in the economy. The unemployment rate represents labour market conditions. Therefore, we specify changes in nominal wages in the traded sector by the expression:

$$WT = F_1(\overset{+}{P_{-1}}, \overset{+}{QT}/LET, \overset{-}{U}) \qquad (25.1)$$

where

- WT = nominal wage index in the traded sector
- P_{-1} = domestic price level lagged one period (a proxy for price expectation)
- U = the unemployment rate
- QT = real value added in the traded sector
- LET = employment in the traded sector.

The sign on the variables indicates expected direction of influence on the dependent variable.

We anticipate strong wage impulses to pass from the traded to the nontraded sector, as the former is considered the leading sector in the wage determination process. Therefore, in addition to the above variables explaining wage movements in the nontraded sector, we include changes in the traded sector wage. Nominal wage change in the nontraded sector is expressed functionally as:

$$WN = F_2(\overset{+}{P_{-1}}, \overset{+}{QN}/LEN \overset{-}{U}, \overset{+}{WT}) \qquad (25.2)$$

where

- WN = nominal wage index in the nontraded sector
- QN = real value added in the nontraded sector
- LEN = employment in the non-traded sector.

Recent research into the determinants of employment in the Caribbean supports the traditional orthodoxy that real output and real wages are the main determinants of employment in the region (Farrell, 1980, p. 97). Farrell also identifies other factors such as the extent of capital availability, production technology, organizational technology and infrastructure as some of the other crucial factors that determine employment in the Caribbean.[1] While these factors are important, the real output variable which serves as a proxy for effective demand for the product cannot be ignored (Boamah, 1981). However, problems of multicollinearity do not permit the use of both the capital stock and the real output variables in one equation. In this chapter, the specification of the employment demand function follows the traditional view, with employment determined by real output, real wage and employment in the previous period. For the two sectors we have:

$$\text{LET} = F_3(\overset{+}{\text{QT}}, \overset{-}{\text{WT}}/\text{PT}, \overset{+}{\text{LET}}_{-1}) \qquad (25.3)$$

and

$$\text{LEN} = F_4(\overset{+}{\text{QN}}, \overset{-}{\text{WT}}/\text{PT}, \overset{+}{\text{LEN}}_{-1}) \qquad (25.4)$$

where

PT, PN = the indices of prices in the traded and nontraded sectors.

The lagged endogenous variables are included because firms usually make only partial adjustment towards their desired labour requirements in each period.

While economic factors may be important in determining labour supply within sectors, in the aggregate they play secondary role to noneconomic factors. We assume that it is demographic factors that principally determine aggregate labour supply. That variable is therefore considered exogenous in the model. Equation 25.10 establishes the equilibrium condition in the labour market. It is postulated on the implied assumption that labour demand is always satisfied.[2]

For all the commodities traded by Barbados in the international markets the smallness of its participation means that it is a price taker in these markets. Therefore, the price received for tradable goods and services is exogenous and real output in the sector is mainly determined by supply conditions in the market. Output of tradables is expected to be sensitive to the prevailing market prices, the cost of variable factors of production, and the level of real investment in capital and intermediate goods. The latter is proxied by real imports of machinery and intermediate products. Hence, for the output of tradable goods and services, we have:

$$\text{QT} = F_5(\overset{+}{\text{PT}}, \overset{-}{\text{RL}}, \overset{-}{\text{ULCT}}, \overset{+}{\text{MKI}}) \qquad (25.5)$$

where

RL = average interest rate on loans
ULCT = unit labour cost in the traded sector

---

[1] In an earlier work, we tested and confirmed this empirically by representing all the factors identified by a composite capital stock variable (see Boamah, 1985). Refer to Boamah (1984) for the derivation of the capital stock variable.
[2] Given the pool of unemployed labour in the system, the assumption that labour demand is always satisfied may be justified.

MKI = real value of imports of capital and intermediate goods.

Nontradables, by definition, are not traded externally. Firms in the sector sell to private residents and the government in the domestic market. Excess supply or demand in this market will require price adjustments to reachieve equilibrium. Therefore, supply and demand factors determine the quantity demanded and supplied. Demand for nontradables is assumed to be dependent on real domestic expenditure, the relative price of nontradables (to tradables) and a trend variable that proxies other exogenous influences such as population changes. That is:

$$QN = F_6(\overset{+}{DE}, \overset{-}{PN/PT}, \overset{+}{T}) \qquad (25.6)$$

where

DE = real domestic expenditures
T = a trend variable.

The supply price that firms in the nontraded sector expect to get for their products is a direct function of factor costs and the going price level in the traded sector, which determines what nontradable firms have to pay for imported materials (Worrell and Holder, 1985). There is also a direct relationship between what firms receive for their products and what they produce. The functional relationship for the supply price of nontradables is therefore expressed as:

$$PN = F_7(\overset{+}{QN}, \overset{-}{ULCN}, \overset{+}{RL}, \overset{+}{PT}) \qquad (25.7)$$

where

ULCN = unit labour cost in the nontradable sector.

We complete the discussion of the commodity market with a behavioural relationship for total private domestic expenditure. Real government expenditure is assumed exogenous. Real private domestic expenditure is determined by real disposable income, the cost of loan capital and the previous level of private expenditure, a proxy for habit persistence. Thus we have:

$$PDE = F_8(\overset{+}{YD}, \overset{-}{RL}, \overset{+}{PDE_{-1}}) \qquad (25.8)$$

where

PDE = real private domestic expenditure, and
YD = real personal disposable income.

The following identities complete the model:

$$P = PT(QT/Q) + PN(QN/Q) \qquad (25.9)$$
$$U = 100(LS - LET - LEN)/LS \qquad (25.10)$$
$$DE = PDE + G \qquad (25.11)$$

and

$$Q = QT + QN \qquad (25.12)$$

In Equations 25.9–25.12

Q = aggregate real value added
G = real government expenditure
LS = aggregate labour supply

It is hoped that the above model, consisting of eight behavioural relationships and four identities, explains conditions in the real sector of the Barbados economy. To make the presentation clearer, we summarize the model below

$$WT = F_1(P_{-1}, QT/LET, U) \tag{25.13}$$
$$WN = F_2(P_{-1}, QN/LEN, U, WT) \tag{25.14}$$
$$LET = F_3(QT, WT/PT, LET_{-1}) \tag{25.15}$$
$$LEN = F_4(QN, WN/PN, LEN_{-1}) \tag{25.16}$$
$$QT = F_5(PT, RL, ULCT, MKI) \tag{25.17}$$
$$QN = F_6(DE, PN/PT, T) \tag{25.18}$$
$$PN = F_7(QN, ULCN, RL, PT) \tag{25.19}$$
$$PDE = F_8(YD, RL, PDE_{-1}) \tag{25.20}$$
$$P = PT(QT/Q) + PN(QN/Q) \tag{25.21}$$
$$U = 100(LS - LET - LEN)/LS \tag{25.22}$$
$$DE = PDE + G \tag{25.23}$$
$$Q = QT + QN \tag{25.24}$$

Equation 25.13, 25.14, 25.16 and 25.22 summarize conditions in the labour market while Equation 25.17, 25.18, 25.19 and 25.24 describe conditions in the commodity market. Equations 25.20 and 25.23 provide a link with the aggregate demand sector in the economy. The model is determinate with twelve equations explaining the twelve endogenous variables, namely: WT, WN, LET, LEN, QT, QN, PN, PDE, DE, P, U and Q.

Before discussing the empirical estimates and the results of the simulation and forecasting of the endogenous variables, we provide a brief description of the workings of the economy which should give an indication of possible policy implications of the model. Suppose there is an exogenous increase in the price of tradable goods (PT): the real wage in the sector would decline and we expect the demand for labour in the traded sector to increase. Output in the traded sector should increase. An increase in PT would also tend to raise PN and hence employment and output in the nontraded sector would also increase. Meanwhile, an increase in both PT and PN should raise the domestic price level. The rising price level in association with the falling rate of unemployment in both sectors would give rise to increased wage demands. The subsequent increase in real wages would tend to moderate the demand for labour in the two sectors. Eventually, the economy may settle at higher levels of PT, WT, QT, PN, QN, WN and P.

25.3 EMPIRICAL ESTIMATES

The above model is basically simultaneous although some of the equations are determined independently. Each of the eight behavioural relationships is overidentified on the basis of both the order and rank conditions. Statistical theory suggests that the ordinary least squares (OLS) estimator may not be appropriate for dealing with simultaneous equation systems but initial exploratory work with the two stage least squares (2SLS) technique did not suggest any changes that would

Table 25.1 Estimates of the equation of the model

(1*) $WT = 90.674 + 0.395P_{-1} - 0.368U - 2.232QT/LET - 9.451DM$
$\quad\quad\ (6.823)\ \ (4.807)\ \ (-0.607) - (-1.351)\ \ (-3.269)$
$R^2 = 0.870\quad D-W = 1.77\quad SE = 6.311$
$F(4, 19) = 31.87\quad AR(1),\ rho = 0.81$

(2*) $WN = 24.243 + 0.113P_{-1} - 0.744U + 0.431WT + 4.875QN/LEN$
$\quad\quad\ (1.974)\ \ (2.051)\ \ (-1.600)\ \ (2.503)\ \ (3.880)$
$R^2 = 0.922\quad D-W = 1.79\quad SE = 5.32$
$F(4, 18) = 52.96\quad AR(1),\ rho = 0.23$

(3*) $LET = 0.014QT + 1.053WT/P + 0.824LET_{-1}$
$\quad\quad\ (2.535)\ \ (1.486)\ \ (11.052)$
$R^2 = 0.827\quad D-W = 1.99\quad SE = 1.29$
$F(3, 20) = 37.74\quad AR(1),\ rho = 0.02$
$h = 0.037$

(4*) $LEN = 26.284 + 0.014QN - 2.364WN/P + 0.535LEN_{-1}$
$\quad\quad\ (2.565)\ \ (1.833)\ \ (-1.881)\ \ (2.486)$
$R^2 = 0.937\quad D-W = 1.90\quad SE = 1.80$
$F(3, 19) = 82.49\quad AR(1),\ rho = 0.34$
$h = 0.527$

(5*) $QT = 211.540 + 0.474PT - 0.741ULCT + 0.204PDL\,(MKI, 1, 3, FAR)$
$\quad\quad\ (9.250)\ \ (1.832)\ \ (-5.413)\ \ (1.673)$
$R^2 = 0.951\quad D-W = 2.23\quad SE = 10.48$
$F(3, 18) = 82.04,\ AR(2),\ rho = -0.63$

(6*) $QN = -124.920 + 0.513DE - 4.601PN/PT + 7.584T$
$\quad\quad\ (-2.872)\ \ (6.764)\ \ (-0.100)\ \ (2.349)$
$R^2 = 0.981\quad D-W = 1.94\quad SE = 19.11$
$F(3,19) = 233.71\quad AR(2),\ rho = 0.09$

(7*) $PN = -180.530 + 0.168QN + 3.458RL + 1.115PT + 0.631ULCN$
$\quad\quad\ (-4.858)\ \ (3.259)\ \ (1.971)\ \ (6.898)\ \ (3.277)$
$R^2 = 0.975\quad D-W = 1.71\quad SE = 9.03$
$F(4, 18) = 143\quad AR(2),\ rho = -0.56$

(8*) $PDE = 104.530 + 0.386YD - 9.616RL + 0.653PDE_{-1}$
$\quad\quad\ (3.219)\ \ (2.574)\ \ (-1.800)\ \ (4.704)$
$R^2 = 0.948\quad D-W = 1.83\quad SE = 29.92$
$F(3, 20) = 139.38\quad h = 0.570$

Note: t-statistics are in parentheses below estimated coefficients;
rho = coefficients of autocorrelation;
SE = standard error of estimate.

invalidate the main conclusions of the study.[3] The estimated model, with the OLS technique, is presented in Table 25.1. But, before discussing the results, we look at the data sources and indicate how specific variables were measured.

25.3.1 Data: definitions, measurements and sources

The data used were obtained from both published and unpublished sources at the Central Bank of Barbados, various issues of Barbados Economic Report and the Abstract of Statistics of the Barbados Statistical Service; and other published sources. The wage indices were derived from Downes (1980) and the Annual Statistical Digest (ASD) of the Central Bank of Barbados. They represent indices of minimum wages paid mainly to manual workers in the respective sectors.

A casual examination of the wage index in the traded sector (WT) revealed cases where the index fell unrealistically sharply. These occurred in the years 1963, 1974 and 1979. A dummy variable DM which took on values of unity for these years and zero otherwise was added to the regressors as an additional variable in Equation 25.1.

By definition, the unit labour cost in a sector (ULC) may be expressed as ULC = wage bill/labour employed. There is no data on the wage bills for the two sectors for the entire reference period. The alternative adopted in this chapter is an index of wage rates adjusted for productivity in the respective sectors. Also, in estimating the employment functions the domestic price index was the deflator for the wage variables because that gave a better fit.

The real domestic expenditure variable (DE) is defined as $DE = C + I$ where C and I represent aggregate real consumption and investment expenditures respectively. Also, the personal disposable income variable (YD) is defined as $YD = Q - TP$, where TP stands for real personal direct taxes. The labour force and employment statistics were estimated and in the following paragraphs we present a detailed description of the methodology used.

(a) Labour force and employment data

Prior to 1975, statistics on the labour force and employment in Barbados were not collected on a continuous basis. We have had to resort to an indirect way to estimate the complete series for the two variables. To do this, we were aided by the available data from the 1965–66 labour force surveys and in the census years of 1960 and 1970.

Information on total population as at 31 December (1958–80) and adult population (aged 15–65) for 1958–69 and 1975–80 was obtained from the Abstract of Statistics (1969, No. 6, Barbados Statistical Service) and various issues

---

[3] We encountered problems correcting for autocorrelation for the two-stage least squares technique with the Economic Analysis Language (EAL) computer package which was used to estimate and simulate the model.

of the Barbados Economic Report. The latter also provided information on the labour force and the percentage of adults in the labour force for the period 1975–80. The missing portion of the adult population (1970–74) was derived on trend. It was based on the average of the ratio of adults in total population for the 5 years preceding 1970 and the 5 years immediately after 1974.

In general, one expects the labour force to be predominantly derived from the adult population. As such the available data on the percentage of the labour force in the adult population (participation rate) provided the base from which the labour force data were derived. The labour force participation rate shows a slight drop from about 72.5% in 1960 to 68.7% in 1966. Between 1966 and 1970 the ratio is fairly stable, moving from 68.7% to 68.1% in 1970. Given the relative stability of the participation rates between the benchmark years one would not expect wide yearly variations in the ratios. The intervening participation rates between 1960 and 1966 were therefore estimated on the trend movement between the two years. The same procedure was adopted to calculate the rates between 1966 and 1970 and thenceforth to 1975 when regular figures became available. The labour force data were subsequently derived from a combination of the participation rates and the series on adult population.

Employment figures were also calculated from the estimates of the proportion of adult population employed, using as benchmarks those years (1960, 1966, 1970 and 1975–80) when data were available. Information indicates that the proportion of adult population employed dropped from 66.9% in 1960 to about 62.2% in 1970, and subsequently to 57.6% in 1980. The lowest ratio occurred in 1975 when 50.9% of the adult population was employed. As was done for the labour force data, the missing employment–adult ratios within the four benchmarks were calculated on trend.

25.3.2 The estimated model

From the estimated model above we find that the wage determination results for the two sectors are mixed. Price expectation (proxied by one period lagged prices) seems to be the main factor driving wages in the traded sector. The unemployment rate and the average productivity variables are all insignificant, with the latter carrying the wrong sign. The results for the nontraded sector, are, however, quite robust with all the variables carrying correctly signed coefficients which are significant at the 5% level, except the unemployment rate which is significant at the 10% level. The results also appear to lend credence to the hypothesis of a wage push arising from the traded sector. Wage settlements in the key tourism and manufacturing industries seem to have strong influence on wage settlements in the nontraded sector. Also the size of the coefficients of the lagged price variables (less than unity) in both sectors seems to confirm the general observation that expected prices are not fully reflected in money wage settlements.[4]

---

[4] The point is that unions do not usually get what they want.

The estimates of the employment functions generally agree with a priori expectations (3*) and (4*). The real output and the real wage variables explain over 82% and 94%, respectively, in the traded and nontraded sectors. All the coefficients for the nontradable function are of the expected signs and are all significant at the 5% level. The coefficient of the real wage variable for the tradable function was, however, not of the expected sign, but it is not significant. Hence there is not enough justification either to reject or accept the original hypothesis.

We realize that because of the lagged dependent variables also appearing as explanatory variables, the Durbin–Watson statistics may not strictly be appropriate. Therefore, the Durbin h-statistics[5] were also computed. In general, there appears to be no serious problem of serial correlation. The original hypothesis for tradable output appears to have been fairly borne out by the results (see 5*). The estimates seem to suggest that supply in the sector increases with rising prices of tradables and the real value of capital and intermediate good's imports which is fitted as a first degree polynomial with a 3-year lag.[6] However, increased wage costs are likely to depress output.

The results for the output of nontradables also appear to be fairly robust although the coefficient of the relative price of nontradables is not significant. The demand for nontradable goods seems extremely responsive to aggregate real domestic expenditure and other exogenous influences proxied by the trend variable. The fit for the nontraded price function is also acceptable, judging from the summary statistics. Over 97% of the variations in the price variable are explained by the real output in the sector, the price of tradables, the cost of loan capital and the unit labour cost variable. As expected the price at which nontraded goods are offered on the market responds positively to increases in demand, increases in the price of tradables, and increases in wages above productivity gains.

The estimates for the private domestic expenditure function also follow a priori expectations. The coefficients of all the variables carry the expected signs and are significant, especially those for the real disposable income and the lagged value of the dependent variable. The interest rate variable, although marginally significant at the 5% level, has an important implication for economic policy. The elasticity of 0.12 implies that a 10% increase in the cost of borrowing is likely to reduce real private expenditure by approximately 10%.

---

[5] The Durbin h-statistic is defined as:

$$h = (1 - \tfrac{1}{2}d)\sqrt{\left[\frac{N}{1 - Nv(b)}\right]}$$

where d is the Durbin–Watson statistic, N is the sample size and $v(b)$ is the estimate of the variance of the coefficient of the lagged endogenous variable. We tested the null hypothesis that the error terms are serially independent against the alternative that the null hypothesis is false. At the 95% level the critical value of h is 1.645 and in both cases the computed h does not exceed the critical value. Therefore we do not reject the null hypothesis that the error terms are serially independent.

[6] The cost of capital variable was dropped as the variable was not significant and its inclusion did not yield acceptable overall fit.

25.3.3 Simulation and forecasting

The summary statistics for the simulation (*ex-post*) are presented in Table 25.2; Table 25.3 reports the *ex-ante* forecast values for key endogenous variables from 1983 to 1990. Judging from these statistics, (especially the Theil U) the model does reasonably well at tracking the endogenous variables. Figures 25.2 to 25.13 provide evidence of the model's tracking performance.

Table 25.2 Simulation statistics

	Average error	Average absolute error	Mean square error	RMSQ	Thiel-U	Standard error
DPE	0.679	30.546	1095.960	33.105	0.021	33.098
QT	1.256	15.572	383.301	19.578	0.043	19.538
DE	−0.458	48.795	3119.098	55.849	0.028	55.847
PN	−2.411	8.126	118.042	10.865	0.043	10.594
Q	1.697	31.320	1589.977	39.875	0.026	39.838
QN	0.440	37.071	1856.631	43.089	0.040	43.086
P	−1.898	6.091	65.831	8.114	0.032	7.888
LEN	−0.705	1.913	6.090	2.468	0.019	2.365
LET	−1.024	1.329	2.668	1.633	0.031	1.272
U	1.711	2.961	12.687	3.562	0.130	3.124
WN	1.684	7.365	78.960	8.886	0.039	8.725
WT	5.369	7.798	102.794	10.139	0.046	8.601

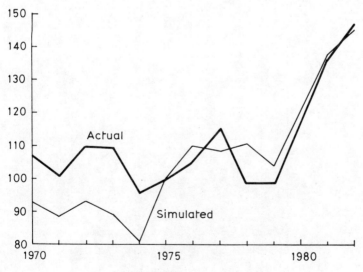

Fig. 25.2 WT (1975 = 100).

Table 25.3 Forecast of major economic aggregates

	1983	1984	1985	1986	1987	1988	1989	1990
Real output								
Overall economy (% change)	−0.1(0.4*)	4.0(3.5*)	−0.9(−0.8*)	2.3	3.0	3.0	3.2	2.5
Traded sector (% change)	5.6(1.5*)	6.3(5.9*)	−2.9	−1.2	−2.4	1.2	2.4	0.6
Nontraded sector (% change)	−2.0(−0.3*)	3.2(3.1*)	−0.1	3.6	5.0	3.7	3.4	3.1
Employment								
Overall economy ('000)	95.3(95.8*)	96.0(93.1*)	98.2	99.4	100.0	100.5	101.0	101.4
Traded sector ('000)	24.4	25.0	25.6	25.7	25.5	25.3	25.2	25.0
Nontraded sector ('000)	70.9	71.0	72.6	73.7	74.7	75.2	75.8	76.4
Unemployment rate (%)	14.5(15.0*)	14.5(17.1*)	13.9(18.7*)	14.2	14.8	15.7	16.6	17.5
Inflation								
GDP deflator (% change)	—	2.9(5.6*)	4.1(4.5*)	7.3	6.6	6.5	5.9	5.8
Nontraded sector prices (% change)	—	3.1(6.5*)	3.6	7.4	6.2	6.4	6.0	5.7

Note: Figures with asterisk in parentheses are actual figures.

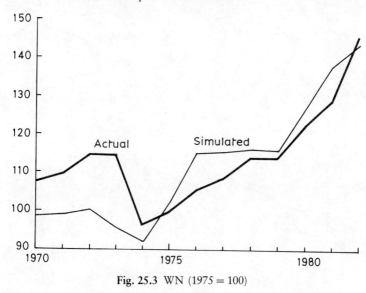

Fig. 25.3 WN (1975 = 100)

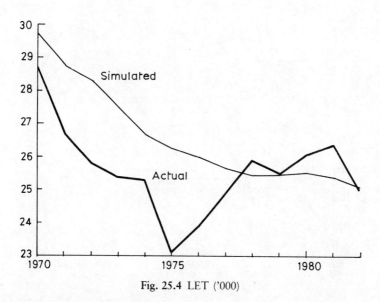

Fig. 25.4 LET ('000)

The model tracks the wage indices for both sectors fairly well in the post-1975 period. The yearly fluctuations have been captured quite accurately. However, in the period between 1970 and 1975, the model consistently underestimates the actual movements of wages. Despite the small values of the Theil inequality coefficients and the root mean square errors (RMSQ), the model falls a bit short a duplicating the

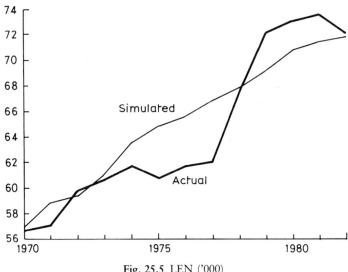

Fig. 25.5 LEN ('000)

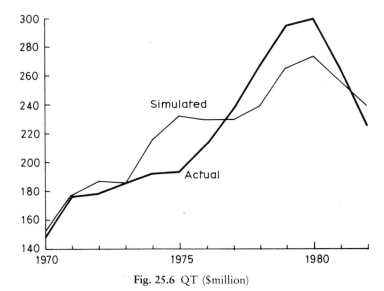

Fig. 25.6 QT ($million)

employment functions in both sectors (Figs 25.4 and 25.5). In the period 1970 to 1978, it consistently overestimates actual employment in the two sectors. The situation is reversed in the post-1978 period when the simulated figures underestimate actual observations. The overestimation of employment in the period to 1978, followed by underestimation implies that the unemployment rate is underestimated

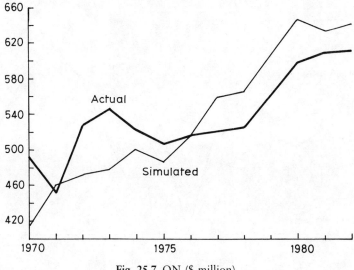

Fig. 25.7 QN ($ million)

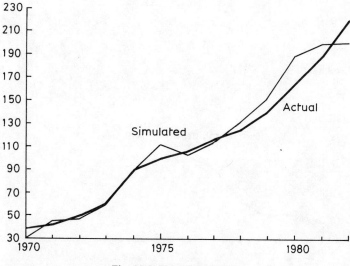

Fig. 25.8 PN (1975 = 100)

in the pre-1978 period and overestimated thereafter. This is quite evident in Fig. 25.12. In general the less than adequate tracking performance of employment may be traced to the fact that much of the data for that variable in both sectors were estimated.

Despite missing some of the yearly fluctuations, the model does reasonably well in tracking the output of tradables (Fig. 25.6). It records the upswing in QT

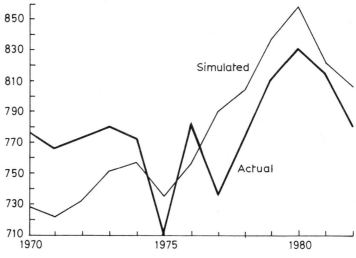

Fig. 25.9 PDE ($ million)

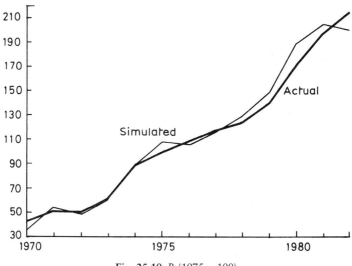

Fig. 25.10 P (1975 = 100)

between 1973 and 1980 and the severe contraction thereafter, although it overpredicts between 1973 and 1977 and underestimates the actual gains in the last 4 years to 1981. As was the case for the output of tradables, the model duplicates the broad pattern of the output of nontradables (Fig. 25.7) but this time it underestimates the actual values in the period 1973–77 and overpredicts in the 1977–82 period. The year-to-year fluctuations have been picked up fairly well.

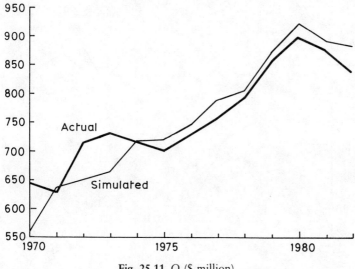

Fig. 25.11 Q ($ million)

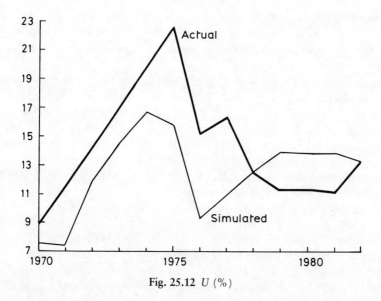

Fig. 25.12 U (%)

The growth pattern of the price of nontradables (PN) has been reasonably tracked by the model (Fig. 25.8). Both the trend and yearly variations of the simulated pattern bear fairly accurate resemblance to actual observations. There was little inflation up to 1973, followed by strong inflationary pressures between

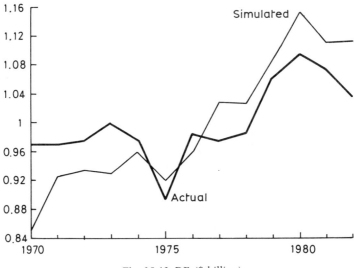

Fig. 25.13 DE ($ billion)

1973 and 1980, arising mainly from the oil price shocks of 1974 and 1979. With the price of the traded sector exogenously determined, the behaviour of the overall price level (P) faithfully duplicates the broad pattern exhibited by the price of nontradables. (Fig. 25.10).

The periodic changes of the real aggregate expenditure variables have generally been reproduced although the model consistently overestimates the actual outturn for the period after 1976 (see Fig. 25.9 and 25.13). The simulation for the overall real output is evidently much better than either of two components, QT and QN. The rapid expansion in real output between 1975 and 1980 and the subsequent depression is portrayed quite accurately (Fig. 25.11).

The outturn for the forecast values is reported in Table 25.A.2 in Appendix A. Table 25.3 also shows the relative movements of some of the important variables in the model. Given the actual values of exogenous variables for 1983–85, the *ex-post* forecasts appear to give a fairly good account of the actual outturn of the key variables for 1983–85 as indicated with asterisks in Table 25.3. The yearly movements of the real output variables are generally well duplicated.

The results suggest that the 1985 downturn for traded goods is likely to continue for 1986 and 1987 while a turnaround is expected for nontraded goods which is predicted to grow by an average of 3.8% in the 5 years to 1990. Overall output is predicted to grow by 3.0% on the average over the same period. The unemployment rate is generally underestimated in the period 1983–85. That pattern is likely to continue throughout the period of the forecast. The inflation figures look realistic, with the GDP deflator and nontraded prices forecast to grow by an average of 6.4% and 6.3%, respectively, between 1986–90.

25.4 SUMMARY AND CONCLUSIONS

With a relatively simple econometric model, the chapter has attempted to investigate statistically some of the factors that determine wages, employment and output in the Barbados economy. The analysis has been undertaken in the traditional framework of a small open economy with broadly identifiable traded and nontraded sectors. Standard tools of statistical inference suggest that the independent variables generally give a reasonably good explanation of the dependent variables.

Expected prices appear to be the main influence on wage determination in the two sectors. Also, there is some evidence of wage push from the traded to the nontraded sector. Employment is principally determined by real output in the respective sectors, and in the case of the nontraded sector, the sectoral real wage also appears to be an important variable.

The price of traded goods and wages adjusted for productivity are important explanatory variables for real output in the sector, while aggregate domestic expenditure and the trend variable largely determine the demand for real output in the nontraded sector. Also, the supply price of nontradables appears to be sensitive to real output in the sector, the price of tradables and the cost of the factors of production.

While the results are interesting in themselves, problems of data quality invite caution on their interpretation. Some refinements in the data, estimating and forecasting techniques may be necessary to improve upon the model's tracking performance. Nevertheless, the outcome of the simulations within sample and the *ex-post* forecasts give a cautiously optimistic indication that the *ex-ante* forecasts can be relied upon as a fairly suitable guide to the short and medium term planning of wages, employment and output in Barbados.

REFERENCES

Ashenfelter, O., Johnson, G. E. and Pencaval, J. H. (1972) Trade unions and the rate of change of money wages in United States manufacturing industry. *Review of Economic Studies*, 39.

Boamah, D. O. (1981) Proposed Specification of an Econometric Model of the Barbadian Economy, Central Bank of Barbados (*mimeo*).

Boamah, D. O. (1984) The stock of fixed capital in Barbados, 1958–1981: some exploratory estimates. *Economic Review*, Central Bank of Barbados, 11, No. 3, December, 8–20.

Boamah, D. O. (1985) Wage formation, employment and output in Barbados. *Social and Economic Studies*, 34, No. 4, December, 199–217.

Brooks, S. and Henry, B. (1983) Re-estimation of the National Institute Model, *National Institute Economic Review*, No. 103, February.

Brown, A. (1980) Employment policy in the open dual economy. *Social and Economic Studies*, 29, No. 4, December, 112–24.

Cumper, G. E. (1959) Employment in Barbados. *Social and Economic Studies*, 8, No. 2, June.

Downes, A. S. (1980) A wage index for Barbados: an exploratory note. *Caribbean Studies*, **20**, No. 2, June, 75–80.

Downes, A. S. and McClean A. W. A. (1982) Wage determination in a small open unionized economy: the case of Barbados, UWI, Cave Hill Campus, November (*mimeo*).

Farrell, Trevor (1980) The root of unemployment. *Social and Economic Studies*, **29**, No. 4, December, 95–111.

Irfan, M. (1982) Wages, employment and trade unions in Pakistan. *The Pakistan Development Review*, **21**, No 1, Spring, 49–72.

Iyoha, M. A. (1978) The relation between employment and growth in developing countries: an econometric analysis. *Social and Economic Studies*, **27**, No. 1, March, 69–84.

Komiya, R. and Yasui, K. (1984) Japan's Macroeconomic performance since the first oil crisis: review and appraisal. *Carnegie–Rochester Conference Series on Public Policy*, **20**, Spring, 67–114.

Ormerod, P. (1982) Rational and non-rational expectations of inflation in wage equations for the United Kingdom. *Economica*, **49**, November, 375–87.

St Cyr, E. (1979) A Note on the Trinidad and Tobago inflationary experience, 1965–1976. *Social and Economic Studies*, **28**, No. 3, September, 618–27.

St Cyr, E. (1981) Wages, prices and the balance of payments: Trinidad and Tobago, 1956–1976. *Social and Economic Studies*, **30**, No. 4, December, 111–33.

Worrell, D. and Holder, C. (1985) A model of price formation for small economies: three Caribbean examples. *Journal of Development Economics*, **18**, December.

APPENDIX A

Table 25.A.1 Forecasting – expected values of exogenous variables 1983–90

	1983*	1984*	1985*	1986	1987	1988	1989	1990
YD^a	721.3	748.4	749.1	777.1	798.1	828.3	844.9	861.8
RL^b	10.5	11.0	10.4	10.7	10.2	9.8	9.8	9.8
PT^c	138.7	143.1	150.3	157.6	165.6	173.9	182.6	191.7
$ULCT^d$	132.3	128.7	135.2	139.2	143.4	147.7	150.7	153.7
MKI^e	359.5	343.5	292.2	307.4	308.9	313.6	318.5	323.1
G^f	253.3	262.8	261.7	271.5	278.9	289.4	295.2	301.1
$ULCN^g$	137.5	133.8	140.5	147.6	154.9	162.7	167.6	172.6
LS^h	111.6	112.2	114.0	115.7	117.5	119.2	121.0	122.8

*Actual values used.

[a] 1986–88: as predicted by the Central Bank econometric model for real output; 89–90: 2.0% annual growth rate.

[b] 1986–88: as predicted by the Central Bank econometric model; 1989–90: assumes rare unchanged from 1988.

[c] Assumes 5% average annual growth rate.

[d] Assumes 3% average annual growth rate for 1986–88; 2% thereafter.

[e] 1986–90: assumes 5% average annual growth of nominal values. Import price pattern as for PT.

[f] Follows the pattern for YD.

[g] Assumes 5% average annual growth rate.

[h] Assumes 1.5% average annual growth rate.

Appendix B

Table 25.A.2 Forecasting results (1983–90)

	1983	1984	1985	1986	1987	1988	1989	1990
DPE	804.0	820.3	823.0	845.2	871.1	895.9	923.9	948.4
QT	244.6	260.1	252.4	249.4	243.5	246.4	252.3	253.7
DE	1057.2	1083.1	1084.8	1160.7	1150.1	1185.3	1219.1	1249.6
PN	217.5	224.2	232.3	249.6	265.0	282.1	299.0	316.2
Q	872.2	907.5	899.1	919.4	946.9	975.5	1006.2	1031.0
QN	627.6	647.4	646.7	670.0	703.4	729.1	753.9	772.2
P	195.4	201.0	209.3	224.6	239.4	254.7	269.8	285.6
LEN	70.9	71.0	72.6	73.7	74.5	75.2	75.8	76.3
LET	24.4	25.0	25.6	25.7	25.5	25.3	25.2	25.0
U	14.5	14.5	13.9	14.2	14.9	15.7	16.6	17.5
WN	148.1	144.0	138.1	141.7	153.8	156.8	157.8	163.7
WT	146.7	139.4	143.5	146.5	152.7	157.8	162.9	168.2

APPENDIX B

Glossary of symbols used

C	Real consumption expenditure
DE	Real domestic expenditure
G	Real government expenditure
I	Aggregate real investment expenditure
LEN	Employment in the non-traded sector
LET	Employment in the traded sector
LS	Aggregate labour supply
MKI	Real value of imports of capital and intermediate goods
P	Domestic price level
PDE	Real private domestic expenditure
PN	Price index in the non-traded sector
PT	Price index in the traded sector
Q	Aggregate real value added
QN	Real value added in the non-traded sector
QT	Real value added in the traded sector
RL	Average interest rate on loans
T	Trend variable
TP	Real personal direct taxes
U	Unemployment rate
ULCN	Unit labour cost in the non-traded sector
ULCT	Unit labour cost in the traded sector
WN	Nominal wage index in the non-traded sector
WT	Nominal wage index in the traded sector
YD	Real personal disposable income

26

Reducing working time for reducing unemployment? A macroeconomic simulation study for the Belgian economy

JOSEPH PLASMANS and
ANNEMIE VANROELEN

In this chapter we investigate the impact of a reduction of working time (RWT) on some main aggregate economic variables, such as (un)employment, production, final demand, profits and the financing shortage of the government. This investigation is performed by simulating the MARIBEL model, which is the current official model of the Belgian Planning Office, over a time period of 5 years (1985–1989).

We found that a single 10% contractual RWT in the private sector in 1985 leads to a 5.6% actual RWT during the same year and to a simultaneous increase of private jobs of 4.2%. In 1989 5.4% more jobs than observed in the base simulation remain. Unemployment becomes 17.5% lower in 1985 and even 19.1% in 1989, involving a reduction of the unemployment rate from 13.0% in the base simulation to 10.7% in 1985 and from 14.7% in the base projection in 1989 to 11.9%. The 10% RWT in 1985 involves a 2.8% decrease of gross purchasing power of the employed in the private sector over the whole period 1984–1989. Hence, there is a trade-off between creation of jobs and buying power, which is investigated in detail in some other simulations, also with respect to the development of profits of private enterprises. The effects of a similar RWT in the government sector are equally investigated.

26.1 INTRODUCTION

In this chapter we investigate, on some main macro-economic variables, the impact of a 10% once-and-for-all reduction of working time (RWT) with the Belgian

MARIBEL model. There is also a comparison with some other national and international models where experiments of RWT have been made. The major point of the chapter concentrates upon the effect of RWT upon employment and unemployment. Also covered is the effect upon production, productivity, profits and competition, and upon the budget deficit of the government. In this chapter RWT is defined as the decrease of total hours worked per employee per year. A RWT over the whole life cycle, by means of increasing the school leaving age or earlier retirement, is kept out of this chapter.

In Section 26.2 some essential relationships of the model that we have used are discussed. In Section 26.3 the simulations are described. In Section 26.4 the effects of a RWT for the employees are investigated, while in Section 26.5 the impact for the employers is pointed out. The results of the government's deficit are reviewed in Section 26.6. In section 26.7 an international comparison is made. Lastly, general conclusions are discussed in Section 26.8

26.2 THE MARIBEL MODEL: AN OVERVIEW

26.2.1 Introduction

The MARIBEL model of the Belgian Planning Office is used twice a year (January–July) for forecasting the Belgian economy for a planning horizon of 5 years.

MARIBEL (Model for Analysis and Rapid Investigation of the Belgian Economy) is a macroeconomic model for the postwar open economy of Belgium. The databank of MARIBEL is called MIRABEL, and contains statistical information from 1953 onwards. The main source of the data is the National Accounts of the National Institute of Statistics. MARIBEL itself is very aggregated for private households and for private firms. The model ignores sectoral or regional subdivisions. The information collected about the public sector is more disaggregated (i.e., local and central government, social security, etc.) In Sections 26.2.2 and 26.2.3 two of the principal relations with respect to RWT, i.e. the production function and the employment function respectively, are described. For the purpose of this chapter, these relationships are the most important. However, when relevant, other functions will be discussed too. Both production and employment functions are those of the private sector only.

26.2.2 The production function of the MARIBEL model

(a) Production

The production model used in MARIBEL considers homogeneous output produced by homogeneous inputs. In the medium run the production technology is described by a Cobb Douglas production function with Hicks' neutral technical progress, showing a unitary elasticity of substitution between capital and labour. Relationships for the (average) labour productivity and the (degree of utilization of the) production capacity of MARIBEL are discussed now.

(b) Productivity

The parameters of the productivity function are estimated by the following equation; written for period $t = 1, 2, \ldots T$ as:

$$\ln \text{QAFEHF}_t = c_1 + c_2 \ln\left(\frac{\text{KF}}{\text{LF}}\right)_t + c_3 \text{TIME} + c_4 \ln \text{QAFEHF}_{t-1} + \varepsilon_t \quad (26.1)$$

where

QAFEHF_t = the (average) labour productivity, constant BF of 1975
 $= 1000 \times (\text{QAFF/LF})_t$
QAFF_t = the volume of production, i.e. the gross domestic product at factor cost of the private sector in constant prices (10^9BF of 1975)
KF_t = the capital stock of the private sector in constant prices (10^9BF of 1975)
LF_t = quantity of labour input in the private sector (10^6 working hours per year)[1]
ε_t = a random error term with expectation 0 and σ_ε^2 variance

An econometric estimation of relationship 26.1 over the period 1954–1981, by ordinary least squares, yields the following point estimates (t-statistics between brackets)

$$\left.\begin{array}{c} c_1 = 1.88 \, (2.2) \\ c_2 = 0.25 \, (1.88) \\ c_3 = 0.002 \, (0.36) \\ c_4 = 0.70 \, (6.3) \\ \text{Durbin Watson statistic} = 1.6 \\ R^2 = 0.99 \end{array}\right\} \quad (26.2)$$

It is noticeable that the coefficient of technological progress, c_3, is very small (0.002) and insignificant. An adjustment of the parameter of technological progress from 0.002 to 0.010 – this would mean an average extra technological progress of 1% per year – seemed to be impossible: estimating this leads to a diverging model.

(c) Production capacity and degree of utilization of capacity

In MARIBEL, the capacity of production is constructed by substituting potential

---

[1] A reduction of LF does not necessarily imply a RWT, since LF is the product of average working time and the number of workers.

inputs for real inputs in the long run CD-production function. So, one obtains from Equation 26.1 the long run production function (26.3) ($c_5 = 0$):

$$\ln \text{QAFF}_t = \frac{c_1}{1 - c_4} - \ln 1000 + \frac{c_2}{1 - c_4} \ln \text{KF}_t + \left(1 - \frac{c_2}{1 - c_4}\right) \ln \text{LF}_t$$

$$+ \frac{c_3}{1 - c_4} \text{TIME} + \frac{\varepsilon_t}{1 - c_4} \qquad (26.3)$$

In Equation 26.3 real input of labour (LF) can be replaced by the potential quantity of labour input (LPF), LPF being defined as

$$\text{LPF}_t := (\text{EOF}_t + \text{UL}_t) \times \text{HPDF}_t \qquad (26.4)$$

Table 26.1 Production, production capacity and degree of utilization of capacity, private sector, Belgium, 1953–1981

Year	Production (QAFF) 10^6BF75	Production capacity (QPF 10^6BF75)	Degree of utilization of capacity (ZQF) (%)
1953	693 776	753 884	92.0
1954	722 846	783 838	92.2
1955	757 781	815 952	92.9
1956	780 655	857 783	91.0
1957	795 004	890 043	89.3
1958	782 577	914 260	85.6
1959	806 443	937 060	86.1
1960	851 587	973 308	87.5
1961	897 546	1 021 861	87.8
1962	946 604	1 082 976	87.4
1963	982 594	1 136 162	86.5
1964	1 057 599	1 184 912	89.3
1965	1 091 898	1 235 166	88.4
1966	1 122 983	1 302 165	86.2
1967	1 167 076	1 368 235	85.3
1968	1 220 163	1 423 340	85.7
1969	1 299 414	1 486 178	87.4
1970	1 384 770	1 560 020	88.8
1971	1 439 034	1 637 301	87.9
1972	1 516 843	1 704 782	89.0
1973	1 615 904	1 781 713	90.7
1974	1 691 692	1 871 907	90.4
1975	1 646 228	1 943 108	84.7
1976	1 751 096	2 002 303	87.5
1977	1 759 497	2 053 455	85.7
1978	1 810 633	2 102 555	86.1
1979	1 852 423	2 153 727	86.0
1980	1 908 779	2 209 404	86.4
1981	1 862 504	2 254 723	82.6

Source: MARIBEL

where:

- EOF_t = total employment in the private sector
- UL_t = unemployment, all categories and
- $HPDF_t$ = the number of per capita contractual working hours per year

If Equations 26.3 and 26.4 are combined this will result in Equation 26.5, which is the potential production function:

$$\ln QPF_t = c'_1 + \frac{c_2}{1-c_4} \ln KF_t + \left(1 - \frac{c_2}{1-c_4}\right) \ln LPF_t$$
$$+ \frac{c_3}{1-c_4} TIME + \eta_t \qquad (26.5)$$

where:

- QPF_t = potential output, i.e. the production capacity
- LPF_t = potential quantity of aggregate labour input in the private sector (10^6 hours/year)

The degree of utilization of capacity ZQF is defined as the ratio between actual production and the production capacity, or

$$ZQF_t := (QAFF/QPF)_t \qquad (26.6)$$

where ZQF_t denotes the degree of utilization of capacity.[2]

In Table 26.1 postwar data for the volume of output, production capacity, and the degree of utilization of capacity are collected. Figures 26.1 and 26.2 represent the same data.

26.2.3 The employment function of the MARIBEL model

(a) Employment

The firms demand a theoretical aggregate number of hours of work. Dividing this by the contractual working time per worker leads to the number of people employed. There are two distorting variables, which are the real wage cost and a tension on the labour market, which can be noted as in Equations 26.7 and 26.8

$$\frac{WFR_t}{PAFF_t} \qquad (26.7)$$

where:

- WFR_t = the rate of wage cost in the private sector (per hour per worker)

---

[2] Notice that underutilization of production capacity is assumed to be proportional to underutilization of potential labour input.

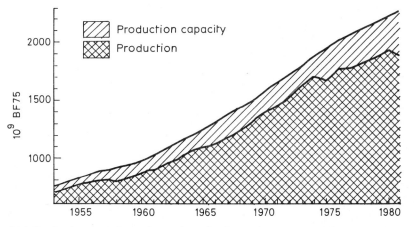

Fig. 26.1 Production capacity and actual production, private sector, Belgium, 1953–1981. Source: MARIBEL (1985)

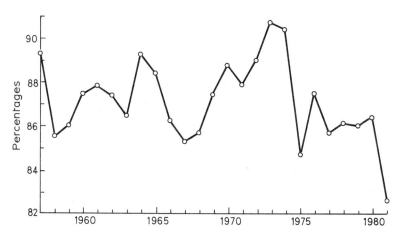

Fig. 26.2 Rate of utilization of production capacity, private sector, Belgium, 1953–1981. Source: MARIBEL (1985)

$PAFF_t$ = the price index (1975 = 1) of gross domestic product at factor cost of the private sector

and

$$\frac{(NA_t - EGO_t) - EOFT_t}{NA_t - EGO_t} \qquad (26.8)$$

where:

NA_t = the active residential population

EGO$_t$ = the total employment in the public sector (including the unemployed occupied by public authorities)
EOFT$_t$ = employment theoretically planned by the enterprises

These two distorting variables produce a gap between the theoretically maximum planned employment and the actual employment. Because of the rigidity on the labour market, firms do not recruit (or dismiss) immediately in times of an increase (or decrease) in production. They do however reorganize the disposable labour (labour hoarding). The short-run employment function can be written as

$$\Delta \ln \text{EOF}_t = \alpha_1 \Delta \ln \text{EOF}_t^* + \alpha_2 (\ln \text{EOF}_{t-1}^* - \ln \text{EOF}_{t-1})$$
$$+ \alpha_3 \ln(\text{WFR/PAFF})_t + \alpha_4 [1 - (\text{LF/HPDF})/(\text{NA} - \text{EGO})]_2 + z_t \quad (26.9)$$

while the planned employment function is assumed to be:

$$\ln \text{EOF}_t^* = \alpha_5 \ln(\text{LF}_t/\text{HPDF}_t) \quad (26.10)$$

where

HPDF$_t$ = the yearly number of contractual working hours
z_t = a random error term standing for omitted variables

The short-run function takes care of the two above-mentioned distorting factors. The estimation of employment function 26.9, where 26.10 has been substituted, by the nonlinear generalized least squares procedure leads to the following results, for the sample period 1954–1984.

$$\alpha_1 = 0.65 \ (6.77)$$
$$\alpha_2 = 0.67 \ (4.58)$$
$$\alpha_5 = 0.99 \ (915) \quad (26.11)$$
$$\alpha_3 = —$$
$$\alpha_4 = 0.42 \ (3.18)$$
Durbin Watson = 1.4

The results can be interpreted as follows:

(1) contractual employment is almost equal to the theoretically planned employment (α_5 is almost 1);
(2) in the short run, there is a considerable difference between actual and theoretically planned employment ($\alpha_1 = 0.65$), which remains for a large part in the long run;
(3) α_3 does not have a value in this estimation; other estimations lead to very low values of this coefficient (0.004 or 0.008) with large variances, which points to multicollinearity. Hence, real wage costs do not have a significant impact on the division of total labour between employment and working times
(4) the indicator of the tension on the labour market is very significant.

It should not be forgotten that the employment functions 26.9 and 26.10 are only valid for the private sector. The employers' employment is not estimated separately.

(b) Contractual and actual working time

As we have seen in (a), the contractual working time is important for determining the aggregate theoretically planned labour input. The actual working time is given by dividing the total quantity of labour by employment (the outcome of the

Table 26.2 Contractual and actual working time, private sector, Belgium, 1953–1985

Year	Contractual WT(HPDF) (hours/year/employee)	Actual WT(HDF) (hours/year/employee)
1953	2336	2140
1954	2335	2147
1955	2332	2150
1956	2298	2114
1957	2274	2042
1958	2274	2025
1959	2262	2018
1960	2236	2045
1961	2227	2018
1962	2225	2049
1963	2223	2066
1964	2209	2043
1965	2157	1995
1966	2129	1973
1967	2102	1942
1968	2093	1950
1969	2061	1933
1970	2024	1877
1971	2008	1849
1972	1975	1788
1973	1926	1758
1974	1887	1702
1975	1859	1584
1976	1847	1600
1977	1840	1579
1978	1836	1575
1979	1827	1564
1980	1792	1550
1981	1790	1503
1982	1784	1502
1983	1775	1503
1984	1766	1494
1985	1762	1474

Source: MARIBEL

employment function). In symbols, this can be written as follows

$$\text{HDF}_t = \frac{\text{LF}_t}{\text{EOF}_t} \qquad (26.12)$$

where HDF_t = the number of actual working hours per year.

In Table 26.2 the development of contractual and actual working time is summarized; Fig. 26.3 gives another presentation of the same data. Contractual working time decreased by 24.6% and actual working time by 31.1% during the period 1953–1985.

(c) Degree of utilization of the potential quantity of labour

This degree of utilization of potential quantity of labour is defined as the real quantity of labour divided by the potential quantity of labour, or

$$\begin{aligned}\text{ZHF}_t &= (\text{LF}/\text{LPF})_t \\ &= (\text{EOF} \times \text{HDF})_t / (\text{EOF} + \text{UL})_t \text{HPDF}_t \end{aligned} \qquad (26.13)$$

where:

ZHF_t = degree of utilization of the potential quantity of labour
LPF_t = potential quantity of labour input in the private sector (10^6 hours/year)
UL_t = unemployment of all categories

In Table 26.3 all the variables of Equation 26.13 are given.

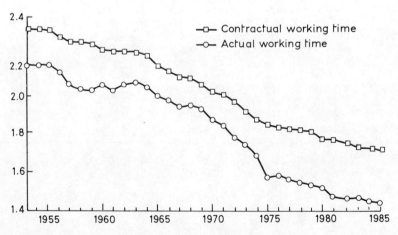

Fig. 26.3 Contractual and actual working time, private sector, Belgium, 1953–1985. Source: MARIBEL (1985)

Table 26.3 Employment, unemployment, quantity of labour, potential quantity of labour and degree of utilization of potential quantity of labour, Belgium, 1953–1985

Year	Employment private sector (EOF) employers + employees	Unemployment (UL)	Quantity of labour (10^6 hours/year), private sector (LF)	Potential quantity of labour (10^6 hours/year), private sector (LPF)	Degree of utilization of potential quantity of labour (ZHF) (%)
1953	3 022 091	191 890	6467.3	7507.9	86.1
1954	3 013 494	173 701	6471.0	7443.3	86.9
1955	3 018 426	104 778	6489.6	7283.3	89.1
1956	3 061 046	95 656	6471.1	7254.1	89.2
1957	3 089 054	82 487	6307.8	7212.1	87.5
1958	3 038 541	121 227	6153.0	7185.3	85.6
1959	2 967 851	119 620	5988.1	6983.9	85.8
1960	2 974 824	103 793	6083.5	6883.8	88.4
1961	3 000 770	80 903	6055.6	6862.9	88.2
1962	3 036 295	70 871	6221.4	6913.4	90.0
1963	3 029 062	57 037	6258.0	6860.4	91.2
1964	3 041 156	47 329	6213.1	6822.5	91.0
1965	3 026 277	60 208	6037.4	6657.5	90.7
1966	3 021 351	60 032	5961.1	6560.3	90.9
1967	3 000 034	82 321	5826.1	6479.1	89.9
1968	2 990 548	99 514	5831.6	6467.5	90.2
1969	3 039 069	78 695	5874.5	6425.7	91.4
1970	3 086 034	68 067	5792.5	6383.9	90.7
1971	3 114 243	66 456	5758.2	6386.8	90.2
1972	3 095 396	85 284	5534.6	6281.8	88.1
1973	3 132 595	90 038	5507.1	6206.8	88.7
1974	3 179 254	97 040	5411.1	6182.4	87.5
1975	3 123 000	177 535	4946.8	6135.7	80.6
1976	3 090 598	238 880	4944.8	6149.5	80.4
1977	3 074 998	273 708	4855.4	6161.6	78.8
1978	3 050 876	299 997	4805.1	6152.2	78.1
1979	3 061 626	316 041	4788.4	6171.0	77.6
1980	3 051 557	336 314	4729.9	6071.1	77.9
1981	2 973 005	433 445	4468.4	6097.5	73.3
1982	2 922 921	511 785	4390.2	6127.5	71.6
1983	2 895 797	545 109	4352.4	6107.6	71.3
1984	2 897 335	546 964	4330.2	6082.6	71.2
1985	2 903 047	549 553	4280.3	6083.5	70.4

Source: MARIBEL

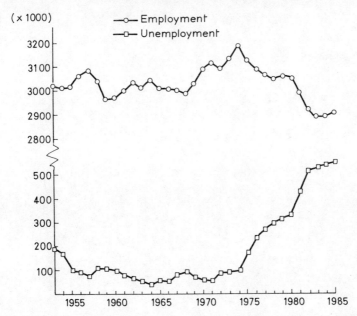

Fig. 26.4 Employment and unemployment, Belgium, 1953–1985. Source: MARIBEL data

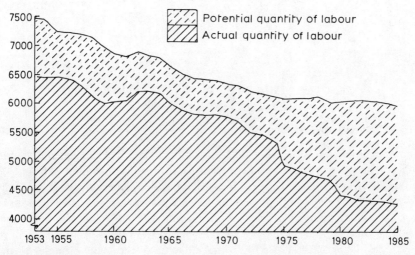

Fig. 26.5 Quantity of labour and potential quantity of labour, Belgium, 1953–1985. Source: MARIBEL (1985)

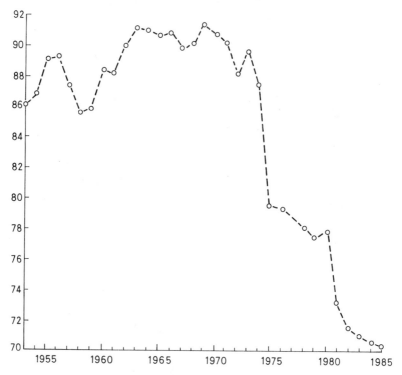

Fig. 26.6 Rate of utilization of potential quantity of labour, Belgium, 1953–1985. Source: MARIBEL data

26.3 THE SIMULATIONS OF RWT, PERFORMED BY THE MARIBEL MODEL

26.3.1 The base projection

The base projection is the May 1985 projection 1985–89 of the Belgian Planning Office. There were two alternative plans in this projection: one with and one without a fiscal decrease. We opted for the latter scenario, because the expenditures side of this projection was not yet built when we started this investigation. In the base projection, there is already implied a small RWT value: the contractual working time is 1766 hours per worker per year in 1985, and 1744 in 1989 (−1.25%). Table 26.4 shows the values of some of the main variables as they are shown in the base projection.

26.3.2 Simulations of RWT

In the eight simulations of RWT we investigate here, we have a reduction of contractual working time of 10% in 1985, which is a single but once-and-for-all reduction. According to the model, this leads to a reduction of actual working time of 5.6% in that first year, which results in a decrease of 2 hours a week.

Simulation 1 models the effect of reducing the contractual working time in the private sector by 10%, all the other endogenous variables of the base projection staying endogenous. The exogenous variables of the base projection remain exogenous. *Simulation 2* repeats the same scenario but looks at the effect of a single real wage increase of 5% per hour. Together with a 5.6% actual RWT, this leads to almost constant real gross wages per year per employee upon the base simulation. In *simulation 3* the real gross wages per employee are at least kept equal compared to the base projection for the next 5 years. This is made by increasing the real hourly wages each year by the same amount as the decrease of the actual working time in the same year. The result of this operation is at least to maintain the gross purchasing power per employee per year. *Simulation 4* is identical to simulation 3, but together with a so-called MARIBEL operation. This means that the employers' labour costs will be reduced by lowering their social security contributions with 50 billion BF. The governmental budget does not reduce, because the operation is financed by increasing the VAT also by 50 billion BF. Whereas the extra real wage increase is 5% in simulation 2, it is just 2.5% in *simulation 5*. At the same time, mandatory employment is imposed. Both the total number of hours worked in the private sector, and the number of employees, are increased by 2.5% with respect to the base simulation. *Simulation 6* is almost identical, but only the number of employees must increase (nothing is said about an increase in the total number of working hours). In *simulation 7*, which is based on simulation 1, RWT is also introduced for the public sector. The contractual RWT is only 5%, whereas in the public sector, the contractual and the actual working time are equal, because of absence of overtime hours. In all the simulations mentioned so far, the supply of labour is supposed not to be influenced by RWT. In *simulation 8*, this assumption is changed. Because of lower real wages per year, it may be necessary for some families to have a second earner (additional workers' effect). Moreover, just because of the RWT, for a lot of people (almost all women), it is more practical to enter labour force now. This results in an assumed increase of labour force by 20% of the extra employment. Wage formation is free (endogenous). The assumption of 20% of extra employment is based on a survey for the Netherlands (Bakhoven and Jansen, 1984). Maybe this rate is too high for Belgium, where the female employment rate is already much higher than in the Netherlands. We also wanted to simulate RWT taking care of the phenomenon of scrapping capital goods: through RWT, firms reorganize and modernize labour. Older, labour intensive, investments will be scrapped earlier and substituted by more capital intensive investments. Simulating this seems impossible with the available version of MARIBEL. A survey of all the simulations is given in Table 26.5.

Table 26.4 The May 1985 projections 1985–89 of the Belgian Planning Office: hypotheses and main results of the base projection

	1984	1985	1986	1987	1988	1989
A. *International context*						
Volume of world trade Δ	7.15	4.75	4.39	5.14	5.21	4.60
World export price in $ Δ	−4.40	−1.45	5.24	4.75	5.27	5.79
World export price in BF Δ	7.97	3.32	3.76	3.27	3.80	4.32
Exchange rate (BF/US $)[a]	57.71	60.51	59.66	58.82	58.00	57.19
German discount rate Δ	4.30	4.00	4.00	3.90	3.90	4.10
Eurodollar discount rate Δ	11.00	10.00	10.00	9.62	11.10	12.00
B. *National context*						
Private cons. 10^9 BF 75	1651.9	1639.2	1630.0	1644.1	1663.9	1687.1
Public cons. 10^9 BF 75	460.1	468.8	464.4	464.7	464.9	465.3
Investments 10^9 BF 75						
(i) Firms	288.0	304.1	309.7	324.1	338.0	352.8
(ii) Government	79.9	78.5	78.4	78.4	78.4	78.5
(iii) Buildings	87.8	93.3	93.9	92.6	94.0	100.0
GNP, market price 10^9 BF 75	2723.2	2767.6	2806.7	2872.8	2947.6	3022.8
Inflation: price of consumption (1975 = 1)	1.773	1.869	1.940	2.003	2.074	2.151
Employment 10^3 persons						
(i) private sector	2897.3	2903.0	2890.1	2881.6	2878.2	2871.7
(ii) public sector	656.7	666.5	669.2	668.5	667.7	667.0
Unemployment 10^3 persons	547.0	549.6	575.3	599.3	616.3	629.6
Net financing shortage of the government[b]	−505.3	−510.8	−510.8	−494.3	−457.0	−410.7
10^9 BF and % GNP	(−11.2)	(−10.7)	(−10.2)	(−9.4)	(−8.2)	(−6.9)

Source: Belgian Planning Office, May 1985.

[a] This value is considered too high: on 31 December 1985, the exchange rate of 1 US $ was 50.36 BF, on 31 January 1986 it was 48.87 BF per US $. This lower dollar rate than the one supposed in MARIBEL, leads to a better competitive position of the Belgian economy, with all the consequences.
[b] January 1986: the net financing shortage of the government is calculated to be 553.4 billion BF in 1985, without the public housing debt (11.7%).

Table 26.5 Survey of all the RWT simulations, with the help of the MARIBEL model (deviations of the base simulation)

Simulation	Contractual working time	Private real wage cost per employee/hour	Employment private sector	Employment public sector	Active population
1	−10% in 1985, private	Endogenous	Endogenous	—	—
2	−10% in 1985, private	+5% in 1985	Endogenous	—	—
3	−10% in 1985, private	+5% in 1985 +↑1985–1989 = ↓ actual working time	Endogenous	—	—
4	−10% in 1985, private	as for 3 above +50 billion BF operation SS and VAT	Endogenous	—	—
5	−10% in 1985, private	+2.5% in 1985	+2.5% mandatory +LF↗2.5% +2.5%	—	—
6	−10% in 1985, private	+2.5% in 1985			
7	−10% in 1985, private −5% in 1985, public	0%	Endogenous	+5%	—
8	−10% in 1985, private	0%	Endogenous	—	+20% Δ employment

Table 26.6 Growth rates of working time, employment and unemployment in the private sector, by a contractual reduction of working time of 10% in 1985 – Belgium

Simulation	Contractual working time		Actual working time				Employment				Unemployment			
	1985	84–89	1985	84–86	86–89	84–89	1985	84–86	86–89	84–89	1985	84–86	86–89	84–89
Base	−0.2	−1.2	−1.1	−2.7	−1.6	−4.3	0.1	−0.5	−1.0	−1.5	0.5	5.2	9.4	15.1
Simulation 1	−10.2	−11.2	−5.6	−8.8	−2.8	−11.4	4.4	5.0	−1.2	3.8	−17.1	−17.8	13.2	−6.9
Simulation 2	−10.2	−11.2	−5.8	−9.6	−3.1	−12.4	4.1	4.2	−2.2	1.9	−16.2	−14.6	17.9	0.6
Simulation 3	−10.2	−11.2	−5.9	−9.8	−4.1	−13.5	4.1	4.1	−3.5	0.5	−16.2	−14.1	24.4	6.9
Simulation 4	−10.2	−11.2	−5.8	−9.5	−4.1	−13.3	4.2	4.4	−3.2	0.0	−16.3	−15.1	23.3	4.7
Simulation 5	−10.2	−11.2	−5.7	−9.1	−3.3	−12.1	8.0	6.7	−3.5	3.0	−32.3	−24.9	28.1	−3.8
Simulation 6	−10.2	−11.2	−7.6	−9.8	−2.3	−11.9	6.8	5.4	−2.4	2.9	−27.0	−19.5	20.3	−3.2
Simulation 7	−10.2	−11.2	−5.4	−8.6	−2.8	−11.1	3.9	4.4	−1.3	3.0	−21.3	−21.3	14.4	−9.9
Simulation 8	−10.2	−11.2	−5.6	−9.0	−2.8	−11.5	4.6	5.3	−1.1	4.2	−14.7	−15.7	12.7	−5.1

26.4 THE EFFECTS OF A RWT FOR THE EMPLOYEES

26.4.1 Employment and unemployment, wages and purchasing power, by the MARIBEL simulations of RWT

(a) Some data and main observations

In this chapter, we focus our interest principally on the impact of a RWT on employment and unemployment. Table 26.6 gives growth rates of employment, unemployment, and working time. A distinction is made between the short run and medium term impact.

It is clear that, over the whole period 1984–89, the actual working time decreases more than the contractual working time, although the extra decrease of the contractual WT in the variants with respect to the base projection results in a slightly extra decrease of actual WT, also with respect to the base projection. In 1985, the effects are vice versa. During that first year, adjustments of the production process are quite impossible; these become possible only in the medium run. Employment increases less than the working time – even the actual one – decreases, except in simulation 5. The employment effect has its largest impact in the short run. This effect has weakened in the medium term, though over the whole period

Table 26.7 Employment and unemployment in the private sector, by a contractual RWT of 10% in 1985, absolute numbers + % of active population (between brackets)

	Variable					
	Employment			Unemployment		
Simulation	1984	1985	1989	1984	1985	1989
Base	2 267 335	2 270 047	2 232 744	546 964	549 553	629 556
	(53.0)	(53.5)	(52.1)	(12.4)	(13.0)	(14.7)
Simulation 1	2 267 335	2 366 158	2 353 011	546 964	453 442	509 289
	(53.6)	(55.8)	(54.9)	(12.4)	(10.7)	(11.9)
Simulation 2	2 267 335	2 361 079	2 311 947	546 964	458 520	550 352
	(53.6)	(55.7)	(54.0)	(12.4)	(10.8)	(12.8)
Simulation 3	2 267 335	2 361 036	2 277 661	546 964	458 564	584 639
	(53.6)	(55.7)	(53.2)	(12.4)	(10.8)	(13.6)
Simulation 4	2 267 335	2 361 600	2 289 895	546 964	457 999	572 404
	(53.6)	(55.7)	(53.4)	(12.4)	(10.8)	(13.4)
Simulation 5	2 267 335	2 449 473	2 336 256	546 964	370 126	526 043
	(53.6)	(57.8)	(54.5)	(12.4)	(8.7)	(12.3)
Simulation 6	2 267 335	2 420 589	2 332 667	546 964	399 011	529 633
	(53.6)	(57.1)	(54.4)	(12.4)	(9.4)	(12.4)
Simulation 7	2 267 335	2 355 878	2 336 105	546 964	430 399	492 861
	(53.6)	(55.5)	(55.5)	(12.4)	(9.4)	(11.5)
Simulation 8	2 267 355	2 371 848	2 361 980	546 964	466 747	519 283
	(53.6)	(55.7)	(54.9)	(12.4)	(12.8)	(12.1)

there is still a positive effect, certainly compared with the base projection.

Table 26.7 summarizes the effect of RWT on employment and unemployment of the employees in the private sector, in absolute numbers. Employment and unemployment are also expressed as percentages of active population.

From Tables 26.6 and 26.7 we can deduce some conclusions concerning the impact upon (un)employment as a consequence of a RWT.

(1) Employment decreases in the base projection over the period 1985–89. That same phenomenon is visible in the projections, but the decrease is not as pronounced.
(2) The simulations where the real wage costs increase are those where the fewest jobs are produced. When the hourly wage cost per employee is kept constant, the employment effect is the largest.
(3) The extra supply of labour leads to an increase in the creation of jobs.
(4) Legal levels of employment produce more jobs only in the short term.
(5) A new MARIBEL operation (see above) leads to a considerable increase in job creation only in the medium term.
(6) A RWT in the public sector accompanied by legal extra increase in jobs, results in a decrease in the gross purchasing power of public servants.

These conclusions will be investigated in more detail in Sections 26.4.1(b)–(g).

(b) The decrease of employment is smaller in the simulations of RWT than in the base projection

It is noticeable that employment declines in all simulations during the period 1986–89. The decrease is larger than in the base projection. This is due to the planned employment function (see Equation 26.10). Employment is only a function of total labour quantity divided by contractual working time in Equation 26.10. A smaller contractual working time goes together with a larger employment level in the short run. In the medium run, a part of the employment is lost; the decrease is larger in the simulations: because of the RWT, the planned employment decreases.

(c) A choice to make: more employment or more purchasing power

Gross purchasing power per employee is defined here as the gross real wage per employee per year. Wealth income is not taken into account. Contributions and taxes are not subtracted, neither are social security benefits added.

The purchasing power of the self-employed and the public servants is not included in Table 26.8. The purchasing power of officials decreases by 4.5% over the period 1984–89, except in simulation 7. There is a RWT for functionaries of 5%, without any real wage increase; the fall in buying power is 9%. The purchasing power of the unemployed people does not change due to a RWT.

From Tables 26.6, 26.7 and 26.8 one can conclude that the results on job creation are larger the less the hourly wages per employee rise. The extra creation of newly

Table 26.8 Growth rates of yearly nominal wages per employee, inflation and gross purchasing power per employee, private sector, by a contractual RWT of 10% in 1985 – Belgium

Simulation	Variable											
	Nominal wage/employee/year				Inflation (consumption prices)				Gross purchasing power/employee/year			
	1985	84–86	86–89	84–89	1985	84–86	86–89	84–89	1985	84–86	86–89	84–89
Base	2.0	3.3	22.0	26.0	5.4	9.4	10.9	21.3	−3.2	−5.6	10.1	3.9
Simulation 1	−1.7	−1.8	22.0	19.8	6.1	10.9	11.1	23.3	−7.3	−11.5	9.8	−2.8
Simulation 2	3.6	3.5	22.6	27.0	6.8	12.5	11.4	25.4	−3.0	−8.0	10.0	1.3
Simulation 3	3.4	6.7	27.4	35.9	6.8	12.9	13.0	27.6	−3.2	−5.5	12.7	6.5
Simulation 4	4.6	7.6	27.0	36.6	8.0	13.9	12.8	28.5	−3.1	−5.5	12.5	6.3
Simulation 5	1.4	1.9	24.0	26.4	6.9	12.7	11.5	25.6	−5.2	−9.6	11.3	0.6
Simulation 6	1.0	0.2	23.2	23.4	6.5	11.7	11.3	24.4	−5.2	−10.3	10.6	−0.8
Simulation 7	−1.4	−1.5	22.2	20.4	6.1	11.1	11.2	23.5	−7.1	−11.3	9.9	−2.5
Simulation 8	−1.8	−2.0	22.0	19.5	6.1	10.9	11.1	23.2	−7.4	−11.7	9.8	−3.0

specified full-time jobs is the largest in the simulation where real gross yearly salary per employee decreases due to the RWT, because of a constant real hourly wage per employee (simulation 1). A RWT of 10% in 1985 produces in that same year 99 000 extra jobs with respect to 1984 (+4.4%). This is a difference of 96 000 (+4.2%) with respect to the base projection. In 1989, the results do decrease in absolute numbers: with respect to 1984, there are 85 000 (+3.8%) extra jobs. Compared with the data for 1989 in the base projection, there are 120 000 (+5.4%) more jobs. Hence, an introduction of a RWT results in a smaller employment decline, so that more newly specified full-time jobs remain.

Unemployment, as a percentage of the active population, decreases from 13.0% in the base projection to 10.7% in simulation 1 for 1985. For 1989, these percentages are respectively 14.7% (base) and 11.9% (simulation 1). This last percentage is lower than in 1984, when unemployment was 12.4%. When the gross real hourly wages of the employees increase during the simulation period 1985–89, raising gross purchasing power, the increase in job creation is smaller. In this case (simulation 3) in 1985 there are 94 000 (+4.1%) jobs more than in 1984, which means a surplus of 91 000 (+4.0%) with respect to the base projection in 1985. For 1989, however, there remain only 10 000 (+0.5%) more people employed in comparison with 1984, or 45 000 (+2.0%) with respect to the base projection in 1989. This leads to an unemployment rate of 10.8% and 13.6% respectively for 1985 and 1989.

If one wants to keep the gross buying power per employee constant, the result will be between simulations 1 and 2. In simulation 2, the hourly real wage increase is equal to the RWT only in the first year. Reducing the number of working hours has in all cases positive results with respect to the base projection, but the effects are smaller when hourly wage costs increase. In this case, the firms try to reorganize labour to increase the productivity of labour. Table 26.9 gives the forecast of

Table 26.9 Growth rate of hourly labour productivity per employee in the private sector, by a contractual RWT of 10% in 1985 – Belgium

	Variable			
	Labour productivity/hour/employee			
Simulation	1985	84–86	86–89	84–89
Base	3.1	6.8	10.7	18.2
Simulation 1	3.4	7.5	11.7	20.1
Simulation 2	3.5	7.9	12.3	21.1
Simulation 3	3.5	8.0	13.1	22.1
Simulation 4	3.5	7.9	12.9	21.8
Simulation 5	0.0	5.0	13.7	19.4
Simulation 6	3.3	7.7	12.0	20.7
Simulation 7	3.3	7.6	11.8	20.3
Simulation 8	3.3	7.5	11.7	20.1

labour productivity per hour per employee in the private sector. The medium run growth rates are higher the more the real wages increase.

(d) Extra supply of labour leading to extra job creation

In simulation 8, a surplus supply of labour due to RWT is assumed. This can be explained by the additional worker effect that exists in the case of RWT, as has already been mentioned in Section 26.3.2. The surplus of workers entering the labour market is supposed to be 20% of the additional employment created by RWT (19 000 people). In this case, an extra 5700 jobs become available in 1985 with respect to simulation 1, or 102 000 (+4.5%) more jobs than in the base projection. Unemployment is 12.8% (13.0%) in the base projection and 10.7% in simulation 1. For 1989, the forecasts are as follows: 95 000 (+4.2%) above the 1984 level, which is an increase of 129 000 (+5.8%) with respect to the base projection and of almost 9000 with respect to simulation 1. It is noticeable that the increase of employment (+5.8%) in this case is larger than the reduction of working time (−5.6%). This is the only simulation where the reoccupation of jobs is larger than the RWT. Unemployment is higher than in simulation 1, also in 1989: it is 12.1% in simulation 8, 11.9% in simulation 1 and 14.7% in the base projection.

The reason for this phenomenon can be found in the MARIBEL employment function. This takes care of the tension on the labour market, which is partially explained by the rise of the active population (see Section 26.2.3).

(e) Legal levels of recruitment just lead to more jobs in the short run

The effect of legal levels of recruitment is investigated in simulation 6: 10% RWT, 2.5% hourly wage increase and 2.5% compensating recruitment. The results are as follows. In 1985, there are 153 000 (+6.8%) jobs more than in 1984, of which 2.5% are due to legal levels of employment. This results in 150 000 (+6.6%) extra jobs compared with the base projection. In 1989, there are only 65 000 (+2.9%) jobs left in comparison with 1984, or 100 000 (+4.5%) more with respect to the base projection. One-third of the surplus is lost in the medium run. Unemployment in 1985 is 9.4% and in 1989 12.4% (= stagnation with respect to 1984).

(f) A new MARIBEL-operation shows how more jobs can be created in the medium run

Some years ago, there was a so-called MARIBEL operation in Belgium. Part of it was a reduction of social security contributions by the employers, which was financed by a rise in VAT. Such an operation is portrayed here. There is a rise in indirect taxes in 1985 by 50 billion BF and a decrease of contributions to social security by the same amount. The intention of this operation is of course to lower the labour costs per employee.

This is done in simulation 4, which is the same as simulation 3 but with the

MARIBEL operation as extra measure. In this simulation, an increase of gross purchasing power appears which will be financed partly by the VAT. In 1985, this results in only 560 jobs more than the 94 000 mentioned in simulation 3. In 1989, there are already 12 000 jobs more than in 1984, which totals at 22 000 (from which 10 000 are already available in simulation 3). So, it is clear that the effect of the simple MARIBEL operation is considerably positive in the medium run.

(g) RWT in the public sector

Simulation 7 treats a RWT in public services. There is no variable in MARIBEL that defines working hours of functionaries. There is no difference between contractual and actual working time in the public sector. Neither overtime nor temporary unemployment are supposed, and the introduction of a contractual RWT is considered to be in force immediately. However, the 5% creation of jobs will be an overestimation: also in the public sector there are some jobs that are not separable. Higher skilful jobs must be done by one person, even if there is a RWT or not.

A reduction of working hours of 5% in 1985, together with 5% legal increase in recruitment, produces 43 000 (+6.6%) jobs, in comparison with 1984. That means 33 000 (+5.0%) more civil servants with respect to the 1985 base projection. In 1989, there are 44 000 (+6.7%) more jobs than in 1984, or 33 000 (+5.0%) more with respect to the base projection. This creation of jobs does remain, in contrast with the jobs created by enforcing the legal levels of recruitment in the private sector (see Section (e) above). This switch can be explained by looking at the wages: in Section (e) there is an increase of purchasing power, which is not the case here. In this simulation, the purchasing power of the public servants decreases by 9.0% during the period 1984–89 (base: 4.5%).

(h) An evaluation of the results

A conclusion of the simulations mentioned here is that the impact on employment of a contractual RWT of 10% in 1985, becomes more positive the more there is a reduction in the hourly wage rate. The growth rate is always less than the RWT.

A single once-and-for-all RWT cannot have the same impact on employment and unemployment as a RWT carried through during 5 successive years. This last scenario has been investigated by the Planning Office (Planbureau, 1985a). The RWT is annually 5% during 5 years, so in total there is a contractual RWT of 25%. This is an extreme situation, but it is interesting for pedagogical reasons. In the case of reduced hourly wages, the effect is cumulative: each year, there are new jobs. In the first year, there are 47 000 extra jobs with respect to the base projection, while in the fifth year there is a surplus of 251 000 jobs.

Some objections can be made regarding the results of all the MARIBEL simulations, as follows:

(1) Within a macroeconomic model such as MARIBEL, without sectoral subdivisions, it is impossible to distinguish between large and small firms, low and high productivity firms, although the size of the firm and the productivity level have an influence on the creation of employment. Small firms, with fewer people doing the same job, have more difficulties for creating a new job. These are rather disposed to make the employees work longer (i.e. overtime work) and will not recruit new workers (Késenne, and Butzen, 1984). While this distinction is not made in MARIBEL, the employment effect will be overestimated as a result.

(2) In the planned employment function of MARIBEL, real wages are not a significant factor. In reality, there are quasi-fixed labour costs, e.g., recruitment costs, vocational training costs, certain social allowances, clothes, canteen, payments for days not worked, etc. Hart (1984) estimates the ratio between quasi-fixed non-wage labour costs and total variable labour costs (i.e. per man-hour costs) at about 20% in the United States and the UK during the late 70s–beginning of the 80s. So, these costs bring the total labour costs to a higher level.

(3) The MARIBEL model does not take a sufficiently high technological change into consideration; only an increase of 0.5% per year has been supposed. MARIBEL contains a homogenous production model and not a vintage production model where technical progress can be estimated endogenously. Because of innovations and reorganizations, firms do increase hourly labour productivity per employee, and therefore need fewer people. There is an increase of labour productivity in the MARIBEL simulations, but this might be underestimated (see Table 26.9). Also this reason makes one believe that the simulated creation of jobs will be exaggerated.

(4) It is probable that there is also an overestimation of the employment effect due to the increase of informal and illegal work. Because of the RWT, people are doing some work at home themselves, instead of demanding it at the market. There is probably also a rise in illegal work because of the fall in wages and the higher costs of overtime work.

In what follows, some comparisons are made with other studies concerning RWT in Belgium.

26.4.2 A comparison with other Belgian investigations concerning RWT

(a) A simulation study of the Belgian Ministry of Economic Affairs

E. Pollefliet, a member of the Ministry of Economic Affairs, has also simulated the effect of RWT (Ministry of Economic Affairs, 1985). His own macroeconomic model, SMOBE (Short-time Model for the Belgian Economy) is used. As is clear from the name, this is a short-run model for the Belgian economy. The main hypotheses of the base projection are summarized in Table 26.10.

We do not investigate all the simulations executed, just the one that makes a comparison with the MARIBEL model possible. In that study, only the short-run effects (1984–86) of a 10% RWT are investigated, but no distinction is made between contractual and actual working time. Simulation 2 of the SMOBE model can be compared with simulation 2 of MARIBEL. In both cases, the wage per year per employee is kept constant. Table 26.11 gives some points of comparison between these two simulations.

It is remarkable that the development of the variables always goes in the same direction, but that the size can differ so much. The difference is considerable, especially concerning unemployment. This is due in a great part to deviations in the labour productivity, which is much higher in SMOBE, probably caused by the vintage character of this model since scrapping is explicitly taken into account. More than half of the initial employment effect (namely 48 700 jobs) is cancelled after 3 years. Only when the growth of labour productivity is very high will there be

Table 26.10 The SMOBE model, hypotheses of the base projection: growth rates – Belgium

	1984	1985	1986
Gross national product	2.2	1.5	1.5
Employment, private sector	−0.3	−0.2	−0.1
Unemployment, % active population	0.7	1.3	1.1
Labour productivity	2.32	2.03	2.05
Real wage cost/hour	−0.26	1.04	1.74
Gross real wage/employee	−0.45	0.14	0.77
Profit rate	1.50	1.55	1.68

Source: Ministry of Economic Affairs (1985, appendix).

Table 26.11 RWT of 10%, yearly wage per employee kept constant: a comparison of MARIBEL with SMOBE for the short term (percentage deviations from the reference simulation)

	1984–1986 MARIBEL	1984–1986 SMOBE
RWT	−10	−10
Employment	+4	+2
Unemployment (10^3)	−108.4	−42.2
Inflation	+2.9	+3.9
Hourly wage cost	+8.4	+10.5
Production	−2.7	−2.0
Labour productivity	+1.1	+6.3
Imports	+0.8	+1.0
Exports	−0.9	−3.2

Source: MARIBEL simulations and SMOBE simulations.

Table 26.12 The effects of a RWT by means of elasticities according to Drèze (1980)

	Weight (1)	Direct effect (2)	Indirect effects — Production loss (3)	Indirect effects — Increase of wages (4)	Total effect $(1) \times (2 + 3 + 4)$
Production loss	β	0	$-\eta_{N\bar{Y}|B}$	$(1-\alpha)\eta_{NW|B}$	$\beta[-\eta_{N\bar{Y}|B} + (1-\alpha)\eta_{NW|B}] +$
No production loss	$1-\beta$	η_{NY}	0	$(1-\alpha)\eta_{NY|B}\eta_{NY}$	$(1-\beta)[\eta_{NY} + (1-\alpha)\eta_{NW|B}\eta_{NY}]$ $= -\eta_{NT|B}$

Source: Drèze (1980), p. 13.
N = employment; Y = production; $\bar{Y}$ = production capacity; W = real wages; T = working time; M = imports of goods and services; X = exports of goods and services; η_{NW} = elasticity of N with respect to W; $\eta_{NW|B}$ = conditional elasticity of N with respect to W, given the balance of trades restriction.

no further scrapping. Calculations by the SMOBE model predict that the labour productivity must increase by 12% in the first year and 8% in the second. If not, labour will be replaced by capital and employment will decrease. Of course, these requirements are very difficult to achieve. Moreover, if they were achieved, there would be high production and the resulting problem of selling the products. There is no doubt about it: the SMOBE simulation study is more pessimistic when looking at the employment effects of RWT. We prefer the MARIBEL simulations, even taking into account the objections noted earlier. The growth rate of labour productivity might be too low in MARIBEL but it seems much too high in SMOBE.

(b) Drèze's investigation of RWT

J. Drèze (1980) takes as point of departure the situation of a small open economy with underemployment and a negative balance of payments. The impact of RWT on employment is investigated with the help of elasticities. The determining factors are (a) the measure of wage cut (α) and (b) the proportion of production loss (β). Drèze recognizes the role of a changing productivity as well, but he does not use it.

Table 26.12 shows how the total effect of RWT on employment is composed. It computes the opposite of the 'employment elasticity of a RWT', given the non-negative balances of trades.

The employment elasticity of a RWT has the following outlook (sum of total effects):

$$-\eta_{NT|B} = (1-\beta)\eta_{NY} - \beta\eta_{N\bar{Y}|B} + (1-\alpha)\eta_{NW|B}[\beta + (1-\beta)\eta_{NY}], \quad (26.14)$$

where

$$\eta_{N\bar{Y}|B} := \frac{\eta_{X\bar{Y}}(1-\eta_{MX}) - \eta_{M\bar{Y}}}{\eta_{MY}}\eta_{NY}, \quad (26.15)$$

with η_{MX} = the import elasticity of exports.

The total effect of a RWT on employment ($-\eta_{NT|B}$) depends on the relative part of (firms with) production loss (β), in comparison with the cases without any spare production ($1-\beta$). A fall of production will take place especially in those firms that have just one shift: reducing working time reduces production time and, hence, production itself if there is no extra productivity increase. In that case, there will be no extra employment. Drèze also takes into account the indirect impact of RWT on employment by adjusting the production capacity ($\bar{Y}$) and a rise of real wage costs (W). The (opposite) employment elasticity according to Equation 26.14 is computed now for various subperiods of the postwar Belgian economy. In Table 26.13 point estimates of the elasticities are given for the short run (Drèze's data, short run SR), for the period 1966–76 (Drèze's data, medium run MR), for the period 1973–83 (MARIBEL data, medium run MR), and for the period 1953–83 (MARIBEL data, long run LR). Interpreting these elasticities, it is noticed that the underlying sample periods are different. Moreover, Drèze's calculations are made only for the manufacturing sector, while the MARIBEL computations contain the whole private sector.

The effects of a RWT for the employers

Table 26.13 Elasticities by Drèze and MARIBEL

	Drèze (manufacturing sector)	MARIBEL (private sector)		
	SR	MR 66–76	MR 73–83	LR 53–83
η_{NY}	0.3	0.9	0.135	0.162
$\eta_{NW|B}$	−0.2	−1.8	−4.688	−0.036
$\eta_{X\bar{Y}}$	0.77	0.77	1.961	1.820
η_{MX}	0.4	0.4	0.83	0.950
η_{MY}	1.53	1.24	1.837	1.940
$\eta_{M\bar{Y}}$	−0.29	−0.29	2.044	1.729
$\eta_{N\bar{Y}|B}$	0.15	0.54	−0.099	−0.137

Source: Drèze (1980); own calculations with the MARIBEL databank.
SR = short run; MR = medium run; LR = long run.

There are some remarkable differences between the medium run elasticities by Drèze and MARIBEL. The most important deviation concerns the elasticity of imports with respect to production capacity ($\eta_{M\bar{Y}}$). Even the sign is different. Drèze supposes that imports decrease when production capacity increases, because of a substitution of imported goods by self-produced goods (the substitution effect). From the MARIBEL databank, imports increase more proportionally than an expansion of production would suggest. This can be explained by the increasing importing needs of primary and intermediary products. This deviation of $\eta_{M\bar{Y}}$ also has an impact on $\eta_{N\bar{Y}|B}$, which uses this elasticity. The long run elasticities follow in the same directions, although they can be larger or smaller. The most important difference is in $\eta_{NW|B}$, A change in real wages, given the balance of payments restriction, has a very small effect on employment when averaging over 20 years.

For alternative values of α (wage decrease) and β (production loss), the size of the employment elasticity of RWT is calculated. The advantage of this approach is that it makes clear what size of wage decrease is necessary to have a positive impact on employment of a RWT.

Table 26.14 gives a synopsis of the values of $-\eta_{NT|B}$, for Drèze and for MARIBEL, for various values of α and β.

From these calculations, it can be concluded that a reduction of real wages has a very small impact on employment in the short run. In the medium run, it is clear that a fall of real wages is indispensable to have a positive employment effect. If the increase of costs must be carried by the employers, the medium run employment effect is negative. This is also true for the medium run MARIBEL version even with half of the wage increase carried by employees. In the long run, all values of $-\eta_{NT|B}$ are almost equal. Neither the fall in wages nor the loss of production seem to have much impact on employment in the long run.

Drèze's theory (only positive employment effect of RWT when real wage cut) remains for the medium run only, when applied on more recent Belgian data,

Table 26.14 The employment of RWT for Belgium, by Drèze and MARIBEL

	\multicolumn{12}{c}{$-\eta_{NT	B}$}
	Drèze SR			Drèze MR(66–76)			MARIBEL MR(73–83)			MARIBEL LR(53–83)		
	α			α			α			α		
β	0	1/2	1	0	1/2	1	0	1/2	1	0	1/2	1
0	0.24	0.27	0.30	−0.72	0.09	0.90	−0.498	−0.181	0.135	0.156	0.159	0.162
1/4	0.09	0.14	0.19	−1.12	−0.29	0.54	−1.521	−0.697	0.126	0.142	0.149	0.156
1/2	−0.05	0.01	0.08	−1.53	−0.67	0.18	−2.543	−1.213	0.117	0.129	0.139	0.150

Source: Drèze (1980, p, 19); calculations with MARIBEL.
SR = short run; MR = medium run; LR = long run.

although there exist considerable differences in underlying elasticities. In the long run, the effects are almost vanishing which is not in accordance with the MARIBEL simulations discussed above.

26.5 THE EFFECTS OF A RWT FOR THE EMPLOYERS

In Section 26.4 of this chapter, we investigated the impact of a RWT on (un)employment and the financial position of employees. In this section, we try to point out the changing position of the firms as a consequence of the RWT. We investigate the evolution of gross wage costs and labour productivity, already mentioned in Section 26.4. Investments, production and production capacity might change also. As a consequence, profits may alter. Last, but not least, we look at the competititve position of the Belgian firms with respect to the world market. Table 26.15 summarizes the impact of RWT upon hourly labour costs and hourly labour productivity per employee, which is shown in growth rates. The third column of Table 26.15 gives the growth rates of real output.

It is clear that the hourly wage cost and the hourly labour productivity increase the most in simulations 3 and 4. The expected high hourly labour productivity per employee is not realized in the first year as the reorganization of labour takes some time. Therefore, the growth rates for 1985 are almost the same for all simulations. It is only in the period 1986–89 that there is a considerable difference among the simulations. Hence, firms seem to respond only slowly to an increase in the wage costs.

Another important consequence of RWT is the decline of real production, as can be seen in Table 26.15. The growth rate is always smaller than without RWT, but it is even smaller in the simulations with a wage increase. Table 26.16 shows the real value of production and production capacity, and the ratio between these two variables, which is the degree of utilization of production capacity.

Production decreases in all simulations with respect to the base projection, especially in those where real labour costs increase. The same conclusion is possible for production capacity. The use of production capacity is higher in all the

Table 26.15 Growth rates of nominal hourly labour costs, hourly labour productivity and real output by a conventional RWT of 10% in 1985, private sector, in Belgium

Simulation	Nominal labour costs/hour				Labour productivity/hour				Real output			
	1985	84–86	86–89	84–89	1985	84–86	86–89	84–89	1985	84–86	86–89	84–89
Base	3.4	6.2	24.0	31.6	3.1	6.8	10.7	18.2	1.9	3.6	8.3	12.2
Simulation 1	4.1	7.7	25.5	35.2	3.4	7.5	11.7	20.1	1.1	2.0	7.8	10.0
Simulation 2	10.0	14.6	26.0	45.0	3.5	7.9	12.3	21.1	0.7	0.9	7.0	8.0
Simulation 3	9.9	18.3	32.8	57.2	3.5	8.0	13.1	22.1	0.7	0.7	5.6	6.3
Simulation 4	11.1	18.9	32.4	57.5	3.5	7.9	12.9	21.8	0.7	1.1	5.7	6.8
Simulation 5	7.5	12.1	28.3	43.8	0.0	5.0	13.7	19.4	0.4	0.6	7.1	7.8
Simulation 6	7.1	11.1	26.1	40.1	3.3	7.7	12.0	20.7	0.8	1.4	7.5	9.0
Simulation 7	4.2	7.8	25.7	35.5	3.3	7.6	11.8	20.3	1.0	1.9	7.8	9.8
Simulation 8	4.1	7.6	25.5	35.1	3.3	7.5	11.7	20.1	1.1	2.1	7.8	10.1

Source: MARIBEL simulations.

Table 26.16 Production and production capacity in the private sector, in billion constant BF of 1975; degree of utilization of production capacity (%); due to conventional RWT of 10% in 1985 – Belgium

	Production			Production capacity			Degree of utilization of production capacity (%)		
	1984	1985	1989	1984	1985	1989	1984	1985	1989
Base	1954.4	1991.5	2192.1	2424.3	2491.0	2808.7	80.6	79.9	78.0
Simulation 1	1945.4	1975.4	2149.0	2424.3	2409.6	2705.4	80.6	81.9	79.4
Simulation 2	1954.4	1967.6	2111.1	2424.3	2408.5	2683.5	80.6	81.6	78.6
Simulation 3	1954.4	1967.7	2077.0	2424.4	2408.5	2669.4	80.6	81.7	77.8
Simulation 4	1954.4	1968.8	2087.8	2424.3	2408.6	2675.5	80.6	81.7	78.0
Simulation 5	1954.4	1962.9	2105.9	2424.3	2407.8	2680.4	80.6	81.5	78.5
Simulation 6	1954.4	1969.7	2129.9	2424.3	2408.8	2694.6	80.6	81.8	79.0
Simulation 7	1954.4	1973.3	2145.4	2424.3	2402.1	2696.5	80.6	82.1	79.5
Simulation 8	1954.4	1976.5	2151.3	2424.3	2413.8	2710.7	80.6	81.8	79.3

Source: MARIBEL simulations.

Table 26.17 Quantity and potential quantity of labour input in private sector, 10^6 hours per year; degree of utilization of potential quantity of labour; by a conventional RWT of 10% in 1985 – Belgium

	Quantity of labour input			Quantity of potential labour input			Degree of utilization potential quantity of labour		
	1984	1985	1989	1984	1985	1989	1984	1985	1989
Base	4330.2	4280.3	4109.0	6082.6	6083.5	6106.3	71.2	70.4	67.3
Simulation 1	4330.2	4234.5	3964.5	6082.6	5473.8	5494.3	71.2	77.4	72.2
Simulation 2	4330.2	4212.7	3860.5	6082.6	5473.8	5494.3	71.2	77.0	70.3
Simulation 3	4330.2	4212.6	3769.0	6082.6	5473.8	5494.3	71.2	77.0	68.6
Simulation 4	4330.2	4215.0	3797.9	6082.6	5473.8	5494.3	71.2	77.0	69.1
Simulation 5	4330.2	4344.8	3907.5	6082.6	5473.8	5494.3	71.2	79.4	71.1
Simulation 6	4330.2	4218.6	3910.7	6082.6	5473.8	5494.3	71.2	77.1	71.2
Simulation 7	4330.2	4228.5	3952.2	6082.6	5420.9	5441.9	71.2	78.0	72.6
Simulation 8	4330.2	4237.3	3969.5	6082.6	5503.9	5524.0	71.2	77.0	71.9

Source: MARIBEL simulations.

simulations, except in simulation 3. The capacity utilization is generally higher the lower the labour costs.

Although in the base projection the degree of utilization decreases during the forecast period, the same is not entirely true for the simulations. In the year of the introduction of the RWT, there is a higher rate in all the simulations with respect to 1984. The production capacity is computed for the production function, by replacing the actual employment by the potential labour input, since capital services are approximated by the capital stock in the production function (see also Section 26.2.2(c) for the production capacity). Hence, the underutilization of production capacity is proportional to the underutilization of the labour input (Table 26.17).

The effects of a RWT for the employers

Potential labour input diminishes with respect to the base projection because of the RWT and the non-complete reallocation. Real labour input decreases less than potential labour input. The decrease of the latter is due to the RWT. The reduction of real labour input is lower the smaller the rise in wages: when wages increase, labour is substituted by capital.

When production is declining, private consumption develops in the same direction, as can be seen in Table 26.18. To be complete, the evolution of public consumption is also shown in this table.

In Table 26.18 private consumption of employed and unemployed is given. This explains why, in 1989, total consumption is smaller than in the base projection, also in simulation 3. In that simulation, only a growth of the purchasing power of workers is supposed.

The decrease in sales creates a problem for the firms. Looking at their profit rates, the problem becomes even worse. As shown in Table 26.19, profits reduce in all the

Table 26.18 Volume of private and public consumption, by a contractual RWT of 10% in 1985, billion BF 1975 prices, for Belgium

	Private consumption			Public consumption		
	1984	1985	1989	1984	1985	1989
Base	1651.876	1639.245	1687.709	460.094	468.825	465.278
Simulation 1	1651.876	1629.385	1659.300	460.094	468.737	465.254
Simulation 2	1651.876	1641.370	1661.753	460.094	468.723	465.286
Simulation 3	1651.876	1641.139	1670.966	460.094	468.716	465.337
Simulation 4	1651.876	1636.903	1666.798	460.094	468.685	465.342
Simulation 5	1651.876	1639.018	1658.584	460.094	468.646	465.291
Simulation 6	1651.876	1632.662	1660.150	460.094	468.686	465.264
Simulation 7	1651.876	1626.946	1655.625	460.094	468.711	465.250
Simulation 8	1651.876	1630.302	1661.033	460.094	468.735	465.252

Source: MARIBEL simulations.

Table 26.19 Growth rates of profits as percentage of GNP, by a contractual RWT of 10% in 1985 – Belgium

	1985	1984–86	1986–89	1984–89
Base	11.3	30.6	−6.2	22.5
Simulation 1	10.8	30.4	−7.7	20.4
Simulation 2	−18.3	1.3	−11.7	−10.5
Simulation 3	−17.7	−15.3	−39.0	−48.3
Simulation 4	−13.9	−10.4	−35.9	−42.6
Simulation 5	−25.2	−1.3	−12.9	−14.0
Simulation 6	−20.6	16.4	−10.1	4.7
Simulation 7	12.5	32.1	−7.6	22.0
Simulation 8	10.7	30.3	−6.0	20.3

Source: MARIBEL simulations.

simulations of RWT with respect to the base simulation, but enormously in the simulations without wage cut. The investments follow this evolution (see Table 26.20, where public investments and investments in houses are also included), although not so spectacularly. The fall in profits oppresses the future!

So far, we have investigated the impact of a RWT on the aggregate firms, without any distinction with respect to size, turnover, sector, etc. Such a distinction is impossible by the MARIBEL model, but has to be taken into consideration evaluating the possibilities of a RWT and the competitive position of a specific firm.

In what follows, we investigate the competitive position of the Belgian economy, resulting in the development of the balance of trades. Table 26.21 outlines the imports and the exports of goods and services in quantities. The balance of trades remains positive in all the simulations, with or without a RWT.

Table 26.20 Volumes of private and public investments, investments in housing sector, billion BF constant 1975 prices, by a contractual RWT of 10% in 1985, for Belgium

	Private investments			Public investments			Housing sector		
	1984	1985	1989	1984	1985	1989	1984	1985	1989
Base	288.0	304.1	352.8	79.9	78.5	78.5	87.8	93.3	99.8
Simulation 1	288.0	301.0	346.8	79.9	77.8	77.9	87.8	93.2	103.0
Simulation 2	288.0	299.5	336.3	79.9	76.5	76.5	87.8	93.4	104.0
Simulation 3	288.0	299.5	326.9	79.9	76.5	76.5	87.8	93.4	107.9
Simulation 4	288.0	299.6	329.6	79.9	75.9	76.0	87.8	93.3	108.2
Simulation 5	288.0	298.5	335.2	79.9	76.2	76.2	87.8	93.3	104.0
Simulation 6	288.0	299.9	341.7	79.9	77.1	77.1	87.8	93.2	103.7
Simulation 7	288.0	300.6	346.4	79.9	77.8	77.8	87.8	93.2	103.4
Simulation 8	288.0	301.2	347.1	79.9	77.8	77.8	87.8	93.2	102.8

Source: MARIBEL simulations.

Table 26.21 Volume of imports and exports of goods and services and balance of trades, billion real BF, 1975 prices, by a contractual RWT of 10% in 1985, for Belgium

	Imports			Exports			Balance of trades		
	1984	1985	1989	1984	1985	1989	1984	1985	1989
Base	1599.8	1670.2	2000.9	1789.4	1885.5	2347.5	189.6	215.3	346.6
Simulation 1	1599.8	1669.6	2017.4	1789.4	1884.2	2345.1	189.6	214.6	327.2
Simulation 2	1599.8	1673.7	2032.4	1789.4	1872.0	2327.5	189.6	198.3	295.1
Simulation 3	1599.8	1673.5	2049.7	1789.4	1872.2	2306.9	189.6	198.7	257.2
Simulation 4	1599.8	1671.0	2041.1	1789.4	1875.3	2311.6	189.6	204.3	270.5
Simulation 5	1599.8	1671.5	2031.3	1789.4	1869.3	2325.6	189.6	197.8	294.3
Simulation 6	1599.8	1670.0	2025.1	1789.4	1877.9	2336.2	189.6	207.9	311.1
Simulation 7	1599.8	1668.7	2018.4	1789.4	1884.2	2345.1	189.6	215.5	326.7
Simulation 8	1599.8	1669.9	2017.5	1789.4	1884.2	2345.1	189.6	214.3	327.6

Source: MARIBEL simulations.

Table 26.22 Volumes of final demands, billion BF in 1975 prices, by a contractual RWT of 10% in 1985 for Belgium

	1984	1985	1986	1987	1988	1989
Base	2761.2	2808.8	2843.6	2904.8	2974.5	3043.5
Simulation 1	2761.2	2790.9	2810.0	2866.1	2931.1	2995.6
Simulation 2	2761.2	2782.2	2786.1	2835.4	2894.6	2953.4
Simulation 3	2761.2	2782.4	2781.2	2818.5	2866.8	2915.5
Simulation 4	2761.2	2783.5	2790.2	2829.2	2877.2	2927.5
Simulation 4	2761.2	2776.9	2780.0	2831.3	2889.9	2947.7
Simulation 6	2761.2	2784.5	2797.3	2850.9	2912.9	2974.4
Simulation 7	2761.2	2788.4	2806.7	2862.8	2927.4	2991.5
Simulation 8	2761.2	2792.0	2811.8	2868.1	2933.4	2998.1

Source: MARIBEL simulations.

Nevertheless, the balance of trades worsens with respect to the base projection in all the simulations of RWT. Imports increase heavily (also in the base simulation!) over time. The exports side of the balance of trades is worsening too: in all the simulations, exports are smaller than in the base projection. In total, the balance of trades is best in the simulations with wage cut, but is always worse than in the base projection. As a kind of summary of this paragraph, the final demands are shown in Table 26.22. Final demands are defined here as the sum of consumption, investments and exports minus imports. This variable too is worse in the simulations without wage cut than in those with reduced wages. The situation is in all simulations worse than in the base projection.

The conclusions of this section can be formulated as follows. Production and production capacity decrease by RWT, while the degree of utilization of capital increases. The total demand of labour is smaller, especially when wages rise. Labour productivity increases most in the simulations with rising wages, but this increase is delayed over time. Also in the simulations with endogenous wage formation, hourly labour productivity is higher with respect to the base projection.

The profits decrease, especially in the case of higher purchasing power. This leads to decreasing investments. In the same time, consumption is lower, and there are more imports and less exports. The final demands decrease with respect to the base projection.

The increase of the wage costs may be underestimated, because of social welfare costs, recruiting and training costs, and all other non-wage labour costs, which are associated with an employee and not with the working time. This phenomenon can make the situation worse, also in the simulation with constant yearly wages per employee.

There is not a lot known about these extra costs of labour (see Hart (1984) for more explanation of non-wage labour costs). It is impossible to calculate those costs within a macroeconomic model such as MARIBEL, because these costs are almost unknown and very different for several kinds of firms. One has to

26.6 THE EFFECTS OF A RWT FOR THE GOVERNMENT

A RWT has an impact on the governmental budget. In Tables 26.23 to 26.25 the development of taxes (direct and indirect) and the transfers of unemployment are pointed out. The taxes of households are supposed to rise, and unemployment benefits to decrease, because of the RWT.

Direct tax income of the central government increases when the pay of employees rises. If the yearly wage per employee decreases, the taxes remain. This is due to the

Table 26.23 Direct taxes of families and firms, billion current BF, by a contractual RWT of 10% in 1985, for Belgium

	Direct taxes of families			Direct taxes of firms		
	1984	1985	1989	1984	1985	1989
Base	660.0	704.5	986.2	119.1	136.5	182.0
Simulation 1	660.0	704.5	985.1	119.1	136.3	180.6
Simulation 2	660.0	732.6	1007.9	119.1	125.1	165.8
Simulation 3	660.0	732.1	1045.9	119.1	125.4	142.3
Simulation 4	660.0	738.2	1054.6	119.1	127.3	147.2
Simulation 5	660.0	737.0	1009.0	119.1	121.8	163.9
Simulation 6	660.0	717.7	996.4	119.1	130.7	173.6
Simulation 7	660.0	702.6	983.4	119.1	136.8	181.2
Simulation 8	660.0	704.6	985.2	119.1	136.3	180.6

Source: MARIBEL simulations.

Table 26.24 Indirect taxes of central government, billion current BF, by a contractual RWT of 10% in 1985, for Belgium

	1984	1985	1989
Base	519.5	566.8	690.4
Simulation 1	519.5	567.4	692.4
Simulation 2	519.5	555.4	678.4
Simulation 3	519.5	555.3	691.4
Simulation 4	519.5	610.9	770.8
Simulation 5	519.5	555.4	678.4
Simulation 6	519.5	552.1	674.2
Simulation 7	519.5	549.7	669.6
Simulation 8	519.5	550.2	669.7

Source: MARIBEL simulations.

Table 26.25 Social security transfers to wage earners: unemployment, billion current BF, by a contractual RWT of 10% in 1985 for Belgium

	1984	1985	1989
Base	165.6	172.2	212.3
Simulation 1	165.6	151.8	198.3
Simulation 2	165.6	153.7	201.6
Simulation 3	165.6	153.7	213.3
Simulation 4	165.6	154.8	211.4
Simulation 5	165.6	136.2	196.8
Simulation 6	165.6	142.3	195.6
Simulation 7	165.6	147.3	186.1
Simulation 8	165.6	154.4	191.3

Source: MARIBEL simulations.

more people employed, who now have to pay taxes as well.[3] Part of the rise of tax income is due to the higher inflation in the simulations with respect to the base projection and to the indexation of tax rates.

Direct taxes of firms are lower (up to 22% in the case of purchasing power of employees increasing) with respect to the base projection. The fall is very small in the simulations with hourly wage costs almost constant. The decrease of direct taxes of firms depends on the lower profit rates. Table 26.24 gives the values of indirect taxes, which are also a part of the governmental income. It is clear that indirect taxes are on a lower level in the simulations than in the reference projection, except in simulation 4.

In simulation 4, indirect taxes are higher because of the MARIBEL operation. The lower level of indirect taxes in the other simulations might be surprising, especially in those cases where the purchasing power per employee increases. But it is already known that private consumption falls as a consequence of RWT, so indirect taxes (above all: value added taxes), develop in the same direction.

Roughly, tax income is higher in the simulations with wage-rise and almost equal to the base projection in simulations without wage-increase. On the expenditures' side of the governmental account, it is clear that the unemployment benefits change as a consequence of RWT because of more people being employed.

As expected, transfers to unemployed diminish due to the fall of unemployment rates. The benefits to unemployed people are higher the more the unemployment rate increases. This rate increases when wages increase. Most people assume that the fall in benefits, together with an increase of taxes, lead unconditionally to an improvement of the financing shortage of the government.

Table 26.26 summarizes the net financing shortage of the government. This

---

[3] Meanwhile, fiscal laws on tax reduction have been voted in parliament, so that future direct tax income could be overestimated.

Table 26.26 Net financing shortage of the government in billion current BF and as a percentage of GDP (between brackets), by a contractual RWT of 10% in 1985 for Belgium

	1984	1985	1986	1987	1988	1989
Base	−505.34	−510.82	−510.78	−494.30	−457.04	−410.68
	(−11.21)	(−10.70)	(−10.22)	(−9.40)	(−8.20)	(−6.93)
Simulation 1	−505.34	−498.93	−507.25	−492.47	−458.86	−417.40
	(−11.21)	(−10.43)	(−10.12)	(−9.34)	(−8.22)	(−7.05)
Simulation 2	−505.34	−451.41	−472.92	−463.55	−436.46	−404.10
	(−11.21)	(−9.40)	(−9.39)	(−8.76)	(−7.81)	(−6.83)
Simulation 3	−505.34	−452.11	−444.59	−429.04	−401.25	−372.61
	(−11.21)	(−9.41)	(−8.80)	(−8.07)	(−7.14)	(−6.28)
Simulation 4	−505.34	−453.25	−428.88	−408.96	−378.39	−344.30
	(−11.21)	(−9.33)	(−8.37)	(−7.58)	(−6.64)	(−5.72)
Simulation 5	−505.34	−424.21	−458.93	−450.34	−429.94	−394.16
	(−11.21)	(−8.84)	(−9.11)	(−8.51)	(−7.60)	(−6.66)
Simulation 6	−505.34	−464.41	−486.27	−475.74	−446.12	−409.32
	(−11.21)	(−9.70)	(−9.68)	(−9.01)	(−7.98)	(−6.92)
Simulation 7	−505.34	−498.68	−508.67	−493.83	−460.60	−419.53
	(−11.21)	(−10.43)	(−10.15)	(−9.37)	(−8.25)	(−7.09)
Simulation 8	−505.34	−500.34	−507.59	−492.46	−458.63	−416.84
	(−11.21)	(−10.46)	(−10.13)	(−9.34)	(−8.22)	(−7.04)

Source: MARIBEL Projections.

variable worsens in the simulations with wage reduction with respect to the base simulation, in contrast to what is often supposed (de Neubourg and Kok, 1984, p. 27). In contrast, the shortage is smaller in the cases of wage-rise. Naturally, this is a consequence of RWT without a wage-cut that cannot be neglected. In 1985, the shortage is in all simulations lower than in the base projection. In 1989, the shortage is 404 billion BF in the simulations with growing purchasing power. The financial position of the government is the best (or: the least bad) when the purchasing power is the higher and the employment effect the smaller. This is an important management implication: for the government, RWT without wage reduction is more interesting for its budget.

The simulations with the best impact on employment worsen the financing shortage. So, the government is confronted with a dilemma: what is the most important to achieve, reducing unemployment or ameliorating the financial position of the state?

26.7 RWT COMPARISONS WITH OTHER (OECD) ECONOMIES

In a large number of OECD member countries, the RWT is primarily linked to the long run improvement of welfare; in particular, RWT is generally not a part of an employment- or job-creating policy. This can be verified from Table 26.27 and Fig. 26.7 from the average number of hours worked per employee per year.

The substantial drop in labour input per employee over the last century becomes

RWT comparisons with other (OECD) economies

Table 26.27 Average annual number of hours worked per employee in the total economy: period 1890–1982

	1890	1913	1929	1950	1970	1979	1982
Austria	2760	2580	2281	1976	1848	1660	—
Belgium	2789	2605	2272	2283	1986	1747	1502
Canada	2789	2605	2399	1967	1805	1730	1720
France	2770	2588	2297	1989	1888	1727	1700
Germany	2765	2584	2284	2316	1907	1719	1640
Italy	2714	2536	2228	1997	1768	—	1650
Japan	2770	2588	2364	2272	2252	2129	2080
Sweden	2770	2588	2283	1951	1660	1451	1430
United Kingdom	2807	2624	2286	1958	1735	1617	1625
United States	2789	2605	2342	1867	1707	1607	1610

Sources: Austria, Canada, Sweden and 1970 figures for all countries: Maddison (1982) and Belgium, France, Germany, Italy, Japan, UK and USA: Maddison (1983). Figures for 1982: Commisseriat Général du Han.

clear from Table 26.27 and Fig. 26.7, showing that the average number of hours worked per employee in 1970 amounted to only about 60% of that in 1870. This downward trend was even more pronounced during the post-war decades in most European countries (except Belgium and France for the Second World War).

The average reduction in annual hours worked per employee during the seventies largely exceeded previous decade averages (except for Canada, where annual hours worked changed less during the seventies than during the fifties and the sixties, and for Germany and the UK, where the reduction progressed at the same (high) rate throughout the post-war period). This becomes clear from Table 26.28, showing the average percentage annual changes in hours worked per employee during the period 1960–1981.

It follows from Table 26.28 that the above-mentioned OECD countries showed an average annual RWT of 0.3% during the 1950s, 0.7% during the 1960s, 1.1% during the 1970s and 0.5% during 1979–81. However, there were substantial differences between countries, and the variations between countries and between periods reflected short-term structural and cyclical changes and increases in part-time female employment in addition to long-term trends towards shorter standard hours, e.g. the shortest working year in Sweden which had already been the case for a long time (introduction of parents' holidays during the 50s, considerable part-time female employment and 3% unemployment in 1980).

It should be stressed, however, that some OECD member countries have taken the view that a policy of RWT is a powerful and effective weapon in the fight against unemployment (e.g. Sweden). In general, *employers* are against a RWT for any purpose other than furthering the long-term improvement of welfare. They fear its effect on production costs, on their competitive position and, consequently, on employment itself. On the other hand, they often favour adjustments aimed at improving flexible working arrangements (part-time work, working time savings,

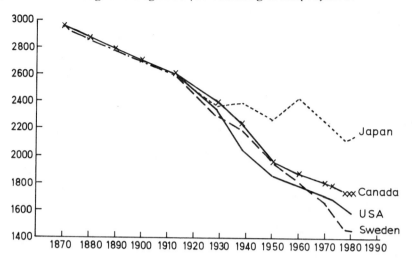
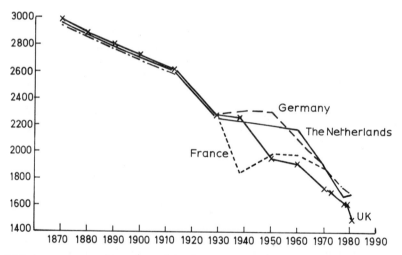

Fig. 26.7 Average annual number of hours worked per employee 1870–1980, OECD economies. Source: as for Table 26.27 plus the Netherlands: Maddison (1983)

etc.). *Unions*, on the contrary, are generally determined to press on with claims for RWT, and, sometimes, their demands are substantial and insistent.

Since, in general, there is considerable uncertainty as to the actual impact of RWT on employment and unemployment, on consumption, investment, production growth, balance of payments, inflation, governmental financing shortage, etc., it is instructive to compare some recent European national econometric models on their various impacts of a RWT policy. The studies of van Ginneken (1984), de Neubourg (1984) and de Neubourg and Kok (1984) will be very helpful in this respect.

Table 26.28 Average annual changes of hours worked per person in the employment of the total economy (%)

	1960–70	1970–73	1973–76	1976–79	1979–81
Belgium	−0.9	−2.2	−3.1	−0.8	−2.0
Canada	−0.8	−0.5	−0.7	−0.6	−0.6
Finland	−0.4	−1.1	−0.6	−0.3	−1.1
France	−0.5	−1.0	−1.1	−0.8	−0.2
Germany	−0.9	−1.5	−0.8	−1.2	−1.2
Japan	−0.8	−1.8	−1.6	0.3	−0.3
Italy	−0.6	−2.2	−0.8	−0.1	−0.1
Netherlands	−1.4	−2.0	−1.7	−1.9	−0.9
Norway	−1.1	−2.4	−1.1	−1.5	−0.6
Sweden	−0.9	−1.7	−0.7	−1.6	−0.7
United Kingdom	−0.1	−0.3	−1.1	−0.9	−2.9
United States	−0.5	−0.2	−1.1	−0.2	−0.6

Source: OECD (1983).

In general, an analysis of a RWT with the help of macroeconomic models permits the study of direct as well as indirect effects of a RWT on various interesting aggregate economic variables, but it should be stressed that, since (almost) all large econometric models have been estimated under the assumption of a constant (parameter) structure, the behavioural relationships represent a somewhat average behaviour during a sample period, and not (in principle) during a period of recession, for which RWT policies will be simulated.

The macroeconomic models utilized for this RWT simulation comparison are:

MARIBEL	for Belgium (Planning Office (1984) and this chapter (1986));
DMS	for France (Oudiz et al. (1979));
HENIZE	for the Federal Republic of Germany (Henize (1981)),
VINTAF II and FREIA	for the Netherlands (Central Planning Bureau (1979) and (1983)) and
TREASURY	for the United Kingdom (Allen (1980)).

In Table 26.29 the impact of a RWT on employment (measured by the employment elasticity, i.e. the (opposite of the) percentage change of the actual number of employees owing to a 1% RWT), on the (absolute) change of the unemployment percentage $\tilde{u}n$, on the (relative) change of private consumption C^p and private output growth Y^p, on the real hourly labour costs in the private sector w^p/p_c, on consumer prices p_c, on the balance of payments (as a percentage of gross domestic product) and on the government budget (also as a percentage of gross domestic product) is summarized for the various above-named econometric models. This impact has been measured for the starting year (of introduction of RWT) and the final year of the simulation forecasting period. Distinction is made

Table 26.29 Percentage changes with respect to the reference simulation

Model	Hypotheses		Year of impact	Employment elasticity $-\eta_{NT}$
MARIBEL (1985) (10% RWT: once in 1985)	(i)	Proportional wage reduction; no assumption about production capacity (simulation 1)	1985 1989	0.423 0.539
	(ii)	Full wage compensation; no assumption about production capacity (simulation 3)	1985 1989	0.401 0.201
DMS (1979) (2.5% RWT per year, trend reduction, 1982–1986)	(i)	Proportional wage reduction, unchanged capacity	1986	0.582
	(ii)	Full wage compensation, unchanged capacity	1986	0.633
HENIZE (1981) (5% RWT: once, plus 1% trend reduction)	(i)	Proportional wage reduction; no assumption regarding capacity	1981	0.680
	(ii)	Full wage compensation, (tax reduction); no assumption regarding capacity	1981	0.840
VINTAF II (1979) (2.5% RWT per year, 1979–1983)	(i)	Proportional wage reduction, reduction of capacity	1983 1988	0.067 0.168
	(ii)	Full wage compensation, reduction of capacity	1983 1988	−0.135 −0.193
	(iii)	Proportional wage reduction, unchanged capacity	1983 1988	0.118 0.193
	(iv)	Full wage compensation, unchanged capacity	1983 1988	−0.084 −0.177
FREIA (1983) (2.5% RWT per year, 1983–1986)	(i)	Proportional wage reduction, reduction of capacity	1986	0.125
	(ii)	Proportional wage reduction, unchanged capacity	1986	0.488
TREASURY (1981) (5% RWT: once)	(i)	Proportional wage reduction, accommodative monetary policy	1981	0.280
	(ii)	Full wage compensation, accommodative monetary policy	1981	0.160

[a] Gross domestic product at factor cost of the private enterprises.
[b] In percentage of gross national product.
[c] Production price.

according to the requirement of a strict proportional per capita wage reduction or to the case of a full wage compensation and to the case of reduction of productive capacity (decrease of the number of machine-operating hours, being proportional to the RWT) or to capacity maintenance.

It should be noted that the Dutch VINTAF II model is only for the market sector (hence, the government sector is excluded), that the French DMS model is concentrated only on the private nonagricultural and nonfinancial enterprises and that the HENIZE, FREIA, MARIBEL and TREASURY models treat the total economy (private + public sector). Hence, by the impact on employment is meant the impact on employment in the correspondingly treated part of the economy, e.g., for FREIA, the impact on the total number of Dutch employees.

RWT comparisons with other (OECD) economies

Δũn (as % of the active population)	ΔC^p (vol.)	ΔY^p (vol.)	Δ(w^p/p_c)	Real income per employee (purchasing power)	Δp^c	Balance of payments (as % of NNI)	Government deficit (as % of NNI)
−2.3	−0.06	−0.81[a]	0.05	−4.09	0.68	0[b]	−0.27[b]
−2.8	−1.68	−1.97[a]	0.32	0.11	−0.00	−0.46[b]	+0.12[b]
−2.2	−0.01	−1.19[a]	4.79	0.05	1.41	−0.62[b]	−1.29[b]
−1.1	−1.00	−5.25[a]	1.28	0.72	0.29	−3.36[b]	−0.65[b]
−2.2	−1.5	−0.2	0.2	−1.6	−0.2[c]	2.7[d]	−1.8[d]
−2.0	−0.1	−0.3	1.8	−0.4	1.7[c]	2.0[d]	−1.2[d]
—	—	—	—	—	—	—	—
—	—	—	—	—	—	—	—
−0.4	−1.8	−1.9	0.1	−2.0	0.9	1.6	0.0
−1.1	−1.2	−1.1	−0.2	−1.3	0.4	2.0	−1.0
1.0	−0.8	−2.3	0.9	−0.5	1.9	−0.7	+1.5
1.2	−0.6	−1.5	0.2	−0.5	1.1	−1.4	+2.6
−0.8	−1.7	−1.7	0.2	−2.0	0.7	1.5	0.0
−1.2	−1.1	−1.0	−1.0	−1.3	0.4	2.1	+0.9
0.6	−0.7	−2.1	1.0	−0.5	1.8	−0.8	+1.5
1.1	−0.6	−1.4	0.2	−0.4	1.0	−1.4	+2.8
−0.8	−1.0	−1.5	0.5	−1.5	1.5	0.7	−0.3
−4.0	−0.7	−0.5	0.7	−0.5	0.0	1.0	−0.5
−1.0	—	0.3	—	−0.4	−0.3	−0.1[d]	−0.1[d]
−0.6	—	−0.3	—	0.9	2.6	0.1[d]	+0.4[d]

[d] In percentage of gross domestic product.
η_{NT} = employment elasticity.
NNI = net national income.

Table 26.29 implies widely differing effects of RWT, sometimes with opposite signs, among macroeconometric models of the various countries. Three main reasons for these observed differences are:

(1) the varying model hypotheses regarding the anticipated capacity – and productivity effects;
(2) the strongly differing specification of the macroeconometric models (e.g. linear vs. nonlinear);
(3) the varying sampling periods and spatial situations of the underlying models.

In the sequel we discuss the impact of a RWT on (un)employment, the volumes of private consumption and output, real gross hourly wage cost, real income per employee (purchasing power), inflation (measured by the percentage price change of private consumption goods), current balance of payments and governmental budget (as percentage of NNI) from the results in Table 26.29.

26.7.1 RWT effects on employment

A RWT raises employment almost everywhere (except for the full-wage compensation scenario in the Dutch VINTAF model) and the employment elasticity is always less than one, so that the percentage increase in jobs is less than the percentage decrease in the number of working hours.

Immediately striking, indeed, are the negative employment effects from a substantial RWT (12.5% in 5 years) under full-wage compensation in the Dutch VINTAF II model: a scrapping condition in real wages only for this clay–clay model leads to the outpricing of labour.

Wage compensation tends to reduce the employment effects, although the two models that introduce wage compensation by means of a tax reduction (the French DMS model and the German HENIZE model) yield better results compared with the situation where wages are reduced proportionally to the RWT.

Employment effects of a RWT differ widely: the most pessimistic (positive) estimate of the employment elasticity is obtained in the VINTAF II variant in the short run: 1% RWT raises employment by only 0.067% during the initial year 1983 (and even has a negative impact when there is full-wage compensation); the most optimistic result is obtained by the German HENIZE model: 1% RWT raises employment by 0.84%.

Assuming no loss of productive capacity and no rise in labour productivity produces the largest employment effect, but it should be stressed that, generally, some of the excess capacity will be left unused due to the RWT (principally in the smaller firms).

The Belgian MARIBEL model ranks in third place regarding the employment effects of a RWT (after the German HENIZE model and the French DMS model and just before the Dutch FREIA model); note that the full-wage compensation variant of MARIBEL (our simulation 3) entails low medium run employment effects, while the medium run employment effects of the proportional wage reduction variant of this model (our simulation 1) are larger than the short run effects. The RWT impact on the unemployment percentage is considerable in Belgium.

26.7.2 RWT effects on the volumes of private consumption and output

The volumes of private consumption and (private) output decrease in nearly all the cases, though with different percentages. In general, private consumption decreases less under full wage consumption (because of the higher private income) than under proportional wage reduction, while the reverse is true for the volume of output.

The growth of the private goods value added volume or of GNP is slower under

RWT than in the base simulation, especially for MARIBEL (larger negative medium run effects!) and VINTAF II (considerable negative short run effects). The estimates of the French and British models are less pessimistic, but nearly all models stress the fact that it is necessary to sacrifice some economic growth for new employment under a RWT.

26.7.3 RWT effects on the real hourly labour costs

Real hourly wages tend to increase more under RWT than without RWT (base simulation); obviously, these increases become more important when the salaries can be maintained (wage compensation). Maintenance of production capacity does not seem to play any role.

There is a large increase of real hourly wages in the full-wage compensation variant of MARIBEL (+4.8% in the short run, +1.3% in the medium run) which is accompanied by an increasing purchasing power per wage-earner.

26.7.4 RWT effects on the real income per employee

The method of determination of income is very different from model to model.

The per capita real income is generally lower than in the base simulation, if full-wage compensation occurs, except for the full-wage compensation variants of the Belgian MARIBEL model and the British TREASURY model.

26.7.5 RWT effects on inflation

Almost all models show an important increase in inflation under RWT, principally under wage compensation.

The inflation also has a tendency to grow under the hypothesis of production capacity reduction, since unit capital costs are then growing.

26.7.6 RWT effects on the current account balance of payments

The current account balance of payments position is ameliorated by a RWT in most models except for MARIBEL and VINTAF II.

The maintenance of salaries tends to stimulate private domestic expenses to the disadvantage of exports. But the balance of payments (trades) may keep positive because of the decrease of imports. The deterioration of the current account balance of payments in the full-wage compensation variants of the Belgian MARIBEL model and the Dutch VINTAF model originates from the large elasticity of imports with respect to (Belgian and Dutch) income.

Notice that a RWT leads to a deterioration of the Belgian balance of trades with respect to that of the base simulation in MARIBEL (which has, however, a positive value of 1.06% of GNP in 1985 and of 4.59% of GNP in 1989).

26.7.7 RWT effects on the governmental budget

The government deficit is lowered when the RWT policy is accompanied by a proportional wage cut, because it produces a significant unemployment decrease and, hence, a decrease of social security allowances. There is only one notable exception, i.e. the Belgian 1989 deficit is estimated somewhat higher than the corresponding base simulation deficit of MARIBEL (deficit of 410.7 billion BF in 1989 or 6.9% of GNP in the base simulation vs. 417.4 billion BF or 7.1% of GNP in simulation 1).

Under full wage compensation, a RWT entails a lower governmental deficit for the Belgian MARIBEL model (our simulation 3) and the French DMS model than in the reference simulation, and a larger public deficit for the Dutch VINTAF II model and the British TREASURY model.

26.8 GENERAL CONCLUSIONS

In this chapter we have investigated the effect of a reduction of working time (RWT) on macroeconomic aggregates. A single once-and-for-all contractual RWT of 10%, in the private sector in Belgium, in 1985, has been simulated for various accommodating policies. This was done with the help of the MARIBEL model, which is the official forecasting model of the Belgian Planning Office. The reference simulations are part of the May 1985 predictions of the Planning Office.

(1) The actual RWT is always smaller than the contractual RWT. A 10% reduction of contractual working time, with respect to the base projection (= without RWT) in 1985, leads to a 4.3% actual extra RWT in that same year (also compared with the base projection), when hourly wages remain constant. If real hourly wages increase because of the RWT (simulation 2), the actual RWT is 4.6% more than in the base projection for 1985.

(2) The 10% surplus contractual and 4.3% extra actual RWT, creates 4.4% more jobs in the private sector in 1985 with respect to 1984, or 4.2% more jobs than in the base simulation, all the other exogenous variables holding constant. This appears if the RWT is accompanied by a real constant hourly wage rate, which means that the real yearly wage per employee decreases with the amount of the RWT. In 1989, the increase in private jobs is 3.8% with respect to 1984, which means 5.4% more jobs compared with the base projection. This results in a decline of the unemployment rates. Without any extra RWT, unemployment, as a percentage of the active population, increases from 12.4% in 1984 over 13.0% in 1985 to 14.7% in 1989. The introduction of the RWT as mentioned here, leads to an unemployment rate of 10.7% in 1985 and 11.9% in 1989.

(3) When the RWT is accompanied by a 5% real hourly wage increase with respect to the base simulation in 1985, the impact on job-creation in the private sector is smaller (simulation 2). In 1985, the increase is of 4.1% with

General conclusions

respect to 1984 and of 4.0% compared with the base simulation. In 1989, 2.0% more jobs than in 1984 remain, which means 3.5% more jobs than in the base projection. The impact is larger with respect to the base simulation than with respect to 1984 because of the underlying employment shrinkage in the base projection. The unemployment rate is 10.8% in 1985 and 12.8% in 1989 in simulation 2.

(4) Increasing real hourly wages per employee in the private sector by the amount of the actual RWT each year, leads to an at least constant labour income per private wage earner per year. The creation of jobs is smaller in this simulation (simulation 3). Compared with 1984, there are 4.1% more jobs in 1985 but only 0.5% more people employed in the private sector in 1989. This means 4.0% extra jobs with respect to the base simulation in 1985 and 2.0% in 1989. Hence, the RWT does not create many jobs over the forecast period but rather stops the decrease in private employment appearing without the RWT. There is no full reallocation of potential jobs at all under this income policy. There are 10.8% people of the active population unemployed in 1985 and 13.6% in 1989.

(5) As can be concluded, the employment effect is smaller the more the real wage income increases. The effect is larger in the short run than in the medium run. The decline in jobs in the medium run is larger when real wages increase. Compared with the situation without a RWT policy, there is less decline in private jobs.

(6) An exogenous contractual level of recruitment leads to more jobs only in the short run; in the medium run the effect almost disappears.

(7) A reduction of wage costs, by reducing the social security contributions of employers and simultaneously a compensating increase in value added taxes (a simplified MARIBEL operation), creates significantly more jobs only in the medium term.

(8) An extra supply of labour (20% of extra employment), due to the RWT, leads to more employment and a larger actual RWT per employee: there are 4.5% extra jobs compared with the base projection in 1985, or a surplus of 0.2% compared with simulation 1.

(9) The level of employment increases more when the purchasing power declines. A situation can be imagined with constant buying power over the simulation period (but a decrease during the first year), which lies between simulations 1 and 2. The employment effect will be some weighted mean of the results of simulations 1 and 2. Simulation 3 (actual RWT = increase in real hourly wage) seems unrealistic in the actual situation: it leads to an increasing purchasing power per employee (+6.5% in 6 years).

(10) In all the situations of a RWT, the total labour cost is at a higher level than without any RWT. The larger these costs, the more profits and investments decrease.

(11) The gross value added of the private sector declines in all the simulations, especially when the RWT is accompanied by an hourly real wage increase.

Hourly labour productivity per employee increases more in that case, which explains the low reallocation of potential jobs. The hourly per capita labour productivity increase is larger than in the base projection in all RWT simulations, but is the largest in simulation 3. While consumption and production decrease, and labour productivity increases, the potential working time is not completely reallocated.

(12) The competitive position of the Belgian economy worsens due to the RWT, especially when accompanied by a real hourly wage increase. The imports are larger, and the exports smaller. The balance of trades is less good, although remaining positive, compared with the base projection. While the balance of trades is 4.59% of GNP in 1989 in the base projection, it is 4.13% in simulation 1 and only 1.23% in simulation 3.

(13) The Belgian governmental budget deficit worsens with respect to the base projection, if the RWT is accompanied by a declining purchasing power per private wage earner. The deficit declines over time and with respect to the base projection, when the RWT is introduced with an hourly real wage increase. The Belgian government seems to be confronted with a dilemma. Either it opts for an employment policy, or it chooses to lead a public deficit recovery policy. In the first case, the extra employment is accompanied by real lower wages per wage earner. For reducing the public deficit, purchasing power has to increase, meaning a smaller employment effect.

(14) When comparing the MARIBEL simulations with simulations of macro-economic models of other OECD economies, we observed that there are no exceptional effects of a RWT in the MARIBEL model. The strength of the simulations based on these country econometric models however is somewhat limited because these simulations do not consider the effects of a similar RWT policy measure taken in another (OECD) country. For this purpose, an integrated multi-country econometric model (e.g. Comet and Hermes for the EEC) should be necessary.

ACKNOWLEDGEMENT

This chapter forms part of a project financed by the Belgian National Science Foundation, under FKFO 20092.85. The writing of this chapter was made possible because of collaboration with the Belgian Planning Office, where we are very grateful to R. Maldague, R. de Falleur, M. Englert, M. Vanden Boer and J. Verlinden.

REFERENCES

Allen, R. (1980) *The economic effects of a shorter working week*, Government Economic Service Workshop Paper, No. 33, London.

Bakhoven, A. and Jansen, C. (1984) Het belang van veronderstellingen bij de berekening van

de macro-economische effecten van herverdeling van arbeid, *Kwantitatieve Methoden*, 5, No. 14, 79–96.
Central Planning Bureau (1979) *Centraal Economisch Plan 1979*, Staatsuitgeverij, The Hague.
Central Planning Bureau (Centraal Planbureau) (1983) *Freia, een macro-economisch model voor de middellange termijn*, Monografie 25, 's Gravenhage.
Central Planning Bureau (Centraal Planbureau) (1985) *Centraal Economisch Plan 1985*, 's Gravenhage, 276 pp.
Cuvillier, R. (1984) *The reduction of working time*, ILO, Geneva, 150 pp.
de Bruyne, G. and de Grauwe, P. (1985) Herverdeling van het werk en makro-economisch evenwicht, *Leuvense Economische Standpunten*, Leuven, No. 36, 21 pp.
de Neuborg C. (1984) What a difference a day makes: figures and estimates on work-time reduction, *Research Memorandum from Institute of Economic Research*, Groningen, No. 141, 44 pp.
de Neuborg C. and Kok L. (1984) *Arbeidstijdverkorting*, Aula, Het Spectrum, Utrecht/Antwerp, 104 pp.
Drèze, J. (1979) Salaires, emploi et durée du travail, *Recherches Economiques de Louvain*, 45 No. 1, Louvain-la-Neuve, 17–34.
Drèze, J. (1980) Réduction progressive des heures et partage du travail, *Rapport pour la 3e Commission du 4e Congres des Economistes Belges de Langue Française*, Mons, November
Drèze, J. and Modigliani, F. (1981) The trade-off between real wages and employment in an open economy (Belgium), *European Economic Review*, 15, 1–40.
Drèze, J. and Sneessens, H. (1985) *What, if anything, we have learned from the rise of unemployment in Belgium, 1974–1983*, Louvain-la-Neuve, 45 pp.
Hart, R. (1984) *The Economics of Non-Wage Labour Costs*, Allen and Unwin, London, 173 pp.
Henize, J. (1981) Can a shorter work week reduce unemployment? A German simulation study, *Simulation*, Nov, pp. 145–66.
Hoger Instituut voor de Arbeid (1986) *Met een stempel door het leven? Werkloosheid en tewerkstelling van jongeren: beschrijving, gevolgen en perspectieven*, Leuven, 435 pp.
Késenne, S. and Butzen, P. (1984) Arbeidsduurverkorting, kostenstructuur van de onderneming en tewerkstelling, *SESO-report*, UFSIA, Antwerp, 15 pp.
Madisson, A. (1982) *Phases of Capitalist Development*, Oxford University Press, Oxford.
Madisson, A. (1983) Comparative analysis of the productivity situation in the advanced capitalist countries, in *International Comparisons of Productivity and Causes of the Slowdown* (ed. J. Kendrick) American Institute, Washington DC.
Martens, B. (1983) Arbeidsduurverkorting in België, in: *Werkelegenheid voor de jaren tachtig*, 16de Vlaams Wetenschappelijk Economisch Congres, Gent, pp. 561–74.
Ministry of Economic Affairs (Ministerie van Economische Zaken), E. Pollefliet (1985) *Dossier Smobe: arbeidsduurverkorting met 10%; enkele alternatieve simulaties*, Brussels, 61 pp.
National Instituut voor de Statistiek, *Nationale Rekeningen*, Brussels, annually.
National Instituut voor de Statistiek, *Sociale Statistieken*, Brussels.
OECD (1983) *Employment Outlook*, Paris.
Oudiz, G., Raoul, E. and Sterdyniak, H. (1979) Réduire la durée du travail, quelles conséquences? *Economie et Statistique*, Paris, pp. 3–17.
Planning Office (Planbureau) (1984) *Maribel*, Brussels, 330 pp.
Planbureau (Englert, M. and Vanden Boer, M.) (1985a) *Arbeidsduurverkorting gesimuleerd door Maribel*, Brussels, 43 pp.
Planbureau (1985b) *Vooruitzichten 1985–1989 van mei 1985*, Brussels, 103 pp.
Planbureau (Englert, M. and Verlinden, J.), (1985c) *Enkele simulaties van arbeidsduurvermindering*, Brussels, 5 pp.
Plasmans, J. and Vanroelen, A. (1985) Arbeidsduurverkorting: een mogelijke oplossing voor

(jeugd)werklooshied?, *SESO-rapport*, UFSIA, Antwerp, No. 183, 43 pp.
Plasmans, J. and Vanroelen, A. (1986a) Arbeidsduurverkorting: een mogelijke oplossing voor jeugdwerklooshied?, in: *Met een stempel door her leven? Werkloosheid en tewerkstelling van jongeren: beschrijving, gevolgen en perspectieven*, HIVA, Leuven, pp. 265–307.
Plasmans, J. and Vanroelen, A. (1986b) Arbeidsduurverkorting en tewerkstelling in België. Een Macro-economische simulatiestudie, *Economisch en Sociaal Tijdschrift*, 1986 en *SESO-beleidsnota*, 1986/12, UFSIA, Antwerp. 18 pp.
Rijksdienst voor Arbeidsvoorziening, *Jaarverslag*, Brussels, annually.
Rijksdienst voor Arbeidsvoorziening, *Maandelijks Bulletin*, Brussels, monthly.
Vanden Boer, M., and Pollefliet, E. (1983) Arbeidstijdverkorting: een poging tot objectivering, *Economisch en Sociaal Tijdschrift*, 2, 119–33.
van Ginneken, W. (1984) La réduction de la semaine de travail et l'emploi. Comparaison entre sept modèles macro-économiques européens, *Revue Internationale de Travail*, **123**, No. 1, 37–56.

27

An econometric model for the determination of banking system excess reserves

JOSÉ LUIS ESCRIVÁ and ANTONI ESPASA

In this chapter, a demand equation for the excess reserves held by banks and saving banks in Spain is presented and estimated. The equation is derived from a profit maximization model by banking agents in a state of uncertainty and where the rational expectations hypothesis is imposed.

ARIMA methods are used to determine the mechanisms of the formation of the expectations. The procedures to estimate the market risks are also described. The estimated model is broken down in order to obtain the desired short and long term excess bank reserves.

The time unit of the model is a period of about ten days. Each month is divided into three periods and the variables are measured as the average daily values for each period.

27.1 AN OVERVIEW

The first study of modelling the demand for excess reserves[1] by the banks in Spain was by Pérez (1976), in which the author derives this demand from a theoretical model where the banks act as profit-maximizing agents in a climate of uncertainty, and he formulates a partial adjustment demand model. Mauleón (1984) takes up the problem of this type of demand for a later period (1978–1982) and develops a model of general adjustment.

This chapter continues along the same lines as these earlier ones, and a model is proposed in which the rational behaviour of the agents is incorporated into the analysis by means of the direct modelling of their expectations.

---

[1]The reserves of a bank are the part of a bank's assets held in cash or on deposit at the Bank of Spain. Excess reserves are the amount of reserves held by a bank above the level established by the reserve requirement regulations.

In Pérez (1976) and Mauleón (1984) the starting point is a simplified bank balance sheet, from which the demand for excess bank reserves arises from the determination of their desired level, after considering the situation of the markets for their assets and liabilities. Therefore, they formulate a problem of optimization of the volume of inventories in a state of uncertainty, in which the banks try to minimize the expected losses through maintaining excess reserves. This approach is also the basis for the present chapter. See also Morrison (1966) and Frost (1967).

In a context of rational expectations, the ratio of excess reserves which a given financial institution wishes to maintain at any moment t is an optimum level derived as a function of all the information available up to the moment $t-1$. Consequently the ratio of excess reserves[2] actually maintained is the desired ratio corrected by the effect of a series of unexpected disturbances and by the existence of adjustment costs to attain this ratio.

The optimum ratio of excess reserves is the result of forecasts, which in a rational expectations context are the best that can be made, and which the financial institution carries out on those variables relevant to the financing or to the future investment of excess reserves, as well as its appreciation of the risk which is being induced by the level of uncertainty existent in the system.

The model described below is an application of this simple context of expectations to the Spanish banking system taken as a whole. The level of excess bank reserves is determined fundamentally in the interbank market and there exists considerable support, both theoretical and empirical, for maintaining the hypothesis of rational expectations in this type of market; see for instance Smith (1976).

The expected value of excess bank reserves, ESB_t^e, given all the information available up to time $t-1$, Ω_{t-1}, can be expressed as follows:

$$ESB_t^e = E(ESB_t/\Omega_{t-1}) = \sum_{j=1}^{n} \alpha_j x_{tj}^e + \sum_{l=1}^{h} \beta_l z_{(t-1)l} + \eta_t^e, \qquad (27.1)$$

where $\eta_t^e = \eta_t - a_t$ and

$$\eta_t = \frac{\theta(L)}{\phi(L)} a_t.$$

$\theta(L)$ and $\phi(L)$ being polynomials of finite order on the lag operator L, and a_t is generated by a white noise process.

In Equation 27.1 the n variables x_t^e are the expected components of the corresponding economic variables which are relevant in the determination of the excess bank reserves; the h variables z_l capture the risk effect; η_t^e is a residual component the nature of which is mainly seasonal.[3]

---

[2] This is defined as the ratio between excess reserves and the deposits and the liabilities of a bank which are affected by the reserve requirements regulations.

[3] With the η_t^e tem in Equation 27.1 we are assuming that the expected value of excess bank reserves also depends on a set of variables that cannot be quantifiable and their effect is relegated to a residual term. This residual term can be dynamic if with it we are going to approximate the expected value of the mentioned non-quantifiable variables.

An overview

The level of excess bank reserves can be broken down into two parts, one expected and the other not:

$$\text{ESB}_t = \text{ESB}_t^e + \text{ESB}_t^{ne} \qquad (27.2)$$

in such a way that the unexpected component takes the following representation:

$$\text{ESB}_t^{ne} = \sum_{j=1}^{n} \gamma_j x_{tj}^{ne} + \varepsilon_t \qquad (27.3)$$

where x_{tj}^{ne} represents the unexpected component of the jth economic variable and ε_t is white noise. If what is obtained in 27.1 and 27.3 is inserted into 27.2, we can express ESB as a function of the expected and unexpected economic variables, of the risk and of a residual element:

$$\text{ESB}_t = \sum_{j=1}^{n} \alpha_j x_{tj}^e + \sum_{j=1}^{n} \gamma_j x_{tj}^{ne} + \sum_{l=1}^{h} \beta_l z_{(t-1)l} + \eta_t^*. \qquad (27.4)$$

The n x_t^e are variables which are not observed and must be estimated from mechanisms for generating expectations. The construction of models in which the rational expectations on the exogenous variables are obtained by vector ARIMA processes, as a general rule, is backed up in recent econometric literature; see Wallis (1980) and Stockton and Glassman (1985). This being the case we see that:

$$x_{tj} = \frac{\theta_j(L)}{\phi_j(L)} a_{tj}, \quad j = 1, \ldots n, \qquad (27.5)$$

and then

$$x_{tj}^e = E(x_{tj}/\Omega_{t-1}) = x_{tj} - a_{tj}.$$

and therefore the unexpected component is

$$x_{tj}^{ne} = a_{tj}.$$

The dynamic specification in Equation 27.4 is correct as long as there are no delays in the availability of information on the variables, as each lag effect of variable x_{tj} on the excess reserves is incorporated in the expected value of x_{tj}, through Equations 27.5. However, in practice we have temporary delays in getting the information, and costs of adjustment to the desired levels, both of which make for dynamic responses. Therefore, it would seem to be more realistic to rewrite Equation 27.4 thus:

$$\text{ESB}_t = \sum_{j=1}^{n} \alpha_j(L) x_{tj}^e + \sum_{j=1}^{n} \gamma_j(L) x_t^{ne} + \sum_{l=1}^{n} \beta_l(L) z_{(t-1)l} + \eta_t^*. \qquad (27.6)$$

This model has a latent problem of simultaneity. The x_j variables have been considered as exogenous and this may not be the case. The determination of the effects of the expected and unexpected components and the possible dynamic responses in a simultaneous context proves to be complicated and we have opted for the following strategy: in this chapter to estimate these effects and dynamic structures assuming a recursive model for the determination of bank reserves, interest rates and excess bank reserves, so that in a subsequent publication, we can broach the topic of simultaneous determination of these variables.

It should be noted that at the ten-day aggregation level at which we are going to work the problem of simultaneity is of less importance than the problem of specification of the dynamic responses to movement of nonobservable variables, the problem which this chapter will focus on.

In Section 27.6 we are assuming that the conditions for identification are satisfied.

27.2 THE FORMATION OF EXPECTATIONS FOR THE RELEVANT EXPLANATORY VARIABLES

To explain the excess reserve ratio, we have incorporated explicitly as explanatory variables the bank reserves and variables which could be proxies for the risk. In doing this we have tried to concentrate on those variables which have been considered decisive in the process of formation of expectations by the banks in their cash management. This makes it possible to define a compact and easily understood model, which could, it is hoped, at the same time be useful in forecasting.

Specifically, it has been assumed that the agents, in addition to their estimate of the risk, form their expectations for three basic variables: the interest rate on operations for one day and one month, and the bank reserves.

The interest rates on overnight operations on the interbank market may be considered as the opportunity cost of maintaining reserves. This, then, is the relevant interest rate, bearing in mind that it is for this period for which the majority of the operations on the market are carried out. In fact, lending over one day has been the normal outlet for the temporary excess reserves of the banks in the sample period considered.

The link between overnight interest rate and excess reserves can be interpreted as follows. If one bank hasn't reached the level of required reserves, it will borrow from the Central Bank the amount of reserves needed of fulfil these requirements. The Central Bank usually fixes the cost of these loans taking as reference the price of money in the interbank market, and the one-month interest rate is a good indicator of this price. So, this rate is a proxy variable of the penalty rate in a excess reserves demand function.

Therefore, we expect the excess reserves to have a negative relationship with the overnight interbank interest rate but a positive one with the one-month interbank interest rate.

Reserves is a third variable on which the banks form expectations. In fact one of the main concerns of the banks consists of anticipating how the monetary authority will operate in the open market, and this leads the agents to form expectations on what the global supply of reserves will be as a mean in each ten-day period. These expectations are already embodied in the cash management of the banks and only the unexpected component of the supply of cash reserves has an effect on excess reserves.

The mechanisms of generation of expectations on these three variables could be approximated by vector autoregressive integrated moving average (ARIMA) models, but since the interventions on these variables are important we have used univariate ARIMA models plus interventions to estimate the expected values, hoping that the differences with respect to a multivariate approach will not substantially change the results of this chapter. This implies that we are considering these variables as exogenous in our model. This is a restriction that we will try to relax in a future publication. For cash reserves we are not using the original series, but an adjusted one, which, is a more reliable indicator of the way in which the central bank has operated (see Frost, 1977 and Mauleón et al., 1985). For the breakdown of cash reserves in expected and unexpected components we use the results obtained in Escrivá et al. (1986).

In Table 27.1 we give the univariate models from which we obtain the expected

Table 27.1 Univariate models for interbank interest rates: sample 2.10.1979–3.12.1984

Univariate stochastic structure	Overnight interest rate $\Delta R1D_t =$ $(1 - \theta_1 L - \theta_7 L^7)(1 - \theta_{18} L^{18}) a_t$		One-month interest rate $\Delta R1D_t = \dfrac{(1 - \theta_3 L^3 - \theta_4 L^4)(1 - \theta_7 L^7)(1 - \theta_{18} L^{18})}{(1 - \phi_1 L)} a_t$	
D3AB82	−0.11	(6.65)	−0.014	(2.96)
D2DI83	−0.08	(4.78)	−0.008	(1.61)
D1EN84	−0.13	(7.25)	−0.021	(3.82)
D2EN84	−0.12	(8.32)	−0.015	(2.63)
ϕ_1			0.21	(2.95)
θ_1	0.46	(7.45)		
θ_3			−0.16	(2.13)
θ_4			0.23	(3.12)
θ_7	0.24	(4.01)	0.29	(3.80)
θ_{18}	−0.24	(3.50)	−0.11	(1.60)
RSS	0.0757	d.f. = 186	0.0070	d.f. = 185
$\hat{\sigma}$	0.020		0.007	
Box–Ljung statistic	$Q(14) = 8.1$	$Q(26) = 21.6$	$Q(14) = 21$	$Q(26) = 34.1$
R_1^2	0.72		0.94	
R_2^2	0.56		0.44	

R_1^2 is the R^2 coefficient for the original series.
R_1^2 is the R^2 coefficient for the stationary series.
The values in brackets after the estimates coefficients are their corresponding t statistics.

interest rates. The four impulse dummies that the models include refer to the same dates as the dummies used by Escrivá (1986) in a univariate model to explain the adjusted cash reserves variable, and they capture the disturbances produced in the market by substantial institutional changes. As we shall see, these same dummies are required in the model for the excess reserve ratio, and we shall discuss the causes that motivated them. Temporary dependence on the interest rates, which appears in models of Table 27.1 was previously referred to in Espasa (1982).

27.3 THE ESTIMATION OF RISK

According to Burger (1971, pp. 55–9), the uncertainty that makes the banks maintain excess reserves is caused by three fundamental factors, which can be represented by the following three variables:

(1) the variability of interest rates
(2) the variability in the flow of assets between the banks and the rest of the economic agents
(3) the expectation that the banks have of the variability of the flow of deposits.

Of these three potential sources of risk the third one has no importance at the aggregate level of this analysis; therefore the construction of variables that estimate the risk factor is limited to the first and second, which, in principle, should have some relevance for the determination of excess reserves.

The variability of the interest rates has been estimated by a moving sample variance. In constructing this we have to decide about the following aspects:

(1) which is the most representative interest rate
(2) the level of time disaggregation at which we are going to work
(3) the transformation, if any, that we are going to apply to the interest rates
(4) the order of the moving sample variance.

For the first point we have chosen the interbank interest rates one week to maturity. For the second point we have decided to work at the maximum level of disaggregation and we have used daily data. The variability of interest is the corresponding to the nonanticipated component of the interest rates and we have estimated this by the variability of the residuals in a univariate model for the logarithmic transformation of the interest rates. At a first approximation we can take the daily series of the interbank weekly interest rates as following a random walk, and therefore the moving sample variance has been calculated on the first differences of the logarithmic transformation of the daily data of the weekly interest rates. This random walk hypothesis seems to be at variance with the results obtained by aggregating the weekly interest rates in means for nonoverlapping period of ten days. The model estimated for this ten-day series is

$$\Delta \text{RIS}_t = (1 - \theta_1 L - \theta_7 L^7)(1 - \theta_{18} L^{18})a_t.$$

Nevertheless in this ten-day period model the moving average of order 1 can be induced by aggregation made on the daily data, and the moving averages of order 7 and 18 are difficult to detect in the daily data due to the presence of midweek holidays. Thus, even though a more elaborate model on the daily data could reject the random walk hypothesis, for the purpose of the present study we can take it as a reasonable one.

In this market information is assimilated quite rapidly, and it has been assumed that the agents evaluate the risk taking as a reference the most immediate past. Therefore we have tried with sample moving variances of periods between five and thirty days, and we have chosen the one which produces the best adjustment for the whole model. This is the sample moving variance of ten days.

With reference to the second source of potential risk, the flow of assets which is specially important in determining the behaviour of the agents on the interbank market is the daily supply of cash reserves by the central bank. A lower variability of this supply will reduce the uncertainty in this market and, therefore, will induce the banks to maintain a lower level of excess reserves. As in the case of the interest rates, only the variability of the unexpected component can have an influence on the risk. A measure of this variability is a sample moving variance of the residuals of a univariate model for the adjusted reserves. For this variable we do not have available daily data, but only data for the mean of each ten-day nonoverlapping period. The best fit was obtained with a moving variance of twelve ten-day periods.

In the model for the excess reserve ratio the moving variance of the adjusted reserves is significant when it appears with a lag of two periods. One could expect this variable to act on the model with only one lag. The divergence between the result obtained and the a priori belief may, in part, be due to the temporary aggregation employed. In any case we have considered an alternative way of measuring the variability of the adjusted reserves.

It may be thought that there is a certain component of risk induced by the seasonal behaviour of the central bank in its supply of cash reserves, if the central bank does not fully adjust its supply to the seasonal movements of the demand. If this occurs it will lead to an adjustment in the market through prices (interest rates) or through quantities (excess reserves) or through a combination of both. In the last two cases we expected to find identifiable seasonal behaviour of the excess reserves, and in fact we did.

Thus in accordance with the foregoing, we have also tried with a sample moving variance of the stationary transformation of the adjusted cash reserves. The results obtained with these two alternative measures of the variability of reserves are reported in Section 27.5.

In addition to the risk caused by the variability in the interest rates and cash reserves, the model makes it possible to obtain, indirectly, a third measurement of uncertainty. The model includes as explanatory variables two expected interest rates, $R1M_t^e$ and $R1D_t^e$, with opposite signs, so that their overall effectiveness in the model (CTI_t^e) can be expressed as

$$CTI_t^e = \alpha_0 R1M_t^e - \alpha_1 R1D_t^e \qquad (27.7)$$

where α_0 and $-\alpha_1$ are the parameters which would be estimated for both variables. If we take α_0 as being greater than α_1 then expression 27.7 may be restated as

$$\text{CTI}_t^e = \alpha_1(\text{R1M}_t^e - \text{R1D}_t^e) + (\alpha_0 - \alpha_1)\,\text{R1M}_t^e \tag{27.8}$$

so that the contribution of the interest rates will consist of the effect of the spread between interest rates, on the one hand, and a positive effect of the monthly interest rate, on the other. The spread measures the risk of lending at one month, as opposed to one day.

27.4 GLOBAL SPECIFICATION OF THE MODEL

According to the expectations formed by the agents on the three basic variables mentioned in Section 27.2, the model can be expressed in a simplified way as follows

$$\text{ESB}_t = C + \xi_t + \alpha_0 \text{R1M}_t^e + \alpha_1 \text{R1D}_t^e + \alpha_2 \text{AC}^{ne} + \eta_t \tag{27.9}$$

where ESB_t is the excess reserve variable; C is a constant justified both on the existence of a fixed risk and the weaknesses and adjustment costs existing in the market; ξ_t measures the variable risk attributable both to the variability of interest rates and the action of the monetary authorities; R1M_t^e and R1D_t^e are the expected values of one-month and overnight interest rates respectively; AC^{ne} is the unexpected reserves component, reflecting unforeseen behaviour by the central bank on the market; η_t is the seasonal disturbance which can be broken down into two components: a predictable one, capturing the effects of seasonal patterns and omitted variables, and a nonpredictable one, a_t, so that

$$\eta_t = \hat{\eta}_t + a_t.$$

Replacing the expected interest rates by expressions more consistent with the desired levels,[4] R1M_t^{e*} and R1D_t^{e*}, and using them as proxies of the expected interest rates in the long run, the interest-rate contribution to the model can be restated as

$$\alpha_0 \text{R1M}_t^e + \alpha_1 \text{R1D}_t^e = \alpha_0 \text{R1M}_t^{e*} + \alpha_1 \text{R1D}_t^{e*} + r_t^*, \tag{27.10}$$

where r_t^* is the residual incurred when shifting from expected levels in the short run to expected levels in the long run.

From expressions 27.9 and 27.10 a desired level of excess reserves in the short run can be derived

$$\text{ESB}_t^d = C + \xi_t + \alpha_0 \text{R1M}_t^{e*} + \alpha_1 \text{R1D}_t^{e*} + \hat{\eta}_t \tag{27.11}$$

Assuming that $\hat{\eta}_t$ captures basically the seasonal structure of the excess reserves, a kind of seasonally-adjusted desired level could be defined by eliminating $\hat{\eta}_t$ in Equation 27.11.

In Equation 27.11 ESB_t^d is composed of a rigid minimum level imposed by

---

[4] The calculation of R1M_t^{e*} and R1D_t^{e*} is as shown in Section 27.6.

market weaknesses C, plus an assessment of variable risk ξ_t, plus the impact of the expectations formed on interest rates, plus, we might say, some seasonal requirements $\hat{\eta}_t$. ESB_t^d can also be expressed alternatively as

$$\text{ESB}_t^d = \text{ESB}_t - \alpha_2 \text{AC}_t^{ne} - a_t - r_t^*. \qquad (27.12)$$

e.g., the short-term desired level (ESB_t^d) equals the final excess reserve level ESB_t, deducting the effect of the contemporary monetary innovation (AC^{ne}), the effect of a random shock arising at time t (a_t) and the deviations between the expected values of interest rates in the long and short run (r_t^*).

The variable risk, ξ_t, can in turn broken down into

$$\xi_t = \xi_t^i + \xi_t^c \qquad (27.13)$$

where ξ_t^i is the risk associated with interest-rate variability and ξ_t^c the risk associated with reserve variability.

Combining Equations 27.9 and 27.13, the so-called desired level of excess reserves, in the long run, can be derived. Assuming $\alpha_0 = -\alpha_1$ we can write:

$$\text{ESB}_t = C + \xi_t^i + \xi_t^c + \alpha_0(\text{R1M}_t^e - \text{R1D}_t^e) + \alpha_2 \text{AC}^{ne} + \eta_t. \qquad (27.14)$$

In a traditional theory of interest-rate structure, the spread between expected one-month and one-day interest rates, ρ^e, measures the risk premium associated with longer terms. In a sense, specification 27.14 implies incorporating an additional method of risk estimation by the agents via the above-mentioned spread into the model. As AC^{ne} is a variable with zero mean (as defined in its construction) and η_t a seasonal disturbance, the effects of which fade out in the long run, we can actually consider a long-term desired level of excess reserves determined by risk only. Within the variable risk, only that associated with cash assets variability can be considered by agents in their long-term expectations; the risk attributable to interest-rate variability captures rather the uncertainty linked to temporary events and is more properly associated with the short term. The desired component will therefore consist of a term associated with the fixed risk, C, which every imperfect market carries with it; of a term associated with the variable risk due to the action of the monetary authorities (ξ_t^c), and finally of a third one which goes along with the risk premium (ρ^e) implied by lending at more than one day. Taking it a step further, it can be expected that, in calm market conditions, the expectations on one-month and overnight interest rates tend to converge, so that it will eventually be the same. In this case, the risk premium for long-term (one month) transactions would be zero and the long-term desired excess reserves would be given by $c + \xi_t^c$.

In Section 27.6, we shall discuss the consistency of the hypothesis developed to derive the desired levels of short-term and long-term excess reserves in the light of the results obtained from the estimation.

27.5 ESTIMATION OF THE MODEL

The kinds of models estimated according to the points put forward in the previous sections can be specified as follows

Table 27.2 Summary of estimated models

Variable	Lag	Parameter	Model I	Model II	Model III	Model IV	Model V
Constant		C	0.041 (2.71)	0.0551 (4.53)	0.0427 (4.78)	0.0555 (6.75)	0.0406 (3.14)
D3AB82	0	ω_0	0.1474 (11.12)	0.1554 (11.09)	0.1468 (10.91)	0.1471 (10.98)	0.1471 (11.38)
D2DI83	0	ω_1	0.1717 (13.49)	0.1677 (12.25)	0.1719 (13.53)	0.1714 (13.36)	0.1664 (13.22)
S1EN84	0	ω_2	0.3826 (24.65)	0.3808 (23.44)	0.3819 (24.29)	0.3804 (24.42)	0.3762 (24.65)
	1	ω_3	−0.3968 (25.02)	−0.3963 (24.45)	−0.3934 (24.36)	−0.3994 (25.19)	−0.3924 (25.17)
D2EN84	0	ω_4	1.0105 (64.17)	0.9994 (64.26)	1.0031 (64.22)	1.0123 (64.11)	1.0098 (64.62)
	1	ω_5	0.0657 (4.63)	0.0472 (3.17)	0.0644 (4.49)	0.0669 (4.70)	0.0640 (4.58)
D2MA84	0	ω_6	0.1102 (7.86)	0.1296 (9.03)	0.1112 (7.88)	0.1129 (8.03)	0.1070 (8.11)
	1	ω_7	0.0718 (5.16)	0.0835 (5.78)	0.0707 (5.01)	0.0733 (5.24)	0.0678 (5.16)
D1OC84	0	ω_8	0.0941 (7.14)	0.0960 (6.89)	0.0930 (7.02)	0.0943 (7.10)	0.0892 (7.12)
R1M^e	0	ω_9	0.4316 (5.19)	0.3971 (5.08)	0.4518 (5.13)	—	0.4421 (5.22)
R1D^e	0	ω_{10}	−0.3599 (6.30)	−0.3721 (6.46)	−0.3771 (6.47)	—	−0.3760 (6.61)
DTI	0	ω_{20}	—	—	—	0.3591 (6.21)	—
AC^{ne}	0	ω_{11}	2.7889 (13.44)	2.5845 (11.97)	2.7441 (13.11)	2.7730 (13.19)	3.0578 (14.17)
	1	d_1	0.2946 (4.15)	—	0.2864 (3.92)	0.2719 (3.82)	—
VTI	1	ω_{12}	0.0011 (2.13)	0.0012 (2.18)	0.0011 (2.30)	0.0014 (2.46)	—
VACNE	1	ω_{13}	—	35.4090 (2.41)	—	—	—
	2	ω_{14}	44.1654 (2.79)	—	—	49.5818 (3.31)	49.8597 (3.17)
	37	ω_{15}	—	−52.4674 (0.81)	—	—	—

Estimation of the model

VACE		1					
ΔLAC^e	ω_{16}	0	—	—	—	—
	ω_{17}	1	—	—	14.2369 (1.87)	—	1.3554 (5.60)
	ω_{18}	1	—	—	—	—	0.3919 (1.65)
	ω_{19}	2	—	—	—	—	0.6581 (2.97)
MA	θ_{36}	36	0.2341 (2.84)	0.2727 (3.37)	0.2184 (2.60)	0.2373 (2.88)	0.1580 (1.79)
AR	ϕ_1	1	0.2190 (2.74)	0.1672 (2.03)	0.2289 (2.86)	0.2184 (2.72)	0.2416 (3.09)
AR	ϕ_3	3	0.2086 (2.58)	0.0904 (1.11)	0.2261 (2.79)	0.1934 (2.39)	0.2108 (2.74)
AR	ϕ_{36}	36	0.7607 (12.98)	0.7095 (11.77)	0.7362 (12.12)	0.7529 (12.79)	0.6049 (8.08)
RSS(d.f.)			0.03107(171)	0.03144(176)	0.03181(171)	0.03131(171)	0.02839(174)
$\hat{\sigma}$ (nonadjusted by degrees of freedom)			0.01347	0.01379	0.01364	0.01353	0.01277
Statistic Box–Peerce–Ljung			$Q(14) = 9.5$ $Q(26) = 18.5$ $Q(36) = 31.7$	$Q(14) = 12.8$ $Q(26) = 20.7$ $Q(36) = 32.7$	$Q(14) = 11.5$ $Q(26) = 20.2$ $Q(36) = 32.3$	$Q(14) = 9.6$ $Q(26) = 18.2$ $Q(36) = 31.4$	$Q(14) = 8.8$ $Q(26) = 17.4$ $Q(36) = 28.7$
$R^2 = 1 - \dfrac{\hat{\sigma}^2}{\text{VAR(ESB)}}$			0.973 0.754^a	0.972	0.973	0.973	0.976

^{a}This R^2 coefficient corresponds to the ESB series corrected of all interventions.

$$\begin{aligned}ESB_t = &\ C_0 + \omega_0 D3AB82 + \omega_1 D2DI83 + (\omega_2 + \omega_3)\, S1EN84 \\ &+ (\omega_4 + \omega_5 L)\, D2EN84 + (\omega_6 + \omega_7 L)\, D2MA84 \\ &+ \omega_8 D1OC84 + \omega_9 R1M_e^e + \omega_{10} R1D_t^e \\ &+ \frac{\omega_{11}}{(1 - d_1 L)} AC_t^{ne} + \omega_{12} L\, VTI_t + \omega_{14} L^2\, VACNE \\ &+ \frac{(1 - \theta_{36} L^{36})}{(1 - \phi_1 L - \phi_3 L^3)(1 - \phi_{36} L^{36})} a_t \end{aligned}$$

The estimated parameters corresponding to this specification are shown in Table 27.2 under model I; this table also shows together with the parameter values, their statistical t in brackets. Figure 27.1 shows the correlogram of residuals. The explanatory variables of the model are explained below.

The model incorporates six dummy variables in order to capture the impact of disturbances in the money market which create a temporary, unusually high rise in excess reserves. A mere glance at the excess reserve graph (Fig. 27.4) makes it possible to detect these anomalies and assess their magnitude. D3AB82 is an impulse dummy variable capturing the effect of money market certificates (CRMs)[5] in the last ten days of April 1982. D2DI83, D2EN84, and S1EN84 are three dummy variables (the first two, impulses, and the third one, step) capturing the impact on excess reserves of the changes in the regulation on reserve requirements in December 1983 and January 1984. These three variables represent altogether a number of impulses in the second 10-day period of December and in the three periods of January, as well as a negative step from the second 10-day period of

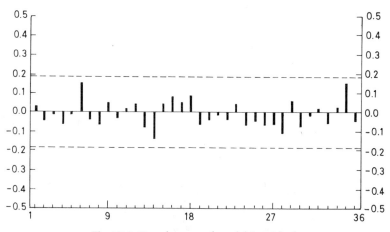

Fig. 27.1 Correlogram of model I residuals

---

[5] Money market certificates were the financial assets used by the Bank of Spain to implement its open market policy between April 1982 and January 1984. From that date they were replaced by Treasury Bills.

January, which evidences that excess reserves, after the strong disturbances experienced at the beginning of 1984, stood at a lower level than in the rest of the sample periods. Thus, while the constant estimated in the model for the period 1980 to 1983 is 0.041, it drops to 0.027 from January 1984. Such a significant fall in the minimum risk level, which is how the constant must be interpreted, can be explained as follows: all the legal changes which occurred in December 1983 and January 1984 led to the ultimate disappearance of the agents' doubts about the new regulations applicable to reserve requirements and also led to the setting up of a new framework for performance on the reserve market, replacing CRMs by treasury bills. Thus the uncertainty was significantly reduced. D2MA84 and D1OC84 are two impulse dummy variables corresponding to the second 10-day period of March and to the first 10-day period of October respectively; both capture unusual increases in excess reserves on those dates, due mainly to temporary downfalls of one-day interest rates on the interbank market.

Escrivá *et al.* (1986) suggest calculating the unexpected reserve component as the innovation of a univariate ARIMA model for reserves. The model also includes four impulse dummies, associated with the two institutional changes mentioned above, and their effect is not included in the unexpected component.

To construct the risk variable attributable to the actions of the monetary authorities – denoted by VACNE in the model – the effects of the four impulse dummies are added to the residuals of the model to calculate the moving variances. This was done because this variable was meant to measure risk, and therefore the uncertainty created on the market by the two institutional changes could not be disregarded.

The unexpected reserve component is incorporated into the model under a rational distributed lag structure. The response of excess reserves to a monetary shock is very quick, 71% of the effect being focused on the current ten-day period and 21% on the following one; beyond the third ten-day period, AC^{ne} practically stops producing any effect; e.g., faced with an unexpected reserves supply, the agents are forced to maintain a higher level of excess reserves within the ten-day period over which the disturbance occurs and, to a much lesser extent, in the two following ten-day periods.

As AC^{ne} is derived from the innovation of a univariate model where the variable is expressed in logarithms[6] and ESB is defined as the ratio between excess reserves and computable liabilities, the resulting overall gain, 3.95, cannot be interpreted in terms of constant elasticity.

The expected interest rate variables, $R1M^e$ and $R1D^e$, have already been discussed and VTI is the risk variable associated with the variability of the interest rate which has been described in Section 27.3.

The stochastic structure estimated for the disturbance of the model determining the excess reserves captures three effects:

---

[6] This implies that the AC^{ne} variable used could be interpreted as being expressed as a decimal fraction of the total reserves.

Table 27.3 Roots of autoregressive polynomial

Roots	Modulus	Argument in degrees	Periods	36 periods
1.4657	1.4657	0d	0.00	0.00
−0.73287−1.6533 i	1.8085	−113d	−3.16	−11.39
−0.73287+1.6533 i	1.8085	113d	3.16	11.39

(1) a seasonal dependence in the excess reserves which is approximated by means of an autoregressive-moving averages structure:
(2) cyclical dependence, with an approximate periodicity of three periods of ten days, incorporated by complex roots of the autoregressive third-order polynomial, AR(3), (see Table 27.3); and
(3) a first-order dependence captured by the positive root of the AR(3) factor.

The existence of the last two types of temporal dependence justify the fact that the banks operate on the market at two levels: with a ten-day treasury management plan to meet the reserve requirements and with a simultaneous more extensive monthly planning connected with monthly bank balance-sheets.

The only discrepancy between the initial theoretical approach and the variables of the estimated model I stems from the fact that the variance of the unexpected reserve component (VACNE) affects excess reserves with a two-period lag whereas it should be expected to affect with a one lag period. In model II of Table 27.2 we apply the filter $(\omega_{13}L + \omega_{15}L^{37})$ to the variable VACNE, but the coefficient ω_{15} does not result statistically significant and, on the other hand, the overall adjustment deteriorates and an exceedingly high value of constant C appears, which is difficult to interpret. In model III (see Table 27.2), VACNE is replaced by a sample moving variance of the stationary transformation of reserves, which can be called VACE. The parameter obtained, ω_{16}, is not significant and moreover, as VACE captures part of the seasonal component of excess reserves, it is difficult to perform with this model a breakdown of the fit separating the merely seasonal effects from those induced by risk. Therefore, no satisfactory alternative to model I could be developed.

In model IV the restriction $\omega_9 = -\omega_{10}$ is imposed and the interest rate effect is estimated as $\omega_{20}(R1M^e - R1D^e)$. The maximum likelihood ratio test does not reject model IV in favour of model I. However, with the restriction that the interest rates have the same coefficient but with different sign, the interest rates have a smaller overall positive contribution which entails a higher compensation of the constant risk (C) and of the variable risks (VTI and VACNE). Model IV versus model I originates an approximation between the levels of desired long-run and short-run excess reserves.

Since targets are set by the central bank in terms of reserves, a new version of the model has been estimated incorporating as advanced indicator the expected

Model breakdown and calculation of short and long-run levels 623

component of reserves exhibited in Table 27.2 as model V. As new information on reserves[7] is available, and in order to forecast excess reserves, model V can be used, which adds to model I the rate of change of the expected component of the adjusted reserves, ΔLAC^e, resulting from the breakdown provided by Escrivá et al. (1986). This variable is entered both contemporaneously and with two lags. With model V, the residual variance is reduced by 10.2% with respect to model I. Two features stand out when comparing the parametric structure of both specifications:

(1) In model V, the distributed rational lag structure of the unexpected component of reserves disappears, hence maintaining exclusively a contemporaneous relation. This was to be expected taking into account that the unexpected component is embodied in the expected component after one period of time.
(2) Likewise, it appears clear in model V that the parameters of the seasonal ARIMA structure drop significantly. This evidences that part of the seasonality of excess reserves is induced by the reserves supply.

Model V should be simply considered as being useful for forecasting purposes since, with the indicator ΔLAC^{ne} included in the model, the effects of certain institutional factors, which are difficult to quantify and which model I relegates to a residual term, may be calculated with a greater approximation. Thus, in model I, the gain of the residual filter is 5.6, while that of model V is 3.5.

27.6 MODEL BREAKDOWN AND CALCULATION OF DESIRED SHORT AND LONG-RUN LEVELS

This section deals with the breakdown of the excess reserves in the contribution of the various explanatory variables. According to the result of this breakdown and taking Section 27.4 as a starting point, the levels of short and long-run excess reserves desired by agents may be obtained.

The contribution of the various economic variables is calculated as follows

$$CR1M_t^e = 0.4316\ R1M_t^e \qquad (27.15)$$

$$CR1D_t^e = 0.3599\ R1D_t^e \qquad (27.16)$$

$$CAC^{ne} = \frac{2.7889}{(1 - 0.2946L)} AC_t^{ne} \qquad (27.17)$$

$$CVTI = 0.0011L\ VTI_t \qquad (27.18)$$

$$CVAC = 44.1654L^2\ VAC_t \qquad (27.19)$$

where $CR1M_t^e$, $CR1D_t^e$, CAC_t^{ne}, $CVTI_t$ and $CVAC_t$ are the respective contributions of the expected one-month interest rate, of the expected overnight interest rate, of

---
[7] Reserves are controlled by the central bank and are known the same day, whereas excess reserves are obtained from the reserve requirement reports compiled by banking institutions and are known after about ten days.

the nonexpected component of reserves, of the interest rate variance and of the reserves variance.

As regards the contribution of the deterministic part of the model, it is possible to differentiate between the part affecting the excess reserve level and the part which captures exclusively transitory disturbances. The contribution of the step variable S1EN84 may be broken down as follows

$$(0.3826 - 0.3968L)\ \text{S1EN84} = 0.3826\ \text{D1EN84} - 0.0142\ \text{S2EN84},$$

where D1EN84 is an impulse variable in the first ten-day period of January 1984 and S2EN84 is a step variable from the second ten-day period of the same year. This step plus the constant estimated in the model provides the minimum level of excess reserves (CNIV_t)

$$\text{CNIV}_t = 0.0410 - 0.0141\ \text{S2EN84}.$$

The contribution of impulses (CIMP) is given by

$$\begin{aligned}\text{CIMP} = &\ 0.1474\ \text{D3AB82} + 0.1717\ \text{D2DI83} \\ &+ (0.3826 + 1.0105L + 0.0657L^2)\ \text{D1EN84} \\ &+ (0.1102 + 0.0718L)\ \text{D2MA84} + 0.0941\ \text{D1OC84}\end{aligned}$$

The contribution of disturbance (CPER), made up of both the contemporary and the lagged contributions of innovations, can be expressed as

$$\text{CPER}_t = \frac{(1 - 0.2341L^{36})}{(1 - 0.2190L - 0.2086L^3)(1 - 0.7607L^{36})} a_t.$$

This expression may also be defined as

$$\text{CPER}_t = \text{CPERS}_t + a_t$$

where CPERS_t is the predictable component, i.e., the part known given the information up to time $t-1$ and a_t is the innovations or unpredictable component.

The behaviour of excess reserves is fully explained by the sum of all the previously defined components

$$\begin{aligned}\text{ESB}_t = &\ \text{CNIV}_t + \text{CIMP}_t + \text{CR1M}_t^e + \text{CR1D}_t^e + \text{CAC}^{ne} + \text{CVTI}_t \\ &+ \text{CVAC}_t + \text{CPERS}_t + a_t\end{aligned}$$

All these components are shown in Appendix B (Table B1), whereas Appendix A records the relevant charts. From this information we can derive some results in relation to the differential contribution of the explanatory variables of excess reserves.

The constant level (CNIV_t) has a great significance within the model as it sets the minimum level on which excess reserves are determined if the contribution of the stochastic explanatory variables is zero. At the same time the value of CNIV_t allows us to differentiate two periods within the sample used (before and after January

Model breakdown and calculation of short and long-run levels

1984) and to associate the different 'minimum levels' experienced by the excess reserves over each of the two periods with a structural change fostered by the regulation amendments affecting the financial institutions and the interbank market between December and January 1984.

The variable $CIMP_t$ is significant to the extent that it allows us to quantify the magnitude of the unusual rise of excess cash reserves in nine specific ten-day periods.

The contribution of moving variances of interest rates ($CVTI_t$) and reserves ($CVAV_t$) could be considered as residual, as they explain a very small part of excess reserve behaviour. As far as interest rates variance is concerned (see Fig. A2 in Appendix A), its average contribution is minimal and only at specific points of the sample, where interest-rate oscillations reach a high level, they have a significant role in the explanation of temporary increases of excess reserves.

The contribution of the reserve variance is also minimal, except for the months following the disturbances which occurred in April 1982 and especially in December and January 1984 (see Figure A3 of Appendix A). This variable therefore plays a double part within the model, capturing on the one hand the existence of a minimum risk, fairly constant, associated with the relative smoothness of the monetary authority's action in their supply of reserves, and on the other hand two important uncertainty effects linked to two institutional changes made by the central bank.

With a view to the breakdown of contributions into short and long term, trying to isolate the two effects indicated above is particularly convenient. The operation is possible because the VAC_t variable can be broken down into the part of the variance corresponding only to the innovations of the univariate reserve model, (VAC_t^r), and into the part due to the impulses of this model (VAC_t^i) which capture the effects of the losses of monetary control associated with the above-mentioned changes. In terms of contribution to the excess reserve, $CVAC_t$ is broken down into

$$CVAC_t = CVAC_t^r + CVAC_t^i.$$

In $CVAC_t^r$, four periods can be distinguished according to the risk level induced by the conducting of monetary policy on the part of the central bank:

- 1980–1981: more and more stable implementation of monetary policy
- 1982: stabilization of the reduction in reserve variance
- 1983: increasingly erratic central bank monetary intervention
- 1984: again progressive reduction in the variance of the unexpected component of reserves down to 1982 levels.

The most important explanatory variables in the determination of excess-reserve formation are the two interest rates and the 'monetary innovations' (see Figs A4, A5, and A6 of Appendix A). The latter explains by itself the strong fluctuation in excess reserve over particular periods of time.

As far as interest rates are concerned, both contribute significantly, one-month rate performing positively, one-day rate negatively. However, overall interest-rate contribution can be reformulated as

$$0.4316\ R1M_t^e - 0.3599\ R1D_t^e = 0.0733\ R1M_t^e + 0.3599\ (R1M_t^e - R1D_t^e),$$

so as to separate the positive contribution of one-month interest rate from the differential effect between both rates, which as indicated in Sections 3 and 4, may be associated with a measure of uncertainty,

To determine each of these contributions the following terms will be used

$$\overline{CR1M_t^e} = 0.0733\ R1M_t^e$$
$$CDTI_t^e = 0.3599\ (R1M_t^e - R1D_t^e).$$

According to this, the contribution of economic variables may be grouped as follows

$$CVEC_t = CAC_t^{ne} + \overline{CR1M_t^e} + CRID_t,$$

where $CRIS_t$ is the contribution of the three risk components

$$CRIS_t = CVAC_t + CVTI_t + CDTI_t^e.$$

Figure 27.2 exhibits the three $CVEC_t$ components simultaneously in order to visually assess their differential contributions to excess reserve explanation.

The total disturbance contribution (CPER) is shown in Fig. A7 of Appendix A, and Figs A8 and A9 display the predictable (CPERS) and unpredictable (a_t) components respectively. CPERS makes two questions evident:

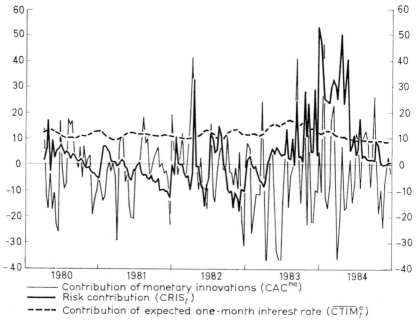

Fig. 27.2 Contribution of economic variables to excess reserve explanation

(1) the existence of a seasonal component and
(2) the absence of a variable in the model, capable of capturing that smooth secular decrease in the excess reserve levels which must be recorded by the forecastable component of the residuals. This decrease, associated no doubt with the slow but progressive improvement of both the interbank market and the agents acting on it, is difficult to model by means of economic variables.

As far as the unpredictable component (a_t) is concerned, it must be pointed out that notwithstanding the set of effects estimated by the model, it makes evident that bank excess reserves are difficult to forecast. The set of random and unpredictable

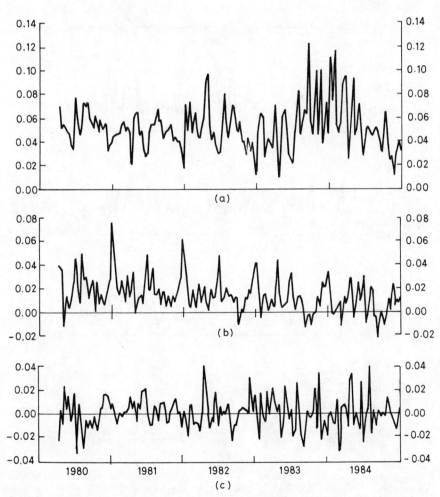

Fig. 27.3 Breakdown of bank excess reserves: (a) contribution of the economic variables and constant term ($CVEC_t = CNIV_t$); (b) predictable disturbance contribution ($CPERS_t$); (c) innovation contribution (a_t)

shocks affecting the market are an important component that impede forecast of excess reserves at very narrow intervals.

Figure 27.3 shows the contribution of economic variables plus the constant $(CNIV_t + CVEC_t)$, the contribution of the predictable disturbance $(PERS_t)$ and the innovations (a_t), in three different sequences on the same scale.

Figure 27.4 was designed in order to assess how excess reserve is progressively smoothed as its more erratic components are drawn out.

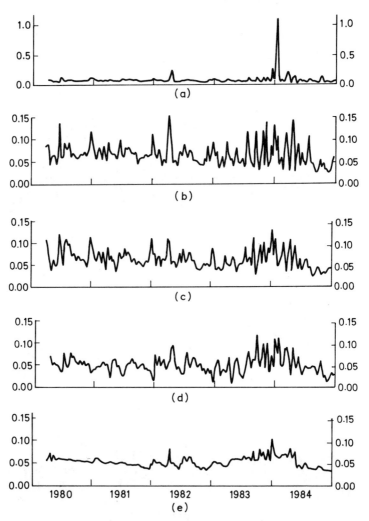

Fig. 27.4 (a) Bank excess reserves (original series); (b) original series adjusted for impulse interventions; (c) series adjusted for impulse interventions and innovations; (d) series adjusted for impulse interventions, innovations and predictable disturbance; (e) series adjusted for impulse interventions, innovations, predictable disturbance and 'monetary innovations'

Model breakdown and calculation of short and long-run levels 629

Figure 27.4(a) exhibits the original excess reserve series, clearly dominated by a number of outliers which impairs observation of its normal behaviour. This behaviour is made evident in Fig. 27.4(b) as the estimated impact of impulse interventions is removed. If innovations are extracted from the intervention-adjusted series the profile shown in Fig. 27.4(c) is obtained; these series provide the one period ahead forecast that the model would have supplied. A greater reduction of excess reserve variance will be obtained by deducting not only innovations but the whole disturbance (Fig. 27.4(d)). If, taking a further step, the effect of 'monetary innovations' is deducted we obtain Fig. 27.4(e), which shows the excess reserve profile once their most erratic components have been eliminated. This excess reserve level is supposed to bring us closer to the so-called desired level expected in the short run by the agents ($ESBCP_t$), which is defined by the sum of the following components

$$ESBCP_t = CNIV_t + \overline{CR1M_t^e} + CDTI_t^e + CVTI_t + CVAC_t^r + CVAC_t^i.$$

However, $ESBCP_t$ has too many oscillations owing to the contribution of the interest rates that, even though expressed in expected values, fluctuate excessively. Apparently it makes no sense to think that agents will use these forecasts of the interest rates when defining their desired level of excess reserves, even though in the short run, but rather a set of forecasts corresponding, in every moment, to a medium term projection of the interest rates. The univariate models presented in Section 27.2 were used, and forecasts, both for overnight and one-month interest rates, 25 periods ahead from the different starting points were generated, so as to eliminate the above-mentioned fluctuations. The 25-period horizon was chosen because by then the prediction converges to a constant in the overnight interest rate and the prediction of the one-month interest rate is almost a constant from it. Thus we can state

$$R1M_t^{e*} = \widehat{R1M}_{t+25}$$
$$R1D_t^{e*} = \widehat{R1D}_{t+25},$$

where $R1M_t^{e*}$ and $R1D_t^{e*}$ are the series of interest rates expected by the agents in the long run. Table B2 in Appendix B incorporates a printout of these variables and of the spread existing between them. Thus, replacing $R1D_t^e$ and $R1M_t^e$ by $R1D_t^{e*}$ and $R1M_t^{e*}$, respectively, when calculating the contributions of the interest rates we obtain the desired level of excess reserves in the short run ($ESBCP_t^*$); therefore

$$ESBCP_t^* = CNIV_t + \overline{CR1M_t^{e*}} + CVTI_t + CVAC_t^r + CVAC_t^i + CDTI_t^{e*}.$$

In $ESBCP_t^*$ a number of components clearly contribute in the short run, while others provide information in the long run. As stated above, $CVTI_t$ approximates a variable and nonconsolidated risk; $CVAC_t^i$ is associated with point uncertainties which only have effect over some periods of time, in which reserves show a high volatility on some specific days, and $CDTI_t^{e*}$ is the risk premium associated with lending over longer terms, which, as discussed earlier, can be considered zero in the limit. Thus, the desired level of excess reserves in the long run (ESBLD$_t^*$) can be expressed as the sum of three components

$$\text{ESBLD}_t^* = \text{CNIV}_t + \overline{\text{CR1M}_t^{e*}} + \text{CVAC}_{t}^r;$$

thus excess reserves in the long run are comprised of a minimum structural risk level (CNIV$_t$) plus the contribution of one-month interest rate expected in the long run,

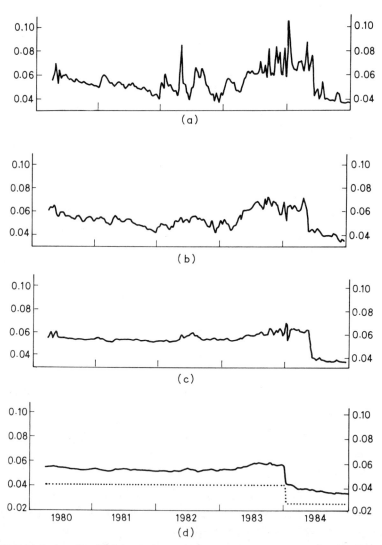

Fig. 27.5 (a) Series adjusted for impulse interventions, innovations, predictable disturbance and monetary innovations; (b) short-term desired level; (c) short-term desired level less effect of interest rate spread; (d) long-term desired level

as an indicator of the cost of future financement (CR1M$_t^{e*}$), plus a fairly stable variable risk (CVAC$_t^r$) linked to monetary authority action as far as its supply of reserves is concerned.

Figure 27.5 shows the excess reserve breakdown continuing from Fig. 27.4 in an attempt to represent the components of the excess reserves that have an economic interpretation.

Figure 27.5(a) provides the desired level of excess reserves in the short run calculated from the expected short-run interest rates (ESBCP$_t$); 27.5(b) exhibits this desired level expressed by the expected interest rates in the long run (ESBCP$_t^*$); in 27.5(c) the contribution of expected interest-rate spread in the long run is eliminated (CDTI$_t^{e*}$); eventually 27.5(d) represents the desired level in the long run. (ESBLP$_t$) simultaneously with the constant minimum structural level (CNIV$_t$), the latter as a dotted line.

This chart allows us to detect to what extent a progressive elimination of the different short-term and risk elements leads to a desired excess reserve level in the long run clearly associated with the structural, more stable elements out of all those determining excess reserves.

It must be emphasized that the deduction carried out on the desired long-run level excludes the own interest rate (RDS$_t$). This apparently surprising outcome coincides with that of Mauleon (1984) and can be justified as follows: the excess-reserve own interest rate, one-day to maturity, is extremely volatile, hardly

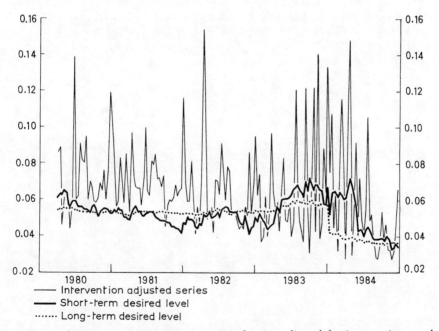

Fig. 27.6 Banking system excess reserves: original series adjusted for interventions, and desired levels in the short and long term

Table 27.4 Various transformations of bank excess reserves[a]

Observation no. and date	ESBCI (1)	MMESBCI (2)	ESBPB (3)	MMESBPB (4)	ESBPA (5)	MMESBPA (6)	ESBCP (7)	MMESBCP (8)	ESBLP (9)	MMESBLP (10)
1 8004.1	0.0860		0.1079		0.0941		0.0610		0.0545	
2 8004.2	0.0890		0.0856		0.0963		0.0632		0.0549	
3 8004.3	0.0460	0.0737	0.0558	0.0871	0.0727	0.0877	0.0619	0.0620	0.0546	0.0547
4 8005.1	0.0610		0.0379		0.0394		0.0650		0.0556	
5 8005.2	0.0840		0.0595		0.0768		0.0632		0.0551	
6 8005.3	0.0630	0.0627	0.0485	0.0486	0.0592	0.0585	0.0572	0.0618	0.0549	0.0552
7 8006.1	0.0440		0.0504		0.0728		0.0557		0.0546	
8 8006.2	0.0590		0.0594		0.0854		0.0588		0.0549	
9 8006.3	0.1390	0.0807	0.1226	0.0775	0.1056	0.0879	0.0588	0.0578	0.0545	0.0547
10 8007.1	0.0600		0.0951		0.0967		0.0589		0.0540	
11 8007.2	0.0820		0.0540		0.0635		0.0554		0.0537	
12 8007.3	0.0910	0.0710	0.0986	0.0826	0.1058	0.886	0.0558	0.0567	0.0536	0.0538
13 8008.1	0.0820		0.1121		0.0943		0.0558		0.0533	
14 8008.2	0.0800		0.0964		0.0613		0.0546		0.0533	
15 8008.3	0.0950	0.0857	0.1013	0.1033	0.0846	0.0867	0.0546	0.0550	0.0535	0.0534
16 8009.1	0.0600		0.0714		0.0873		0.0535		0.0532	
17 8009.2	0.0700		0.0761		0.0731		0.0531		0.0528	
18 8009.3	0.0670	0.0657	0.0785	0.0753	0.0796	0.0734	0.0560	0.0542	0.0533	0.0531
19 8010.1	0.0590		0.0612		0.0547		0.0569		0.0533	
20 8010.2	0.0580		0.0727		0.0779		0.0541		0.0526	
21 8010.3	0.0600	0.0590	0.0646	0.0862	0.0577	0.0634	0.0506	0.0536	0.0521	0.0526
22 8011.1	0.0690		0.0644		0.0650		0.0531		0.0524	
23 8011.2	0.0650		0.0614		0.0500		0.0542		0.0526	
24 8011.3	0.0760	0.0700	0.0606	0.0621	0.0572	0.0604	0.0549	0.0541	0.0530	0.0527
25 8012.1	0.0600		0.0446		0.0636		0.0548		0.0532	
26 8012.2	0.0770		0.0625		0.0758		0.0540		0.0533	
27 8012.3	0.1190	0.0853	0.1148	0.0739	0.1247	0.0880	0.0520	0.0536	0.0534	0.0533

28 8101.1	0.1070		0.0984		0.1021	0.0520		0.0538	
29 8101.2	0.0840	0.0823	0.0825		0.0912	0.0544		0.0539	0.0536
30 8101.3	0.0560		0.0651	0.0820	0.0792	0.0531	0.0909	0.0530	
31 8102.1	0.0590		0.0605		0.0731	0.0551		0.0527	
32 8102.2	0.0830	0.0707	0.0832		0.0657	0.0536		0.0523	0.0524
33 8102.3	0.0700		0.0736	0.0725	0.0728	0.0541	0.0772	0.0522	
34 8103.1	0.0580		0.0573		0.0602	0.0500		0.0516	
35 8103.2	0.0650	0.0700	0.0837		0.0830	0.0501		0.0517	0.0516
36 8103.3	0.0670		0.0620	0.0876	0.0664	0.0487	0.0899	0.0514	
37 8104.1	0.0530		0.0396		0.0685	0.0485		0.0511	
38 8104.2	0.0970	0.0743	0.0930		0.0848	0.0517		0.0516	0.0516
39 8104.3	0.0730		0.0630	0.0852	0.0522	0.0549	0.0656	0.0522	
40 8105.1	0.0870		0.0739		0.0634	0.0565		0.0532	
41 8105.2	0.0660	0.0583	0.0572		0.0641	0.0524		0.0530	0.0532
42 8105.3	0.0660		0.0630	0.0847	0.0680	0.0537	0.0845	0.0535	
43 8106.1	0.0570		0.0389		0.0552	0.0512		0.0534	
44 8106.2	0.0710	0.0760	0.0524		0.0731	0.0507		0.0534	0.0534
45 8106.3	0.1000		0.0800	0.0571	0.1009	0.0522	0.0784	0.0534	
46 8107.1	0.0650		0.0679		0.0698	0.0525		0.0534	
47 8107.2	0.0620	0.0693	0.0721		0.0710	0.0532		0.0534	0.0534
48 8107.3	0.0810		0.0905	0.0768	0.0905	0.0530	0.0771	0.0535	
49 8108.1	0.0790		0.0730		0.0847	0.0528		0.0535	
50 8108.2	0.0650	0.0783	0.0819		0.0634	0.0498		0.0524	0.0531
51 8108.3	0.0710		0.0768	0.0772	0.0686	0.0495	0.0655	0.0533	
52 8109.1	0.0720		0.0618		0.0535	0.0482		0.0534	
53 8109.2	0.0670	0.0707	0.0592		0.0641	0.0483		0.0534	0.0533
54 8109.3	0.0730		0.0840	0.0823	0.0651	0.0483	0.0609	0.0533	
55 8110.1	0.0450		0.0538		0.0510	0.0476		0.0530	
56 8110.2	0.0560	0.0537	0.0634		0.0566	0.0471		0.0526	0.0528
57 8110.3	0.0600		0.0582	0.0564	0.0516	0.0472	0.0537	0.0526	
58 8111.1	0.0590		0.0547		0.0561	0.0449		0.0526	
59 8111.2	0.0610	0.0627	0.0515		0.0509	0.0450		0.0526	0.0527
60 8111.3	0.0660		0.0559	0.0540	0.0592	0.0442	0.0554	0.0527	
61 8112.1	0.0620		0.0618		0.0640	0.0437		0.0526	

634 The determination of banking system excess reserves

Table 27.4 (Contd.)

Observation no. and date	ESBCI (1)	MMESBCI (2)	ESBPB (3)	MMESBPB (4)	ESBPA (5)	MMESBPA (6)	ESBCP (7)	MMESBCP (8)	ESBLP (9)	MMESBLP (10)
62 8112.2	0.0650		0.0643		0.0709		0.0428		0.0522	
63 8112.3	0.0660	0.0643	0.0779	0.0680	0.1012	0.0767	0.0412	0.0426	0.0521	0.0523
64 8201.1	0.1160		0.1147		0.0951		0.0468		0.0523	
65 8201.2	0.0700	0.0817	0.0776	0.0903	0.0760	0.0797	0.0460	0.0478	0.0524	0.0524
66 8201.3	0.0590		0.0767		0.0678		0.0504		0.0526	
67 8202.1	0.0590		0.0522		0.0562		0.0469		0.0524	
68 8202.2	0.0810		0.0743		0.0711		0.0492		0.0525	
69 8202.3	0.0620	0.0673	0.0715	0.0660	0.0595	0.0623	0.0486	0.0489	0.0524	0.0524
70 8203.1	0.0410		0.0487		0.0524		0.0451		0.0520	
71 8203.2	0.0560		0.0655		0.0690		0.0441		0.0517	
72 8203.3	0.0470	0.0480	0.0632	0.0591	0.0560	0.0591	0.0470	0.0450	0.0523	0.0520
73 8204.1	0.0710		0.0740		0.0558		0.0455		0.0522	
74 8204.2	0.1540		0.1132		0.0718		0.0514		0.0531	
75 8204.3	0.1096	0.1115	0.1035	0.0969	0.0918	0.0731	0.0504	0.0491	0.0526	0.0526
76 8205.1	0.0490		0.0561		0.0537		0.0507		0.0527	
77 8205.2	0.0500		0.0622		0.0718		0.0538		0.0528	
78 8205.3	0.0560	0.0517	0.0599	0.0594	0.0572	0.0609	0.0497	0.0514	0.0530	0.0528
79 8206.1	0.0420		0.0473		0.0564		0.0506		0.0536	
80 8206.2	0.0710		0.0543		0.0633		0.0510		0.0548	
81 8206.3	0.0700	0.0610	0.0789	0.0602	0.0961	0.0719	0.0531	0.0516	0.0553	0.0546
82 8207.1	0.0830		0.0570		0.0571		0.0503		0.0546	
83 8207.2	0.0910		0.0919		0.0761		0.0544		0.0537	
84 8207.3	0.0740	0.0760	0.0726	0.0738	0.0842	0.0725	0.0556	0.0534	0.0539	0.0541
85 8208.1	0.0600		0.0584		0.0729		0.0562		0.0541	
86 8208.2	0.0780		0.0691		0.0716		0.0556		0.0531	
87 8208.3	0.0750	0.0710	0.0845	0.0707	0.0802	0.0749	0.0530	0.0549	0.0521	0.0531
88 8209.1	0.0530		0.0764		0.0698		0.0541		0.0521	
89 8209.2	0.0540		0.0634		0.0650		0.0528		0.0527	
90 8209.3	0.0520	0.0530	0.0609	0.0669	0.0635	0.0681	0.0538	0.0535	0.0535	0.0528

Model breakdown and calculation of short and long-run levels 635

91 8210.1	0.0490			0.0343		0.0524		0.0540	
92 8210.2	0.0440		0.0469	0.0448		0.0480		0.0538	
93 8210.3	0.0450	0.0463	0.0419	0.0452	0.0436	0.0466	0.0490	0.0539	0.0539
94 8211.1	0.0440		0.0419	0.0498		0.0463		0.0543	
95 8211.2	0.0560		0.0418	0.0555		0.0518		0.0551	
96 8211.3	0.0510	0.0503	0.0542	0.0601	0.0489	0.0463	0.0481	0.0540	0.0545
97 8212.1	0.0860		0.0506	0.0520		0.0410		0.0531	
98 8212.2	0.0520		0.0565	0.0724		0.0446		0.0529	
99 8212.3	0.0680	0.0687	0.0512	0.0842	0.0535	0.0460	0.0439	0.0530	0.0530
			0.0526						
100 8301.1	0.0950		0.0913	0.0918		0.0506		0.0542	
101 8301.2	0.0630		0.0855	0.0769		0.0480		0.0536	
102 8301.3	0.0740	0.0773	0.0570	0.0508	0.0779	0.0525	0.0505	0.0544	0.0541
103 8302.1	0.0370		0.0414	0.0879		0.0505		0.0541	
104 8302.2	0.0400		0.0485	0.0871		0.0482		0.0541	
105 8302.3	0.0620	0.0463	0.0439	0.0550	0.0446	0.0468	0.0485	0.0543	0.0542
106 8303.1	0.0400		0.0421	0.0533		0.0435		0.0537	
107 8303.2	0.0540		0.0452	0.0589		0.0455		0.0539	
108 8303.3	0.0970	0.0637	0.0755	0.0509	0.0543	0.0486	0.0452	0.0541	0.0539
109 8304.1	0.0840		0.0558	0.0619		0.0472		0.0542	
110 8304.2	0.0550		0.0540	0.0938		0.0532		0.0558	
111 8304.3	0.0590	0.0593	0.0517	0.0713	0.0538	0.0570	0.0525	0.0556	0.0552
112 8305.1	0.0430		0.0655	0.0623		0.0530		0.0554	
113 8305.2	0.0650		0.0717	0.0678		0.0604		0.0572	
114 8305.3	0.0830	0.0637	0.0596	0.0689	0.0656	0.0609	0.0581	0.0575	0.0567
115 8306.1	0.0490		0.0391	0.0720		0.0608		0.0568	
116 8306.2	0.0480		0.0547	0.0909		0.0608		0.0571	
117 8306.3	0.0500	0.0517	0.0561	0.0956	0.0500	0.0613	0.0610	0.0580	0.0573
118 8307.1	0.0400		0.0588	0.0733		0.0624		0.0585	
119 8307.2	0.0400		0.0638	0.0635		0.0658		0.0598	
120 8307.3	0.1210	0.0697	0.0960	0.0852	0.0729	0.0669	0.0650	0.0586	0.0590
121 8308.1	0.0660		0.0584	0.0750		0.0687		0.0598	
122 8308.2	0.0480		0.0693	0.0735		0.0656		0.0595	

636 *The determination of banking system excess reserves*

Table 27.4 (*Contd.*)

Observation no. and date	ESBCI (1)	MMESBCI (2)	ESBPB (3)	MMESBPB (4)	ESBPA (5)	MMESBPA (6)	ESBCP (7)	MMESBCP (8)	ESBLP (9)	MMESBLP (10)
123 8308.3	0.0420	0.0520	0.0715	0.0684	0.0728	0.0738	0.0640	0.0661	0.0589	0.0594
124 8309.1	0.0370		0.0525		0.0466		0.0615		0.0587	
125 8309.2	0.1220		0.1195		0.0783		0.0703		0.0586	
126 8309.3	0.0581	0.0703	0.0581	0.0761	0.0574	0.0608	0.0649	0.0836	0.0582	0.0585
127 8310.1	0.0310		0.0347		0.0495		0.0723		0.0601	
128 8310.2	0.0680		0.0672		0.0586		0.0687		0.0606	
129 8310.3	0.1220	0.0737	0.1011	0.0677	0.0855	0.0845	0.0854	0.0688	0.0581	0.0596
130 8311.1	0.0430		0.0572		0.0802		0.0687		0.0586	
131 8311.2	0.1410		0.1059		0.0854		0.0684		0.0576	
132 8311.3	0.0390	0.0743	0.0626	0.0752	0.0847	0.0834	0.0657	0.0676	0.0579	0.0580
133 8312.1	0.0460		0.0603		0.0826		0.0659		0.0580	
134 8312.2	0.0863		0.0962		0.1058		0.0577		0.0582	
135 8312.3	0.0640	0.0661	0.0824	0.0796	0.0931	0.0938	0.0575	0.0804	0.0579	0.0580
136 8401.1	0.1334		0.1353		0.1303		0.0875		0.0576	
137 8401.2	0.0755		0.0752		0.0858		0.0522		0.0424	
138 8401.3	0.1170	0.1053	0.1160	0.1088	0.0692	0.0951	0.0842	0.0613	0.0421	0.0474
139 8402.1	0.0410		0.0546		0.0668		0.0652		0.0418	
140 8402.2	0.0630		0.0520		0.0688		0.0651		0.0417	
141 8402.3	0.0327	0.0467	0.0652	0.0572	0.0732	0.0696	0.0616	0.0640	0.0410	0.0414
142 8403.1	0.0820		0.0781		0.0603		0.0850		0.0408	
143 8403.2	0.1158		0.1087		0.0856		0.0627		0.0393	
144 8403.3	0.0742	0.0907	0.0699	0.0856	0.0764	0.0741	0.0605	0.0627	0.0388	0.0398
145 8404.1	0.0280		0.0408		0.0773		0.0811		0.0397	
146 8404.2	0.1130		0.0838		0.1061		0.0883		0.0389	
147 8404.3	0.1480	0.0963	0.1144	0.0797	0.1073	0.0969	0.0720	0.0865	0.0394	0.0390
148 8405.1	0.0300		0.0400		0.0573		0.0681		0.0398	
149 8405.2	0.0540		0.0582		0.0731		0.0626		0.0394	
150 8405.3	0.0920	0.0587	0.0993	0.0652	0.1040	0.0782	0.0584	0.0624	0.0381	0.0391

Model breakdown and calculation of short and long-run levels

		ESBCI	MMESBCI	ESBPB	MMESBPB	ESBPA	MMESBPA	ESBCP	MMESBCP	ESBLP	MMESBLP
151	8406.1	0.0410		0.0550		0.0529		0.0428		0.0384	
152	8406.2	0.0720		0.0452		0.0598		0.0441		0.0387	
153	8406.3	0.0490	0.0540	0.0732	0.0578	0.0807	0.0845	0.0437	0.0435	0.0382	0.0384
154	8407.1	0.0410		0.0442		0.0320		0.0408		0.0371	
155	8407.2	0.0530		0.0489		0.0392		0.0442		0.0373	
156	8407.3	0.1060	0.0687	0.0669	0.0534	0.0761	0.0491	0.0455	0.0434	0.0378	0.0374
157	8408.1	0.0490		0.0673		0.0595		0.0431		0.0372	
158	8408.2	0.0520		0.0489		0.0378		0.0404		0.0368	
159	8408.3	0.0370	0.0460	0.0479	0.0547	0.0348	0.0440	0.0385	0.0406	0.0366	0.0389
160	8409.1	0.0290		0.0271		0.0197		0.0387		0.0365	
161	8409.2	0.0280		0.0301		0.0388		0.0388		0.0364	
162	8409.3	0.0360	0.0310	0.0396	0.0323	0.0358	0.0314	0.0391	0.0369	0.0367	0.0365
163	8410.1	0.0339		0.0539		0.0275		0.0395		0.0368	
164	8410.2	0.0180		0.0405		0.0452		0.0378		0.0359	
165	8410.3	0.0470	0.0483	0.0354	0.0433	0.0541	0.0423	0.0411	0.0395	0.0365	0.0364
166	8411.1	0.0340		0.0304		0.0401		0.0400		0.0370	
167	8411.2	0.0337		0.0359		0.0601		0.0381		0.0370	
168	8411.3	0.0284	0.0313	0.0340	0.0335	0.0413	0.0471	0.0332	0.0371	0.0365	0.0368
169	8412.1	0.0320		0.0456		0.0483		0.0346		0.0368	
170	8412.2	0.0450		0.0487		0.0456		0.0365		0.0369	
171	8412.3	0.0660	0.0477	0.0487	0.0477	0.0520	0.0480	0.0349	0.0353	0.0364	0.0367

(1) ESBCI   series adjusted for impulse interventions of banking system excess reserves
(2) MMESBCI   monthly average of ESBCI
(3) ESBPB   forecast one period ahead of excess reserves.
(4) MMESBPB   monthly average of ESBPB
(5) ESBPA   forecast one period ahead of excess reserves not knowing 'monetary innovations'
(6) MMESBPA   monthly average of ESBPA
(7) ESBCP   desired level of excess reserves in the short run
(8) MMESBCP   monthly average of ESBCP
(9) ESBLP   desired level of excess reserves in the long run
(10) MMESBLP   monthly average of ESBLP

[a] The monthly averages exhibited together with the ten-period observations are arithmetic averages and not legal averages, weighted by the number of days of the period.

reflecting the market consolidated trends; the agents therefore cannot refer to it for long-term planning. One-month interest rate, by contrast is much more stable, so that the agents can use its expected value in the long run as an indicator of financing in a near future.

Figure 27.6 shows simultaneously the original excess reserve series adjusted for the impulse interventions (thin line), the desired level in the short run (thick line) and the desired level in the long run (dotted line). This figure enables us to assess the magnitude of the different shocks affecting the excess reserve performance with the consequent systematic shift from desired levels. The periods when the desired level in the short run falls beneath the desired level in the long run are linked to situations in which expected overnight interest rate stands above expected one-month interest rate.

Table 27.4 shows the excess reserve series adjusted for interventions (columns 1 and 2), the forecast provided by the model (columns 3 and 4), the forecast the agents could make without knowing the 'monetary innovation' (5 and 6), and the desired level both in the short and long run (columns 7 to 10).

Using this model enables monetary authorities to forecast excess reserves and to evaluate the impact of the 'monetary innovation', as well as their desired levels in the short and long run. Excess reserve forecasts can be of great help in order to predict the computable liabilities of the banking system; calculation of expected values in the short and long run is very important for planning monetary authority action.

ACKNOWLEDGEMENTS

This chapter is part of a larger research project, 'An econometric model for monetary control' in which José Pérez, M. Cruz Manzano and Julia Salaverría are also coauthors. The suggestions and comments of José Pérez were highly valuable in the final version of this chapter. We are also grateful to M. Cruz Manzano for her comments on different questions faced in the chapter. In any case, we remain solely responsible for the mistakes made in this work. Typewriting was performed by Coral Aldea, whose patient attitude throughout this work we have greatly appreciated.

REFERENCES

Burger, A. E. (1971) *The Money Supply Process*, Wadsworth Publishing Company, Belmont.
Escriva, J. L., Espasa, A., Pérez, J. and Salaverria, J. (1986) A short-term econometric model for the Spanish Monetary Policy. Bank of Spain (unpublished paper).
Espasa, A. (1982) Problemas y enfoques en la predicción de tipos de interés., working paper 8214, Research Department, Bank of Spain.
Frost, D. A. (1967) *Banks Demand for Excess Reserves*, University of California, Los Angeles.
Frost, D. A. (1977) Short-run fluctuations in the money multiplier and monetary control. *Journal of Money, Credit and Banking*.

Mauleón, I. (1984) La demanda de activos de caja en el sistema bancario en el periodo 1978–1982: un estudio empirico, Economic Studies 36, Bank of Spain.

Mauleón, I., Pérez, J. and Sanz, B (1985) Los activos de caja y la oferta de dinero, unpublished paper, Research Department, Bank of Spain.

Morrison, G. R. (1966) *Liquidity Preferences for Commercial Banks*, University of Chicago Press, Chicago.

Pérez, J. (1976) Sistema bancario: demanda de activos rentables y creación de depósitos bancarios, 1963–1972, unpublished paper, Research Department, Bank of Spain.

Smith, C. W. (1976) Option pricing: a review. *Journal of Financial Economics*, 3, January/March.

Wallis, K. P. (1980) Econometric implications of the rational expectations hypothesis. *Econometrica*, 48, No. 1, January.

APPENDIX A
SUPPLEMENTARY DIAGRAMS

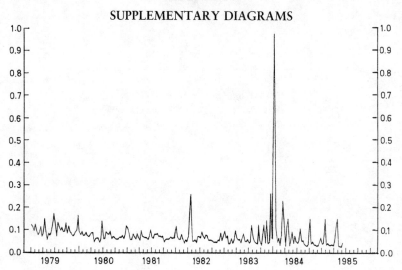

Fig. 27.A1 Excess reserves

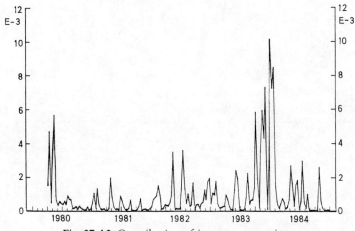

Fig. 27.A2 Contribution of interest rate variance

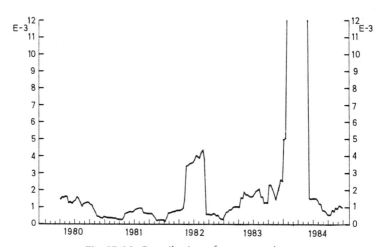

Fig. 27.A3 Contribution of reserve variance

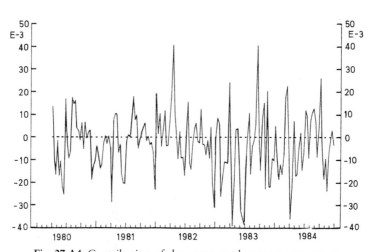

Fig. 27.A4 Contribution of the unexpected reserves component

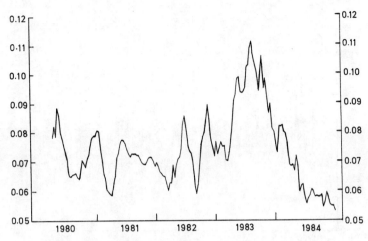

Fig. 27.A5 Contribution of one-month interest rate

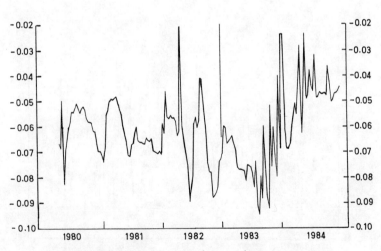

Fig. 27.A6 Contribution of overnight interest rate

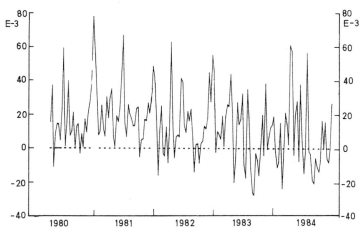

Fig. 27.A7 Disturbance contribution to excess reserve explanation

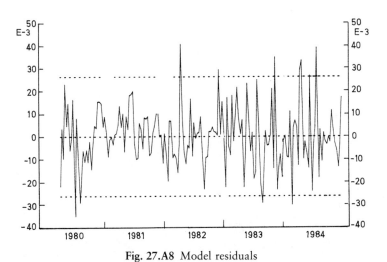

Fig. 27.A8 Model residuals

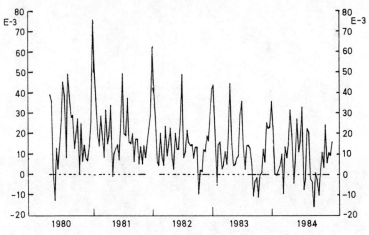

Fig. 27.A9 Predictable disturbance

APPENDIX B
SUPPLEMENTARY TABLES

(see overleaf, pages 644–651)

Table 27.B1 Model breakdown into explanatory variable contributions

Observation no. and date	ESB_t	$CIMP_t$	$CNIV_t$	$CRDS_t^e$	$CR1M_t^e$	$CVAC_t$	$CVTI_t$	CAC_t^{ne}	$CPERS_t$	a_t
1 8004.1	0.08600	0.00000	0.04110	−0.06368	0.07485	0.00146	0.00154	0.01388	0.03879	−0.02193
2 8004.2	0.08900	0.00000	0.04110	−0.06558	0.07856	0.00159	0.00469	−0.01068	0.03590	0.00342
3 8004.3	0.04600	0.00000	0.04110	−0.04740	0.07516	0.00155	0.00046	−0.01689	0.00185	−0.00982
4 8005.1	0.06100	0.00000	0.04110	−0.07892	0.08467	0.00163	0.00365	−0.00149	−0.01273	0.02309
5 8005.2	0.06400	0.00000	0.04110	−0.06613	0.08181	0.00159	0.00565	−0.01734	0.01278	0.00453
6 8005.3	0.06300	0.00000	0.04110	−0.06262	0.07620	0.00124	0.00063	−0.01075	0.00268	0.01453
7 8006.1	0.04400	0.00000	0.04110	−0.05819	0.07451	0.00133	0.00031	−0.02241	0.01375	−0.00639
8 8006.2	0.05900	0.00000	0.04110	−0.05546	0.07172	0.00119	0.00056	−0.02600	0.02632	−0.00044
9 8006.3	0.13900	0.00000	0.04110	−0.05162	0.06910	0.00133	0.00044	0.01705	0.04525	0.01635
10 8007.1	0.06000	0.00000	0.04110	−0.05180	0.06720	0.00137	0.00036	−0.00154	0.03841	−0.03512
11 8007.2	0.06200	0.00000	0.04110	−0.05022	0.06218	0.00159	0.00059	−0.00955	0.00826	0.00805
12 8007.3	0.09100	0.00000	0.04110	−0.04841	0.06200	0.00150	0.00035	−0.00713	0.04922	−0.00764
13 8008.1	0.08200	0.00000	0.04110	−0.05020	0.06267	0.00128	0.00090	0.01788	0.03851	−0.03014
14 8008.2	0.08000	0.00000	0.04110	−0.05212	0.06294	0.00102	0.00068	0.01509	0.02770	−0.01640
15 8008.3	0.09500	0.00000	0.04110	−0.05044	0.06333	0.00119	0.00065	0.01673	0.02878	−0.00634
16 8009.1	0.06000	0.00000	0.04110	−0.04960	0.06180	0.00124	0.00014	0.00407	0.01262	−0.01137
17 8009.2	0.07000	0.00000	0.04110	−0.05087	0.06119	0.00128	0.00016	0.00295	0.02027	−0.00607
18 8009.3	0.06700	0.00000	0.04110	−0.05407	0.06409	0.00128	0.00028	−0.00128	0.02709	−0.01150
19 8010.1	0.05900	0.00000	0.04110	−0.05532	0.06745	0.00115	0.00009	0.00650	0.00022	−0.00219
20 8010.2	0.05800	0.00000	0.04110	−0.05552	0.06628	0.00110	0.00029	−0.00518	0.02464	−0.01472
21 8010.3	0.06000	0.00000	0.04110	−0.05584	0.06508	0.00088	0.00016	0.00690	0.00632	−0.00461
22 8011.1	0.06900	0.00000	0.04110	−0.05890	0.06771	0.00071	0.00011	−0.00060	0.01426	0.00462
23 8011.2	0.06500	0.00000	0.04110	−0.05916	0.06910	0.00044	0.00005	0.00244	0.00744	0.00360
24 8011.3	0.07600	0.00000	0.04110	−0.06309	0.07205	0.00044	0.00006	0.00339	0.00662	0.01544
25 8012.1	0.06000	0.00000	0.04110	−0.06647	0.07476	0.00035	0.00025	−0.01906	0.01362	0.01544
26 8012.2	0.07700	0.00000	0.04110	−0.06656	0.07546	0.00031	0.00002	−0.01329	0.02547	0.01449
27 8012.3	0.11900	0.00000	0.04110	−0.06819	0.07546	0.00044	0.00015	−0.00989	0.07569	0.00424
28 8101.1	0.10700	0.00000	0.04110	−0.07057	0.07710	0.00040	0.00043	−0.00379	0.05369	0.00864

Appendix B

29	8101.2	0.08400	0.00000	0.04110	−0.06520	0.07675	0.00035	0.00105	−0.00871	0.03716	0.00150
30	8101.3	0.05600	0.00000	0.04110	−0.05276	0.07094	0.00035	0.00014	−0.01410	0.01941	−0.00909
31	8102.1	0.05900	0.00000	0.04110	−0.05126	0.06775	0.00035	0.00133	−0.01260	0.01386	−0.00153
32	8102.2	0.08300	0.00000	0.04110	−0.04772	0.06329	0.00035	0.00036	−0.00253	0.02834	−0.00020
33	8102.3	0.07000	0.00000	0.04110	−0.04678	0.06087	0.00031	0.00008	0.00085	0.01721	−0.00364
34	8103.1	0.05800	0.00000	0.04110	−0.04726	0.05758	0.00035	0.00008	−0.00289	0.00833	0.00071
35	8103.2	0.08500	0.00000	0.04110	−0.04652	0.05670	0.00027	0.00014	−0.00064	0.03136	0.00132
36	8103.3	0.06700	0.00000	0.04110	−0.04584	0.05598	0.00027	0.00000	−0.00447	0.01492	0.00503
37	8104.1	0.05300	0.00000	0.04110	−0.04759	0.05583	0.00027	0.00002	−0.02896	0.01890	0.01344
38	8104.2	0.09700	0.00000	0.04110	−0.05023	0.05996	0.00022	0.00015	0.00809	0.03372	0.00398
39	8104.3	0.07300	0.00000	0.04110	−0.05261	0.06252	0.00035	0.00197	0.01075	−0.00111	0.01002
40	8105.1	0.06700	0.00000	0.04110	−0.05716	0.06815	0.00062	0.00087	0.01048	0.00983	−0.00688
41	8105.2	0.06600	0.00000	0.04110	−0.06028	0.06971	0.00066	0.00023	−0.00687	0.01266	0.00879
42	8105.3	0.06600	0.00000	0.04110	−0.06336	0.07291	0.00071	0.00005	−0.00308	0.01463	0.00305
43	8106.1	0.05700	0.00000	0.04110	−0.06798	0.07408	0.00075	0.00015	−0.01625	0.00707	0.01809
44	8106.2	0.07100	0.00000	0.04110	−0.06838	0.07381	0.00071	0.00002	−0.02064	0.02580	0.01858
45	8106.3	0.10000	0.00000	0.04110	−0.06395	0.07277	0.00075	0.00088	−0.02098	0.04940	0.02004
46	8107.1	0.06500	0.00000	0.04110	−0.06347	0.07094	0.00084	0.00064	−0.00188	0.01975	−0.00292
47	8107.2	0.06200	0.00000	0.04110	−0.05944	0.06944	0.00088	0.00025	0.00109	0.01881	−0.01014
48	8107.3	0.08100	0.00000	0.04110	−0.05724	0.06833	0.00093	0.00003	0.00003	0.03731	−0.00949
49	8108.1	0.07900	0.00000	0.04110	−0.06246	0.06948	0.00093	0.00005	0.00837	0.01558	0.00595
50	8108.2	0.08500	0.00000	0.04110	−0.06313	0.06933	0.00093	0.00017	0.01850	0.01497	0.00313
51	8108.3	0.07100	0.00000	0.04110	−0.06306	0.06939	0.00066	0.00065	0.00799	0.01984	−0.00557
52	8109.1	0.07200	0.00000	0.04110	−0.06352	0.06901	0.00066	0.00004	0.01016	0.00616	−0.00838
53	8109.2	0.06700	0.00000	0.04110	−0.06372	0.06890	0.00062	0.00001	−0.00487	0.01718	0.00778
54	8109.3	0.07300	0.00000	0.04110	−0.06131	0.06764	0.00062	0.00001	−0.00114	0.01707	0.00900
55	8110.1	0.04500	0.00000	0.04110	−0.06233	0.06659	0.00062	0.00006	0.00256	0.00497	−0.00856
56	8110.2	0.05600	0.00000	0.04110	−0.06287	0.06601	0.00062	0.00025	0.00477	0.01349	−0.00736
57	8110.3	0.06000	0.00000	0.04110	−0.06158	0.06586	0.00049	0.00073	0.00653	0.00504	0.00184
58	8111.1	0.05900	0.00000	0.04110	−0.06619	0.06676	0.00035	0.00002	−0.00144	0.01407	0.00432
59	8111.2	0.06100	0.00000	0.04110	−0.06662	0.06804	0.00018	0.00011	0.00056	0.00814	0.00949
60	8111.3	0.06800	0.00000	0.04110	−0.06699	0.06841	0.00022	0.00011	−0.00331	0.01635	0.01211
61	8112.1	0.06200	0.00000	0.04110	−0.06748	0.06817	0.00022	0.00007	−0.00221	0.02189	0.00025

Table 27.B1 (Contd.)

Observation no. and date	ESB_t	$CIMP_t$	$CNIV_t$	$CRDS_t^e$	$CRIM_t^e$	$CVAC_t$	$CVTI_t$	CAC_t^{ne}	$CPERS_t$	a_t
62 8112.2	0.06500	0.00000	0.04110	−0.06639	0.06698	0.00022	0.00004	−0.00661	0.02894	0.00073
63 8112.3	0.06600	0.00000	0.04110	−0.06771	0.06531	0.00022	−0.00018	−0.02324	0.06209	−0.01195
64 8201.1	0.11600	0.00000	0.04110	−0.05592	0.06654	0.00013	0.00023	0.01956	0.04306	0.00130
65 8201.2	0.07000	0.00000	0.04110	−0.05972	0.06533	0.00035	0.00068	0.00161	0.02823	−0.00759
66 8201.3	0.05900	0.00000	0.04110	−0.04396	0.06341	0.00066	0.00080	0.01086	0.00583	−0.01971
67 8202.1	0.05900	0.00000	0.04110	−0.05369	0.06288	0.00066	0.00090	−0.00402	0.00434	0.00684
68 8202.2	0.08100	0.00000	0.04110	−0.05435	0.06214	0.00071	0.00152	0.00318	0.02002	0.00669
69 8202.3	0.06200	0.00000	0.04110	−0.05299	0.06189	0.00075	0.00093	0.01202	0.00780	−0.00950
70 8203.1	0.04100	0.00000	0.04110	−0.05397	0.06001	0.00075	0.00011	−0.00374	0.00445	−0.00771
71 8203.2	0.05600	0.00000	0.04110	−0.05384	0.05735	0.00080	0.00015	−0.00349	0.02344	−0.00951
72 8203.3	0.04700	0.00000	0.04110	−0.05530	0.05989	0.00080	0.00037	0.00726	0.00910	−0.01623
73 8204.1	0.07100	0.00000	0.04110	−0.06064	0.05981	0.00080	0.00032	0.01821	0.01445	−0.00305
74 8204.2	0.15400	0.00000	0.04110	−0.05892	0.06575	0.00084	0.00029	0.04155	0.02258	0.04082
75 8204.3	0.25700	0.14740	0.04110	−0.01972	0.06162	0.00093	0.00052	0.01167	0.00739	0.00610
76 8205.1	0.04900	0.00000	0.04110	−0.05782	0.06552	0.00141	0.00088	0.00246	0.00256	−0.00713
77 8205.2	0.05000	0.00000	0.04110	−0.06388	0.06769	0.00340	0.00349	−0.00962	0.02001	−0.01219
78 8205.3	0.05600	0.00000	0.04110	−0.06800	0.06835	0.00345	0.00005	0.00264	0.01227	−0.00386
79 8206.1	0.04200	0.00000	0.04110	−0.07166	0.07092	0.00353	0.00014	−0.00917	0.01238	−0.00525
80 8206.2	0.07100	0.00000	0.04110	−0.08544	0.07892	0.00358	0.00010	−0.00894	0.02501	0.01668
81 8206.3	0.07000	0.00000	0.04110	−0.07929	0.08172	0.00362	0.00013	−0.01719	0.04882	−0.00892
82 8207.1	0.06300	0.00000	0.04110	−0.07603	0.07886	0.00375	0.00130	−0.00018	0.00815	0.00604
83 8207.2	0.09100	0.00000	0.04110	−0.05625	0.07268	0.00402	0.00359	0.01588	0.01092	−0.00094
84 8207.3	0.07400	0.00000	0.04110	−0.05359	0.06998	0.00393	0.00131	−0.01158	0.02143	0.00142
85 8208.1	0.06000	0.00000	0.04110	−0.05763	0.06948	0.00384	0.00043	−0.01448	0.01564	0.00162
86 8208.2	0.07800	0.00000	0.04110	−0.05518	0.06648	0.00420	0.00102	−0.00245	0.01394	0.00889
87 8208.3	0.07500	0.00000	0.04110	−0.03914	0.05918	0.00433	0.00025	0.00430	0.01450	−0.00952
88 8209.1	0.05300	0.00000	0.04110	−0.03971	0.05625	0.00380	0.00035	0.00664	0.00800	−0.02342

89 8209.2	0.05400	0.00000	0.04110	−0.05095	0.05933	0.00057	0.00168	−0.00162	0.01326	−0.00937	
90 8209.3	0.05200	0.00000	0.04110	−0.05784	0.06640	0.00057	0.00005	−0.00259	0.01319	−0.00889	
91 8210.1	0.04900	0.00000	0.04110	−0.07068	0.07264	0.00057	0.00035	0.01263	−0.00965	0.00205	
92 8210.2	0.04400	0.00000	0.04110	−0.07436	0.07498	0.00053	0.00040	−0.00287	0.00216	0.00206	
93 8210.3	0.04600	0.00000	0.04110	−0.07444	0.07632	0.00062	0.00015	−0.00334	0.00147	0.00412	
94 8211.1	0.04400	0.00000	0.04110	−0.08397	0.07958	0.00062	0.00045	−0.00795	0.01204	0.00215	
95 8211.2	0.05600	0.00000	0.04110	−0.08318	0.08546	0.00049	0.00050	−0.00132	0.01114	0.00181	
96 8211.3	0.05100	0.00000	0.04110	−0.08193	0.08033	0.00053	0.00125	−0.00945	0.01879	0.00038	
97 8212.1	0.08600	0.00000	0.04110	−0.08014	0.07379	0.00035	0.00066	0.00452	0.01628	0.02945	
98 8212.2	0.05200	0.00000	0.04110	−0.07041	0.07109	0.00027	0.00175	−0.02118	0.02863	0.00076	
99 8212.3	0.06800	0.00000	0.04110	−0.06948	0.06849	0.00027	0.00194	−0.03160	0.04192	0.01537	
100 8301.1	0.09500	0.00000	0.04110	−0.06716	0.07331	0.00049	0.00043	−0.00055	0.04366	0.00373	
101 8301.2	0.06300	0.00000	0.04110	−0.05701	0.06919	0.00071	0.00105	0.00860	0.02189	−0.02253	
102 8301.3	0.07400	0.00000	0.04110	−0.05777	0.07129	0.00075	0.00094	−0.00617	−0.00552	0.01705	
103 8302.1	0.03700	0.00000	0.04110	−0.06362	0.07322	0.00084	0.00176	−0.02651	0.01459	−0.00437	
104 8302.2	0.04000	0.00000	0.04110	−0.06269	0.07173	0.00084	0.00040	−0.01862	0.01572	−0.00848	
105 8302.3	0.06200	0.00000	0.04110	−0.06176	0.07188	0.00097	0.00017	−0.01112	0.00267	0.01809	
106 8303.1	0.04000	0.00000	0.04110	−0.06073	0.06724	0.00102	0.00020	−0.01122	0.00451	−0.00212	
107 8303.2	0.05400	0.00000	0.04110	−0.06352	0.06691	0.00102	0.00025	−0.01172	0.01116	0.00880	
108 8303.3	0.09700	0.00000	0.04110	−0.06642	0.06994	0.00102	0.00029	0.02453	0.00503	0.02152	
109 8304.1	0.06400	0.00000	0.04110	−0.07239	0.07385	0.00097	0.00096	−0.00608	0.01741	0.00818	
110 8304.2	0.05500	0.00000	0.04110	−0.07345	0.07960	0.00155	0.00069	−0.03979	0.04433	0.00097	
111 8304.3	0.05900	0.00000	0.04110	−0.07342	0.08703	0.00150	0.00022	−0.01961	0.01486	0.00733	
112 8305.1	0.04300	0.00000	0.04110	−0.07367	0.08852	0.00190	0.00008	0.00327	0.00434	−0.02254	
113 8305.2	0.06500	0.00000	0.04110	−0.07393	0.09391	0.00172	0.00005	0.00389	0.00496	−0.00670	
114 8305.3	0.08300	0.00000	0.04110	−0.07774	0.09452	0.00172	0.00121	−0.00928	0.00811	0.02336	
115 8306.1	0.04900	0.00000	0.04110	−0.07161	0.08935	0.00163	0.00241	−0.03285	0.00907	0.00989	
116 8306.2	0.04800	0.00000	0.04110	−0.07200	0.08910	0.00159	0.00192	−0.03618	0.02917	−0.00669	
117 8306.3	0.05800	0.00000	0.04110	−0.07262	0.08978	0.00163	0.00002	−0.03944	0.03565	0.00187	
118 8307.1	0.04000	0.00000	0.04110	−0.07445	0.09107	0.00177	0.00006	−0.01453	0.01378	−0.01879	
119 8307.2	0.04800	0.00000	0.04110	−0.08001	0.09807	0.00194	0.00001	0.00030	0.00243	−0.01585	
120 8307.3	0.12100	0.00000	0.04110	−0.07030	0.09834	0.00199	0.00013	0.01082	0.01394	0.02498	

648 The determination of banking system excess reserves

Table 27.B1 (Contd.)

Observation no. and date	ESB_t	$CIMP_t$	$CNIV_t$	$CRDS_t^e$	$CRIM_t^e$	$CVAC_t$	$CVTI_t$	CAC_t^{ne}	$CPERS_t$	a_t
121 8308.1	0.06600	0.00000	0.04110	−0.08719	0.10409	0.00208	0.00075	−0.01657	0.01417	0.00757
122 8308.2	0.04800	0.00000	0.04110	−0.09029	0.10608	0.00159	0.00224	−0.00422	0.01279	−0.02129
123 8308.3	0.04200	0.00000	0.04110	−0.07569	0.10054	0.00159	0.00037	−0.00123	0.00484	−0.02952
124 8309.1	0.03700	0.00000	0.04110	−0.08421	0.09854	0.00124	0.00064	0.00586	−0.01069	−0.01548
125 8309.2	0.12200	0.00000	0.04110	−0.05706	0.09571	0.00128	0.00066	0.04125	−0.00341	0.00247
126 8309.3	0.05200	0.00000	0.04110	−0.07458	0.08975	0.00124	0.00134	−0.00126	−0.00144	−0.00414
127 8310.1	0.03100	0.00000	0.04110	−0.08427	0.09577	0.00230	0.00584	−0.01479	−0.01121	−0.00372
128 8310.2	0.06800	0.00000	0.04110	−0.08800	0.10134	0.00230	0.00191	0.00869	−0.00009	0.00075
129 8310.3	0.12200	0.00000	0.04110	−0.04926	0.09060	0.00203	0.00016	0.01554	0.00088	0.02094
130 8311.1	0.04300	0.00000	0.04110	−0.07237	0.09391	0.00181	0.00355	−0.02302	0.01223	−0.01421
131 8311.2	0.14100	0.00000	0.04110	−0.05742	0.08825	0.00141	0.00593	0.02056	0.00610	0.03508
132 8311.3	0.03900	0.00000	0.04110	−0.07004	0.08250	0.00186	0.00430	−0.02213	0.02502	−0.02360
133 8312.1	0.04600	0.00000	0.04110	−0.07605	0.08568	0.00221	0.00729	−0.02229	0.02235	−0.01429
134 8312.2	0.26000	0.17170	0.04110	−0.03822	0.07734	0.00261	0.00008	−0.00954	0.02286	−0.00791
135 8312.3	0.06400	0.00000	0.04110	−0.06556	0.07647	0.00252	0.00293	−0.01068	0.03561	−0.01839
136 8401.1	0.51600	0.38260	0.04110	−0.02277	0.07309	0.00495	0.01013	0.00500	0.02383	−0.00193
137 8401.2	1.08600	1.01050	0.02690	−0.02287	0.06935	0.00504	0.00724	−0.01063	0.00015	0.00034
138 8401.3	0.10700	0.06057	0.02690	−0.05743	0.07830	0.01352	0.00844	−0.02363	−0.00051	−0.00905
139 8402.1	0.04500	0.00000	0.02690	−0.06555	0.07809	0.02411	0.00144	−0.01225	0.00183	−0.00956
140 8402.2	0.06300	0.00000	0.02690	−0.06570	0.07872	0.02420	0.00038	−0.01684	0.00430	0.01104
141 8402.3	0.03200	0.00000	0.02690	−0.06395	0.07631	0.02394	0.00014	−0.00805	0.00989	−0.03317
142 8403.1	0.08200	0.00000	0.02690	−0.05800	0.07583	0.02464	0.00040	0.01782	−0.00948	0.00387
143 8403.2	0.22600	0.11020	0.02690	−0.05329	0.07277	0.02495	0.00073	0.02312	0.01355	0.00706
144 8403.3	0.14600	0.07180	0.02690	−0.04844	0.06578	0.02350	0.00051	−0.00650	0.00819	0.00425
145 8404.1	0.02800	0.00000	0.02690	−0.05263	0.06511	0.02354	0.00012	−0.03652	0.01430	−0.01283
146 8404.2	0.11300	0.00000	0.02690	−0.04098	0.06573	0.02279	0.00021	−0.02230	0.03142	0.02922
147 8404.3	0.14800	0.00000	0.02690	−0.02696	0.06362	0.02385	0.00065	0.00704	0.01928	0.03363
148 8405.1	0.03000	0.00000	0.02690	−0.05970	0.06860	0.02328	0.00267	−0.01731	−0.00443	−0.01001

Appendix B

		ESB$_t$	CIMP$_t$	CNIV$_t$	CRDS$_t^e$	CR1M$_t^e$	CVAC$_t$	CVTI$_t$	CAC$_t^{me}$	CPERS$_t$	a_t
149	8405.2	0.05400	0.00000	0.02690	−0.04975	0.06583	0.02288	0.00165	−0.01684	0.00558	−0.00224
150	8405.3	0.09200	0.00000	0.02690	−0.02249	0.05685	0.01550	0.00027	−0.00472	0.02701	−0.00733
151	8406.1	0.04100	0.00000	0.02690	−0.04694	0.05888	0.00150	0.00155	0.00208	0.01105	−0.01402
152	8406.2	0.07200	0.00000	0.02690	−0.04552	0.05932	0.00146	0.00183	−0.01465	0.01583	0.02684
153	8406.3	0.04900	0.00000	0.02690	−0.03609	0.05547	0.00150	0.00007	−0.00745	0.03282	−0.02422
154	8407.1	0.04100	0.00000	0.02690	−0.04242	0.05292	0.00150	0.00055	0.01217	−0.00741	−0.00320
155	8407.2	0.05300	0.00000	0.02690	−0.04410	0.05462	0.00150	0.00295	0.00972	−0.00265	0.00406
156	8407.3	0.10600	0.00000	0.02690	−0.03049	0.05565	0.00141	0.00046	−0.00916	0.02215	0.03908
157	8408.1	0.04900	0.00000	0.02690	−0.04663	0.05765	0.00119	0.00014	0.00774	0.02027	−0.01827
158	8408.2	0.05200	0.00000	0.02690	−0.04612	0.05712	0.00119	0.00100	0.01101	−0.00224	0.00315
159	8408.3	0.03700	0.00000	0.02690	−0.04447	0.05518	0.00080	0.00002	0.01317	−0.00365	−0.01094
160	8409.1	0.02900	0.00000	0.02690	−0.04516	0.05557	0.00080	0.00005	0.00740	−0.01840	0.00185
161	8409.2	0.02800	0.00000	0.02690	−0.04505	0.05516	0.00075	0.00004	−0.00869	0.00098	−0.00208
162	8409.3	0.03600	0.00000	0.02690	−0.04468	0.05528	0.00057	0.00004	0.00400	−0.00248	−0.00362
163	8510.1	0.14800	0.09410	0.02690	−0.04580	0.05600	0.00053	0.00002	0.02644	−0.01020	0.00000
164	8410.2	0.03800	0.00000	0.02690	−0.03466	0.05185	0.00057	0.00004	−0.00474	0.00050	−0.00246
165	8410.3	0.04700	0.00000	0.02690	−0.04036	0.05339	0.00084	0.00258	−0.01871	0.01077	0.01160
166	8411.1	0.03300	0.00000	0.02690	−0.04813	0.05669	0.00075	0.00053	−0.00963	0.00332	0.00256
167	8411.2	0.03300	0.00000	0.02690	−0.04732	0.05515	0.00097	0.00020	−0.02420	0.02421	−0.00290
168	8411.3	0.0280	0.0000	0.0269	−0.0449	0.0527	0.0009	0.0000	−0.0072	0.0056	−0.0060
169	8412.1	0.0320	0.0000	0.0269	−0.0446	0.0522	0.0011	0.0000	−0.0007	0.0107	−0.0136
170	8412.2	0.0450	0.0000	0.0269	−0.0439	0.0524	0.0011	0.0001	0.0031	0.0092	−0.0037
171	8412.3	0.0660	0.0000	0.0269	−0.0424	0.0507	0.0010	0.0000	−0.0033	0.0158	0.0173

ESB$_t$ bank excess reserves, original series
CIMP$_t$ impulse contribution
CNIV$_t$ constant contribution
CRDS$_t^e$ expected (overnight) interest rate contribution
CR1M$_t^e$ expected one-month interest rate contribution
CVAC$_t$ reserve variance contribution
CVTI$_t$ interest rate variance contribution
CAC$_t^{me}$ contribution of the monetary innovations
CPERS$_t$ predictable disturbance contribution
a_t innovation contribution

650 The determination of banking system excess reserves

Table 27.B2 Long-term expected interest rates

Observation no and date	$R1M^{e*}$ (1)	RDS^{e*} (3)	Diff. (3) = 1 − 2	Observation no. and date	$R1M^{e*}$ (1)	RDS^{e*} (2)	Diff. (3) = 1 − 2	Observation no. and date	$R1M^{e*}$ (1)	RDS^{e*} (2)	Diff. (3) = 1 − 2
1 8004.1	0.1646	0.1505	0.0141	64 8201.1	0.1554	0.1714	−0.0160	127 8310.1	0.2327	0.2150	0.0177
2 8004.2	0.1677	0.1571	0.0106	65 8201.2	0.1526	0.1722	−0.0196	128 8310.2	0.2438	0.2274	0.0163
3 8004.3	0.1642	0.1446	0.0196	66 8201.3	0.1490	0.1568	−0.0078	129 8310.3	0.2147	0.1959	0.0188
4 8005.1	0.1829	0.1677	0.0152	67 8202.1	0.1466	0.1584	−0.0119	130 8311.1	0.2206	0.2028	0.0179
5 8005.2	0.1747	0.1683	0.0064	68 8202.2	0.1464	0.1593	−0.0129	131 8311.2	0.2061	0.1917	0.0144
6 8005.3	0.1737	0.1687	0.0050	69 8202.3	0.1452	0.1579	−0.0127	132 8311.3	0.2032	0.1923	0.0109
7 8006.1	0.1680	0.1655	0.0025	70 8203.1	0.1396	0.1587	−0.0191	133 8312.1	0.2039	0.2020	0.0019
8 8006.2	0.1731	0.1631	0.0100	71 8203.2	0.1351	0.1563	−0.0212	134 8312.2	0.1979	0.1987	−0.0008
9 8006.3	0.1690	0.1584	0.0107	72 8203.3	0.1422	0.1574	−0.0153	135 8312.3	0.1916	0.1992	−0.0076
10 8007.1	0.1635	0.1516	0.0119	73 8204.1	0.1406	0.1595	−0.0189	136 8401.1	0.1877	0.1938	−0.0060
11 8007.2	0.1549	0.1521	0.0028	74 8204.2	0.1538	0.1591	−0.0052	137 8401.2	0.1891	0.1905	−0.0014
12 8007.3	0.1573	0.1530	0.0043	75 8204.3	0.1508	0.1590	−0.0082	138 8401.3	0.1847	0.1789	0.0059
13 8008.1	0.1565	0.1531	0.0035	76 8205.1	0.1482	0.1574	−0.0092	139 8402.1	0.1780	0.1780	0.0000
14 8008.2	0.1562	0.1546	0.0016	77 8205.2	0.1492	0.1630	−0.0137	140 8402.2	0.1808	0.1790	0.0017
15 8008.3	0.1541	0.1525	0.0016	78 8205.3	0.1528	0.1691	−0.0163	141 8402.3	0.1752	0.1806	−0.0055
16 8009.1	0.1486	0.1477	0.0010	79 8206.1	0.1601	0.1759	−0.0158	142 8403.1	0.1748	0.1734	0.0014
17 8009.2	0.1440	0.1434	0.0006	80 8206.2	0.1761	0.1939	−0.0178	143 8403.2	0.1656	0.1706	−0.0050
18 8009.3	0.1509	0.1440	0.0069	81 8206.3	0.1820	0.1953	−0.0133	144 8403.3	0.1550	0.1593	−0.0043
19 8010.1	0.1553	0.1459	0.0094	82 8207.1	0.1726	0.1954	−0.0227	145 8404.1	0.1523	0.1534	−0.0011
20 8010.2	0.1492	0.1466	0.0025	83 8207.2	0.1581	0.1738	−0.0157	146 8404.2	0.1528	0.1377	0.0151
21 8010.3	0.1460	0.1516	−0.0057	84 8207.3	0.1602	0.1665	−0.0063	147 8404.3	0.1528	0.1259	0.0269
22 8011.1	0.1506	0.1496	0.0010	85 8208.1	0.1640	0.1666	−0.0026	148 8405.1	0.1591	0.1483	0.0108
23 8011.2	0.1550	0.1509	0.0042	86 8208.2	0.1597	0.1657	−0.0060	149 8405.2	0.1515	0.1505	0.0009
24 8011.3	0.1602	0.1551	0.0051	87 8208.3	0.1450	0.1536	−0.0087	150 8405.3	0.1320	0.1203	0.0117
25 8012.1	0.1653	0.1618	0.0034	88 8209.1	0.1445	0.1488	−0.0043	151 8406.1	0.1353	0.1266	0.0086
26 8012.2	0.1677	0.1662	0.0015	89 8209.2	0.1532	0.1581	−0.0048	152 8406.2	0.1401	0.1293	0.0108
27 8012.3	0.1667	0.1711	−0.0045	90 8209.3	0.1645	0.1637	0.0008	153 8406.3	0.1332	0.1174	0.0159
28 8101.1	0.1713	0.1773	−0.0060	91 8210.1	0.1710	0.1762	−0.0052	154 8407.1	0.1228	0.1149	0.0079
29 8101.2	0.1732	0.1747	−0.0015	92 8210.2	0.1691	0.1862	−0.0171	155 8407.2	0.1311	0.1215	0.0096

32 8102.2	0.1514	0.1487	0.0027	95 8211.2	0.1894	0.2007	−0.0113	158 8408.2	0.1264	0.1203	0.0061
33 8102.3	0.1515	0.1467	0.0048	96 8211.3	0.1755	0.2010	−0.0256	159 8408.3	0.1252	0.1203	0.0049
34 8103.1	0.1430	0.1481	−0.0051	97 8212.1	0.1642	0.1999	−0.0357	160 8409.1	0.1254	0.1200	0.0055
35 8103.2	0.1445	0.1495	−0.0050	98 8212.2	0.1610	0.1890	−0.0279	161 8409.2	0.1247	0.1187	0.0060
36 8103.3	0.1407	0.1483	−0.0076	99 8212.3	0.1622	0.1871	−0.0249	162 8409.3	0.1282	0.1216	0.0066
37 8104.1	0.1381	0.1459	−0.0078	100 8301.1	0.1747	0.1851	−0.0104	163 8410.1	0.1292	0.1214	0.0078
38 8104.2	0.1453	0.1459	−0.0006	101 8301.2	0.1631	0.1812	−0.0181	164 8410.2	0.1190	0.1141	0.0049
39 8104.3	0.1481	0.1459	0.0023	102 8301.3	0.1727	0.1800	−0.0073	165 8410.3	0.1238	0.1185	0.0054
40 8105.1	0.1580	0.1509	0.0071	103 8302.1	0.1699	0.1850	−0.0151	166 8411.1	0.1288	0.1217	0.0071
41 8105.2	0.1541	0.1560	−0.0018	104 8302.2	0.1692	0.1867	−0.0175	167 8411.2	0.1265	0.1239	0.0026
42 8105.3	0.1614	0.1607	0.0008	105 8302.3	0.1670	0.1878	−0.0207	168 8411.3	0.1204	0.1295	−0.0091
43 8106.1	0.1599	0.1661	−0.0062	106 8303.1	0.1583	0.1867	−0.0284	169 8412.1	0.1211	0.1269	−0.0058
44 8106.2	0.1608	0.1683	−0.0075	107 8303.2	0.1621	0.1857	−0.0236	170 8412.2	0.1237	0.1248	−0.0011
45 8106.3	0.1588	0.1641	−0.0053	108 8303.3	0.1647	0.1857	−0.0210	171 8412.3	0.1184	0.1227	−0.0043
46 8107.1	0.1573	0.1611	−0.0038	109 8304.1	0.1677	0.1894	0.0584				
47 8107.2	0.1572	0.1580	−0.0008	110 8304.2	0.1824	0.1913	0.0089				
48 8107.3	0.1576	0.1586	−0.0010	111 8304.3	0.1832	0.1802	0.1174				
49 8108.1	0.1629	0.1657	−0.0027	112 8305.1	0.1719	0.1787	0.1045				
50 8108.2	0.1496	0.1582	−0.0086	113 8305.2	0.2007	0.1918	0.0089				
51 8108.3	0.1626	0.1754	−0.0128	114 8305.3	0.2024	0.1960	0.0064				
52 8109.1	0.1617	0.1762	−0.0145	115 8306.1	0.1967	0.1923	0.0043				
53 8109.2	0.1621	0.1762	−0.0141	116 8306.2	0.2002	0.1950	0.0052				
54 8109.3	0.1597	0.1733	−0.0136	117 8306.3	0.2075	0.1973	0.0102				
55 8110.1	0.1582	0.1728	−0.0146	118 8307.1	0.2129	0.2012	0.0118				
56 8110.2	0.1571	0.1742	−0.0171	119 8307.2	0.2274	0.2096	0.0178				
57 8110.3	0.1570	0.1747	−0.0176	120 8307.3	0.2244	0.2035	0.0210				
58 8111.1	0.1572	0.1791	−0.0219	121 8308.1	0.2387	0.2175	0.0212				
59 8111.2	0.1594	0.1812	−0.0218	122 8308.2	0.2366	0.2264	0.0102				
60 8111.3	0.1581	0.1821	−0.0239	123 8308.3	0.2283	0.2156	0.0127				
61 8112.1	0.1566	0.1814	−0.0248	124 8309.1	0.2265	0.2201	0.0064				
62 8112.2	0.1534	0.1799	−0.0265	125 8309.2	0.2263	0.1955	0.0308				
63 8112.3	0.1512	0.1820	−0.0308	126 8309.3	0.2190	0.2037	0.0154				

R1M** Long-term expected interest rate in one-month transactions
RDS** Long-term expected interest rate in one-day transactions

28

Forecasting versus policy analysis with the ORANI model

PETER B. DIXON, BRIAN R. PARMENTER
and MARK HORRIDGE

28.1 INTRODUCTION

ORANI is a detailed general equilibrium model of the Australian economy.[1] It has been applied many times by economists in universities, government departments and business in analyses of the effects on industries, occupational groups and regions of changes in policy variables (e.g., taxes and subsidies) and in other aspects of the economic environment (e.g., world commodity prices). These applications have been comparative static, i.e. they have been concerned with questions of how different the economy would be with and without the changes under investigation. They have not been concerned with forecasting the future state of the economy.

More recently we have experimented with the model for forecasting. In a pilot exercise (Dixon, 1986), ORANI was used to forecast growth rates of industry outputs in Australia for the period 1985–1990. This revealed some problems inherent in the use of computable general equilibrium models for forecasting.

In Section 28.2, we describe the difference between comparative static analysis and forecasting, in general terms, and with specific reference to the ORANI model. We also discuss some computational difficulties which arise in applying ORANI to forecasting. In Section 28.3 we present some numerical examples to supplement the theoretical material in Section 28.2. Finally, Section 28.4 contains some brief concluding comments on the strengths and limitations of ORANI as a forecasting device.

---

[1] The model, which is fully described in Dixon, Parmenter, Sutton and Vincent (1982), hereinafter DPSV (1982), was developed at the IMPACT Project. IMPACT is a joint research endeavour of several agencies of the Australian government in collaboration with the University of Melbourne, La Trobe University and the Australian National University. It has been directed since its inception in 1975 by Alan A. Powell, Ritchie Professor of Research in Economics at the University of Melbourne. For a recent overview of the Project, see Powell (1985). For an overview of ORANI applications, see Parmenter and Meagher (1985). Parmenter and Meagher also provide useful comments on forecasting versus policy analysis.

28.2 THE DIFFERENCE BETWEEN COMPARATIVE STATICS AND FORECASTING

The difference between comparative static analysis and forecasting is illustrated in Fig. 28.1. This depicts two paths for the variable V_1, derived from an economic model. The control path, AB, shows what would happen in the absence of a policy change under consideration while the shocked path, AC, shows what would happen if the policy change were implemented. Comparative static analysis is concerned with the gap between C and B. This gap measures the effect of the policy change after t years. In forecasting, on the other hand, we are concerned with whether variable V_1 is going to reach point C or point B or some other point. That is, the focus is on the gap between C or B and A.

In the context of policy questions, the most appropriate model-based input is often a comparative static analysis. However, there are other situations in which this is not adequate. Businesspeople, for example, require forecasts of industry outputs, employment and other variables to assist them to make investment decisions. With a view to meeting some of their requirements, the Institute of Applied Economic and Social Research is currently devoting resources to the application of the ORANI model to forecasting.

For comparative statics, a version of the model containing no dynamic mechanisms is sufficient. This can be represented as

$$\mathbf{V}_1(t) = g(\mathbf{V}_2(t)), \tag{28.1}$$

where

g is the solution function relating the vector of endogenous variables at time t ($\mathbf{V}_1(t)$) to the exogenous variables at time t ($\mathbf{V}_2(t)$).

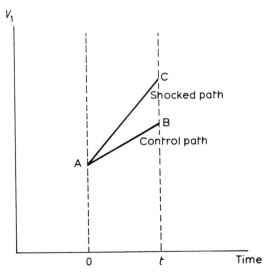

Fig. 28.1 The difference between comparative statics and forecasting

The difference between comparative statics and forecasting

In comparative static exercises with this model, we compare two values of $V_1(t)$, the control value, $V_1^C(t)$ (e.g., point B in Fig. 28.1) and the shocked value, $V_1^S(t)$ (e.g., point C in Fig. 28.1). Results are of the form

$$\begin{aligned} C(V_2^C(t), S) &= V_1^S(t) - V_1^C(t) \\ &= g(V_2^C(t) + S) - g(V_2^C(t)), \end{aligned} \qquad (28.2)$$

where

$C(V_2^C(t), S)$ is the vector of effects on the endogenous variables at time t of deviating the exogenous variables by the vector S from their control vector, $V_2^C(t)$.

Equation 28.2 involves no dynamic considerations. However, Fig. 28.1 suggests that dynamic considerations might be involved in the generation of a control solution $(V_1^C(t), V_2^C(t))$. The figure indicates that the control state of the economy (represented by the point B) has evolved from its current state (point A).

To date in comparative static exercises we have not paid much attention to the characteristics of the control solution. Because of delays in the publication of economic statistics, especially input–output tables, it is very difficult to obtain a picture of even the current state of the economy at the level of detail required by ORANI. Usually the data bases for such models refer to some period in recent history.[2] We have generally assumed that the historical data or a simple scaling up of them are adequate control solutions for comparative-static exercises. These procedures may not however provide a plausible forecast of the state of the economy for a future year t in the absence of the policy shocks to be analysed. Experience with sensitivity analyses (e.g. Dixon et al., 1986; Bruce, 1985) indicates that the comparative-static results are not very sensitive to details of the control solution. Intuitively, the question is whether the effects of a tariff reform (say) in the Australian economy of the late 1970s are likely to be very different from the effects of a tariff reform in the Australian economy of the late 1980s.

In contrast, for forecasting, considerably more care must be taken in generating control solutions. The reason is that forecasts explicitly entail comparisons between states of the economy at different points in time (usually the present and some future period). The inclusion of some dynamic mechanisms is, therefore, inevitable. Forecasting versions of ORANI with which we are experimenting can be represented as:

$$V_1(t) = f(V_2(t), Z(0), t), \qquad (28.3)$$

where

$Z(0)$ is a subvector of variables at time 0, and

f is the solution function relating the endogenous variables at time t ($V_1(t)$), to

---

[2] The current ORANI data base refers to 1977–78 but contains data on the agricultural sector averaged over a number of years. Hence, the data contain a representation of agriculture which is typical of recent history.

the exogenous variables at time t ($\mathbf{V}_2(t)$) and to the initial conditions represented by $\mathbf{Z}(0)$.

From Equation 28.3, we see that forecasting versions of ORANI contain only very limited dynamic mechanisms. The values of endogenous variables at time t are explained purely by the values of exogenous variables at time t and by the values of variables at time 0. The dynamic mechanisms which are included are capital-accumulation relationships.[3] A simplified form of these is [4]

$$K(t) = K(0) + \int_{s=0}^{t} I(s)\,ds \qquad (28.4)$$

and

$$I(s) = I(0)e^{i(t)s}, \qquad (28.5)$$

where

$K(t)$ and $K(0)$ are capital stocks at times t and 0 respectively,
$I(t)$ and $I(0)$ are investment at times t and 0, and
$i(t)$ is average annual rate of growth of investment over the period 0 to t.

It is the assumption that investment grows smoothly at the rate $i(t)$ that allows us to simplify dramatically the dynamics of our model. It avoids the necessity of explicitly including any variables relating to periods between 0 and t. On substituting Equation 28.5 into 28.4 we obtain

$$K(t) = K(0) + I(0)\left(\frac{e^{i(t)t} - 1}{i(t)}\right), \quad \text{if } i(t) \neq 0$$

$$= K(0) + I(0)t, \quad \text{if } i(t) = 0. \qquad (28.6)$$

Using Equation 28.3 in forecasting, we compare one value of the endogenous vector at time t (our forecast value, $\mathbf{V}_1^F(t)$) with the initial value of this vector, $\mathbf{V}_1(0)$. Results of forecasting exercises are usually reported as growth rates for the period 0 to t calculated as simple transformations of the vector

$$\begin{aligned}
F(t, \mathbf{V}_2^C(t) + \mathbf{G}, \mathbf{V}(0)) &= \mathbf{V}_1^F(t) - \mathbf{V}_1(0) \\
&= [\mathbf{V}_1^F(t) - \mathbf{V}_1^C(t)] + [\mathbf{V}_1^C(t) - \mathbf{V}_1(0)] \\
&= [f(\mathbf{V}_2^C(t) + \mathbf{G}, \mathbf{Z}(0), t) - f(\mathbf{V}_2^C(t), \mathbf{Z}(0), t)] \\
&\quad + [f(\mathbf{V}_2^C(t), \mathbf{Z}(0), t) - \mathbf{V}_1(0)],
\end{aligned} \qquad (28.7)$$

where

$\mathbf{G}$ is the vector of deviations that we expect (forecast) in the exogenous variables over the period 0 to t away from their control values, $\mathbf{V}_2^C(t)$.

---

[3] Earlier versions of ORANI, including that described in DPSV (1982), had no explicit accumulation relationships. The most authoritative paper on these relationships in ORANI is Horridge (1985).
[4] In Equations 28.4 and 28.5 we ignore depreciation, we fail to append industry subscripts and we make no distinction between foreign and domestic ownership of capital stocks. While these complications are included in ORANI, see Horridge (1985) and Dixon et al. (1984), they are not essential at this stage.

The difference between comparative statics and forecasting

In applying Equation 28.7 we must supply observations on the initial values of variables ($V_1(0)$ and $Z(0)$) and a scenario for the exogenous variables in period t ($V_2^C(t) + G$). The second line of Equation 28.7 suggests a convenient decomposition of the forecast vector (F) into:

(1) the difference between the forecast values of the endogenous variables at time t ($V_1^F(t)$) and their control values at time t ($V_1^C(t)$), and
(2) the difference between the control values at time t and the initial values ($V_1(0)$).

Note that the first part of this decomposition is a purely comparative static computation, i.e., it involves a comparison between alternative values of variables at the same point in time, namely period t. The dynamic aspects of our forecasts are thus confined to the second part of the decomposition, i.e., to the computation of a valid control solution for the model for time t.

In our pilot forecasting exercise (Dixon, 1986), we assumed that the control values for the exogenous variables were the initial values, and that the second term in our decomposition was zero. These were also the assumptions implicit in the theoretical discussion by Parmenter and Meagher (1985). Underlying this approach is the false assumption that if $V_2^C(t) = V_2(0)$, then $V_1^C(t) = V_1(0)$. We should expect, however, that with $V_2^C(t) = V_2(0)$,

$$V_1(0) \neq V_1^C(t) = f(V_2(0), Z(0), t), \quad t > 0. \tag{28.8}$$

Even in the absence of changes in the exogenous variables, year t will differ from year 0 because of capital accumulation. This is obvious in Equations 28.5 and 28.6. If $I(0) > 0$ and we set $I(t) = I(0)$, then we find from Equation 28.5 that $i(t) = 0$. By substituting into Equation 28.6, we obtain $K(t) > K(0)$ for $t > 0$. In other words, if $I(0) > 0$ it is impossible to have

$$(K(t), I(t), i(t)) = (K(0), I(0), i(0)). \tag{28.9}$$

More generally, the initial conditions, represented as $Z(0)$ in Equation 28.3, will normally rule out the possibility of a stationary state. Not only are the initial values of the variables usually an implausible representation of any hypothetical state of the economy in a future year t but for the dynamic forecasting model (Equation 28.3) (although not for the static model (Equation 28.1)) they do not usually constitute a valid solution for year t.

In principle, it is straightforward to calculate a legitimate value for $V_1^C(t)$ via Equation 28.8. The forecast can then be completed by computing the first term in the decomposition in Equation 28.7 setting G according to forecasts of differences between the exogenous variables at time t and their values at time 0. In practice, however, evaluation of the control solution is complicated because in performing calculations with ORANI we adopt linear approximations, i.e., we do not have explicit forms for the functions g or f.

For example, in performing the computations required in evaluating the right-hand side of Equation 28.2 and the first term on the right-hand side of Equation 28.7

we adopt approximations of the form

$$g(\mathbf{V}_2^C(t) + \mathbf{S}) - g(\mathbf{V}_2^C(t)) = B(\mathbf{V}^C(t))\mathbf{S} \tag{28.10}$$

and

$$f(\mathbf{V}_2^C(t) + \mathbf{G}, \mathbf{Z}(0), t) - f(\mathbf{V}_2^C(t), \mathbf{Z}(0), t) = D(\mathbf{V}^C(t), \mathbf{Z}(0), t)\mathbf{G}, \tag{28.11}$$

where

> B and D are matrices of partial derivatives of g and f with respect to the exogenous variables $(\mathbf{V}_2)$, evaluated at the points $(\mathbf{V}^C(t))$ and $(\mathbf{V}^C(t), \mathbf{Z}(0), t)$ respectively.

Note that computations 28.10 and 28.11 require a valid control solution around which to perform the linearization. The matrices B and D are computed as functions of both the endogenous and exogenous variables $(\mathbf{V}^C(t) = (\mathbf{V}_1^C(t), \mathbf{V}_2^C(t)))$ where the $\mathbf{V}^C(t)$ satisfy Equations 28.1 or 28.3.

The reasons for adopting these approximations are explained in DPSV (1982, especially section 8) and Dixon (1985). Here we simply note that

(1) The approximations have proved satisfactory in a wide range of ORANI applications.
(2) The matrices B and D are easily evaluated from input–output data and econometric estimates of substitution elasticities.
(3) The forms of g and f are underivable in any practical way so that Equations 28.2 and 28.7 are not directly usable in computations.
(4) In the rare cases in which the approximations in Equation 28.10 and 28.11 are inadequate, more accurate solutions are available, although with considerably increased computational difficulty. The approach which we have taken in these cases is to compute the effects of the entire shock (e.g., $\mathbf{S}$ in Equation 28.10 or $\mathbf{G}$ in Equation 28.11) in a series of linear computations of the effects of smaller shocks. Between each pair of such computations the matrix of partial derivatives is re-evaluated using data updated to include the effects of previous shocks. For example, instead of using Equation 28.11 we might compute the effects of the shock $\mathbf{G}$ via n computations in each of which a shock of size $\mathbf{G}/n$ is imposed. First we compute

$$\Delta \mathbf{V}_1^1(t) = D(\mathbf{V}^C(t), \mathbf{Z}(0), t)(\mathbf{G}/n),$$

then

$$\Delta \mathbf{V}_1^2(t) = D(\mathbf{V}_1^C(t) + \Delta \mathbf{V}_1^1(t), \mathbf{V}_2^C(t) + \mathbf{G}/n, \mathbf{Z}(0), t)(\mathbf{G}/n),$$

etc. Details of the multistep procedure are given in DPSV (1982, ch. 5 and section 47). For ORANI it is shown that a 2-step solution combined with an extrapolation rule produces values for the endogenous variables very close to exact, nonlinear solutions.

We have now devised a method of deriving a valid control solution $(\mathbf{V}^C(t))$ for the forecasting model in year t which has two essential properties:

(1) it continues to rely primarily on linear approximations such as Equations 28.10

and 28.11, i.e., it does not require an explicit form for underivable components of the function f; and
(2) it allows us, if necessary, still to employ the multistep solution method for making computations free of linearization errors.

Our method is based on the observation that we can separate the model into two components, namely, the accumulation equations and the rest, the latter being a strictly static system. Hence, a valid control solution to the model can be represented as

$$K^C(t) = f^{**}(I^C(t), \mathbf{Z}(0), t), \tag{28.12a}$$

$$\mathbf{V}_1^C(t) = f^*(\tilde{\mathbf{V}}_2^C(t), K^C(t), I^C(t)). \tag{28.12b}$$

Note that in Equation 28.12b, both the capital stock at time t ($K(t)$) and investment at time t ($I(t)$) are treated as exogenous variables. $\tilde{\mathbf{V}}_2^C(t)$ is defined as the vector of control values of exogenous variables excluding $K(t)$ and $I(t)$. It remains true that it is impractical to derive an explicit form for f^*.[5] This is not the case for f^{**}, however. By postulating a control value for $I(t)$ and given the initial conditions ($\mathbf{Z}(0)$) we can easily derive a valid control solution for $K(t)$ using Equation 28.12a. We can then use an approximation to Equation (28.12b) to compute control values for the endogenous variables $\mathbf{V}_1(t)$ using the previously calculated control values for $K(t)$ and $I(t)$, and using initial values for the control values of the remaining exogenous variables, $\tilde{\mathbf{V}}_2(t)$. That is, we compute

$$\Delta \mathbf{V}_1(t) \equiv \mathbf{V}_1^C(t) - \mathbf{V}_1(0), \tag{28.13}$$

via

$$\Delta \mathbf{V}_1(t) = f^*(\tilde{\mathbf{V}}_2(0), K^C(t), I^C(t)) - f^*(\tilde{\mathbf{V}}_2(0), K(0), I(0))$$

$$= D^*(\mathbf{V}(0)) \begin{bmatrix} 0 \\ \Delta K(t) \\ \Delta I(t) \end{bmatrix}, \tag{28.14}$$

where D^* is the matrix of partial derivatives of f^* with respect to the exogenous variables $\tilde{\mathbf{V}}_2$, K and I. $\Delta K(t)$ and $\Delta I(t)$ are deviations of the control values of these variables (obtained via Equation 28.12a) from their initial values. Note that D^* is evaluated at the initial values of *all* the variables. Because f^* is strictly static, i.e., it excludes the accumuation relationships, $\mathbf{V}^C(t) = V(0)$ is a valid control solution to f^* for any period t.[6] Finally, we use the solution to Equation 28.14 to update the initial values of the endogenous variables, yielding

$$\mathbf{V}_1^C(t) = \mathbf{V}_1(0) + \Delta \mathbf{V}_1(t). \tag{28.15}$$

Equations 28.14 and 28.15 give only an approximation to the control values of $\mathbf{V}_1(t)$

---

[5] The function f^* in our forecasting version of ORANI is very similar to the function g in our comparative static version.
[6] Control solutions to f^* are thus exactly analogous to control solutions for g in model Equation 28.1.

which are consistent with our control values for $K(t)$ and $I(t)$ (and $\tilde{V}_2(t)$). However, if more accuracy is required the multistep solution method outlined above (point (4)) can be applied.

To summarize, we have described a two-part method for deriving a valid control solution to our entire forecasting model for period t. The solution is

$$\mathbf{V}^C(t) = (\mathbf{V}_1^C(t), \tilde{V}_2(0), K^C(t), I^C(t)),$$

where $K^C(t)$ and $I^C(t)$ are calculated directly from Equation 28.12a and $\mathbf{V}_1^C(t)$ is computed from Equations 28.14 and 28.15. Note that the solution to 28.14 and 28.15 is closely related to the second term in our decomposition of the forecasting Equation 28.7.[7]

We can now use Equation 28.11 to compute the first term in the decomposition of Equation 28.7. Moreover, in making this calculation there is nothing to stop us reassigning variables between the exogenous and endogenous sets. Our method for evaluating $\mathbf{V}^C(t)$ involved a computation (Equation 28.14) in which $K(t)$ and $I(t)$ were exogenous, although one of these variables can be thought of as endogenized by Equation 28.12a. In solving Equation 28.11, however, we would usually include both $K(t)$ and $I(t)$ as endogenous variables. The inclusion of the accumulation relationship in f is sufficient to endogenize one of these variables. The other would usually be replaced on the exogenous list by the rates of return on capital, about which an exogenous forecast would be made.

The only outstanding problems in implementing Equation 28.11 concern the shock vector, $\mathbf{G}$. In forecasting, in contrast to pure comparative static exercises, considerable effort must be devoted to forecasting the exogenous variables. In the forecasting exercise reported by Dixon (1986) the setting of $\mathbf{G}$ involved:

(1) the consideration of data on tariff and quota changes for the period 1967 to 1983 and the assessment of announced government intentions concerning protection for steel, textiles, clothing, footwear and automobiles over the next five years;
(2) the consideration of data on world price movements for the period 1967 to 1983 and the assessment of how future price trends are likely to differ from past trends in view of likely changes in energy markets;
(3) the use of input–output calculations applied to price projections to forecast world-wide technological change by industry, and the assessment of how technological developments in Australian industries are likely to deviate from world-wide experience;
(4) the projection of the size of the labour force and the number of households to 1990;
(5) the assessment of likely changes between 1985 and 1990 in the extent of excess capacity in different Australian industries, and;

---

[7] The difference is that the variables treated as endogenous in Equation 28.14 may not coincide precisely with the list of endogenous variables in the complete forecast. Nevertheless, the solutions to Equations 28.14, 28.15 and 28.12a contain all the information necessary to construct the second term in Equation 28.7.

(6) the assessment of likely future changes in the cost of capital to Australian businesspeople.

In formulating our vector **G** of forecast *changes* in the exogenous variables, we must be careful to deviate from the control values of these variables. Similarly, in interpreting the changes in the endogenous variables projected by Equation 28.11 we must recognize that they are deviations from their control values, not their initial values.

28.3 A NUMERICAL EXAMPLE

A very simple model for numerical illustration of our forecasting procedure is the capital-accumulation Equation 28.6 combined with

$$Q(t) = K(t) + I(t). \tag{28.16}$$

We can interpret Equation 28.16 as explaining aggregate employment ($Q(t)$). At time t, one unit of labour is employed in conjunction with each unit of capital for current production and one unit is employed per unit of capital creation. However, the interpretation of Equation 28.16 is not important, nor is its exact form. For our purpose, the role of Equation 28.16 is to represent the rest of ORANI, i.e., the nonaccumulation relationships, f^*. Note that Equation 28.16 is purely static, i.e., choosing $Q(t)$ as the endogenous variable, it has the same form as Equation 28.12b. Equation 28.6, on the other hand, has the form of Equation 28.12a.

We first transform the equations to distinguish the average annual rates of growth of the variables over the period 0 to t. We define all growth rates as in Equation 28.5 and rewrite Equations 28.16 and 28.6 as[8]

$$e^{q(t)t} = \frac{K(0)}{Q(0)} e^{k(t)t} + \frac{I(0)}{Q(0)} e^{i(t)t} \tag{28.16$'$}$$

and

$$e^{k(t)t} = 1 + \frac{I(0)}{K(0)} \left(\frac{e^{i(t)t} - 1}{i(t)} \right), \quad i(t) \neq 0 \tag{28.6$'$}$$

$$= 1 + \frac{I(0)}{K(0)} t, \quad i(t) = 0$$

Hypothetical base-period data for the model are:

$$(K(0), I(0), Q(0)) = (100, 4, 104). \tag{28.17}$$

Note that these data satisfy Equation 28.16 for $t = 0$.

Imagine that we wish to use Equations 28.16$'$ and 28.6$'$ to make forecasts of the

---

[8] The conversion to growth-rate form is the 'simple transformation' of the forecast vector referred to on p. 656.

average annual rates of growth of employment, capital and investment over a five-year period. The first task is to find a valid control solution for year 5. A 'no-change' scenario

$$(\bar{q}(5), \bar{k}(5), \bar{i}(5)) = (0, 0, 0), \tag{28.18}$$

would satisfy Equation 28.16′ but not 28.6′. Our procedure is to solve Equation 28.6′ directly for a valid control. Two obvious choices are

(i) 'no change' in investment

$$(k(5), i(5)) = (0.0365, 0.0), \tag{28.19}$$

(ii) 'balanced growth'

$$(k(5), i(5)) = (0.04, 0.04). \tag{28.20}$$

As we shall see, our solutions are not very sensitive to the choice between (i) and (ii). To complete the control solution we require a value $q(5)$ compatible with our chosen controls $k(5)$ and $i(5)$. In deriving this we assume that Equation 28.16′ is unavailable. Instead we must use a first-order linear approximation to it, the linearization being performed around the no-change values of the variables (Equation 28.18) The linearized equation is

$$q(t) - \bar{q}(t) = \left[\frac{K(0)e^{\bar{k}(t)t}}{Q(0)e^{\bar{q}(t)t}}, \frac{I(0)e^{\bar{i}(t)t}}{Q(0)e^{\bar{q}(t)t}} \right] \left[\begin{array}{c} k(t) - \bar{k}(t) \\ i(t) - \bar{i}(t) \end{array} \right]. \tag{28.21}$$

Equation 28.21 is analogous to Equation 28.14, with the qualification that our example is in terms of changes in the average rates of change of the variables, not changes in their levels. Using Equations 28.17 and 28.18, Equation 28.21 reduces to

$$q(5) = (0.9615, 0.0385) \binom{k(5)}{i(5)}, \tag{28.22}$$

which yields

$$q(5) = 0.0351, \text{ given Equation 28.19}$$
$$= 0.04, \text{ given Equation 28.20}.$$

Hence, we have derived two alternative valid control solutions for Equations 28.16′ and 28.6′. They are

$$(\bar{q}(5), \bar{k}(5), \bar{i}(5)) = (0.0351, 0.0365, 0.0) \tag{28.23}$$

and

$$(\bar{q}(5), \bar{k}(5), \bar{i}(5)) = (0.04, 0.04, 0.04). \tag{28.24}$$

These are valid control solutions in the sense that, given the initial conditions (Equation 28.17) they satisfy Equations 28.16′ and 28.6′ simultaneously.

To complete our forecast we require a first-order linear approximation to the entire model 28.16′ and 28.6′ evaluated at one of our valid control solutions. Linearizing Equation 28.6′ we obtain

$$(te^{\bar{k}(t)t})(k(t) - \bar{k}(t)) = \gamma(i(t) - \bar{i}(t)), \tag{28.25}$$

where

$$\gamma = \frac{I(0)}{K(0)} \left[\frac{e^{\bar{i}(t)t}(\bar{i}(t)t - 1) + 1}{\bar{i}(t)^2} \right], \quad \bar{i}(t) \neq 0 \qquad (28.26)$$

$$= \frac{t^2 I(0)}{2K(0)}, \quad \bar{i}(t) = 0.$$

Combining Equations 28.21 and 28.25, the complete linear system is

$$\begin{bmatrix} 1, & -\dfrac{K(0)e^{\bar{k}(t)t}}{Q(0)e^{\bar{q}(t)t}}, & -\dfrac{I(0)e^{\bar{i}(t)t}}{Q(0)e^{\bar{q}(t)t}} \\ 0, & te^{\bar{k}(t)t}, & -\gamma \end{bmatrix} \begin{bmatrix} q(t) - \bar{q}(t) \\ k(t) - \bar{k}(t) \\ i(t) - \bar{i}(t) \end{bmatrix} = 0. \qquad (28.27)$$

We use Equation 28.27 to compute the comparative static component of our forecast. Evaluating Equation 28.27 at one of our valid control solutions (say Equation 28.24), using initial conditions (Equation 28.17),[9] gives

$$\begin{bmatrix} 1, & -0.9615, & -0.0385 \\ 0, & 6.1070, & -0.5719 \end{bmatrix} \begin{bmatrix} q(5) - 0.04 \\ k(5) - 0.04 \\ i(5) - 0.04 \end{bmatrix} = 0. \qquad (28.28)$$

Assume that $q(t)$ is the exogenous variable, then, from Equation 28.28

$$\begin{bmatrix} k(5) - 0.04 \\ i(5) - 0.04 \end{bmatrix} = - \begin{bmatrix} 0.9615, & -0.0385 \\ 6.1070, & -0.5716 \end{bmatrix}^{-1} \begin{bmatrix} 1 \\ 0 \end{bmatrix} [q(5) - 0.04], \qquad (28.29)$$

or

$$\begin{pmatrix} k(5) \\ i(5) \end{pmatrix} = \begin{pmatrix} 0.04 \\ 0.04 \end{pmatrix} + \begin{pmatrix} 0.7285 \\ 7.7796 \end{pmatrix}(q(5) - 0.04). \qquad (28.30)$$

Next, assume that our exogenous forecast of $q(5)$ is 0.05, then via Equation 28.30 our forecasts for $k(5)$ and $i(5)$ are

$$\begin{pmatrix} k(5) \\ i(5) \end{pmatrix} = \begin{pmatrix} 0.0473 \\ 0.1180 \end{pmatrix}.$$

These values are listed in Table 28.1 together with forecasts derived by solving Equations 28.16′ and 28.6′ directly and via Equation 28.27 using Equations 28.23 and 28.18 as control solutions.

The first row of the table gives exact solutions generated by solving Equations 28.16′ and 28.6′ directly. Note that answers derived from multistep solution of Equation 28.27 (fifth row) reproduce the exact solutions without error. Single-step

---

[9] Note that the coefficients in the first row of Equation 28.27 could also be written as

$$\left[1, \frac{K^C(t)}{Q^C(t)}, \frac{I^C(t)}{Q^C(t)} \right].$$

This form makes it apparent that the coefficients can be calculated from updated data. For example, when Equation 28.24 is used the control data are

$$(Q^C(t), K^C(t), I^C(t)) = (127.03, 122.14, 4.89).$$

Table 28.1 Alternative forecasts of $k(5)$ and $i(5)$ given $q(5) = 0.05$

Forecasts	Endogenous variables	
	$k(5)$	$i(5)$
Via exact solution of Equations 28.16' and 28.6'	0.047	0.109
Via single-step solutions of Equation 28.27		
Evaluated at Equation 28.23: investment constant in control solution	0.048	0.132
Evaluated at Equation 28.24: balanced growth control solution	0.047	0.118
Evaluated at Equation 28.18: invalid 'no change' control solution	0.037	0.371
Via multistep solution Equation 28.27 evaluated at Equation 28.24: balanced growth control solution	0.047	0.109

solutions of Equation 28.27 using either of our valid control solutions (second and third rows) produce acceptable approximations to the exact solutions. Use of Equation 28.27 under the false assumption that no change in all variables is a valid control solution for year 5 (fourth row) does not yield an acceptable approximation.

28.4 CONCLUDING REMARKS

The strength of general equilibrium models is their ability to handle interindustry linkages. Industries are linked through input–output relationships and also through their competition in factor markets. The importance of these linkages has long been recognized in policy analysis. For example, when we consider the effects of increases in tariffs we must account for cost increases flowing from sectors receiving greater protection to other sectors, particularly export sectors, via interindustry transactions and labour markets. It is not surprising, therefore, that frequent use is made in policy debates of comparative static simulations from general equilibrium models.

Our experience with the ORANI model suggests that interindustry linkages are also of vital importance in forecasting. For example, in Dixon (1986), our exogenous scenario included slow growth in foreign demand for Australian agricultural products. Via the ORANI model we derived forecasts of slow growth in agricultural outputs. The model also indicated that growth in the mining sector would be rapid. This is because a poor performance in agriculture will improve competitive conditions for our mineral sector by leading to a deterioration in our real exchange rate. Thus, in forecasting prospects for the mining sector it was necessary to consider carefully prospects for other sectors of the economy.

Although general equilibrium models have much to offer as forecasting devices,

Concluding remarks

their potential cannot be realized fully without considerable effort. As we saw in Section 28.2, the workload involved in a general equilibrium forecast is much greater than that required for a comparative static calculation. Unlike the situation in comparative statics, in forecasting we must make a detailed assessment of the likely future course of a large number of exogenous variables.

Another problem which increases the workload in forecasting relative to that in comparative statics is delays in the collection and publication of statistics. For example, input–output tables which are the main data input for general equilibrium models, are available for Australia from the Australian Bureau of Statistics only up to 1979. If we are trying to forecast industry growth rates from 1985 to 1990 then it is clearly of importance to know the details of the situation in 1985, particularly for industries such as construction and agriculture which are subject to strong cyclical or climatic influences. For example, if we failed to recognize that capital creation was unusually high in 1985, then we would be in danger of overestimating growth prospects for construction industries to 1990. In terms of Equation 28.27, up-to-date information on the state of the economy is required in the evaluation of $I(0)/K(0)$. If this is set too low, then our forecast growth rate for investment, $i(t)$, is likely to be too high. This suggests that despite difficulties associated with publication lags, the data bases of general equilibrium models should be updated before these models are used in forecasting.

In Section 28.2, we saw that forecasting with ORANI requires a slightly different computational approach from that used in comparative statics. This is because of difficulties in generating a valid control solution for year t. Fortunately, it appears that the computational problems associated with forecasting can be overcome without significant disturbance to the ORANI computing codes. This can be done by (a) generating a control solution for ORANI's accumulation relationships in their nonlinear form and (b) using this solution as exogenous shocks in a linear computation of control values for variables not included in the accumulation equations.

Finally, it is worth mentioning two issues which traditionally are of central interest to forecasters: validation and dynamics. As with most general equilibrium models, ORANI has not been subjected to detailed validation. One of the advantages of adapting the model to forecasting applications is that validation will then be possible. The Institute is currently undertaking a project in which the performance of the model in tracking developments of the 1970s will be assessed.

In comparison with other forecasting techniques, e.g., Box–Jenkins methods, in the general equilibrium approach comparatively little attention is given to dynamics. This is a problem not only in forecasting but also in comparative statics. The nature of the difficulty can be seen by examination of either Equation 28.2 or 28.7. In these equations it is not clear at what time the shocks S and G are applied. Normally in comparative statics we assume that the shock S is applied at time zero and is sustained up to time t. In forecasting we often assume that the shock G develops smoothly over the period zero to t. It should be emphasized, however, that the theoretical structure of a general equilibrium model provides no guidance as to the appropriate interpretation of the evolution of the exogenous shock.

In work at the IMPACT Project, Cooper et al.[10] have made a substantial contribution to overcoming the dynamic limitations of general equilibrium models. They have shown how given paths for exogenous shocks over the period zero to t can be aggregated into the appropriate shocks **S** and **G** in Equations 28.2 and 28.7 to be applied in a comparative static computation for time t. In forming **S** and **G**, their method gives greater weight to shocks occurring early in the period zero to t, rather than later. Shocks (e.g., tariff changes) occurring close to t have little time in which to influence outputs, prices, employment, etc. It is likely that adoption of Cooper et al.'s methods will lead to substantial improvements in both the forecasting and comparative static capabilities of the ORANI model.

REFERENCES

Bruce, Ian (1985) The sensitivity of ORANI 78 projections to the data base used, *Preliminary Working Paper* OP-53, July, IMPACT Project, University of Melbourne.

Cooper, Russel J. and McLaren, Keith (1983) The ORANI–MACRO interface: an illustrative exposition. *Economic Record*, 59 (June), 166–179.

Cooper, Russel, J., McLaren, Keith R. and Powell, Alan A. (1985) Short-run macroeconomic closure in applied general equilibrium modelling: experience from ORANI and agenda for further research, in *New Developments in Applied General Equilibrium Analysis*, (eds John Piggott and John Whalley), Cambridge University Press.

Dixon, P. B. (1985) The solution procedure for the ORANI model explained by a simple example, in *New Mathematical Advances in Economic Dynamics* (eds D. Batten and P. Lesse), Croom Helm, London and Sydney, pp. 119–29.

Dixon, P. B. (1986) Prospects for Australian industries: 1985 to 1990. *Australian Economic Review*, No. 73, 1st quarter.

Dixon, P. B., Parmenter, B. R., Sutton, J. and Vincent, D. P. (1982) *ORANI: A Multisectoral Model of the Australian Economy*, North-Holland. (Cited in the text as DPSV (1982).)

Dixon, P. B., Parmenter, B. R. and Rimmer, R. J. (1984) Extending the ORANI model of the Australian economy: adding foreign investment to a miniature version, in *Applied General Equilibrium Analysis* (eds H. E. Scarf and J. B. Shoven), Cambridge University Press, New York, pp. 485–533.

Dixon, P. B., Parmenter, B. R. and Rimmer, R. J. (1986) ORANI projections of the short run effects of a 50 per cent across-the-board cut in protection using alternative data bases, in *General Equilibrium Trade Policy Modelling* (eds T. N. Srinivasan and J. Whalley), MIT Press, Cambridge, MA.

Horridge, M. (1985) The long-run closure of ORANI: first implementation, *Preliminary Working Paper* OP-50, February, pp. 87 + 16 + 39, IMPACT Project, University of Melbourne.

Parmenter, B. R. and Meagher, G. A. (1985) Policy analysis using a computable general equilibrium model: a review of experience at the IMPACT Project. *Australian Economic Review*, No. 69, 1st Quarter, 3–15.

Powell, Alan A. (1985) A brief account of activities over the period 1st February 1982 to 28th February 1985, with a prospectus for further developments, *Report* R-05, IMPACT Project, University of Melbourne, pp. 67.

---

[10] See, for example, Cooper and McLaren (1983) and Cooper et al. (1985).

29

An applied general equilibrium model of the United States economy

JOHN V. COLIAS

In this chapter a multisector applied general equilibrium model in the Johansen class is created for the United States economy. The nonlinear model is linearized and solved for percentage changes of the variables. The development of a complete modelling system permits flexibility in equation specification and data aggregation.

A GAMS (General Algebraic Modelling System) program manipulates and aggregates data of 184 input–output and 128 final demand sectors, and calculates sectoral value added. Capacity for rapid reaggregation of data to any desired sectoral breakdown improves the flexibility of applied general equilibrium modelling.

The Johansen-style model is used to analyse the impact on key economic variables of the accelerated depreciation and investment tax credit provisions for the first year of the Economic Recovery Tax Act of 1981, compared with what would have occurred with a neutral capital income tax. The model investigates the impact on production, employment, capital utilization, exports, imports, and relative returns to labour and capital. Of this model's nine sectors – farming, extractive, construction, household, transportation, communications, services, high technology, and smokestack – the high technology sector suffers the greatest decline in output.

29.1 THE JOHANSEN-STYLE APPLIED GENERAL EQUILIBRIUM MODEL

29.1.1 Review and overview

Recently, Dixon *et al.* (1982) extended the Johansen (1960) method of stating and solving a computable general equilibrium (CGE) model. The resulting ORANI model of the Australian economy has impressive detail and flexibility of use. Whereas Johansen's technique of linearizing a nonlinear model introduces linearization error, Dixon et al. (1982, pp. 51–60, 199–251) employ a multistep solution algorithm which eliminates these errors. Shoven and Whalley (1984, p. 1021) report that Lans Bovenburg and Wouter Keller use a similar iterative solution procedure to eliminate linearization errors in Keller's (1980) CGE model of the Netherlands.

Johansen's model (1960, pp. 3–23) studied the benefits and reasonability of balanced and unbalanced growth strategies for Norway. Black and Taylor (1974, pp. 37–8) used the Johansen approach to analyse the effects on resource allocation of trade liberalization in Chile. Staelin (1976, pp. 39–40) investigated tariff impacts in a noncompetitive economy, the Ivory Coast. ORANI of Australia (Dixon et al. 1982, pp. 344–53) simulated both industry policies related to tariffs, taxes, and subsidies on imports, exports, and sales; and macroeconomic strategies regarding changes in aggregate consumption, investment, and government spending.

This chapter reports the development of a Johansen-style CGE model for the United States economy and its application to taxation policy.

The INTERSAGE (Intermediate size Sectoral Applied General Equilibrium) model of the United States economy consists of a data program (Colias, 1984 and Colias, 1985, pp. 196–245) and the Johansen-style CGE model (Colias, 1985, pp. 338–46). Both are written in GAMS (general algebraic modelling system), a higher level modelling language (Kendrick and Meeraus, 1985). The Johansen method of presenting and solving a CGE model (Dixon et al., 1982) and GAMS enhance each other's flexibility in use (Kendrick, 1984). The resulting GAMS/CGE model (Colias, 1985, pp. 347–452) is straightforward to implement on the computer and permits easy reaggregation and respecification of equations.

The development of this modelling system complements the extensive efforts to apply general equilibrium models to the US economy by Ballard et al. (1985), Jorgenson (1984), Slemrod (1983), Shoven and Whalley (1972), and others.

29.1.2 The model

(a) Assumptions

The equations of the US model may be grouped in the following categories: (1) household demand; (2) export demand; (3) commodity supply; (4) input demand; (5) zero pure profit equations; (6) price equations; (7) market clearing equations; (8) personal income, expenditure, and savings; (9) trade balance equations.

These equations incorporate the following assumptions (not a comprehensive list):

(1) utility maximization in consumption
(2) cost minimization in production
(3) zero pure profits

(4) market clearing
(5) unitary elasticities of substitution in consumption among types of commodities
(6) zero substitutability in production among intermediate and factor composites
(7) unitary elasticities of substitution between the domestic and imported version of each commodity and between labour and capital
(8) ownership of all capital and labour by a single representative household
(9) complete mobility of labour and capital among industries
(10) fixed total supplies of labour and capital, and
(11) an inverse response of exports to the domestic price.

Assumptions (1)–(4) describe a competitive world. However, the model does not simulate a purely competitive environment since (a) the economy does not have a large number of buyers and sellers – the INTERSAGE model has only eleven buyers, namely, the domestic household, nine industries, and the rest of the world, (b) demand curves are not infinitely elastic, and (c) total supplies of all goods are exogenous.

Margins, that is, commodities which facilitate the flow of final goods from producers to users (Dixon *et al.*, 1982, p. 106), are treated as regular commodities in INTERSAGE. Dixon *et al.* (1982, pp. 106–8) cite potential problems resulting from this approach. For example, with retail trade entering the household utility function, retail may be substituted for food, an unrealistic result. A distinct treatment of margins should be added to the model in order to remedy this problem. A detailed treatment of margins may be found in Chapter 3 of Dixon *et al.* (1982).

(b) Equations

In this subsection, the household and input demand equations (both percentage change and structural) will be presented. The presentation should clarify the construction of the INTERSAGE model. For a complete listing of the structural and percentage change equations plus the GAMS model, see Appendices B and C of Colias (1985).

As a Johansen model, the INTERSAGE model consists of percentage change equations derived from a set of theoretical counterparts. For example, the household demand equations derive from household utility maximization. In contrast to the percentage change equations, the structural equations are stated with variables in levels rather than percentage changes. In the equations reported below, the variables with a tilde above them are in percentage changes.

The variables and parameters are defined for each equation of the model as needed. In the notation for variables and parameters, superscripts are part of the variable name and subscripts are indices. For example, for x_i^k which equals the demand for capital goods by industry i, the k superscript defines the type of demand, and the i subscript is an index for industries. The key sets of the model appear in Table 29.1.

Table 29.1 Sets of the INTERSAGE model

Set	Definition
$c \in C$	Set of commodities
$c \in C_i$	Set of commodities produced by industry $i \in I$
$i \in I$	Set of industries
$i \in I_c$	Set of industries which produce commodity $c \in C$
$s \in S$	Set of sources $\equiv \{\text{domestic, imported}\}$
$s \in SD$	Set of domestic sources $\equiv \{\text{domestic}\}$
$s \in SF$	Set of foreign sources $\equiv \{\text{imported}\}$
$\phi \in \Phi$	Set of factors $\equiv \{\text{labour, capital}\}$

(i) HOUSEHOLD DEMAND

The percentage-change household demand equations appear as follows:

$$\tilde{c}_{c's'} = \tilde{y}^e + (\varepsilon_{c'}^s - \varepsilon^c)\left(\sum_{s \in S} s_{c's}^c \tilde{p}_{c's}^c\right) - \varepsilon_{c'}^s \tilde{p}_{c's'}^c$$

$$+ (\varepsilon^c - 1)\left[\sum_{c \in C} s_c^A \left(\sum_{s \in S} s_{cs}^c \tilde{p}_{cs}^c\right)\right] \quad c' \in C, \ s' \in S \quad (29.1)$$

where

c_{cs} = household consumption of commodity c from source s
y^e = total expenditure (aggregate consumption)
p_{cs}^c = price paid by consumers for commodity c from source s
s_c^A = ratio of household expenditure on commodity c to household expenditure on all commodities
s_{cs}^c = ratio of household expenditure on commodity c from source s to household expenditure on commodity c regardless of source
ε^c = the elasticity of substitution in consumption among commodities
ε_c^s = the elasticity of substitution in consumption among sources

Equation 29.1 derives from a classical programming problem (Equation 29.2) of utility maximization.

Choose c_{cs} to maximize

$$u = \left[\sum_{c \in C} (v_c^A)^{1/\varepsilon^c} (c_c^A)^{((\varepsilon^c - 1)/\varepsilon^c)}\right]^{(\varepsilon^c/(\varepsilon^c - 1))} \quad (29.2)$$

subject to

$$c_c^A = \left[\sum_{s \in S} (v_{cs})^{(1/\varepsilon_c^s)} (c_{cs})^{((\varepsilon_c^s - 1)/\varepsilon_c^s)}\right]^{(\varepsilon_c^s/(\varepsilon_c^s - 1))}$$

$$\sum_{c \in C} \sum_{s \in S} p_{cs}^c c_{cs} = y^e$$

The Johansen-style applied general equilibrium model

where

u = total utility
c_c^A = a quantity aggregator function for consumption of commodity c
v_c^A = a distribution parameter in consumption
v_{cs} = a distribution parameter in consumption

The utility function is a nested CES function which allows substitution, on the one hand, among commodities, and on the other hand between the domestic and imported version of each commodity. The degree of substitutability as specified by the elasticities, ε^c and ε_c^s in Equation 29.1 should be estimated. General equilibrium modellers usually search the literature for these values. For the tax application of INTERSAGE which follows, ε^c and ε_c^s are assumed equal to one. Base period budget shares, s_c^A and s_{cs}^c in Equation 29.1, are calculated from the base period data set.

(ii) INPUT DEMAND

The percentage-change input demand equations of the INTERSAGE model have the following form:

$$\tilde{x}_i^l = \tilde{q}_i^l - \alpha_i^k \tilde{p}^l + \alpha_i^k \tilde{p}_i^k \quad i \in I \tag{29.3}$$

where

x_i^l = the use of labour by industry i
p^l = the price of labour (wage rate)
p_i^k = the price (cost) of capital services in industry i
α_i^l, α_i^k = the ratio of production expenditure by industry i on labour or capital, respectively, to expenditure on both labour and capital ($\alpha_i^l + \alpha_i^k = 1, i \in I$)

$$\tilde{x}_i^k = \tilde{q}_i^l + \alpha_i^l \tilde{p}^l - \alpha_i^l \tilde{p}_i^k \quad i \in I \tag{29.4}$$

where

x_1^k = the use of capital by industry i
α_i^l, α_i^k = ratio of production expenditure by industry i on labour or capital, respectively, to expenditure on both labour and capital ($\alpha_i^l + \alpha_i^k = 1, i \in I$)

$$\tilde{x}_{c\kappa i}^l = \tilde{q}_i^l - \alpha_{c\lambda i}^i \tilde{p}_{c\kappa}^l + \alpha_{c\lambda i} \tilde{p}_{c\lambda}^i \quad \begin{array}{l} \kappa \in \{\text{domestic}\}, \quad c \in C \\ \lambda \in \{\text{imported}\}, \quad i \in I \end{array} \tag{29.5}$$

where

x_{csi}^i = use of intermediate commodity c from source s in industry i
p_{cs}^i = price of intermediate commodity c from source s
α_{csi}^i = ratio of expenditure by industry i on commodity c from source s to expenditure by the same industry on commodity c from all sources

$$\left(\sum_{s \in S} \alpha_{csi}^i = 1, c \in C, i \in I \right)$$

$$\tilde{x}_{c\lambda i}^i = \tilde{q}_i^i - \alpha_{c\kappa i}^i \tilde{p}_{c\lambda}^i + \alpha_{c\kappa i} \tilde{p}_{c\kappa}^i \quad \begin{array}{l} \kappa \in \{\text{domestic}\}, \quad c \in C \\ \lambda \in \{\text{imported}\}, \quad i \in I \end{array} \tag{29.6}$$

The percentage-change input demand Equations 29.3–29.6 derive from cost minimization subject to a technology or production constraint (Equation 29.7).

For each $i \in I$, choose x^i_{csi}, x^l_i, and x^k_i to minimize

$$\sum_{c \in C} \sum_{s \in S} [p^i_{cs} x^i_{csi} + p^l x^l_i + p^k_i x^k_i] \tag{33.7}$$

subject to

$$q^l_i = \min \{f^A_i / a^f_i ; x^{iA}_{ci} / a_{ci} | c \in C\}$$

$$f^A_i = x^{|\alpha^l_i|}_i x^{k\alpha^k_i}_i$$

= a quantity aggregate variable for factors employed by industry i

$$x^{iA}_{ci} = \prod_{s \in S} x^i_{csi} \alpha^i_{csi} \quad c \in C$$

= a quantity aggregate variable for use of intermediate commodity c in industry i

where

f^A_i = the total use of factors by industry i

x^{iA}_{ci} = the use of intermediate commodity c from all sources in industry i

a^f_i = ratio of factor inputs to the activity level in industry i

a_{ci} = ratio of the use of intermediate commodity c (domestic and imported) to the activity level in industry i

As in Dixon et al. (1982, p. 22), a nested Leontief and Cobb–Douglas function describes the production technology. Commodities, capital, and labour combine to produce commodities. The nested Leontief–Cobb–Douglas function permits substitution in production between the domestic and imported version of each good but not among types of commodities. Substitution also occurs between labour and capital.

The elasticities of substitution in production have the value of one since the functions are Cobb–Douglas in their second levels. The share parameters, α^l_i, α^k_i, and α^i_{csi} are all calculated from the base period data.

29.2 THE GAMS DATA PROGRAM

CGE models require a consistent base period data set. Consistency means:

(1) Demands equal supplies for all commodities.
(2) Nonpositive profits are made in all industries.
(3) All domestic agents (including the government) have demands that satisfy their budget constraints.
(4) The economy is in zero external sector balance (Mansur and Whalley, 1984, p. 91).

Since the US national income accounts assume equilibrium only in the aggregate, not in individual markets, a large amount of data tranformation is needed to create a microconsistent data set. The GAMS data program accomplishes this task.

The program requires data collection for gross output by commodity, GNP as a total, GNP disaggregated by both commodity and use, and for intermediate flows (input–output table). Value added data need not be recorded since it is calculated as a residual. Labour share data must be found, however, to split value added between labour and capital. The data is transformed in five general steps:

(1) Scale disaggregated GNP data to make it consistent with aggregate GNP.
(2) Calculate value added as a residual and then aggregate value added data.
(3) Find the returns to labour and capital.
(4) Adjust the input–output matrix to ensure consistency with gross output, final sales, and value added data – use the iterative procedure introduced by Richard A. Stone (1962) and formalized by Michael Bacharach (1965).
(5) Aggregate the input–output and final sales data.
(6) Transform and rearrange the data to prepare it for use in the CGE model.

A listing of the GAMS program which performs steps (1)–(6) may be found in Collias (1984 and 1985, pp. 194–245). In the tax application which follows, the GAMS data program aggregated 184 input–output sectors into 9.

29.3 AN APPLICATION OF THE UNITED STATES MODEL

In this section, the INTERSAGE model analyses the impact on the United States economy of acceleration of depreciation for tax purposes and of increased investment tax credits. The neutrality of these tax policies as investment incentives has been an area of active debate. The analysis in Colias (1985) investigates a related, but different, aspect of these tax policies; namely, their distributional impact on key economic variables such as output, employment, capital use, relative returns to labour and capital, exports, and imports. Three of these – output, employment, and exports – are discussed below. The sectoral analysis of key economic variables adds to Fullerton and Henderson's (1983) welfare analysis of the Accelerated Cost Recovery System.

29.3.1 Expected theoretical results

Two major theoretical results of the Harberger (1962) type of analysis are the output effect and the factor substitution effect (Mieszkowski, 1967, pp. 252–3). Both of these effects operate also in the INTERSAGE model. In addition, the presence of international trade in INTERSAGE incorporates trade effects. Each of these effects will be discussed.

The output effect impacts the relative returns of labour and capital. For example, in both the Harberger and INTERSAGE models, an increase in the effective capital income tax rate on X raises the price of X relative to that of Y, causing households to demand more Y and less X. Hence, the output of Y rises relative to X. If the taxed industry is the labour (capital) intensive one, then more labour (capital) is released than can be absorbed (at given prices) by the untaxed industry. Thus, the return to capital (labour) rises relative to the return to labour (capital) (Boadway, 1979, p. 309).

The factor substitution effect also has an impact on the relative returns to labour and capital. The factor substitution effect is the substitution of labour for capital in the taxed industry when a capital income tax is imposed. The capital intensity, as measured by capital/labour, declines in the taxed and rises in the untaxed industry. The changes in the capital intensities cause the return to capital to fall relative to the return to labour (Boadway, 1979, p. 309).

If the taxed industry is capital intensive, then the impacts of the output and factor substitution effects on the relative returns to capital and labour reinforce each other. However, if the taxed industry is labour intensive, then the impacts of the two effects on the relative factor returns oppose each other, and the resulting change in the return to capital over the return to labour is an empirical question (Boadway, 1979, 308–10).

With international trade, an increase in the price of the good from the taxed industry diminishes export demand, and hence, reduces the output of the taxed good. The increased price of the taxed good causes domestic industries to substitute imported for domestic goods. This also reduces the output of the taxed good.

These arguments lead to the conclusion that the presence in INTERSAGE of international trade reinforces the output effect.

29.3.2 Marginal and average effective tax rates

According to Fullerton (1984, p. 25), two kinds of corporate effective tax rates may be defined. The average effective corporate tax rate is 'observed corporate taxes divided by "correctly measured" corporate income'. The marginal effective corporate tax rate is the 'the expected real pre-tax rate of return on a marginal investment, minus the real after-tax return to the corporation'. The analyses with the INTERSAGE model employ average rates.

Marginal and average effective corporate rates usually are not equal. Fullerton and Henderson (1983, p. 23) calculated average and marginal effective rates for eighteen industries, for different time periods. They found a correlation coefficient between these two vectors of between zero and 0.3.

Fullerton (1984, p. 30) correctly states that average effective rates measure 'cash flows and distributional burdens' while marginal effective rates reveal 'incentives to use new capital'. Thus, the use of percentage changes in average effective rates to simulate the impact of ERTA (1981) produces simulation results caused by distributional changes, not by investment.

If investment did result from ERTA (1981), it had a distributional impact by increasing the return to capital owners. However, an increased return to capital owners need not signify a successful use of accelerated depreciation and investment credits in stimulating investment. For example, firms may have sought an immediate reduction of their income taxes by building structures and leasing them out, rather than investing in goods-producing capital.

Since the overall impact of capital income tax policy on sectoral output and similar economic variables depends upon distributional changes, it is average, not marginal rates, that are used in the analysis with INTERSAGE.

29.3.3 Analysis of the Economic Recovery Tax Act (ERTA) of 1981

(a) Some preliminaries

Since percentage changes in average effective tax rates also incorporate the effects of past revisions of the tax code, in order to isolate the impact of ERTA (1981), the *Tax Notes* (Horst, 1982, p. 348) method of calculating the average effective corporate tax rate is modified to find the percentage change in the effective rate which would have resulted if all determinants of the effective rate, other than increased depreciation and investment credits, did not change. The procedure for calculating percentage changes in these effective rates appears in Colias (1985, pp. 464–7).

Calculations are made for nine sectors. Two of the nine, high technology and smokestack, were chosen because of their prominence in political debate. The remaining seven sectors were chosen to include other leading sectors of the economy. The resulting percentage changes in effective tax rates are reported in Table 29.2. The nine sectors are ordered from most to least benefited by ERTA (1981). Corporate tax data for the calculations is derived from returns of active corporations in the Internal Revenue Service's (1983 and 1984) Statistics of Income for 1980 and 1981. The returns for 58 sectors of the economy were aggregated into nine sectors.

Two of the nine sectors, high technology and smokestack, come from Hulten and Robertson (1984, p. 332), who define the high technology sectors based on criteria

Table 29.2 Percentage changes in average effective corporate income tax rates due to accelerated depreciation and investment credit provisions of ERTA (1981)

Industry	Tax rate
Transportation	−217.4
Communications	−52.8
Smokestack	−41.0
Services	−25.1
Farming and products	−23.1
Construction	−22.3
Extractive and products	−18.6
High technology	−17.2
Household products	−5.6

suggested by the Tomaskovic-Devey and Miller (1983, p. 58) study. High technology includes:

(1) chemicals (SIC 28)
(2) nonelectrical machinery (SIC 35)
(3) electrical machinery (SIC 36)
(4) transportation equipment, except motor vehicles (SIC 37, except 371)
(5) instruments (SIC 38).

Smoke stack industries include:

(1) textiles (SIC 22)
(2) paper (SIC 26)
(3) petroleum refining (SIC 29)
(4) primary metals (SIC 33)
(5) fabricated metal products (SIC 34)
(6) motor vehicles (SIC 371).

The remaining seven sectors consist of other leading sectors of the economy – see Colias (1985, p. 287).

(b) Results for the accelerated depreciation and investment tax credit provisions of ERTA (1981)

The accelerated depreciation and investment tax credit provisions of the Economic Recovery Tax Act of 1981, all other things being equal, would have caused the percentage changes in effective rates of tax on capital income reported in Table 29.2.

In order to analyse what changes occurred in the US economy as a result of the accelerated depreciation and investment tax credit provisions of ERTA (1981) which would not have occurred under a neutral tax – where a neutral tax refers to an equal percentage reduction of all average effective capital income tax rates – a nine sector version of a United States general equilibrium model simulates the replacement of a neutral lump-sum tax by the tax changes reported in Table 29.2. To isolate only the changes due to the use of ERTA (1981) accelerated depreciation allowances and investment tax credits, the ERTA (1981) provisions and lump-sum tax are assumed to generate equal amounts of government revenue.

The results to be discussed are (1) production, (2) employment, and (3) exports. A more complete analysis of the results is found in Colias (1985, pp. 295–310).

(i) PRODUCTION

The percentage changes in outputs provides an overall indication of the impact of the non-neutral tax strategy (Table 29.3). Transportation outscored all other sectors with a positive growth in output of 2.63%. The farming, communications, and smokestack sectors grew relatively modestly, registering growth rates of

Table 29.3 Percentage changes in output due to accelerated depreciation and investment credit provisions of ERTA (1981)

Industry	Output
Transportation	2.63
Farming and products	0.42
Communications	0.32
Smokestack	0.14
Construction	−0.01
Extractive and products	−0.08
Services	−0.14
Household products	−0.44
High technology	−0.78

0.42%, 0.32%, and 0.14%, respectively. The construction, extraction, and services sectors all declined slightly with negative growth percentages of 0.01%, 0.08%, and 0.14%, respectively. The household industry declined significantly by 0.44%. High technology industries fared worst of all with a 0.78% fall in output.

The impact on output reflects the relative benefits from the incentive policies of accelerated depreciation and increased investment tax credits. Those sectors which benefited most from these two aspects of ERTA (1981) experienced a decrease in average effective tax rate on capital income relative to other sectors. A relative decrease in average effective tax rate implies an output effect according to the following mechanism. The relative decrease in tax rate causes a decline in the price of commodities of the benefited sector relative to other goods: households purchase more of the commodities relative to others.

However, growth in output depends not only on changes in household demand but also on shifts in exports and purchases of domestic intermediates. Exports increase more for those commodities with prices declining relative to prices of other goods. The relative decrease in prices depends upon both relative tax benefits and changing supply conditions. Purchases of domestic intermediates follow output changes modified by substitution between domestic and imported commodities. All of these causal forces operate simultaneously to determine the overall percentage changes in sectoral outputs.

The output of high technology industries declined relative to all others not because it received a lesser benefit from accelerated depreciation and the investment tax credit – since the household industry received an even smaller tax benefit (Table 29.2 – but because of the low tax benefit combined with (1) a relative large drop in exports and (2) a relatively great decrease in imports of high technology intermediates.

The high technology sector experienced the largest percentage decrease of exports, 1.71%, and largest increase of imports, ranging from a 0.84% increase of high technology imports by the high technology sector to a 4.24% increase by the transportation industries.

Table 29.4 Percentage changes in employment due to accelerated depreciation and investment credit provisions of ERTA (1981)

Industry	Employment
Extractive and products	1.79
Household products	1.20
Farming and products	0.88
High technology	0.81
Smokestack	0.44
Services	0.36
Construction	0.10
Communications	−0.38
Transportation	−8.86

(ii) EMPLOYMENT

The results for employment appear in Table 29.4. If labour were mobile and the labour market competitive, then most industries would have hired more people under ERTA (1981) than under a neutral tax plan. Only communications and transportation employment would have decreased. These wage earners would have been retrained to be hired in other industries. Since the labour market does not actually clear in the short run, the results may be interpreted as long-run outcomes. Alternatively, the relative percentage changes in employment may be understood as pressures in the labour market which were felt during 1981.

The results for employment reflect the combined factor substitution and output effects which have opposing impacts on employment. For example, consider the extractive industry. It received less benefit from ERTA (1981) than did all others except high technology and household (Table 29.2). The implied increase of capital price in extractive relative to capital prices in most other industries caused extractive product prices to rise relative to other product prices. Consumers decreased their purchases of extractive commodities and increased their purchases of most others. As extractive industries declined, they fired labour. Hence, output effect decreased employment. The increase in the ratio of capital to labour prices, 6.6% (higher than for any industry except household), caused a substitution of labour for capital. Thus, the factor substitution effect opposed the output effect.

The combined output and substitution effects of the accelerated depreciation and investment credit provisions of ERTA (1981) resulted in greater increases in employment for the extractive than for any other industry.

(iii) EXPORTS

The use of the non-neutral tax plan, caused a 7.63% growth (Table 29.5) in exports of transportation (airline services is an example). All the other exports declined. High technology suffered the largest drop in exports, 1.71%.

Table 29.5 Percentage changes in exports due to accelerated depreciations and investment credit provisions of ERTA (1981)

Commodity	Quantity
Transportation	7.63
Farming and products	−0.31
Communications	−0.31
Smokestack	−0.48
Extractive and products	−0.86
Construction	−0.95
Services	−1.08
Household products	−1.31
High technology	−1.71

(c) Conclusions

By changing the sign of the INTERSAGE results, some policy conclusions would result. A tax policy incorporating the reverse of the depreciation and investment credit provisions of ERTA (1981) would cause the high technology, household, service, extraction, and construction industries all to increase in output. The smokestack sector, communications, and farming would decline moderately, and transportation would suffer the greatest reduction in output. The relatively strong performance of high technology under such a tax plan would result from a combination of (1) a relatively large tax benefit and (2) a large percentage increase in exports and decrease in imports.

If labour were mobile, the transportation and communications sectors would increase their work forces. Employment would decrease in all other sectors. The greatest decrease would occur in the extractive industries because of a strong factor substitution effect. Employment would decrease more in the high technology than in the smokestack sector, and more in the smokestack industries than in services.

29.4 EXTENSIONS

The application of the Johansen-type INTERSAGE model to general equilibrium analysis of United States tax policy marks only the beginning of needed model development. The credibility of results would improve with (a) six extensions stated below and (b) the addition of dynamic and monetary phenomena to the analysis.

The United States INTERSAGE model should be extended to incorporate:

(1) the multistep solution procedure of Dixon *et al.* (1982, pp. 51–60, 199–251) to eliminate linearization errors

(2) reliable estimates of elasticities of substitution in both demand and production – see Shoven and Whalley (1984, pp. 1030–1, 1042–43)
(3) better system techniques for simultaneous estimation of equations – see Jorgenson (1984)
(4) a sectoral accelerator theory of investment – see Colias (1985, pp. 107–20)
(5) multiple consumers – see Colias (1985, pp. 58–83)
(6) more recent input–output data.

Researchers have already been incorporating both dynamic and monetary elements into applied general equilibrium analysis. Shoven and Whalley (1984, pp. 1022–33) survey several models possessing dynamic features. Ballard et al. (1985) and Bovenburg and Keller (1983) both incorporate time by solving for a sequence of single-period equilibria (Shoven and Whalley, 1984, pp. 1029–30). Auerbach and Kotlikoff (1983, p. 465) have incorporated intertemporal utility functions. Savings – which may be interpreted as future consumption – has been modelled by all of the above in their multiperiod framework. Savings appears in single-period models by Piggott and Whalley (Fullerton et al., 1984, p. 394), Serra-Puche (1984), and Fullerton and Gordon (1983). Keller (1980) and Ballentine and Thirsk (Fullerton et al., 1984, p. 372) state their models in percentage-change form. Dixon et al., (1982, p. 118) interface the ORANI general equilibrium model with a dynamic macro model. Colias (1985, pp. 58–83) suggests a method for endogenizing investment in the Johansen-type model.

29.5 CONCLUSION

A complete Johansen-style computable general equilibrium model has been formulated from data input to model solution, all in the GAMS programming language. The ease with which modifications can be made in aggregation and model specification, made possible by combining the Johansen approach with GAMS, renders the task of building complex CGE models more tractable.

A nine sector analysis of the accelerated depreciation and investment tax credit provisions of the US Economic Recovery Tax Act of 1981 replacing a lump-sum tax of equal yield reveals ERTA (1981)'s impact on prices, outputs, and demands which would not have been observed under a lump-sum tax (without price distortions) of equal yield. Although high technology is more capital intensive than certain of the other sectors, its output declines relative to all of them, a result due to the increased imports and decreased exports of high technology.

These analyses add to the extensive efforts to apply general equilibrium models to the US economy by Ballard et al., (1985), Jorgenson (1984), Slemrod (1983), Shoven and Whalley (1972), and others. Furthermore, they provide a basis for future applications of more complicated Johansen models to the US economy.

ACKNOWLEDGEMENTS

The IC^2 Institute of the University of Texas at Austin, Texas in the USA funded the research reported in this paper. I thank David A. Kendrick who provided valuable guidance and assistance.

I also thank George Kozmetsky, Randolph M. Lyon, W. W. Rostow and Daniel T. Slesnick.

REFERENCES

Auerbach, Alan J. and Kotlikoff, Laurence J. (1983) National savings, economic welfare, and the structure of taxation, in *Behavioral Simulation Methods in Tax Policy Analysis* (ed. Martin Feldstein), University of Chicago Press, Chicago.

Bacharach, Michael (1965) Estimating non-negative matrices from marginal data. *International Economic Review*, 6(3), 294–310.

Ballard, Charles L., Fullerton, Don, Shoven, John B. and Whalley, John (1985) *A General Equilibrium Model for Tax Policy Evaluation*, University of Chicago Press, Chicago.

Black, Stephen L. and Taylor, Lance (1974) Practical general equilibrium estimation of resource pulls under trade liberalization. *Journal of International Economics*, 4, No. 1, April, 37–58.

Boadway, Robin W. (1979) *Public Sector Economics*, Winthrop Publishers, Cambridge.

Bovenburg, Lans and Keller, Wouter J. (1983) Dynamics in applied general equilibrium models. Netherlands Central Bureau of Statistics, Voorburg, Internal Rep., May 1983d.

Colias John V. (1984) Aggregation of input–output, value added, and final sales data: a GAMS data program explained with matrix algebra, working paper 84–08–3, The IC^2 Institute, The University of Texas at Austin, Austin, Texas.

Colias, John V. (1985) *A Sectoral Applied General Equilibrium Model of the United States Economy*, University Microfilms International, Ann Arbor.

Dixon, Peter B., Parmenter, B. R., Sutton, John and Vincent, D. P. (1982) *ORANI: A Multisectoral Model of the Australian Economy*, Contributions to Economic Analysis, No. 142. North-Holland, Amsterdam.

Fullerton, Don (1984) which effective tax rate? *National Tax Journal*, 37, No. 1, March, 23–41.

Fullerton, Don and Gordon, Roger H. (1983) A reexamination of tax distortions in general equilibrium models, in *Behavioral Simulation Methods in Tax Policy Analysis* (ed. Martin Feldstein), University of Chicago Press, Chicago.

Fullerton, Don and Henderson, Yolanda K. (1983) Long run effects of the accelerated cost recovery system, working paper 828, National Bureau of Economic Research. Cambridge, Massachusetts.

Fullerton, Don, Henderson, Yolanda, K. and Shoven, John B. (1984) A comparison of methodologies in empirical general equilibrium models of taxation, in *Applied General Equilibrium Analysis* (eds Herbert E. Scarf and John B. Shoven), Cambridge University Press, Cambridge.

Harberger, Arnold C. (1982) The incidence of the corporation income tax. *Journal of Political Economy*, 70, No. 3, June, 215–40.

Horst, Thomas (1982) Some issues in the calculation and uses of effective corporate income tax rates. *Tax Notes*, 17, No. 5, November, 347–51.

Hulten, Charles R. and Robertson, James W. (1984) The taxation of high technology industries. *National Tax Journal*, 37, No. 3, September, 327–45.

Internal Revenue Service (1983) Statistics of income – 1980 corporation income tax returns; US Government Printing Office, Washington, DC.

Internal Revenue Service (1984) Statistics of income – 1981 corporation income tax returns, US Government Printing Office, Washington, DC.

Johansen, Leif (1960) *A Multi-Sectoral Study of Economic Growth*, North-Holland Publishing Company, Austerdam.

Jorgenson, Dale W. (1984) Econometric methods for applied general equilibrium analysis, in *Applied General Equilibrium Analysis* (eds Herbert E. Scarf and John B. Shoven), Cambridge University Press, Cambridge.

Keller, Wouter J. (1980) *Tax Incidence: A General Equilibrium Approach*. Contributions to Economic Analysis, No. 134, North-Holland, Amsterdam.

Kendrick, David A. (1984) Style in multisectoral modelling, in *Applied Decision Analysis and Economic Behaviour* (ed. A. J. Hughes Hallett), Martinus Nijhoff, Dordrecht.

Kendrick, David A. and Meeraus, Alexander (1985) *GAMS: An Introduction*, Development Research Department, The World Bank, Washington, DC.

Mansur, Ahsan and Whalley, John (1984) Numerical specification of applied general equilibrium models: estimation, calibration, and data, in *Applied General Equilibrium Analysis* (eds Herbert E. Scarf and John B. Shoven). Cambridge University Press, Cambridge.

Mieszkowski, Peter M. (1967) On the theory of tax incidence. *Journal of Political Economy*, 75, No. 3, June, 250–62.

Serra-Puche, Jaime (1984) A general equilibrium model for the Mexican economy, in *Applied General Equilibrium Analysis*, (eds Herbert E. Scarf and John B. Shoven), Cambridge University Press, Cambridge.

Shoven, John B. and Whalley, John (1972) A general equilibrium calculation of the effects of differential taxation of income from capital in the US *Journal of Public Economic*, 1, No. 3/4, November, 281–321.

Shoven, John B. and Whalley, John (1984) Applied general equilibrium models of taxation and international trade: an introduction and survey. *Journal of Economic Literature*, 22, No. 3, September, 1007–51.

Slemrod, Joel (1983) A general equilibrium model of taxation with endogenous financial behavior, in *Behavioral Simulation Methods in Tax Policy Analysis* (ed. Martin Feldstein), University of Chicago Press, Chicago.

Staelin, Charles P. (1976) A general-equilibrium model of tariffs in a noncompetitive economy. *Journal of International Economics*, 6, No. 2, February, 39–63.

Stone, R. and Brown, J. A. C. (1962) A computable model of economic growth (A programme for growth 1) Chapman and Hall, London.

Tomaskovic-Devey, Donald and Miller, S. M. (1983) Can high-tech provide the jobs? *Challenge*, 26, No. 2, May–June, 57–63.

30

A quarterly econometric model for the Spanish economy

IGNACIO MAULEÓN

This chapter presents a compact and aggregative model of the Spanish economy developed under two constraints: (a) data availability, and (b) search for the appropriate dynamic specification. The sampling period spans 1974 to 1983, and all equations have been tested against several possible misspecifications. This calls for a small model that aims at capturing the salient features of the Spanish economy in a stylized way. The model describes mainly the monetary side of the economy because of the first constraint.

The financial side analyses the money and credit markets. Short and long-term rates of interest are determined in the credit market and the crowding-out effect of the public deficit fits the model naturally. The long-term rate is explicitly modelled and the usual reduced form type of approach embodied in the term-structure equation is dropped. Imperfect substitution among foreign and domestic assets is assumed. The money supply reaction function is explicitly taken into account. The rational expectations hypothesis does not play an important role in the Spanish economy and real demand shocks have a lasting effect on output. However, money increases rapidly lead to higher prices. The labour market includes the conventional demand and supply of labour functions, and a wage equation of the phillips type. The cost of the working (borrowed) capital is considered so that the interest rate enters the demand for labour equation. Finally, the exchange rate is explained by a reaction function that is meant to describe a managed floating exchange rate system, where the Bank of Spain's foreign currency reserves have an important role in the short run, but PPP holds in the long run.

30.1 INTRODUCTION

The model of the Spanish economy presented here is based on quarterly data estimated over the period 1974–1983. The main body of the model was first presented in Mauleón and Pérez (1984), and this chapter reports on the empirical results for the whole model, with an extension to the labour market. The model basically develops the monetary side of the economy. This simply reflects the lack of reliable quarterly data for most real series. This model is the first attempt that has brought together several quarterly econometric equations in a medium-sized model. Earlier attempts at the Bank of Spain in the mid-seventies include work by Sánchez (1978), Pérez (1975, 1977), Rodriguez (1978), Albarracin (1978), and Bonilla (1978), and the model of this chapter tries to bridge the gap with them.

From the econometric point of view, the model has been developed under two binding constraints: (a) data availability, (b) careful search for appropriate specification. The data constraint implies that the model is basically a monetary one. Special attention has been paid to the specification of the equations in several respects: dynamic specification, variables included, residual autocorrelation, simultaneity, normality of residuals and absence of outliers, stability and homoscedasticity, are the main points under study in every equation of the system. The model is made up of fifteen behavioural equations grouped into four sectors (foreign, money and credit markets, goods and labour market), and eight identities. The effective exchange rate is modelled as a reaction function that drives it around long-run PPP while allowing for short-term deviations in response to changes in the level of required reserves. The main feature of the money and credit markets is the modelling of interest rates in the credit market. Imperfect substitution among assets is assumed and the financial crowding-out effect of the public deficit fits the model naturally. The supply of money is explained by means of a reaction function that aims at capturing the behaviour of the Bank of Spain. Short-run deviations of the annual money growth target are allowed to fit swings of the exchange rate and the rate of inflation. The goods and labour market are standard, except for the introduction of money in the price equation, and the interest rate in the demand for labour. A more detailed explanation of these points is offered later in the chapter.

Although the model is of a fairly tentative kind, due partly to the lack of reliable data for some series, it may be the starting point of more ambitious projects. In particular it can provide a framework for future work in some statistical series, and it can also help organize future empirical research.

30.2 AN OVERVIEW OF THE MODEL

In this section the basic analytic structure of the model is introduced, leaving the detailed empirical specification to the next paragraph. The model is a medium-sized one, and as can be seen, it is mainly the monetary block that has been specified with care. This reflects partly the lack of reliable quarterly statistics of the real sector. The salient features of the model have already been pointed out in the introduction, and it is made up of the following set of equations:

An overview of the model

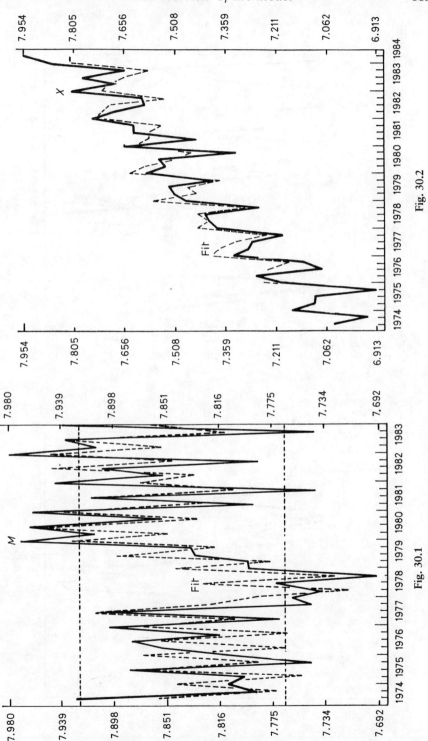

Fig. 30.1

Fig. 30.2

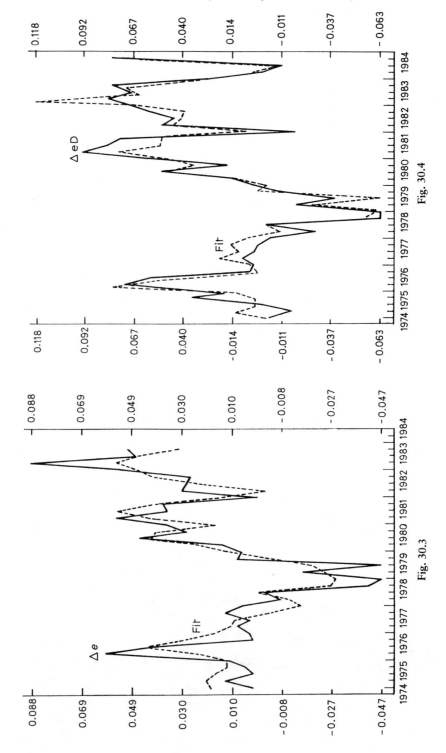

Fig. 30.3

Fig. 30.4

An overview of the model

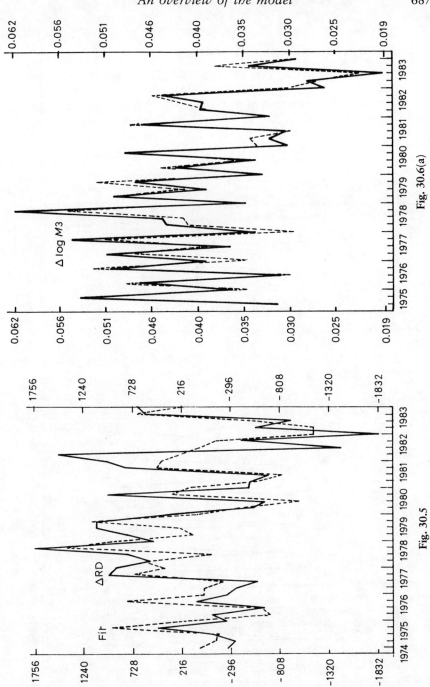

Fig. 30.5

Fig. 30.6(a)

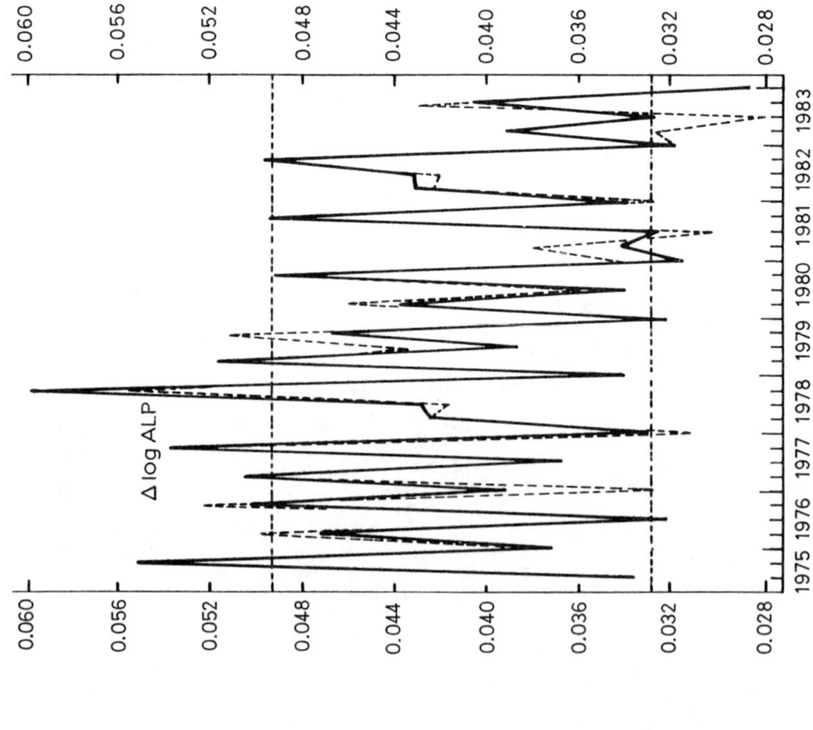

688   *Econometric model for the Spanish economy*

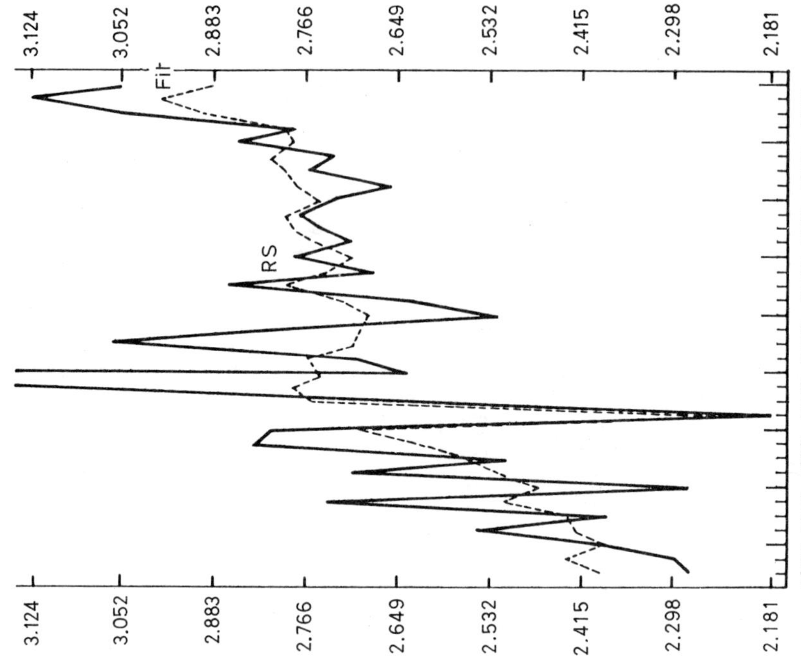

Fig. 30.7

Fig. 30.6(b)

An overview of the model

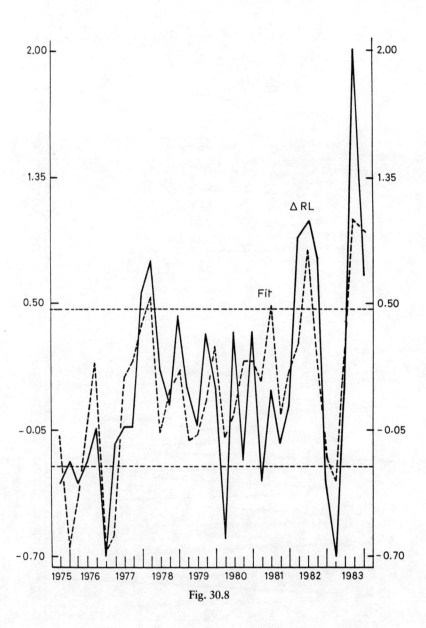

Fig. 30.8

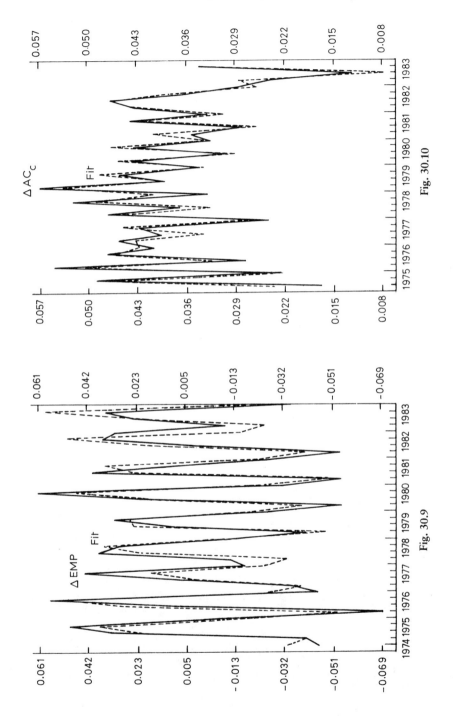

Fig. 30.10

Fig. 30.9

An overview of the model

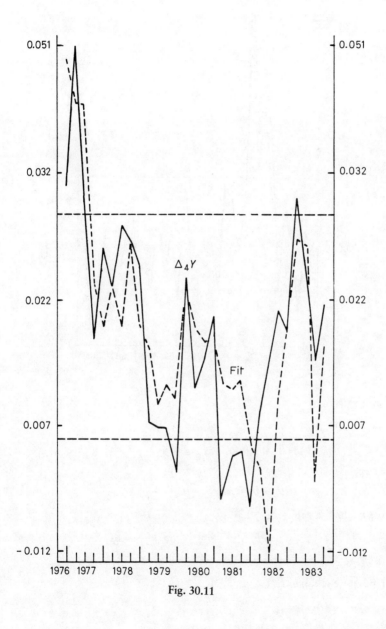

Fig. 30.11

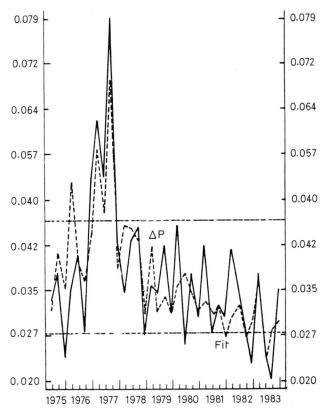

Fig. 30.12

(a) Foreign sector

Imports (Fig. 30.1):

$$M = a_0 + a_1(\text{FD}) + a_2[(e + p_w - p)^e - (e + p_w - p)_{-1}] \quad (30.1)$$

Exports (Fig. 30.2)

$$X = b_0 + b_1 \text{IM} + b_2(e + p_w - p) \quad (30.2)$$

Effective exchange rate (Fig. 30.3):

$$\Delta e = c_0 + c_1 \Delta(p - p_w) - c_2(e + p_w - p)$$
$$- c_3 \Delta \text{RD} + c_4(R_w - \text{RS}) \quad (30.3)$$

Dollar/peseta exchange rate (Fig. 30.4):

$$\Delta e\text{D} = d_0 + d_1 \Delta e \quad (30.4)$$

Bank of Spain's dollar reserves (Fig. 30.5)

$$\Delta \text{RD} = f_0 + f_1 \Delta e \quad (30.5)$$

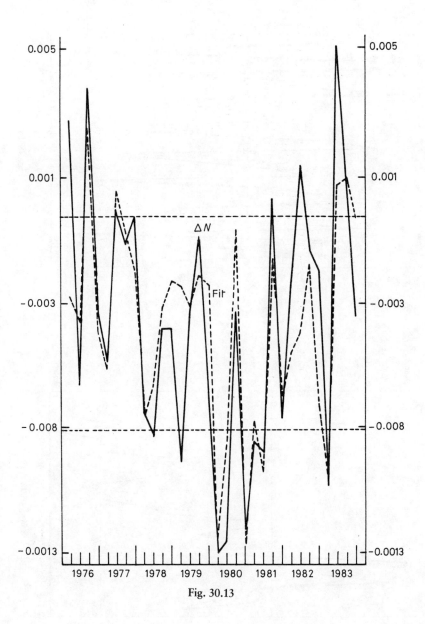
Fig. 30.13

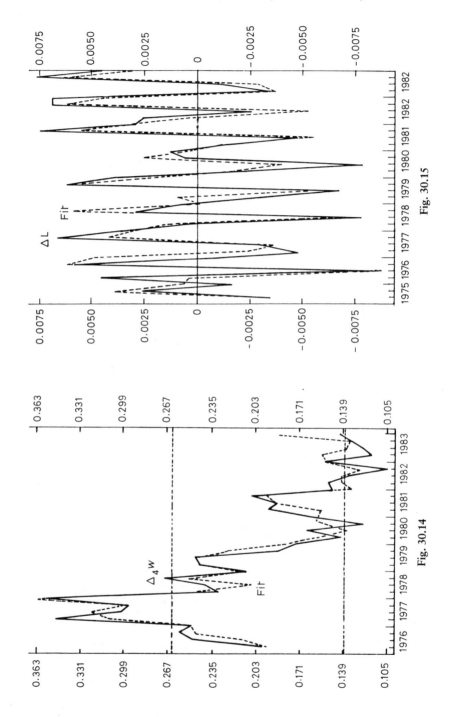

Fig. 30.15

Fig. 30.14

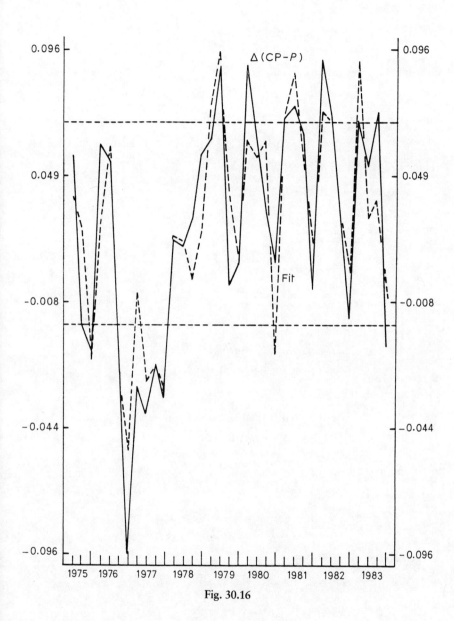

Fig. 30.16

(b) Money and credit markets

Supply of money (Fig. 30.6):

$$\Delta M3 = g_0 + g_1 \Delta AC_c + g_2 \Delta^2 RS \tag{30.6}$$
$$\Delta ALP = g'_0 + g'_1 \Delta AC_c + g'_2 \Delta^2 RS \tag{30.6'}$$

Short-term interest rate (Fig. 30.7):

$$RS = h_0 + h_1(CP - CT) + h_2(P + Y - ALP) \tag{30.7}$$

Long-term interest rate (Fig. 30.8):

$$\Delta RL = j_0 + j_1 \Delta(CP - CT) + j_2 \Delta(P + Y - ALP) \tag{30.8}$$

Demand for currency (Fig. 30.9):

$$\Delta(EMP - P) = k_0 + k_1 \Delta Y - k_2 \Delta P \tag{30.9}$$

Supply of adjusted bank reserves (Fig. 30.10):

$$\Delta AC_c = l_0 + l_1 \Delta AC_c^* \\ - l_2 \Delta e - l_3 \Delta P \tag{30.10}$$

(c) Goods market

Demand for goods (Fig. 30.11):

$$\Delta_4 Y = n_0 + n_1 \Delta_4 X - n_2 \Delta RL \\ + n_3 \Delta(e + p_w - p) + n_4 \Delta_4 (G - P) \tag{30.11}$$

Price equation (Fig. 30.12):

$$\Delta P = q_0 + q_1 \Delta M3 + q_2 \Delta(e + p_w) \\ + q_3 \Delta Y + q_4 \Delta(w + N - Y) \tag{30.12}$$

(d) Labour market

Demand for labour (Fig. 30.13):

$$\Delta N = r_0 - r_1 \Delta(w - p) - r_2 \Delta(e + p_w - p) \\ - r_3 \Delta RL \tag{30.13}$$

Wage equation (Fig. 30.14):

$$\Delta_4 w = s_0 + s_1 \Delta M3 + s_2 \Delta Y \\ + s_3 \Delta(e + p_w) + s_4 \Delta_4 U \\ + s_5 \Delta(Y - N) \tag{30.14}$$

Supply of labour (Fig. 30.15):

$$\Delta L = r_0 - r_1 \Delta Y - r_2 \Delta U \tag{30.15}$$

(e) Identities and definitions

(variables with a '+' as r.h. superindex are in levels)
Unemployment rate definition:
$$U = 1 - N^+/L^+ \tag{30.16}$$

Adjusted and unadjusted reserves linkage:
$$\Delta AC_c = \Delta AC - (PC^+/AC^+)\Delta q \tag{30.17}$$
$$PC^+ = M3^+ - EMP^+ \tag{30.18}$$

Monetary base definition:
$$BM^+ = EMP^+ + AC^+ \tag{30.19}$$

Balance sheet of the Bank of Spain:
$$BM^+ = RSP^+ + R^+ \tag{30.20}$$

Total domestic credit to the economy:
$$CT^+ = M3^+ - R^+ + DP^+ \tag{30.21}$$

Government budget constraint:
$$\Delta CP^+ = \Delta RSP^+ + \Delta DP^+ \tag{30.22}$$
$$= DNF + DP_{-1} \cdot RL_{-1} \tag{30.23}$$

(f) Feed back equation (Fig. 30.16)

$$\Delta(CP - p) = m_0 - m_1 \Delta Y + m_2 \Delta RL \tag{30.24}$$

(All variables are measured in natural logarithms except interest rates, the Bank of Spain's dollar reserves and the rate of unemployment, which are in levels. A list with the definition of all variables is provided in Appendix A.)

The first block of equations focuses mainly around Equation 30.3, which is meant to represent a managed floating exchange rate system, where the Bank of Spain's dollar reserves have an impact on the exchange rate through a reaction function. Dollar reserves enter the equation with a first difference as a direct reflection of the Bank of Spain's reserves policy, which is addressed to offset, to a certain extent, sudden swings in the real exchange rate. This enables the authorities to solve, in the short run, potential conflicts between the objectives of monetary and exchange rate policies. Finally, the authorities try to drive the exchange rate around an equilibrium level based on a long-run purchasing power parity.

Equation 30.5 is a reduced form type of equation, required in order to close the model. In a more developed version of the model, it should be replaced by Equations 30.1, 30.2 and the two following identities,

$$\Delta R = \Delta CF + X \cdot p_x - M \cdot p_w \cdot e \tag{30.25}$$
$$R = eD \cdot RD \tag{30.26}$$

Equation 30.25 is the balance of payments identity (simplified), where CF is foreign credit in domestic currency. Capital mobility has been heavily regulated in the past, and that is very likely the reason why estimation of an equation for CF did not yield any reasonable result.

A second block of equations (30.6–30.10) represents the money and credit markets. In this model, the bank credit market plays a key role in the transmission of monetary impulses to the exchange rate, output and prices. This is a consequence of the high level of intermediation in the Spanish financial markets. The intermediate money target AC_c^*, real output Y, and the price level P, enter as explanatory variables to determine short and long-term interest rates (RS, RL), the adjusted bank reserves (AC_c), and money ($M3$) or liquidity balances (ALP). Within the range of the money growth yearly target announced by the authorities at the end of every year, the Bank of Spain pays attention, in the short run, to other variables so as to smooth out large fluctuations in monetary variables, especially interest and exchange rates (Equation 30.10). The role of the policy mix in determining domestic interest rates should be emphasized. By controlling bank reserves, the monetary authorities affect the supply of credit to the nonbanking domestic sector (Equations 30.6 and 30.21). But the credit finally available to firms and households (CT − CP) will be the amount remaining after the absorption of the public sector (CP). Public sector pressure on bank credit markets has adopted different forms in the last decade but its joint effects on interest rates and credit availability to the private sector have been quite similar. This is because the general government finances a very high portion of its accumulated deficit in the monetary and banking system (around 80% in 1983).

The next two sets of equations are a simplified real block. First, the goods market endogenizes output and the price level. Real total domestic demand is made to depend on the long-term interest rate, real general government spending, the real exchange rate and real exports. The price equation can be thought of as a mark-up equation on unit costs, where the size of the mark-up responds to demand pressure. Finally, producers drive prices to the desired level, when the credit conditions are favourable, so that money, as a proxy for credit availability, enters the price equation.

The labour market is made up of the three usual equations: demand for and supply of labour, and the augmented Phillips type of wage equation. The demand for labour depends on the nominal long term rate of interest. This is a consequence of the fact that the Spanish stock market is fairly narrow so that firms have to rest heavily upon bank credit fo finance the working capital needed to operate their businesses. As a result, the level of financial intermediation is very high as has been mentioned previously.

The wage equation includes a target for real wage growth measured by average productivity. The dynamics of the equation are somewhat messy: the variables $M3$, Y, $e + p_w$, are just the long-run determinants of prices and it is in this way that in long-run equilibrium the coefficient of prices in Equation 30.14 is nearly one.

Finally, the supply of labour is of a standard type. Increases in the rate of unemployment are likely to decrease the perceived probability of finding a job, and have, therefore, a negative impact on the number of people who look for one. Increases in income make leisure more attractive, thereby decreasing the supply of labour. In contrast, real wage increases make leisure more costly and should drive up the labour supply. However, as the next section discusses, no such effect has been found in the Spanish economy.

A final block of equations deals with the problem of relating several variables by means of identities and definitions. Equations 30.17 to 30.21 deal with the money and credit block. They are of a fairly standard type except for Equation 30.17, which relates the adjusted and unadjusted bank reserves. This equation is fully detailed in Frost (1977), and Mauleón et al. (1986). Basically, what it does is to summarize all monetary policy measures (supply of bank reserves, AC, and changes in the coefficient of required reserves, q), in a single indicator, adjusted bank reserves, AC_c. This is the only way of measuring the stance of monetary policy, especially when there are many banks subject to different coefficients for several different reasons (for example, some credit institutions undergoing difficulties are exempt from fulfilling the required reserve coefficient).

The government budget constraint (Equation 30.22) is presented in a very simplified way. It is included for the sake of coherency although it does not play any important role in the model, except in simulations of the monetary and fiscal policy mix. The total deficit is precisely the change in the domestic credit demanded by the general government (ΔCP). It can be broken down into the direct appeal to the Bank of Spain (RSP), and the change in the stock of public debt outstanding (DP). By definition, the total deficit must equal the fiscal deficit plus the financial cost of the stock of outstanding public debt (Equation 30.23). The nonfinancial deficit is partly exogenous, but to a great extent it is just a byproduct of the low real growth rates of the economy (for example, social security transfers to unemployed, and early retirements). We should then have a feedback equation of the type

$$\Delta(DNF - p) = \alpha_0 - \alpha_1 \Delta Y \tag{30.27}$$

In the absence of reliable data, Equation 30.27 can be a proxy equation.

30.3 EMPIRICAL RESULTS

This section presents the empirical results obtained in estimating the model presented in the preceding section. The dynamic specification in every equation has been studied carefully, that is, the appropriate lags of each variable have been selected, neglecting the nonsignificant coefficients. Insignificant variables that might in principle enter a given equation have also been dropped. Every equation has been tested against a battery of possible misspecifications, which include at least a stability test, a serial correlation test, and a test of the overall lag specification.

Heteroscedasticity and simultaneity tests have also been included when required. The errors of some equations have also been tested for normality.

Partly because complete data series are not available for the past decade and partly because it seemed sensible to select a period where the basic relationships could be reasonably supposed to be stable, the sampling period considered for estimation has been 1974–1983. The model (Equations 30.1–30.15) has been estimated, then, for the period spanning 1974 to 1983 with quarterly data. Most equations have been estimated by ordinary least squares (OLS) after having tested and rejected the simultaneity hypothesis. In some, an autoregressive structure in the error has been added, when the autocorrelation tests indicated that it was appropriate. Finally, some equations have been estimated jointly by maximum likelihood methods when simultaneity did matter. A more detailed account of the empirical results by equation is given next.

(a) Imports (non energy)

$$M = 4.83 + 0.093\,(D2 + D4) + 0.98\,\text{FD}$$
$$(2.5)\quad (7.2)\qquad\qquad (3.6)$$
$$0.88\,[(e + p_w - p)^e - (e + p_w - p)_{-1}] + \hat{u}$$
$$(3.4)$$
$$\hat{u} = 0.28\hat{u}_{-1} + \hat{\varepsilon}$$
$$(1.7)$$
$$T = 39(74.3 - 83.4),\quad \hat{\sigma} = 0.05,\quad R^2 = 0.72$$
$$C_1(4) = 3.3,\quad C_2(8) = 14,\quad C_3(3) = 1.0 \qquad (30.28)$$

Imports depend on final demand (FD) with near unitary elasticity. They also depend on anticipated relative price changes (i.e. if a depreciation of the domestic currency is expected, that leads to an increase in current imports). Energy imports are not modelled because they are directly under government control.

(b) Exports

$$X = 3.3\,(D2 + D3) + 3.4\,(D1 + D4) + 1.3\,\text{IM}$$
$$(5.6)\qquad\qquad (5.8)\qquad\qquad (20.4)$$
$$0.44\,(e + p_w - p)_{-2}$$
$$(3.5)$$
$$T = 40\,(74.3 - 84.2),\quad \hat{\sigma} = 0.074,\quad R^2 = 0.93$$
$$C_1(6) = 12,\quad C_2(15) = 15,\quad C_3(4) = 3.2 \qquad (30.29)$$

Real exports depend basically on and index of world trade (IM), which accounts for almost the whole of the R^2. Relative prices do not have a big impact and the price elasticity of the equation is lower than those usually reported for other economies.

(c) Effective exchange rate

$$\Delta e = \underset{(3.5)}{-0.63} + \underset{(1.8)}{0.43\Delta(P - P_w)} + \underset{(2.2)}{0.26\Delta e_{-1}}$$
$$\underset{(3.7)}{-0.14(e + p_w - p)_{-4}} - \underset{(2.8)}{0.003(RS - R_w)_{-6}}$$
$$\underset{(2.8)}{-0.000007\Delta(RD + RD_{-1})}$$

$$T = 37(74.4 - 83.4), \hat{\sigma} = 0.016, R^2 = 0.74$$
$$C_1(4) = 11.7, C_3(4) = 2.1, C_4(1) = 0.8, C_5(4) = 3.7$$
$$C_6(2) = 1.5, C_7(1) = 0.0006\,(RD), C_7(1) = 0.0006\,(P) \quad (30.30)$$

This equation is the Bank of Spain's feedback reaction function that describes a managed floating system for the effective exchange rate. In the very long run, purchasing power parity holds, but the exchange rate is subject in the short run to large swings in response to interest rate differentials and changes in the level of foreign reserves. Inflation differentials also have a short-run impact which is far below unity.

The fit as measured by the R^2 is reasonably high when compared with other estimations that also take as dependent variable the rate of change of the exchange rate. The stability test is somewhat too high, but the parameter estimates are rather stable. This is likely to reflect a bigger residual variance in the last part of the sample.

(d) Peseta–dollar exchange rate

$$\Delta eD = \underset{(15.0)}{1.3\Delta e} + \underset{(2.1)}{0.18\Delta e_{-4}}$$

$$T = 40(74.4 - 84.3), \quad \hat{\sigma} = 0.015, \quad R^2 = 0.86$$
$$C_1(4) = 1.0, \quad C_3(4) = 2.3, \quad C_4(1) = 2.1, \quad C_5(4) = 3.8$$
$$C_6(2) = 0.3, \quad C_7(1) = 2.6 \quad (30.31)$$

This is a reduced form type of equation required to give an estimate of the balance of payments in domestic currency (i.e. $R = eD*RD$). Both exchange rates have approximately the same mean, although the 'effective exchange rate' is more stable than the 'dollar peseta rate', as might have been expected. Otherwise, the performance of the equation is fairly satisfactory, according to the criteria given above.

(e) Bank of Spain's dollar reserves:

$$\Delta RD = \underset{(2.7)}{0.36\Delta RD_{-1}} - \underset{(2.4)}{0.3\Delta RD_{-3}} - \underset{(3.1)}{10092\Delta e} + \underset{(5.0)}{969D3}$$

$$T = 37(74.4 - 83.4), \quad \hat{\sigma} = 574, \quad R^2 = 0.55$$
$$C_1(4) = 2.2, \quad C_2(9) = 3.2, \quad C_3(4) = 0.26, \quad C_4(1) = 0.001$$
$$C_5(4) = 3.7, \quad C_6(2) = 0.05, \quad C_7(1) = 0.12 \quad (30.32)$$

This equation is required to close the foreign sector. In a more developed model it should be replaced by the balance of payments identity, and behavioural equations for imports, exports, and capital movements. The equation is a reduced form expression of a very simple kind and that is the likely reason for the low R^2. Nevertheless, all tests are passed easily.

(f) Supply of money

$$\Delta M3 = \underset{(6.3)}{0.008 D1} - \underset{(2.0)}{0.004 D3} + \underset{(3.0)}{0.24 \Delta M_{-1}} + \underset{(10.0)}{0.63 \Delta AC_c}$$
$$+ \underset{(2.6)}{0.12 \Delta AC_{c-4}} + \underset{(2.2)}{0.002 \Delta^2 RS}$$

$$T = 35(75.2 - 83.4), \quad \hat{\sigma} = 0.003, \quad R^2 = 0.93$$
$$C_1(8) = 4.7, \quad C_2(10) = 14, \quad C_3(4) = 8 \text{(IV test)} \tag{30.33}$$

The money supply function is derived from the money multiplier of the adjusted bank reserves. This is the target variable that the Spanish monetary authorities try to control, rather than the monetary base. This choice has been made because the relationship between M3 and bank reserves is more stable than the monetary base multiplier. The use of an adjusted series takes into account changes in the coefficient of required reserves and the number of banks subject to it.

The money supply equation performs very well from a statistical viewpoint except for the autocorrelation test that is a bit high. But this is probably due to temporal aggregation since monthly regressions were very similar, and displayed no autocorrelation. The long run elasticity of money with respect to adjusted bank reserves is unity as expected, and the response of the money supply very fast (Mauleón *et al.* 1986). The short-term interest rate enters the equation with a positive sign as it should, reflecting its negative relationship with the excess of bank reserves. The money supply equation has been estimated jointly with the short-term interest rate equation by a maximum likelihood method. Dealing with simultaneity adequately in this case was essential to estimate correctly the interest rate effect in the supply of money.

Alternatively, an equation explaining a broader monetary aggregate is the following,

$$\Delta ALP = \underset{(6.9)}{0.004} + \underset{(6.3)}{0.0056 SAL} + \underset{(5.6)}{0.42 \Delta ALP_{-1}}$$
$$+ \underset{(8.7)}{0.46 \Delta AC_c} + \underset{(3.0)}{0.12 \Delta AC_{c-4}}$$
$$+ \underset{(2.5)}{0.0034 \Delta^2 IM}$$

$$T = 35(75.2 - 83.4), \quad \hat{\sigma} = 0.003, \quad R^2 = 0.9$$
$$C_1(8) = 13, \quad C_2(8) = 6.4 \tag{30.34}$$

This equation may be a substitute for the M3 equation. This is because there have been recent changes in the financial sector that have led the authorities to target this later variable, rather than M3. The main reason behind the financial innovations that have taken place in Spain has been the need to finance an increasing public deficit with short-term treasury bills.

(g) Interest rate equations

$$RS = 200 + 4.6(CP - P) + 45(P + Y - M)_{-1} - 6.7D21$$
$$(4.0)(4.0)(3.5)\phantom{(P + Y - M)_{-1} - }(4.3)$$
$$T = 35(75.2 - 83.4), \quad \hat{\sigma} = 1.8, \quad R^2 = 0.68$$
$$C_1(8) = 4.7, \quad C_2(8) = 14, \quad C_3(4) = 4.2 \tag{30.35}$$

$$\Delta RL = 0.36\Delta RL_{-1} - 31\Delta(M - P - Y)_{-1} + 7\Delta(CP - P)$$
$$(2.7)\phantom{\Delta RL_{-1} - }(4.7)\phantom{\Delta(M - P - Y)_{-1} + }(4.6)$$
$$+ 0.05\Delta(RS_{-1} + RS_{-2}) + 19\Delta Y_{-1} + 0.27SRL + \hat{u}$$
$$(2.5)\phantom{\Delta(RS_{-1} + RS_{-2}) + }(2.9)\phantom{\Delta Y_{-1} + }(2.6)$$
$$\hat{u} = 0.6\hat{u}_{-3} + \hat{\varepsilon}$$
$$\phantom{\hat{u} = }(3.0)$$
$$T = 34(75.3 - 83.4), \quad \hat{\sigma} = 0.8, \quad R^2 = 0.59$$
$$C_1(4) = 9.0, \quad C_2(14) = 5.6, \quad C_3(3) = 3.4 \tag{30.36}$$

The long-term interest rate is determined in the credit market, being directly affected therefore by the credit demanded by the public sector (CP). The short-term rate enters as an explanatory variable through the optimizing credit behaviour of private banks. Since there are reasons to believe that the market does not adjust quickly enough, the interest rate moves to adjust this disequilibrium and is then measured in first differences (Mauleón and Pérez, 1984). This temporary unadjustment translates into the market for excess reserves and has an impact on short-term rates. This rate is then determined by similar variables (as a matter of fact, the Bank of Spain currently defines an indicator of the pressure in the credit market by the difference between short and long-term rates).

The amount of credit demanded by the private sector is a direct function of the usual variables, (Y, M, P), so that the interest rate goes up with Y, P and down with M (or total credit available). Given a fixed volume of credit available to the economy, the amount available for the private sector is constrained by the credit that goes to the public sector. There are then two types of variables in the model that drive up interest rates, as modelled in Equation 30.2. These two interest rate equations take the place of the usual money demand function in most models. If the demand for money is to take into account the financial effects of the public deficit, it must include wealth as an argument, and the public bonds as one of its components. But in Spain, until 1983, most of the public deficit had been financed with short-term debt held by private banks. Therefore, it competes directly with other assets in the banks' portfolios. Altogether, is seems more natural to model interest rates directly in the credit market, than the demand for money. The interest rate equations in this chapter are somewhat similar to that of Feldstein and Eckstein (1970), already well known in the literature.

The short-term interest rate equation has been modelled in levels and includes the expected explanatory variables. No lagged dependent variable enters the equation reflecting the fact that adjustments in the money market take place fast. There are three outliers that call for a different treatment. Two of them are due to institutional alterations of the market, and since the model is static, the solution from the estimation viewpoint is simply to drop them from the sample. The third has a sizeable impact on excess reserves and thereby on money supply. Then, it cannot be omitted from the sample, so that the econometric solution was just to pick it up with a dummy variable in the interest rate equation.

The estimation results for the long-term interest rate equation are as expected although the fit is somewhat low. The equation is not very stable but this might be due to a possible outlier in 84(3). The main problem in this case was to choose a representative rate. Spanish financial markets are narrow and have been heavily regulated in the past. Finally, the choice had to be made by exclusion (more details are given in Mauleón and Pérez (1984)). The serial correlation test detects negative correlation at all lags although this is not very significant. This might be a result of overdifferencing, but estimation in levels persistently gave a DW near zero and a lagged dependent coefficient close to unity when included. Therefore, it was decided to model the equation in differences.

(h) Demand for currency:

$$\Delta(\text{EMP} - p) = -0.021 D1 + 0.042 D2 + 0.056 D3$$
$$(2.6) \quad (5.9) \quad (7.2)$$
$$- 0.64 \Delta p + 0.6 \Delta(Y + Y_{-2})$$
$$(5.0) \quad (2.4)$$

$T = 37(74.4 - 83.4)$, $\hat{\sigma} = 0.016$, $R^2 = 0.82$
$C_1(4) = 4.2$, $C_2(18) = 15$, $C_3(4) = 1.8$, $C_4(1) = 0.1$
$C_5(4) = 1.7$, $C_6(2) = 2.2$, $C_7(1) = 0.2(P)$ (30.37)

The demand for real currency balances has a strong seasonal component, as might have been guessed from casual observation. The income elasticity is close to unity, and the public seems to be fairly sensitive to the opportunity cost of holding money, as measured by the rate of price inflation. As a result, a given increase in the monetization of the deficit will be considerably more inflationary. This is because the increased rate of inflation, by reducing the demand for currency, leads to a direct increase in bank reserves and therefore in the total supply of liquidity through the usual money multiplier mechanism embodied in Equation 30.6.

(i) Supply of adjusted bank reserves

$$\Delta AC_c = 0.016(D1 + D3) + 0.02A38 - 0.32\Delta AC_{c-4}$$
$$\quad\quad (11.0) \quad\quad\quad (5.3) \quad\quad (5.7)$$

$$0.31\Delta AC_c^* + 0.06\Delta DAC_c^*$$
$$(14.0) \quad\quad (12.5)$$

$$-0.22\Delta p_{-1} - 0.064\Delta(e + e_{-1})$$
$$(3.3) \quad\quad (5.5)$$

$$T = 34(75.2 - 83.3), \quad \hat{\sigma} = 0.004, \quad R^2 = 0.9$$
$$C_1(4) = 3.6, \quad C_3(4) = 5.8, \quad C_4(1) = 0.57, \quad C_5(4) = 1.3$$
$$C_6(2) = 0.8, \quad C_7(1) = 0.02 \tag{30.38}$$

The supply of adjusted bank reserves has a seasonal component (D1, D3), one outlier for observation 38 (A38), and depends on the annual growth target (AC_c^*, DAC_c^*), and on the changes of the effective exchange rate and the rate of inflation.

Since 1977 the Bank of Spain fixes at the beginning of every year an annual target for the growth rate of M3, that is made public. This target is implemented through a daily control of the supply of adjusted bank reserves. In fact, the elasticity of M3 with respect to adjusted bank reserves is unity (Mauleón et al., 1986), and that is why finally AC_c^* can be taken to be the M3 growth target. This yearly growth target is distributed evenly among the four quarters of each year, in order to obtain the final data for AC_c^*. Before 1977, the simple mean of the dependent variable for each year, is taken as explanatory (DAC_c^*). The long-run coefficient of ΔAC_c^* is 0.24 which makes 0.96 in a year, that is, almost unity as one would expect it to be. In the short run, the supply of money is restricted when the domestic currency depreciates. Thus, it is an intervention of a stabilizing kind. The supply of money is also restricted when inflation accelerates. Since the annual target for inflation is not available for the whole of the sample period, the actual rate of inflation has been used as an explanatory variable. However, it is very reasonable to assume that the proper variable should be $\Delta(p - p^*)$, Δp^* being the inflation target.

(j) Demand for goods

$$\Delta_4 Y = 0.73\Delta_4 Y_{-1} + 0.05\Delta_4 X_{-2} - 0.01\Delta RL$$
$$(5.0) \quad\quad\quad (2.0) \quad\quad\quad (2.7)$$

$$+ 0.12\Delta(PM - P)_{-2} + 0.046\Delta_4(G - P)_{-4} + \hat{u}$$
$$(2.7) \quad\quad\quad\quad (2.1)$$

$$\hat{u} = 0.32\hat{u}_{-3} + \hat{\varepsilon}$$
$$(1.6)$$

$$T = 29(76.4 - 83.4), \quad R^2 = 0.6, \quad \hat{\sigma} = 0.01$$
$$C_1(8) = 8, \quad C_2(15) = 22, \quad C_3(3) = 2.8 \tag{30.39}$$

The output equation is a classical IS curve. Imports enter through lagged y's and the real exchange rate measured by PM − P. Then, we are not in a fixed output (or growth of) model, and exogenous demand shocks have a positive and permanent impact on output.

We do not have quarterly data in Spain for most real variables and the GDP is no exception. The data used here have been interpolated at the research department of the Bank of Spain. This is very likely the explanation for the somewhat odd within-year behaviour of the series. Use of the filter $(1-L^4)$ rather than $(1-L)$, smooths this problem since $(1-L^4) = (1-L)(1+L^1+L^2+L^3)$. The GDP equation has been estimated by full information maximum likelihood methods, jointly with the interest rate equation. This was essential to detect the impact of interest rates on output, as simultaneity bias was strong in this case. Since this implied losing four extra observations at the beginning of the sample, the results reported on the interest rate equation come from a single least squares fit for this last variable over the longer sample. Since the elasticity of employment with respect to real wages is less than unity, a slowdown in the rate of inflation coupled with slowly adjusting labour contracts (or at least nominal wages), implies an increase in real total labour income. Thus, consumption and output are likely to increase. On the other hand, inflation increases relative price distortions and therefore uncertainty, reducing imvestment and output. Therefore, the rate of inflation is a potential explanatory variable in an output equation, although in the present case it was not significant, due probably to the quality of the data. One of the explanatory variables in this equation is the nominal rather than the real interest rate. If it is the real interest rate that enters the equation, using the nominal rate leaves the inflation rate out with a positive contribution to output. This effect may be making up for the consumption effect discussed above, and provides an explanation both for the significance of the nominal interest rate and the insignificant coefficient of the rate of inflation in the output equation.

(k) Price equation

$$\Delta P = -0.017 D1 - 0.023(D2+D3) + 0.78 \Delta M + 0.18 \Delta PM$$
$$\quad\quad\quad (3.1) \quad\quad (4.1) \quad\quad\quad\quad (10.0) \quad\quad (4.3)$$
$$+ 0.43(Y-Y_{-3})_{-2} + 0.23(CU-CU_{-2})_{-2}/3$$
$$\quad (3.6) \quad\quad\quad\quad\quad (4.2)$$
$$T = 35(75.2 - 83.4), \quad \hat{\sigma} = 0.007, \quad R^2 = 0.65$$
$$C_1(4) = 3.9, \quad C_2(18) \doteq 3.2, \quad C_3(4) = 1.4 \quad\quad\quad\quad (30.40)$$

The price equation can be thought of as a flexible mark-up equation on unit costs that is sensitive to demand pressure, as measured by output, that is $p^* = \delta(Y)$ CU. But producers do not put up prices immediately: rather, they do it when the market conditions allow it, that is, when credit is available to potential buyers so that higher prices can be paid for. Then, $\dot{p} - p^* = F(M)$, so that finally, we have M, Y, and unit costs explaining prices. A price equation with money appearing strongly, such as that presented in this chapter, is usually regarded as a clear sign of an economy working at full capacity. But estimation results, and plain evidence, show that idea to be at odds with the Spanish reality. The explanation offered here is just a possibility among others, perhaps not the best, but surely not the worst, and tries to reconcile two undeniable facts: (1) demand shocks have a positive and lasting effect on output; (2) money has a strong and quick impact on prices, and almost null on output (these two facts are common to many other inflationary economies).

Most coefficients are fairly significant and have the expected size and sign, and the equation passes all tests very easily. The first motivation behind the introduction of 'money' in the price equation has been that previous empirical work has shown a strong relationship between money and prices. On the other hand, prices (and wages) cannot increase indefinitely given a fixed amount of money. The standard model only admits the feedback in a fairly cumbersome and weak way through interest rates, output, and employment. The introduction of money dampens considerably the wage–price spiral facing an exogenous shock, unless money accommodates the process. This is because the estimated coefficients of wages and prices are smaller in this latter case. Finally, feedback equations from prices and wages to money did not give any indication that causality was running in this direction.

(l) Demand for labour

$$\Delta N = 0.38 \Delta N_{-2} - 0.033 \Delta(PM - P) - 0.029 \Delta((PM - P)_{-2} + (PM - P)_{-3})$$
$$(3.8) \qquad (2.2) \qquad (3.5)$$
$$- 0.006 \Delta RL_{-2} - 0.027 \Delta_4 (W - P)_{-3}$$
$$(5.3) \qquad (3.0)$$

$$T = 32(76.1 - 83.4), \quad \hat{\sigma} = 0.003, \quad R^2 = 0.65$$
$$C_1(4) = 4.3, \quad C_2(14) = 3.1, \quad C_3(4) = 3.4 \tag{30.41}$$

The demand for labour is fairly standard except for the interest rate effect. Since the volume of short-term credit required by a firm moves along with inflation, it is the nominal rate that must appear in the equation. One apparently missing variable is the stock of capital. But since the equation has been estimated in first differences, and the interest rate is an important determinant of investment, this variable is also taken into account. (There are no reliable quarterly data for capital or investment, so that this is again another reason to use the interest rate as an explanatory variable.)

Employment is made to depend on real wages, real exchange rate, and a nominal interest rate. All coefficients have the expected sign, the tests are passed easily, and although the fit is not impressive it is not too bad either. One can find some justification for modelling the equation either in differences or in levels. Since unemployment, like interest rates, has been trending over the sample period, it was decided to model it in differences. Estimation in levels is also risky since it can yield totally distorted long-run elasticities (see also Mauleón 1985c). It is not uncommon to find in the literature labour demand functions estimated in levels, with near unity long-run elasticity of employment to real wages. However, a simple analysis reveals the sheer impossibility of this result for the Spanish case: unemployment has risen by 15 percentage points over the decade 74–83, but real wages have risen 60%. Therefore, the unit elasticity could only be plausible if there was another very strong positive impact on employment over this period. But all remaining effects in the equations reported are usually negative.

The nominal interest rate appears very strongly in the equation, as a proxy for financial costs and investment as discussed before. Some omitted variables are taxes, unemployment benefits, costs of firing and hiring, and mismatching effects. There are no quarterly data available for them, but they are not likely to have an important weight in the Spanish case, except for the employer's contribution to the social security. But this should be added to wages in order to obtain labour costs. Since this variable has risen faster than wages, and the increase in unemployment to be explained is fixed, the most likely effect of taking into account this tax would be to reduce the labour cost coefficient in the demand for labour.

(m) Wage equation

$$\Delta_4 W = 0.97 \Delta_4 M + 1.8 \Delta_4 (Y - N)_{-3} - 3.1 \Delta_4 U_{-2}$$
$$(6.5) \qquad (4.5) \qquad (3.5)$$
$$+ 0.26 \Delta PM_{-1} + 1.1 (Y - Y_{-5})_{-2} + \hat{u}$$
$$(1.8) \qquad (2.8)$$
$$\hat{u} = 0.47 \hat{u}_{-1} + \hat{\varepsilon}$$
$$(2.6)$$
$$T = 32(76.1 - 83.4), \quad \hat{\sigma} = 0.022, \quad R^2 = 0.88$$
$$C_1(4) = 9, \quad C_2(18) = 8.7, \quad C_3(3) = 5.3 \tag{30.42}$$

The wage equation is an augmented Phillips type of function, where a target for real wage growth is taken explicitly into account (as measured by average productivity). Price expectations are assumed to be rational in the long run. In the short run, expectations are modelled in such a way that the long-run determinants of prices enter the wage equation but with unrestricted lags. In this way, the model allows for greater dynamic flexibility. The reason for proceeding in this way has been that introducing directly in the equation prices, current and lagged, did not work acceptably.

The wage equation includes the long-run determinants of prices with unrestricted lags (M, Y, PM), and the static solution is close to long-run rationality. Total wages, rather than wages net of the employees' contribution to the social security, are modelled. But taking net wages, the results were fairly similar, so that total wages were finally selected for the sake of coherency with the remainder of the model.

(n) Supply of labour

$$\Delta L = -0.006 D2 + 0.009 D3 + 0.008 D4 + 0.52 \Delta L_{-2} - 0.33 \Delta L_{-4}$$
$$ (5.4) \qquad (7.0) \qquad (4.2) \qquad (3.2) \qquad (2.2)$$
$$-0.2 \Delta Y - 0.38 \Delta U_{-1}$$
$$(3.2) \qquad (2.8)$$

$T = 35(75.2 - 83.4)$, $\hat{\sigma} = 0.0025$, $R^2 = 0.8$
$C_1(4) = 1.5$, $C_2(11) = 9.3$, $C_3(4) = 4.2$, $C_4(1) = 0.8$
$C_5(4) = 3.5$, $C_6(2) = 1.1$ (30.43)

The supply of labour has a strong seasonal component ($D2$, $D3$, $D4$), and depends negatively on current income, and the lagged rate of unemployment as expected. Although one would expect to find a positive correlation with real wages, no such effect has been found in the estimation process.

(o) Bank reserves linkage

$$\Delta AC_c = \Delta AC - (PC^+/AC^+) \Delta q \qquad (30.44)$$

This equation is required if it is wished to run simulations in which the coefficient of required reserves, q, is changed. The explicit derivation of this equation jointly with a detailed explanation can be found in Mauleón *et al.* (1986).

(p) Feedback equation

$$\Delta(CP - P) = 0.46 \Delta(CP - P)_{-1} - 2.8 \Delta y_{-1}$$
$$ (5.2) \qquad\qquad\qquad (5.2)$$
$$- 0.95 \Delta y_{-3} + 0.025 \Delta RL_{-3} + 0.056 D1 + 0.049 D2$$
$$(1.9) \qquad\quad (2.6) \qquad\quad (5.9) \qquad (4.8)$$

$T = 35(75.2 - 83.4)$, $\hat{\sigma} = 0.025$, $R^2 = 0.78$
$C_1(4) = 6$, $C_2(8) = 2.6$, $C_3(4) = 4.8$ (30.45)

This feedback equation reflects basically the negative impact on the deficit of a slowdown in economic activity, and the cumulative effects of the stock of outstanding public debt through the interest rate. The status of the equation is somewhat unclear, being a byproduct of the lack of data. As pointed out in the previous section, this equation should be replaced by Equations 30.13 and 30.27, if the proper data set was available.

30.4 CONCLUSIONS

A reasonable way of assessing an econometric model is based on the analysis of the long-run solution and short-run simulations. Because some of the series used in this model are somewhat unreliable, it is not safe to draw detailed conclusions from the joint study of the whole model. Rather, it may be wiser to use, for the time being, only the safest parts of the model, namely the equations for exports, the effective exchange rate, the money supply, the supply of bank reserves, the demand for currency, prices and wages. With an improved data base, and perhaps with more research and more refined techniques, it is hoped that the whole model will become operational in the near future. A few comments on the properties of the model are worth pointing out, though. First, as far as short-run responses are concerned, the model is fast to adjust. Dynamic equations modelled in levels, often yield unrealistically long adjustment lags. This has been avoided here by modelling equations in growth rates wherever it was required. As for long-run properties, the model produces a highly inflationary path, when the public deficit is monetized. If it is not, it crowds out private activity through a reduction in domestic credit available to the private sector, which drives up interest rates in the credit market.

APPENDIX A

List of variables

(Interest rates, the rate of unemployment, and the Bank of Spain's dollar reserves are defined in levels. All remaining variables are taken in natural logarithms.)

ALP	broad monetary aggregate (M3 + very liquid short-term assets, mainly short-term treasury bills)
AC_c	adjusted bank reserves
AC_c^*	money growth target
AC	unadjusted bank reserves
BM	monetary base
CU	unit labour costs (W^*N/Y)
CP	domestic credit to the public sector
CT	total domestic credit to the economy
DNF	nonfinancial deficit
DP	outstanding short term public debt (there is hardly any long term public debt)
e	effective exchange rate (an increase means a depreciation in the peseta)
eD	peseta dollar exchange rate
FD	final demand (GDP, plus imports)
EMP	currency in public hands
G	central government expenditure (nominal terms)
IM	index of world imports
L	supply of labour (total labour force)
M	nonenergy imports (real terms)
M3	M3 (deposits plus currency)
N	employment (total)
P	consumer price index
PC	eligible liabilities
PM	imports price index ($e + p_w$)

Appendix A

q	required reserves coefficient
P_x	exports price index
P_w	world consumer price index
R	Bank of Spain's foreign reserves (denominated in domestic currency)
RL	long-term yield on private industrial bonds
RS	one-month interbank rate
RSP	public sector borrowing from the Bank of Spain
R_w	short-term world rate
RD	bank of Spain's dollar reserves
U	rate of unemployment
X	real exports
Y	GDP
W	quarterly wage rate
$D1, D2, D3, D4$	quarterly seasonal dummies

Testing procedures

Several criteria have been used to check the validity of the equations of the model, although not all tests have been tried on every equation. The criteria used are the following:

(a) absence of serial correlation
(b) no significant variables and lags omitted
(c) stability
(d) no outliers
(e) homoscedasticity of the errors
(f) normality of the errors
(g) no simultaneity

The serial correlation tests can be evaluated by means of an auxiliary regression of the errors on its own lagged values. The stability tests are of the predictive ability type, and are asymptotically equivalent to the Chow test. Two types of heteroscedasticity in the errors have been tested: (1) the error variance is assumed to be proportional to the mean of the dependent variable; (2) the ARCH type of heteroscedasticity, as suggested by Engle. The normality of the errors is tested by means of the Bera–Jarque test. The simultaneity test of this chapter is the Hausman test. All these tests are very nicely presented in a unified way in Pagan and Hall (1983).

The tests are denoted in the main body of the chapter as follows:

$C_1(.)$ stability test
$C_2(.)$ overall lag specification
$C_3(.)$ serial correlation test
$C_4(.)$ heteroscedasticity test
$C_5(.)$ heteroscedasticity (ARCH type)
$C_6(.)$ normality of the errors
$C_7(.)$ absence of simultaneity

The number of degrees of freedom are given in brackets and all tests are asymptotically distributed as a χ^2.

REFERENCES

Albarracin, J. (1978) La función de inversión bajo una tecnologia Putty–Clay: un intento de estimación para la economia española. *Estudios Económicos*, 12, Banco de España (Bank of Spain).

Bonilla, J. M. (1978) Funciones de importación y exportación para la economia española. *Estudios Económicos*, 14, Banco de España (Bank of Spain).

Feldstein, M, and Eckstein, O. (1970) The fundamental determinants of the interest rate. *Review of Economics and Statistics*.

Frost, E. (1977) Short run fluctuations in the money multiplier and monetary control. *Journal of Money Credit and Banking*.

Mauleón, I. (1985a) Una función de exportaciones para la economia española, Documento de Trabajo 8507, Banco de España (Bank of Spain).

Mauleón I. (1985b) Análisis econométrico de las importaciones españolas, Documento interno, Servicio de Estudios, Banco de España (Bank of Spain).

Mauleón, I. (1985c) The public deficit and the labour market in Spain: some links and implications (presented at the 5th World Congress of the Econometric Society, Boston).

Mauleón, I. and Pérez, J. (1984) Interest rates: determinants and consequences for macroeconomic performance in Spain, in *Nominal and Real Interest Rates: Determinants and Influences*, Bank for International Settlements, Basle.

Mauleón, I., Pérez, J. and Sanz, B. (1986) Los activos de caja y la oferta de dinero. *Estudios Economicos*, 40, Banco de España (Bank of Spain).

Pagan, A. and Hall, A. (1983) Diagnostic tests as residual analysis. *Econometric Reviews*.

Pérez, J. (1975) Un modelo para el sector financiero de la economia española. *Estudios Económicos*, 6, Banco de España (Bank of Spain).

Pérez, J. (1977) El tipo de rendimiento de las obligacionses y la demanda de depósitos. *Estudios Economicos*, 9, Banco de España (Bank of Spain).

Rodriguez, J. (1978) Una estimación de la función de inversión en viviendas en España. *Estudios Economicos*, 13, Banco de España (Bank of Spain).

Sánchez, A. (1977) Relaciones econométricas sobre precios y salarios. *Estudios Económicos*, 8, Banco de España (Bank of Spain).

31

Macroeconomic policies and adjustment in Yugoslavia: some counterfactual simulations

FAHRETTIN YAGCI and STEVEN KAMIN

31.1 INTRODUCTION

After a period of successful growth in the 1970s, Yugoslavia was hit very hard by a severe balance of payments crisis at the end of the 1970s that has overshadowed the economic scene since. What were the causes of the crisis? Could it have been averted through the adoption of alternative policies in the 1970s? What can be said about the adjustment policies pursued since 1980 in the light of the lessons of the 1970s? These questions are addressed in this chapter with the help of some counterfactual policy simulations for the period 1973–79, using a macroeconomic model developed for the Yugoslav economy.

The simulations yielded three main conclusions. Firstly, the payments crisis was attributable to a deterioration in the policy environment, particularly with respect to the exchange rate, wages and interest rates, although external shocks and/or excessive expansion of domestic demand have played an aggravating role. Secondly, the crisis could have been avoided with alternative macroeconomic policies. Thirdly, Yugoslavia could have achieved external balance and satisfactory growth in investment and output since 1980 had the policy environment been reformed promptly and significantly.

Traditionally, excessive expansion of demand has been regarded as a prime culprit in foreign exchange problems. Inappropriate expenditure-switching policies have also been pointed to, although their importance has been considered to be only secondary. As a consequence, stabilization programs have emphasized the importance of reducing expenditures in responding to balance of payments difficulties. Under such programs, however, excessive reductions in expenditures with insufficient expenditure-switching and structural adjustment might result in a substantial sacrifice of employment and growth that would lead to social unrest and abandonment of the program. By focusing specifically on some expenditure-switching policies, this chapter also addresses this issue. In the context of Yugoslavia's experience, it analyzes the role of the exchange rate policy in keeping the balance of payments in check without sacrificing employment and growth.

The rest of the chapter is structured as follows. Section 31.2 provides a brief description of the economic policies and growth in the 1970s. The core of the model is sketched in Section 31.3, while Section 31.4 evaluates alternative growth paths the economy would have taken under different exchange rate, wage, and interest rate policies than those pursued in 1973–79. Section 31.5 summarizes the main findings of the chapter.

31.2 FROM GROWTH TO CRISIS

Yugoslavia chose to maintain a policy of high growth after the 1973 oil price increase and during the consequent turbulence in the world economy. The rate

Table 31.1 Yugoslavia – main macroeconomic data

	1966–72	1973–79	1980–85
Growth rates in real terms (%)			
GSP	5.9	6.1	0.9
Domestic absorption	5.7	6.4	−3.1
Gross fixed investment	6.0	8.0	−7.8
Total consumption	5.7	5.6	−0.9
Employment[a]	2.0	4.2	2.5
Labour productivity[a]	3.8	1.8	−0.2
Net income per worker[a]	5.1	3.4	−6.7
Exports of goods	7.0	2.9	4.0
Imports of goods	10.8	7.5	−7.5
Implicit GSP deflator	11.2	17.8	45.2
Money supply (M1)	4.9	10.0	−8.0
Share in GSP (%)			
Total investment	38.1	42.2	43.1
Gross fixed investment	30.5	34.1	27.5
Changes in stocks	7.6	8.1	15.6
Domestic savings	37.0	39.8	43.1
Current account balance	−1.1	−2.4	−0.5
Others			
Real interest rate on saving deposits (%)	n.a.	−9.6	−18.8
Maximum real lending rate (%)	n.a.	−7.8	−10.8

[a] In the social sector.
Source: Federal Statistical Office, *Statistical Yearbook*, various issues; and National Bank of Yugoslavia, *Quarterly Bulletin*, various issues.

of growth of gross social product (GSP)[1] amounted to 6.1% on average in the 1973–79 period, slightly higher than that of the preceding seven years (Table 31.1). This growth was driven by an acceleration of capital expenditures: fixed investment rose to a rate of 8.0% in 1973–79 from the 6.0% of 1966–72. In the same period, the ratio of total investment to GSP increased from 30.5% to 34.1%.

Yugoslavia's ambitious investment drive, which was aimed at moderating the unemployment problem, was not, however, matched by a corresponding increase in savings. The rate of domestic savings rose from 37.0% in 1966–72 to only 39.8% in 1973–79. The rising level of investment therefore had to be financed by foreign savings.

The composition of aggregate demand changed substantially in the 1970s. The share of exports of goods and non-factor services in aggregate demand fell from 24.2% in 1973 to 17.7% in 1979, while the share of imports of goods and non-factor services remained constant at 29.0%. As a result, the share of the current account deficit in GSP rose from 1.1% in 1966–72 to 2.4% in 1973–79, reaching a peak of 6.0% in 1979.

The increased liquidity in the international capital markets made borrowing a convenient substitute for improving the policy environment. However, by the end of the 1970s, the country was facing a severe balance of payments crisis, total debt having increased from $4.3 billion in 1973 to $13.7 billion in 1979. The structure of the terms of the debt also changed substantially: the average maturity and grace period fell from 21.4 and 4.2 years to 14.9 and 3.5 years, respectively, in the same period. Creditworthiness eroded, and external financing dried up.

Since 1980, although severe austerity measures have been introduced, the macroeconomic policies have not been reformed significantly. Consequently, the country has become bogged down in a long period of stagflation.

What were the causes of the payments crisis? This question has been raised repeatedly by interested parties within and outside Yugoslavia, since a correct identification of the main reasons is imperative to designing an effective program of adjustment to restore stability and growth.

Contrary to the general perception, the deterioration in the terms of trade induced by the first oil shock does not appear to have been a significant contributor to the crisis. The terms of trade worsened by an annual average rate of only 0.5% in 1973–79, which is substantially lower than the annual average deterioration of 1.2% in all non-oil developing countries in the same period. Moreover, as Table 31.2 indicates, after falling 13.0 percentage points in 1974, the terms of trade gradually improved. This moderate deterioration may be explained by the fact that the share of oil in total imports was only 3.9% in 1973. In addition, the bulk of Yugoslavia's exports was comprised of manufactured products for which prices were rising.

---

[1] GSP is the counterpart of GDP in the Yugoslav accounting system. It yields somewhat lower figures than GDP does as it excludes part of services such as education, health, public administration, and banking and insurance.

Table 31.2 Yugoslavia – some economic indicators (% unless otherwise indicated)

	Terms of trade	Index of real exchange rate[a]	Index of labour productivity	Index of real income per worker	Inflation	Real interest rate on savings deposits	Real maximum lending rate	Growth rate of fixed investment	Ratio of current account balance to GSP
1972	100.0	100.0	100.0	100.0	15.2	-8.9	-7.2	3.1	2.8
1973	102.0	84.8	102.6	98.9	18.9	-10.7	-9.9	2.9	2.5
1974	89.0	70.4	106.2	109.4	22.5	-14.3	-13.5	9.1	-4.6
1975	93.0	72.2	104.4	108.4	19.2	-11.0	-9.2	9.7	-3.4
1976	94.0	72.1	105.2	113.6	13.4	-5.2	-3.4	8.1	0.7
1977	93.0	66.8	108.3	115.7	14.7	-6.7	-4.7	9.4	-3.6
1978	96.0	63.7	110.7	124.1	14.9	-6.8	-3.9	10.5	-2.3
1979	95.0	58.4	113.7	126.2	20.7	-12.6	-9.7	6.4	-6.0
1980	94.0	63.5	112.6	114.6	30.3	-22.1	-18.3	-5.9	-4.0
1981	93.0	70.4	111.0	109.3	40.2	-32.0	-28.2	-9.8	-0.9
1982	97.0	81.4	109.0	105.2	31.8	-22.3	-9.8	-5.5	-0.7
1983	95.0	111.0	105.5	93.6	40.8	-21.7	-3.5	-9.7	0.4
1984	96.0	124.1	105.8	86.8	52.6	-10.1	-1.0	-8.2	1.1
1985	97.0	127.2	104.9	83.3	74.4	-4.4	-4.4	-9.5	1.2

[a] Nominal exchange rate multiplied by the ratio of GNP deflator of USA to GSP deflator of Yugoslavia.
Source: Federal Statistical Office, *Statistical Yearbook*, various issues; National Bank of Yugoslavia, *Quarterly Bulletin*, various issues; IMF, *International Financial Statistics*, various issues.

Policy weaknesses and distortions

The short-lived world recession in 1974/75 was also not a main suspect, given that the build-up of the crisis occurred over a longer period and continued after the recession ended.

These considerations suggest that the payments difficulties that emerged in 1979 were attributable more to policy-induced domestic shocks than to exogenous ones. On the domestic front, the acceleration of fixed investment obviously contributed to the deterioration of external payments. A less ambitious investment drive would have resulted in a more favorable balance of payments, although it would also have generated lower growth and employment. Besides, as the data indicate, the destabilizing element was not the level of demand but its composition: the rate of net exports to aggregate demand rose from -4.8% in 1973 to -11.3% in 1979. Behind this significant switch away from exportables was the substantial worsening of the policy environment in general and the rapid appreciation of the exchange rate in particular. Indeed, one of the main points coming out of this chapter is that the payments crisis of 1979 may be ascribed to the accumulated effects of misconceived exchange rate, wage, and interest rate policies.

31.3 POLICY WEAKNESSES AND DISTORTIONS

The nominal exchange rate was almost constant from 1972 to 1979, whereas in the same period domestic prices tripled. Consequently, the real exchange rate appreciated continuously, averaging 7.7% a year in 1973–79, and the index of the real exchange rate fell from 100.0 in 1972 to 58.4 in 1979 (Table 31.2). This dramatic loss of competitiveness contributed substantially to the fall in the growth rate of exports from 7.0% in 1966–72 to 2.9% in 1973–79 (Table 31.1).

Distortions in the factor markets are observed in both the labour and the capital markets. The average rate of growth of labour productivity in the social sector was 1.8% in 1973–79. However, real wages in the same sector grew at an average rate of 3.4% in that period, an indication of a substantial margin between productivity and wage payments. On the cost side, wage payments in excess of productivity exerted pressure on prices, a situation that led to lower competitiveness. On the demand side, higher wages shifted the distribution of income in favour of social sector wage earners and possibly affected the rate of domestic savings adversely.

Real interest rates on both deposits and loans were grossly negative (Table 31.2). The low rates for deposits discouraged financial savings and the repatriation of workers' remittances. The low rates for lending gave rise to excess demand for credit and necessitated credit rationing, which interfered with its efficient allocation.

Other policy weaknesses of a more structural nature compounded the effects of the deteriorating macroeconomic environment in the 1970s. Although discussed here briefly, these weaknesses are not included in the counterfactual simulations because their impact on the performance of the economy is harder to quantify.

The effective rate of protection is an appropriate indicator for measuring trade-related distortions in the product market. There is evidence that the variation in effective protection among subsectors was high. The average tariff rate (tariff and all tariff-like changes) was around 15.0% in 1973–79. The rates varied between 0.0 and 35.0%; in general, however, those for intermediates were substantially lower than the rates for final products. In addition, the use of quantitative restrictions was widespread. All these factors point to higher effective rates and significant variation across products.

Until recently, most prices have been determined administratively in Yugoslavia by self-management agreements among the parties involved. This rigid price system, coupled with the high and differential levels of protection, created substantial distortions in the product market and affected efficiency adversely.

Solidarity among the socially owned firms led to widespread socialization of financial losses and enabled firms to operate under very soft budget constraints.[2] Because of the lack of financial discipline, economic agents were not held accountable for their actions, and non-viable enterprises could continue to operate. In addition, investment decisions were not guided by appropriate criteria for project selection. Consequently, the flow of resources into the most profitable activities was seriously constrained.

Factor mobility among the regions was very limited, including capital, labour and foreign exchange. The reason is that firms in Yugoslavia are regionally based, and they tend to establish vertically integrated production units within their own regions to obtain the maximum possible benefits from intraregional contacts (securing credit, obtaining licenses, etc.). The resultant regionalism has hindered the dissemination of technical and managerial skills, created excess capacity, constrained specialization and the division of labour, produced large interregional differences in productivity and income, and severely affected overall efficiency and productivity.

31.4 THE MODEL

The model abstracts from sectoral details and input-output relationships in order to focus sharply on interactions among the major macroeconomic performance indicators: output, trade and current account balances, prices, savings, and investment; and the major macro policy instruments: exchange rate, wages, and interest rates. Three essential elements dominate the structure of the model.

Firstly, only one type of goods is assumed to be produced domestically, and it is either exported abroad or marketed internally, depending on the relative prices of exports and domestically sold output. Consumers, firms, and the government have a demand for both these goods and for final imports, depending on income and relative prices. In the 'flex-price' closure of the model, national output and

---

[2] Under these circumstances the elasticity of response to changes in incentives is likely to be low.

the prices of domestically marketed goods are determined so as to satisfy simultaneously the country's foreign exchange constraint and the condition of zero excess demand in the market for domestically marketed output. In the 'fix-price' formulation, either the foreign exchange constraint or the zero excess demand condition is relaxed, depending on the assumption made regarding Yugoslavia's access to foreign credit.

Secondly, output is assumed to depend strongly on the use of imported intermediate goods. In cases where foreign capital inflows are not accommodative, this condition gives the balance of trade and payments exceptionally important roles in determining output.

Thirdly, the model incorporates a number of institutional features of the Yugoslav economy in the 1973–79 period that depart strongly from conventional assumptions about price, wage, and interest rate flexibility. In Yugoslavia, employment and wages are both (by and large) institutionally determined social objectives, and so both are modeled as exogenous policy instruments. Interest rates and the allocation of credit to firms are determined administratively as well, and so they are also modeled as exogenous policy instruments rather than as the outcome of supply and demand in the financial markets. Finally, it was decided that the practice of price determination in the Yugoslav self-managed enterprise system in the 1970s is best approached through a mark-up model rather than through a framework of competitive market pricing.[3] The model's major components are summarized and interpreted below.

31.4.1 Production

The economy produces only one type of goods, denoted Q, using labour (L), capital (K), and an imported intermediate type of goods (N). We assume the production function to be of the Leontief fixed-proportion type, separable into each of the three inputs

$$Q = Q(L, K, N) = \min[l(L), k(K), n(N)] \qquad (31.1)$$

Because of the growing problem of unemployment in Yugoslavia, it is assumed that the labour constraint is never binding. It is also assumed that the capital constraint is potentially binding, so that at all times

$$Q \leqslant QP = k(K), \qquad (31.2)$$

where QP stands for potential output. However, once installed, the capital stock is fixed, so that production may at times plausibly fail to reach the capital constraint. Finally, because imported intermediates are both scarce and variable, the relationship between output and imported intermediates is considered to be binding at all times. Thus,

$$Q = nN \qquad (31.3)$$

---

[3] With the liberalization of some prices in the last two years, price behaviour has changed substantially. The suggested mark-up model is obviously not appropriate to explain the current inflation.

Real income, Y, is defined as

$$Y = Q - N \qquad (31.4)$$

31.4.2 Supply behaviour

Given total output, the share that suppliers export abroad depends on the ratio of the domestic price, P, to the currency-converted, subsidy-adjusted export price, $PE = ER*(1+su)*PEW$, where ER represents the exchange rate (in dinars per dollar), su the subsidy rate, and PEW the exogenous world price of the goods. Accordingly, real exports, E, are determined as

$$E = E(PE/P, Y) \quad \delta E/\delta(PE/P) > 0, \quad \delta E/\delta Y > 0 \qquad (31.5)$$

Similarly, the supply of domestically marketed output is determined as

$$\begin{aligned} SQ &= Y - E = Y - E(PE/P, Y) = SQ(PE/P, Y) \\ \delta SQ/\delta(PE/P) &< 0, \quad \delta SQ/\delta Y > 0 \end{aligned} \qquad (31.6)$$

31.4.3 Demand Behaviour

World demand for Yugoslavia's goods is considered to be perfectly elastic at the world price, PEW. As Equation 31.3 indicates, domestic demand for imported intermediates, N, is responsive only to output, not to relative prices. Domestic demand for the imported final goods, FM, and domestically produced and sold goods, DQ, depend on both their relative prices and total absorption, or expenditure A

$$FM = FM(PM/P, A) \quad \delta FM/\delta(PM/P) < 0 \quad \delta FM/\delta A > 0 \qquad (31.7)$$

$$DQ = A - FM = A - FM(PM/P, A) = DQ(PM/P, A) \qquad (31.8)$$
$$\delta DQ/\delta(PM/P) > 0, \quad \delta DQ/\delta A > 0$$

where $PM = PMW*(1+tf)*ER$ represents the domestic currency price of imported final goods, tf the tariff rate, and PMW the exogenous world price of the final imports. Total absorption in the model depends on the endogenous consumption of households, HC, as well as on a number of exogenous expenditure components, those being government consumption, GC, government investment, GI, and economic sector investment, EI:

$$A = HC + EI + GC + GI \qquad (31.9)$$

Household consumption, HC, in turn depends (through a number of distributional equations) on income, Y, as well as on the real interest rate r:

$$HC = HC(Y, r) \quad \delta HC/\delta Y > 0, \delta HC/\delta r < 0 \qquad (31.10)$$

The real interest rate is defined as

$$r = i - \dot{P}, \qquad (31.11)$$

where $\dot{P}$ represents the rate of growth of (actual) prices and i is a weighted average of nominal interest rates for different deposit categories (a dot on a variable indicates a percentage change).

Total absorption, imports, and the demand for domestic goods can now be summarized

$$A = A[Y, r(i, \dot{P}), EI, GC, GI] \tag{31.12}$$

$$FM = FM[PM/P, A(Y, r(i, \dot{P}), EI, GC, GI)] \tag{31.13}$$

$$DQ = DQ[PM/P, A(Y, r(i, \dot{P}), EI, GC, GI)]. \tag{31.14}$$

31.4.4 Price Behaviour

Under the Yugoslav system of self-management, the price determination of the 1970s appears to be approached best through an administered-price model, with firms charging a mark-up over prime costs in setting their prices. In this model, domestic prices, P, are determined as a multiplicative mark-up over wages, intermediate input costs, and working capital costs, inclusive of indirect taxes

$$P = (1 + mr)(1 + t)[W*L/Q + r_1*K_W/Q + PN*N/Q], \tag{31.15}$$

where

- mr = mark-up rate
- t = indirect tax rate
- W = nominal wage rate
- L = employment
- r_1 = lending rate
- K_W = working capital
- N = imported intermediate input
- PN = domestic currency price of imported intermediate input

With mr and t kept constant, this relationship is converted into the estimated equation in growth rates actually used in the model:

$$\dot{P} = sw*\dot{W} + sk*\dot{r}_1 + sn*\dot{PN}, \tag{31.16}$$

where the estimated coefficients, sw, sk, and sn are interpreted as the shares of labour, working capital, and imported intermediates in prime cost, respectively.

31.4.5 Key equilibrium relationships

The core of the model consists of three relationships between output and relative prices. First, the demand for imports, $N + FM$, along with net services payments, NFI, must equal the flow supply of foreign exchange provided by exports, E, reserve drawdowns, R, and net capital inflows, CI:

$$E + CI + R = N + FM + NFI \tag{31.17}$$

or

$$E(P/PE, Y) + CI + R = N(Y) + FM[P/PM, A(Y, r, EI, GC, GI)] + NFI. \tag{31.18}$$

For given CI, R, and NFI, increases in income, Y (and hence output), raise both exports and imports. Our data confirm that the marginal propensity to import out of income exceeds the positive response of exports to increases in output, so that imports expand more than exports in the event of a rise in Y. Accordingly, if the foreign exchange constraint is to hold when income grows, relative prices, P/PM and P/PE, must fall to discourage imports and encourage exports. With RP representing either P/PE or P/PM, since both PE and PM are independent of Y, the locus of (RP, Y) combinations that represent points of *external balance* are depicted as curve EB in Fig. 31.1.

The second key relationship in the model equates the demand for domestically marketed output, DQ, with the supply of domestically marketed output, SQ:

$$SQ(PE/P, Y) = DQ[PM/P, A(Y, r, EI, GC, GI)]. \tag{31.19}$$

Increases in income increase both the supply and demand for domestically marketed output. Our data indicate that the marginal propensity to purchase domestically produced goods out of income is exceeded by the marginal propensity to supply them. This pattern corresponds to a stable system in a basic Keynesian framework. Accordingly, increases in income are associated with the development of excess supplies of domestically marketed goods at the same relative price; the price of domestic goods relative to final imports must therefore drop to raise domestic demand, lower domestic supplies, and re-establish *internal balance*. In Fig. 31.1, IB represents the locus of (RP, Y) points at which internal balance holds, all else being equal.

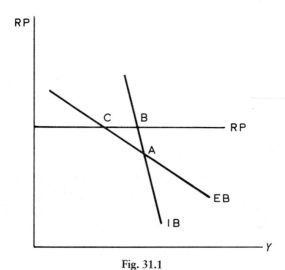

Fig. 31.1

The model

The third key relationship in the model is the price determination equation already described above. Since none of the determinants of the domestic price, P, and hence the relative price, RP, is related to the level of income or output, the RP curve depicting this condition in Fig. 31.1 is a horizontal line.

31.4.6 Model solution under different closures

Obviously, all three of the relationships described in Fig. 31.1 cannot be binding at once. Depending on the type of closure deemed appropriate, either the external balance condition, EB, the internal balance condition, IB, or the price relationship, RP, must be relaxed.

In a setting with flexible prices and instantaneously clearing markets, movements in production costs, demand for goods, and output would be reflected in real income and relative prices rather than in the aggregate level of domestic prices. In that case, the RP relationship would be eliminated, so that equilibrium would occur at point A in Fig. 31.1 the point at which both the foreign exchange constraint is met and the supply of domestically marketed output equals its demand. As the flexible price scenario appears to be especially inappropriate to Yugoslavia's economy in the 1970s, the RP relationship is retained, and either the internal balance or external balance condition is residualized.

The choice of closure for the model is influenced most by Yugoslavia's access to foreign credit. In the 1970s, Yugoslavia had relatively easy access to foreign credit, and combined extensive stimulation of demand with substantial borrowing overseas. As the capital account was essentially accommodative of the pressures on the current account, the external balance condition did not bind. This closure is achieved in our model by endogenizing the level of net capital inflows, CI. Equilibrium under this closure occurs at point B in Fig. 31.1, with the fixed level of relative prices, RP, determining output along the IB curve. Note that at point B, the relative price is above the EB curve. This result means that the level of exports are lower, and the level of imports and the trade deficit higher than the levels that would be consistent with the foreign exchange constraint as shown for some given amount of net capital inflows.

After 1979, Yugoslavia's prior accumulation of foreign debt, coupled with the generally less favourable environment for international commercial lending, made further external borrowing untenable. Yugoslavia's foreign exchange constraint thus became binding, so that the appropriate model closure provides for an exogenously specified level of net capital inflows. Yugoslavia was forced to take corresponding steps to restrain the level of domestic absorption. In our model, the internal balance, IB, constraint is relaxed by residualizing fixed investment, EI, in total absorption.

$$A = HC(Y, r) + EI + GC + GI. \tag{31.20}$$

EI is predetermined when the EB constraint is relaxed, but is residualized under the binding EB closure, so that the IB constraint is then relaxed. Equilibrium occurs at point C in Fig. 31.1 in this closure.

31.5 ALTERNATIVE POLICIES FOR YUGOSLAVIA IN THE 1970s

The framework elaborated above vividly highlights the impact of Yugoslavia's macroeconomic policies on its current account balance and foreign debt over the course of the 1970s. Figure 31.2 depicts the Yugoslav economy as it was initially in the decade, (at a point such as A in 1972), on the IB curve on account of the accommodating capital inflows. High investment, low interest rate, and high wage growth policies stimulated total expenditures and shifting of the IB curve to the right and the EB curve downward. At the same time, high wage growth caused the real exchange rate to appreciate over time. Hence, the Yugoslav economy moved from A in 1972 to a point such as B in 1979 that is considerably farther removed from the locus of external balance.

Theoretically, the Yugoslav authorities could have offset the undesirable effects of their investment program with sufficiently stringent exchange rate, wage, and interest rate policies. It is not clear a priori whether 'reasonable', 'non-extreme' actions in these macro policy areas could have substantially moderated the growth of Yugoslavia's foreign debt. This section describes a variety of counterfactual simulations intended to assess whether reasonable, alternative macro policies could have enabled Yugoslavia to pursue its investment program of the 1970s without incurring the external financial burden it experienced at the onset of the 1980s. The model is solved, in every case, under the open balance of payments closure, a choice that reflects Yugoslavia's easy access to foreign credit during the period. The levels of both private and government investment are kept at their historical levels for the period, as are employment, all world prices, workers' remittances, and reserve movements. Nominal exchange rate, wage, and interest

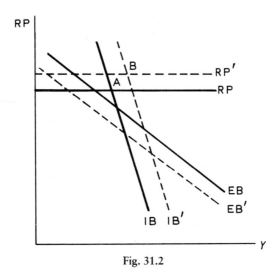

Fig. 31.2

Table 31.3 Summary of simulation results

	Actual		S1		S2		S3		S4	
	Average 1973–79	1979	Average 1973–79	1979	Average 1973–79	1979	Average 1973–79	1979	Average 1973–79	1979
Growth rate (%)										
GSP	6.1	7.0	6.5	7.2	5.5	3.9	6.2	7.2	6.2	6.4
Household consumption	5.5	5.8	5.1	5.7	4.4	0.8	5.0	5.7	4.6	4.0
Exports	2.9	4.5	5.4	5.1	1.7	−1.4	4.1	5.7	5.3	4.6
Imports	7.5	18.1	6.9	17.6	7.5	17.0	7.1	17.6	6.7	16.8
Inflation	17.7	20.7	18.5	20.9	18.8	26.3	16.1	19.2	22.4	40.8
Ratios (%)										
Current account/GSP	−2.4	−6.0	0.1	−1.1	−2.6	−7.0	−0.1	−3.6	0.2	−1.0
Debt service ratio	22.2	27.2	19.8	20.6	23.2	30.2	20.5	23.4	19.8	20.7
Household income/GSP	63.7	62.2	60.3	59.1	61.4	60.9	62.0	60.6	60.2	58.9
Household saving rate	14.5	14.1	14.7	14.2	14.9	14.9	14.7	14.3	14.9	14.8
Domestic saving rate	35.4	37.2	36.7	40.0	35.7	38.2	36.3	38.7	36.9	40.5
Others ($ billion)										
Gross capital inflow	2.6	3.9	1.4	0.8	2.7	4.6	1.9	2.2	1.4	0.7
Total external debt	8.0	13.8	5.6	6.0	8.5	15.2	6.4	9.1	5.6	5.8

rate policies are the instruments that are changed in the alternative counterfactual scenarios described below.[4]

Scenario 1 (S1): constant real exchange rate

The real exchange rate (measured in terms of the exchange rate-adjusted domestic price level and the US GNP deflator) appreciated at a rate of 7.7% a year in the 1973–79 period. By holding the real exchange rate constant at its 1973/74 value and keeping all other exogenous variables unchanged, this scenario simulates the effects of an active exchange rate policy. The nominal exchange rate is allowed to adjust to keep the real exchange rate at the specified level.

Scenario 2 (S2): higher real interest rates

Ex post facto real interest rates were highly negative in the 1973–79 period. S2 simulates the effects of higher rates for both loans and deposits. The nominal rates are gradually increased until the *ex post facto* real rates become positive in 1979.

Scenario 3 (S3): lower wage growth

Real wages in Yugoslavia grew faster than productivity in 1973–79. S3 is designed to simulate the effects of a nominal wage rate set to equate the growth of the real wage to that of labour productivity over the period.

A summary of the simulation results is provided in Table 31.3. The model has been calibrated to equate its base case solution for the period 1973–79 with actual historical values for those years.[5]

31.5.1 S1: Constant real exchange rate

The simulation results for S1 indicate that a constant real exchange rate policy could have substantially moderated the growth of Yugoslavia's foreign debt over the course of the 1970s. The ratio of the current account to GSP registers an average 0.1% for 1973–79 in the S1 simulation, as compared to Yugoslavia's actual experience of −2.4%. As a result, the simulated total foreign debt figure for Yugoslavia in 1979 is $6.0 billion, as compared to the actual $13.8 billion, while the simulated debt service ratio in that year is 20.6% as compared to the actual figure of 27.2%.

---

[4] The coefficients of the behavioural equations are also kept constant across the scenarios. It might be argued that each scenario represents a different policy regime and that the coefficients (and the values of some of the exogenous variables) might have changed under each scenario. This argument might be correct, but there is no way to estimate the hypothetical coefficients that would have prevailed under each policy regime.
[5] Most of the behavioural equations in the model are estimated econometrically.

Under S1, both expenditure-switching and expenditure-reduction effects cause Yugoslavia's external imbalance to decline. Expenditure-switching results from the increases in the prices for exports and imports (measured in domestic currency) relative to domestic prices, a trend that encourages exports and discourages imports at every level of output and absorption. In 1973–79, Yugoslavia's exports and imports actually increased at an annual average of 2.9% and 7.5%, respectively. Under the constant real exchange rate scenario, the rates of growth of real exports and imports are 5.4% and 6.9%, respectively. These figures indicate substantial expenditure-switching.

The improved current account performance in S1 also reflects reductions in domestic expenditures. Real private consumption grows at a 5.1% rate between 1973 and 1979, compared with Yugoslavia's actual performance of 5.5%. The decline in real consumption growth, in turn, reflects the impact of heightened inflation on real wage growth over this period. By increasing the domestic currency cost of imported inputs, the constant real exchange rate policy raises Yugoslavia's average rate of inflation in 1973–79 from 17.7% (actual) to 18.5% (S1). Because the path of nominal wages is fixed in S1, the growth of real wages falls. As may be seen in Table 31.3, this effect lowers the share of household income in GSP from 63.7% (actual) to 60.3% (S1) over the period and therefore helps increase the rate of domestic savings from 35.4% to 36.7%. The shifts in the distribution of income that accompany the rise in inflation more than offset the effects of the lower real interest rates in depressing the savings rate. In the face of the drop in domestic demand, the improved performance of output is explained by substantially greater net exports.

31.5.2 S2: Higher real interest rates

A higher path of real interest rates (for both deposits and loans) over the 1970s could have benefitted the Yugoslav economy in two ways. First, it might have increased the rate of savings and hence improved the current account. Second, higher interest rates might have acted to screen out less productive investment projects and hence have improved the efficiency of investments; this effect is not incorporated in the Yugoslavia model, which is more narrowly macroeconomic in focus, but it should be kept in mind.

Notwithstanding the theoretical desirability of higher interest rates, the simulation results for S2 indicate that a policy of high interest rates would not only have produced stagflation, but would actually have worsened the current account and debt accumulation during the 1970s. These effects result largely from the impact of higher lending rates on the prime costs of production and hence domestic prices. The rate of inflation for 1979 rises from an actual level of 20.7% to 26.3% in S2, compared with 20.9% in Scenario 1. While the average rates of inflation for the 1973–79 period are comparable for S2 (18.8%) and S1 (18.5%), the average real exchange rate appreciates considerably more in S2, since the path of the nominal exchange rate is kept fixed at historic levels. As a result, export growth falters significantly under S2. This situation reinforces the impact of lower real wages (caused by the higher inflation) on consumer spending, and the growth in total output falls to an average of 5.5% for the period, compared to the actual 6.1%. Increases in household savings resulting from higher real deposit rates also contribute to the fall-off in expenditures.

The cumulative effects of the loss of competitiveness are considerable. Average export growth for the period falls to 1.7%, while import growth amounts to 7.5%. As a consequence, the current account balance rises to an average of -2.6% over the period, and total debt increases to $15.2 billion in 1979, which is $1.4 billion more than the actual debt.

Under the current specification of the model, the only favourable effect that higher interest rates have is a slightly greater rate of domestic savings than that realized in 1973–79. The improved savings are attributable to both increased deposit rates and income redistribution toward enterprises.

Two points are worth mentioning. First, S2 confirms that if efficiency gains are not considered and if prices are determined mainly by costs and a constant mark-up, then an isolated active interest rate policy might be stagflationary. Second, as will be seen in the discussion of S4, a compensatory devaluation to neutralize the inflationary effects of the higher interest rates could eliminate the undesirable trade effects of the financial reforms.

31.5.3 S3: Lower wage growth

In this scenario, the growth of nominal wages is set so as to equate the growth of real wages with the growth of the endogenously determined output/labour ratio. In response to the slower nominal wage growth under S3, inflation averages only 16.1% for the period, compared to the actual 17.7%. The lower rates of inflation in turn slow down the overvaluation of the real exchange rate and hence demonstrably improve Yugoslavia's external balance. In S3, Yugoslavia's current account balance averages only -0.1% of GSP, compared to the historical average of -2.4% for 1973–79, while accumulated foreign debt is $9.1 billion in 1979 rather than the actual $13.8 billion.

These gains in external balance in Scenario 3 are not achieved without cost. Because of the direct impact of the policy of low wage growth on household income, the growth of real private consumption drops to 5.0% annually for 1973–79, lower than both Yugoslavia's actual 5.5% growth and S1's 5.1%. The resultant decrease in total domestic expenditures just offsets the increase in activity originating in the export sector, so that, output growth is left about unchanged from its actual level. With output unchanged and the balance of trade improved, the rate of domestic savings also improves under S3.

31.5.4 S4: A Combined scenario (S4)

S4 represents a policy package: a combination of S1, 2, and 3. Under S4, the response of the external sector to the inflationary impact of the policy of high interest rates (S2) and the deflationary impact of the policy of low wage growth (S3) are essentially nullified by the policy of an active real exchange rate (S1). Both S2 and S3 have largely contractionary effects on demand and output, but they are more than offset by the impact of the real exchange rate in boosting exports and discouraging imports. Real private consumption grows at a rate of only 4.6% in 1973-79 in S4, compared to Yugoslavia's historical average of 5.5%.

The contribution of S2 and S3 to the outcome of the policy package is evident in the improved domestic saving rate in particular: the rate of domestic savings under S4 is 40.5% in 1979, compared to the actual 37.2%. This improvement is the result of both the higher rates for deposits and the redistribution of income toward enterprises.

The current account responds well to both the expenditure-switching and expenditure-reduction policies: it moves from the actual deficit of 2.4% to a simulated surplus of 0.2% over the period. The simulated level of foreign debt in 1979 amounts to $5.8 billion, compared to the actual level of $13.8 billion. As a consequence, the debt service ratio in 1979 falls from the actual rate of 27.2% to 20.7%.

31.6 CONCLUSIONS

The results of our simulation experiments strongly indicate that the explosive growth of Yugoslavia's foreign debt in the 1970s and the consequent severe balance of payments crisis were not inevitable. Nor would Yugoslavia have been forced to sacrifice its investment program and high growth and employment targets in order to achieve external balance. If Yugoslavia had avoided the deterioration in its policy environment then, despite unfavourable external conditions, its foreign indebtedness at the end of the decade would have been less than half the actual value. The alternative policies, moreover, are not 'extreme', 'unreasonable', or 'restrictive' in any sense. Stable real exchange rates, positive real interest rates, and the equating of real wage growth with productivity growth are conditions associated with any normal, well-functioning economy.

Note also that the favourable response of the economy to the alternative policies simulated with the model are not the result of using unrealistically high elasticities (the econometrically estimated elasticities used are more on the conservative side, with the elasticity of export supply and import demand with respect to relative prices very close to unity). Rather, it is attributable to the correction of some of the policy mistakes made in macroeconomic management.

A detailed evaluation of the recent economic policies would fall outside the scope of this paper. However, the simulation results provide some basis for assessing the main adjustment strategy Yugoslavia has adopted since 1980. Failure to take corrective action promptly forced Yugoslavia to rely exclusively on reductions in expenditures to control the payments crisis. Since 1980, fixed investment and real income in the social sector have declined by 7.8% and 6.7%

a year, respectively. The growth of output has stagnated around 0.9% as a reaction to the protracted contraction in demand (Table 31.1). Policy reforms have been considerably delayed. Compared to its 1972 value, the exchange rate was overvalued in the first three years of adjustment (Table 31.2), real interest rates are still negative, and the intended reforms related to financial discipline, investment criteria, and foreign exchange and credit allocation are not yet in place, six years after the outbreak of the crisis.

The simulation experiments for 1973–79 indicate that external balance could had been achieved without sacrificing investment and growth if the right policies had been in place. The inference is that Yugoslavia could have combined external balance with satisfactory growth of investment and output if the policy environment had been reformed promptly and significantly after 1979.

Finally, our results confirm the usefulness of simulation models and scenario analyses in evaluating the interactive effects of separate policy instruments in combined policy packages. Note for example, that taken in isolation, the higher real interest rate path tested in Scenario 2 produced some unfavourable results. However, when combined with an active real exchange rate policy, its deleterious impact on real exchange rates was muted, and it was freed to raise savings rates and, to a certain extent, to lower the share of household income.

ACKNOWLEDGEMENTS

The authors would like to thank Wafik Grais and Mansour Farsad for many valuable discussions, EPDCO staff for useful comments, Barbara Ossowicka for the help with the data and computer work, and Vicky Sugui for skilful typing. The chapter expresses the authors' views alone and should not be interpreted as reflecting the opinions of the World Bank, the Board of Governors of the Federal Reserve System, their staff, on the above individuals. The research described in this chapter was performed while Steven Kamin was consulting for the World Bank.

Index

Page numbers followed by n refer to notes; page numbers in *italics* refer to tables or figures.

Abel, A. B.   108n, 109
Accelerated cost recovery system (ACRS)   147
Adams, F. G.   165, 166n, 169
Adjustment costs
  data   132–3
  double-cost hypothesis   128–9, *137*
  empirical methodology   129–32
  empirical results   134–8
  estimation procedures   133–4
  in APEX equations   125–9
  in portfolio demand functions   125–9
  in UK financial markets   120–40
  lagged portfolio shares in   128
  optimal marginal adjustment model   126, 128
  partial adjustment models   125n
  see also APEX equations
Agri-economic models   283–331
  beef cattle model   308
  cohort population model   285, 306–8
  EEC-EGG   299–304
  egg production model   310–11
  heterogenous data   284–5
  hog livestock model   308–9, *310*
  Kalman filter   287–94
  populations models   285, 306–8
  price models   286
  stochastic aspects   287
  use of
    detailed example   299–304
    forecasting   299
    microcomputers and software   299–300, 304–5
    optimal control   299
    statistics filtering   294–6, *297*

  understanding reality   296, *297*, *298*, 299
  unknown parameters   294
Albarracian, J.   684
Allen, R.   599
Anderson, B. D. O.   287
Anderson, G. J.   120
Anderson, O. *et al.*   167
Anderson, R.   159n
Anderson, R. *et al.*   158, 159n
Ando, A.   144n, 145n
Andreano, R. L.   504
Anglard, Patrick   283–331
Anticipations model   165–82
  A-variable values   166–8
  assessment of inventories   171, *172*
  constant term adjustment   173
  construction of   169–72
  exports   177
  flow chart   *170*
  forecasting   172–3
    evaluation of   173, *174–6*, 177–8
    from 30 December 1985   179–81
    of steady growth   179–81
  imports   177
  investment   177
  private consumption   173
  production in   171–2, 173
Aoki, M.   283, 287, 291
APEX equations   120, 121–5, *134*, 138–9
  see also Adjustment costs
Arbitrage   410, 412
ARIMA method   609, 611, 613, 621
Arrow, K. T.   220
Artus, P.   185
Aschauer, D. A.   206n

Ashenfelter, O. et al. 540
Asset price expectation formation equations, see APEX equations
Asset pricing, see APEX equations; Financial markets
Asset stocks 125–6
  autocorrelation in 121
Association of South East Asian Nations 21
Assymmetric case spillover 78
Aubin, J. P. 81n
Auerbach, A. J. 476n, 680
Aukrust, O. 394
Australia, see ORANI model

Babic, M., 510
Bacharach, M. 673
Backus, D. 119
Bacon, R. 331
Baillie, R. T. 410n
Bajt, A. 513
Bakjoven, A. 571
Balance of payments 16
  Japanese 27
  RWT effects 603
Balance of trade, MARIBEL and 590–1
Balance-budget policy, VICTOR model 368, 369, 370
Ballard, C. L. et al. 668, 680
Banking excess reserves 609–51
  ARIMA method 609, 611, 613, 621
  estimation of model 617, 618–19, 620–3
  expectations
    formation variables 612–14
    rational 609–10
  global specification of model 616–17
  interest rate variations 612, 614, 625–6, 639, 641
  model breakdown 623–38, 644–51
  money markets and 620
  overview 609–12
  reserves 612, 640
  risk estimation 612, 613, 614–16
  short and long-run levels 623–38
  VACE 622
  VACNE 621, 622
Barbados
  commodity market 542–3
  empirical estimates 545–56
    data 546–7
    estimated model 547–8
    labour force data 546–7

employment function 540, 541–4, 547–8
short-term forecasting 539–59
simulation and forecasting 549–56
wages function 540, 541–4, 547–8
Barro, R. J. 146n, 204, 334
Barro–Ricardo equivalence proposition 203, 204, 206, 213, 216
Barten, A. P. 75–6
Bartlett, B. 44
Basar, T. 81, 82n
Basevi, G. et al. 74
Beef cattle model 308
Beenstock, M. 327–52
Beenstock, M. et al. 328, 337, 348
Begg, D. 74
Belgium
  RWT, see MARIBEL model
  TVP model 193, 194, 195–8
Bera–Jarque test 711
Berg, L. 229
Berger, Kjell 457–71
Bergsten, C. Fred et al. 74
Bergstrom, A. R. 250
Bergstrom–Wymer model 250
Bernanke, B. S. 104n, 105n
Berndt, E. R. 249
Berndt–Wood model 249
Bewley, R. A. 133
Biart, M. 169n
BITNET–EARNIT system 8
Black, F. 119
Black, S. L. 668
Black economy, Netherlands 364–5
Blanchard, O. et al. 491
Blanchard, O. J. 108n,109, 119
Blinder, A. S. 159n, 162, 206
Blomquist, N. S. 353n
Blumle, G. 206n
Blundell, R. W. 120
Boamah, Daniel O. 539–59
Boardway, R. W. 674
Bodkin, R. G. 415
Bohm, B. 133
Bole, V. 511
Bonilla, J. M. 684
Bonn economic summit 72, 74
Boothe, P. M. 434
Boskin, M. J. 146n
Bouey, Governor 407n 408n
Bovenburg, L. 668, 680
Box–Jenkins methods 665

Brainard, W. C.   119, 125
Brandsma, A. S.   38, 79
Branson, W. H.   85
Brayton, Flint   141–64
Bremer, Stuart A.   12
Brennan, M. J.   105n, 110
Bresnahan, T. F.   80
British Gas Corporation   347, 348
Britton, A.   213
Brooks, S.   540
Brown, A.   539
Brown, O. V.   353n
Bruce, I.   655
Bruckmann, Gerhart   12
Budgets
 balanced, *see* VICTOR model
 consolidation, *see* Freiburger and Tübinger econometric model
 RWT effects   603
Buitelaar, P.   355n 356n
Buiter, W. H.   53, 59
Business cycle   222
 policy   474
 RWI model   498–504
Butter, F. A. G. den   474n
Butzen, P.   581

Canadian foreign exchange model   407–55
 estimated model   445–55
 exchange rate as equilibrating instrument   409
 foreign assets increased   416, *417–19*, 420
 forward exchange rate   410, 411
 interest parity assumption   411, *412, 413*, 414
 interest rate disturbances   420–9
  Canadian rate shock   425, *426–8*, 429
  Canadian YTB held   425
  foreign shocks   420, *421–3*, 424–5
 market efficiency   *414, 415*
 monetary disturbances   429–40
  domestic   434, *435–9*, 440
  high-powered money growth   429, *430–3*, 434
 policy simulations   416–40
 statistical estimates   441–5
Canzoneri, M. E.   53, 59, 72, 78
Capital
 accumulation, Yugoslavian   519–20
 allowances   333
 market, CUBS model   332–4
 net flows   409
 stock analysis, conservation in, *see* Conservation
Capital Asset Pricing Model (CAPM)   107
Cappelen, A.   457–71
Cattier, J.   185
Causality tests   110–11, 112–13
 Granger   112–13
Central Planning Bureau (Hague)   356, 360, 474, 599
Centrally Planned Economies   2, 4
CGE model   667–72
Chappelen, A.   393, 395
Chase model   165
Chick, V.   500, 504,
China   19
 LINK model and   4
Chirinko, R.   183
Chow test   393, 711
City University Business School model, *see* CUBS model
Clark, Peter B.   141–64
Co-ordination, *see* Policy co-ordination
Cobb–Douglas production function   55, 76, 148, 158, 247, 249, 250, 562, 672
Cochrane–Orcutt technique   442
Coen, R. M.   163
Cohen, D.   149n
Cohen, K.   166n
Cohort population model   306–8
 controlled population   307–8
 free population   307
Colias, J. V.   667–82
COMET model, *see* Policy co-ordination
Comparative advantage   88–90
Conservation   245–64
 central heating fuel consumption   251, 254, *255–6*, 257
 energy-economy models   246–7, *248, 249–50*
 gasoline demand in US   250, *251*, 252–4
 US airlines   251, 257, *258–9*
 US iron and steel industry   251, *260–1*
Consumption
 Freiburger and Tübinger econometric model   206, *207, 208*, 209
 life-cycle model of   144
 MPS model   144–6, 162
 private   39, *43*

Consumption *contd.*
  A-model 173
  RASMUS 2b model 266, 269
  RWT effects 602
  VICTOR model 364–5
  Swedish model 229, 230
Cooper, R. J. *et al.* 666
Cooper, R. N. 72, 73, 91, 95
Co-operative policies, *see* Policy co-ordination
Corden, W. M. 38, 73
Corporate bond rate 154
Cost-push model, *see* Yugoslavian inflation
Council of Economic Experts (CEE) 491, 502
Courchene, T. 425n
CPE, *see* Centrally Planned Economies
Cross, R. 362
Crowding out 213, 275
  US fiscal shock 57
CUBS model 327–52
  empirical analysis, 337–51
  gas price increases 347, 348
  government employment increase 349, 350
  model properties 337
  National Insurance contributions 330, 331, 337, 338–9, 346–7
  public sector finance 350, 351
  social capital stock 348
  Social Security benefits 339, 340
    pensions and lower retirement age 343, 344
    SERPS 346, 347
  taxation
    cut in indirect 344, 345, 346
    Laffer curve 341, 342, 343
    personal 340, 341
  theory
    aggregate demand and fiscal policy 334
    aggregate supply 329–30
    capital market 332–4
    general disequilibrium 336–7
    general equilibrium 334, 335, 336
    labour market 330–2
Cukierman, A. 105n
Cumper, G. E. 539
Currie, D. 72
CUSUM stability tests 495

DaCuhna, N. 82

Data Resources Inc. (DRI) model 162, 165, 463
Davidson, P. 220, 498
Debt
  debt paradox hypothesis 204, 205–8, 210
  developing countries 15
  foreign indebtedness 16
  government, direct taxation reduction and 48, 49
  LINK model 3
    Mexico 5, 6–7
  stocks of 151n
  see also Deficits; United States, budget deficit
Deficits
  financing 85–6, 212
  West Germany 209, 210, 211
Demand
  aggregate
    CUBS model 334
    RASMUS 2b model 266
  behaviour, Yugoslavian adjustment model 720–1
  equations 34
  for goods, Spanish model 696, 705–6
  for labour, Spanish model 696, 707–8
  functions, TVP model
    aggregate constrained demand 191
    aggregate effective demand 188–93
    aggregate notional demand 191
    constrained demand 190
    effective demand 187, 188
    notional demand 187, 190–1
    sales dominated demand 187, 190–1
  MCM 55
Depreciation 147
  accelerated 474, 676–9
  MPS model 154, 162n
D'Estaing, Giscard 74
Deutsch, Karl 12
Deutsche Bundesbank 210
Developing countries, debt problems 15
Deviliers, M. 171n
Dickens, R. 121
Dickens, W. T. 120
Dicks-Mireaux, L. A. 38, 45
Disarmament 15, 17–18
Discount rate policies in RASMUS 2b model 274–5
Disequilibrium, general 336–7
Dixon, P. B. 653–66

Index

Dixon, P. B. *et al.* 667, 668, 669, 672, 679, 680
DMS model 599, 600–1, 602, 603
Doblin, C. P. 245, 247
Dolenc, M. 520
Dollar, US, *see* Exchange rate
Dow, J. C. R. 38, 45
Downes, A. S. 539, 546
Draper, D. A. G. 362n
Dreze, J. H. 44, 183, 184, 583, 584, 585–6
Driehuis, W. 356n 473–90
Drollas, L. P. 249, 250
Drollas–Greenman model 249, 250
Duggal, V. G. 165
Durbin–Watson statistic 358n
Dwyer, G. P., Jr. 204, 213
Dymanic soft systems approach 13–14

Eastman, H. C. 385
Eckstein, O. 39, 498, 500, 704
Eckstein, O. *et al.* 165
Economic Recovery Tax Act 141–64, 667
  empirical estimates 152–3
  INTERSAGE model 674, 675, 676, 677–9
  long-run consequences of 151–2
  M1 growth rate held constant 159, 160, 161–3
  see also MPS model
Edison, Hali J. 53–70
Egg production model 310–11
  EEC-EGG 299–304
Eisner, R. 183
Ekstedt, Hasse 219–43
Eltis, W. 331, 341
Employment
  consumption and 229
  CUBS model 330–2, 349, 350
  government deficit and 214, 215
  INTERSAGE model 678
  labour market equations 34
  private sector 38, 39, 40
  RWT effects 600–2
  see also MARIBEL model
  see also Unemployment
Endogenously time-varying parameter model, *see* TVP model
Energy-economy models 3, 246–7, 248, 249–50
  see also Conservation
Environmental issues

in FUGI model 13
  see also Conservation
Enzler, J. J. 159n 161n
EPA model 78n
Equilibrium, general, *see* General equilibrium models
Equity prices, Swedish model 240, *241*, 242
Escriva, J. L. 609–51
Escriva, J. L. *et al.* 613, 621
Espasa, Antoni 609–51
ETA-MACRO model 247
European Economic Community
  deficit financing 85–6
  policy co-ordination with US 71–101
Euthoven, A. C. A. 240
Evans, M. K. 166, 167, 168, 357n, 372
Evans, P. 104n, 204
Exchange rate 16
  as equilibrating instrument 409
  Canadian dollar, *see* Canadian foreign exchange model
  CUBS model 337
  dollar (US) 55
    depreciation and LINK model 4, 7–8
  MPS and 149–50
  forward 410, 411
  in LINK model 3
  in MCM 56
  Spanish model 701
  Yugoslavian adjustment model 726–7
  Yugoslavian inflation and 518
Expectations 498, 500
  rational *131*, *136*, 206, 609–10
Expenditure
  domestic *41*
  FUGI model 15
  government 76
    in RASMUS 2b model 267, 272, *273*
  multipliers 78n, 79
  real 55
  military 15, 17–18
Exports 39, *41*, 45
  A-model 177
  in MCM 55
  INTERSAGE model 678
  price deflator *3*, 6, 15
  prices 43
  Spanish model 700
  Swedish model 234, *235*, 236
  VICTOR model 361–2
  see also Imports; World trade

736 Index

Fair model 169
Farrell, T. 539, 541
Fautz, W. 494
Federal Planning Office (Yugoslavia) 520, 521
Federal Statistical Office (Yugoslavia) 521, 534
Feedback equation 697, 709
Feldstein, M. 74, 96, 704
Financial markets 16
  adjustment costs and mean-variance analysis 120–40
  flows in Sweden 219–43
  perfect efficiency 414, 415
Fiscal policy
  CUBS model 334
  neutrality 205–8, 210, 212
Fischer, S. 113
Flanagan, R. J. et al. 494, 495, 502
Forecasting
  Box–Jenkins methods 665
  in Barbados, see Barbados
  see also ORANI model
Forrester, J. 12
Forward exchange rate 410, 411
Foster, J. 167
Frankel, J. A. 120
FREIA model 355n 474, 476–8, 599, 600–1, 602, 603
Freiburger and Tübinger econometric model 204, 205n, 212–14
  consumption functions 206, 207, 208
  results 214–17
Friedland, B. 292, 293
Friedman, B. N. 126, 128
Friedrich, D. et al. 504
Friend, I. 146n
Friend, J. 165, 166n 169
Fromm, G. 39
Frost, D. A. 610, 613
Frost, E, 699
FUGI model 11–30
  alternative policies
    global disarmament 17–18
    interest rate lowering 17
    Japan opening markets 18
    R and D increased 18
  alternative projection 21–5
  baseline projection 19, 21
  baseline scenario 17
  dymanic soft systems approach 13–14
  GDPs 15, 20, 22, 23

Japanese economy, impacts of alternative policies 25–9
model structure 14–17
Full Information Maximum Likelihood (FIML) 133
Fullerton, D. 673, 674
Fullerton, D. et al. 680
Future of Global Interdependence model, see FUGI model

GAMS program 667, 668, 672–3
Ganderberger, O. 204
Gartner, M. 494
Gas prices, UK 347, 348
GDP, see Gross Domestic Product
Gendreau, Francoise 283–331
General equilibrium models
  CUBS model 334, 335, 336
  Johansen-style 667–72
  US economy, see INTERSAGE model
  see also ORANI model
Gerard, Marcel 183–201
Germany, West, see West Germany
Giersch, H. 495
Glassman, . 611
GLOBUS model 12
Gordon, R. H. 184, 680
Gramm–Rudman Act 54, 57–8, 68
  see also Policy co-ordination
Gramm–Rudman–Hollings legislation 4
Granger causality tests 112–13
Gray, J. 53, 59
Gray, J. A. 72, 78
Green, C. J. 120–40
Greenman Jonathon V. 245–64
Gregory, M. 167
Gregory, P. R. 249
Griffin, J. M. 249
Gross Domestic Product 112
  FUGI model 15, 20, 22, 23, 25
  investment as percentage of 38–9
Gross National Product
  ERTA and 152, 153, 154
  government deficit and 214, 215, 216
  oil price fall and 462, 463
Guetzkow, Harold 12
Gurley, J. 226

Haberley, G. 146n
Hahn, F. H. 220
Haitovsky, Y. 169
Hall, A. 711

Hall, R. 193
Hamada, K. 53, 59, 72, 73
Hansen, G. 206n
Hansson, I. 353n
Harberger, A. C. 673
Harildstad, A. 396
Harris, R. 385
Harsayni, J. C. 94
Hart, R. 581, 591
Hartog, H. den 44, 354, 355, 474n
Hartog, H. den et al. 356, 358
Hasbrouck, J. 146n
Hausman test 711
Hecheltjen, P. 168, 169
Heilemann, U. 213, 491–506
Helliwell, J. F. 434
Hemming, R. 347
Henderson, Y. K. 673, 674
Henize, J. 599
HENIZE model 599, 600–1, 602, 603
Henry, B. 540
HERMES model 606
Hickman, Bert G. 163
Hickman–Coen Annual Growth (HCAG) model 163
Hicks, J. 231
Hicks' neutral technical progress 562
Hildreth–Lu method 172
Hoel, Michael 381–405
Hog livestock model 308–9, *310*
Holder, C. 543
Holt, C. A. 80
Hooper, P. 149n, 150n
Horridge, M. 653–66
Horst, T. 675
Horton, G. 463
Hotelling, H. 247
Hudson, E. A. 247, 249
Hudson–Jorgenson model 247, 249
Hughes Hallet, A. J. 71–101
Hulten, C. R. 675

Ichikawa, Kaoru 313–326
IMPACT project 653n, 666
Imports 39, 42
 A-model 177
 in MCM 55
 Spanish model 700
 Swedish model 236, 237
 see also Exports; World trade
Industrialization rate 19
Inflation 38, *41*
 expected, in MCM 55
 oil price and 462, 463
 LINK model 6
 RWT effects 603
 see also Yugoslavian inflation
Information exchanging, see Policy co-ordination
Interest rate 78
 assest demand and 120
 banking excess reserves 612, *613*, 614, 625–6, *639*, 641
 co-ordinated lowering 17
 disturbances in Canadian model 420–9
 fiscal policy and 213
 FUGI model 15
 interest parity assumption 411, *412*, *413*, 414
 MCM 55
 MPS model 150–1, 157, 158
 RASMUS 2b model 271, 274–5
 response to US protectionism 8–9
 Spanish model 606, 703–4
 Yugoslavian adjustment model 727–8
 Yugoslavian inflation and 519
INTERSAGE model 667–82
 assumptions 668–9
 Economic Recovery Tax Act 667, 674, 675, 676, 677–9
  accelerated depreciation 676–9
  investment tax credit 675, 676–9
 effective tax rates 674–5
 employment 678
 equations 669–72
  household demand 670–1
  input demand 671–2
 expected theoretical results 673–4
 exports 678
 extensions 679–80
 GAMS program 667, 668, 672–3
 margins 669
 production 676–7
 sets *670*
Inverted Haavelmo effect 45, 370
Investment
 A-model 177
 equations 38–9, 55
 financing 184
 in durable equipment 111
 marginal q and 105, 107–9
 market volatility 104
 MPS model 147–9, 157, 161, 162
 non-residential fixed 111

Investment contd.
  oil price shocks and   103–4
  one year lagged net investment
    ratio   38–9
  private, West Germany   209
  ratio   45
  residential   *154*, 155
  risk and   103–17
    causality tests   110–11, 112–13
    multivariate regressions   110–11, 113
    nondiversifiable   105
    option prices   106–7
    stock returns variance   110–11, *114–15*
    rule   106
  spending determinants   105–7
  subsidies   358n, 473–90
    direct output effect   477
    FREIA model   474, 476–8
    indirect output effect   478
    obsolescence effect   478
    policy-makers views on   474–8
    SECMON, *see* SECMON
    VINTAF model   474
  Swedish model   231, *232–3*, 234
    in buildings   231, *232*
    in machinery   *233*, 234
  tax credit   148n 277, 474, 675, 676–9
  TVP model, *see* TVP model
  VICTOR model   357–9
Irfan, M.   540
Isolationist policies   85–6, 87
Iyoha, M. A.   539

Jackson, P. M.   353n
Jahnke, W.   501, 504
Jansen, C.   574
Jansen, E. S.   393, 394, 395
Japan
  FUGI model
    impacts of alternative policies   25–9
    opening markets   18
  US–Japan trade imbalance   27, 28
  see also Policy co-ordination
Jazwinskii, A. H.   287, 289, 290, 292
Jensen, M. C.   119
Johansen, L.   72, 96, 667–8
Johansen-style model   667–72
Johnson, L.   161n
Johnston, J.   38
Jorgenson, D. W.   193, 247, 249, 668, 680

Kalai, E.   94

Kaldor, N.   222, 239, 475
Kalman filter   185, 287–94
  extended filtering   292–4, 296
    constant biases   293
    not constant biases   293–4
  noise assumptions   288–9
  optimal control methods   291–2
  parameter identification   289
  unstable system identification   289–90
  use of   293–6, 297
Kamin, S.   713–30
Katona, G.   167, 168
Kay, J,   347
Kay, J. A.   247
Keller, W. J.   668, 680
Kendrick, D. A.   668
Kennedy–Johnson tax cuts   366n
Kesenne, S.   581
Keyfitz, M.   306
Keynes, J. M.   504
Klein, L. R.   1–10, 12, 31, 74, 166n, 169, 366n
Klein, L. R. *et al.*   4, 12
Kline, D. M.   458
Kménta, J.   354
Knivve, A.   479n
Knobl, A.   168
Knoester A.   31–51, 356n, 360n, 362, 370
Knudsen, Vidar   457–71
Kok, L.   598
Komiya, R.   541
Kopcke, R. W.   105, 108n
Koromzay, V. *et al.*   83
Kotlikoff, L. J.   680
Kracun, Davorin   507–38
Kragh, B.   226
Kronenberger, S.   204
Kuh, E.   502
Kuipers, S. K.   354, 355
Kuipers, S. K. *et al.*   354, 355

Labour market, *see* Employment
Laffer curve   341, *342*, 343, 366
Lagrange multiplier tests   393
Lags   336
  correlation   321
  optimal   322
  regression   321
  see also Optimally lagged models
Lambelet, J-C.   204
Lambert, J. P.   184, 186
Layard, P. R.   329
Least Developed Countries (LDCs)   4

Leibfritz, W. 210
Leibundgut, B. 284
Leontief, W. *et al.* 12
Leontief function 672
Levine, P. L. 72
Lewington, P. 327–52
Likelihood tests 133–4
  Quandt log 495
Lindbeck, A. 360
LINK model 1–10, 12, 462–3, 464
  1986 baseline case 3–5
  dollar depreciation and 4, 7–8
  oil price changes and 3–4, 6–7
  protectionism and 8–10
Longworth, D. 408n
Lorentsen, L. 457
LSE methodology 392
Lucas critique 79, 82, 185, 210
Lucas, R. E. 184, 185, 210
Ludeke, D. 504
Ludeke, D. *et al.* 205n 206, 212
Luenberger, D. C. 306

McAleer, M. *et al.* 392
McDonald, R. 105n, 106n, 107, 110
McLean, A. W. A. 539
McMahon, P. C. 410n
McNees, S. K. 165, 172
Main Economic Indicators
  (OECD) 314–15
Majd, S. 105n, 110
Malcomson, J. 478n
Malinvaud, E. 32, 44, 73, 96, 473n
Manne, A. S. 247
Mansur, A. 672
MARIBEL model 561–607
  Dreze's model compared 583, 584, 585–6
  employees, effect of RWT 574–86
    extra job creation 578, 579–80
  employers, effect of RWT 586–92
    balance of trade 590–1
  employment function
    employment 565, 567–8
    labour utilization 569, 570
    working time 568, 569
  government, effect of RWT 580, 592–5, 596
  production function
    capacity utilization 563, 564, 565, 566, 588
    production 562–3, 564, 566, 588

production capacity 563, 564, 565, 566, 588
productivity 563
results evaluation 580–1
RWT comparisons with other countries 596, 597–8, 599, 600–1, 602–3
simulations of RWT 571, 572–3, 574
SMOBE model compared 581, 582, 584
Markovitz, H. M. 119
Marston, R. C. 53, 59
Martiensen, J. 504
Marwah, Kanta 407–55
Mauleon, I. 609, 610, 613, 631, 683–712
Mauleon, I. *et al.* 699, 702
Mauskopf, Eileen 143n, 145n
Meadows, D. H. *et al.* 12, 247
Meagher, G. A. 653n
Mean-variance efficiency
  in UK financial markets 120–40
  test results *136*
  test schema *131*
Meeraus, A. 668
Mehra, R. K. 289
Mellon tax cuts 366n
Merton, R. C. 113
Mesarovic, M. 12
Meut, P. A. 185
Mexico, debts in LINK model 3, 5, 6–7
Meyer, W. 505
Mieszkowski, P. M. 673
Miller, J. 204, 206
Miller, M. 53, 59, 72
Miller, S. M. 676
Minsky, H. P. 222n
Mirlees, J. 247
MIT–Penn–Social Science Research Council model, *see* MPS model
Mizutani, Kazuo 321
MODAG A 457
Model for Analysis and Rapid Investigation of the Belgian Economy, *see* MARIBEL model
Models, *see individual entries*
Modigliani, F. 44, 144n, 166n
Molitor, B. 502
Monetary policy
  Canadian model 429–40
  feedback rule 159
  M1 held constant 159, *160*, 161–3
money supply
  FUGI model 15

Monetary policy *contd.*
  Spanish model 606, 702
  US and EEC 76–7
Money, in MCM 55, 57
Monte Carlo method 133
Moore, J. B. 287
Morrison, G. R. 610
Motorola surcharge 8, 9
MPS model 142, 143–52
  consumption and 162
  consumption specification in 144–6
  cycling behaviour 161–2
  depreciation 154, 162n
  ERTA/TEFRA consequences 151–2
  exports 155
  external repercussions 149–51
  federal deficit 162
  housing 154, 155
  interest rates in 157, 158
  investments and 147–9, 157, 161, 162
  output devoted to consumption 153, 154, 156n
  potential output 147–9
  production function and, putty-clay 157–8
  savings 155, 156, 162
  unemployment rate constant 153–9
  see also Economic Recovery Tax Act
Muller, F. *et al.* 354, 355
Multicountry Model (MCM) 53, 54–70, 78n
Multipliers
  COMET model 77–9, 91, 98, 99
  government expenditure 78n, 79
  in MCM 56
  negative balanced-budget 32, 45
  policy co-ordination 77–9, 91, 98, 99
Munch, H. J. 495, 498
Mundell, R. A. 73, 89
Mundell–Fleming model 149n

Naggl, Walter 165–82
Nash equilibrium 59–60, 69, 70, 93–4
  open loop 81n, 82
Nash, J. F. 93
National Bank of Slovenia 519, 521
National Insurance contributions 330, 331, 337, 338–9, 346–7
Natzmer, W. von 203–218
Neese, J. 502
Nerb, G. 166
Netherlands
  investment subsidies 473–90

  see also VICTOR model; SECMON; Supply-side policies
Neubourg, C. de 598
New York City rescue 366n
Newly industrialized countries (NICs) 19, 21
Nickell, S. J. 125, 329, 478n
Norway
  effect of oil price fall 457–71
    international repercussions 462–3
    high price oil scenario 464, 465
    low price oil scenario 465, 466–8, 469
    restrictive fiscal policy 469, 470, 471
  see also RIKMOD
Nymoen, Ragnar 381–405

Oberhauser, A. 204, 205
Official development assistance (ODA) 16, 18
Oil prices
  effect on Norway 457–71
  investment behaviour and 103–4
  sensitivity of LINK model 3–4, 6–7
  see also World oil market model (WOM)
Oligopolies 144
Olson, M. 362
Onishi, Akira 11–30
OPEC 458, 459–60
Optimally lagged models 313–26
  analysis principals 313–14
  method 325–6
  procedure for building model 326
  theory 321–5
  verification 314–21
Option pricing 106–7
ORANI model 653–66, 667–8, 680
  comparative statics and forecasting 654, 655–61
  numerical example 661–4
  valid control solution 658–60
Ormerod, P. 540
Oudiz, G. 53, 58, 59, 60, 69, 72, 73, 78n, 96
Oudiz, G. *et al.* 599

Pagan, A. 711
Palsson, Halldor P. 407–55
Parmenter, B. R. 653–66
Partial adjustment models 125n
Pen, J. 370
Pensions, *see* Social security, benefits
Perez, J. 609, 610, 684, 703

Pestel, E. 12
Pfajfar, L. 520
Phillips, A. W. 38, 45
Phillips curve 38, 45, 162, 363, 367, 476, 494
Phillips–Lepsey–Friedman approach 55
Pindyck, R. S. 103–17, 249, 250
Planbureau 580
Plantes, M. K. 119
Plasmans, J. 561–607
Polak, E. 82
Policy co-ordination
  COMET model 71–101, 606
    bargaining strategies 92–5
    co-operative policies 86, 87, 88
    competitive strategies 86, 87
    empirical results 83–8
    expected target values 89, 90–1
    externalities 91
    gains from co-ordination 91–2, 93
    incentives to co-operate 88–90
    instrument adjustments 87, 97
    isolationist policies 85–6, 87
  empirical simulations 61–8
    foreign monetary
      accommodation 63, 64, 65
    initial effects 61, 62, 63
    joint accommodation 67, 68
    US monetary accommodation 65, 66
  multicountry model 53–70
  multipliers 77–9, 91, 98, 99
  non-cooperative policy with
    anticipation 79–82
  spillovers 72, 73, 77–9, 91, 99
  world recession and 73–5
  see also Gramm–Rudman Act
Pollefleit, E. 581
Population models 285, 306–8
Portfolio choice, see also Financial markets
Portfolio demand functions 125–9
Poser, G. 168, 169
Powell, A. A. 653n
Powell, S. 463
Poznanski, K. Z. 260
Praet, P. 169n
Prices 25
  energy, and conservation 245–6
  equations 34
    Spanish model 606, 706–7
  equity, constant price index for 220, 221
  export price deflators 3, 6, 15
  FUGI model 15
  MCM 55
  models 286
  option pricing 106–7
  Producer Price Index 112
  RASMUS 2b model 267
  Swedish model 239, *240*
  wholesale price index 24, 25
  Yugoslavian adjustment model and 721
Producer behaviour, see RIKMOD
Production
  anticipations model 171–2, 173
  capacity, VICTOR model 359–61, 370, *371*, *372*
  INTERSAGE model 676–7
  medium term in Sweden 219–22, 226, 227–36
  Yugoslavian adjustment model 719–20
Production function 148–9, 501
  clay–clay vintage 360
  Cobb–Douglas 562
  Leontief–Cobb–Douglas 672
  MARIBEL model 562–6, 588
  MPS model 157–8
  putty–clay assumption 144, 148–9, 157–8
Protectionism 8–10, 718
Prywes, M. 249
Public sector equations 34–5
Public sector finance 350, 351
Purchasing power parity 337

Quandt log-likelihood test 495

Ramsey, J. B. 354
RASMUS 2b model 265–82
  dynamic properties 272, *273*–6, 277
  forecasting performance 268, *269*–71, 272
  overall structure of model 265–8
  policy options 277–81
  taxation in 267, 275, 276, 277, 278, 279, *280*, 281
Rational expectations 206, 609–10
  test results *136*
  test schema *131*
Rault, A. 283–331
Recession 73–5
Reduction in working time, see MARIBEL model
Rees, H. J. B. 96

Regressions 495
 multivariate 110–11, 113
 Yule–Walker method 227
 see also APEX equations
Research and development 18
Reserves
 in MCM 55
 Spanish model 696, 697, 705, 709
 see also Banking excess reserves
Richardian equivalence theorem 145, 146n
Ridder, P. B. de 358, 474n
Rieder, B. 133
RIKMOD 381–405
 empirical properties 392–9
  price equations 392–5
 employment adjustment costs 390–2, 395–6
 estimated results 400–5
 sectoral production function 385, 386, 387
 simulation properties 396–9
 theoretical considerations 382–92
 with inventories 390–2, 395–6
 without inventories 383–9
Risk
 aversion coefficient 120, 138
 banking excess reserves 612, 613, 614–16
 see also Investment, risk and
Roberts, Ch. C. 168
Robertson, J. W. 675
Rodriguez, J. 684
Roland, Kjell 457–71
Roll, R. 119
Rosanna, R. J. 392
Rose, A. K. 408n
Ross, S. 119n
Rotemberg, J. J. 85, 104
Roth, T. 44
Rouhani, R. 306
Rutten, F. W. 356n, 362n, 370n
RWI Business-cycle model 498–504

Sachs, J. 53, 58, 59, 60, 69, 72, 73, 78n, 96
Sachverstandigenrat 491, 502
St Cyr, E. 539
Sakuri, M. 72, 73
Salmon, M. 53, 59, 72
Sanchez, A. 684
Santomero, A. 495

Sargent, T. J. 121
Saudi Arabia 458
 export price deflator 3, 6
Savings
 Freiburger and Tübinger econometric model 206n
 MPS model 146n, 155, 156, 162
Say's law 371–2
Scher, W. 205
Schipper, L. et al. 255
Schmidt, Helmut 74
Schmidt, R. 494, 495
Scholes, M. 119
Schwartz, E. S. 105n, 110
Seater, J. J. 495
SECMON 474, 478–81
 empirical results 481–2, 486–90
 outline of model 484–6
Sector models, Netherlands, see VICTOR model
Sekulic, M. 510
SERPS 346, 347
Serra-Puche, J. 680
Sharpe, W. P. 119
Shaw, E. 226
Short-time Model for the Belgian Economy, see SMOBE model
Shoven, J. B. 668, 680
Sieband, Jon C. 265–82
Siegel, D. 105n, 106n, 107, 110
Sievert, O. 204, 210
Slemrod, J. 668, 680
Smith, C. W. 610
Smith, D. 306
Smith, G. 119
SMOBE model 581, 582, 584
Smorodinski, M. 94
Sneessens, H. 183, 184
Social Accounting and Auditing Service (Yugoslavia) 520, 521
Social capital stock 348–9
Social security 205n
 benefits 339, 340
  lowering retirement age 343, 344
  SERPS 346, 347
 contributions 45
 RASMUS 2b model 267
 transfers 593
Spain
 banking excess reserves, see Banking excess reserves
 quarterly econometric model 683–712

Index

bank reserves 696, 697, 705, 709
constraints 684
demand for currency 696, 704
demand for goods 696, 705–6
demand for labour 696, 707–8
dollar reserves 696, 701–2
empirical results 699–709
exchange rates 701
exports, 700
feedback equation 697, 709
imports 700
interest rates 696, 703–4
money supply 696, 702
overview 684, 685–95, 696–9
price equation 696, 706–7
supply of labour 696, 709
testing procedures 711–12
wage equation 696, 708
Spillover effects 72, 73, 77–9, 91, 99
 begger-thy-neighbour case 78
 locomotive case 78
Stability tests 495, 711
Staelin, C. P. 668
State-earnings-related pension scheme, *see* SERPS
Statistics filtering 294–6, 297
 see also Kalman filter
Stein, J. L. 205n, 212n
Steindel, Charles 144n
Stevens, G. V. G. *et al.* 149n
Stiglitz, J. 247
Stock returns variance 110–11, *114–15*
Stockton, - 611
Stolen, N. M. 394
Stone, R. A. 673
Strigel, W. H. 166
Stuart, C. 353n
Stykolt, S. 385
Summers, L. H. 156n
Supply
 aggregate, in CUBS model 329–30
 behaviour in Yugoslavian adjustment model 720
 equation 33–4
 of labour in Spanish model 696, 709
Supply-side policies 31–51
 estimated equations 32, 38–9, *40–3*, 44
 general model 32–8
 shifting forward taxation 31–2, 38, 39, 45
 values for fixed weights 36

see also FREIA model; RIKMOD; VICTOR model
Sutton, J. 653n
Swank, Job 265–82
Sweden
 economic growth
  aim and theoretical approach 219–25
  analytical model 225–6
  empirical findings 226–7
  financial flows and 219–43
 flow chart of economy 224, *225*
 prices 239, *240*
 production potential and actual 219, 220, 227–36
  consumption and 229, *230*
  exports 26, *234*, *235*
  imports *236*, *237*
  investment and 231, *232–3*, *234*
  wages *236*, *237*, *238*
Symansky, S. 149n, 150n
Synchronization 58
System Dynamics model 12

Tatom, J. A. 104n
Taubman, P. 165, 166n, 169
Tax Equity and Fiscal Responsibility Act, *see* Economic Recovery Tax Act
Taxation
 CUBS model
  cut in indirect 344, *345*, 346
  personal 340, *341*
 direct
  higher indirect taxes and 47, *48*, 49
  international policy co-ordination 46, 47
  issuing government debt *48*, 49
  public spending cuts and 44–5, *46*, 47
 INTERSAGE model 674–5
 Laffer curve 341, *342*, *343*, 366
 MARIBEL model and 592, *593*
 RASMUS 2b model 267, 275, *276*, *277*, *278*, *279*, *280*, 281
 shifting forward 31–2, 38, 39, 45
 VICTOR model 366, *367*, 368
 see also Fiscal policy
Taylor expansions 186, 187
Taylor, J. B. 72, 96
Taylor, L. 668
Tewes, T. 167
Theil, H. 167

Throop, A. W.   157
Timbrell, M.   38
Tinbergen, J.   31, 59
Tinter, G.   133
Tjan, H. S.   44, 354, 355, 358, 474n
Tobin, J.   31, 119, 125, 146n, 222n, 225, 231
Tomaskovic-Devey, D.   676
Trade, *see* World trade
Treasury model (UK)   599, *600–1*, 602, 603
Treyz, G. I.   169
Tryon, Ralph   53–70
Tse, E.   306
Turnovsky, S. J.   73
TVP model   183–201
  application
    estimation   193–5
    statistics   195–8
  general framework   184–5
  theory
    aggregation over individuals   188–93
    demand functions   186–93
    individual level   186–9
    minimization process   *188*

Ulph, A. M.   247
Ulph, D.   80
Uncertainty, *see* Risk
Unemployment   4, 45
  ERTA and   153–9
  LINK model   6
  rate   142n, 153–9
  see also Employment
United Kingdom, *see* CUBS model; Supply-side policies; Treasury model (UK)
United Nations World model   12
United States
  airlines fuel consumption   251, 257, 258–9
  budget deficit   155, 156n, 158–9, 162
    adjustment options, *see* RASMUS 2b model
    financing   85–6
    reduction   4
    see also Gramm–Rudman Act
  dollar rates, *see* Exchange rate
  Federal Reserve Board   15
    see also MPS model; Multicountry Model (MCM)
  gasoline demand in   250, 251, 252–4

general equilibrium model, *see* INTERSAGE model
iron and steel industry, conservation and   251, *260–1*
policy co-ordination with EEC   71–101
RASMUS 2b model, *see* RASMUS 2b model
US–Japan trade imbalance   27, 28
see also Policy co-ordination; Supply-side policies
Utilization
  capacity
    RIKMOD   395
    Swedish   222, 227, 231
  index   76
  rate   38, 39, 44, 45, 205, 216, 371
    in MCM   55
    in RASMUS 2b model   266, *270*

Van de Klundert, Th.   358
Van den Noord, P. J.   354, 355, 473–90
Van der Windt, N.   32, 38, 45, 76, 356n, 362
Van der Windt, N. *et al.*   265
Van Drimmelel, W.   363
Van Duijn, J. J.   353n, 356
Van Ginneken, W.   598
Van Hulst, N.   363
Van Riet, A.   358n
Van Schail, A. B. T. M.   354, 355
Van Sinderen, J.   38, 358n, 360n
Van Zon, A. H.   354, 355
Vanden Berghe, C.   183–201
Vanroelen, Annemie   561–607
Vatthauer, M.   505
Verbruggen, J. P.   353–81
Verdoorn, P. J.   475
Verdoorn's law   475, 476
VICTOR model   353–81
  empirical results
    black economy   364–5
    exports   361–2
    investment   357–9
    private consumption   364–5
    production capacity   359–61
    wage rate   363–4
  simulations   365–75
    balanced-budget policy   368, 369, 370
    capacity increase   370, *371, 372*
    tax reduction   366, *367*, 368
    wage moderation   *373–4*, 375

Index

supply-side elements 356–7
Vincent, D. P. 653n
VINTAF model 474
   VINTAF II model 599, *600–1*, 602, 603
Vukman, J. 520

Wages
   bargaining 38, 45, 47, 362
      in West Germany, see West Germany
   cost level neutral 500, 502, 503
   equations 34
      estimated 38, 39
      MCM 55
      nominal 42
      Spanish model 696, 709
   RASMUS 2b model 267, 270
   real 31–2, 39, 44
   Swedish model 236, *237*, *238*
   VICTOR model 362–4
      moderation in 373–4, 375
   wage-leader hypothesis 362
Wallis, K. P. 611
Weiserbs, D. 185
Werin, L. 240
West Germany
   budget consolidation
      deficit 203–5
      see also Freiburger and Tübinger econometric model
   consumption function 206, *207*, *208*, 209
   debt paradox hypothesis 204, 205–8, 210
   deficit 209, 210, *211*
   economic situation 208–11
   fiscal deficit
      causal relationships 212, *213*
      see also budget consolidation
   fiscal neutrality, 205–8, 210, 212
   wage bargaining 491–506
      cost level neutral wages 500, 502, 503
      determinants of 493, *494–5*, *496–7*
      macroeconomic consequences 498, *499*, 500–2, *503*, 504
   see also Policy co-ordination; Supply-side policies
Westberg, Lars 219–43
Weyant, J. P. 458
Whalley, J. 668, 672, 680
Wharton Econometric Forecasting Associates (WEFA) model 2, 162, 165
Wharton index, modified 393
Wille, E. 204
Wills, H. 138
Wolffram operator 256
Wolffram, R. 256
Woll, A. et al. 495
Wolters, J. 168
WOM, see World oil market model (WOM)
Wood, D. O. 249
World Energy Conference 247
World Multilevel model 12
World oil market model (WOM) 457–71
   high price scenario 458, *459*, *460*
      Norwegian economy and 464, 465
   international macroeconomic repercussions 462–3
   low price scenario 458, *459*, *460*, 461
      Norwegian economy and 465, 466–8, 469
   temporary or long term fall 460–2
World recession 73–5
World trade 22, *24*
   US–Japan trade imbalance 27, *28*
   see also Exports; Imports
Worrell, D. 543
Wymer, Professor 250

Yagci, F. 713–30
Yasui, K. 541
Yugoslavia
   adjustment model 718–21, 722, 723
      demand behaviour 720–1
      equilibrium relationships 721, 722, 723
      exchange rates 726–7
      interest rates 727–8
      price behaviour 721
      production 719–20
      simulations 724–5, 726–9
      solution under different closures 723
      supply behaviour 720
      wage growth 728
   inflation 507–38
   inflation, see Yugoslavian inflation
   macroeconomics and adjustment 713–30
   policies and growth 714, 715, 716, 717
   policy weaknesses and distortions 717–18

Yugoslavia *contd.*
  protection 718
Yugoslavian inflation
  adjustment stage of model, starting point 516
  cost and price growth mechanisms 511–20
  decomposition procedure 508–11
  depreciation and 515–16, 518
  exchange rate policy 518
  higher tax burdens 515
  imbalance between sectors 513–15
  input–output model 508
  input–output table of economy 520–1, 522–30
  model results 531–2, 533, 534–5, 536
  planned inflation 512–13
  price structure 508–11
Yule–Walker autoregressive method 227

Zahn, P. 495
Zarnowitz, V. 165
Zeuthen, F. 45
Zizmond, E. 511